Synopsis of Neuropsychiatry

Synopsis of Neuropsychiatry

Editors

Barry S. Fogel, M.D.
Clinical Professor of Psychiatry
Harvard Medical School
Boston, Massachusetts;
Adjunct Professor of Community Health
Brown Medical School
Providence, Rhode Island;
Neuropsychiatrist and Behavioral Neurologist
Brigham Behavioral Neurology Group
Brigham and Women's Hospital
Boston, Massachusetts

Randolph B. Schiffer, M.D.
Director, Department of Neuropsychiatry
Texas Tech Health Science Center
Lubbock, Texas

Stephen M. Rao, Ph.D., A.B.P.P.
Professor of Neurology
Co-Director, Section of Neuropsychology
Medical College of Wisconsin
Milwaukee, Wisconsin

Philadelphia • Baltimore • New York • London
Buenos Aires • Hong Kong • Sydney • Tokyo

Acquisitions Editor: Charles W. Mitchell
Developmental Editor: Joanne Husovski
Production Editor: Jeff Somers
Manufacturing Manager: Kevin Watt
Compositor: Lippincott Williams & Wilkins Desktop Division
Printer: Maple Press

227 East Washington Square
Philadelphia, PA 19106-3780 USA
LWW.com

Printed in the USA

Library of Congress Cataloging-in-Publication Data

Synopsis of neuropsychiatry / editors, Barry S. Fogel, Randolph B. Schiffer, Stephen M. Rao.
p. cm.
Based on: Neuropsychiatry / editors, Barry S. Fogel, Randolph B. Schiffer ; associate editor, Stephen M. Rao. 1996.
Includes bibliographical references and index.
ISBN 0-683-30699-5 (alk. paper)
1. Neuropsychiatry. I. Fogel, Barry S. II. Rao, Stephen M. III. Neuropsychiatry.
[DNLM: 1. Neuropsychology. 2. Mental Disorders—diagnosis. 3 Mental Disorders—therapy. 4. Neurophysiology. 5. Psychiatry. WL 103.5 S993 1999]
RC341.S962 1999
616.8—dc21
DNLM/DLC
For Library of Congress 99-28397
CIP

---

10 9 8 7 6 5 4 3 2 1

To John T. Montford, David R. Smith, and Joel Kupersmith,
who had the foresight to create the first integrated
Department of Neuropsychiatry, at Texas Tech University,
in 1998.

Contents

Section III. Syndromes and Disorders

Contributing Authors

John R. Absher, M.D.
Assistant Professor of Neurology
Associate in Psychiatry and Biobehavioral Medicine
Bowman Gray School of Medicine
Medical Center Boulevard
Winston-Salem, North Carolina

Robert Ader, Ph.D., M.D.
George L. Engel Professor of Psychological Medicine
Director, Center for Psychoneuroimmunology Research
University of Rochester School of Medicine and Dentistry
Rochester, New York

Michael J. Aminoff, M.D., F.R.C.P.
Professor of Neurology
University of California, San Francisco
San Francisco, California

David M. Bear, M.D.
Professor of Psychiatry
University of Massachusetts Medical School
Chairman, Department of Psychiatry
The Medical Center of Central Massachusetts
Worcester, Massachusetts

Denise L. Bellinger, Ph.D.
Assistant Professor of Neurobiology and Anatomy
University of Rochester School of Medicine and Dentistry
Rochester, New York

D. Frank Benson, M.D.
Professor of Neurology, Emeritus
University of California, Los Angeles, School of Medicine
Los Angeles, California

Dawn Bowers, Ph.D.
Associate Professor of Neurology and Clinical and Health Psychology
University of Florida College of Medicine
Gainesville, Florida

J. Douglas Bremner, M.D.
Assistant Professor of Psychiatry
Yale University School of Medicine
New Haven, Connecticut
Research Scientist
National Center for Post-Traumatic Stress Disorder
Veterans Administration Medical Center
West Haven, Connecticut

David Caplan, M.D., Ph.D.
Associate Professor of Neurology
Harvard Medical School
Neuropsychology Laboratory
Massachusetts General Hospital
Boston, Massachusetts

Dennis S. Charney, M.D.
Professor of Psychiatry
Associate Chairman for Research
Yale University School of Medicine
New Haven, Connecticut
Chief, Psychiatry Service
Veterans Administration Medical Center
West Haven, Connecticut

Peggy Compton, R.N., Ph.D.
Assistant Clinical Professor of Acute Care Nursing
University of California, Los Angeles, School of Nursing
Research Scientist
The Los Angeles Addiction Treatment Research Center
Los Angeles, California

Yeates Conwell, M.D.
Associate Professor of Psychiatry and Oncology
Director, Laboratory of Suicide Studies
University of Rochester School of Medicine and Dentistry
Rochester, New York

Jody Corey-Bloom, M.D., Ph.D.
Assistant Professor of Neurology
University of California, San Diego, School of Medicine
Attending Neurologist
Veterans Administration Medical Center
San Diego, California

Dean C. Delis, Ph.D., A.B.P.P
Professor of Psychiatry
University of California, San Diego, School of Medicine
Director, Psycyhological Assessment Unit
Veterans Administration Medical Center
San Diego, California

Beatriz M. DeMoranville, M.D.
Clinical Instructor in Medicine
Brown University School of Medicine
Chief, Endocrinology Section
Veterans Administration Medical Center
Providence, Rhode Island

Orrin Devinsky, M.D.
Professor of Neurology
New York University School of Medicine
Chief, Department of Neurology
NYU Medical Center
Hospital for Joint Diseases
New York, New York

George H. Dooneief, M.D., M.P.H.
Assistant Professor of Neurology
College of Physicians and Surgeons of Columbia University
Assistant Attending Neurologist
Columbia-Presbyterian Medical Center
Columbia University, Sergievsky Center
New York, New York

Maureen P. Dymek
Doctoral Candidate in Clinical Psychiatry
University of Alabama at Birmingham
Birmingham, Alabama

Deborah Fein, Ph.D.
Professor of Psychology
University of Connecticut School of Medicine
Department of Psychology
Storrs, Connecticut

Robert G. Feldman, M.D.
Professor and Chairman
Department of Neurology
Boston University School of Medicine
Boston, Massachusetts

David L. Felten, M.D., Ph.D.
Kilian J. and Caroline F. Schmitt Professor and Chair
Department of Neurobiology and Anatomy
University of Rochester School of Medicine and Dentistry
Rochester, New York

Howard L. Fields, M.D., Ph.D.
Professor of Neurology, Physiology, and Psychiatry
University of California, San Francisco
Department of Neurology
San Francisco, California

Christopher M. Filley, M.D.
Associate Professor of Neurology and Psychiatry
Director, Behavioral Neurology Section
University of Colorado Health Sciences Center
Denver, Colorado

David A. Fishbain, M.D., F.A.P.A.
Professor of Psychiatry and Neurological Surgery
University of Miami School of Medicine
Miami, Florida
Comprehensive Pain and Rehabilitation Center
South Shore Hospital and Medical Center
South Miami Beach, Florida

Barry S. Fogel, M.D.
Clinical Professor of Psychiatry
Harvard Medical School
Boston, Massachusetts;
Adjunct Professor of Community Health
Brown Medical School
Providence, Rhode Island;
Neuropsychiatrist and Behavioral Neurologist
Brigham Behavioral Neurology Group
Brigham and Women's Hospital
Boston, Massachusetts

Joaquin M. Fuster, M.D., Ph.D.
Professor of Psychiatry
Neuropsychiatric Institute and Hospital
Los Angeles, California

Douglas Galasko, M.D.
Assistant Professor of Neuroscience
University of California, San Diego, School of Medicine
Staff Physician in Neurology
Veterans Administration Medical Center
San Diego, California

Andrew W. Goddard, M.D.
Assistant Professor of Psychiatry
Yale University School of Medicine
New Haven, Connecticut

D. Gregory Gorman, M.D.
Associate Clinical Professor of Neurology
University of New Mexico School of Medicine
Neurobehavioral Program
Department of Neurology
Lovelace Medical Center
Department of Neurology
Albuquerque, New Mexico

Jordan Grafman, Ph.D.
Chief, Cognitive Neuroscience Section
Medical Neurology Branch
National Institute of Neurological Disorders and Stroke
National Institutes of Health
Bethesda, Maryland

Eric Granholm, Ph.D.
Assistant Professor of Psychiatry
University of California, San Diego, School of Medicine
Staff Psychologist
Veterans Administration Medical Center
San Diego, California

Steven J. Grant, M.A.
Senior Staff Fellow
Intramural Research Program
National Institute on Drug Abuse
National Institutes of Health
Bethesda, Maryland

Lee Anne Green, M.A.
Doctoral Candidate in Clinical Psychology
Research Assistant in Psychology
University of Connecticut
Department of Psychology
Storrs, Connecticut

Kenneth M. Heilman, M.D.
The James E. Rooks Jr. Professor of Neurology
Department of Neurology
University of Florida College of Medicine
Gainesville, Florida

Robin E. Henderson, Ph.D.
Assistant Professor of Orthopedics (Rehabilitation Medicine) and Psychiatry
University of Rochester Medical Center
Rochester, New York

Ivor Jackson, M.D.
Professor of Medicine
Chief, Division of Endocrinology
Brown University School of Medicine
Director, Division of Endocrinology
Rhode Island Hospital
Department of Endocrinology
Providence, Rhode Island

Dilip V. Jeste, M.D.
Professor of Psychiatry and Neuroscience
University of California, San Diego, School of Medicine
Director, Geriatric Psychiatry
Clinical Research Center
Veterans Administration Medical Center
San Diego, California

Stephen Joy, M.A.
Doctoral Candidate in Clinical Psychology
University of Connecticut
Storrs, Connecticut
Instructor in Psychology
Albertus Magnus College
New Haven, Connecticut

Daniel K. Kido, M.D.
Professor of Radiology
Chief, Neuroradiology Section
Mallinckrodt Institute of Radiology
Department of Radiology
St. Louis, Missouri

Michael D. Kopelman, Ph.D., F.R.C.Psych.
Reader in Neuropsychiatry
Neuropsychiatry and Memory Disorders Clinic
United Medical and Dental Schools of Guy's and St. Thomas' Hospitals
St. Thomas' Hospital
Academic Unit of Psychiatry
London, United Kingdom

Walter Ling, M.D.
Professor and Chief, Substance Abuse Program
University of California, Los Angeles, School of Medicine
Director, Los Angeles Addiction Treatment Research Center
Associate Chief of Psychiatry for Substance Abuse
West Los Angeles Veternas Administration Medical Center
Los Angeles, California

Edythe D. London, Ph.D.
Adjunct Associate Professor of Pharmacology and Experimental Therapeutics
University of Maryland School of Medicine
Associate Professor of Radiology
Johns Hopkins School of Medicine
Baltimore, Maryland
Chief, Neuroimaging and Drug Action Section
National Institute on Drug Abuse
National Institutes of Health
Bethesda, Maryland

John A. Lucas, Ph.D.
Assistant Professor of Psychiatry
Mayo Medical School
Rochester, Minnesota
Senior Associate Consultant
Mayo Clinic Jacksonville
Jacksonville, Florida

Kelley S. Madden, Ph.D.
Scientist in Neurobiology and Anatomy
Department of Neurobiology and Anatomy
University of Rochester Medical Center
Rochester, New York

Karen S. Marder, M.D., M.P.H.
Assistant Professor of Neurology
College of Physicians and Surgeons of Columbia University
Columbia University
Sergievsky Center
New York, New York

Robert J. Morecraft, Ph.D.
Assistant Professor of Anatomy and Structural Biology
Department of Anatomy and Structure Biology
University of South Dakota
Vermillion, South Dakota

Michael J. Morgan, Ph.D.
Lecturer in Psychology
University College of Swanson
Department of Psychology
Singleton Park
Swanson South, Wales
United Kingdom

Jonathan Mueller, M.D.
Associate Clinical Professor of Psychiatry
University of California, San Francisco, School of Medicine
Director, San Francisco Neuropsychiatric Associates
San Francisco, California

Linda M. Nagy, M.D.
Assistant Professor of Psychiatry
Yale University School of Medicine
New Haven, Connecticut
Research Director, Outpatient Division
National Center for Post-Traumatic Stress Disorder
Veterans Administration Medical Center
West Haven, Connecticut

John B. Penney, Jr., M.D.
Professor of Neurology
Harvard Medical School
Neurologist
Massachusetts General Hospital
Neurology Service
Boston, Massachusetts

Susan P. Proctor, D.Sc.
Research Assistant Professor of Neurology and Environmental Health
Boston University Schools of Medicine and Public Health
Boston Veterans Administration Medical Center
Boston, Massachusetts

Henry J. Ralston III, M.D.
Professor of Anatomy
W.M. Keck Foundation Center for Integrative Neuroscience
Department of Anatomy
University of California, San Francisco, School of Medicine
San Francisco, California

Stephen M. Rao, Ph.D., A.B.P.P
Professor of Neurology
Co-Director, Section of Neuropsychology
The Medical College of Wisconsin
Departments of Neurology and Psychology
Milwaukee, Wisconsin

John J. Ratey, M.D.
Assistant Professor of Psychiatry
Harvard Medical School
Boston, Massachusetts
Medfield State Hospital
Medfield, Massachusetts

Richard Rawson, Ph.D.
Executive Director, Matrix Center
Deputy Director
University of California, Los Angeles
Alcoholism and Addiction Medicine Services
Los Angeles, California

Alison Reeve, M.D.
Assistant Professor of Psychiatry and Neurology
Mental Health Programs
The University of New Mexico School of Medicine
Department of Psychiatry
Albuquerque, New Mexico

Howard A. Ring, M.D.
Senior Lecturer in Psychiatry
London Hospital Medical College
Honorary Senior Lecturer
Institute of Neurology, Queen Square
Royal London Hospital
Academic Department of Psychiatry
London
United Kingdom

Lynn C. Robertson, Ph.D.
Professor of Neurology and Psychiatry
University of California, Davis, School of Medicine
Davis, California
Research Career Scientist
Veterans Administration Medical Center
Neurology Service
Martinez, California

Mary M. Robertson, M.B.Ch.B., M.D., D.P.M., F.R.C.Psych.
Reader in Neuropsychiatry
University College London Medical School
Consultant Neuropsychiatrist
National Hospital for Neurology and Neurosurgery
Middlesex Hospital
Department of Psychiatry UCLMS
London
United Kingdom

Robert G. Robinson, M.D.
Professor and Chairman
Department of Psychiatry
University of Iowa College of Medicine
University of Iowa Hospitals and Clinics
Iowa City, Iowa

Andres Salazar, M.D.
Professor of Neurology
Uniformed Services University of the Health Sciences
Director, Defense and Veterans Head Injury Program
Walter Reed Medical Center
Army Head Injury Study
Washington, D.C.

Stephen P. Salloway, M.D.
Assistant Professor of Clinical Neurosciences and Psychiatry and Human Behavior
Brown University School of Medicine
Director of Neurology
Butler Hospital
Providence, Rhode Island

Mary Sano, Ph.D.
Assistant Professor of Neurology
College of Physicians and Surgeons of Columbia University
Columbia University
Sergievsky Center
New York, New York

Jeffrey L. Saver, M.D.
Assistant Professor of Clinical Neurology
University of California, Los Angeles, School of Medicine
Attending Neurologist
University of California, Los Angeles Center for Health Sciences
Neurological Services
Los Angeles, California

Randall T. Schapiro, M.D.
Clinical Professor of Neurology
University of Minnesota Medical School
Medical Director of Rehabilitation
The Fairview MS Center
Fairview Riverside Medical Center
Minneapolis, Minnesota

Randolph B. Schiffer, M.D.
Director
Department of Neuropsychiatry
Texas Tech Health Science Center
Lubbock, Texas

Bettina Schmitz, M.D.
Psychiatrische Klinik und Poliklinik
Klinikum Benjamin Franklin
Free University
Berlin, Germany

Robert Taylor Segraves, M.D., Ph.D.
Professor of Psychiatry
Case Western Reserve University
Metro Health Medical Center
Cleveland, Ohio

Katerina Semendeferi, Ph.D.
Postdoctoral Associate in Neurology and Anatomy
Department of Anatomy
University of Iowa
Iowa City, Iowa

Yvette I. Sheline, M.D.
Assistant Professor of Psychiatry and Radiology
Department of Psychiatry
Washington University School of Medicine
St. Louis, Missouri

Steven M. Southwick, M.D.
Associate Professor of Psychiatry
Yale University School of Medicine
New Haven, Connecticut
Director, National Center for Post-Traumatic Stress Disorder
Clinical Neurosciences Division
Veterans Administration Medical Center
West Haven, Connecticut

James F. Toole, M.D.
Teagle Professor of Neurology
Professor of Public Health Sciences
Director, Stroke Center
Department of Neurology
Bowman Gray School of Medicine
Winston-Salem, North Carolina

Javier I. Travella, M.D.
Research Fellow
Department of Psychiatry
University of Iowa Hospitals and Clinics
Iowa City, Iowa

Michael R. Trimble, M.D., F.R.C.P., F.R.C.Psych.
Professor of Behavioral Neurology
University Department of Clinical Neurology
Institute of Neurology
The National Hospital
Queen Square
London
United Kingdom

Gary W. Van Hoesen, Ph.D.
Professor of Anatomy and Neurology
Department of Anatomy
University of Iowa College of Medicine
Iowa City, Iowa

Stanley Walens, Ph.D.
Research Associate
Department of Psychiatry
University of California, San Diego, School of Medicine
San Diego, California

Lynn Waterhouse, Ph.D.
Director, Child Behavior Study
Trenton State College
Trenton, New Jersey

Charles E. Wells, M.D.
Clinical Professor of Psychiatry and Neurology
Vanderbilt University School of Medicine
Nashville, Tennessee

Donald R. Wesson, M.D.
Medical Director and Scientific Director
MPI Treatment Services
Summit Medical Center
Oakland, California

Roberta F. White, Ph.D., A.B.P.P
Professor of Neurology (Neuropsychology) and Environmental Health
Boston University School of Medicine and School of Public Health
Director, Clinical Neuropsychology
Research Director, Boston Environmental Hazards Center
Veterans Affairs Medical Center
Boston, Massachusetts

Peter J. Whitehouse, M.D., Ph.D.
Professor of Neurology
Case Western Reserve University
Director, Alzheimer Center
University Hospitals of Cleveland
Cleveland, Ohio

Tony M. Wong, Ph.D.
Assistant Professor of Physical Medicine and Rehabilitation
University of Rochester School of Medicine and Dentistry
Director of Neuropsychiatry
St. Mary's Hospital
Brain Injury Rehabilitation Program
Rochester, New York

Jessica Yakeley, M.A., M.B.B.Chir., M.R.C.P.
Registrar in Psychiatry
Maudsley Hospital and Institute of Psychiatry
London
United Kingdom

Rachel Yehuda, Ph.D.
Assistant Professor of Psychiatry
Yale University School of Medicine
New Haven, Connecticut
Scientific Director, National Center Post-Traumatic Stress Disorder
Veterans Administration Medical Center
West Haven, Connecticut

Stephen R. Zukin, M.D.
Professor of Psychiatry and Neuroscience
Albert Einstein College of Medicine of Yeshiva University
Bronx, New York
Director, Division of Clinical and Services Research
Division of Clinical and Services Research
National Institute on Drug Abuse
Rockville, Maryland

Preface

Since the appearance of its parent volume, *Neuropsychiatry*, in 1996, we have been considering the release of *Synopsis of Neuropsychiatry*. Our aim has been to provide a concise summary of the larger book, reformatted and reorganized so as to appeal to busy clinicians of several disciplines. We have combined chapters where appropriate, and we have reduced the density of the basic neuroscience coverage in this text. Reference lists have been truncated, and we have emphasized review articles and citations of clinical relevance. We have retained the therapy sections of the parent volume and have updated them in the light of certain major new advances in the field of neuropsychiatry since 1996. The scope of the condensation has been great, with a reduction to 35% of the original text length.

The field of neuropsychiatry remains a work in progress at the time of this writing. Monthly advances march forward apace, from the molecular biology of neurologic disease to the development of psychotherapeutic treatment algorithms for substance abuse. No one can fully grasp the magnitude of this revolution in the behavioral sciences, and no one can be sure where it will take us. At some institutions, divisions of neurology, psychiatry, and neuropsychology are being combined into new neuropsychiatry or behavioral science departments. Laboratory divisions of traditional anatomy and physiology are commonly merging into neuroscience departments that operate scientifically at an entirely new molecular genetic level. As the century ends, we are reminded of the poignant cycles in human affairs which saw a similar revolution in medical school structure and organization at the beginning of this century. Departments of Natural Theology and Rhetoric gave way to the new divisions of Psychiatry, Anatomy, and Biology, making possible the many advances that were to come. We will hope for as much from the century that lies ahead, and we will be pleased if this *Synopsis of Neuropsychiatry* can make some small contribution.

SECTION I

Assessment and Treatment

1
Neuropsychiatric Assessment

Jonathan Mueller and Barry S. Fogel

OVERVIEW

The neuropsychiatric examination is a specialized evaluation that comprises a complete psychiatric evaluation, a neurologic history and examination, detailed assessment of cognitive mental status, and the integration of findings to form inferences about altered brain function and its relationship to changes in mood, thinking, or behavior. Its points of departure are the adult psychiatric evaluation, as described in the American Psychiatric Association's Practice Guidelines for the Psychiatric Evaluation of Adults (1); the neurologic examination, as described comprehensively by DeJong (2); and the neurobehavioral mental status examination, as described by Strub and Black (3), Trzepacz and Baker (4), and Weintraub and Mesulam (5).

This chapter surveys the components of the neuropsychiatric examination, emphasizing parts that are virtually always done and mentioning points that regularly have been of clinical value to the authors. A comprehensive discussion of examination techniques is beyond the scope of this chapter; readers unfamiliar with the references cited are encouraged to consult them.

INDICATIONS AND CONSIDERATIONS

The neuropsychiatric examination is designed to provide a descriptive assessment of both subjective and objective dimensions of patients' behavioral disorders. The examination is designed to identify and localize regional or systemic deficits in brain function. The examination is conducted to assist in the diagnosis of specific neurobehavioral syndromes, which are described subsequently.

EXAMINATION AND ASSESSMENT

Mental Status Examination

Level of Consciousness

In the state of alert wakefulness, patients respond promptly and appropriately to auditory, visual, and tactile stimulation. When a patient's level of alertness or capacity for interaction fluctuates, the examiner should describe the quality of interactions as precisely as possible. The degree of arousal or level of consciousness is evaluated by examining the patient's spontaneous behavior and responsiveness to environmental stimuli.

Patients who respond only to vigorous and repeated stimulation are described as *stuporous.* Whenever a disturbance in the level of consciousness exists, it must be realized that the remainder of the neurologic examination may be affected and that the validity of the data elicited may be partly diminished.

Depth of stupor or coma, especially in the traumatic context, can be quantified with the Glasgow Coma Scale (GCS) (6). The GCS quantifies reactivity to verbal and physical stimulation in terms of a 15-point scale, subdivided into eye-opening responses (score: 1 to 4), motor responses (score: 1 to 6), and verbal responses (score: 1 to 5). On this scale, a score of 3 represents deep coma, whereas a score of 15 represents alert wakefulness.

Behavior

General patterns of observed behavior should be described. Dysfunction of the basal ganglia is suggested by movements that are rapid, jerky, ticlike, or choreiform (i.e., hyperkinesis) or impoverished in frequency and amplitude (i.e., hypokinesis). Patients with parkinsonism, stuporous conditions, and drug toxicity may move and think slowly. Anxiety may manifest as repetitive tapping of the feet, turning of the head, or any other number of stereotyped movements. Hyperkinetic movements should also be noted and described, as should strange or bizarre movement patterns, such as catatonia.

Speech

Spontaneous speech can be characterized in terms of its (a) flow, (b) volume, (c) pressure, (d) rhythm, and (e) intonation.

Disturbances of speech as a motor act, as distinguished from language disorders, are referred to as *dysarthria.* Dysarthria can be spastic, flaccid, ataxic, hypokinetic, or hyperkinetic. In practice, however, distinctions among types of dysarthria are somewhat difficult to make because patients often suffer from mixed dysarthrias that are characterized, for instance, by both spastic and ataxic features.

Cognitive Status Examination

Orientation

The patient is first asked about name, place, and time. Orientation to location may be assessed through inquiry about the city, type of building, address, name of building (if any), and floor or office number.

Orientation to time is assessed with reference to year, month, date, day, and time of day. When patients give an inaccurate response, it is important to record their precise response rather than simply note that their answer was incorrect.

Attention

Attention span is most typically assessed by asking the patient to repeat a string of digits given at a rate of one per second. In presenting a digit string, it is important to avoid any grouping of the numbers that makes the task easier. Most healthy people can repeat a six-digit string. It is important to test the upper and lower limits of a patient's attention span. Patients who are unable to repeat five or six digits should be presented a series of shorter digit strings (beginning with two or three digits and increasing in length) to determine how long a sequence they can retain. On the other hand, patients who easily repeat a six-digit sequence should be requested to repeat progressively longer sequences until they are unable to do so accurately.

A second important measure of attention is the patient's ability to register the list of words presented for the standard verbal memory task. After the patient has been given a list of words, he or she is asked to repeat them immediately. This allows the examiner to establish that the patient has in fact registered the words. If the patient does not register all the words, additional presentations are made, and the number of trials required to register the list of words is recorded.

Cognitive tests that provide a broader view of attention capacities include the following:

Digit-symbol test

Cancellation tasks: a page covered with letters or digits is presented to the patient who is asked to strike out a particular digit or letter whenever it occurs

Trails, or the *trail-making test:* the patient's task is to track in sequence a set of numbers or letters alternating with numbers

Stroop test: the patient begins by naming several different colors and is then asked to identify the color of the ink in which each of a list of color names is printed and then to name the color of the ink in which the names of different colors are printed, where the names do not correspond to the color of the ink

Corsi's block-tapping test: a series of numbered blocks arrayed on a board are tapped in a sequence that the patient is asked to imitate (7)

Language

In contrast to disorders of speech, such as mutism and dysarthria, in which writing is preserved, disorders of language represent central problems in the reception and manipulation of linguistic symbols. Disturbances of writing usually parallel those of spoken language in aphasic conditions. Fig. 1-1 enables clinicians to classify aphasia rapidly by assessing fluency, comprehension, and repetition.

Adequate assessment of language entails consideration of (a) fluency, (b) comprehension, (c) repetition, (d) naming, (e) reading, (f) writing, and (g) spelling (8).

Fluency (Spontaneous Speech)

Fluency refers to the rate, rhythm, and degree of effort in producing spontaneous speech. An alternative way to assess fluency is to request that a patient describe a physical scene (such as the cookie theft picture from the Boston Diagnostic Aphasia Examination) (7). In describing spontaneous speech, the examiner records word-finding pauses and paraphasic errors (substitution of inappropriate syllables or entire words). Although test batteries such as the Boston Diagnostic Aphasia Examination provide quantitative scoring of the degree of fluency, most neurologists and psychiatrists describe speech in all-or-nothing terms as either fluent or nonfluent.

Comprehension

Because the capacity to repeat can be preserved in some patients with impaired comprehension, comprehension should be tested separately from repetition.

A common method of assessing comprehension is to have the patient perform one-, two-, or three-step tasks. Those who correctly

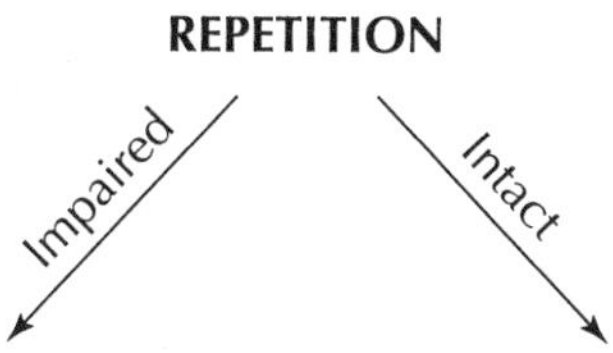

REPETITION

Impaired ↙ ↘ Intact

Perisylvian Syndromes	Fluency	Comprehension	Nonperisylvian Syndromes
1. Broca's aphasia	–	+	1. Anomic aphasias
2. Wernicke's aphasia	+	–	2. Transcortical aphasias
3. Conduction aphasia	+	+	Motor
4. Global aphasia	–	–	Sensory
			Mixed ("isolation of the speech area")
			3. Subcortical aphasias
			Basal ganglia/internal capsule
			Thalamus
			Marie's quadrilateral space

FIG. 1-1. Classification of aphasic syndromes. (From Mueller J, Flynn FG, Fields HL. Brain and behavior. In: Goldman HH, ed. *Review of general psychiatry.* 4th ed. Norwalk, CT: Appleton & Lange, 1995:58.)

perform three-step tasks may be considered to have intact comprehension. Careful documentation of such inability to comprehend multistep instructions is of great value to hospital staff and family members.

Repetition

Repetition is impaired by all four of the major aphasic disorders that result from damage to the perisylvian speech areas of the language hemisphere: Broca's, Wernicke's, conduction aphasia, and global aphasia (see Fig. 1-1). Thus, any patient who can accurately repeat phrases of significant length either has no language disorder or has one of the nonperisylvian aphasias: anomic, subcortical, or transcortical (see Fig. 1-1). Repetition can be assessed for single words, short phrases, and lengthy sentences.

Naming

Most patients with aphasic disorders have word-finding difficulty (dysnomia). Therefore, patients with bizarre speech can be initially screened for aphasia by being asked to name objects. A simple object to use for this task is a pen; the patient is asked to name the point, cap, and clip.

Reading

Most patients with aphasic disorders have reading problems. Clinicians who wish to familiarize themselves with a variety of reading disturbances are referred to the monograph by Benson, *Aphasia, Alexia, and Agraphia* (8) and to standard neuropsychological texts (9).

Patients may be asked to read individual words, sentences, or paragraphs of varying difficulty. It is important to realize that reading aloud is different from reading for comprehension. The fact that a patient can read a passage aloud does not indicate that he or she is able to understand it, much less remember it.

Writing

Patients with true aphasic disorders usually exhibit disturbances in both speech and writing. Because it may be difficult for examiners to transcribe the speech of aphasic patients, it is helpful to obtain samples of the patient's writing. Sentences, either written to dictation or following a theme suggested by the examiner, are requested.

Spelling

Orthographic errors are common following insults that produce aphasia. It is important, however, for the clinician to establish that the patient's current difficulties with spelling represent change from an established baseline.

Memory

In patients whose memory disturbance is secondary to gross brain dysfunction, a distinction is made between isolated memory disturbance (pure amnestic syndrome) and a more widespread disturbance of intellectual function that includes impaired memory (either a dementia syndrome or a confusional state).

Short-term verbal memory is typically assessed at the bedside by asking the patient to recall a list of four words. The examiner must establish that the patient actually registers the words initially; otherwise, it is impossible to distinguish a memory disturbance from an attention problem. It is helpful for the examiner to employ a graded approach in the assessment of memory. If, for instance, 3 points are given for each word recalled spontaneously, 2 points for a word recalled after a category prompt, and 1 point if the word is chosen from a list of three or four alternatives, the test has a 12-point range. This is superior to recording simply the number of words recalled because it offers qualitative data on any memory problem. Its greater detail also enables the clinician to form a clearer opinion of the progression (or remission) of any memory disturbance when the patient is reexamined on another occasion (3).

Assessment of memory is a crucial element in all neuropsychological evaluations; thus, it is not surprising that a large number of tests for verbal memory have been developed (10, 11). Among the more popular tests is the Logical Memory Paragraphs from the Wechsler Memory Scale (12). In this test, patients are read two paragraphs, each containing 25 idea units, and are asked to repeat as much of the two stories as possible. Following a 30- or 60-minute interval, patients are again asked to repeat whatever they recall from each of the two stories. This allows the examiner to look at the decay of memory over time.

Another popular test is the Rey Auditory Verbal Learning Test (13). This test consists of a list of 15 words that are read to the patient five times. After each reading, the patient is asked to recall as many words as possible. The Rey test allows the examiner to look at the patient's learning curve when confronted with a "supraspan" list of words. After the five trials, a second list of 15 words is read, and the patient is again asked to recall as many words as possible. After this "interference" trial, the patient is asked to recall as many words as possible from the original list. Finally, the patient is read a list of 30 words, half of which were on the first list. He or she is asked to say "yes" after each word that appeared on the initial list.

A second type of memory testing is that which assesses visuospatial recall. The examiner may place several objects around the patient's room at the start of the interview, and after 5, 10, or 15 minutes, inquire how many objects were hidden, what they were, and where they were placed. Another approach is to ask a patient to study a drawn or printed figure and then reproduce it shortly afterward.

Remote memory, which refers to a patient's recall of past events, may be difficult unless the examiner is familiar with the patient's biography. An alternate way of exploring remote memory is to ask the patient about famous people or historical events. Patients with dementia or with amnesia due to acquired brain disease may accurately recall events from the distant past but not from the more immediate past.

Visuospatial Skills

Visuospatial skills usually are assessed by paper-and-pencil drawing exercises. Patients are asked to draw circles, triangles, and three-dimensional cubes, a person, a bicycle, a house, or the face of a clock with the hands placed at a particular time. The clock drawing in particular has established value as a screening test for dementia (14).

In addition to drawing tasks, patients may be asked to manipulate either tokens or three-dimensional blocks to make a series of designs of increasing complexity. The examiner records the time required to complete each design.

Most techniques for assessing visuospatial skills rely on motor activity of the patient. An alternative approach is to ask the patient to identify a particular geometric figure among a series of figures oriented in different planes. Such tasks require mental manipulation of figures without physical effort. They may be useful, therefore, in examining patients who have difficulty using their upper extremities.

Calculations

Patients are asked to perform mental computations. The speed and ease with which they perform these tasks is noted. Some examiners give patients practical questions, such as the following one: "If you went to the store with a dollar bill and bought three items for five cents, and four items for two cents, how much change would you receive?"

Reasoning

The capacity to reason may be examined in several ways. Practical judgment, appreciation of abstract similarities, and interpretations of proverbs have all been used. The examiner should be aware, however, that the "practical judgment" that a patient appears to exercise in a hypothetical situation may not carry over into a real-life situation. This is particularly true in cases of frontal lobe dysfunction.

Thought Process

The organization, coherence, and style of a patient's thought process are noted. The examiner records a patient's capacity to respond to questions in a focused and coherent fashion, noting any tendencies toward *circumstantiality* (in which excessive and digressive details are given) or *tangentiality* (responses that veer off from the question without returning to it). The nature of a patient's associative processes in response to questions is reflected in both the degree to which relevant information is organized and the manner in which historical information is conveyed.

Content

The themes or subject matter of a patient's conversation is recorded. Traumatized patients may have recurring intrusive preoccupations with sounds, images, or feelings associated with the trauma. Psychotic patients may be preoccupied with recurring hallucinations. Other themes of a patient's conversation may be anger, guilt, diminished self-esteem, desire for closeness, or fear of intimacy.

Insight

Patients vary tremendously with regard to their awareness of, and degree of concern about, problems in their life. Anosognosia describes the striking lack of awareness that is observed in patients with right-hemisphere strokes. Critchley's book, *The Parietal Lobe* (15), describes a wide range of this and similar neglect syndromes.

Mood and Affect

Affect refers to the outward display of felt inner emotion, whereas *mood* refers to the state of inner emotionality. In general, mood may be thought of as an internal emotional climate. For many people, mood states last for days or weeks, whereas for others, including rapidly cycling bipolar patients, patients with borderline personalities, normal adolescents, and patients with limbic epilepsy, mood may fluctuate rapidly and dramatically (16).

Feelings of anxiety can occur in many different guises. Heart palpitations, diaphoresis, shortness of breath, dizziness, and feelings of impending doom may all reflect an underlying anxiety or panic disorder. Many factors, such as anxiety, insomnia, exhaustion, and medications, can affect patients' mental states and, therefore, their test performance.

Neurologic Examination

The detail with which the examination is done varies according to clinical circumstances. When conducting the examination, the detailed attention given to each part should be in accord with the diagnostic hypotheses that have evolved by that point in the clinical encounter. The examiner should not hesitate to return to earlier parts of the examination, such as mental status or visual fields, if later findings change the likelihood and probable location of a focal neurologic deficit.

Cranial Nerves

Cranial Nerve I

Disturbance of the sense of smell may occur in a wide variety of neurologic disorders. Shearing of the olfactory filaments as they traverse the cribriform plate occurs frequently with closed head injury and can take place without loss of consciousness. Complete anosmia should be distinguished from partial loss of olfactory capacity (hyposmia).

The olfactory nerves should be assessed with substances that do not irritate the trigeminal nerves, such as vanilla or coffee. Scratch-and-sniff cards are available for systematic, validated olfactory screening using standardized stimuli.

Cranial Nerve II

Examination of the optic nerve entails examination of the optic fundus (retina), visual fields, visual acuity, and the pupil. Funduscopic examination allows consideration of the retinal vascularity and the optic disc. Visual acuity can be assessed with a pocket screener or a wall chart. It is important to indicate whether acuity is assessed with patients wearing corrective lenses or glasses. In the absence of corrective lenses, a pinhole can be used to correct for refractive errors. Visual fields can be tested with small red and white objects on the tip of a pointer; cotton swabs on long sticks will serve if standard test objects are not at hand. Each eye is examined independently. Pupillary reactivity to direct and consensual light is recorded, as is the near reflex (accommodation and convergence).

Cranial Nerves III, IV, and VI

Extraocular motility is examined by observing the patient's upward, downward, lateral, and angular gaze. The examiner looks closely for nystagmus, diplopia, or limitation of gaze in any direction.

Cranial Nerve V

Facial sensation over each of the three branches of the trigeminal nerve is tested with light touch and pinprick, and each side is compared with the other. The corneal reflex (afferent limb by way of cranial nerve V; efferent limb by way of cranial nerve VII) should be tested with a wisp of cotton. Its absence can be the sole evidence of trigeminal nerve dysfunction; its presence is evidence against facial anesthesia.

Cranial Nerve VII

Evidence of facial asymmetry is noted both at rest and during facial expression. In peripheral facial nerve lesions, the entire hemiface is involved, whereas in a central lesion, the forehead is spared. Central lesions involving the right face may also be associated with aphasia. Temporal lobe lesions can produce symmetry on spontaneous emotional expression with intact facial movements in response to commands.

Cranial Nerve VIII

Hearing may be assessed at the bedside with whispering, tuning forks, or, ideally, an audioscope. The examiner also performs the Weber test, in which a 512-Hz tuning fork is placed on the vertex and the patient asked whether the sound is heard equally well in both ears. In conductive hearing loss, the sound is louder on the defective side. In sensorineural hearing loss, the sound is louder on the less impaired side. Finally, bone conduction is compared with air conduction by placing the base of the vibrating tuning first on the mastoid process behind the ear and asking the patient to compare the strength of that sound with the sound of the tuning fork in the air. In conductive hearing loss, bone conduction is greater than air conduction.

The vestibular component of cranial nerve VIII may be tested with the Nylen-Barany maneuver, which stimulates the vestibular apparatus. The patient is initially seated and then moved rapidly into a supine posture with head tilted 45 degrees backward below the level of the table and 45 degrees to the left. The patient resumes the seated posture, and the maneuver is repeated, moving the head backward and 45 degrees to the right. The examiner notes whether nystagmus or symptoms of vertigo are produced by this maneuver and compares the intensity and duration of nystagmus produced by turning of the head toward the left and the right sides.

Cranial Nerves IX and X

The afferent limb of the gag reflex is by way of cranial nerve IX, and the efferent limb is by way of cranial nerve X. Swallowing is a complex act that can be impaired even if the gag reflex is intact. Timed swallowing of liquid is a sensitive bedside test for dysphagia. A rate of less than 10 mL per second is strongly associated with objective demonstration of dysphagia on cineesophogram.

Cranial Nerve XI

Function of the accessory nerve is tested by palpation of the sternocleidomastoid muscle as the patient turns his or her head to the left and right alternately while being opposed by the hand of the examiner. Weakness of the *left* sternocleidomastoid impairs head turning to the *right* and vice versa. Shrugging or elevation of the shoulders is accomplished through activation of the trapezius muscle. The examiner requests the patient to elevate his or her shoulders while the examiner palpates the involved muscle.

Cranial Nerve XII

The patient is asked to protrude the tongue and move it from side to side. In unilateral paralysis of cranial nerve XII, the tongue is observed to deviate toward the side of the weakness.

Motor Examination

Muscle Mass

Muscle mass is observed for any asymmetry of bulk and for any focal atrophy due to peripheral lesions or disuse.

Muscle Tone

Muscle tone is tested by passively moving the patient's limbs. The examiner compares tone in both extremities and notes the presence or absence of increased tone. Increased tone may be characterized as having a jackknife quality typical of upper motor neuron disease, a paratonic quality (a ratchetlike, intermittent tone seen in frontal lobe disease), cogwheeling (parkinsonian), or a negativistic quality (the rigid resistance of catatonia). Absence or flaccidity of muscle tone is indicative of lower motor neuron disease, and decreased tone is associated with cerebellar lesions. Tone varies considerably among healthy people.

Muscle Strength

Muscle strength is graded from 0 to 5, as follows:

0 = no movement
1 = trace movement
2 = movement with gravity in a horizontal plane
3 = movement against gravity
4 = movement against gravity and against applied force
5 = normal strength

The examiner compares left versus right, proximal versus distal, and upper versus lower extremity strength.

Reflexes

Reflexes are graded from 0 to 4, with a score of 2 being average. Briskness of reflexes varies in healthy people, and areflexia that is not asymmetric and is not associated with weakness or other motor or sensory problems is not necessarily pathologic.

Pathologic Reflexes

In healthy adults, stimulation of the plantar surface of the foot produces either no response or plantar flexion (a downgoing toe). In patients with upper motor neuron pathology (i.e., a pyramidal tract lesion), the great toe may dorsiflex (i.e., extend), and the toes fan out. An upgoing toe with or without fanning of the other toes is referred to as *Babinski's sign.* Babinski's sign in an adult usually implies structural disease of the central nervous system or a toxic-metabolic encephalopathy.

A number of "primitive" reflexes (i.e., grasp, suck, snout, root) may be seen in infants during the first year of life. Reappearance of these reflexes suggests frontal lobe disorders. The occurrence of primitive reflexes in elderly people, however, is of relatively low diagnostic specificity.

Coordination

Coordination (cerebellar) testing consists of finger-nose-finger, heel-knee-shin, and rapid alternating movements. Weakness can interfere with coordination testing. Spasticity can cause clumsiness but not the dysmetria typical of cerebellar lesions.

Station

Is the patient able to stand comfortably with feet together? Is there evidence of swaying when the patient is asked to close his or her eyes (positive Romberg's sign)? Can the patient stand on one foot?

Gait

The examiner should observe several components in a patient's gait: How wide is the stride? How broad-based is the gait? Are movements made in a shuffling, tentative fashion? Is there a foot-drop with a steppage gait? Is turning done smoothly or en bloc? Is there asymmetry of arm swing? The time taken to walk a fixed distance can be used as a quantitative measure of a patient's gait.

Abnormal Movements

Is there a paucity or an excess of movements (hypokinesia or hyperkinesia)? Are there spontaneous dyskinesias (such as those of Huntington's disease or tardive dyskinesia)? Is there a tremor visible at rest (parkinsonian tremor), postural tremor (familial or essential tremor), or tremor that appears only with intention (cerebellar tremor)?

Soft Signs

The neurologic and psychiatric literature has been inconsistent regarding whether the traditional neurologic examination is of assistance in the evaluation of the various neuropsychiatric syndromes.

"Soft" neurologic signs include the following:

Fine motor incoordination: awkwardness at such tasks as handwriting, finger-nose-finger and finger pursuit in the absence of frank cerebellar dysmetria

Dysrhythmia: a lack of smooth transitions between different motor tasks

Mirror movements: the contralateral overflow of motor activity in homologous muscle groups

Synkinesis: ipsilateral overflow of extraneous associated movements when the patient is asked to perform an activity involving a specific set of muscle groups

Both mirror movements and synkinesis can be elicited by asking the patient to perform activities involving discrete muscle groups, such as the following: (a) tap one foot; (b) alternately tap heel and toe; (c) pat one's thigh; (d) flip-flop one hand on one's thigh; (e) tap one's thumb and index finger repeatedly; (f) touch one's index, middle, ring, and little fingers to one's thumb.

Sensation

The examiner tests primary sensory perception by examining the patient's awareness of pinprick, light touch, position, and vibration. Parietal sensory testing consists of examining a patient's capacity to recognize numbers or letters written on the palm or on the sole of the foot while the patient is not looking (graphesthesia), the ability to recognize an object from manipulating or palpating it (stereognosis), and the ability to detect sensory stimuli applied at the same time in different anatomic regions (double-simultaneous stimulation).

CLINICAL ISSUES

Abnormalities of Orientation and Attention

Loss of orientation to location may be seen in a variety of contexts, most commonly in confusional states, dementing processes, and amnestic disorders—all of which prevent patients from updating themselves on their current location (i.e., remembering transitions from one place to another). Patients with visual-perceptual problems may also be unable to use visual cues to orient themselves.

Disorders of attention are the hallmark of confusional states (either quiet confusional states or agitated confusional states—sometimes referred to as *delirium*). Confusional states most commonly occur in the context of toxic, infectious, and metabolic encephalopathy and usually respond to treatment of the causative medical condition. Patients with preexisting dementia syndromes are at particular risk for the development of superimposed confusional states (sometimes called *beclouded dementia*) in response to a wide variety of problems, such as drug toxicity, pulmonary and urinary tract infections, anesthesia, and sleep deprivation.

Abnormalities of Linguistic Prosody

Prosody, the emotional coloring of speech, is a manifestation of affect. Alteration in prosody can reflect either a change or dysfunction of mechanisms of affective expression. Loss of emotional coloring in speech can occur in a number of disorders, including major depression, degenerative illnesses affecting the basal ganglia, drug-induced akinesia, damage to the right frontal lobe, or Broca-type aphasia secondary to left frontal damage. Both the patient and caregivers should be asked whether any change has been noted in the patient's style of speaking.

Abnormalities of Affective Regulation

Significant discrepancies between inner emotion and outward affect occur frequently in patients with brain diseases affecting the frontal lobes or their connections. Affective states can be altered by brain diseases involving the right hemisphere. Pseudobulbar palsy refers to a state in which upper motor neuron regulation of the brain-stem nuclei subserving emotional expression is disrupted. Such patients may complain of excessive and sometimes even explosive affective display in response to trivial emotional stimuli. They may describe "emotional incontinence," either sobbing uncontrollably when feeling only somewhat sad, or laughing in a grotesquely loud and disinhibited fashion in response to something that may strike them as only mildly amusing. Occasionally, such patients become tearful when finding something humorous. At other times, they may complain of laughing when they experience sadness.

Affect and thought content (as expressed verbally) are normally congruous; marked discrepancy suggests mental illness or anatomic or physiologic disconnection of limbic areas associated with emotion from the neocortical and brain-stem areas associated with emotional expression. Patients with schizophrenia or with bilateral frontal lobe damage may smile strangely when discussing suicide or discuss feelings of extreme agitation with little or no facial expression. In addition to the appropriateness of affect to content of thought, the examiner pays particular attention to the range, intensity, and lability of affect. Emotion is reflected not just in facial expression but also in shifts of posture, gesticulation, flow, and coloring of speech.

Abnormalities of Visuospatial Function

The classic syndromes of visuospatial disability are those associated with parietal lobe damage, particularly that of the nondominant hemisphere. The drawings of these patients often omit major elements of the figure being copied, with particular difficulty manifested with the left side of the drawing. On the other hand, damage to the dominant parietal lobe may produce some difficulty reproducing some aspects of figures, but the deficits tend to be less coarse and may be confined to difficulty with accurate depiction of the inner details of complex figures.

Damage to the inferior parietal lobule of the dominant hemisphere may also produce a constellation of cognitive difficulties known as Gerstmann's syndrome. This consists of the following tetrad of findings: acalculia, agraphia, right-left confusion, and finger agnosia. Damage to this region, however, seldom produces a pure Gerstmann's syndrome. Some degree of language difficulty, usually a receptive aphasia with word-finding difficulty, typically accompanies damage to the left inferior parietal lobule.

THE DIAGNOSIS OF DEMENTIA: FALSE-NEGATIVE AND FALSE-POSITIVE TEST RESULTS

It is not uncommon for a patient with superior premorbid intelligence to have significant decrements in intellectual functioning but still perform within the normal range on cognitive screening tests designed to detect major performance deficits or on intelligence tests and school entrance examinations, such as the SAT, GMAT, LSAT, GRE, and MCAT, and grades from high school, college, or graduate school (17).

Many of the briefer bedside instruments for cognitive assessment, such as the Mini-Mental State Examination and the Cognitive Capacity Screening Examination, approach cognition as univariate (i.e., global) and have such low cutoff scores for the diagnosis of dementia that impairment must be very significant before detection occurs. This leads to false negative assessments in patients with restricted deficits, particularly of right-hemisphere or frontal lobe functions (18, 19).

False-positive assessments of cognitive decline tend to occur in two situations: when educational limitations or cultural factors are not taken into account and when long-standing, stable learning disabilities are misinterpreted as more recently acquired deficits. Restricted difficulties with spatial orientation, sentence structure, or arithmetic, for instance, commonly reflect learning disabilities that may not have been previously defined.

The diagnosis of a learning disability is particularly likely to be missed or delayed if the patient has dropped out of school without any formal evaluation of academic difficulties or if the patient had above average intelligence and was able to pass courses despite difficulties. Directly asking the patient whether there were special areas of difficulty in school, regardless of the grades obtained, may yield clues. Examples of schoolwork or narrative evaluations by teachers are sometimes available.

FRONTAL LOBE LESIONS

Frontal lobe lesions are compatible with normal performance in many areas of higher intellectual functioning. To assess frontal lobe deficits, it is necessary to employ tests that allow one to examine a patient's capacity for sustaining and shifting attention, for self-monitoring, and for responding to challenges in the face of shifting environmental cues and contingencies. Because examiners often perform their assessment in a structured environment and provide patients with explicit instructions on how to proceed on tests, they actually create an artificial world in which the examiner "takes over the role of the patient's frontal lobes." When patients with frontal lobe damage are observed outside the test situation, their real-life behavior often differs considerably from their verbal responses to hypothetical situations.

THE NEUROLOGIC EXAMINATION OF PATIENTS SUSPECTED OF HYSTERIA, MALINGERING, AND OTHER PSYCHOGENIC DISORDERS

DeJong's encyclopedic text, *The Neurologic Examination*, devotes 18 pages to signs that may be seen in patients with hysteria, malingering, and other psychogenic conditions (2). Although the detection of nonphysiologic findings on a neurologic examination raises strong questions of psychological factors, the presence of these findings cannot be used as proof that the entirety of a patient's presentation is feigned, hysterical, or psychological in origin.

Inconsistency of motor performance is typical of psychogenic illness. On testing of strength, the examiner may discover that a patient exerts effort but only for a brief period. Others fail to exert a full effort. When this occurs, the examiner should establish whether pain is a factor and determine whether the patient is able, even for a brief period, to give a maximal effort.

When asked to turn the head first to one side, then the other, the patient with hysterical hemiplegia may mount less effort in turning his or her head toward the allegedly weak side. If the examiner palpates the contralateral sternocleidomastoid muscle, he or she will feel the lack of tension (i.e., effort) on turning toward the allegedly paretic side.

Hoover's sign may also be helpful in patients with either hysterical hemiplegia or hysterical monoplegia. The patient is asked to lie supine and then raise first the strong leg, then the weak leg. As the patient pushes up with the intact leg, the examiner offers resistance to this raising, while at the same time positioning his or her hand beneath the heel of the allegedly paretic leg. If both legs are in fact intact, the examiner can feel downward pressure from the allegedly paretic leg as the patient attempts to raise the intact leg against the resistance of the examiner. The examiner next places his or her hand beneath the heel of the intact leg and requests the patient to lift the allegedly paretic limb against resistance. Failure to exert an effort to raise the allegedly paretic limb may be detected by the absence of downward pressure in the intact leg.

Patients who present with neurologically based weakness have reflexes that are increased with pyramidal tract lesions or decreased with lower motor neuron lesions. Examiners may test cooperation by palpating both agonist and antagonist muscle groups. For instance, when asked to extend the knee, there should be some increase in muscle tone within the quadriceps accompanied by relaxation within the hamstrings. If the hamstrings tighten and the quadriceps remain flaccid, this is inconsistent with full cooperation. In cases of hysterical hemiplegia, some patients may, in response to an unanticipated noxious stimulus, move the allegedly paretic limb.

When patients complain of sensory deficits, the pattern of their alleged alteration in sensation should be noted. Nondermatomal distributions may be claimed for alleged sensory loss. Patients who claim anesthesia may be instructed to respond with a "yes" when they feel a pinprick and a "no" when they do not feel it. If a patient responds "no" when an allegedly anesthetic region is touched, he or she demonstrates perception of the stimulus. If a patient complains of blindness, any reaction to visual threat (blinking or head aversion) should be noted, as should pupillary reaction to light.

Sensory complaints that split the midline in a precise fashion tend to be of nonorganic origin, particularly when they involve both the face and trunk.

SUMMARY

The neuropsychiatric examination remains the mainstay of clinical practice in neuropsychiatry. The examination documents functional performance at a certain point in time and permits the generation of hypotheses that govern further diagnostic evaluations as well as treatment. The clinical examination can assist in providing answers to the following questions:

- Is there a disturbance of brain function that is clinically discernible? In a patient with known brain injury or lesion, what functional consequences are present?
- What diagnostic hypotheses should be considered in a given patient? What further investigations with imaging, electrophysiologic, or psychometric tools might be appropriate?
- Do we need to consider the commitment of resource-intensive or invasive technology to the particular clinical problem before us?

REFERENCES

1. American Psychiatric Association Work Group on Practice Guidelines for the Psychiatric Evaluation of Adults. Practice guidelines for the psychiatric evaluation of adults. *Am J Psychiatry* 1995;152[Suppl 11]:65—-80.
2. DeJong R. *The neurologic examination.* 4th ed. New York: Harper & Row, 1979.
3. Strub RL, Black FW. *The mental status examination in neurology.* 3rd ed. Philadelphia, PA: FA Davis, 1993.
4. Trzepacz PT, Baker RW. *The psychiatric mental status examination.* New York: Oxford University Press, 1993.
5. Weintraub S, Mesulam M-M. Mental state assessment of young and elderly adults in behavioral neurology. In: Mesulam M-M. *Principles of behavioral neurology.* Philadelphia, PA: FA Davis, 1985:71–124.
6. Jennett B, Bond M. Assessment of outcome after severe brain damage: a practical scale. *Lancet* 1975;1: 480–481.

7. Lezak M. Verbal functions and language skills. In: *Neuropsychological assessment.* 3rd ed. New York: Oxford University Press, 1995:523–558.
8. Benson DF. *Aphasia, alexia, and agraphia.* New York: Churchill Livingstone, 1979.
9. Friedman R, Ween JE, Albert ML. Alexia. In: Heilman KM, Valenstein E, eds. *Clinical neuropsychology.* New York: Oxford University Press, 1993:37–62.
10. Delis DC, Kramer JH, Kaplan E, Ober BA. *California Verbal Learning Test:* adult version. San Antonio, TX: The Psychological Corporation, 1987.
11. Buschke H, Fuld PA. Evaluation of storage, retention, and retrieval in disordered memory and learning. *Neurology* 1974;11:1019–1025.
12. Wechsler D. *Wechsler Memory Scale:* revised manual. San Antonio, TX: The Psychological Corporation, 1987.
13. Lezak M. Memory I: tests. In: *Neuropsychological assessment.* 3rd ed. New York: Oxford University Press, 1995:429–498.
14. Pan GD, Stern Y, Sano M, Mayeux R. Clock-drawing in neurological disorders. *Behavioral Neurology* 1989;2: 39–48.
15. Critchley M. *The parietal lobes.* London: Hafner, 1953.
16. Heilman KM, Bowers D, Valenstein E. Emotional disorders associated with neurological diseases. In: Heilman KM, Valenstein E, eds. *Clinical neuropsychology.* 3rd ed. New York: Oxford University Press, 1993: 461–498.
17. Wechsler D. *Wechster Adult Intelligence Scale:* revised manual. New York: The Psychological Corporation, 1981.
18. Folstein M, Folstein S, McHugh PR. Mini-Mental State: A practical method for grading the cognitive state of the patient for the clinician. *J Psychiatr Res* 1975;12: 189–198.
19. Jacobs JW, Bernard MR, Delgado A, Strain JJ. Screening for organic mental syndromes in the medically ill. *Ann Intern Med* 1977;86:40–47.

2

Neuropsychological Evaluation

Stephen M. Rao

OVERVIEW

The use of psychological tests to study brain dysfunction took hold in the decades following World War II, with the pioneering experimental studies of focal brain damage by Hans-Lukas Teuber, Brenda Milner, Alexander Luria, Arthur Benton, Henry Hècaen, and Ward Halstead. These psychological investigations, based on accidents of nature (strokes and tumors), warfare (penetrating head wounds), and surgery (cortical excision of epileptic foci), provided important new insights regarding cerebral localization and lateralization. These early studies formed the basis for the scientific field of human neuropsychology.

Clinical neuropsychology as a professional specialty is a fairly recent development, originating in the 1970s (1). Whereas *human neuropsychology* is multidisciplinary, involving specialists in experimental, cognitive, and clinical psychology, neurology, psychiatry, linguistics, speech pathology, and neuroscience, *clinical neuropsychology* is typically practiced by clinical psychologists who are charged with applying the scientific knowledge derived from human neuropsychological research to the evaluation and treatment of suspected brain dysfunction. For more detailed discussions of the points raised in this chapter, the reader is referred to textbooks pertaining to clinical neuropsychological assessment (2) and the scientific field of human neuropsychology (3). Cognitive and behavioral interventions of brain dysfunction are covered in detail elsewhere (4–6).

INDICATIONS AND CONSIDERATIONS

The reasons for performing a clinical neuropsychological assessment have broadened over the years. Since the advent of high-resolution structural brain imaging techniques, the emphasis of neuropsychological assessment has shifted from lesion localization to a more comprehensive characterization of the patient's cognitive and emotional status. Neuropsychological testing has been used to accomplish the following:

- Establish baseline measures to monitor changes in time associated with progressive cerebral diseases (neoplasms and demyelinating and dementing conditions) or the recovery from acute brain disorders (traumatic head injury, stroke)
- Measure outcome in clinical efficacy trials involving surgical, pharmacologic, and behavioral interventions
- Predict changes in quality of life, for example, employment and social adjustment
- Formulate rehabilitation and remedial interventions, including education and vocational planning
- Provide information to patients and family members coping with cognitive and emotional changes associated with brain dysfunction
- Establish forensic opinions regarding the effects of brain dysfunction in civil and criminal cases, including issues of competency

EXAMINATION AND ASSESSMENT

The clinical neuropsychological examination may consist of a fixed or flexible battery of neuropsychological tests. Examples of fixed batteries include the Halstead-Reitan Test Battery (7) and the Luria-Nebraska Battery (8). The standard battery typically takes 3 to 8 hours to administer and provides a comprehensive assessment of cognitive, perceptual, linguistic, and sensory-motor skills. The same tests are given to all patients regardless of the referral question. An alternative approach, advocated initially by Luria (8) and described by Lezak (2), calls for flexibility in the selection of neuropsychological tests because the battery is specifically tailored to the referral question and is influenced by the interview and prior medical and psychosocial records. This approach also emphasizes a qualitative analysis of test performance, as best exemplified by the Boston process approach (9).

SPECIFIC NEUROPSYCHOLOGICAL TESTING PROCEDURES

In this section, citations to standardized neuropsychological tests can be obtained from Lezak's textbook (2).

Intelligence

Neuropsychological assessments typically begin with a measurement of intellectual functioning. The most common test used in clinical practice is the Wechsler Adult Intelligence Scale revised test (WAIS-R), which consists of six verbal and five performance subtests. Three intelligence quotients, a Full-Scale, Verbal, and Performance IQ, are derived. A major advantage of this test is its large, stratified, normative database, which corrects for age differences.

Comparisons of tested IQ values and premorbid IQ estimates (from demographic variables or tests like the National Adult Reading Test) can be achieved to obtain a gross measure of the degree of cognitive deterioration. A major disadvantage of this test is that it was designed to provide a measure of a patient's general ability levels. Consequently, performance on each subtest is influenced by multiple cognitive operations. In addition to comparing estimated and actual IQ test scores, clinicians compare the Verbal and Performance IQ scores as a measure of lateralized brain dysfunction. Clinicians may also interpret both intersubtest and intrasubtest scatter as indicators of brain dysfunction.

Conceptual Reasoning and Executive Functions

Neuropsychological assessments have traditionally included measurement of nonverbal, abstract thinking or concept formation. The two most common tasks used in this regard are the Wisconsin Card Sorting Test (WCST) and the Category Test. The WCST, in particular, is used by more than 70% of practicing neuropsychologists. Both measures are sensitive to cerebral dysfunction in neurologic and psychiatric disorders. The WCST is viewed as a measure particularly sensitive to focal frontal lobe dysfunction. A greater number of WCST perseverative errors are observed in patients with frontal lobe lesions than in those with nonfrontal lesions, although this relationship has not been uniformly observed.

Patients with executive function disorders exhibit perseveration (an inability to stop a sequence of actions once begun), a loss of initiative or intention to act, an inability to generate plans, a tendency to act impulsively, and problems incorporating feedback in modifying their behavior. A wide variety of tasks have been used to tap this diverse cluster of cognitive abilities, including the Tower of Hanoi and Tower of London tests, verbal and design fluency tasks, Porteus Maze Test, and repetitive line drawings. Executive functions can also involve the ability to plan, follow, arrange, or recall the temporal order or sequence of events and the ability to evaluate the accuracy of one's own performance.

Attention

Attention is frequently impaired in neurologic disorders and may be the primary area of cognitive dysfunction in psychiatric disorders, such as schizophrenia. Attention can be segregated into focused, sustained, and divided processes. *Focused attention* refers to the process of searching for and locating target stimuli. *Sustained attention,* sometimes used interchangeably with the term *vigilance,* refers to the monitoring of target stimuli over an extended duration. *Divided attention* refers to the ability to perform two tasks simultaneously.

Most neuropsychological tests of attention used in clinical practice assess one or more of these attentional components. These tests include the Trail Making Test, the Stroop Color-Word Interference Test, the Symbol Digit Modalities Test, and the Paced Auditory Serial Addition Test. Some attentional tasks are designed to assess unilateral spatial neglect. Two of the more common of these tasks are the Letter Cancellation and Line Bisection tasks.

Memory

Ample evidence exists for multiple memory systems in the human brain. *Primary* or *working memory* is defined as the information processing system dedicated to the temporary storage and manipulation of information. Information held in primary memory is lost to conscious awareness if not immediately rehearsed. This system can be assessed by using tasks that measure the amount of information that can be briefly held in short-term store, such as the Digit Span subtest of the WAIS-R, or the rate of forgetting from short-term storage using the Brown-Peterson task.

Secondary or *long-term memory* represents a larger-capacity, more permanent store of newly acquired information that has been consolidated from primary memory. Both recent and remote personal information and historical events are considered part of secondary memory. Tasks that assess secondary memory typically ask the patient to recall or recognize units of information that exceed the capacity of primary memory (i.e., at least 8 or 9 units). Standardized secondary memory tasks include the Wechsler Memory Scale revised test, the Rey Auditory-Verbal Learning Test, the California Verbal Learning Test, the Buschke Selective Reminding Test, the Rey-Osterrieth Complex Figure Test, and the 7/24 Spatial Recall Test.

Language

The assessment of aphasia typically involves an evaluation of spontaneous speech, repetition, comprehension, naming, reading, and writing. Several comprehensive aphasia batteries have been developed over the years, including the Boston Diagnostic Aphasia Examination, the Western Aphasia Battery, the Multilingual Aphasia Examination, the Illinois Test of Psycholinguistic Ability, and the Porch Index of Communicative Ability. Reitan's Aphasia Screening Test provides a less thorough language assessment but can be administered rapidly. More specific language tasks include the Token Test to assess verbal comprehension of commands of increasing difficulty, the Boston Naming Test to assess the ability to name pictured objects, the Peabody Picture Vocabulary Test—Revised to assess auditory comprehension of picture names, and the Controlled Oral Word Association Test to assess spontaneous generation of words beginning with a given letter.

Perception

An assessment of higher perceptual processes is commonly included in a comprehensive neuropsychological evaluation. *Agnosia* is classically defined as a failure to recognize a percept that cannot be accounted for by defects in elementary sensory function. The correct interpretation of perceptual tasks, therefore, cannot be made unless more primary sensory data are collected using screening tests of visual, auditory, tactile, and olfactory function either during the neuropsychological examination (e.g., visual field assessment through bedside confrontation, two-point tactile stimulation) or from specialized diagnostic procedures performed by other specialties (e.g., formal perimetric visual field testing, audiometric testing). Double simultaneous stimulation can frequently identify patients experiencing a mild, residual neglect that may not be observed during unilateral presentations.

The most common tests for assessing visuospatial perception include the Benton Facial Recognition Test, the Benton Line Orientation Test, the Hidden Figures Test, and the Hooper Visual Organization Test. The Halstead-Reitan Test Battery contains several tests of tactile (Finger Localization Test, Tactile Form Perception, Fingertip Number-Writing Perception, Tactual Performance Test) and auditory (Seashore Rhythm Test, Speech-Sounds Perception Test) perception.

Praxis and Motor Dexterity

Apraxia is a disorder of skilled movements that cannot be accounted for by primary motor dysfunction (diminished strength, speed, and coordination), sensory loss, impaired language comprehension, or inattention to commands. Testing for apraxia involves an assessment of the motor system at the highest level of programming. Specialized screening batteries have been developed for assessing limb apraxia and include the production of gestures to command or imitation. Constructional apraxia is assessed by asking patients to copy designs from a two- or three-dimensional model or to generate common objects spontaneously, such as a house, a clock, or a daisy. More commonly used tests include the Bender Gestalt Test, the Rey-Osterrieth Complex Figure Test, the Benton Visual Retention Test, the Block Design subtest of the WAIS-R, and the Benton Test of Three-Dimensional Constructional Praxis. Neuropsychological assessments also include tests of strength (Hand Dynamometer Test), speed (Finger Tapping Test), and coordination (Grooved Pegboard Test, Purdue Pegboard Test) involving the upper extremities.

Academic Achievement

A common complaint of patients is an impairment in their ability to read, spell, or calculate as well as they had before the onset of brain damage. Although academic achievement tests have been developed specifically for assessing developmental learning disorders, they can also be useful in evaluating and understanding acquired cognitive deficits in the academic skill areas. The most commonly used tests in neuropsychological assessments include the Woodcock-Johnson Psycho-Educational Battery–Revised and the Wide Range Achievement Test 3.

Personality and Socioadaptive Functions

Personality change is one of the most common and debilitating symptoms associated with acquired brain disorders. Such changes can occur as a direct result of the brain damage or disease, as a psychological reaction to experiencing a chronic injury or disease, as premorbid personality characteristics, or as a combination of these factors. Common interpersonal problems resulting directly from brain dysfunction include loss of impulse control, insensitivity, emotional lability, irritability, loss of self-awareness, lack of initiative, euphoria, and an inability to profit from experience. Patients with brain damage can also experience emotional reactions to their disability, including anger, anxiety, denial, dependency, repression, and depression.

The development of psychometric instruments for assessing personality change in neurologic disorders has lagged behind the development of instruments for assessing cognition. Commonly used self-report instruments include the Minnesota Multiphasic Personality Inventory, the Beck Depression Inventory, the Zung Depression Scale, the Geriatric Depression Inventory, and the State-Trait Anxiety Inventory. Self-rating scales may lack validity in patients who experience gross changes in personal insight and self-awareness. An alternative method of assessing personality change is to use relative ratings, which can also be compared with self-reported symptoms. One such instrument, the Katz Adjustment Scale, has been used with success in evaluating interpersonal problems of patients with neurologic deficits from the perspective of a close relative or friend.

METHODS OF INTERPRETATION

Neuropsychological assessment typically involves an integration of interview data, medical and school records when available, and neuropsychological test data. Determination of the presence, type, and degree of brain dysfunction is based on several inferential methods. The most common methods include measurement of the patient's level of performance, the appearance of pathognomonic signs of brain dysfunction, testing for lateralized brain dysfunction, and a profile analysis that is consistent with known pathology.

Level of Performance

The most common method for inferring brain impairment from neuropsychological testing is based on a deficit model, which assumes that cognitive impairment has occurred when a discrepancy is observed between the patient's level of test performance and his or her estimated ability level before the onset of brain dysfunction. This method also assumes that the patient's premorbid ability level can be accurately predicted. School records, when available, may yield pertinent data, such as group-administered intellectual and academic achievement test scores. More commonly, the patient's highest level of functioning is inferred indirectly through demographic data (e.g., education, occupation). Clinicians may also use specialized tests of premorbid intellectual ability, such as the National Adult Reading Test.

Pathognomonic Signs

An additional method of inferring cognitive decline is based on detection of qualitative changes that are characteristic of specific types of brain dysfunction, particularly in the acute state. Such changes have an extremely low base rate in the normal population. Examples include the appearance of left-sided neglect, motor preservations, confabulation, and paraphasic speech.

Laterality

The third inferential method involves testing of abilities that imply dysfunction in one cerebral hemisphere relative to the other. Most neuropsychological examinations include measurements of upper extremity motor strength, speed, and coordination as a means of comparing right and left body sides. Assessments of visual, auditory, and tactile perceptual asymmetries are performed for similar purposes. Clinicians also test more complex perceptual and memory processes with verbal and visual test stimuli as a means of inferring lateralized dysfunction.

Profile Analysis

This interpretive approach posits that a neurobehavioral syndrome would demonstrate a fairly specific pattern or profile of deficits on a comprehensive neuropsychological test battery. It is assumed that there is consistency in the expression of cognitive functions within an individual. When test scores are converted to a similar metric (standard scores, T scores, percentiles) and plotted, one can readily observe a patient's strengths and weaknesses.

ISSUES AFFECTING INTERPRETATION

Test Reliability and Validity

Reliability can be assessed by examining a test's consistency over time or its internal consistency (i.e., Do all test items assess the same psychological construct?). Whereas tests with high levels of reliability may be insensitive in detecting brain dysfunction, the converse is not true: Unreliable measures cannot be valid measures of brain dysfunction. An important question for all neuropsychological instruments is whether the test is capable of discriminating between healthy individuals and patients with presumed brain dysfunction (test validity). Although a test may demonstrate a statistically significant group difference based on mean scores, one may also ask whether the instrument has *clinical* utility (i.e., Can the test be used in making discriminations on a case-by-case basis?). Test validity can also involve an evaluation to determine whether the neuropsychological test measures the psychological construct (e.g., recent memory) that it is purported to evaluate.

Normative Standards

Estimates of cognitive decline require the use of normative data to determine the ranking of the patient's test performance relative to an appropriate peer group. Neuropsychological tests are frequently influenced by age, education, race, socioeconomic status, and, less frequently, sex. The effects of these demographic factors can be complex and nonuniform either across or within cognitive domains. The ideal neuropsychological test, therefore, has normative data derived from a large (preferably nationwide), stratified, and randomized standardization sample. Several textbooks provide norms for neuropsychological tests (2, 10–12).

Motivational Factors

The outcome of neuropsychological testing can be influenced by the degree of cooperation and effort put forth by the patient. The validity of neuropsychological test performance may be called into question in patients seeking financial compensation for injuries. Precise prevalence estimates of malingering on neuropsychological testing are unknown. In recent years, neuropsychologists have improved the accuracy of detecting malingering by applying specialized testing procedures.

Affective Disorders

Patients with major affective disorders may exhibit potentially reversible cognitive decline on neuropsychological testing. Depressed patients perform poorly on memory tests requiring high degrees of mental effort, such as on free recall tests, but perform normally on memory tests that are more automatic and require less effort, as in recognition testing formats or incidental learning tasks. Patients with major affective disorders may also be experiencing a form of subcortical dementia, which may correlate to changes on structural and functional neuroimaging. Depression may also lead individuals to overreport or overperceive cognitive and functional difficulties.

Medication Effects

A variety of prescription medications can affect cognitive functioning and thereby alter the interpretation of neuropsychological testing. Examples include antidepressants with significant anticholinergic side effects (e.g., amitriptyline), anticonvulsants (particularly phenobarbital), anxiolytics (particularly the benzodiazepines), and antipsychotics. In addition, various drugs of abuse (e.g., alcohol, stimulants, cocaine, heroine, cannabinol, nicotine) can have short- and long-term effects on cognitive test performance. Typically, these effects are most often observed on measures of attention, memory, information processing speed, and fine motor dexterity.

Sensorimotor Dysfunction

Many neuropsychological tests that assess higher cognitive functions make the assumption that patients have adequate primary sensory and motor functions to perform the task. Many neurologic and nonneurologic conditions produce sensorimotor impairment from disease or trauma, with the greatest impact occurring outside the cerebral hemispheres. Such lesions should have little or no effect on cognitive processes but can reduce performance scores on cognitive tests, particularly those that demand motor speed and dexterity or fine visual acuity. Clinicians attempt to select neuropsychological tests that provide unambiguous test findings.

Practice Effects

Repeated neuropsychological testing results in improvements in test scores in both normal and brain-damaged populations. *Practice effects,* as they have come to be known, can present challenges in test interpretation. They may result from the explicit and conscious recollection of the identical test stimuli, suggesting the use of alternate test forms of equivalent difficulty. Unfortunately, recent findings suggest that practice effects occur as the result of procedural-implicit memory, which may be more resistant to decay over time and is not controlled by alternate test forms.

Fatigue Effects and Duration of Testing

Lack of sleep and fatigue can also decrease attention and concentration and thus secondarily affect various other cognitive functions. Patients may experience fatigue and reduced motivation at the end of a lengthy neuropsychological testing session. These effects may potentially influence the validity of such tests. To counteract this problem, most neuropsychologists administer tests that are most likely to be influenced by fatigue (e.g., attention and memory) at the beginning of the test session.

Alternatively, several investigators have developed screening examinations for quickly assessing cognitive functions at the bedside (see Chapter 1). The most commonly used screening instrument, the Mini-Mental State Examination (MMSE), is brief (5 to 10 minutes), can be administered by a healthcare professional without psychological training, and yields a single score that minimizes interpretative skills. Unfortunately, the MMSE is insensitive to most of the milder forms of cognitive dysfunction or to dementing disorders affecting primarily subcortical structures. The Mattis Dementia Rating Scale (13) addresses many of the criticisms of the MMSE but takes longer to administer and requires greater psychological expertise to administer and interpret.

Preexisting Conditions

Preexisting intellectual weaknesses or learning disabilities may also account for lowered test scores. Research suggests that learning difficulties or disabilities may persist in part, or wholly, into adulthood. Abuse of alcohol or other psychoactive substances also need to be considered. Short of permanent and severe brain dysfunction, alcohol abuse can lead to less obvious but permanent effects or to significant dysfunction that may last for years.

ROLE OF TESTING IN FORENSIC NEUROPSYCHOLOGY

The involvement of neuropsychologists in the legal arena appears to occur most frequently in personal injury and workers' compensation cases. Practitioners do participate in a range of other cases, such as those involving an individual's competence to manage his or her own affairs, criminal responsibility, and child custody. Many cases involve head injury or claims of head injury, or, increasingly, exposure to neurotoxins.

Attorneys may adopt at least five basic strategies for challenging neuropsychological evidence:

1. Raising questions about the expert's credentials
2. Pointing out flaws in the conduct of the examination, and in particular, errors of omission and commission
3. Establishing the presence of bias; errors of omission and commission serve as one potentially potent means for doing so
4. Challenging the underlying scientific bases of the expert's work

5. Showing that the expert's assumptions and conclusions are faulty; for example, the attorney may introduce contrary facts or attempt to establish flaws in the expert's reasoning

A common problem in neuropsychological testimony is the overestimation of prior functioning owing to sole reliance on a plaintiff's self-report. Other bases for overdiagnosis include insufficient appreciation of normal variation in functioning or the fact that most people have at least some cognitive weaknesses, failure to recognize the extent of overlap between normal and abnormal populations, and the use of inappropriate performance or normative standards.

Each of these factors can also work in the opposite direction. Common errors of omission also include the failure to assess for malingering adequately, incomplete consideration of alternative explanations of etiology, and, perhaps most important, failure to obtain and review collateral information. Collateral sources, such as prior school and work records and interviews with people who knew the person well both before and after injury, can help in addressing four issues that are almost always germane in legal cases:

1. Past functioning
2. Current functioning
3. Cause of the presenting difficulties
4. Examiner's honesty and accuracy as an informant

QUALIFICATIONS TO PRACTICE CLINICAL NEUROPSYCHOLOGY

The practice of clinical neuropsychology is fairly new in the healthcare delivery system. Before 1980, training in neuropsychology was limited to a few doctoral programs and clinical internships. It was not uncommon for clinical psychologists in clinical practice to take brief workshops on neuropsychological assessment and, without a comprehensive knowledge of the behavioral neurosciences or clinical neurology, begin assessing patients with neuropsychological test procedures.

Recognizing the need to identify qualified neuropsychological practitioners, the American Board of Professional Psychology in 1983, in conjunction with the newly formed American Board of Clinical Neuropsychology, identified clinical neuropsychology as a specialty area and developed an examination process to evaluate the training, knowledge, and skills of psychologists specializing in neuropsychology. At present, there are more than 300 board-certified clinical neuropsychologists in North America.

SUMMARY

The neuropsychological examination is an important method for understanding the functional abnormalities of patients with a variety of neuropsychiatric disorders. The clinical neuropsychological examination provides evidence to support the existence of acquired brain damage and describes the pattern and severity of cognitive and emotional impairment in a patient with suspected brain dysfunction. With repeat neuropsychological testing, it is possible to assess changes (i.e., worsening, improvement) in performance that may indicate concomitant alterations in brain status. Recent studies have demonstrated that patterns of neuropsychological test performance can be predictive of a patient's everyday functioning (e.g., work, personal hygiene, driving).

REFERENCES

1. Benton A. Evolution of a clinical specialty. *Clin Neuropsychologist* 1987;1:5–8.
2. Lezak MD. *Neuropsychological assessment.* 3rd ed. New York: Oxford University Press, 1995.
3. Kolb B, Whishaw IQ. *Fundamentals of human neuropsychology.* 3rd ed. New York: WH Freeman, 1990.
4. Meier MJ, Benton AL, Diller LD. *Neuropsychological rehabilitation.* New York: Guilford Press, 1987.
5. Wilson BA. *Rehabilitation of memory.* New York: Guilford Press, 1987.
6. Sohlberg MM, Mateer CA. *Introduction to cognitive rehabilitation: theory and practice.* New York: Guilford Publications, 1989.
7. Reitan RM, Davison LA. *Clinical neuropsychology: current status and applications.* New York: Wiley, 1974.
8. Luria AR. *Higher cortical functions in man.* New York: Basic Books, 1966.
9. Milberg WP, Hebben N, Kaplan E. The Boston process

approach to neuropsychological assessment. In: Grant I, Adams KM, eds. *Neuropsychological assessment of neuropsychiatric disorders.* New York: Oxford University Press, 1986:65–86.

10. Heaton RK, Grant I, Matthews CG. *Comprehensive norms for an expanded Halstead-Reitan Battery: demographic corrections, research findings, and clinical applications.* Odessa, FL: Psychological Assessment Resources, 1991.
11. Spreen O, Strauss E. *A compendium of neuropsychological tests: administration, norms, and commentary.* New York: Oxford University Press, 1991.
12. Beardsley JV, Matthews CG, Cleeland CS, Harley JP. *Experimental T-score norms on the Wisconsin Neuropsychological Test Battery.* Madison, WI: University of Wisconsin Center for Health Science, 1978.
13. Mattis S. Mental status examination for organic mental syndrome in the elderly patient. In: Bellak L, Karasu TB, eds. *Geriatric psychiatry.* New York: Grune & Stratton, 1976.

3

Diagnostic Testing

Daniel K. Kido, Yvette I. Sheline, and Alison Reeve

NEUROIMAGING

OVERVIEW

The purpose of noninvasive neuroradiologic imaging techniques, such as computed tomography (CT) and magnetic resonance imaging (MRI), is to display lesions of the central nervous system (CNS) in patients with corresponding signs and symptoms. The purpose of this chapter is to demonstrate how the brain can be segmented into lobes by identifying key sulci and fissures and thus to facilitate the correlation of neurologic changes with specific locations within those lobes. We begin by demonstrating how sulci and fissures present in images of the whole brain can be used to segment it into lobes. In turn, we show how whole brain information about sulci and fissures can be applied to individual MRI and CT slices to segment them into lobes. Finally, we review how MRI and CT scans should be requested from a radiologist if they are to be studied in detail.

INDICATIONS

Neuroimaging must be used in neuropsychiatry as one element in a planned diagnostic process. Decisions concerning neuroimaging should follow the clinical assessment. When particular expense or inconvenience to the patient is entailed, consultation with neuroimaging colleagues is indicated.

EXAMINATION AND ASSESSMENT

Surface Anatomy of the Cerebral Hemispheres

Two-dimensional images of the lateral, medial, and superior surfaces of whole brains are familiar to all physicians. Relating information about gyri and sulci from the whole brain to specific axial CT and MRI scans is difficult, however, because these techniques slice the brain into less familiar planes. This difficulty is most evident on MRI scans taken in the axial and sagittal planes. In an attempt to overcome this barrier, we take a set of sagittal MRI scans and reconstruct them into a set of two-dimensional textbook-type whole brain images (Fig. 3-1). Review of the major sulci and fissures on these images can then be used to segment the brain into lobes.

The images of the whole brain shown in Fig. 3-1 were obtained using a software program called VOXEL-MAN (Siemens, Erlangen, Germany). Similar whole brain images can be produced from MRI scans by using programs such as Analyze (Mayo Foundation for Medical Education and Research, Rochester, MN) and 3D VIEWNIX (Medical Image Processing Group, Philadelphia, PA). Later, in the asymmetry section, these whole brain images are transformed into two-dimensional surface maps to allow comparison of each side of the brain. In the section of this chapter on sectional anatomy, the information regarding the location of sulci and fissures in

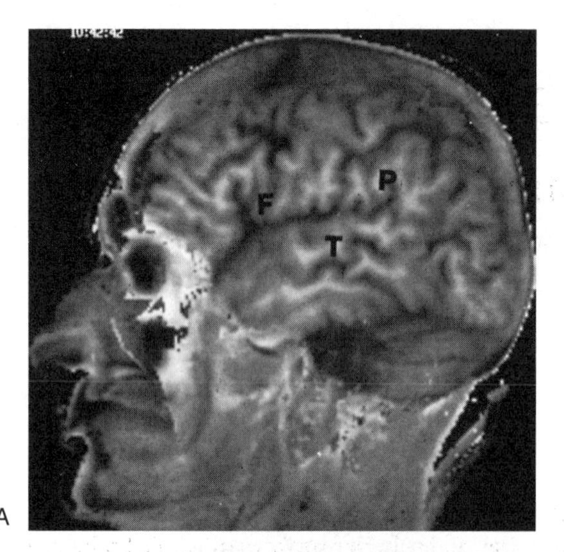

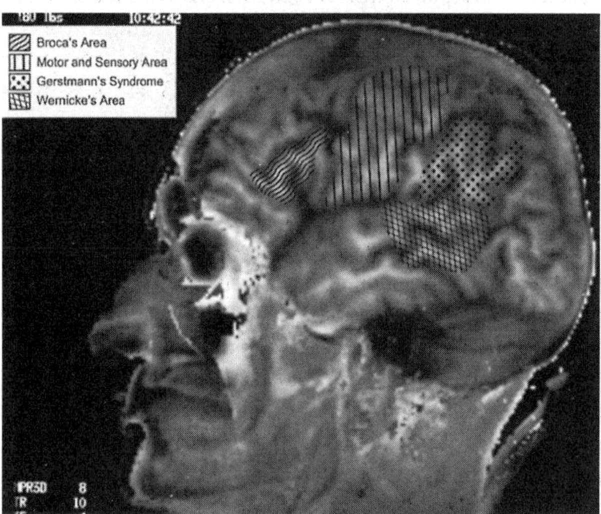

FIG. 3-1. Lateral view of a reformatted whole brain. **A:** The sylvian fissure separates the frontal (F) and parietal (P) lobes from the temporal (T) lobe. **B:** Broca's area *(wavy lines)*, motor and sensory cortex *(vertical lines)*, Wernicke's area *(cross-hatches)*, and inferior parietal lobule *(small circles)* are shown.

the whole brain, which follows, is applied to individual axial, coronal, or sagittal scans of the brain to demonstrate how these slices can be segmented into lobes by using internal landmarks.

Lateral Surface of the Cerebral Hemispheres (Whole Brain)

The most prominent surface marking on the lateral surface of the reformatted whole brain is a deep horizontal depression (sylvian fissure) about two thirds of the way between the top and the bottom of the image (1). The sylvian fissure separates the frontal lobe and anterior portion of the parietal lobe from the anterior two thirds of the temporal lobe (see Fig. 3-1*A*). The sylvian fissure terminates posteriorly in the supramarginal gyrus of the inferior parietal lobule.

Superior to the posterior portion of the sylvian fissure is another horizontal sulcus, the intraparietal sulcus. The intraparietal sulcus is located above both the supramarginal and angular gyri and separates these gyri from the superior portion of the parietal lobe.

Medial Surface of the Cerebral Hemispheres

The posterior boundary of the somatosensory cortex, on the medial surface of each hemisphere, is formed by the marginal ramus of the cingulate sulcus (Fig. 3-2). The marginal ramus is formed when the cingulate sulcus terminates above the body of the corpus callosum and divides into the horizontally directed subparietal sulcus and the coronally oriented marginal ramus of the cingulate sulcus. The cingulate sulcus forms an arch around the corpus callosum.

The marginal ramus forms not only the posterior border of the somatosensory cortex (the paracentral lobule) but also the anterior border of the precuneus. The precuneus is the medial surface of the parietal lobe, posterior to the sensory cortex. The inferior surface of the precuneus is outlined by the horizontally oriented extension of the cingulate sulcus, whereas its posterior border is formed by the parieto-occipital fissure (see Fig. 3-2). The parieto-occipital fissure forms the anterior border of the cuneus (i.e., occipital lobe). The parieto-occipital fissure joins the more inferiorly located calcarine sulcus behind the splenium of the corpus callosum.

Superior Surface of the Cerebral Hemisphere (Whole Brain)

Sectional Anatomy

Axial. The description of sulci and fissures in the previous section on the whole brain can be used to divide individual axial scans into lobes (segments). Four axial scans are described in detail to illustrate how these axial scans can be segmented. Axial sections close to the vertex can be almost directly correlated to the superior view of the whole brain. Thus, information regarding sulci in the frontal lobe can be used to segment axial slices above the lateral ventricle into a frontal and a parietal lobe. On an axial slice close to the vertex, the superior frontal sulcus can be traced backward to its junction with the precentral sulcus, as it can be on the whole brain images (Fig. 3-3). In turn, the central sulcus is located just behind the precentral sulcus. The superior frontal sulcus is sagittally oriented, whereas the precentral and central sulci are coronally oriented. The central sulcus is identifiable because gray matter in the motor cortex is two times the thickness of gray matter in the sensory cortex. The full course of the central sulcus can be identified by tracing the central

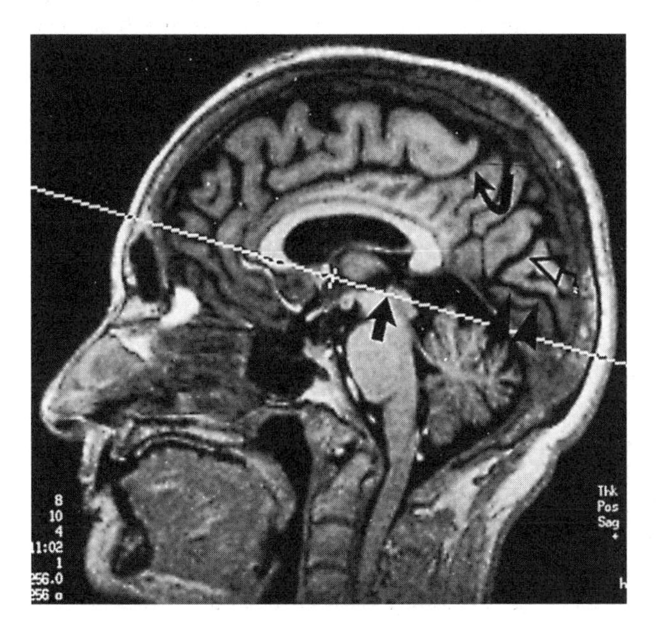

FIG. 3-2. Sagittal section through the medial surface of a brain. A horizontal line is drawn through the bicommissural plane. A small vertical line is located just behind the anterior commissure. Posterior commissure *(arrow)*, marginal ramus of the cingulate sulcus *(curved arrow)*, parieto-occipital fissure *(open arrow)*, and calcarine sulcus *(arrowheads)* are also shown.

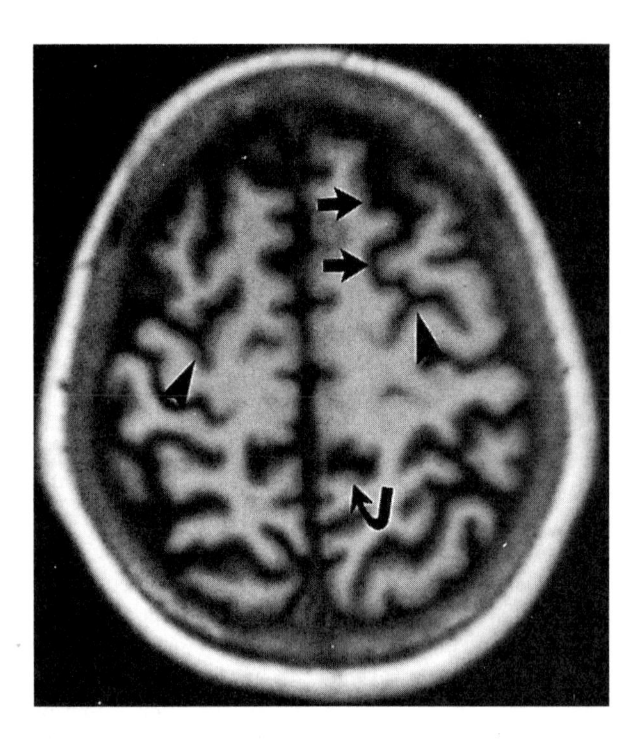

FIG. 3-3. Vertex scan (axial). The left superior frontal sulcus *(arrow)* is better visualized than the right. However, the junction between the superior frontal and precentral sulcus *(arrowheads)* can be identified bilaterally. The marginal ramus of the cingulate sulcus *(curved arrow)* is also shown.

sulcus from slice to slice until it terminates just above the sylvian fissure.

On a slightly more inferior slice through the centrum ovale, the parieto-occipital fissure is located on either side of the interhemispheric fissure behind the marginal ramus of the cingulate sulcus (Fig. 3-4). By extending a horizontal line from the parieto-occipital fissure to the lateral surface of the brain, the parietal lobe can be arbitrarily segmented from the occipital lobe.

Further inferiorly, on an axial slice through the posterior portion of the sylvian fissure, the frontal and parietal lobes can be separated from the temporal lobe by identifying Heschl's gyrus on the superior surface of the temporal lobe. Identifying Heschl's gyrus on the superior surface of the temporal lobe permits separation of the temporal lobe from the more superiorly located frontal and parietal lobes (Fig. 3-5). Heschl's gyrus is located about at the junction between the middle third and the posterior third of the sylvian fissure and differs from other gyri by being oblique instead of perpendicular to the surface of the brain. This plane is also important because it contains the inferior portion of the subcortical nuclei of the basal ganglia as well as the thalamus (see reference 2 for an atlas resource).

Sagittal. The lobar boundaries described in a previous section (Medial Surface of the Cerebral Hemispheres) are visible in the midline sagittal section. In the midline, a vertical line perpendicular to the anterior commissure runs just behind the anterior border of the paracentral lobule (see Fig. 3-2). Further inferiorly beneath the bicommissural plane, the vertical line runs just behind the anterior wall of the third ventricle (lamina terminalis). A vertical line perpendicular to the posterior commissure runs just behind the posterior border of the paracentral lobule.

The paramedian sections also clearly demonstrate the third ventricle and the structures associated with it. The anterior commissure is located at the superior end of the lamina terminalis at its junction with the rostrum of the corpus callosum. The anterior commissure is thus located anterior to Monro's foramen and the columns of the fornix. The posterior commissure is located in front of and

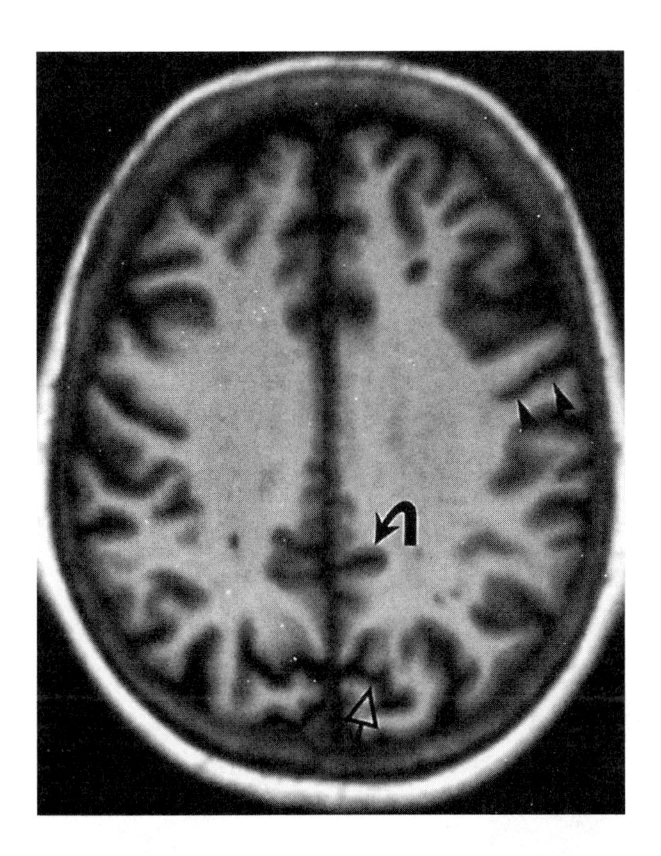

FIG. 3-4. Centrum ovale scan (axial). The marginal ramus of the cingulate sulcus *(curved arrow)* and parieto-occipital fissure *(open arrow)* are located on either side of the interhemispheric fissure. The central sulcus *(arrowheads)* is also shown.

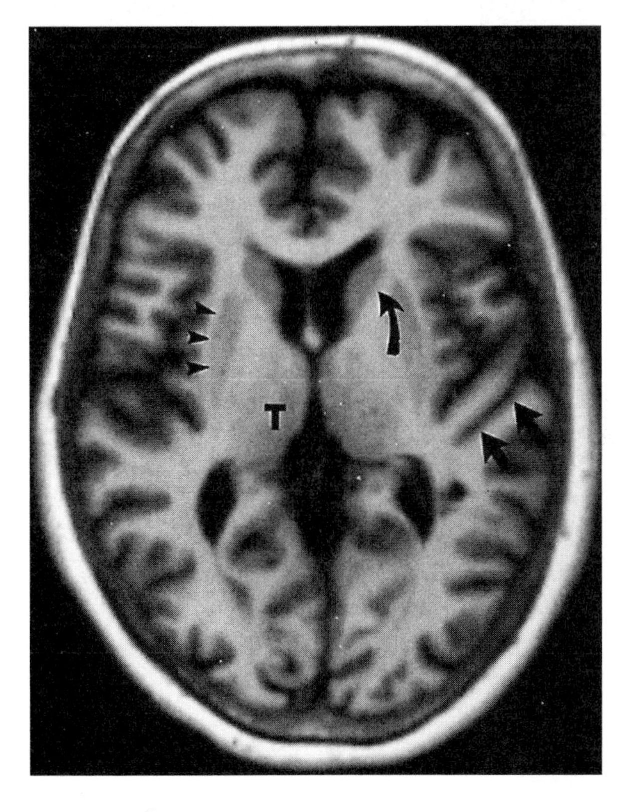

FIG. 3-5. Superior temporal lobe scan (axial). Heschl's gyrus *(arrows)* is located within the sylvan fissure. The head of caudate nucleus *(curved arrow)*, putamen *(arrowheads)*, and thalamus (T) are also shown.

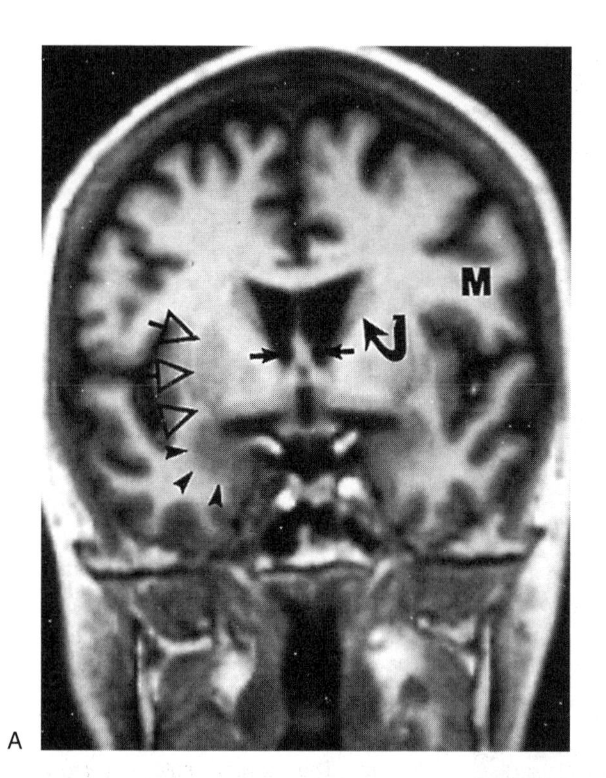

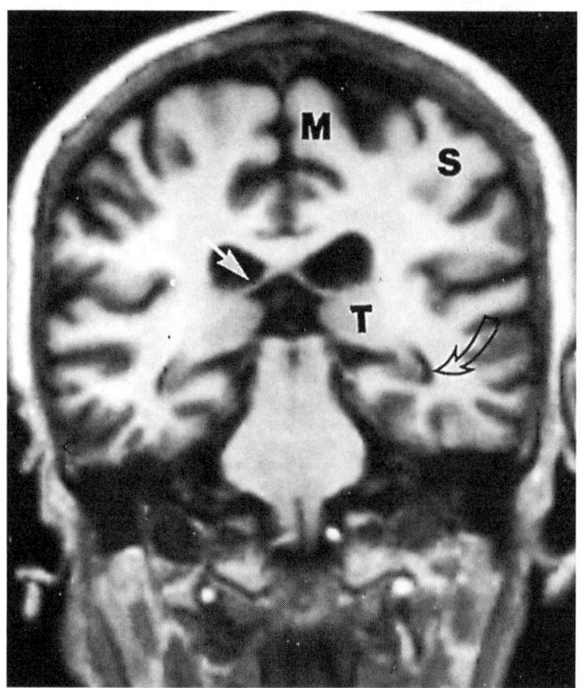

FIG. 3-6. Coronal scans through the anterior **(A)** and the posterior **(B)** commissures. Shown are the precentral gyrus (M); postcentral gyrus (S); amygdala *(arrowheads)*; fornix *(arrows)*; head of caudate nucleus *(curved arrow)*; putamen *(open arrows)*; thalamus (T); and hippocampus *(open curved arrow)* medial to the temporal horn.

beneath the pineal body. Superior to the roof of the third ventricle and frontal horns of the lateral ventricle is the corpus callosum. The cingulate gyrus arches around the periphery of the corpus callosum.

Coronal. Coronal scans between the anterior and posterior commissures show the central sulcus. The inferior end of the central sulcus is located close to the bottom of the coronal scan taken through the anterior commissure, whereas the superior end is located close to the top of the coronal scan taken through the posterior commissure (Fig. 3-6). Thus, as sections progress posteriorly from the anterior to the posterior commissure, the motor cortex moves progressively superiorly from the sylvian fissure to the vertex of the brain.

Coronal images through the anterior commissure but inferior to the bicommissural plane contain the tip of the amygdala (see Fig. 3-6), which extends posteriorly for about 1 cm. Coronal images through the posterior commissure, again inferior to the bicommissural plane, contain the posterior portion of the body of the hippocampus where it begins to diverge (see Fig. 3-6*B*). The hippocampus is easiest to examine just anterior to the plane through the posterior commissure because it has not begun to diverge or become redundant, as it is at its tip (pes hippocampus) (Fig. 3-7).

The hippocampus can be located on coronal scans by following the medial surface of the temporal lobe (parahippocampal gyrus) superiorly until it meets the hippocampal sulcus. The portion of parahippocampal gyrus that forms the inferior lip of hippocampal sulcus is called the *subiculum* (see Fig. 3-7). The superior lip is the *dentate gyrus.* The subicu-

A
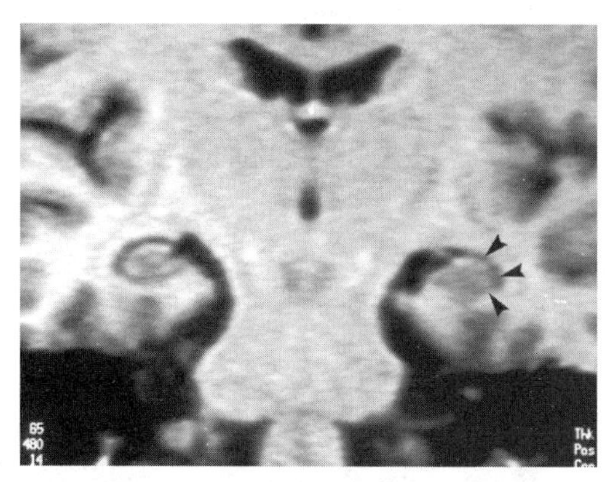

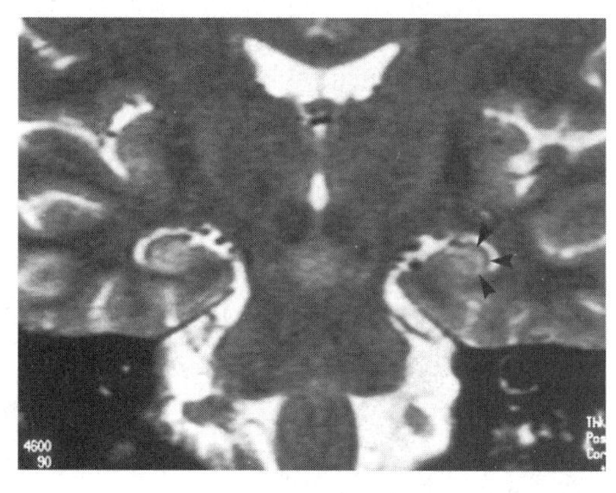
B

C
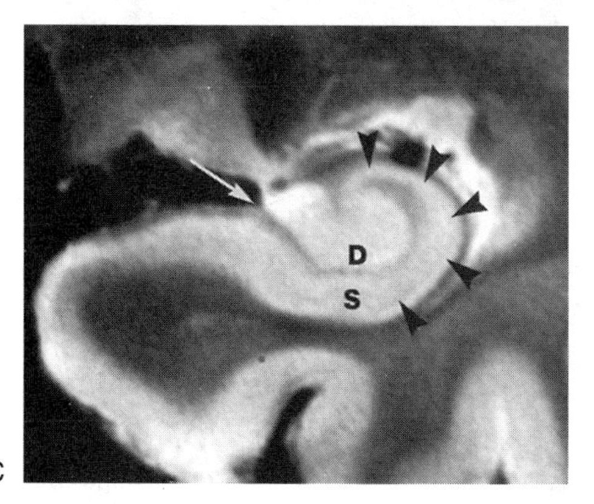

FIG. 3-7. High-resolution coronal image of the hippocampus *(arrowheads)* taken between the anterior and posterior commissures. **A:** T1-weighted gradient-echo image. TR480/TE14; matrix = 144/512; FOV = 24 × 18 cm; slice thickness = 4 mm; FA is 65 degrees; 8 spin-echo acq. **B:** T2-weighted fast spin-echo image taken in a plane corresponding to **A**. TR460/TE90; matrix = 192 × 512; FOV = 26 × 20 cm; slice thickness = 4 mm; 4 spin-echo acq. **C:** Proton density image of a brain specimen (fixed in formaldehyde) through roughly the corresponding left hippocampus. The subiculum (S) and dentate (D) gyri can be separated from each other by the hippocampal fissure *(arrow)*. The hippocampal formation *(arrowheads)* is also shown. TR1500/TE 19; matrix = 256 × 512; slice thickness = 2 mm; 7 acq.

lum is separated from the dentate gyrus located above it by the hippocampal sulcus. The subiculum continues superiorly and medially in a reverse C configuration as the hippocampal formation (Ammon's horn). The subiculum and dentate gyrus differ from the rest of the cortex by having only three layers instead of six. Superior and lateral to the hippocampal formation is the alveus, which contains efferent fibers from the hippocampal region. The outflow fibers from the alveus (fimbria) form a bump along the superior medial border of the temporal lobe. The fimbria merges posteriorly with the fornix.

White Matter

The white matter deep to the gyri on the axial vertex and centrum ovale sections can also be used to locate the precentral and postcentral gyri; this is an alternate way of separating the frontal and parietal lobes. The white matter medullary pattern can be divided into the following six divisions from anterior to posterior: superior frontal gyrus, middle frontal gyrus, precentral gyrus, postcentral gyrus, inferior parietal lobule, and superior parietal lobule (see Fig. 3-4). On the vertex view, the middle frontal gyrus and inferior parietal lobule may not always be visible, whereas the remaining four divisions are always present (see Fig. 3-3). This method of identifying the central sulcus is especially useful in sections through the lower edge of the centrum ovale, where the central sulcus is more shallow and thus more difficult to identify, and in young children, in whom gyri and sulci may be difficult to identify.

The corpus callosum is the main interhemispheric commissure and connects the white matter, which was described in the previous paragraph. The corpus callosum is divided into four sections from anterior to posterior: the genu, which connects the frontal lobes; the body, which connects the motor and somatosensory cortices, as well as the temporal lobes; and the splenium, which connects the occipital lobes (Fig. 3-8). The fourth division is the rostrum, which connects the genu to the lamina terminalis. Linear measurements have been made throughout the genu, body, and splenium of the corpus callosum in the sagittal plane. The average width of the genu is 7.9 mm, the body 3.5 mm, and the splenium 8.4 mm.

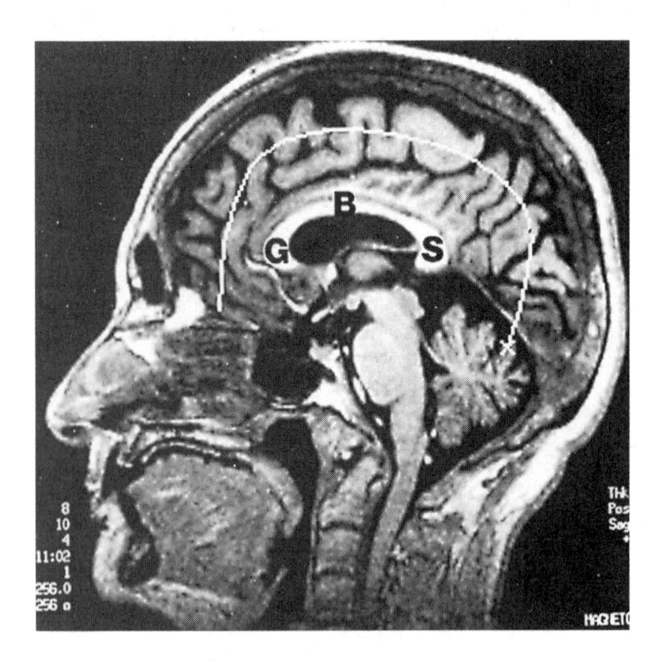

FIG. 3-8. Sagittal section through the midline of the brain (same as Fig. 3-2). The genu (G), body (B), and splenium (S) of the corpus callosum are also shown.

Several smaller interhemispheric commissures exist on the bicommissural plane: The anterior commissure connects the olfactory bulbs and the inferior and middle temporal gyri; the posterior commissure connects the thalami and some structures in the midbrain (see Fig. 3-2). The most unusual interhemispheric connection is the fornix. The body of the fornix is located just beneath the body of the corpus callosum. Interhemispheric commissural fibers cross the midline to connect the bodies of the fornix (see Fig. 3-6*B*). The medial part of the body of the fornix is connected to the corpus callosum close to the midline and the lateral part to the superior surface of the thalamus. The body of the fornix in this region thus forms part of the floor of the lateral ventricles. The anterior extension of the fornix (columns of the fornix) descends just anterior to Monro's foramen and connects it to the mammillary bodies (see Fig. 3-6*A*). The posteroinferior extension of fornix, called the *crus* of the fornix, connects the body of the fornix to the fimbria. The crus of the fornix is present on the coronal images taken behind the posterior commissure (see reference 3 for an overview of white matter pathways and their imaging).

Deep (Subcortical) Nuclei

The high contrast that MRI provides between gray and white matter permits direct visualization of many deep nuclei. In the axial plane through the posterior sylvian fissure, the subcortical nuclei in the basal ganglia (caudate nucleus, putamen, and globus pallidus) and the thalamus are imaged (see Fig. 3-5). The caudate nucleus can be separated from the putamen and globus pallidus because of the intervening V-shaped internal capsule (white matter). The caudate nucleus is located medial to the anterior limb of the internal capsule, and the thalamus is located medial to the posterior limb of the internal capsule. In contrast, the putamen and globus pallidus lie lateral to the internal capsule. The putamen, which is located lateral to the globus pallidus, can also be differentiated from the globus pallidus by the amount of paramagnetic material it contains. The globus pallidus contains more paramagnetic material than the putamen, and thus appears darker on T2-weighted images.

The thalamus extends anteriorly from the interventricular foramen posteriorly to the posterior commissure (see Fig. 3-5). It is bordered laterally by the posterior limb of the internal capsule and medially by the third ventricle. The myelinated fibers (internal medullary lamina), which subdivide the thalamus into an anterior, a medial, and a lateral group, are not currently visible by MRI.

The amygdala, another subcortical gray matter collection, is located in the temporal lobe just below the bicommissural plane (see Fig. 3-6*A*). The amygdala is located superior and medial to the tip of the temporal horn and just lateral to the uncus. It is a large oval mass that is difficult to measure because its lateral borders merge into the adjacent white matter.

Lateral Ventricles

The lateral ventricles are divided from anterior to posterior into the frontal horns, body (cella media), atrium (trigone), and occipital horns. The temporal horns are connected to the body of the lateral ventricles and occipital horns by the atria. These structures have been measured in a variety of ways to determine indirectly when cerebral atrophy exists. Ventricular casts of the lateral ventricles indicate that the ventricular volume is about 16 mL, whereas the volume of the third ventricle is less than 2 mL. Ventricular volumes greater than 30 mL suggest that there is generalized cerebral atrophy. The size of the ventricular system has been related to the brain by an index termed the *ventricular—-brain ratio*.

Enlargement of the width of the frontal horns (more than 3.5 cm) reflects loss of brain parenchyma in the frontal lobes. This measurement has been refined by relating it to the maximal width of either the brain or the cranial cavity (ratio, 0.16 to 0.29). Maximum frontal horn ratios greater than 40% or less than 18% suggest abnormal ventricles. Enlargement of the maximal width of the cella media of the lateral ventricles reflects loss of brain tissue in the paracentral lobule, precentral gyrus, and postcentral gyrus. The maximal width of the cella media has been related to the width of the cranium (cella media ratio, 0.29). Measurements of the atria and occipital horns are difficult to reproduce because of their variability (see reference 4 for an early review of ventricular measurements on CT imaging).

Functional Imaging

Nuclear Studies. Functional studies that examine regional cerebral blood flow and cerebral metabolism have been helpful in identifying patients with psychiatric illnesses. Positron emission tomography (PET) has been used to study cerebral glucose metabolism after the injection of [^{18}F] fluoro-2-deoxy-d-glucose (FDG). Regional cerebral blood flow has been studied by PET after the injection of oxygen $^{-}$15. Similar information has been obtained with single-photon emission computed tomography (SPECT) after the injection of ^{123}I and ^{99}Tc tracers.

Magnetic Resonance Imaging. In addition to detailed information about brain anatomy, MRI can now provide information about brain function. Neuronal activity that induces small changes in brain microvasculature can be detected with the appropriate imaging sequences and hardware. It is believed that the most common functional technique (blood-oxygen level–dependent effect) relies on a difference in the amount of paramagnetic oxyhemoglobin in the microvasculature at the activated site, which in turn leads to a change in T2 signal intensity. Studies measure signal difference between task performance and resting state. To date, it has only been possible to detect large changes in brain activity, such as those induced in the primary visual cortex by photic stimulation or in the primary motor cortex by hand or finger squeeze (Fig. 3-9). Other studies are also being performed to investigate the brain's response to auditory and sensory stimulation. In functional MRI experiments, just as in PET or SPECT, the observed changes occur with a latency of several seconds, reflecting a time delay between the neuronal activity and the measured MRI response. MRI changes are thus a derivative of net neuronal activity rather than a direct measure of neuronal activity, which operates in the millisecond range.

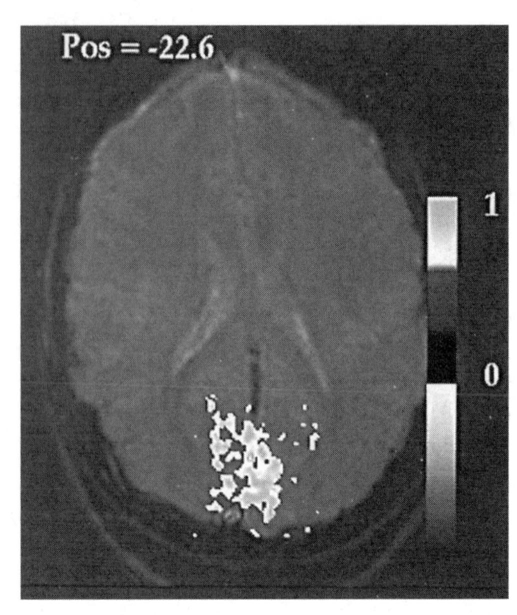

FIG. 3-9. Functional magnetic resonance imaging changes in the occipital lobe during visual stimulation.

The new imaging techniques and equipment used to obtain functional MRI scans have also been used to study cerebral blood flow (perfusion) and blood volume noninvasively. Determining the reserve in these parameters, through the administration of acetazolamide, could be helpful in evaluating patients with stroke as well as dementia. Cerebral blood flow and volume can also be examined after injection of several MRI contrast agents by dynamically studying the susceptibility effect of these agents as they pass through the brain. The advantage of using a contrast agent is that it can potentially produce greater signal changes than can the noncontrast techniques. Unfortunately, however, because the contrast agent must be injected, such experiments can be repeated only a limited number of times.

Diffusion-weighted MRI has been used to examine regional shifts in water content after brain injury and cerebral ischemia. To date, there are no published accounts of diffusion-weighted imaging being performed in patients with psychiatric illnesses (see reference 5 for a description of principles underlying contrast infusion imaging).

Magnetic resonance spectroscopy (MRS) is a technique that can be used to quantify biochemical differences in specific brain regions because the different chemical forms of an element create peaks at characteristic frequencies. Proton MRS measures neurons through the *N*-acetylaspartate (NAA) peak (Fig. 3-10). Additional resonances that are frequently observed at 1.5 T include glutamine/glutamate (Glu/Glx), creatine/phosphocreatine (Cr), choline, (Cho), and myoinositol (myo-Ins). Under pathologic conditions, lactate may also be detected. Phosphorus MRS measures the energy metabolites phosphocreatine (PCr), inorganic phosphate (Pi), and adenosine triphosphate (ATP) as well as phosphomonoesters (AMP) and phosphodiesters (ADP). Like functional MRI, the strength of MRS is its noninvasive nature with no known side effects. The major limitations of MRS performed at the relatively low field strength used in clinical scanners are the poor separation between the metabolite peaks and the poor signal-to-noise ratio compared with those obtained with high field strength laboratory instruments.

In proton MRS, the NAA peak is believed to be an index of neuronal health; a decrease in the NAA-to-Cr ratio indicates either neuronal dysfunction or its loss within a volume. A decrease in the NAA-to-Cr ratio has been observed in demented patients with Alzheimer's disease and in human immunodeficiency virus (HIV)-positive patients with dementia. In Alzheimer's disease, the decrease occurs initially in the temporal lobes and then generally throughout the entire brain. In schizophrenic patients, a decrease in

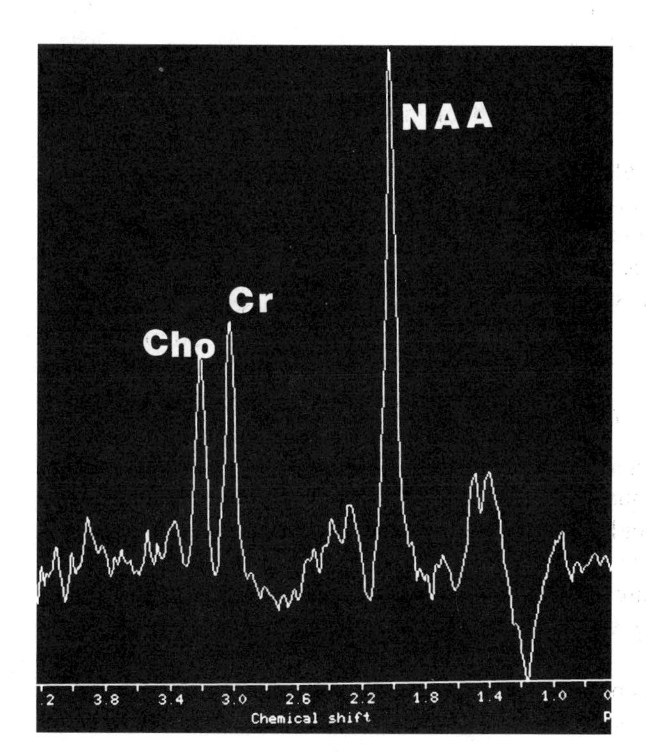

FIG. 3-10. Normal proton spectroscopy. Cho, choline; Cr, creatine/phosphocreatine; NAA, *N*-acetylaspartate.

the NAA-to-Cr ratio has been observed in the temporal and frontal lobes. Patients with major depression have demonstrated elevated choline peaks in the brain. Similar changes have been demonstrated in the basal ganglia not only of depressed patients but also of those with bipolar disorder in the depressed state. The pharmacokinetics of psychotropic agents can be investigated in vivo using ^{19}F MRS. The naturally occurring lithium isotope ^{7}Li is MRS sensitive and can be used to compare serum and brain lithium.

CLINICAL ISSUES

Intraparietal Sulcus

Lesions to this area of the brain in the dominant hemisphere may result in Gerstmann's syndrome: digital agnosia, impaired ability to calculate and write, and right-left disorientation.

Hippocampus

Decrease in hippocampal volume is frequently present in patients who have Alzheimer's disease. In patients with schizophrenia, this decrease in size occurs particularly on the left, where there is associated decrease in size of the amygdala and parahippocampal gyri. Atrophy and gliosis have been reported in some patients with psychomotor seizures. Recently, a right-sided hippocampal volume decrease has been reported in patients with posttraumatic stress disorder.

Subcortical White Matter Hyperintensities

White matter hyperintensities (unidentified bright objects, leukoaraiosis) occur frequently in middle aged and older adults but should be regarded as abnormal in people younger than 45 years. These focal hyperintensities are present on proton and T2-weighted images but not on T1-weighted images.

Although the significance of these lesions is unknown, their frequency and severity appears to increase in patients with cardiovascular risk factors and cerebrovascular symptoms. Gerard and Weisberg reported that patients with cardiovascular risk factors have four times the incidence of white matter hyperintensities (6). Furthermore, patients with both cerebrovascular risk factors and cerebral symptoms have a 10 times higher incidence of white matter hyperintensities. Similar lesions can also be seen in patients with multiple sclerosis (MS), radiation necrosis, vasculitis, or multiple infarcts. White matter lesions have been reported in higher numbers of psychiatric patients compared with healthy controls, and particularly in older patients with depression.

Caudate

The width of the caudate nucleus has been measured on axial scans. Widths of less than 9 mm suggest that there is atrophy, as in Huntington's disease. Similarly, putamen widths of less than 10 mm indicate that they may also be atrophic. Volumetric measurements should be able to detect atrophy in these structures more accurately.

Basal Ganglia

Although the nuclei in the basal ganglia are primarily associated with the extrapyramidal system, they are also important to intellectual function. Dysfunction of these nuclei, as reflected in their atrophy or the abnormal deposition of paramagnetic material, may result in slowing of mental processes, incoordination, inability to solve complex problems, and memory failure, as seen in patients with Parkinson's disease and Huntington's disease. Cognitive impairment of this nature has led to the concept of *subcortical dementia* (dementia without aphasia and apraxia) (see reference 7 for a review of basal ganglia functional issues).

Thalamus

Lesions in the posterior portion of the lateral nucleus can result in contralateral hemianesthesia, transient hemiparesis, choreathetosis, and pain, as well as mild hemiataxia and astereognosis (Dejerine-Roussy syndrome). Neuropsychological dysfunction is common in the presence of lesions affecting the anterior subdivision, whereas a change in consciousness, amnesia, and even confabulation can result from a lesion in the dorsomedial group.

Amygdala

The amygdala is important because it is a key structure in the limbic system associated with emotional control.

Ventricular Enlargement

Enlargement of the ventricles has been demonstrated in schizophrenic patients in both CT and MRI studies. Most studies have found ventricular enlargement to occur early in the course of schizophrenia and have found the degree of enlargement to correlate with severity of symptoms. Some studies have also found an increase in the ventricular–brain ratio in patients with depression, particularly in late-life depression, when there may be an association with dementing illnesses. Evidence for ventricular enlargement in bipolar illness is less conclusive.

Functional Imaging

Functional studies are helpful in differentiating the various forms of dementia. Patients with Alzheimer's disease, the most common dementing illness, have temporal-parietal cortex hypometabolism, which is correlated with behavioral deficits and clinical course, demonstrated on PET and SPECT studies. In contrast, patients with Pick's disease have predominantly frontal hypometabolism, and patients with Huntington's disease and Parkinson's disease have striatal hypometabolism, although depressed Parkinson's patients may also have frontal hypometabolism. Another characteristic that can be used to distinguish the dementias is the presence of focal changes. Multiinfarct dementia can be distinguished from Alzheimer's disease by the presence of asymmetric, scattered perfusion defects.

Other psychiatric disorders also show regional changes in metabolism relative to healthy controls. Schizophrenic patients, for example, have often been shown to have frontal hypometabolism. Although this was initially thought to be confounded by neuroleptic treatment effects, subsequent investigations in never-medicated schizophrenic patients indicated a difference in metabolic rate independent of medication exposure. In addition, decreases in striatal perfusion and increases in medial temporal lobe perfusion have been demonstrated.

Patients with unipolar major depression and those with bipolar depression in the depressed or mixed state have also demonstrated frontal hypometabolism. When bipolar depressed patients become manic, this metabolic rate increases. Depressed patients also appear to have increased limbic metabolism, as well as frontal hypometabolism, perhaps reflecting the involvement of the functional circuit in which these structures are encompassed.

Magnetic Resonance Spectroscopy

The number of MRS studies being performed on neuropsychiatric patients is increasing. Schizophrenic patients have been found to have a decrease in phosphomonoesters (AMP) in the frontal lobes, presumably reflecting alterations in cell membrane phospholipid metabolism. Changes in AMP and ADP have also been reported in patients with dementia of the Alzheimer's type and in HIV-positive dementia patients.

Summary

MRI and other forms of functional imaging are in phases of rapid scientific evolution. Refinements of current clinical applications will be continuous, and new advances are certain. These basic imaging tools are likely to remain part of neuropsychiatric diagnosis in the coming years.

CLINICAL NEUROPHYSIOLOGY

Overview

Electrophysiologic techniques have contributed steadily over the years to our understanding of neuropsychiatric diseases. It is likely that technologically based modifications of basic electrophysiologic tools, such as electroencephalography (EEG), will make new diagnostic contributions in the future. In this section, we review some of the basic aspects, as well as clinical applications of EEG and evoked responses.

Examination and Assessment

Electroencephalography

Electroencephalography (EEG) is a standardized method of recording the electrical activity of the brain at the scalp.

The EEG measures a summation of electrical activity, as recorded as a difference in electrical potential between two active recording electrodes. It is not a direct measure of brain electrical activity. Rather, brain activity is measured indirectly as a difference between electrical potentials received at two recording electrodes. The choice of placement of electrodes and the properties of material between the source (the brain) and the electrodes affect the type, intensity, and specificity of the activity measured.

Factors That Influence the Electroencephalographic Record

The electrical signals of the neurons of the brain are relatively weak. Therefore, the signal coming from the electrode must be amplified. Amplifiers are used to increase the voltage difference between the input and output. They are characterized in terms of sensitivity and gain.

Sensitivity refers to the ratio of the input voltage to the output pen deflections in the standard EEG recording. A typical ratio is 7 μV/mm. A higher numeric value for sensitivity reflects less amplification; that is, a sensitivity of 10 indicates 10 μV/mm. EEG machines have an array of amplifiers set up to allow individual sensitivity adjustment for each channel or general adjustment for all channels.

Gain is the ratio of the voltage obtained at the output of the amplifier to the voltage applied at the input. Often described in terms of volts and microvolts, gain is sometimes expressed in terms of decibels, or 20 times the logarithm in base 10 of the gain. For example, a gain of 10 equals 20 dB; a gain of 1,000,000 equals 120 dB. In clinical EEG, sensitivity is usually used to describe the amplification because the gain is not directly measured from the output of the amplifier.

Electrical signals coming through a recording channel have a characteristic frequency. The band width of the signal is defined by the frequency range within which the signal is contained. Some information will be lost if the frequency range of the recording channels is narrower than the frequency range of the EEG signal. Likewise, extraneous information will be included if the frequency range of the recording channels is wider than the band width of the EEG signal. Different types of filters are used to maximize the EEG signal and minimize noise and signals of noncerebral origin.

In practice, EEG recording channels can have high-pass, low-pass, or band-pass filters. High-pass filters are designed to let higher frequencies pass through and to block lower frequencies, such as those below 1 Hz. Low-pass filters are designed to let low-frequency activity through and to filter high-frequency activity. Band-pass filters are designed around a range of frequency to filter on both the high and low ends of the frequency spectrum.

Many filters induce a phase difference between the output and input signals. This phase difference can create a time difference between different tracings. The choice of filters affects the recording achieved. When the output is to a paper record, the timing of signals may be crucial in interpretation of the clinical data. In a computerized EEG data collection system, it may be more important to allow more "noise" through if necessary to preserve phase relationships.

Montage and Reference Electrodes

Recordings over the head are usually done following the international 10-20 system. This localizes the electrodes proportionately (10% and 20% spacings) between certain landmarks: inion, nasion, and preauricular points (Fig. 3-11). By convention, the electrodes over the left side of the head are numbered odd, and those on the right side of the head are even. Z stands for zero or midline. The letters of the electrode placement reflect relative position over the head: Fp (frontopolar), F (frontal), C (central), P (parietal), T (temporal), O (occipital), A (auricular), M (mastoid).

The number of electrodes applied varies depending on the *montage* (series or patterns in which the electrodes are connected to each other), the population being studied, the type of activity considered important to detect, and the number of channels and amplifiers available. The American EEG Society recommends that the full 21 electrodes of the 10-20 system be used and that a record be obtained in both bipolar and referential montages. Standard clinical laboratories may have from 8 to 32 channel EEG machines with which to make recordings. The number of montages recorded differs depending on the number of channels available to make a simultaneous recording.

The montages used in EEG are of two general types. The first is *referential,* the second *bipolar.* A common form of referential montage is to have a series of electrodes attached to the same common reference, often the earlobe or mastoid. The printout can then be alternated between left and right, which aids the detection of hemispheric asymmetry. Another form of commonly used reference is that of the linked ears. It was thought that this would make a neutral reference; however, the ear lobe tends to be easily "contaminated" by electrical activity of the heart. One attempt to avoid this difficulty involves putting one electrode over the right sternoclavicular junction and another on the spine of the first thoracic vertebrae. These two electrodes are connected together through a 20,000 μ-variable resistor that can change in its resistance, such that the electrocardiogram is canceled or essentially canceled. A different type of reference is a common average reference, created by a computerized analysis system, after the recording is completed. These various attempts to make a neutral reference must always be checked carefully for artifacts. In addition, it is important that the details by which the reference was chosen are carefully documented.

Bipolar montages are particularly useful to compare the polarity of signals and to detect the so-called phase reversal that occurs around an area of high-amplitude signal change. To be useful, bipolar montages should cover relatively large areas of the head so that interpretation of the direction (or polarity) can be made in a reasonable manner. There is a general rule that higher voltages are seen as the interelectrode distance increases.

Some typical EEG tracings and standard frequency activities are listed in Table 3-1. The disorders or states that particularly elicit a frequency are not exclusive but rather characteristic for a frequency type.

The type of disorder being studied influences the recording strategy. Intermittent, infrequent activity is the most difficult to catch. At times, correlating patient behavior with changes in EEG activity is essential. Simultaneous video and EEG recording is one way to differentiate clinical and electrical events that are occurring in relation to one another from events that are occurring completely independently from each other. The video camera can be directed to encompass whole body activity or to focus on certain areas of interest. Portable systems allow the equipment to be set up in any hospital room and can be used with scalp electrodes or implanted electrodes and with cabled or radiotransmitted connections to amplifiers.

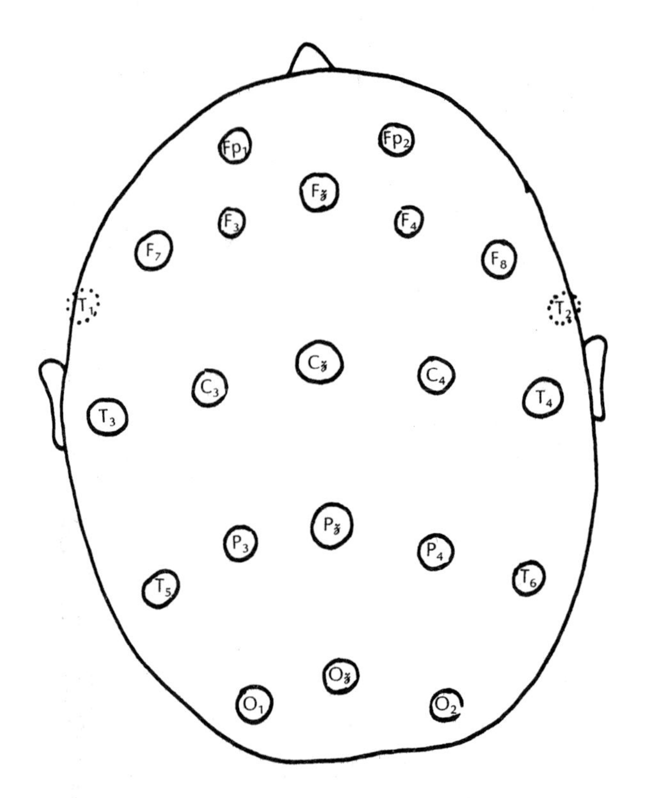

Fp: fronto-polar
F: frontal
C: central
P: parietal
T: temporal
O: occipital
ODD: left side of midline
z: midline
EVEN: right side of midline

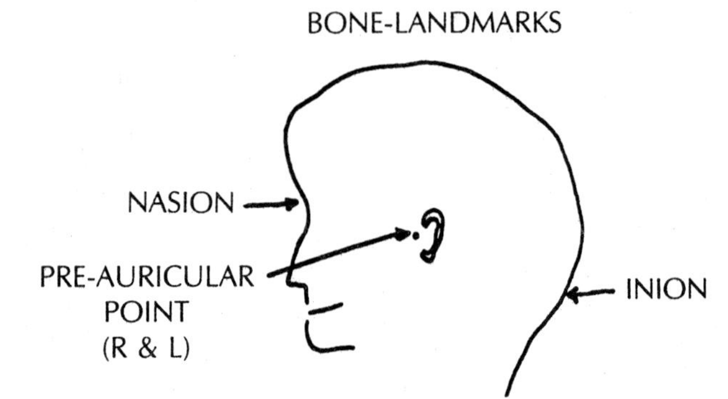

This system is a proportionate placement of electrodes (at 10% and 20% increments) between anterior-posterior/right-left landmarks.

FIG. 3-11. International 10-20 system of electrode placement.

Other forms of telemetry are useful for long-term monitoring of EEG activity during the patient's normal daily routines. Here, the video aspect is not so important. The purpose of these recordings is to obtain long-term records of infrequently occurring activity (such as sharp waves or spikes). Patients are asked to keep a diary of their activity, especially noting any sensory changes that occur and the time of day and duration of symptoms. Patients who are candidates for surgical removal of an epileptic focus need to be carefully assessed for the stability of the focus and the degree of disability it is producing.

TABLE 3-1. *Electroencephalograph Frequency Band and Its Characteristics*

Name	Frequency (H_2)	Location	Factors
β	13–25	Frontal	Increased by benzodiazepines and barbiturates
α	8–12	Occipital	Awake, resting state with eyes closed (meditative); does not slow appreciably with increased age
θ	5–8	Central	Awake, task oriented; may be generalized in drowsiness
δ	1–4	Frontal and central	Generalized in deep sleep; seen in encephalopathy, coma, and toxic states

EEG FREQUENCY EXAMPLES

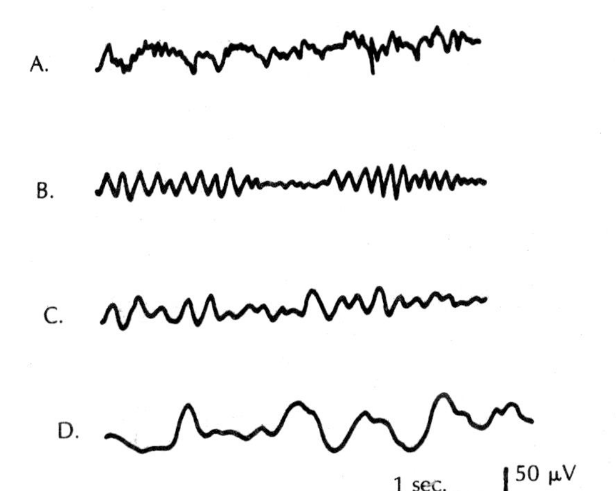

A. Beta: 20-25 Hz.; B. Alpha: 9-10 Hz.;
C. Theta: 5-6 Hz.; D. Delta: 2-2.5 Hz.

EEG recordings that have been collected over hours to days can be analyzed by computer to yield a summary of frequency and amplitude patterns over small chunks of time (e.g., 2, 4, or 10 seconds). If there is an area of interest, one can return to the original (expanded) data set to reread the activity more carefully. During certain procedures, such as during carotid endarterectomy, on-line analysis of EEG activity in 2-second bits can provide immediate and clearly visible change in brain activity to the surgical team. Immediate feedback about decrements in brain activity can guide the vigor of the procedure and help to improve outcome for the patient.

Sleep Studies

Sleep monitoring is of two types. The first, obtained frequently during routine EEG, is brief, is often drug-induced, and may not require any special adaptation of technique. The second is to monitor through all phases and types of sleep, usually over a period longer than 8 hours. Again, maintaining adequate electrode-to-head contact is essential. This may be done by an electrode cap, needle electrodes, or implanted (cortical or depth) electrodes, and some form of cable-to-transmitter box arrangement, rather than direct connection to amplifiers. Evaluation for sleep apnea and other respiratory compromise require monitoring oxygenation, heart rate, and peripheral muscle tone in addition to brain activity. Disorders of excessive daytime drowsiness are important to evaluate because of the associated psychological malfunction they produce.

Interpretation

Over the decades, the paper presentation of the EEG record has become more standardized. Usually, this tracing is read by the naked eye. This leads to an emphasis on recognition of familiar patterns and deviation from those patterns.

The advent of CT mapping of the EEG has made analysis of the record much easier in many ways. The advantage of computerized, quantitative EEG is that the display can be rearranged easily, after the recording session is over. Information can be gleaned from the record months and years after the recording session. Areas of activity can be highlighted, with gray scales or color scales, to make it easy to see temporal and regional changes in the record. Many different programs are available—some for dedicated systems and some that run on a personal computer.

Understanding the Sources of the Electroencephalogram

The electrical potential at any specific point in the brain is the sum of potentials generated by underlying cellular activity. Biochemical activity at the cellular level generates ionic currents that create the electrical potentials and currents. The electrical and magnetic fields associated with these potentials and currents follow well-known physical laws. For EEG, the field potential of a group or population of neurons reflects the sum of the contributions of a collection of individual neurons. Asynchronous and irregular firing of neurons cannot be accurately detected by EEG electrodes distant from the source. Therefore, the scalp EEG reflects activity from regularly arranged neuronal sources, firing in more or less synchronous patterns. The contribution from radially and tangentially oriented currents is equally appreciated at the EEG electrode. The pyramidal neurons of layers IV and V of the cortex are often activated synchronously, producing brain activity measured at the scalp.

The EEG, by providing information about multiple sources of activity in several different montages, may permit deductions about the underlying source of activity. Large contiguous areas of activity, as opposed to many discrete areas that are near each other, however, are difficult to distinguish from each other unless a great number of electrodes placed over the scalp. Different algorithms for dipole localization use different approaches. Some make assumptions regarding what activity is "background" at all electrodes and what other activity represents "true" activity of interest. Others make assumptions about the timing or orientation of the EEG signal of interest. For a review of resolution issues in EEG, see reference 8.

Evoked Potentials and Fields

The evoked response in EEG is the activity linked to a specific stimulus, as distinguished from background activity. Multiple presentations of the stimulus are given, and the activity is summed and averaged. Better noise cancellation can be obtained with more repetitions of the stimulus, but signal can be last as the brain habituates to a repeated stimulus. When the evoked responses under study are due to cortical activity, the ability for any brain to adapt to the stimulus is great.

The summation (averaging) period in an evoked response study usually is selected to start just before the stimulus and to continue for some time longer than the response being studied. The number of repetitions for cortical evoked responses is usually somewhere between 50 and 200 and for subcortical responses, between 1000 and 4000 repetitions (see Fig. 3-12).

Sensory Modality

Traditional evoked responses have used three sensory modalities: visual, somatosensory, and auditory. There are standardized norms for early, middle, and late responses. In the visual domain, several standardized types of stimuli are used for eliciting these responses (e.g., checkerboard, flash, central versus peripheral field). Somatosensory potentials can be evoked by a painful stimulus that is generally an electric shock just above threshold and sufficient to cause a muscle twitch, for example, a thumb twitch. Auditory stimuli include broad-band clicks presented in rapid succession for brain-stem—evoked responses, or se-

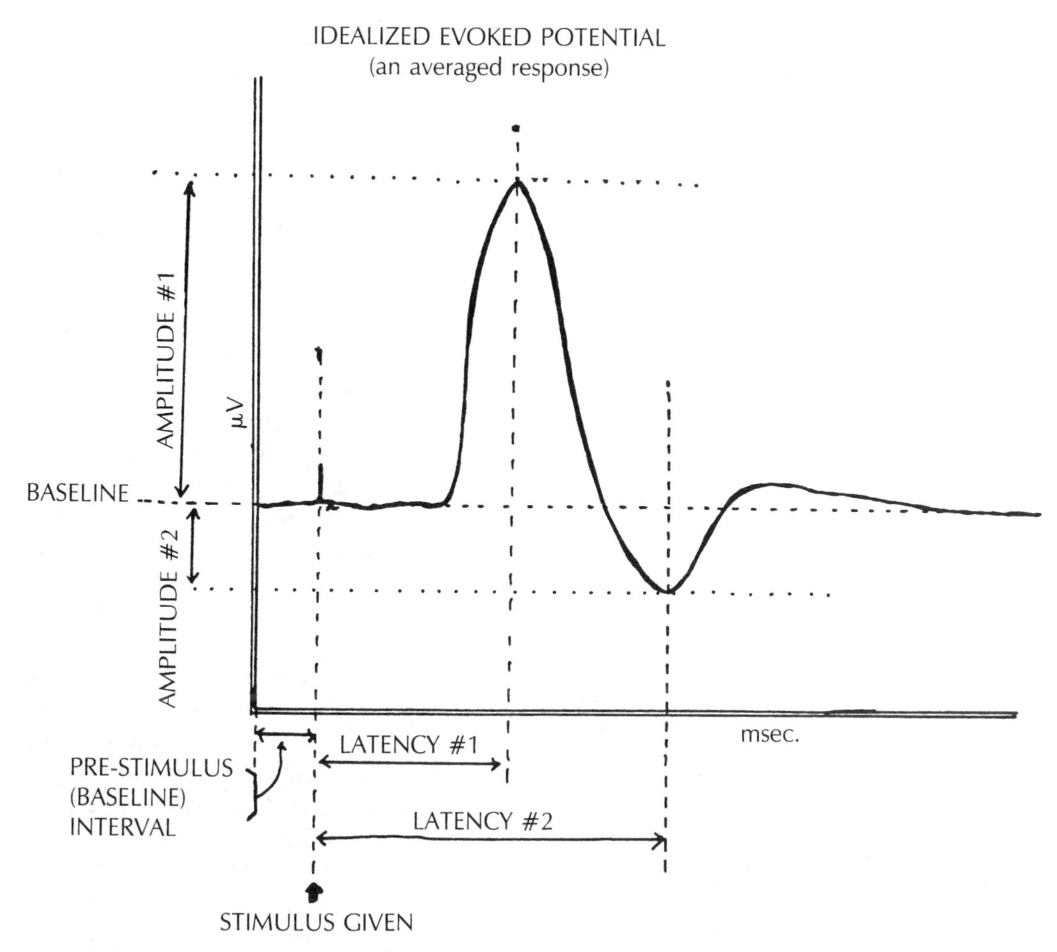

FIG. 3-12. Sample (idealized) evoked potential illustrating general points of reference.

ries of tone pips or more complicated sound presentations for later cortical responses.

Considerable research effort has focused on potential charges associated with various cognitive tasks. These tasks are sometimes cross-modal, involving auditory and visual stimuli presented simultaneously. They can differ within a category, such as variations in visual stimuli between color stimuli and written word stimuli that do not correspond to the color, or visual stimuli of words that sound similar when spoken but of which some are nonsense.

The rate of stimulus presentation in each domain is crucial for producing a maximal response. For example, when auditory tones are presented at rates faster than one per second, the amplitude of the response is significantly decreased. If, on the other hand, the interstimulus interval is closer to 4 or 6 seconds, various components associated with the early cortical response can begin to be isolated.

The quality of the stimulus, the instrumentation used to generate it, and the specific timing characteristics must all be documented so that repeat studies can be done. Subtle differences between different types of equipment can produce minor differences in the stimulus, particularly in visual and auditory systems, such that the response may either be enhanced or reduced and therefore not replicate previous findings.

Brain-stem auditory evoked potentials are a sensitive test of CNS myelination and of acquired demyelination. The interpeak latency from peak 1 to 5 is a sensitive measure of the degree of myelination. It is considerably longer in newborns and shortens with increasing age, stabilizing at about 3 years of life.

Findings of Questionable Significance

A number of findings are seen on EEG recordings, particularly clinical EEG recordings, that are of questionable significance. Incidental sharp waves, spikes, slow waves, exaggerated transitional phenomena (e.g., excessive posthyperventilation high-voltage activity), and low-grade dysrhythmia can all occur without diagnostic specificity. It is noteworthy that patients with neuropsychiatric disorders appear to have a greater incidence of these types of findings in a setting in which there are no clearly distinct accompanying behavioral changes.

The caveat to this discussion is that these same incidental findings can occasionally be seen in healthy subjects without symptoms. The challenge is to determine whether there is meaning to any disruption in the well-modulated EEG. If one is inclined to ignore this finding in a healthy subject, how much can be made of the same finding in someone who has a neuropsychiatric illness, without a clearly defined behavioral abnormality?

CLINICAL ISSUES

Schizophrenia

"Nonspecific" dysrhythmia and slow waves are frequently reported in patients with diagnoses of schizophrenia. Typically, these nonspecific findings are not lateralized. Until the last decade, there has been widespread consensus among psychiatrists and neurologists that the EEG was essentially useless for the detection of clinically relevant abnormalities in patients with schizophrenia, apart from demonstrating seizure foci in patients with psychosis and epilepsy. The quantification inherent in evoked responses, with the requirement to repeat stimuli until a significant signal-to-noise ratio is achieved, and the development of computerized recording systems slowly began to change both what was being measured and what questions that were asked.

Evoked potentials show many group tendencies for acute and chronic schizophrenia. Somatosensory responses tend to have higher overall amplitude before 100 msec, but lower amplitude after 100 msec. The distribution of P30 and N60 responses were found to be more posterior in patients with schizophrenia than in healthy, depressed, or nonpsychotic subjects (9). The amplitude of auditory responses is usually lower after 50 msec than in healthy subjects, and average latency is often shorter.

Many components of auditory evoked responses, especially at 100 msec, are influenced by attention. In much of the literature before 1980, and even since then, the level of attention of patients with schizophrenia was not well controlled or documented. Thus, we cannot be sure whether the lower amplitude of averaged responses is due to variation in latencies, increase in random background activity, variation in arousal or attention to a task, or an inherent inability to activate as large a group of neurons or to elicit as large a response. Visual evoked responses tend to have somewhat *greater* amplitude for patients with chronic schizophrenia than healthy subjects, with less after-rhythm. The P300, in both auditory and visual domains, tends to have low amplitude and be less responsive to uncertainty in the stimulus presentation.

Another hypothesis to explain the symptoms of schizophrenia is that of impaired filtering of sensory information, or altered sensory gating. Using paired auditory clicks, presented 1 second apart, with 5 to 8 seconds between pairs (ISI), some observers have demonstrated that healthy subjects and their relatives have a reduced amplitude of response (P50) to the second click. Patients with schizophrenia have no reduction in their response to the second click; their first-degree relatives are between the patients and the healthy subjects in size of response to the second click. It may be that schizophrenia impairs the ability of the individual to dampen an automatic processing response at primary auditory cortex when there is no new information contained in the signal. For more details, review reference 10.

Sleep recordings reveal persistent abnormality of sleep structure in treated and untreated patients with schizophrenia. Although delayed sleep onset, more arousals, and more early morning awakenings are present in never-treated, untreated, and treated patients, these findings are not specific for schizophrenia. Rather, they most likely contribute to impaired arousal, excessive sleepiness during the day, and a tendency to be disorganized.

Antipsychotic medications have been shown to cause intermittent slow waves of large amplitude (on EEG) over central and frontal regions. These patterns may have contributed to some interpretations of records as low-grade dysrhythmia. On the other hand, patients who have not been taking antipsychotic medications or who have never been exposed to antipsychotic medications have also been found to have nonspecific changes in the EEG. In nonpsychotic subjects, low doses of antipsychotic drugs do not cause gross changes in EEG patterns at visual inspection.

Another class of drug frequently administered to patients with schizophrenia is benzodiazepines. Known to increase fast (β) activity, benzodiazepines can confound the interpretation of fast activity in patients with schizophrenia.

The standard clinical EEG will continue to be useful as a screening tool to make sure a treatable, definable organic etiology is not causing the impairments in function (e.g., tumor, seizure disorder). EEG can be expected to become more sophisticated and to provide meaningful information about rapidity of response to different classes of drugs, predicting what patients will respond to specific classes of drugs. The EEG in combination will be most useful as a tool for developing an understanding of sequential processing of information during periods of exacerbation and remission.

Parkinson's Disease

In Parkinson's disease, the EEG remains normal in the early stages of the disease. Certain evoked potentials may be abnormal. For example, audiospinal facilitation is significantly reduced in amplitude during the 75- to 150-msec period after conditioning stimulation using the soleus H-reflex test. This is improved by treatment with levodopa but not anticholinergic agents. This represents an example of dysfunction of the reticular nuclei in Parkinson's disease.

Patients with Parkinson's disease and dementia demonstrate EEG abnormalities, including slowing of the α-rhythm to below 7 Hz.

Epilepsy

EEG is widely used to document the existence of seizures or the presence of a focal electrical abnormality that is causing behavioral disturbances. For partial seizures, EEG has limited sensitivity because the seizure events do not occur continuously (the EEG can be normal interictally) and because the electrical activity may not be generated in or near the cortical surface (not all activity from depth is recorded by the scalp EEG). The clinical history of alterations in behavior, perception, and cognition may support a diagnosis of partial seizures even when the EEG is normal.

The degree of accuracy of localization from the traditional noninvasive EEG is limited, however. A significant proportion of patients with true epilepsy have no interictal sharp waves on scalp EEG. These same patients may have significant interictal seizure activity on subdural electrode array recordings.

Pseudoseizures

The differentiation of seizures from pseudoseizures is a particular use of the clinical EEG. Video monitoring of the onset of behavioral changes with the EEG can document either absence or presence of electrical activity accounting for epileptiform behavior. Epileptiform activity on the EEG can be missed because of motor activity, especially of the head musculature. Recording can be done with cortical and depth electrodes as well, which increases the likelihood of capturing an active source. Mesial temporal lobe epileptic foci are the most difficult to document accurately and can cause many unusual sensory complaints and behaviors.

It is often the case that epileptic seizures and pseudoseizures are intermixed in the same patient. For a review of EEG utility in the diagnosis of epilepsy, see reference 11.

Depression

With regard to depressive spectrum illness and EEG, sleep studies have been the most consistently useful either for diagnosis or for follow-up of clinical course. Biologic markers of abnormalities of sleep are found in people prone to this disorder compared with controls. These markers may include decreased sleep efficacy and decreased amount of slow-wave sleep during non-REM sleep. Frequent awakenings may also occur.

Compared with people who are not depressed, patients with depression often show increases in δ and τ activity over the right posterior temporal regions and bilaterally increased β activity over the frontal areas.

An overview of these data suggest that EEG is not useful for the diagnosis of depression and other affective disorders but may help on occasion to discriminate affective disorder from other neuropsychiatric disorders, or to suggest treatment possibilities.

Evoked responses have not as yet proved clinically useful in the diagnosis of the affective disorders (12). There may be some slight differences with regard to waves associated with attention function, but there is no reason to believe that such differences are specific to depression.

Electroencephalography and Drug Effects

The EEG can be a useful tool in the measurement of drug effects on brain function. What is most clear from all research to date is that there is a lack of drug-specific characteristic findings for all individuals under all conditions. Evoked potentials can provide a mechanism by which aspects of the drug can be highlighted differentially. For example, lithium highlights the early components of the evoked potential compared with antidepressants, neuroleptics, and other medications. (for a recent review, see reference 13). The purposes of performing pharmacologic EEG are summarized as follows:

- To study drugs of abuse
- To classify psychoactive compounds
- To estimate the therapeutic indications for, or responses to, certain classes of drugs
- To develop bioavailability and pharmacodynamic correlations
- To classify subgroups of psychiatric patients
- To anticipate responders to certain types of treatment

Certain categories of drug produce predictable responses; for example, benzodiazepines increase fast (β) activity. Under certain conditions, there appears to be a decrease in synchronization, which can appear as varying lateralization within certain frequency bands. Perhaps the best approach is to combine resting state and some form of activation condition specific to both the disease state being studied and the presumed mechanism of action for the class of drug being studied (Table 3-2) .

TABLE 3-2. *Effect of Psychoactive Substances on Electroencephalography Frequencies*

	δ	θ	α	β
Benzodiazepines	↓			¶
Antipsychotics	¶	¶		
Tricyclics	¶			(¶)
Opioids	¶		¶	
Phenobarbital	¶		↓	(¶)
Methylphenidate				↓
Nicotine			↓	
Alcohol			¶	

Adapted from ref. 13, with permission.

Violence

There is an increased rate of neurologic impairment in patients who are persistently violent compared with those who are transiently violent or nonviolent. Traditional EEG has not been very useful in distinguishing violence-prone inpatients in psychiatric settings. There have been some reports of increased amounts of slow activity over frontotemporal areas in habitually aggressive forensic offenders (14).

Aging and Dementia

Alzheimer's disease of mild to moderate severity is reported to produce increased slowing on the EEG (15). The slow activity tends to be found primarily over frontal and midline recording locations, especially in the δ and τ bands (in 0- to 2-Hz and 4- to 6-Hz ranges). SPECT scanning may occasionally show an association between the temporoparietal cerebellar ratio (T-P/C) on the SPECT scan and the peak frequencies reported from the EEG leads.

Depressed patients may sometimes be differentiated from demented patients in that they had more α_2 activity with less δ and τ activity. In general, quantitative EEG differences between cognitively impaired elderly patients and matched controls are maximal in frontal and temporal regions.

The visual evoked responses to checkerboard pattern-reversal stimulation shows a curvilinear relationship between P100 latency and age over the life span. The shortest latency occurs during middle adulthood, with increasing latencies found in patients older than 60 years of age. There have been several reports of reduced contrast sensitivity with increasing age over 40 to 50 years. Other groups have found that the evoked potential latency of elderly subjects became comparable to younger subjects when the stimulus contrast was increased to an appropriate degree (having accommodated for physical changes to the lens and eyeball that accompany increasing age).

Mental Retardation and Down's Syndrome

This heterogeneous group of disorders has a wide clinical spectrum, from mild to very severe intellectual deficits. There is little in the EEG that can specifically diagnose the etiology of a particular type of mental retardation. Generally, the EEG shows diffuse patterns of slowing and, at times, dysrhythmia similar to a dementia pattern, usually with preserved α-rhythm. Depending on the severity of the illness, there is a range in abnormal findings from mild retardation with relatively well-organized EEG to severe retardation with disorganization of the EEG and large slow waves, sometimes associated with spikes.

EEG abnormalities consistent with a diagnosis of dementia were seen in patients older than 35 years of age who had epilepsy and Down's syndrome. Patients under the age of 35 years with or without epilepsy showed no EEG abnormalities.

Stroke

EEG can be useful in the evaluation of recovery from stroke. While imaging techniques such as CT and MRI permit assessing lesion size, neither technique measures which tissue remains functional or has the potential to recover function. Physiologic measures can assist in the assessment of recovery and in quantifying the degree of alteration in normal function that has resulted from the stroke, such as motor function, sleep patterns, and alerting or sensory perceptual changes.

Neurologic examination combined with EEG (mapping) permits localization of the lesion and compares favorably to CT scan in some patients. Candidates for endarterectomy have been studied preoperatively and intraoperatively with EEG to assess the risk for intraoperative or postoperative stroke. EEG changes during surgery give immediate and accurate feedback to the surgeon on the state of perfusion to the brain. This feedback is important in reducing rates of intraoperative stroke. EEG mapping may also provide more information on lateralization of deficit than routine EEG and may correlate better with subcortical lacunar infarction than standard EEG (16).

EEG recordings have been used in some cases of transient ischemic attacks, usually after resolution of the episode (because of the short duration of symptoms in most patients).

These recent data suggest that EEG recordings can be used to assess the volume of brain at night during an ischemic episode and thereby provide a means for evaluating the efficacy of neuroprotective strategies. At present, neuropsychological tests have their greatest utility in cases of clinical stroke with discrepant findings or anatomic brain imaging.

Traumatic Brain Injury

Somatosensory evoked potentials (SSEPs) have been reported to be useful predictors of good recovery in patients who were in coma due to traumatic brain injury (TBI) (17).

Seizures are a frequent sequela of TBI, either from scarring or from an excitatory lesion. As with other sources of seizure, patients with TBI may have psychogenic seizures. The risk associated with treating pseudoseizures with antiepileptic drugs is that cognitive processes may be impeded even further, while reinforcing the sick role. The EEG is especially useful in making the differential diagnosis between the two types of seizure behavior.

EEG is often used in the intensive care unit to monitor the stability of patients in the acute postinjury phase of illness. Because of the limitations of space and demands for other equipment, EEG has often been limited to single-channel recording. Even a single channel can provide documentation of paroxysmal events, general arousal (particularly change in arousal), and drug effects.

Pain

The coherence of the EEG can be affected by pain. Specific SSEP values can be altered by chronic pain, with longer latencies the usual finding. Arkhipova and colleagues reported on patients studied with EEG before and after surgery for deafferentation chronic pain (18). EEG improvement after surgery included increased interhemispheric and intrahemispheric coherence, a broader distribution of frequencies at rest, decreased bilateral synchronous τ activity, a shorter P220 to P300 SSEP, and more readily detectable visual evoked responses.

Migraine

Researchers have used EEG extensively to attempt to characterize better the brain activity during migraine and to attempt to identify people who are prone to having migraine headache (19).

Basilar migraine is characterized by symptoms referable to dysfunction of the brain stem. EEG studies show increased δ activity, generalized, which resolves after symptoms subsided. Both before and after an attack, the EEG is normal. In cases in which symptoms have a slightly longer duration, the EEG can provide useful confirmation of transient alteration in brain function. Transient occipital spike–wave complexes during basilar migraine attacks have also been recorded. There is a higher incidence of EEG abnormalities during headache in migraine sufferers with aura than in those without aura. Children with migraine with aura have an increased τ-to-α ratio in posterior temporal and occipital areas than those without aura, those with tension headache, and healthy, age-matched controls.

Migraine is sometimes associated with epilepsy, most often the complex partial type. Increased background slow activity is seen particularly in patients with epilepsy and classic migraine (20). Occipital spike–wave complexes have been recorded during migraine attacks in children who also had seizures.

Evoked potential measurements as a diagnostic tool for migraine has not been particularly useful. The extreme photosensitivity usually present during the headache attack precludes valid testing.

In routine evaluation of chronic headache, clinical judgment cannot be replaced by neurophysiologic testing. The low rate of nonspecific abnormalities in routine EEG argues against the routine ordering of these tests in the absence of some specific finding in the clinical history or neurologic examination.

Multiple Sclerosis

EEG shows diffuse or focal patterns of slowing, depending on the area affected and severity of disease in MS. Early in the course of illness, abnormal findings return to normative values during remission. Impaired mental function can be associated with generalized slowing and, more rarely, epileptiform discharges (21).

Brain-stem auditory evoked potentials are frequently abnormal. Prolongation of the peaks I to V interval indicates brain-stem dysfunction. In contrast to patients with metabolic encephalopathy, patients with MS typically have a normal wave I. Brain-stem auditory evoked potentials are most useful in patients who have a history consistent with MS and do not have clinically evident brainstem lesions.

Abnormal SSEPs are reported in most patients with MS (22). The most frequently ordered, confirmatory test for MS is the visual evoked potential during an initial episode of acute onset of loss of visual acuity. Waveforms are prolonged overall, as well as at specific latencies corresponding to sites of demyelinating plaques. Repeated SSEPs, however, do not correlate with the course of the disease. Topographic or quantitative EEG allows for more sensitive monitoring of CNS changes and alterations in evoked potentials by this method have been found to correlate with changes in clinical state.

Autism

The developmental disorders, including autism, pervasive developmental disorder, and Asperger's syndrome, do not have pathognomonic EEG findings. Minor generalized EEG abnormalities are not uncommon. In a group of children with autism, Rothenberger found that the power spectrum of the EEG could reflect the degree of "change" or "nonchange" in children as he followed them longitudinally (23).

Human Immunodeficiency Virus and Acquired Immunodeficiency Syndrome

Electrophysiology can be an important component in the evaluation of the effects of HIV in the CNS because anatomic imaging techniques are not sensitive enough to follow disease progression in many cases. Noninvasive EEG measures, from resting state coherence and power spectrum analysis to specific evoked potentials, are being used in HIV-positive patients who do not have the full-blown clinical disease. As the research continues, one can expect more specific knowledge about the progression of the disease in the CNS (24).

Other Neuropsychiatric Disorders

Somatosensory evoked potentials (SSEPs) and somatosensory evoked fields can be useful diagnostic tools when there is a focal lesion, such as a brain tumor. The response delay generally is longer over the side or site of the lesion.

Children and adolescents with mitochondrial encephalomyopathies demonstrate slowing of the α-rhythm and epileptiform discharges on EEG.

SUMMARY

The field of EEG and magnetoencephalography (MEG) is evolving rapidly. Rather than overwhelm the reader with minutiae of the differences and similarities of the two techniques, the emphasis has been on fundamentals of each technique and applications that are applicable to the study of neuropsychiatric disorders. There is a long-standing realization among users of these physiologic techniques that individualized applications are relevant and further our understanding of the expression of disorders in the brain.

Although specific findings in certain disorders have been identified, the greatest importance of EEG and MEG may remain as techniques for recording brain *function*. Integration of source modeling with other anatomically based imaging studies will support the validity of these physiologic measurements. The particular value of EEG and MEG to the neuropsychiatrist is that they remain directly tied to neuronal function, normal and abnormal, without reliance on uptake of labeled substances, such as are required by metabolic or blood flow measures. With the development of whole brain MEG systems, we are entering a decade in which noninvasive study of brain function is becoming a reality. The reader is encouraged to integrate the findings from a variety of technical approaches to brain study. The information collated from different types of studies (i.e., anatomic imaging, functional measures with varying time courses from milliseconds to hours) will increase our appreciation of normal and disordered brain function.

Focused application to clinical and research questions will be the most rewarding approaches to the use of EEG or MEG. In this manner, there can be continued refinement of our understanding of the complex interactions that characterize the communication between neurons in the brain. As with early hopes in EEG studies, the likelihood that a single, readily identifiable pattern of activity (EEG or MEG) will characterize any neuropsychiatric disorder is almost nil. In most disorders, we are considering difficulties that wax and wane, can vary in their response to pharmacologic treatment, and are affected by the state of the individual (e.g., stressed versus relaxed). Careful study of physiologic changes (aided by developing technology) that fluctuate in meaningful patterns from minute to minute, even from millisecond to millisecond, are needed to develop understanding of CNS function in neuropsychiatric disorders.

References

1. Gado M, Hanaway J, Frank R. Functional anatomy of the cerebral cortex by computed tomography. *J Comput Assist Tomogr* 1979;3:1–19.
2. Talairach J, Tournoux P. *Co-planar stereotaxic atlas of the human brain.* New York: Thieme Medical Publishers, 1988.
3. Curnes JT, Burger PC, Djang WT, Boyko OB. MR imaging of compact white matter pathways. *Am J Neuroradiol* 1988;9:1061–1068.
4. Hahn FJY, Rim K. Frontal ventricular dimensions on normal computed tomography. *Am J Roentgenol* 1976; 126:593–596.
5. Rosen BR, Belliveau JW, Vevea JM, Brady TJ. Perfusion imaging with NMR contrast agents. *Magn Reson Med* 1990;14:249–265.
6. Gerard G, Weisberg LA. MRI periventricular lesions in adults. *Neurology* 1986;36:998–1001.
7. McHugh PR. The basal ganglia: the region, the integration of its systems and implications for psychiatry and neurology. In: Franks AJ, Ironside JW, Mindham RHS, et al, eds. *Function and dysfunction in the basal ganglia.* Manchester, England: Manchester University, 1990.
8. Gevins A. High resolution EEG [Review]. *Brain Topogr* 1993;5:321–325.
9. Roth WT, Pfefferbaum A, Horvath TB, Berger PA, Kopell BS. P3 reduction in auditory evoked potentials of schizophrenics. *Electroencephalogr Clin Neurophysiol* 1980;49:497–505.
10. Adler LE, Waldo MC, Freedman R. Neurophysiologic studies of sensory gating in schizophrenia: comparison of auditory and visual responses. *Biol Psychiatry* 1985; 20:1284–1296.
11. Niedermeyer E, Lopes de Silva F, eds. *Electroencephalography: basic principle, clinical applications, and related fields.* Baltimore, MD: Urban & Schwarzenberg, 1987.
12. Burkhart MA, Thomas DG. Event-related potential measures of attention in moderately depressed subjects. *Electromyogr Clin Neurophysiol* 1993;88:42–50.
13. Sannita WG. Quantitative EEG in human neuropharmacology. *Acta Neurol* 1990;12:389–409.
14. Williams D. Neural factors related to habitual aggression: consideration of differences between those habitual aggressives and others who have committed crimes of violence. *Brain* 1969;92:503–520.
15. Martin-Loeches M, Gil P, Rubia FJ. Two-Hz-wide EEG bands in Alzheimer's disease. *Biol Psychiatry* 1993;33: 153–159.
16. Kappelle LJ, van Huffelen AC, van Gijn J. Is the EEG really normal in lacunar stroke? *J Neurol Neurosurg Psychiatry* 1990;53:63–66.
17. Hume AL, Cant BR. Central somatosensory conduction after head injury. *Ann Neurol* 1981;10:411–419.
18. Arkhipova NA, Shevelev IN, Grokhovskii NP. The spatial organization of bioelectrical activity during the treatment of the chronic pain syndrome [title translated from Russian]. *Zh Vyssh Nerv Deiat Im I P Pavlova* 1993;43:730–737.
19. Sand T. EEG in migraine: a review of the literature. *Funct Neurol* 1991;6:7-22.
20. Marks DA, Ehrenberg BL. Migraine-related seizures in adults with epilepsy, with EEG correlation. *Neurology* 1993;43:2476–2483.
21. Levic ZM. Electroencephalographic studies in multiple sclerosis: specific changes in benign multiple sclerosis. *Electroencephalogr Clin Neurophysiol* 1978;44: 471–478.
22. Chiappa KH. Pattern shift visual brainstem auditory and short-latency somatosensory evoked potentials in multiple sclerosis [Review]. *Neurology* 1980;30: 110–123.
23. Rothenberger A. Research on autism: different aspects of changes observed during the development period] [title translated from French]. *Encephale* 1992;18:217–223.
24. Syndulko K, Singer EJ, Nogales-Gaete J, Conrad A, Schmid P. Laboratory evaluations in HIV-1-associated cognitive/motor complex. *Psychiatr Clin North Am* 1994;17:91–123.

4
Treatment

Barry S. Fogel, Randall T. Schapiro, and Tony M. Wong

OVERVIEW

This chapter discusses drug therapy and rehabilitation therapy in neuropsychiatry. Advice is offered on the selection and monitoring of drug therapy for patients with neuropsychiatric disorders, including the decision to use drug treatment, the choice of target symptoms and syndromes, methods of monitoring treatment response and adverse effects, use of drug combinations, and concerns regarding side effects and drug interactions.

Rehabilitation therapies currently have a limited empirical database concerning efficacy, but the outlines of present practice standards at our medical centers are provided. For a review of general principles concerning neuropsychiatric pharmacology, see references 1 and 2.

DRUG THERAPY

Antidepressant Drugs

Efficacy

The effectiveness of antidepressant drugs is best established for the treatment of primary major depression, but a relatively small number of randomized controlled trials (RCTs) support their efficacy for major depression secondary to common neurologic diseases (3). Successful trials have been reported for nortriptyline, citalopram, and trazodone in the treatment of poststroke depression. Nortriptyline, amitriptyline, and citalopram have been shown to help pathologic crying after stroke. In Alzheimer's disease, numerous case series and uncontrolled studies suggest that usual antidepressant drug treatments are effective. Desipramine has been reported to be superior to placebo in the treatment of major depression associated with multiple sclerosis (MS). Four RCTs support the efficacy of tricyclic antidepressants for major depression associated with Parkinson's disease (PD). The benefit of selegiline in depressed patients and in parkinsonism is supported by a controlled study. An open trial supported a combined antidepressant and antiparkinsonian effect of bupropion. RCTs of antidepressants for major depression after traumatic brain injury (TBI) and for depression accompanying multiinfarct dementia have not been carried out.

Apart from major depression, the best established indication for antidepressants is the treatment of obsessive-compulsive disorder (OCD). Here, clomipramine, fluoxetine, and fluvoxamine have proved effective by RCTs sufficient to satisfy the U.S. Food and Drug Administration (FDA).

Other Neuropsychiatric Applications

During the past several years, a number of reports have appeared regarding the application of antidepressant drugs to neuropsychiatric problems other than major depression and OCD. These applications have included the following:

- Dysthymia in patients with mental retardation
- Adult attention deficit disorder
- Self-injurious behavior in developmentally disabled people

- Chronic fatigue syndrome (CFS) and fibromyalgia
- Chronic nonmalignant pain
- Miscellaneous disorders of impulse control
- Emotional lability, including pathologic laughing and crying, in patients with brain injury
- Posttraumatic stress disorder (PTSD)
- Tourette's syndrome with attention deficits

Neurotoxicity

The reader is referred to the comprehensive review of Stoudemire and Fogel (4).

Selective Serotonin Reuptake Inhibitors

The selective serotonin reuptake inhibitors (SSRIs) have emerged as the antidepressants of first choice. At high doses, these drugs can produce an apathetic frontal lobe syndrome. SSRIs also increase neuromuscular excitability and can produce or aggravate tremor or myoclonus. Among the SSRIs, fluoxetine is most likely to cause nervousness or restlessness. All of the SSRIs can cause headaches.

Tricyclic Antidepressants

Tricyclic antidepressants (TCAs) can cause hallucinations or delirium, particularly in elderly patients and those with dementia. Like the SSRIs, TCAs can cause or aggravate tremor or myoclonus, but, with the exception of the antidepressant-neuroleptic amoxapine, they are rarely associated with extrapyramidal reactions, such as akathisia and acute dyskinesia. TCAs can have discrete effects on memory in the absence of delirium.

Monoamine Oxidase Inhibitors

The monoamine oxidase inhibitors (MAOIs) increase neuromuscular excitability. In usual dosages, they do not cause delirium, cognitive impairment, or extrapyramidal reactions.

Bupropion has the greatest propensity to induce seizures of all of the currently marketed antidepressants. In patients without neurologic risk factors for seizures who are taking a full therapeutic dose, the risk is about 0.4%. Accordingly, patients with risk factors for seizures should be taking a prophylactic antiepileptic drug (AED) if bupropion is to be prescribed. Bupropion is a stimulating antidepressant and can produce agitation or a delirious psychosis in occasional vulnerable patients.

Nefazodone has a relatively benign neurologic side-effect profile. Sedation and ataxia are the most common limiting neurologic side effects. However, nefazodone is a fairly weak antidepressant and must be pushed to the upper end of its therapeutic range (i.e., 400 to 600 mg/day) to have equivalent efficacy to TCAs or conventional SSRIs. At this higher dosage, sedation or ataxia can pose significant problems in patients with preexisting brain disease.

Venlafaxine has a neurologic side-effect profile resembling that of the SSRIs. In addition, it is associated with hypertension: The rates are 3% to 5% on dosages of less than 200 mg/day, 7% on dosages of 201 to 300 mg/day, and 13% on dosages of more than 300 mg/day. Paroxysmal sweats, sometimes drenching the patient's bed or clothing, can also be seen. Mirtazapine is a newly approved tetracyclic antidepressant that increases the release both of norepinephrine and serotonin.

Drug–Drug Interactions

Concern about drug interactions is greatest with the SSRIs and with nefazodone, both of which are potent inhibitors of the hepatic cytochrome P-450 system. Fluvoxamine inhibits several cytochrome isoenzymes; it has clinically significant effects to slow the metabolism and increase the blood level of TCAs, alprazolam, diazepam, theophylline, propranolol, warfarin, methadone, and carbamazepine. When given together with nefazodone, carbamazepine dosage should be reduced and blood levels monitored frequently during dosage adjustments (5).

Of particular concern in neuropsychiatry is the interaction of antidepressant drugs and

AEDs. The concurrent administration of an SSRI and carbamazepine may raise carbamazepine levels. The concurrent administration of carbamazepine, phenytoin, or primidone with a TCA may lower tricyclic levels.

Pharmacodynamic drug interactions involving antidepressant drugs are usually related to additive effects of the antidepressant and other drugs on neurotransmitter receptors, specifically α-adrenergic, serotonergic, and cholinergic receptors. Thus, hypotension can occur when TCAs or trazodone are combined with antihypertensive α-adrenergic blockers. The *serotonin syndrome*—a state of delirium, tremulous rigidity, and autonomic instability—can occur when an SSRI is combined with an MAOI or with clomipramine. An anticholinergic delirium can occur when TCAs are combined with other drugs that have central anticholinergic effects, such as first-generation antihistamines.

Choice of Antidepressant

Table 4-1 summarizes some of the issues relevant to choosing and initiating antidepressant therapy in a neuropsychiatric patient.

Patients with significant agitation and anxiety are more likely to tolerate sertraline, paroxetine, or nefazodone than bupropion or fluoxetine. Patients with apathy and retardation may benefit from the stimulating effects of bupropion or fluoxetine.

Among anxious patients, those with obsessions, compulsions, or premorbid obsessive-compulsive personality should receive SSRIs. Those with predominant anxiety and insomnia might be treated with nefazodone. Patients with compulsive behaviors, such as paraphilia and compulsive gambling, should be treated with SSRIs.

MAOIs generally should be regarded as second-line agents. Consideration of an MAOI as first-line therapy would be given for patients with severe phobic symptoms.

Patients with a history of poor response to antidepressant drugs, with rapid cycling of mood induced by antidepressants, or with a mixture of depressive symptoms and paroxysmal symptoms typical of partial seizures should initially be treated with mood-stabilizing AEDs, such as carbamazepine.

Dosage Titration

Starting doses of antidepressants in patients with gross brain disease should be low enough to test the patient for unusual sensitivity to the drug:

TCAs (excluding protriptyline): 10 mg q.h.s.
Fluoxetine and paroxetine: 5 mg q.d.
Sertraline: 25 mg q.d.
Nefazodone: 50 mg q.h.s.
Bupropion: 75 mg q.d.
Mirtazapine: 15 mg q.d.
MAOIs: tranylcypromine 10 mg every morning and noon
Phenelzine: 15 mg b.i.d.
Protriptyline: 5 mg q.d.

If the patient tolerates the test dose of the drug, the daily dose may be increased every 2 to 3 days until limiting side effects develop, a conventional therapeutic dosage is reached, or, in the case of the secondary amine tricyclics, a therapeutic blood level is attained. Through the course, if the patient experiences significant improvement in symptoms, the dose is held at that level for at least 1 week. If the patient is continuing to improve, the dose is not increased further. If a plateau has been reached, and there are persistent symptoms, dose increases are resumed. The increment of dose increase is about the same as the initial dose, with the provision that bupropion and the MAOIs are given on a t.i.d. schedule and venlafaxine and nefazodone on a b.i.d. schedule.

Continuation doses may be increased if a patient does not experience significant side effects from an SSRI, venlafaxine, nefazodone, or an MAOI:

Fluoxetine and paroxetine: 80 mg per day
Sertraline: 300 mg per day
Nefazodone: 600 mg per day
Mirtazapine: 45 mg per day
MAOIs: tranylcypromine 90 mg per day
Phenelzine: 90 mg per day

TABLE 4-1. *Issues Relevant to Choosing and Instituting an Antidepressant for Neuropsychiatric Patients*

Drug	Benefits	Precautions	Relevant Controlled Trials	Initial Daily Dosage
Amitriptyline	Sedation, analgesia cardiac effects; orthostatic hypotension	Strongly anticholinergic; quinidine-like poststroke depression; pain; headache	Pathologic laughing and crying; given at bedtime	10–25 mg, usually
Nortriptyline	Meaningful blood levels; less hypotension, sedation, and anticholinergic effect than amitriptyline	Quinidine-like cardiac effects; anticholinergic effects	Pathologic laughing and crying; poststroke depression level 50–150 ng/mL)	10–25 mg, usually given at bedtime (target blood level so-iso ng/mL
Imipramine	Antipanic and antiphobic actions; less sedating than amitriptyline	Quinidine-like cardiac effects; orthostatic hypotension; anticholinergic effects	Depression in Parkinson's disease	10–25 mg
Desipramine	Least anticholinergic of tricyclic antidepressants; can be stimulating	Quinidine-like cardiac effects; orthostatic hypotension anticholinergic effects	Depression in multiple sclerosis; depression in Parkinson's disease	10–25 mg (target blood level >125 ng/mL).
Clomipramine	Potent antiobsessional effects	As for amitriptyline, plus myoclonus	None	12.5–25 mg
Protriptyline	Potent stimulant; alternative to direct stimulants	As for amitriptyline	None; case reports suggest utility as a stimulant in TBI patients	5 mg
Fluoxetine	Helpful for dysphoria and irritability; can be activating; effective for OCD at higher dosages	Akathisia, agitation, anxiety may appear on treatment initiation; apathy at very high doses: drug interactions due to P450 inhibition	None; numerous case series suggest efficacy for depression due to general medical diseases, including neurologic diseases	5–10 mg
Fluvoxamine	Specific for OCD	Potent enzyme inhibitor with significant drug interactions; sedation, myoclonus, GI side effects from hyperserotonergic state all possible	None	25–50 mg
Paroxetine	Less activating than fluoxetine; may cause less sedation and fewer GI symptoms than sertraline; chief benefits are relief of dysphoria and irritability	Drug interactions from P450 inhibition; nausea; potential for sedation	None	5–10 mg
Sertraline	Helpful for dysphoria and irritability; may enhance executive cognitive function	Sedation, lower GI side effects; drug interactions from P450 inhibition	None	12.5–25 mg
Trazodone	Sedating and anxiolytic; may help agitation in dementia	Orthostatic hypotension, excessive sedation, priapism risk in men	Poststroke depression (improved rehabilitation outcome)	25 mg, usually at bedtime
Nefazodone	Anxiolytic, less sedating than trazodone; promotes sleep without disrupting sleep architecture	Sedation (orthostatic hypotension rare and priapism not seen); drug interactions due to inhibition of P450 3A4	None	50 mg at bedtime, or 50 b.i.d.
Bupropion	Stimulating, antiapathy, mildly antiparkinsonian, no cardiotoxicity	Risk of seizures, especially with single doses >150 mg or total daily dosage >450 mg; mild anticholinergic effects; insomnia; weight loss possible	Depression in Parkinson's disease (levodopa requirement reduced)	75 mg b.i.d.
Tranylcypromine	Stimulating, antiapathy, no cardiotoxicity, broad spectrum of antidepressant efficacy	Orthostatic hypotension, insomnia, risk of drug interactions or hypertensive crisis from tyramine-rich foods	None; case series suggest utility in subcortical vascular dementia with apathetic depression	10 mg morning and noon
Phenelzine	Antianxiety and antiphobic effects; analgesic and antimigraine effects	As for tranylcypromine; more likely to be sedating; may cause major weight gain	Migraine	15 mg morning and noon
Selegiline	Well-tolerated; stimulating and antiparkinsonian effects	Risk of drug interactions or drug-food interactions with dose >10 mg/day; can cause nausea	Early Parkinson's disease—helps motor symptoms, mood, cognition	5 mg b.i.d.

OCD, obsessive-compulsive disorder; GI, gastrointestinal; TBI, traumatic brain injury.

Bupropion is not increased beyond 450 mg per day (150 mg t.i.d.) because there is unacceptable risk of seizures at higher dosages. TCAs may, on occasion, be increased beyond 300 mg per day for desipramine or imipramine or 150 mg per day for nortriptyline. When this is done, a tricyclic serum level and an electrocardiogram should be done to establish the safety of going to a higher dosage.

Duration of Therapy

When antidepressants are used to treat primary major depression, maximal treatment response takes 4 to 8 weeks. Therapy, if successful, is continued for at least 6 months at the dosage used to induce a remission of depression. It is not known whether these durations apply equally to major depression secondary to gross brain diseases.

For a review of depressive syndromes in neurologic disease, see reference 6.

Antipsychotic Drugs

Efficacy

Antipsychotic drugs, or neuroleptics, are determined to be effective for the treatment of schizophrenia and of mood disorders with delusions. They have also been proved useful for psychotic symptoms and behavioral disturbances in dementia and in mental retardation. No particular antipsychotic drug has shown consistent superiority to any other in the treatment of psychotic complications of dementia.

The efficacy of haloperidol and of pimozide for reducing tics in Tourette's syndrome has been established by clinical trials.

Neurotoxicity

Chronic neurologic toxicity includes tardive dyskinesia, tardive dystonia, tardive akathisia, tardive dysmentia, and dopamine supersensitivity psychosis. Tardive dyskinesia characteristically involves the buccal, oral, and lingual muscles but may also include choreoathetosis of the upper extremities or dyskinesia of the larynx and pharynx. Tardive dystonia, like acute dystonia, is usually axial. Tardive akathisia resembles acute akathisia. Tardive dysmentia is a chronic disturbance of attention, concentration, and executive cognitive function that usually accompanies a tardive movement disorder. Dopamine supersensitivity psychosis is a state of agitation, confusion, and affective instability that emerges after neuroleptics are withdrawn after long-term use.

All of the neuroleptic drugs can also cause or provoke seizures. However, neuroleptic-induced seizures are most common with the atypical agent clozapine, for which the incidence rate reaches 10% at doses of 900 mg per day or higher. The agent least likely to cause seizures is not definitively established, although molindone and fluphenazine performed better than other antipsychotic drugs in an in vitro model.

Drug–Drug Interactions

The neuroleptics are metabolized by oxidative enzymes in the liver. Specific cytochromes are responsible for the metabolism of each specific drug. Most of the neuroleptic drugs have many active metabolites, complicating the interpretation of blood levels.

Carbamazepine, a potent enzyme inducer, lowers blood levels of haloperidol. On the other hand, phenothiazine and thioxanthene neuroleptics can raise carbamazepine levels by inhibiting carbamazepine metabolism. Fluvoxamine raises the blood level of clozapine when the two drugs are given together; other SSRIs may have similar but weaker effects.

Neuroleptics, which block dopamine receptors, have additive or synergistic effects when combined with drugs that reduce dopamine release or affect dopamine's interaction with second messengers. Drugs that interact in this way with neuroleptics include carbamazepine, lithium, and calcium-channel blockers.

A second set of pharmacodynamic interactions, which may also be either beneficial or problematic, occurs between the neuroleptics and drugs that potentiate inhibition by the γ-aminobutyric acid (GABA) system, such as *benzodiazepines* and *barbiturates*. Conversely, tranquilizing effects may be enhanced, with less likelihood of tremor or akathisia. On the other hand, sedation, lethargy, quiet delirium, or apathy and akinesia can be produced (7).

Choice of Neuroleptic

Considerations relevant to the choice and initiation of neuroleptic therapy are presented in Table 4-2.

The high-potency agents, of which *haloperidol* is the prototype, have less sedating, anticholinergic, and hypotensive effects but frequently cause extrapyramidal reactions, whereas the low-potency agents, of which thioridazine is the prototype, are associated with a lower incidence of extrapyramidal side effects (i.e., acute neurotoxicity). The mid-potency agents represent a compromise.

The atypical agent *risperidone* is a high-potency D_2 receptor blocker that also has potent $5HT_2$ antagonist effects. Risperidone, particularly at doses of 6 mg per day or less, causes substantially fewer extrapyramidal side effects than haloperidol. The most commonly encountered side effects are weight gain, sedation, orthostatic hypotension, and hyperprolactinemia.

Clozapine, the first atypical neuroleptic available in the United States, is a low-potency agent that is a relatively weak blocker of D_2 receptors, a significant blocker of D_3 and D_4 receptors, and a potent blocker of $5HT_2$ receptors. It has an extremely low rate of extrapyramidal side effects. When a movement disorder occurs, it usually consists of bradykinesia or mild akathisia, rather than rigidity or tremor. Clozapine can be given to patients with severe PD without aggravating their movement disorder and may actually ameliorate tardive dyskinesia and tardive dystonia. Clozapine always alters the electroencephalogram, and it causes seizures at a rate that increases with dosage: The annual incidence is 1% at dosages below 300 mg/day, 2.7% at dosages from 300 to 599 mg/day, and 4.4% at dosages between 600 to 900 mg/day. It has many systemic side effects, including fever, hypotension, anticholinergic effects, weight gain, hypersalivation, and suppression of the bone marrow. Agranulocytosis occurs in about 1% of patients.

Olanzapine and ***quetiapine*** are atypical antipsychotic agents recently approved by the FDA that have fewer extrapyramidal side effects than traditional drugs. There is little or no neuropsychiatric experience with these agents.

When choosing a neuroleptic for neuropsychiatric indications, the following guidelines may be useful:

1. In acute behavioral emergencies in which a neuroleptic is to be used for short-term stabilization or "chemical restraint," haloperidol and risperidone are the best agents. Dosage is titrated every hour, with increments of 0.5 to 2 mg for fragile patients, and 5 to 10 mg for patients who are physically robust. A combination of one of these drugs with lorazepam (0.5 to 1 mg for fragile patients; 2 mg for robust patients) gives a more rapid behavioral response.
2. When the patient's primary problem appears to be overwhelming anxiety with some psychotic features, such as mild paranoia or disorganization, a sedative neuroleptic, such as thioridazine or perphenazine, may be chosen for treatment. Olanzapine also appears effective in such situations. Typical starting dosages of thioridazine would be 10 mg t.i.d. for fragile patients and 25 mg t.i.d. for more robust patients; dose is increased as tolerated, with sedation, hypotension, and anticholinergic effects usually the limiting factors.
3. For psychotic phenomena, such as delusions and hallucinations in people with

TABLE 4-2. *Issues Relevant to Choosing and Initiating a Neuroleptic in Neuropsychiatric Patients*

Drug	Initial Dose (mg)	Usual Max. (mg)	Motor	Anticholinergic	Hypotensive	Sedative	Special Benefits	Special Concerns
Haloperidol	0.5–1.0	5–10	+++	+	±	+	Can give IV; depot form; cardiac safety	Frequent extrapyramidal effects, some malignant
Fluphenazine	0.5–1.0	5–10	+++	+	±	+	Depot form; least effect on seizures	Same as haloperidol
Thiothixene	1.0–2.0	10	++	+	±	+	Somewhat less motor effect than haloperidol	
Perphenazine	2–4	16–24	++	++	±	++	Antianxiety effect; combines well with TCA for psychotic depression	Drug interaction: raises TCA blood levels
Molindone	5–10	50–100	++	0	0	+	Lack of systemic side effects; does not cause weight gain	No IM preparation available; not sedative; poor choice for behavioral crises
Chlorpromazine	10–25	100	+	+++	+++	+++	Marked sedation may be useful in crises; good antiemetic	Systemic side effects risky in frail elders; causes seizures
Thioridazine	10–25	100–150	+	+++	++	++	Lowest motor effect of “typical” antipsychotics	Strong quinidine-like effects on heart; lack of IM preparation
Mesoridazine	10–25	100	+	+++	++	++	Like thioridazine, but available IM	Like thioridazine
Risperidone	0.25–0.5	4–6	±	±	+	++	Rare motor side effects at low doses	High cost; hypotension, . especially if dose raised quickly; no IM prep
Clozapine	12.5–25	200	0	+++	+	+++	Virtually no motor side effects; drug of choice in patients with Parkinson’s disease and psychosis	Weekly blood counts required; many systemic side effects; may cause seizures

TCA, tricyclic antidepressant.

gross brain disease who do not present a behavioral emergency, risperidone is a first-choice agent. Starting doses range from 0.5 mg at bedtime to 1 mg b.i.d. in a more severely disturbed yet physically robust person. The dose is increased if necessary and as tolerated to 3 mg b.i.d.

Clozapine is the drug of choice for psychosis in PD.

Dosage Titration

In nonemergent situations, the therapeutic dosage is initiated from lower initial doses. Dosage is increased weekly until behavioral symptoms are clearly improved; the dosage is then kept at that level for another month to assess its effect on thought disorder before considering further dosage increases.

When extrapyramidal reactions occur, the first consideration is whether dosage reduction or a switch to risperidone is reasonable in the given patient. If these measures are not feasible or have been done, drug treatment of the reaction comes next. First-line drugs for treatment of extrapyramidal reactions are benztropine or trihexyphenidyl for tremor, acute dystonia, or acute dyskinesia; amantadine or bromocriptine for rigidity and akinesia; and propranolol for akathisia.

Drug	Typical starting dosages	Typical maximum dosages:
Benztropine	2 mg b.i.d. to q.i.d.	10 mg per day
Trihexyphenidyl	5 mg b.i.d. to q.i.d.	20 mg per day
Amantadine	100 mg b.i.d. to t.i.d.	300 mg per day
Bromocriptine	2.5 mg q.d. to b.i.d.	30 mg per day
Propranolol	20 mg t.i.d.	360 mg per day

Fragile patients should begin at about half the usual starting dose. Rapid upward dosage titration should be carried out until symptoms are relieved.

Duration of Therapy

A reasonable approach is to attempt tapering after the patient has had stable behavior or well-controlled psychotic symptoms for 3 months. The drug should then be tapered over the next 3 months, with an interruption of the taper if clinically significant symptoms recur.

Antianxiety Drugs

The category of antianxiety drugs reviewed in this section comprises the benzodiazepines and the $5HT_{1A}$ partial agonist drug buspirone. However, many other classes of drugs have been used to treat specific anxiety disorders. The SSRIs are the drugs of choice in patients with OCD, and numerous antidepressants have been effective for treating recurrent panic attacks. The more sedating neuroleptics may be the drugs of choice for anxiety associated with paranoid phenomena and disorganized thinking. Anxiety due to hypomania is appropriately treated with lithium or a mood-stabilizing AED.

Efficacy

The effectiveness of benzodiazepines for symptoms of anxiety and for insomnia has been well established in patients without significant neurologic disease. Buspirone is more effective than placebo for symptoms of generalized anxiety but is less consistently effective than the benzodiazepines (8).

Although there are no RCTs specifically confirming the antianxiety effects of benzodiazepines or buspirone in neurologically ill populations, there is general consensus among clinicians that benzodiazepines are therapeutically effective for treating anxiety in patients with gross brain disease.

Other Neuropsychiatric Indications

Benzodiazepines have been used effectively for the following additional indications:

- Myoclonus, including nocturnal myoclonus
- Parasomnia associated with slow-wave sleep

- Rapid behavioral stabilization of acutely agitated patients
- Catatonia

Neuropsychiatric indications for buspirone supported by recent uncontrolled experience include the following:

- Self-injurious behavior in some adults with mental retardation
- Tardive dyskinesia
- Hostility and irritable "type A" behavior

Neurotoxicity

The neurologic side effects of the benzodiazepines are directly related to their effect on the benzodiazepine receptor to facilitate GABA-related inhibitory neurotransmission. They include cognitive impairment, memory impairment, decreased alertness, and impaired coordination. The more severe forms of these side effects include confusion, amnesia, lethargy, apathy, somnolence, ataxia, and falls. These neurologic side effects are more likely to occur in elderly patients.

Drug–Drug Interactions

The benzodiazepines subdivide into two groups according to their metabolism: Some are metabolized by hepatic oxidation; others are conjugated with glucuronide and are eliminated by the kidney. The benzodiazepines that are oxidized are more often involved in pharmacokinetic interactions because any drug that inhibits the relevant hepatic oxidative enzymes can raise benzodiazepine levels and prolong the drug's half-life. Likewise, agents that induce hepatic oxidative enzymes can reduce blood levels of these benzodiazepines and shorten their effective duration of action. Diazepam, chlordiazepoxide, alprazolam, and clonazepam all undergo oxidative metabolism; oxazepam and lorazepam do not. Common drugs that induce benzodiazepine metabolism include theophylline and carbamazepine, with clinically significant effects of the latter drug on the metabolism of alprazolam and clonazepam. Common drugs that inhibit benzodiazepine metabolism include valproate, SSRIs, nefazodone, erythromycin, and cimetidine.

Choice of Antianxiety Drug

Benzodiazepines are the drugs of first choice in situations of acute anxiety or panic and in situations in which a specific benzodiazepine effect, such as an antiepileptic action, is desired.

When benzodiazepines are to be used for treating insomnia, the important distinction is between use for 1 or 2 days and use for 1 week or more. For very short-term use, long-acting drugs, such as flurazepam, offer the advantage of minimal rebound insomnia after the drug is stopped, and their propensity to accumulate is not relevant. When a drug is to be used for 1 week or more, a drug with a medium half-life, such as temazepam, should be selected.

Dosage Titration

Dosage titration for buspirone must deal with the wide interpatient variability in first-pass metabolism for this drug. For some patients, 5 mg t.i.d. is an adequate dose; for others, 20 mg t.i.d. has little effect. Accordingly, patients should be started on 2.5 mg b.i.d. to t.i.d. if they are fragile and 5 mg t.i.d. if they are not, and then, if there are neither therapeutic effects nor side effects, the dose should be raised every 2 to 3 days until the patient gets relief or develops a limiting side effect, or a dose of 20 mg t.i.d. is reached. If a patient reaches 20 mg t.i.d. without side effects but without benefit, the dose should be maintained for 2 weeks. If there is still no effect, the dosage may be raised to 30 mg t.i.d. for an additional 2 weeks.

When short-acting benzodiazepines are used, patients are started on half of the manufacturer's recommended dosage, given as divided doses. The dose can be raised every other day until the patient gets relief or limiting side effects develop. When long-acting agents are used, 2 weeks should be allowed between dose increases.

Duration of Therapy

Anxiety disorders can have either a relapsing-remitting or a chronic course. After it is established that the disorder has a chronic course, the patient can be kept on antianxiety drugs indefinitely, as long as there are few or no side effects. If the course of the disorder is unknown, or the patient has never had a trial off medication since it was started, gradual withdrawal of medication may be attempted 3 to 6 months after the patient is free of major symptoms of the anxiety disorder. Buspirone should be tapered over about 1 month; benzodiazepines over 2 to 3 months. If anxiety symptoms emerge during the taper, nonpharmacologic treatment should be provided. If an unacceptable level of symptoms recur, dosage should be restored to the minimum that controls the symptoms.

Lithium

Efficacy

Lithium has been shown to be an effective agent for acute mania in 70% to 80% of manic patients. Well-designed double-blind studies have shown it to prevent relapses in about two thirds of patients with bipolar disorder. RCTs have not been extended to the population of patients with mania secondary to gross brain diseases. Other reports support the use of lithium for augmentation of antidepressants in major depression and combined use with neuroleptics in the treatment of schizoaffective disorder (9).

Other Neuropsychiatric Indications

Lithium has been used effectively for the treatment of the following neuropsychiatric disorders:

- Impulsive aggression
- Self-injurious behavior
- Irritability and aggression in patients with Huntington's disease
- Cluster headache prevention

Neurotoxicity

The occurrence of neurologic side effects rises rapidly with lithium levels greater than 1.5 mEq/mL. Neurologic side effects of lithium seen within the usual therapeutic range include impaired concentration or memory, apathy, restlessness, myoclonus, and tremor. Other side effects seen with toxic levels of lithium are confusion, hallucinations, and rigidity resembling that produced by neuroleptics. At extreme levels of lithium (more than 2.5 mEq/L), patients can develop seizures and may develop muscular weakness.

Lithium Versus Alternatives

As a mood-stabilizing agent, the major alternatives to lithium are the mood-stabilizing AEDs, of which carbamazepine and valproate are best established.

Drug–Drug Interactions

Because lithium is eliminated by the kidney in a parallel transport system with sodium, drugs and clinical situations that increase sodium retention also increase lithium retention. Drugs that can raise lithium levels in this way include thiazide diuretics and nonsteroidal antiinflammatory drugs (NSAIDs).

Coadministration of lithium with a neuroleptic increases the risk for an acute extrapyramidal reaction over the risk associated with the neuroleptic alone. Coadministration of lithium with an antidepressant increases the risk for tremor, myoclonus, or cognitive side effects. Coadministration of lithium with an AED increases the risk for confusion, ataxia, sedation, or tremor.

Dosage Titration

The target blood level for dosage titration is 1.2 to 1.4 mEq/L. In elderly, demented, or parkinsonism patients, in whom the risk is higher, the target blood level is 0.8 to 1 mEq/L. A small, fragile patient with normal or perhaps slightly decreased renal function might be started on 300 mg b.i.d. A large, robust patient with known normal renal function might be started on 600 mg t.i.d.

During aggressive dosage titration, blood levels are determined every other day. The patient is also examined for neurologic symptoms, which are correlated with the blood level obtained at the time of the examination.

Duration of Therapy

If the patient has bipolar disorder that is simply altered in its presentation because of brain disease (e.g., mental retardation or dementia), prolonged treatment should be carried out. If the disorder is the consequence of an acute brain insult, such as stroke or TBI, it would be reasonable to attempt to taper the lithium after the patient had time to recover from the insult. If the patient's behavioral syndrome is related to a progressive or degenerative brain disease, a reasonable time to attempt tapering of lithium is when clinically significant progression of the disease has been noted.

Antiepileptic (and Mood-Stabilizing) Drugs

Clonazepam has been broadly accepted as a treatment for anxiety and panic, and both ***carbamazepine*** and ***valproate*** have been accepted as effective in the treatment of mania and as prophylaxis of bipolar disorder. Issues relevant to choosing and initiating lithium or a mood-stabilizing AED are presented in Table 4-3.

Efficacy

The efficacy of carbamazepine for the treatment of acute mania and its comparability to lithium for this purpose have been demonstrated by 11 RCTs. The effectiveness of valproate in treating acute mania is supported by 6 RCTs. None of the controlled studies of carbamazepine or valproate addressed the use of those agents in patients with mood disorders secondary to gross brain disease, epilepsy, or dementia. Both lamotrigine and gabapentin have been reported in small series to be useful as alternatives or adjuncts to carbamazepine and valproate for the treatment of bipolar disorder, particularly atypical or treatment-refractory cases.

Other Neuropsychiatric Indications

The AEDs have been used empirically in neuropsychiatric conditions in which paroxysmal brain activity is part of a hypothesized mechanism of symptom production. Some of the potential indications for AEDs include the following:

- PTSD (carbamazepine and valproate)
- Neuropathic pain (carbamazepine, valproate, and clonazepam)
- Aggressive behavior (carbamazepine and valproate)
- Treatment-resistant depression with partial seizure-like symptoms (carbamazepine)
- Self-injurious behavior in mentally retarded patients (carbamazepine)
- Withdrawal from long-term benzodiazepine therapy (carbamazepine)
- Affective symptoms in people with mental retardation (valproate)

Neurotoxicity

The AEDs, with the exception of felbamate, all share the tendency to cause sedation, ataxia, nystagmus, and other signs of central nervous system (CNS) depression. Because of reported hepatotoxicity, felbamate can be used only in very limited circumstances. In addition, several of the AEDs have characteristic neurologic side effects.

Carbamazepine toxicity often includes diplopia. Carbamazepine is also anticholinergic and can cause typical signs of anticholinergic toxicity in susceptible patients.

Valproate often causes an action tremor. Valproate also raises serum ammonia, and signs of hepatic encephalopathy can develop in patients with occult liver disease treated with this drug. Phenytoin usually produces nystagmus and ataxia at toxic levels. Prolonged exposure to phenytoin at excessive levels can cause permanent cerebellar damage.

TABLE 4-3. *Issues in Choosing and Initiating Lithium or Antiepileptic Drugs*

Drug	Benefits	Precautions	Controlled Trials	Initial Dose
Lithium	Low cost; very well-known side effects and established monitoring schedule	Low therapeutic index; may induce hypothyroidism; causes tremor and may aggravate parkinsonism; synergistic toxicity with neuroleptics	None; case series support efficacy for mania due to gross brain disease, impulsive aggression, and self-injurious behavior	300 mg b.i.d.; usual target blood level about 1.0 mEq/mL
Carbamazepine	Efficacy for rapid cycling; antiepileptic effects due to enzyme induction; rare hematologic side effects require patient warning but not routine monitoring of CBC	Quinidine-like cardiac effects; anticholinergic effects; drug interactions	None; case series support efficacy for mania due to gross brain disease and impulsive aggression	100 mg b.i.d.; slow titration to target blood level of 8 to 12 μg/dL when the drug is used as sole therapy
Valproate	Efficacy for rapid cycling, antianxiety effects; antiepileptic effects	Rare hepatic toxicity requires warning but not routine monitoring of enzymes; can cause pancreatitis or hyperammonemia; weight gain common; may cause tremor	None; case series support efficacy for mania due to gross brain disease, and for mood instability and aggression in mentally retarded and dementia patients	250 mg b.i.d. (125 mg if unusually sensitive to side effects); gradual titration to blood level of 50–100 μg/mL: may load rapidly in case of behavioral emergency due to mania
Clonazepam	Strong antianxiety and antipanic effects; sedative; treats myoclonus	Long half-life predisposes to accumulation and risk of falling; ataxia and sedation the main side effects that limit therapy; some risk of disinhibition	None; case series support efficacy for anxiety, panic, and agitation due to gross brain disease	0.25–0.5 mg, usually at bedtime
Gabapentin	Antianxiety and mood-stabilizing actions suggested by patients' experiences in epilepsy trials; no significant drug interactions	Additive sedation and ataxia with other drugs	No systematic trials of psychotropic activity	100 mg t.i.d.; may raise gradually to maximum of 900 t.i.d., but mechanism of action implies diminishing effect of increases
Lamotrigine	Subjective well-being improved in patients enrolled in epilepsy clinical trials	Interactions with valproate; can cause severe skin rashes, especially if dose raised rapidly	None; anecdotes support efficacy for rapid-cycling mood disorders	50 mg at bedtime or b.i.d.; 25 mg every other day if patient is taking valproate because of increased risk for rash

CBC, complete blood count.

Clonazepam toxicity typically includes sedation and ataxia. An apathetic confusional state can also occur. Lamotrigine and gabapentin tend to have side effects of sedation, dizziness, ataxia, or impaired coordination.

Lamotrigine and gabapentin are new AED's whose utility in neuropsychiatric disorders is yet unknown.

Drug–Drug Interactions

Carbamazepine concentrations are increased by drugs that inhibit the enzymes that metabolize it or that compete with carbamazepine at the active sites of those enzymes. Calcium-channel blockers (verapamil and diltiazem), macrolide antibiotics (erythromycin and clarithromycin), and cimetidine all can raise carbamazepine levels to a clinically significant degree. Carbamazepine can induce enzymes that metabolize other drugs as well as carbamazepine itself. The following are four important examples of drugs metabolized by carbamazepine:

- Warfarin
- Oral contraceptives
- Opiate analgesics
- TCAs

Valproate inhibits both oxidative metabolism and glucuronide conjugation of a number of drugs. With coadministration, lorazepam has a higher level and a longer elimination half-life. Valproate also prolongs the metabolism of lamotrigine, raising its half-life by 15% to 25%. Valproate is highly bound to serum albumin. Aspirin displaces valproate from serum albumin, raising the free drug level, sometimes to the toxic range.

Pharmacodynamic interactions involving the AEDs generally consist of additive CNS depression. However, carbamazepine also decreases dopamine turnover and can intensify extrapyramidal side effects when coadministered with neuroleptics (10).

Choice of Antiepileptic Drug

When choosing an AED for mood stabilization in a neuropsychiatric patient, valproate and carbamazepine can be viewed as equivalent. When an AED is used as an antidepressant, carbamazepine is preferable to valproate because there is more clinical evidence of carbamazepine's antidepressant effects. Alternative AEDs for various neuropsychiatric situations include lamotrigine and gabapentin.

Dosage Titration

The dosages and blood levels appropriate for the treatment of epilepsy are followed when using AEDs as psychotropics. Therapeutic dosages and levels are usually approached gradually. When AEDs are discontinued for reasons other than serious adverse reactions, discontinuation is by gradual taper. A typical, conservative dosage strategy might be as follows:

Carbamazepine. Start with 100 mg q.h.s. If tolerated, increase every 2 to 3 days by 100 mg, building up to an evenly divided t.i.d. schedule (thus, 100 mg b.i.d., 100 mg t.i.d., then 100–100–200 on a t.i.d. schedule, etc.). A predose blood (trough) level would be checked after reaching 200 mg t.i.d. The target level should be 8 to 12 μg/mL. When the target blood level is reached, the dosage would then be held constant for another 2 weeks.

Valproate. Start with 250 mg q.h.s. If tolerated, increase every 2 to 3 days by 250 mg, building up to a t.i.d. schedule (thus, 250 b.i.d., 250 t.i.d., then 250–250–500, etc.). A trough blood level should be checked at 1000 mg per day and weekly thereafter. Dosage increases should continue until a level of 75 to 100 μg/mL is reached.

Lamotrigine. Start with 50 mg q.h.s. If tolerated, increase by 50 mg every week on a b.i.d. schedule, eventually reaching 200 mg b.i.d.

Gabapentin. Start with 100 mg q.h.s. Build up in 100-mg increments every 2 to 3 days, on a t.i.d. schedule, eventually reaching 600 mg t.i.d. After reaching 300 mg t.i.d., increments may be 300 mg.

Duration of Therapy

Considerations regarding the duration of therapy with AEDs as mood stabilizers are identical to those presented earlier for lithium therapy. For a review of AED therapy, see reference 11.

Stimulants

Stimulants, such as dextroamphetamine, methylphenidate, and pemoline, are the well-established drug treatment for attention deficit hyperactivity disorder (ADHD). In sleep medicine, they are the treatment of choice for narcolepsy. In consultation psychiatry, they are used to rapidly mobilize hospitalized medically ill patients with apathy and depressed mood. In neuropsychiatric practice, additional applications draw on the stimulants' capacity to enhance arousal, motivation, and the ability to sustain attention (12).

Other Neuropsychiatric Indications

Poststroke depression. Methylphenidate may be able to produce a remission of depressive symptoms after stroke.

ADHD among patients with developmental disabilities. ADHD symptomatology in patients with mental retardation and other developmental disabilities has improved with the use of stimulants.

Apathy. Methylphenidate or dextroamphetamine can reduce apathy in patients with a variety of causes for apathy, including depression, TBI, and adverse effects of AEDs, antidepressants, or neuroleptics.

Neurotoxicity

In usual therapeutic doses, neurologic side effects of the stimulants include insomnia, irritability, nervousness, headache, and tremor. Acute overdose can produce an agitated confusional state with hyperreflexia and seizures; a manic-like psychosis is also possible. Prolonged use at unusually high doses can produce a paranoid psychosis with features of acute schizophrenia. Withdrawal from long-term use can be associated with headache, apathy, fatigue, and hypersomnia. Some children with ADHD treated with stimulant medications develop tics or dyskinesias.

Drug–Drug Interactions

Methylphenidate inhibits oxidative enzymes that metabolize TCAs, warfarin, and AEDs, including phenobarbital, phenytoin, and primidone. All of these drugs may require downward dosage adjustments to avoid toxicity.

All of the stimulants have pharmacodynamic interactions with psychotropic drugs. They enhance antidepressant effects. Their stimulating and anorexiant effects are antagonized by neuroleptics and by lithium.

Choice of Stimulant

Issues related to the choice of stimulant and to initiating therapy are mentioned in Table 4-4 along with similar details for the dopamine agonist drugs.

Dosage Titration

Initial dosage of dextroamphetamine or methylphenidate is 2.5 mg given in the morning and at noon in a fragile or small person; 5 mg in the morning and at noon in a more robust and larger person. Therapeutic effects are immediate. If there are neither side effects nor an apparent therapeutic response, the dose is increased the next day to 5 mg given in the morning and at noon or 10 mg in the morning and at noon, respectively. Dosage may be increased further in units of 2.5 mg for the small or fragile patient and of 5 mg for the larger and healthier patient.

Pemoline is started at 18.75 mg once a day. The dose is raised by 18.75 mg once a week until the patient improves, limiting side effects develop, or a dose of 112.5 mg per day is reached.

TABLE 4-4. *Issues Relevant to Choosing and Initiating a Dopamine Agonist or Stimulant*

Drug	Benefits	Precautions	Controlled Trials	Initial Daily Dose
Bromocriptine	Antidepressant, antiparkinsonian, antiapathetic effects	Orthostatic hypotension; nausea; confusion; depression or mania; ergot hypersensitivity (pulmonary or retroperitoneal fibrosis)	Many for Parkinson's disease and several for primary major depression; case series support for apathy and abulia from gross brain disease, and for drug-induced parkinsonism	2.5–5 mg; drug is usually given on a b.i.d. schedule
Pergolide	Can be given once a day because of long half-life; may be less expensive because of high potency; otherwise like bromocriptine	As for bromocriptine; some patients however, do better on one than another	Many of Parkinson's disease; case series support for depression; case reports for apathy and abulia	0.05–0.10 mg q.d.
Amantadine	Antiparkisonian, antiapathetic; rapid onset of action	Orthostatic hypotension, confusion, anticholinergic effects	In drug-induced parkinsonism has fewer cognitive side effects than benztropine	50–100 mg b.i.d.
Dextroamphetamine	Improves alertness, attention, motivation	May increase anxiety or induce mania; subject to abuse because of euphoriant effects	For narcolepsy; case series support benefit for apathy and for depression in the medically ill	2.5–5 mg morning and noon
Methylphenidate	Improves alertness, attention, motivation	May increase anxiety or induce mania; subject to abuse because of euphoriant effects	Many of ADHD; one for depression in the medically ill; case series support benefit for apathy	2.5–5 mg morning and noon
Pemoline	Improves alertness, attention, motivation; less subject to abuse than dextroamphetamine and less tightly controlled; longer duration of action permits once-a-day dosing	Onset of action may take weeks; dysphoric reactions more common than with dextroamphetamine; overall efficacy probably lower; may induce mania or anxiety	Many for ADHD; no studies in other neuropsychiatric conditions	18.75 mg q.d.

ADHD, attention deficit hyperactivity disorder.

Duration of Therapy

Treatment for ADHD and other attentional disorders is long term. In children, drug-free intervals should be scheduled to allow for growth unimpeded by the stimulants' anorectic and potential growth-suppressing effects. For adults with attentional disorders, tolerance may become a pharmacologic problem. If an adult on stimulant therapy for ADHD finds that the benefits of stimulants are wearing off, a drug holiday should be considered.

When stimulants are used to treat illness-associated depression, the drugs should be withdrawn once the acute illness has resolved or stabilized. If the patient's depression persists after withdrawal of the stimulant, a standard antidepressant should be initiated. If stimulants are given for a week or more, they should be withdrawn gradually, ideally by no more than 25% every 2 to 3 days.

When stimulants are used to treat apathy, therapy may continue until the underlying cause of the apathy has resolved. If the cause of the apathy is permanent, stimulant therapy may continue indefinitely.

Dopamine Agonists

The dopamine agonist antiparkinsonian drugs *selegiline*, *amantadine*, *bromocriptine*, and *pergolide* have been used for neuropsychiatric indications other than PD. The best established neuropsychiatric indication is bromocriptine for treatment-resistant depression.

Other Neuropsychiatric Indications

Dopamine agonists have other neuropsychiatric indications, including the following:

- Neuroleptic-induced apathy and abulia
- Agitation and assaultiveness during recovery from traumatic coma
- Akinetic mutism
- Hemineglect
- Amnesia due to mediobasal forebrain injury
- Transcortical motor aphasia

Neurotoxicity

The dopamine agonists can cause confusion, agitation, insomnia, hallucinations, paranoid ideation, disorganized thinking, incoherent speech, emotional lability, hypersexuality, hypomanic and manic phenomena, depression, or anxiety. They can also cause involuntary movements, including choreoathetosis, dyskinesia, and tremor.

Drug–Drug Interactions

At doses of selegiline above 10 mg per day, this selective type B MAOI loses its selectivity. When monoamine oxidase is nonselectively inhibited, patients can develop serious hypertension from indirect adrenergic agents or manifest a serotonin syndrome if they take an SSRI.

Choice of Dopamine Agonist

When a dopamine agonist drug is to be used for neuroleptic-induced parkinsonism, amantadine is usually selected first because of extensive experience with its use for this indication. If amantadine at a full dose does not relieve the patient's symptoms, bromocriptine is used next.

Dosage Titration

The starting dose of *amantadine* is 50 mg b.i.d. in a small or fragile person; the starting dose is 100 mg b.i.d. in a larger and more robust person. Dosage increases should be in 50-mg increments in the former case; 100-mg increments in the latter case. Dosage usually should not exceed 100 mg t.i.d. Dosage must be reduced in renal insufficiency.

The starting dose of *selegiline* is 5 mg q.d. This can be increased to 5 mg b.i.d. after 2 days if the initial dose is tolerated.

The starting dose of *bromocriptine* is 1.25 mg b.i.d. in a small or fragile patient and 2.5 mg b.i.d. in a larger and more robust person. Dosage increases are in steps of 1.25 mg in the former type of patient and 2.5 mg in the latter type. There is no absolute maximum dosage for neuropsychiatric indications if the drug is well tolerated; however, the dosage should not be increased beyond 10 mg t.i.d. unless there is clear evidence of benefit.

The starting dose of *pergolide* is 0.05 mg q.d. in a small or fragile person and 0.1 mg q.d. in a larger and more robust person. The dose can be raised in increments of 0.05 to 0.1 mg per day to a total of 5 mg per day.

Duration of Therapy

Duration of therapy for major depression is 6 to 12 months, followed by an attempt to taper the medication. When dopamine agonists are used for drug-induced parkinsonism, efforts should be made periodically (e.g., at 2 and at 4 months) to reduce the dosage (13).

Adrenergic Agents

The adrenergic agents most often used in neuropsychiatry are the β-adrenergic blockers, especially *propranolol* and *clonidine*, a mixed α_1- and α_2-agonist. In addition to their original indication of hypertension, these agents have well-established efficacy for several neuropsychiatric indications and have been tried for numerous other ones.

Efficacy

The effectiveness of propranolol has been established for migraine prophylaxis and for treatment of essential tremor. The effectiveness of clonidine has been established for reducing tics in Tourette's syndrome and decreasing the intensity of withdrawal symptoms after discontinuation of opiates.

Other Neuropsychiatric Indications

Both propranolol and clonidine have been used in a number of conditions in which anxiety or symptoms of anxiety are present or inferred, including the following:

- ADHD; the dose, reached by gradual titration, is 3 to 10 μg/kg/day. Propranolol has been used as an adjunct to stimulants, to reduce impulsive behavior, at a dose of 2 to 8 mg/kg/day.
- Agitation in dementia
- Rage attacks in children, adolescents, and adults with structural brain dysfunction
- Akathisia due to neuroleptics

For a review of aggressive behavior in neuropsychiatric patients, see reference 14.

Neurotoxicity

The most common neurologic adverse effects of propranolol and clonidine are fatigue, sedation, and decreased libido. The fatigue and sedation may combine with other symptoms to generate a depressive syndrome.

Drug–Drug Interactions

Pharmacokinetic interactions with propranolol are related to its oxidative metabolism. Propranolol levels are raised by coadministration of chlorpromazine or cimetidine; they are lowered by coadministration of phenytoin or phenobarbital. Propranolol inhibits the metabolism of theophylline, leading to increased serum drug concentrations at a fixed dose.

The major reported interaction of clonidine is with TCAs. Coadministration decreases the hypotensive effect of clonidine. Pharmacodynamic interactions occur with other antihypertensive drugs (additive or synergistic antihypertensive effect) and with other drugs with sedative or CNS depressant effects (increased lethargy, fatigue, and depressive symptoms).

Decision to Use an Adrenergic Agent

Issues related to the choice and initiation of an adrenergic agent are summarized in Table 4-5.

TABLE 4-5. *Issues Relevant to Choosing and Initiating an Adrenergic Agent*

Drug	Benefits	Precautions	Controlled Trials	Initial Daily Dose
Propanolol	Antianxiety, helpful for migraine, tremor, irritability, impulsive aggression	Sedation; hypotension; can aggravate asthma or heart failure; various drug interactions related to its oxidative metabolism	Migraine, essential tremor; case series support utility for agitation and aggression, and for neuroleptic-induced akathisia	20–40 mg t.i.d.
Nadolol	Less fat-soluble and more β_1 selective; therefore, less sedation and aggravation of asthma; may help impulsive aggression; long half-life so may be taken once a day	As with propranolol	None; open trials support benefit for aggression in mental retardation	40 mg
Clonidine	Major anxiolytic effects, including blocking of anxiety from drug withdrawal—decreases firing of locus ceruleus	Sedation, hypotension, confusion	ADHD, Tourette's syndrome, opiate or nicotine withdrawal; case series support use for impulsive aggression and memory loss in Korsakoff's syndrome	0.1 mg q.d. to t.i.d., depending on indication
Guanfacine	Decreases firing of locus ceruleus; stimulates frontal lobe α_2 receptors to improve executive function; long duration of action permits once-daily dosing	As with clonidine, but less sedation; can cause dyspepsia or nausea	ADHD; anecdotal support for use in impulsive aggression	1 mg q.d.

ADHD, attention deficit hyperactivity disorder.

Opiate Receptor Blockers

Oral opiate receptor blockers, of which naltrexone is the only one currently available in the United States, were introduced as an aid in the treatment of opiate abuse and, subsequently, in the treatment of alcoholism. These drugs may be tried in other forms of habitual behaviors, particularly those that involve some self-inflicted pain or discomfort (15).

Other Neuropsychiatric Indications

Other neuropsychiatric indications for naltrexone include the following:

- Self-injurious behavior in mentally retarded adults
- Autism
- Bulimia
- Tourette's syndrome

For a sample review of these indications, see reference 16.

Neurotoxicity

Patients on naltrexone have reported a wide range of mental symptoms, including restlessness, insomnia, nightmares, confusion, hallucinations, paranoia, and fatigue.

Drug–Drug Interactions

There is only one drug interaction of consequence: the precipitation of an acute and severe withdrawal state if the drug is given to a person still taking narcotics on a regular basis.

Dosage Titration

No titration is needed. A fixed dose of 50 mg per day, 100 mg every other day, or 150 mg every third day should provide adequate opiate receptor blockade.

Duration of Therapy

After blocking opiate receptors, a brief rebound of the behavior that was previously reinforced by endogenous opiate release can occur. Behavioral and environmental therapies to extinguish the unwanted behavior should be instituted, taking advantage of the change in internal reinforcement. An attempt to taper and discontinue naltrexone can be made once the patient has a stable period without significant self-injury.

Conclusion

A rich literature of clinical reports and some controlled trials provide the neuropsychiatrist with an empirical basis for planning drug therapy in a neuropsychiatric practice. Generally accepted uses of available neurotropic agents have been described, along with dosing suggestions and drug interactions.

REHABILITATION THERAPY

The World Health Organization (WHO) has defined an *impairment* as "caused by the underlying disorder resulting in clinical signs and symptoms" (17). *Disability* is defined as "reflecting the personal limitations imposed upon the activities of daily living (ADLs)." *Handicap* reflects the environmental situation that limits the person with a disability from achieving an optimal social role, and it results from a combination of disability, impairment, and a number of other factors, including resources, intelligence, and psychological status.

The scope of rehabilitation therapy relates broadly to function, disability, and handicap.

Clinical Issues

Bringing an impairment closer to its original, premorbid function is called *restorative* rehabilitation. This is the therapy that is attempted in most rehabilitation units. Other patients with chronic neurologic impairments or progressive neurologic disease may also benefit from rehabilitative modalities. This is called *maintenance* rehabilitation.

Cognitive rehabilitation, has become a growing practice in recent years. There are limited outcome data bearing on the question of whether specific cognitive rehabilitation and remediation is more effective than general rehabilitation. For a review, see reference 18.

The most important goal for a rehabilitation program is the development of a life management plan. This plan should be a blueprint that allows a person with a disability to function at the highest possible level. The plan should have input from the therapists involved in the patient's management. Many professionals are involved in the development of this overall strategy, each with a distinct and an overlapping role. The physician oversees the medical process leading to the production of the blueprint. The physical therapist sets goals directing physical improvement in the function of body parts. The occupational therapist emphasizes daily living activities. The speech pathologist works with language, communication, swallowing, and cognition. Nursing reinforces all that is done by others and initiates bladder and bowel training regimens. Neuropsychologists evaluate and generate plans for improving behavior and cognitive deviations. Social workers evaluate the person's general role in society.

Patient Selection for Rehabilitation

In selecting a patient for rehabilitation of a neurologic problem, certain criteria must be fulfilled. Rehabilitation is not easy, and great physical and mental work are necessary for complete success. Thus, the patient must be able to withstand physically the significant time commitment. Most rehabilitation programs require 2 to 4 hours of solid rehabilitative therapies each day. A patient who is not capable of being up that long from bed each day, or who does not have that prospect in the near future, is not appropriate for a formal rehabilitation setting.

Cognition also plays a major role in determining eligibility. Although many neurologic patients have cognitive difficulties, some have such extensive problems that their capacity to learn and store new information is severely impaired. Excluding these patients, thus, becomes essential because the prognosis for effective rehabilitation under these conditions is poor.

It is important to identify alcohol and other drug problems before admission.

The patient must be able to focus attention on the business of rehabilitation. Although progressive neurologic disease may be rehabilitated, there should be some hope that the efforts will result in some period of increased functioning. The judicious use of medication to control impulses or to deal with anger or depression can be considered. Behavioral management programs can also be considered (19).

Indications

Depression

Neurologic illness may lead to frank depression without cognitive or behavioral deficits. This may be reactive or secondary to brain disease. Symptoms of depressive illness are apparent in three areas: affective regulation, somatic concerns, and cognition. Typically, there is withdrawal, poor concentration, and feelings of hopelessness. This may be accompanied by fatigue, constipation, decreased appetite, early-morning awakenings, and decreased libido.

Practice Hints: Depression

- Note changes from previous behavior patterns.
- Compare the patient's mood before and after the neurologic illness to determine the possibility of depression. Sometimes, a behavioral pattern that is not necessarily abnormal by many standards is clearly abnormal for a specific person. Deviation from previous baseline behavior can be an indicator of a depression that could inhibit the rehabilitative process if not treated.
- Obtain psychological testing and evaluation.

Multiple Sclerosis

Loss of motor function may be obvious in people with MS. Less obvious, but very important, may be cognitive dysfunction. In recent years, there has been stronger evidence of cognitive problems in many patients with MS, and in some, these are significant deficits.

Cognitive rehabilitative techniques for MS are widespread but not yet proved in terms of efficacy. Compensatory strategies using notebooks, word associations, tape recorders, and other devices are helpful.

At the outset, a clear explanation of MS is important. It is essential for the patient to understand what MS is and what it is not. It is vital to communicate that a diagnosis of MS is not an automatic prescription for a wheelchair. Education may allow for a certain amount of empowerment for the patient.

After the diagnosis is explained from a neurologic standpoint, adjustment must take place. This is eased by professional psychological help. Group therapy for newly diagnosed patients who do not have significant disabilities is very helpful. If the group of patients has a wide variety of disabilities, however, too much comparison and the shock of seeing the variety of MS-associated problems can detract from the therapeutic atmosphere (20).

Practice Hints: Multiple Sclerosis

- If cognitive decline is suspected, confirm with neuropsychological evaluation.
- If cognitive decline is evident, obtain baseline studies with a goal of teaching compensatory techniques. If full neuropsychological testing is not indicated, perform a screening evaluation. If patient job survival is at hand, full testing is necessary.
- Teach compensatory techniques. Compensatory tools (e.g., computer aids, notebooks) may be useful for many of the disabilities of MS, including the neuropsychiatric ones.
- Continue to stimulate intellectually (do not isolate).
- If cognitive problems surface, there is a tendency for the patient to withdraw. To try to prevent patient withdrawal, mainstream, use day care, and continue to include the patient in day-to-day activities. Denial of the cognitive problems will only cause exasperation for all involved.
- If the patient is depressed, use an appropriate antidepressant and counseling.

Chronic Fatigue Syndrome

Chronic fatigue syndrome (CFS) is an aggregation of symptoms that continues to elude clear diagnosis or case definition. One commonly used definition was developed by the Centers for Disease Control (CDC) in 1988, which involved two major criteria and 14 minor criteria. According to this definition, a patient must meet both of 2 major criteria and either 8 of the 11 symptom criteria or 6 of the 11 symptom criteria and 2 of 3 physical criteria. The two major criteria that must be met are the following:

1. New onset of persistent or relapsing, debilitating fatigue or easy fatigability in a patient with no previous history of similar symptoms, which does not resolve with bed rest, with the fatigue severe enough that the patient's average daily activity is reduced to below 50% of premorbid levels for at least 6 months
2. Other clinical conditions that may produce similar symptoms must be excluded by thorough evaluation.

Although complaints of fatigue are commonly encountered in primary care practice, a recent study suggests that the actual occurrence of CFS when using the CDC criteria may be quite low (0.3% point prevalence).

Practice Hints: Chronic Fatigue Syndrome

- Remain open to all possible causes of this syndrome. Patients with CFS frequently expect the professional to indicate that they have a mental health problem, causing them to move on to another professional, adding the physician to their long list of failures.

- Symptom management is essential to allow for a degree of comfort in an uncomfortable process. Pain should be managed without habit-forming medication. If depression is present, treat it as for other medical problems. Physical therapy may be useful for managing weakness.
- The newer energizing antidepressants are very helpful, even if clinical depression is not obvious. Explain to the patient that use of antidepressants does not indicate that the syndrome is viewed as psychiatric—this is essential to gain acceptance of the therapy by the patient.
- As with other chronic processes, follow-up is necessary. Follow-up shows interest on the part of the professional and reassures the patient.
- Counsel as appropriate. Counseling from a professional may be helpful if done in a supportive, nonpejorative manner.

Cerebrovascular Disease

Higher cortical functions become more important than paresis from a rehabilitation point of view. It is far easier to compensate for a plegic arm or leg than for an agnosia, apraxia, or most other neuropsychological disorders. Impairments in attention/concentration, memory, language, visuomotor/visuoperceptual skills, and the so-called executive functions (e.g., maintenance of mental set, organization, response inhibition) may be present, depending on the location or type of stroke involved. Complicating rehabilitative efforts may be the presence of denial or unawareness of deficit syndromes in some patients.

Various neurobehavioral problems are seen in the stroke patient. For example, patients with an executive dysfunction secondary to a frontal stroke may exhibit impulsivity or behavioral/emotional disinhibition. They may have difficulty controlling their frustration and may give up easily, or they may behave impulsively in ways that are counterproductive to their rehabilitative regimen and ultimately to their well-being (e.g., getting out of bed or out of their wheelchair without help). In some cases, they may lose patience so quickly that they are at risk of harming themselves or others.

Depression frequently accompanies stroke and may be a primary or secondary sequel to the cerebrovascular (CVA) accident. If the disease is diffuse, multifocal, or bilateral, a hyperemotional syndrome results. This usually presents with slurred speech, dysphagia, and gait apraxia.

The CVA rehabilitation process is similar to others, with the involvement of a team of professionals, including a physician, physical therapist, occupational therapist, speech pathologist, neuropsychologist, and nurse. Neurobehavioral abnormalities or tendencies may be recognized by any or all involved.

Because of the impairment of higher cortical function seen after many strokes, neuropsychological assessment should play a crucial role in the rehabilitation of these patients. Although impairments in higher cortical function may often be evident through clinical observation, the neuropsychological evaluation provides some objective documentation of deficits. In addition, subtle deficits that may not be evident by observation or detected by less sensitive procedures are more likely to be discovered in a comprehensive neuropsychological evaluation.

Practice Hints: Cerebrovascular Disease

- Undertake a thorough rehabilitative evaluation, using physical therapy, occupational therapy, and speech therapy professionals.
- Depression disorders can submerge and complicate rehabilitation attempts. Depression must be diagnosed and treated as quickly as possible.
- If there is doubt about the neuropsychological status of the patient, objective testing can be the basis for a more logical rehabilitative plan.

Parkinson's Disease

Parkinson's disease (PD) is a progressive disease of dopamine production that also causes a primary, slowly progressive dementia. Although PD usually occurs in the sixth and seventh decades of life, it may occur earlier. Depression commonly accompanies the progressive process. The most common psychiatric manifestations of PD are as follows:

- Delirium
- Dementia
- Depression

Significant dementia occurs in about 30% of patients. Neuropsychological deficits occur in up to 40% of those tested without obvious mental impairment. Frontal lobe dysfunction appears to be involved to some extent in the decreased cognition, as well as in the depression of patient's with PD.

Physical rehabilitation aims at increasing mobility. Stimulating the mobility of the mind should also be an aim of the rehabilitation team.

Practice Hints: Parkinson's Disease

- Assess ambulation and provide gait training for all patients.
- Treat depression with antidepressants when clinically indicated.
- Observe carefully for signs of cognitive deficits; consider compensatory techniques.

Neuromuscular and Skeletal Pain

Low back pain is among the more costly injuries in the United States. The pain may be intolerable, but often there is little evidence on formal examination to indicate severity of the injury. Workers' compensation laws have encouraged the growth of a whole industry of disability determinations.

Chronic pain is among the more disabling symptoms. It has strong implications in the neuropsychiatry of disability and rehabilitation. Despite major advances in pharmacologic approaches to pain and advances in aggressive surgical procedures, there is no solution for the management of persistent pain.

There are three steps in the rehabilitation of patients with pain:

1. Accurate physical and psychological diagnosis
2. Establishment of reasonable goals
3. Selection and implementation of treatments

The impact of pain on the patient and family must be assessed. Pain may be seen as a solution to a problem rather than the problem itself. It may legitimize unacceptable behavior. Litigation may slow the rehabilitation process considerably.

Practice Hints: Musculoskeletal Pain

- Avoid habit-forming medication.
- Attempt aggressive physical and occupational therapies.
- Eliminate litigation; initiate treatment after litigation is settled.
- Manage depression; depression should be treated as in any other disorder. It may be masked behind the guise of anger.

Conversion Reactions

Chronic diseases and disabilities often lead to disability behavior that appears beyond that expected by the impairment. Thus, the disability and the handicap appear exaggerated. If this exaggeration is feigned, the term *malingering* applies.

The patient who is experiencing a conversion reaction is unaware of the exaggeration of symptoms and findings beyond that expected from the physical impairment. The patient has displaced his or her anxiety or depression toward a physical symptom.

There is no standard rehabilitative approach to this situation. It appears more appropriate to work with the patient physically while trying to help psychiatrically in a more subtle fashion.

There is usually some secondary gain to be obtained from the exaggeration, and this must be limited or corrected to make progress (21).

Practice Hints: Conversion Reaction

- Neuropsychiatric counseling may become possible later in the management of the patient. If psychiatric counseling begins too early, long-term improvement will not occur.
- Behaviorally oriented treatment is typically most successful.

Traumatic Brain Injury

Traumatic brain injury (TBI) is a major problem, with 50,000 to 75,000 people each year in the United States suffering severe head injury.

At the time of injury, the person often looks quite different from a neuropsychiatric perspective than he or she will appear several months later. Although plasticity of the nervous system is well known in children, the extent to which the adult brain is plastic after injury is uncertain.

There are three causes of abnormal behaviors in the TBI patient:

1. Impaired cognition
2. Focal cortical injury
3. Psychological inability to adapt to injury-related stressors

The patient with TBI has problems with higher cortical functions and concentration. Behavior modification techniques, along with psychoactive medication, may be necessary to control the outbursts of anger and inappropriate behavior that may occur. A structured, controlled environment with trained staff is essential.

Alongside the hyperemotionality may be a hypersexuality. This may be expressed with aggressiveness secondary to the brain damage.

Goals set for a typical cognitive rehabilitation program include living and managing at home (daily activities, medication administration, independent living, increased endurance, preparation of light meals); leisure activities (community recreation, fitness programs, group activities, public transport, telephone); community living (management of frustration, social situations, budget, safety, group discussions, driver's license); and prevocational issues (vocational planning, complete multistep activities, increased endurance, punctuality).

Models of cognitive rehabilitation programs fall into three categories:

1. Functional adaptation
2. General stimulation
3. Process-specific approach

The functional approach involves environmental manipulations or compensatory strategies. General stimulation uses tasks that encourage cognitive processing. A process-specific approach is oriented toward target remediation of deficits in specific cognitive areas.

Coma stimulation programs have become more popular even though they have not been shown to change outcomes of patients in coma.

Practice Hints: Traumatic Brain Injury

- Many neuropsychiatric problems in TBI resolve with time. The brain does heal and settle down to some degree with time. Patience is essential in treatment.
- Neuropsychological evaluation is very helpful and should be repeated as necessary.
- Behavior problems are common with TBI. The staff needs to be aware of the concomitant behavior problems and to avoid, if at all possible, methods involving physical restraint.
- The family should be involved in the therapy.

For a review, see reference 22.

Spinal Cord Injury

Spinal cord injury (SCI) affects 7,000 to 10,000 Americans each year, with a prevalence rate of 150,000 to 200,000 cases, and most patients are young adults. A pure SCI spares the brain and the resulting secondary cognitive complications; however, it does not spare the self-image problems that accompany such a drastic injury. The typical pattern of SCI develops in a young male adult who suddenly moves from a strong image to a weak one. In addition to learning to live without the functional use of his legs, arms, or both, he has to change his own self-image. The emotional trauma is immense. The stages of grief have been noted previously and apply here.

Sexuality is particularly an issue for both men and women with SCI. Body image alteration is very prominent, and sexual function is always altered. Alternative ways of expressing sexuality need to be discussed, along with exploration of different ways of approaching sexual performance.

Practice Hints: Spinal Cord Injury

- Neuropsychological testing may reveal concomitant head injury.
- Depression is always present after SCI. Management should be aggressive with all modalities.
- Open discussion and counseling regarding body image are absolutely necessary.
- Aggressive physical and occupational therapy are essential.
- Any rehabilitation program should stress the positive.

Medicolegal Aspects of Disability

In addition to dealing with the challenging clinical issues among patients with disabilities, the rehabilitation professional is often expected to make decisions or determinations on vocational capacity that have legal implications. There are two primary reasons for the legal significance of such assessments:

1. Several forms of income replacement and benefits, such as Social Security disability insurance, workers' compensation, and long-term disability insurance, are based on the disabled person's projected lost earning capacity.
2. State and federal laws prohibit employment discrimination against otherwise qualified people with handicaps.

Because of the inherent difficulties involved in determining disability objectively, it is common that medical professionals differ in their conclusions on a particular patient, leading to an adversarial situation in which a nonmedical professional (i.e., judge) decides which opinion is more valid.

SUMMARY

This chapter has presented merely a cursory overview of some current issues in the neuropsychiatry of rehabilitation. This continues to be a developing specialty in which growth has been reflected in many areas in recent years, including the increase of rehabilitation hospitals and units with a neurobehavioral emphasis; the appearance of additional rehabilitation-oriented professional journals that encompass neuropsychiatric issues; and the development of professional organizations, such as the American Society of Neurorehabilitation. Despite the tremendous advances in interest and in the knowledge base, the efficacy of rehabilitation remains under study, particularly in the area of cognitive remediation and rehabilitation.

REFERENCES

1. Stoudemire A, Fogel BS, Gulley L, Moran MG. Psychopharmacology in the medical patient. In: Stoudemire A, Fogel S, eds. *Psychiatric care of the medical patient.* New York: Oxford University Press, 1993:155–206.
2. Gualteri CT. *Neuropsychiatry and behavioral pharmacology.* New York: Springer-Verlag, 1991.
3. Burke MJ, Preskorn SH. Short-term treatment of mood disorders with standard antidepressants. In: Bloom FE, Kupfer DJ, eds. *Psychopharmacology: the fourth generation of progress.* New York: Raven Press, 1995: 1053–1066.
4. Stoudemire A, Fogel BS. Psychopharmacology in medical patients: an update. In: Stoudemire A, Fogel BS, eds. *Medical psychiatric practice.* Vol 3. Washington, DC: American Psychiatric Press, 1995:79–150.
5. Moller HJ, Fuger J, Kasper S. Efficacy of new generation antidepressants: meta-analysis of imipramine-controlled studies. *Pharmacopsychiatry* 1994;27:215–223.
6. Starkstein SE, Robinson RG. *Depression in neurologic*

disease. Baltimore, MD: Johns Hopkins University Press, 1993:28–49.
7. Javaid JI. Clinical pharmacokinetics of antipsychotics. *J Clin Pharmacol* 1994;34:286–295.
8. DeVane CL, Ware MR, Lydiard RB. Pharmacokinetics, pharmacodynamics, and treatment issues of benzodiazepines: alprazolam, adinazolam, and clonazepam. *Psychopharmacol Bull* 1991;27:463–473.
9. Lenox RH, Manji HK. Lithium. In: Schatzberg AF, Nemeroff CB, eds. *Textbook of psychopharmacology.* Washington, DC: American Psychiatric Press, 1995: 303–358.
10. McElroy SL, Keck PE. Antiepileptic drugs. In: Schatzberg AF, Nemeroff CB, eds. *Textbook of psychopharmacology.* Washington, DC: American Psychiatric Press, 1995:351–376.
11. Levy RH, Mattson RH, Meldrum BS, eds. *Antiepileptic drugs.* 4th ed. New York: Raven Press, 1995:889–896.
12. Wilens TE, Biederman J. The stimulants. *Psychiatric Clin North Am* 1992;15:191–222.
13. Kahn RS, David KL. New developments in dopamine and schizophrenia. In: Floom FE, Kupfer DJ, eds. *Psychopharmacology: the fourth generation of progress.* New York: Raven Press, 1995:1193–1204.
14. Silver JM, Yudofsky SC. Aggressive behavior in patients with neuropsychiatric disorders. *Psychiatr Ann* 1987;17:367–370.
15. Philipp M, Fickinger M. Psychotropic drugs in the management of chronic pain syndromes. *Pharmacopsychiatry* 1993;26:221–234.
16. Campbell M, Anderson LT, Small AM, Adams P, Gonzales NM, Ernst M. Naltrexone in autistic children: behavioral symptoms and attentional learning. *J Am Acad Child Adolesc Psychiatry* 1993;32:1283–1291.
17. World Health Organization. *International classification of impairments, disabilities, and handicaps.* World Health Organization, Geneva 1980.
18. Harley JP, Allen C, Braciszeski TL, et al. Guidelines for cognitive rehabilitation. *Neurorehabilitation* 1992;2: 62–67.
19. Rohe D. Psychologic aspects of rehabilitation. In: DeLisa J, ed. *Rehabilitation medicine principles and practice.* Philadelphia, PA: JB Lippincott, 1988.
20. Schapiro R. *Multiple sclerosis, a rehabilitative approach to management.* New York: Demos, 1991.
21. Sullivan M, Buchanan D. The treatment of conversion disorders in a rehabilitation setting. *Can J Rehabil* 1989;2:3:175–180.
22. Bonke C, Boake C. Traumatic brain injury rehabilitation. *Neurosurg Clin North Am* 1991;2:473–482.

SECTION II

Functional Brain Systems

5

Neurochemistry

John B. Penney, Jr.

This chapter reviews the synaptic organization and neurotransmitter systems that make neurotransmission possible. It continues with a review of the anatomic distribution of these transmitters, modulators, and their receptors. Finally, the distributions of second-messenger systems are briefly described.

SYNAPTIC ORGANIZATION AND NOMENCLATURE

Except for the gases, whose mechanisms of action as neuronal messengers are entirely different, all neurotransmitters require a number of specialized structures at the synapse to convey a message from one neuron to another (Fig. 5-1).

Within the presynaptic terminal are the enzymes necessary for synthesis of the neurotransmitter. Terminals typically have several mitochondria to supply energy. Adjacent to the presynaptic membrane are a number of *synaptic vesicles* in which the neurotransmitter is stored. Several *vesicular transporters* have been described. These are proteins that transport the neurotransmitter into the storage vesicle. Vesicular transporters are specific for the neurotransmitter that is being stored.

Along the presynaptic membrane itself, there are a number of specialized molecules. There are also release points to which the synaptic vesicles bind. Once bound, the vesicles fuse with the membrane and release neurotransmitters. There are also high-affinity *neuronal transporters* that remove the neurotransmitter from the synaptic cleft. These transporters serve two purposes:

1. They terminate the action of the most neurotransmitters by removing them from the synaptic cleft.
2. They spare the presynaptic cell the work of resynthesizing the neurotransmitter by providing a premade source of the chemical.

There may also be *presynaptic receptors* located on the presynaptic terminal membrane. These molecules bind neurotransmitters but do not transport them into the terminal. Instead, binding of the transmitter stimulates the receptor to modulate the activity of the terminal through one of two mechanisms. Presynaptic receptors can influence neurotransmitter release by being coupled to ion channels that change the presynaptic terminal's membrane potential. Because neurotransmitter release is governed by the amount of membrane depolarization that occurs when the impulse that has been conducted along the axon reaches the terminal, depolarizing the presynaptic terminal decreases neurotransmitter release, whereas hyperpolarizing it increases release. Alternatively, presynaptic receptors may be coupled by means of second messengers to either change the rate of neurotransmitter synthesis or to modulate it. *Autoreceptors* bind the neurotransmitter that the presynaptic terminal is releasing. *Heteroreceptors* bind other neurotransmitters.

In the synaptic cleft are enzymes that degrade the neurotransmitter after it has been released from the presynaptic terminal. These enzymes are particularly important for terminating the actions of acetylcholine and the peptides.

The postsynaptic membrane is studded with a high density of neurotransmitter recep-

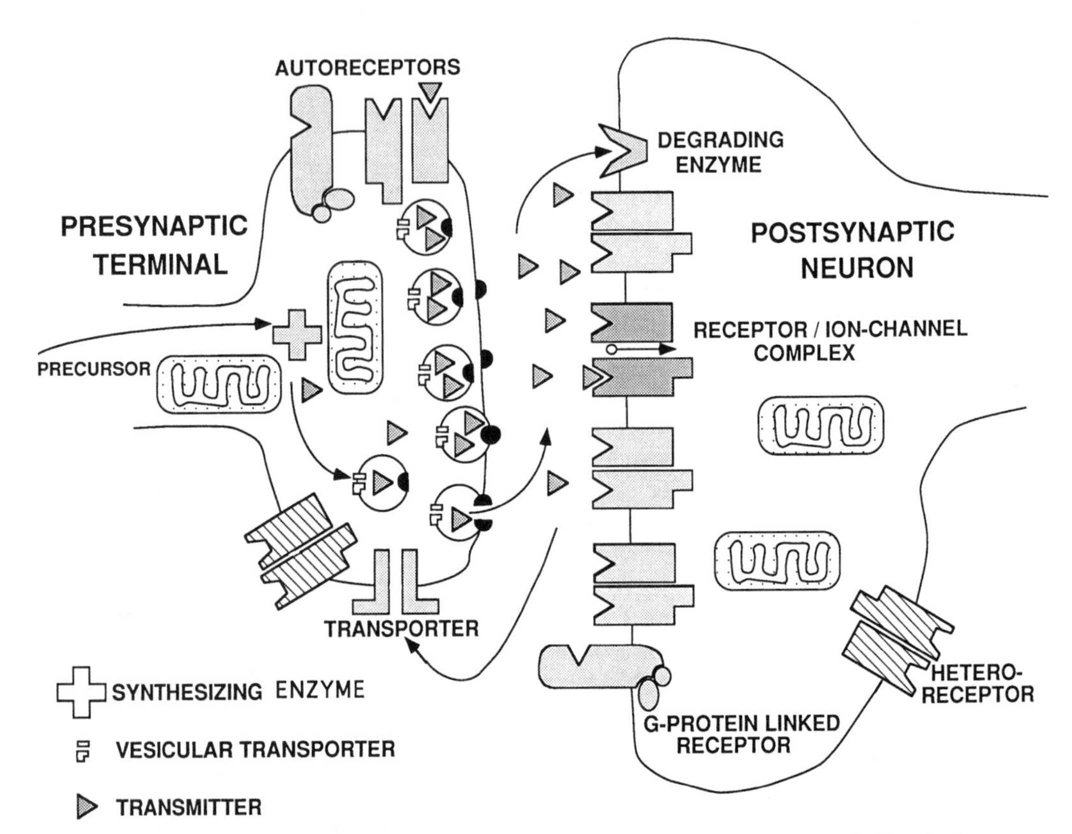

FIG. 5-1. Schematic drawing of the fate of neurotransmitters in synaptic transmission. In the presynaptic terminal, the transmitter is synthesized from a precursor, transported into storage vesicles by a vesicular transporter, and released into the synaptic cleft by depolarization of the presynaptic terminal. Once released, the transmitter diffuses into the synaptic cleft, where it can interact with receptors that have sites to which it can bind. Receptors can be on either the postsynaptic or the presynaptic cell. Autoreceptors on the presynaptic terminal regulate transmitter synthesis and release. Receptors on the postsynaptic membrane initiate changes in the postsynaptic cell that are the ultimate result of transmission at the synapse. Heteroreceptors (receptors for other neurotransmitters) may also regulate synaptic activity. Receptors may be either of the ligand-gated ion channel or the G-protein–coupled type. Ligand-gated ion channels open on binding the transmitter to permit passage of ions through the cell's plasma membrane. (One channel is shown being opened and passing an ion on the postsynaptic cell.) G-protein–coupled receptors initiate second-messenger cascades within the cells on which they reside. Regardless of whether the transmitter binds to a receptor, it eventually is removed from the synapse either by being metabolized by a degrading enzyme, so that it is no longer active or by being transported back into the presynaptic terminal by a high-affinity transporter.

tors. Binding of the neurotransmitter to its postsynaptic receptor, except for the gases, starts the postsynaptic cell response to the transmitter. Receptors either are linked to ion channels that open to initiate the cell response or are linked to guanosine triphosphate–binding proteins (G proteins), which, in turn, activate second-messenger systems within the postsynaptic cell. Numerous types of receptors may be located at a single synapse.

NEUROTRANSMITTER SYSTEMS

The distributions of neurotransmitter systems are presented in a series of schematic drawings showing the neurotransmitter-spe-

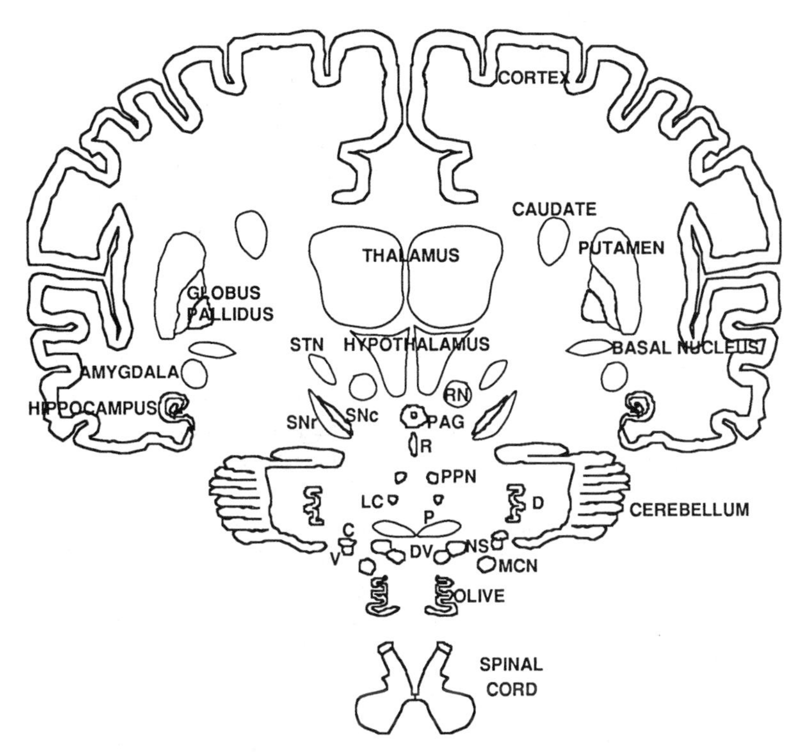

FIG. 5-2. Key to the subsequent neurotransmitter and receptor distribution figures. The names of various structures in the brain are provided here. The subsequent figures are all schematic drawings that show the current concept of the distribution of a neurotransmitter, its receptors, or both. The loci of transmitter-specific neurons are shown as filled circles with their pathways shown as lines projecting from the neurons. The density of receptors is shown as cross-hatching of the various regions. Where the density of a receptor in human brain is unclear, the density has been estimated from reports of other species and from measurements from similar structures. C, cochlear nucleus; D, dentate nucleus of cerebellum; DV, dorsal motor nucleus of the vagus nerve; LC, locus ceruleus; MDFIG, somatic motor cranial nerve nuclei; NS, nucleus of the solitary tract; P, pontine nuclei; PAG, periaqueductal gray; PPN, pedunculopontine nucleus; R, raphe nuclei; RN, red nucleus; SNc, substantia nigra pars compacta; SNr, substantia nigra pars reticulata; STN, subthalamic nucleus; V, vestibular nuclei.

cific pathways and the densities of their receptors in a variety of brain regions. The names of these regions are shown on the accompanying figure (Figs. 5-2). In the subsequent figures, the neurotransmitter specific pathways and densities of receptors are shown.

Amino Acid Neurotransmitters

Amino acids are the neurotransmitters used by most neurons for rapid (millisecond time scale) intracellular communication. Well in excess of 90% of neurons use an amino acid for neurotransmission. Frequently, cells that use an amino acid also use an amine or a peptide. The amino acids used for excitatory neurotransmission are glutamic acid and, possibly, aspartic acid. The inhibitory amino acids are glycine, γ-aminobutyric acid (GABA), and possibly taurine and proline.

Glutamic Acid

Glutamic acid (glutamate) is by far the most important excitatory neurotransmitter in vertebrate brains. Almost all neurons are capable of responding to glutamic acid's effects.

At present, there are no drugs that affect the glutamate-synthesizing enzymes or the glutamate transporters. Glutamate release from the presynaptic terminal is inhibited by several drugs, including lamotrigine and riluzole. Lamotrigine has recently been introduced as an antiepileptic drug; both agents are under study as neuroprotective agents.

Glutamate Pathways

Most of the excitatory pathways in the brain are known to use glutamate as a neurotransmitter (Fig. 5-3). These include the cortical output pathways to subcortical structures, which include the basal ganglia, thalamus, pons, brain stem, and spinal cord. The long intracortical pathways are glutamatergic, as are the pathways within the hippocampal formation. Subthalamic nucleus neurons use glutamate. The main inputs to the cerebellum, the climbing fiber pathway from the inferior olive, and the mossy fibers from the pontine nuclei, use glutamate. The only excitatory neurons in the cerebellar cortex, the cerebellar granule cells, are glutamatergic. Many of the primary sensory fibers use glutamate as one of their transmitters.

Glutamate Pharmacology and Receptors

There are at least six pharmacologic types of mammalian postsynaptic excitatory amino acid receptors. Three of these receptors are linked to ion channels. Each of the ion channel-linked receptors has been named for a prototypic drug that acts specifically at that receptor.

The α-amino-3-hydroxy-5-methylisoxazole-4-propionic acid (AMPA) receptor is a monovalent cation (i.e., sodium) channel that mediates most of the fast, excitatory neurotransmission

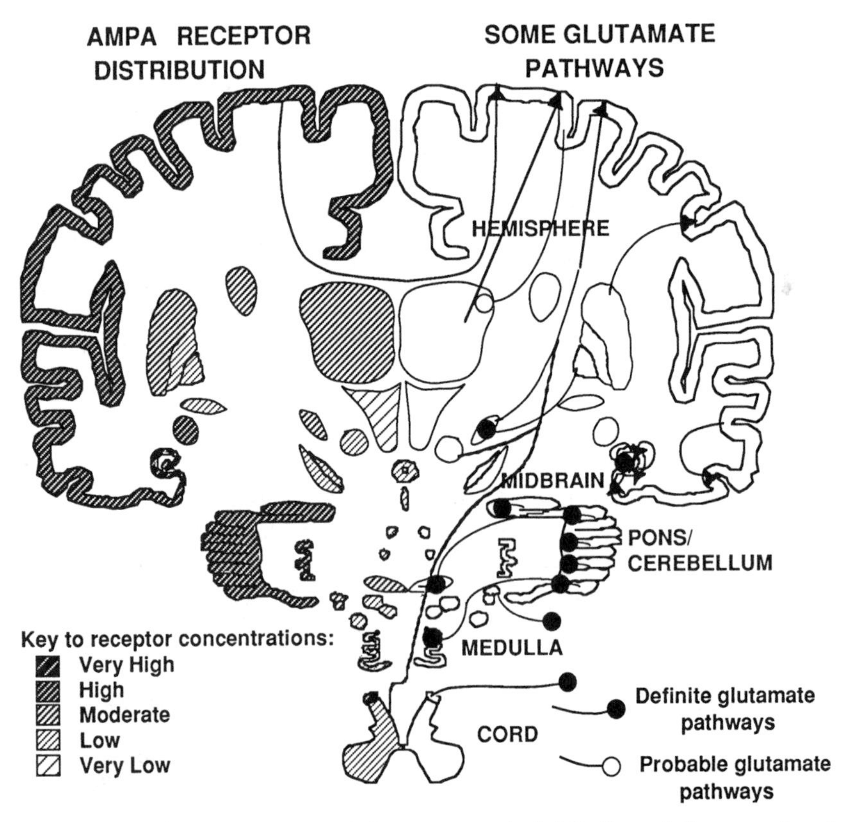

FIG. 5-3. Some glutamate pathways and the presumed distribution of the α-amino3-hydroxy-5-methylisoxazole-4-propionic acid (AMPA) subtype of excitatory amino acid receptors.

in the brain. Several drugs that are antagonists of the AMPA receptor are under development as potential antiepileptics. When glutamate is released from a presynaptic terminal, the excitatory postsynaptic potential that quickly appears at the postsynaptic membrane is due to activation of AMPA receptors. This makes it likely that AMPA receptors are used to convey highly time- and location-specific information along the excitatory pathways. The other glutamate receptors are slower in onset but have longer effects.

AMPA receptors are present throughout the brain. Their distribution is shown in Fig. 5-3. They are densest in the hippocampus; dense in neocortex, on the dendrites of Purkinje's cells in the cerebellum and in the dorsal horn of the spinal cord; moderate in the caudate, putamen, thalamus, and subthalamus; and of lower density in the rest of the brain.

The second of the ion-channel linked receptors—and the most studied—is the *N*-methyl-*D*-aspartate (NMDA) receptor. It is linked to a voltage-gated calcium channel. Activation of the channel is modulated by binding sites for glycine and polyamines on the exterior of the receptor. Drugs are currently being developed that interact with the glutamate-, glycine-, and polyamine-binding sites on the receptor. Under normal membrane potentials, even if glutamate binds to its receptor site on the molecule, the channel is not permeable to calcium because a magnesium ion blocks the channel. When the membrane is somewhat depolarized, however, the channel no longer binds magnesium. Thus, when glutamate binds to its receptor site under depolarized conditions, the channel opens, and calcium enters the neuron. This NMDA receptor—mediated calcium entry is associated with some forms of learning. On the other hand, activation of this receptor during extremely depolarizing situations, such as ischemia and epilepsy, appears to cause neuronal death after a delay of about 24 hours. The distribution of NMDA receptors is shown in Fig. 5-4.

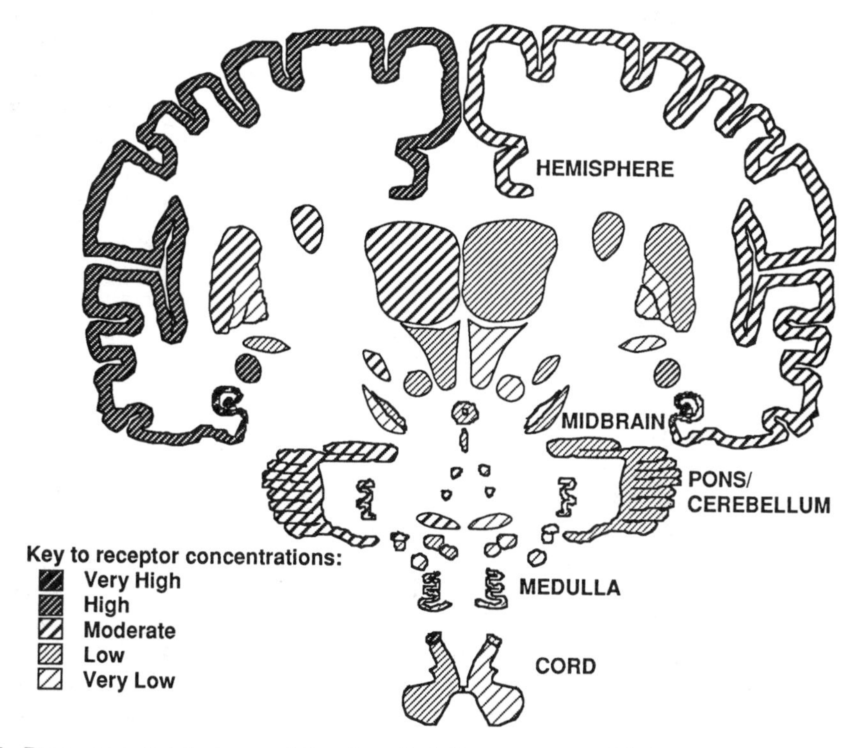

FIG. 5-4. Presumed distributions of the *N*-methyl-*D*-aspartate and kainate subtypes of excitatory amino acid receptors.

The kainate receptor is linked to a monovalent cation channel that does not desensitize as rapidly as the AMPA receptor—linked channels do. Certain naturally occurring toxins that may contaminate shellfish, such as kainate itself and the more potent domoic acid, are able to stimulate this receptor so that it stays open for a long time. This allows sodium to flood into the neuron and kill it through an osmotic overload. The distribution of kainate receptors is shown in Fig. 5-4.

In addition to the three types of ion channel—linked excitatory amino acid receptors, there are also three pharmacologic types of G-protein—-linked receptors that are activated by glutamate. These receptors have been labeled the "metabotropic" glutamate receptors. They have the typical seven membrane-spanning domains of the G-protein—-linked receptors. Type 1 receptors are extremely dense in the Purkinje's cells of the cerebellum, with moderate amounts in the thalamus and lesser amounts in basal ganglia, hippocampus, and cortex. Type 2 receptors are relatively dense in cortex, hippocampus, basal ganglia, and the cerebellum. Type 2 receptors are also present in glial cells.

Glycine

Glycine, like glutamate, is an essential amino acid that is present in all cells. Because the alkaloid poison strychnine acts by blocking glycine receptors, glycine was shown to be a neurotransmitter before glutamate. Glycine has two major actions in the brain outside of its role in intermediary metabolism. The first is as an inhibitory neurotransmitter. Glycine released at most synapses interacts with inhibitory receptors on the postsynaptic neurons. These inhibitory glycine receptors can be blocked by strychnine. There is, however, another role for glycine. The glycine that is normally present in the extracellular fluid binds to a modulatory site on the NMDA type of glutamate receptors. Glycine needs to be bound to this site for the NMDA receptor to be activated. The seizures that are a prominent symptom of the rare, inherited, metabolic disease nonketotic hyperglycinemia are probably caused by excess glycinergic stimulation of this modulatory site on the NMDA receptor. Several drugs that bind to this site are under development as potential neuroprotective agents and antiepileptics.

Glycinergic Neurons

All known glycinergic neurons are inhibitory interneurons. The best known are the inhibitory interneurons of the ventral horn of the spinal cord. These include Renshaw cells, which mediates recurrent inhibition of the immunoglobulin A (IgA) motor neurons and the inhibitory Ia afferent-coupled interneurons that mediate crossed inhibition of the spinal reflexes. The distribution of glycinergic neurons is shown in Fig. 5-5.

Dysfunction of the synapses formed by glycinergic neurons causes a marked increase in the reflexes with spread of innervation from one spinal cord segment to another. Total failure of these synapses causes the sensory input from one contracting muscle to stimulate other muscles to contract. Shortly thereafter, all muscles are maximally contracting owing to positive feedback. This leads to opisthotonus, paralysis, and respiratory failure.

Glycine Pharmacology and Receptors

The pharmacology of glycine is known because of the symptoms (described previously) of failure of glycinergic synapses. Tetanus toxin causes symptoms and death in this way because it blocks the release of glycine from the presynaptic terminal. The rare inherited syndrome hyperekplexia, characterized by excessive startle reactions, is caused by minor mutations in the α-subunit of the glycine receptor.

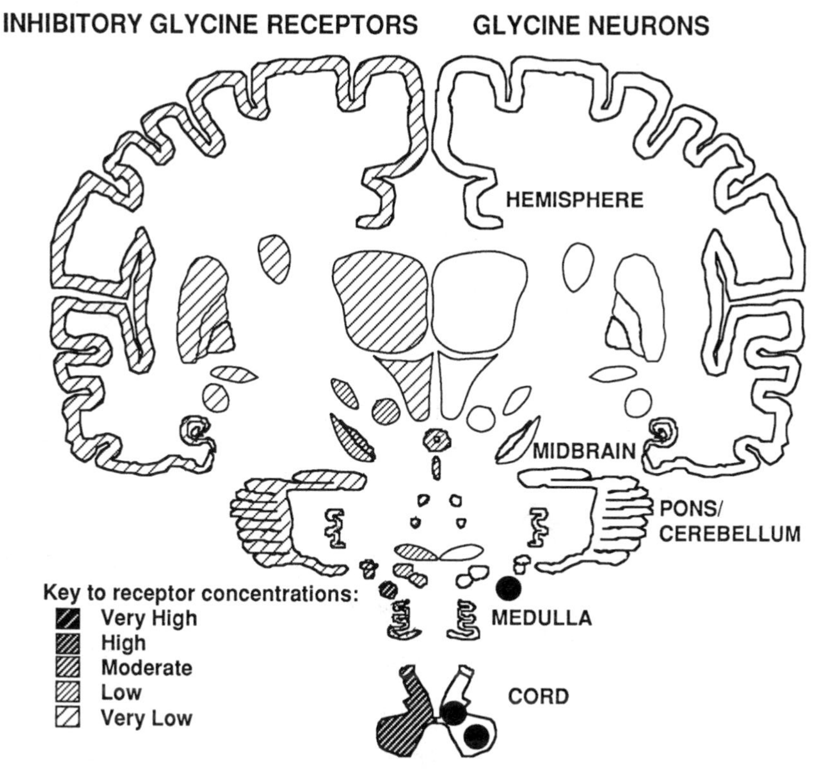

FIG. 5-5. Location of some glycine neurons and the presumed distribution of inhibitory glycine receptors.

γ-Aminobutyric Acid

γ-Aminobutyric acid (GABA) is the predominant, fast-acting, inhibitory neurotransmitter in the nervous system. GABA is synthesized from glutamate by the specific enzyme, glutamic acid decarboxylase (GAD). GAD is found only in neurons that use GABA as a neurotransmitter and serves as a useful immunohistochemical marker for these neurons. GABA is ordinarily removed from the synapse by a high-affinity neuronal transport system and restored in synaptic terminals.

There are several classes of drugs whose mechanism of action is to enhance GABA's actions. These drugs function as sedatives, antiepileptics, and anxiolytics. Some of these act by inhibiting the metabolism of GABA by GABA-transaminase. These include the antiepileptics gabapentin and valproate.

GABA Pathways

GABA neurons serve as the local circuit-inhibitory neurons throughout the central nervous system (CNS) except in the spinal and brain-stem motor nuclei, where glycine serves this purpose. There are also several long projection pathways that use GABA. One is the Purkinje's cell pathway from cerebellar cortex to the deep cerebellar nuclei. Similarly, the only output neurons of the caudate nucleus, putamen, and nucleus accumbens (striatum) are GABA-ergic, medium-sized neurons whose dendrites are covered with synaptic spines (medium spiny neurons). The neurons of the globus pallidus and the pars reticulata of the substantia nigra are large, tonically firing neurons that also use GABA. The distribution of GABA neurons and their pathways is shown in Fig. 5-6.

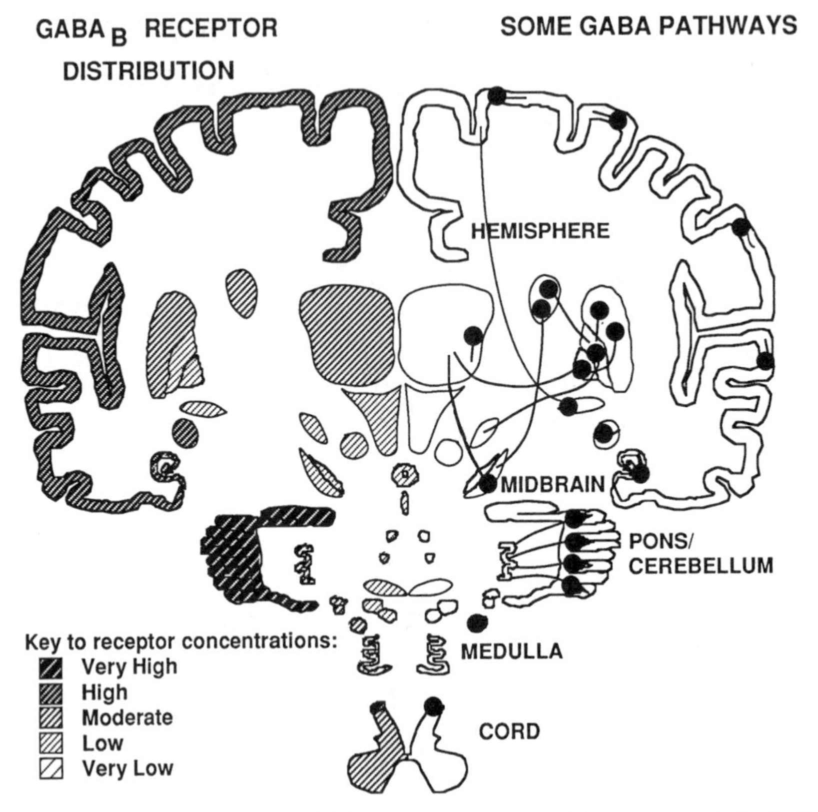

FIG. 5-6. Some γ-aminobutyric acid (GABA) pathways and the presumed distribution of $GABA_B$ receptors.

GABA, Benzodiazepine, and Barbiturate Pharmacology and Receptors

Many drugs—the benzodiazepines, the barbiturates, certain convulsants, and baclofen—act by modulating the activity of one of the two types of postsynaptic GABA receptors.

$GABA_A$ receptors are dense throughout the brain, wherever there are GABA synapses. Benzodiazepines modulate the activity of these receptors through a binding site that is different from the one to which GABA binds. Benzodiazepines cannot open the channel unless GABA is present. The barbiturates bind to a site in the chloride channel and can open the channel without GABA being present. This difference in the actions of the two classes of drugs explains the greater toxicity of the barbiturates.

The other major type of GABA receptor is the $GABA_B$ receptor. $GABA_B$ receptors are G-protein linked. Activation of these receptors results in inhibition of cyclic AMP, an increase in the permeability of potassium channels, and a decrease in the permeability of sodium channels. Binding is very high in the cerebellum and dorsal horn of the spinal cord. High binding is present in the cortex. The very high binding in dorsal horn or spinal cord likely represents receptors that regulate the release of neurotransmitter from primary afferents through presynaptic inhibition.

Acetylcholine

Acetylcholine is the major transmitter of the peripheral motor system, the presynaptic autonomic nervous system, the postsynaptic innervation of sweat glands, and the entire postsynaptic parasympathetic system. It also serves as a neurotransmitter in a number of CNS pathways. Acetylcholine is released from presynaptic terminals and interacts with postsynaptic receptors in a manner very similar to that of amino acids. There is no high-affinity transport system for acetylcholine. Instead, there is an enzyme—acetylcholinesterase—present in the synapse that breaks acetylcholine down into choline and acetate. A number of drugs that are useful for treating myasthenia gravis act by blocking the activity of acetylcholinesterase. By preventing acetylcholine from being metabolized, these drugs prolong acetylcholine's actions at the synapse. Recently, centrally acting cholinesterase inhibitors, such as tetrahydroaminoacridine (tacrine), have been targeted as potentially useful in the treatment of Alzheimer's disease.

Acetylcholine Pathways

There are six major types of acetylcholine neurons within the CNS. The distribution of these neurons is shown in Fig. 5-7.

Within the CNS, there are two major cholinergic projections. One arises in the pedunculopontine nuclei of the midbrain and projects to the medulla, substantia nigra, thalamus, and globus pallidus. These neurons are part of the ascending reticular activating system. These neurons may play an important role in regulating the sleep–wake cycles. A second major CNS projection for acetylcholine is provided by a complex of nuclei, including the medial septal nucleus, the diagonal band of Broca, and the basal nucleus of Meynert. These neurons innervate all the cortical structures, including the neocortex, the hippocampus, and the amygdala. The acetylcholine provided by these neurons appears to be vital for normal learning and memory function. The degeneration of these neurons is one of the major hallmarks of Alzheimer's disease.

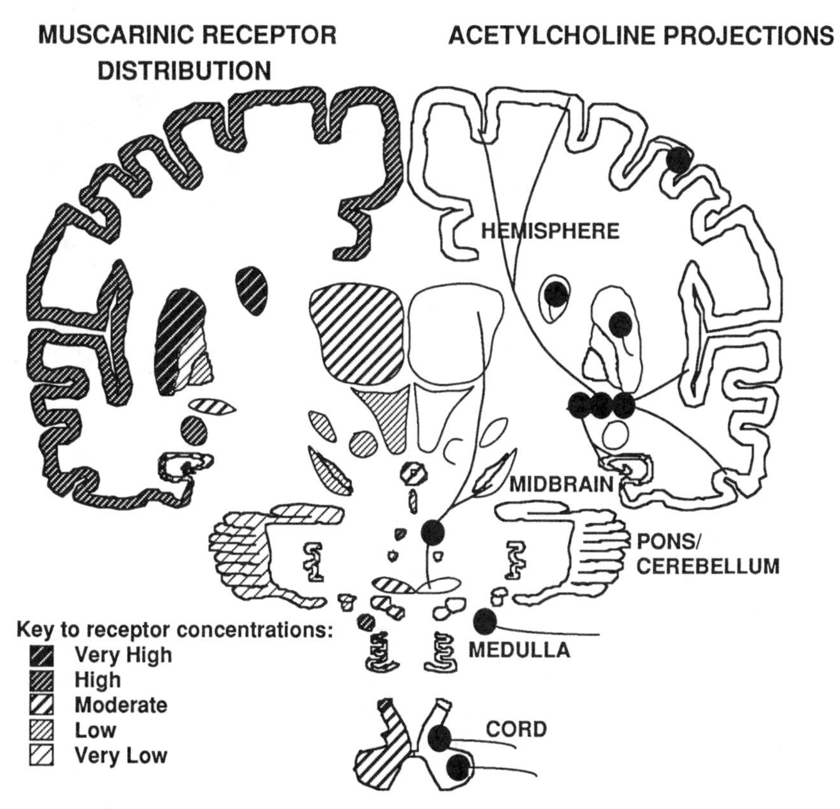

FIG. 5-7. Some acetylcholine pathways and the presumed distribution of muscarinic acetylcholine receptors.

Acetylcholine Receptors

There are two major classes of acetylcholine receptors. The nicotinic receptors at the neuromuscular junction are the prototypic ligand-gated ion channel receptor complex. These are the only receptors that have been purified to homogeneity and that are known to exist in a pentamer conformation. The ion channel opened by the receptor is a sodium channel, and activation of the receptor stimulates an excitatory postsynaptic potential.

The nicotinic receptors located on neurons in the CNS and sympathetic ganglia are pharmacologically different from those located at the neuromuscular junction. They are coded by a different family of genes, and different neurons may express different nicotinic receptors. Many of these receptors are located on presynaptic terminals, where they may govern neurotransmitter release.

The other major type of acetylcholine receptor is the muscarinic receptor. These receptors are present at the synapses where the postsynaptic parasympathetic neurons terminate, such as those in the heart, intestine, and sweat glands. Atropinic (antimuscarinic) drugs produce their tachycardia, constipation, and decreased sweating by blocking these peripheral muscarinic receptors.

Blockade of central muscarinic receptors can profoundly affect learning and memory. Drugs that block these receptors include not only the well-known anticholinergics but also tricyclic antidepressants, antihistamines, and many of the atypical neuroleptics.

Amine Neurotransmitters

Dopamine

Dopamine is an amine neurotransmitter of central importance in motor systems within the brain.

L-Dopa is converted to dopamine by the actions of L-aromatic acid decarboxylase. Dopamine is metabolized by a combination of monoamine oxidase and catechol-*o*-methyltransferase to homovanillic acid. However, the actions of dopamine at dopamine synapses are usually terminated by high-affinity transport back into the presynaptic dopamine terminal by a transporter that is specific for dopamine. The stimulant drugs of abuse, cocaine and the amphetamines, act by blocking the reuptake of dopamine by way of this transporter. In addition, amphetamines can stimulate the release of dopamine.

Dopamine Pathways

Dopamine cell bodies are located in a number of distinct nuclei in the brain stem and hypothalamus. The location of these nuclei and their pathways are shown in Fig. 5-8. The major source of dopamine for the brain is the chain of cells running from the substantia nigra pars compacta through the ventral tegmental area. The nigral neurons project to the caudate nucleus and putamen, where they govern motor activity. Dysfunction of these neurons produces the symptoms of parkinsonism. The ventral tegmental area neurons project to the nucleus accumbens and the entire cortex, particularly the frontal lobe. Dysfunction (or possibly hyperactivity) of these neurons is thought to contribute to the symptoms of Tourette's syndrome and schizophrenia.

Dopamine Receptors

There are two main pharmacologic classes of dopamine receptors. D_1 receptors are linked to the stimulation of cyclic AMP, whereas D_2 receptors are linked to cyclic AMP inhibition. Both receptor subtypes are extremely dense in the caudate nucleus, putamen, and nucleus accumbens. D_1 receptors are located on postsynaptic striatal neurons. They are also located on the terminals of these neurons in the medial global pallidus and substantia nigra pars reticulata. Most striatal D_2 receptors are also on intrinsic striatal neurons, although some exist as autoreceptors on dopamine neuron terminals, where they govern dopamine release. The D_2 receptors are blocked by the classic neuroleptics, such as haloperidol, and the antiemetics, such as metoclopramide. Blockade of these receptors

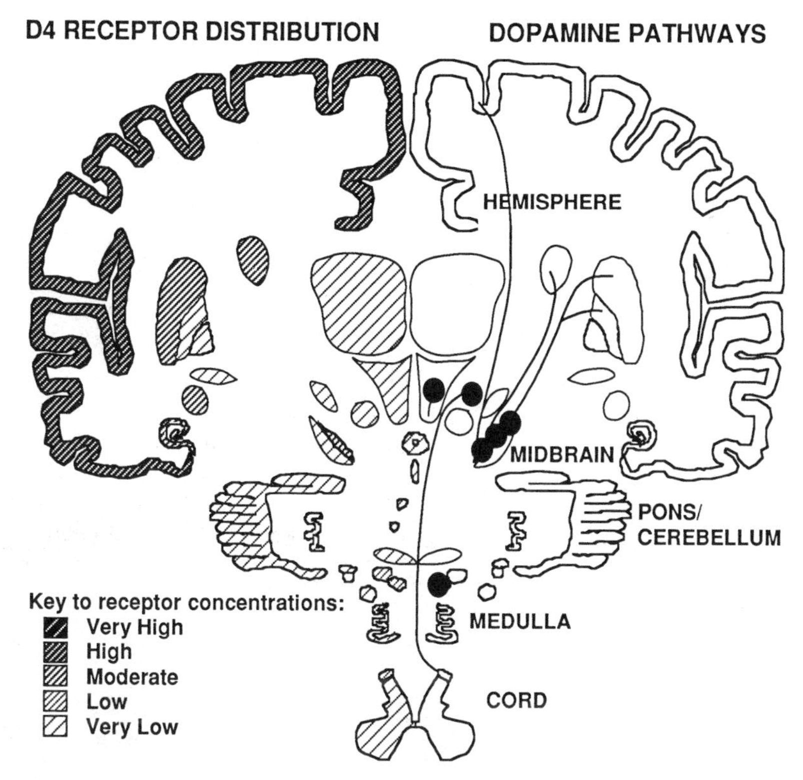

FIG. 5-8. Some dopamine pathways and the presumed distribution of the dopamine D_4 receptor.

produces parkinsonism as a prominent side effect.

There is abundant evidence for interaction between the D_1 and D_2 receptor subtypes. Both are required for normal motor activity. Further studies are needed to measure the amount of overlap between the neurons that express the different pharmacologic subtypes of dopamine receptors.

Norepinephrine

Norepinephrine is synthesized from dopamine by the enzyme dopamine-β-hydroxylase. Like dopamine, norepinephrine is metabolized by monoamine oxidase and catechol-*o*-methyltransferase. Again like dopamine, the major route of inactivation of norepinephrine at the synapse is through a high-affinity transport system into the presynaptic terminal. Paroxetine is a specific inhibitor of the norepinephrine transport system. Tricyclic antidepressants, particularly desipramine, also block norepinephrine reuptake.

Norepinephrine is thought to play major roles in arousal, memory, and affect.

Norepinephrine Pathways

In the peripheral nervous system, norepinephrine is the major transmitter of the postsynaptic sympathetic nervous system. These neurons have their cell bodies in the chain of sympathetic ganglia that lie bilaterally in the paravertebral gutter from the lower cervical to the lumbar regions. The axons of these neurons are distributed to the blood vessels throughout the body, irides, salivary glands, heart, lungs, intestines, and bladder. These neurons play important roles in the regulation of blood pressure, pupillary dilation, cardiac strength and rhythm, pulmonary airway dilation, and intestinal and vesicle mobility.

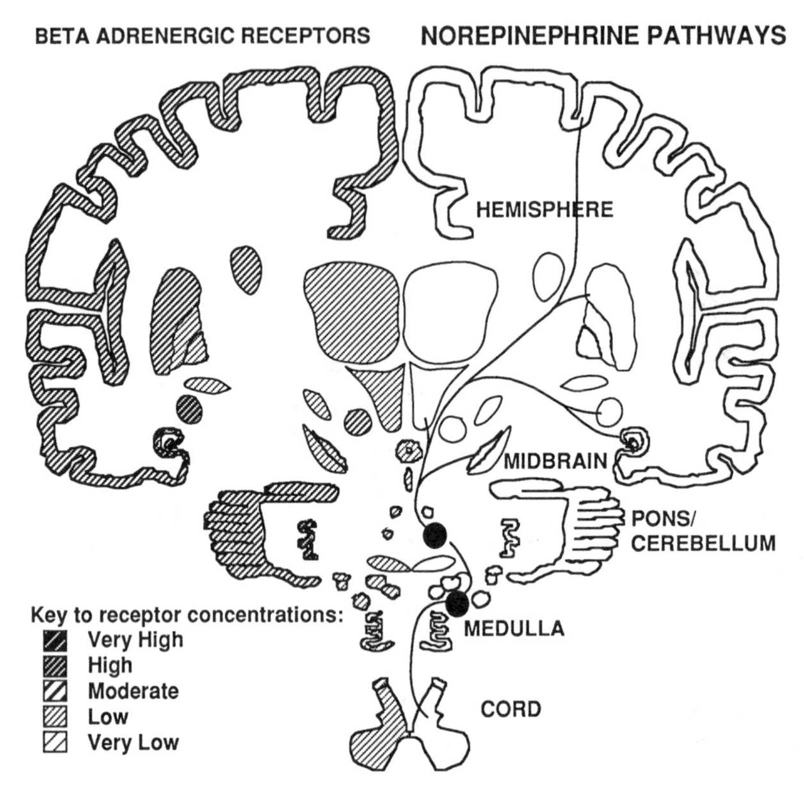

FIG. 5-9. Some norepinephrine pathways and the presumed distribution of β-adrenergic receptors (receptors for norepinephrine and epinephrine).

There are several major norepinephrine-containing cell groups in the mammalian brain. The distribution of these neurons is shown in Fig. 5-9. The locus ceruleus serves as the major source of norepinephrine for the forebrain. This dense cluster of pigmented neurons is located in the dorsal lateral pons. The axons of the ceruleus neurons project in the median forebrain bundle, where they distribute to the hypothalamus, thalamus, basal ganglia, amygdala, hippocampus, and entire neocortex. These projections from the locus ceruleus are thought to play an important role in arousal, memory (being responsible for the hippocampal theta rhythm), and affect.

Norepinephrine Receptors

Norepinephrine receptors are all of the G-protein–coupled type. There are two main pharmacologic subtypes of these receptors, α- and β-receptors. α-Receptors are linked to the inositol phosphate system and to the inhibition of cyclic AMP, whereas β-receptors are linked to stimulation of cyclic AMP. Both receptors are found not only in the brain but also in the periphery. Each major pharmacologic class has several subtypes as well.

α_1-Receptors are located postsynaptically throughout the brain and periphery. They are high in thalamus and hippocampus; moderate in basal ganglia, cerebral, and cerebellar cortices and some brain-stem nuclei; and low elsewhere. α_2-Receptors, on the other hand, are located postsynaptically and as autoreceptors on the synaptic terminals of norepinephrine neurons, where they regulate norepinephrine release. They are particularly abundant in cerebral cortex and locus ceruleus. β-Receptors are also distributed throughout the brain and spinal cord. There is evidence that β-receptors downregulate with antidepressant treatment.

Serotonin

Serotonin is synthesized from tryptophan in a manner entirely analogous to that by which norepinephrine is synthesized from tyrosine. 5-Hydroxytryptophan (5HT) is the immediate precursor for serotonin and has had some success as a drug in the treatment of myoclonus. Serotonin is metabolized to 5-hydroxyindoleacetic acid. Like norepinephrine, serotonin is thought to play a major role in the arousal, sleep, and affective systems within the brain.

Serotonin is mostly inactivated at the synapse by a high-affinity transport system. The specific serotonin transport inhibitors fluoxetine and sertraline are useful antidepressants. Tricyclic antidepressants are also potent serotonin reuptake inhibitors. The usefulness of these drugs has led to the hypothesis that serotonin plays a major role in depression.

Serotonin Pathways

All serotonin pathways have their origin in a series of nuclei that are located on the midline of the brain stem, the raphe nuclei. Serotonin pathways are shown in Fig. 5-10. Within the brain stem, serotonin neurons play a role in arousal and sleep. Ascending pathways from the raphe nuclei distribute to the substantia nigra, the rest of the basal ganglia, the thalamus, hypothalamus, cortex, amygdala, and hippocampi.

Serotonin Receptors

Pharmacologic studies originally divided serotonin receptors into three main subtypes. $5HT_1$ receptors were defined as those having a high affinity for serotonin, whereas $5HT_2$ receptors were defined as those with a high affinity for lysergic acid diethylamide (LSD).

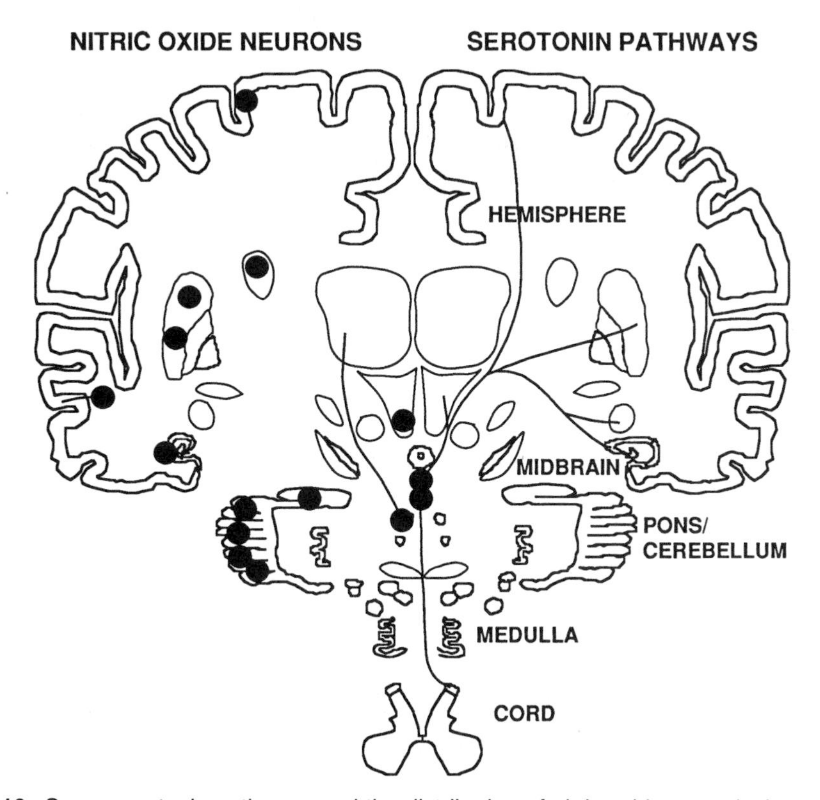

FIG. 5-10. Some serotonin pathways and the distribution of nitric oxide—producing neurons.

Both of these receptors turned out to be G-protein linked. However, the $5HT_3$ receptor, which is found only on peripheral sensory nerve fibers, turned out to be a classic ligand-gated ion channel that permits the passage of cations. A $5HT_4$ receptor that was linked to cyclic AMP stimulation has been described pharmacologically, but the gene or genes for this receptor have not yet been cloned.

Adenosine

The purine adenosine has recently been identified as a neuromodulator in the CNS. Adenosine has a well-established role as a precursor to both DNA and RNA. Thus, enzymes for its synthesis and metabolism are ubiquitously distributed. There is also a high-affinity transport system for adenosine. Specific neurons that use adenosine as their neurotransmitter, however, have not been identified.

Adenosine interacts with two types of postsynaptic receptors. The pharmacology of these receptors is important because the xanthines, such as caffeine, act as antagonists at adenosine receptors. Studies have concluded that adenosine antagonism may be responsible for the major pharmacologic effects of caffeine rather than inhibition of phosphodiesterase. There are two subtypes of adenosine A2 receptors. The $A2_a$ receptor is found only in the striatum, whereas the $A2_b$ receptor is found only the periphery and is not thought to play a role in neurotransmission.

Peptide Neurotransmitters

A number of peptides have been shown to act as neuromodulators within the CNS. Most of these peptides were originally isolated in another bodily system and then shown to be present in brain by purification or demonstration of physiologic action, or (most commonly) their presence has been inferred by immunohistochemical staining. In every case that has been studied, neuropeptides were found to be co-transmitters with some other neuroactive compound, such as GABA or serotonin, being released at the same synapse. The best studied of these neuromodulators are the endogenous opiates and substance P.

Endorphins

The endorphins were all isolated as endogenous compounds that have activity at opiate receptors.

The endorphins are synthesized by cleavage of much larger peptide precursor molecules. The three known precursors are pro-opiomelanocortin, proenkephalin, and prodynorphin.

Endorphin Pathways

Enkephalin neurons serve as local interneurons in the primary sensory receiving areas of the dorsal horn of the spinal cord and the spinal tract of the trigeminal nucleus. It is thought that this is a primary site where the enkephalins suppress pain sensation. There is also dense enkephalinergic innervation of the central gray and periaqueductal gray from the floor of the fourth ventricle through the entire midbrain. Within the forebrain, there are some enkephalinergic neurons in the cortex and hippocampus, although no cell contains a high concentration of this neuromodulator. However, the major source of enkephalin in the forebrain is in the striatum. About half of the spiny neurons of the caudate nucleus and putamen contain enkephalin as a co-neurotransmitter. These neurons send their projections almost exclusively to the lateral segment of the globus pallidus, with recurrent collaterals innervating the striatum. The concentration of enkephalin in the globus pallidus is by far the highest of any place in the brain.

Opiate Receptors

A number of different binding sites for opiates have been described using ligand-binding techniques. The three that are probably most specific for opiate actions are the μ, δ, and κ-receptors. μ-Receptors appear to bind the morphine-like opiate pain medications most specifically. δ-Receptors are most specific for the enkephalins, whereas the κ-receptors are most specific for the dynorphins. The paradox of these binding sites and binding sites for ligands that bind to many of the other peptide transmitters is that the distribution of ligand-binding sites is radically different from the distribution of peptide-containing nerve terminals. Whether the receptors are present in places to which the peptides can diffuse over long distances, whether the peptides are somehow paradoxically present in great concentrations at places where they have no functional role, or whether the presence of high concentrations of peptide causes downregulation of receptor numbers remains to be determined.

Opiate receptors are found selectively throughout the brain, with significant numbers in the dorsal horn of the spinal cord and spinal trigeminal nucleus. There are also receptors present in cerebellum, where there is no enkephalinergic innervation, and in the central gray of the brain stem.

Opiate κ-receptors are located in the dorsal horn or spinal cord and trigeminal nucleus, in layers 1 and 2 of the cortex, caudate nucleus, and putamen, medial dorsal nucleus of thalamus, and amygdala with low binding in the globus pallidus. The δ sites are different from the μ sites in that there are significant numbers of δ receptors in the hippocampal formation, particularly in the dentate gyrus, whereas there are very few δ receptors in the cerebellum.

κ-Receptors are concentrated in the deep layers of the cortex, particularly layers 5 and 6. There is slightly more binding in globus pallidus for κ-receptors than there is for either μ or κ-receptors. There is also a significant concentration of κ-receptors in the basal lateral nucleus of the amygdala. κ-Receptor sites are also found in the granule cell layer of the cerebellum, whereas the μ-receptors appear to be more concentrated in the molecular cell layer.

Tachykinins

Three main tachykinins have been described in the mammalian brain: substance P, neurokinin A, and neuromedin K. By far, the best described of these is substance P. The tachykinins all share the same carboxyterminal amino acid sequence, phenylalanine—an amino acid-glycine-leucine-amino terminal.

Neuromedin K is spliced from the preprotachykinin B gene, which is present in both the CNS and in the periphery. Preprotachykinin A (and therefore substance P and neurokinin A) is mainly expressed in the trigeminal ganglion, dorsal root ganglia, and striatum, whereas preprotachykinin B (and thus neuromedin K) is primarily synthesized in the hypothalamus and in the intestines.

Substance P Distribution

Substance P is a major neurotransmitter of primary afferent nerve terminals. Its primary roles appear to be in pain transmission through the small, unmyelinated, primary afferent nerve fibers. Stimulation of substance P fibers produces burning pain. Capsaicin, the active ingredient in chili peppers, stimulates the release of substance P where it is applied locally to a nerve, such as the skin, or the mucous membranes of the mouth and centrally. Capsaicin applications produce a temporary sensation of burning, followed by relative anesthesia as the substance P in the affected nerve is first released and then depleted.

Within the CNS, substance P neurons are densely concentrated within the nucleus of the solitary tract, moderately around the cranial nerve nuclei, and moderately in the other medullary tegmental nuclei. A dense cluster of substance P neurons resides in the parabrachial nucleus. These neurons innervate other pontine nuclei, as well as the caudal parts of the substantia nigra.

Tachykinin Receptors

The tachykinin receptors are coupled by G proteins to the activation of the phosphatidylinositol system. There are three receptors: NK_1 (formerly called the substance P receptor), NK_2 (formerly called the substance K or substance E receptor), and NK_3 (formerly called the neurokinin B receptor). The NK_1 receptor is most susceptible to substance P. It is found in both nervous system and in the periphery. The NK_2 is most susceptible to neurokinin A and is found only in the periphery. The NK_3 receptor gene is expressed in both the CNS and in the periphery.

There is a large divergence between the location of the neuropeptides and the location of their receptors. For example, neurokinin A is present in the brain, but the receptor is located neither in the brain nor in the spinal cord. Thus, the actual role of these neuromodulators in CNS function remains in doubt because of the lack of correlation between presence of the neuromodulator and presence of its effector.

Somatostatin

Somatostatin is a peptide that was first discovered to play a role in the control of growth hormone secretion. It was also found to be present in the intestinal system, as well as in scattered interneurons throughout the brain. Four subtypes of somatostatin receptors have been described.

Gases

It was recently discovered that at least two diffusible gases are used as intercellular messengers within the CNS and the periphery. Several years ago, such a role was described for nitric oxide. More recently, it has been proposed that carbon monoxide is also a neurotransmitter. The mechanism of action of the gases is completely different from that of the other neurotransmitters and neuromodulators because they are so diffusible. They are not stored in vesicles. There is no release mechanism at the synaptic terminal, and there is no postsynaptic receptor for these gases. The gases are synthesized in presynaptic terminals. They then diffuse freely throughout the tissues, including across the synaptic cleft, where they interact directly with second-messenger systems within the "postsynaptic" neuron.

Nitric Oxide

The first gas discovered to play a role as a neurotransmitter was nitric oxide. It has recently been shown that nitric oxide is the endothelial relaxing factor. It is synthesized by a specific enzyme, nitric oxide synthase, in endothelial cells, and it diffuses from there into the smooth muscle cells that surround the arterioles. There, it stimulates guanylate cyclase, the synthetic enzyme for the second-messenger molecule, cyclic GMP. It is the cyclic GMP that produces relaxation of the arteriolar smooth muscle. This nitric oxide and cyclic GMP–mediated system is vital to the arteriolar relaxation that allows blood to engorge the penis. Thus, nitric oxide is the neurotransmitter responsible for penile erections.

Nitric oxide has been proposed as a potential neurotoxin because exposure of neurons to excess amounts of this compound causes neurons to die. Two proposed mechanisms can possibly make nitric oxide toxic. The simpler theory is that nitric oxide is a free radical and can combine with oxygen to form even more toxic free radicals. The second proposed mechanism is that nitric oxide causes inhibition of glyceraldehyde-3-phosphate dehydrogenase, a vital enzyme in the metabolism of glucose. Thus, excessive stimulation of nitric oxide may constrain neurons from metabolizing glucose, their primary energy source.

Carbon Monoxide

It has recently been found that carbon monoxide can also be produced in the brain by the enzyme hemeoxygenase-2. Like nitric oxide, carbon monoxide is a freely diffusible compound. Once synthesized, it appears to be able to stimulate guanylate cyclase, just as nitric oxide can. Thus, carbon monoxide and nitric oxide both stimulate the same enzyme, and carbon monoxide can greatly potentiate the response to nitric oxide. Hemeoxygenase-

2 is enriched in olfactory structures, pyramidal cells of the hippocampus, and Purkinje's cells of the cerebellum, but it is present throughout the brain.

SECOND MESSENGERS

There are two extremely well-described second-messenger systems within the brain. Both are ordinarily stimulated by G-protein—-coupled receptors. One is the cyclic AMP system, which is stimulated by adenylate cyclase. The other is the phosphatidylinositol cycle in which phosphatidylinositol (bis)-phosphate is cleaved by phospholipase C to inositol triphosphate and diacylglycerol. Diacylglycerol, in turn, stimulates protein kinase C and calcium influx. Ligands have been found that bind with high affinity to adenylate cyclase and to protein kinase C, making it possible to identify the sites in the brain, where these two second-messenger systems are located (Fig. 5-11).

Protein kinase C is also located throughout the brain, but it is not nearly as dense as adenylate cyclase is in the striatum and its terminals. The densest distribution of protein kinase C is in the cell bodies, dendrites, and axons of CA3 and CA1 pyramidal cells of the hippocampus and in the Purkinje's cells of the cerebellum. Moderate levels are found in cortex, striatum, thalamus, substantia nigra, and brain stem. In the spinal cord, protein kinase C is largely located in the substantia gelatinosa.

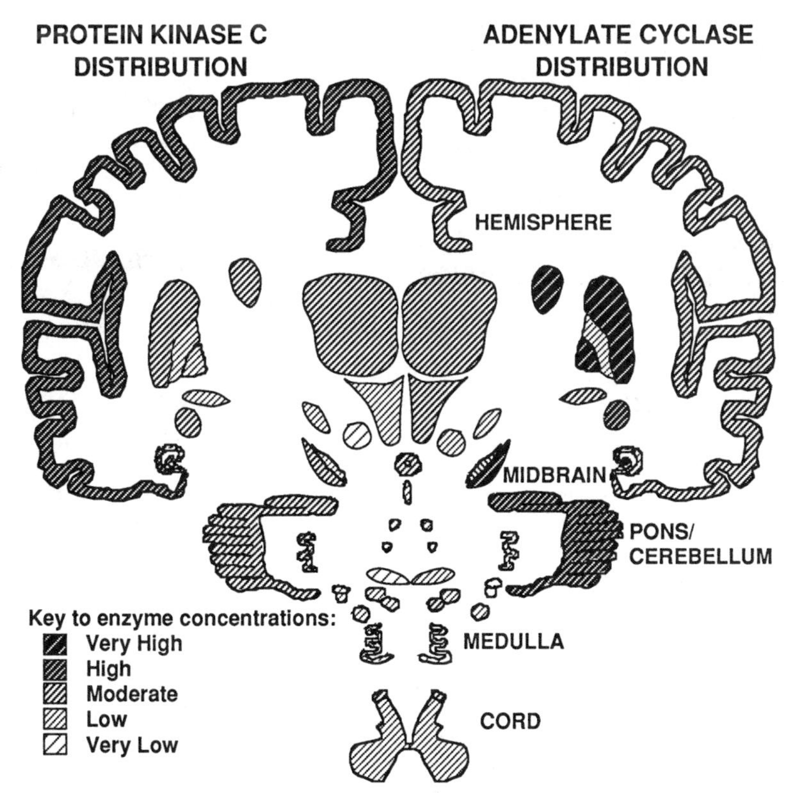

FIG. 5-11. Presumed distribution of two important second-messenger system enzymes: adenylate cyclase, which forms cyclic AMP, and protein kinase C, which is involved in phosphatidylinositol metabolism.

SUMMARY

Neuroscientific research is proceeding steadily with regard to the neurochemical substrates of behavior, cognition, and emotion. Receptor subtypes for each of the transmitter systems discussed in this chapter are expanding in number, and the genetic basis for these receptor subtypes is under investigation in several laboratories. As the genes for these receptors are identified and cloned, the potential for pharmacologic manipulation will become greater. Neuropharmacologic and psychopharmacologic research is focusing on agents that are more specific with regard to receptor subtype interactions. Drugs that block excitatory amino acid transmission may be useful for several applications, including possible therapies for neurodegenerative diseases. New agents that affect GABA and acetylcholine transmission are becoming available for the treatment of Alzheimer's disease, epilepsy, and the anxiety disorders.

ACKNOWLEDGMENTS

I would like to thank Rita Zollo for secretarial assistance. This work was supported by USPHS grants NS19613 and AG08671.

SUGGESTED READINGS

General

Mendelsohn FAO, Paxinos G, eds. *Receptors in the human nervous system.* San Diego, CA: Academic Press, 1991.

Paxinos G, ed. *The human nervous system,* San Diego, CA: Academic Press, 1990.

Watson S, Girdlestone D. TIPS receptor nomenclature supplement. *Trends Pharmacol Sci* 1993(Suppl);14:1–43.

Transporters

Amara SG, Kuhar MJ. Neurotransmitter transporters: recent progress. *Annu Rev Neurosci* 1993;16:73–93.

Kanai Y, Smith CP, Hediger MA. The elusive transporters with a high affinity for glutamate. *Trends Neurol Sci* 1993;16:365–370.

Glutamate Receptors

Collingridge GL, Singer W. Excitatory amino acid receptors and synaptic plasticity. *Trends Pharmacol Sci* 1990;11:290–296.

Meldrum B, Garthwaite J. Excitatory amino acid toxicity and neurodegenerative disease. *Trends Pharmacol Sci* 1990;11:379–387.

Nakanishi S. Molecular diversity of glutamate receptors and implications for brain function. *Science* 1992;258:597–603.

Seeburg PH. The TINS/TIPS lecture: the molecular biology of mammalian glutamate receptor channels. *Trends Neurol Sci* 1993;16:359–364.

GABA Receptors

Burt DR, Kamatchi GL. $GABA_A$ receptor subtypes: from pharmacology to molecular biology. *FASEB J* 1991;5:2916–2923.

Acetylcholine Receptors

Sargent PB. The diversity of neuronal nicotinic acetylcholine receptors. *Annu Rev Neurosci* 1993;16:403443.

Dopamine Receptors

Sibley DR, Monsma FJ. Molecular biology of dopamine receptors. *Trends Pharmacol Sci* 1992;13:61–68.

Norepinephrine Receptors

Kobilka B. Adrenergic receptors as models for G protein-coupled receptors. *Annu Rev Neurosci* 1992;15:87–114.

Serotonin Receptors

Julius D. Molecular biology of serotonin receptors. *Annu Rev Neurosci* 1991;14:335–360.

Humphrey PPA, Hartig P, Hoyer D. A proposed new nomenclature for 5-HT receptors. *Trends Pharmacol Sci* 1993;14:233–236.

Adenosine Receptors

Fastbom J, Pazos A, Probst A, Palacios JM. Adenosine A1 receptors in the human brain: a quantitative autoradiographic study. *Neuroscience* 1987;22:827–839.

Peptide Receptors

Herkenham M. Mismatches between neurotransmitter and

receptor localizations in brain: observations and implications. *Neuroscience* 1987;23:1–38.
Nakanishi S. Mammalian tachykinin receptors. *Annu Rev Neurosci* 1991;14:123–136.

Gases

Snyder SH, Bredt DS. Biological roles of nitric oxide. *Sci Am* 1992;266:68–77.
Verma A, Hirsch DJ, Glatt CE, Ronnett GV, Snyder SH. Carbon monoxide: a putative neural messenger. *Science* 1993;259:381–384.

Second Messengers

Worley PF, Baraban JM, De Souza EB, Snyder SH. Mapping second messenger systems in the brain: differential localizations of adenylate cyclase and protein kinase C. *Proc Nat Acad Sci U S A* 1986;83:4053–4057.

6

Autonomic Nervous System, Limbic System, and Sexual Dysfunction

Michael J. Aminoff, Gary W. Van Hoesen, Robert J. Morecraft, Katerina Semendeferi, and Robert Taylor Seagraves

THE AUTONOMIC NERVOUS SYSTEM

OVERVIEW

The autonomic nervous system is concerned primarily with the innervation of various internal organs, the maintenance of the internal environment, and the regulation of processes that are not usually considered to be under voluntary control. It is traditionally divided into the sympathetic and parasympathetic systems, which have seemingly opposing activities, ensuring the harmonious integration of different functions. The autonomic nervous system is also instrumental in generating the physical responses to emotional stimuli that characterize certain aspects of behavior. Thus, the cardiovascular changes accompanying anger (i.e., an increase in heart rate and blood pressure) or embarrassment (i.e., a cutaneous vasodilation or "blush") depend on autonomic activity. Similarly, the dry mouth of excitement, the excessive sweating that occurs with anxiety, the urinary incontinence associated with intense fear, and the "wide-eyed" appearance of surprise all depend on autonomic activity. Such autonomic responses have to be integrated with any somatic responses necessitated by the emotional stimulus, such as offensive or predatory maneuvers occasioned by anger, or defensive behavior elicited by fear.

ANATOMY

Autonomic Efferent Pathway: Central Autonomic Structures

The autonomic nervous system is represented at many different levels of the central nervous system (CNS), including the *cortex* of the superior frontal gyrus and areas 4 and 6 of the cerebral cortex. Cerebral pathology may lead to disturbances of cardiovascular, pilomotor, sudomotor, or gastric function. For more than a century, it has been known that cortical lesions or stimulation may influence the heart, respiratory rate, and blood pressure. The anterior cingulate cortex is involved in the control of bladder and bowels, and loss of sphincter function accompanies bilateral cingulate lesions. The temporal lobes and amygdalas are also involved in higher-order autonomic regulation. There are profuse connections between those parts of the cerebral cortex involved in autonomic activity and other CNS regions having autonomic functions, but the specific pathways mediating this activity are unknown.

The *hypothalamus* appears to be a major (direct or indirect) relay station for autonomic pathways from the spinal cord, brain stem, and hippocampus and is connected to the premotor frontal cortex. There are also rich efferent connections (direct or indirect) with autonomic neurons in the spinal cord and a close association between the hypothalamus and hypophysis. Hypothalamic stimulation influences cardiovascular, pilomotor, and thermoregulatory function. The hypothalamic region also influences feeding behavior; hypothalamic pathology leads to either hyperphagia and obesity or aphagia and weight loss, depending on the precise site of the lesion.

There are major, often reciprocal, connections between certain *pontine nuclei* and lower brain stem (medullary), forebrain, and hypothalamic structures that are also concerned with cardiovascular and respiratory control. The brain stem has important influences on ventilation. Neuronal activity related to respiration occurs in discrete regions of the upper pons, and electrical stimulation of these regions leads to change in the ventilatory phase. Descending pathways conduct impulses from the brain stem to the preganglionic sympathetic neurons in the intermediolateral cell columns in the thoracic and upper lumbar regions of the cord. The axons of these cells exit with the anterior nerve roots as preganglionic sympathetic fibers passing to adjacent ganglia. Descending spinal pathways also connect with the parasympathetic outflow in the cranial and sacral regions. The spinal autonomic fibers are small in diameter and are probably most profuse in the lateral funiculi, but their precise pathway is unknown. For an anatomic review, see references 1 and 2.

Neurotransmitter and Neuromodulators

The major postganglionic neurotransmitter of the sympathetic system is *norepinephrine*, which is transported along postganglionic fibers to their terminals, where it is stored in dense-core vesicles until released. There are two types of adrenergic receptors—the so-called α- and β-receptors—and subclasses within these two broad categories. *Epinephrine* is released from the adrenal medulla and has important effects on cardiovascular function as well as other responses to stress.

Acetylcholine is the main peripheral neurotransmitter in the parasympathetic system, but it is also released from preganglionic sympathetic fibers and at postganglionic sympathetic nerve terminals to the sweat glands and certain blood vessels in the skeletal muscles as well as from fibers supplying the adrenal medulla. Numerous central pathways are cholinergic. Cholinergic receptors are divided into muscarinic receptors (e.g., in CNS neurons, skeletal postganglionic sympathetic neurons, and smooth and cardiac muscle) and nicotinic receptors (autonomic ganglia, skeletal neuromuscular junctions, and spinal cord).

Neuropeptides are present in autonomic neurons and may function as co-transmitters or modulators of neurotransmission, but their precise role has yet to be clarified. Neuropeptide Y is present in a variety of sympathetic, parasympathetic, and enteric neurons; it is contained, for example, in many nonadrenergic vasomotor neurons in the sympathetic ganglia. Opioid (and other) peptides are similarly widespread in peripheral autonomic neurons and may coexist with neuropeptide Y in the same nerve cells.

Sexual Function

The hypothalamus and limbic regions have an important role in sexual arousal. Electrical stimulation of various hypothalamic areas causes erection, whereas lesions in these regions—and especially of the medial preoptic anterior hypothalamic area—suppress sexual behavior. Bilateral temporal lobectomy in monkeys and humans leads to an increase in sexual activity, as well as other behavioral changes, and a similar clinical picture may occur with other encephalopathies.

Descending pathways traverse the midbrain, lower brain stem, and lateral columns of the cord to the thoracolumbar and sacral regions. Spinal cord lesions may have profound effects on sexual function, depending on the extent and level of the lesion. Many patients with complete cervical lesions have reflex erections, but the pleasurable experience of orgasm is abolished. With more caudal lesions, erectile and ejaculatory failure are common.

The sympathetic and parasympathetic fibers innervate the blood vessels of various pelvic structures, erectile tissues in the penis and clitoris, and smooth muscle of the vagina, uterus, prostate, and seminal vesicles. The mechanisms involved in erection are complex and poorly understood. Cavernosal tissue becomes engorged because of increased blood flow resulting from dilation of cavernosal and helicine arteries. This engorgement leads to compression of emissary veins, and venous drainage is therefore reduced. Detumescence results from contraction of the helicine arteries and the trabecular walls of the lacunar spaces of cavernosal tissue. Blood inflow is thus reduced, and decompression of veins leads to increased drainage. The sympathetic nervous system is primarily responsible for emission. The smooth muscle of the epididymis, vas deferens, seminal vesicles, and prostate contracts, as does the sphincter at the bladder neck to prevent retrograde ejaculation. Ejaculation (i.e., rhythmic seminal transport along, and expulsion from, the urethra) and orgasm are mediated primarily by pudendal somatic efferent and afferent fibers, respectively. The neuropharmacologic basis of these events is unclear, but erection probably involves vasoactive intestinal polypeptide and endothelium-derived releasing factor, which are activated by the parasympathetic system (3).

Clinical Evaluation of Autonomic Function

A variety of noninvasive tests of autonomic function have been developed to test the integrity of parasympathetic (vagal) fibers to the heart and the sympathetic vasomotor and sudomotor fibers. Such tests can be used to evaluate the autonomic nervous system in patients with symptoms suggestive of dysautonomia. The aim is to determine the presence and severity of any autonomic disturbance and the site of the underlying lesion. These studies can also be used to determine the functional integrity of the unmyelinated and small myelinated fibers that compose the peripheral component of the autonomic nervous system in patients with small-fiber peripheral neuropathies.

Heart Rate and Blood Pressure Responses

Measurement of the heart rate response to *deep breathing* is a noninvasive, sensitive, quantitative test that is simple to perform and provides a reliable index of the afferent and efferent parasympathetic (vagal) innervation of the heart. A tachycardia normally occurs during inspiration because of a reduction in cardiac vagal activity (Fig. 6-1). For clinical purposes, an electrocardiograph is used to record the R-R intervals continuously over 60 seconds in the recumbent patient. After a 5-minute rest period, the patient takes six deep breaths over 1 minute. The difference between the longest and shortest R-R intervals is noted. The variation of heart rate with respiration is age dependent but is normally at least 15 beats per minute; values of less than 10 beats per minute are clearly abnormal. In addition, an expiratory-to-inspiratory (E:I) ratio is calculated from the ratio of the mean of the maximum R-R intervals during expiration to the mean of the minimum R-R intervals during inspiration.

The heart rate response to *standing* is another useful, simple, noninvasive test of vagal function. Upon standing from the recumbent position, there is an initial tachycardia followed by a bradycardia that begins after about 20 seconds and stabilizes after about the 30th heartbeat. For clinical purposes, the R-R interval of the electrocardiogram (ECG) can be measured during the performance of the maneuver, and the ratio of the R-R interval at the 30th beat to the 15th beat is determined. This 30:15 ratio depends on vagal function and is age dependent, but in young adults, it normally exceeds 1.04.

The heart rate responses to *passive head-up tilt* can also be measured after the patient has remained supine for 10 minutes. There is normally an increase of between 10 and 30 beats per minute, reflecting a change in both sympathetic and parasympathetic activity, with 60-degree head-up tilt. The systolic and dias-

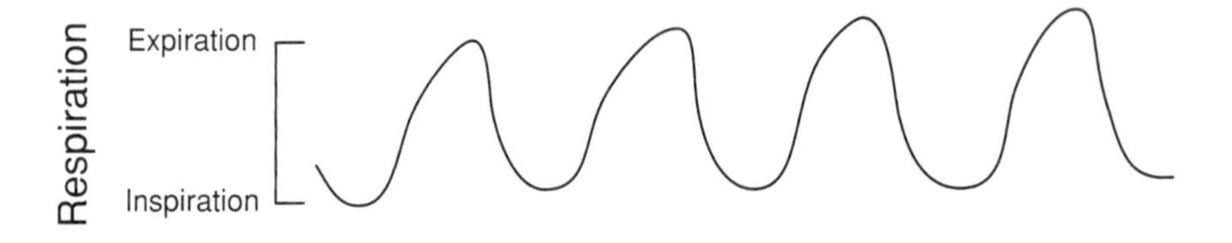

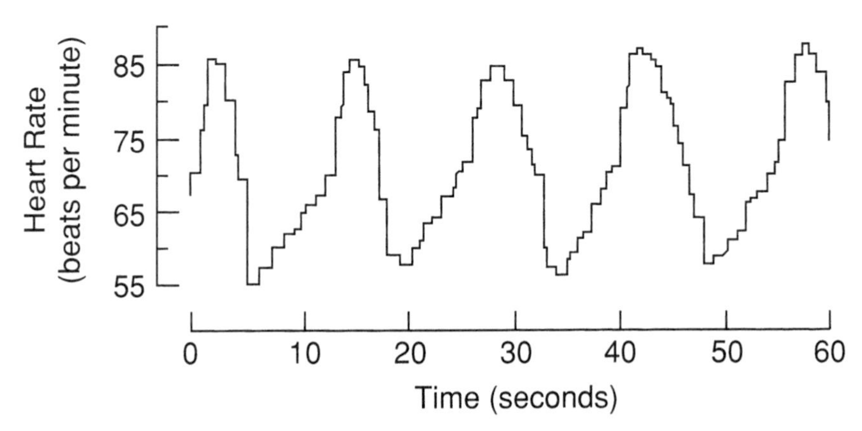

FIG. 6-1. Variation in heart rate with deep breathing in a normal subject.

tolic blood pressures usually fall slightly, but this does not exceed 20 and 10 mm Hg, respectively, in healthy subjects. Recent studies have emphasized the value of maintaining the head-up tilt for up to 60 minutes, particularly in patients in whom vasodepressor syncope is suspected.

The cardiovascular responses to sustained hand grip (30% of maximum for 5 minutes) have also been used as a means of assessing adrenergic function. The sustained muscle contraction produces a rise in heart rate and blood pressure. The diastolic pressure normally rises by at least 15 mm Hg, when the last value recorded before release of the hand grip is compared with the mean value obtained in the 3 minutes before commencing the maneuver. The test, however, is of limited sensitivity and specificity.

The *Valsalva maneuver*, in which the subject performs a forced expiration against a closed glottis, provides important information concerning the integrity of cardiovascular innervation. For clinical purposes, an ECG machine or heart rate monitor can be used to record the heart rate responses (Fig. 6-2). The recumbent subject makes a sustained expiration that is sufficient to maintain a column of mercury at 40 mm for 15 seconds by blowing into a mouthpiece with a calibrated air leak. A Valsalva ratio can then be determined from the shortest interbeat interval (or fastest heart rate) generated during the forced expiration divided into the longest interbeat interval (slowest rate) that occurs after it; this reflects both sympathetic and vagal function. A ratio of 1.1 or less is abnormal and 1.21 or greater normal, but the ratio is age dependent; in subjects younger than 40 years, the Valsalva ratio normally exceeds 1.4. For a review of these tests, see references 4 and 5.

Tests of Cutaneous Vasomotor Function

Cutaneous blood flow can be studied by plethysmography or a laser Doppler flow me-

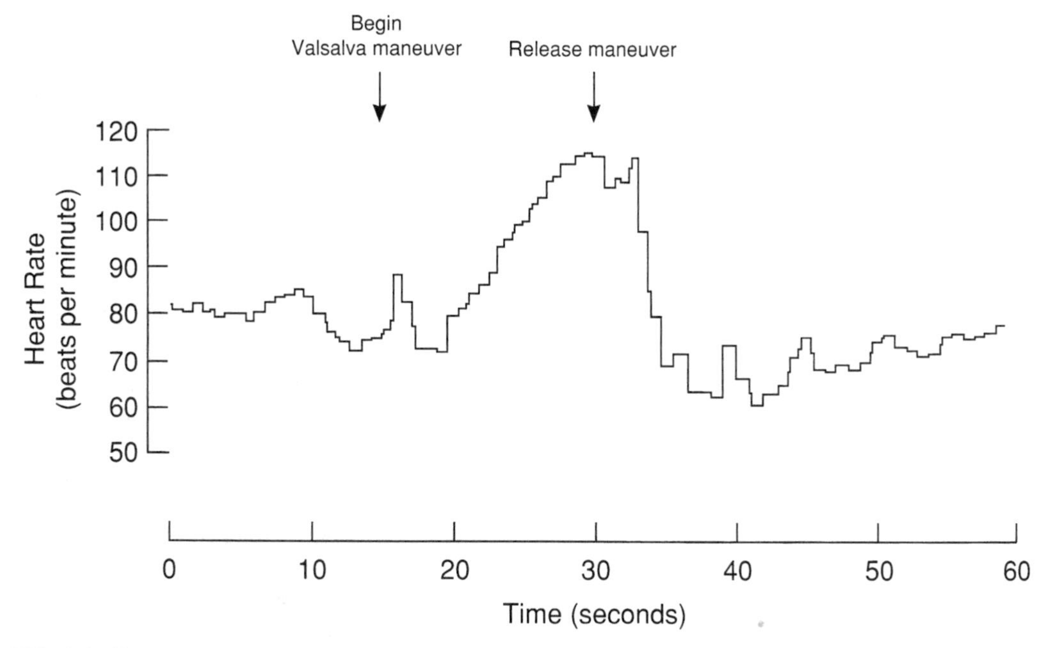

FIG. 6-2. Response to the Valsalva maneuver, recorded with a heart rate monitor in a normal subject. There is a tachycardia during the forced expiratory maneuver and a compensatory bradycardia when the maneuver is released.

ter. The vasomotor responses to various stimuli are evaluated as a measure of adrenergic function. For convenience, digital blood flow is recorded for clinical purposes.

A sudden inspiratory gasp normally produces a digital vasoconstriction as a spinal reflex. The response is lost or impaired in patients with a cord lesion or a disturbance of sympathetic efferent pathways, such as in peripheral neuropathies. A cold stimulus (ice-cold water at 4°C) to the opposite hand also produces a reflex vasoconstriction, as does mental stress.

Sweat Tests

The thermoregulatory sweat test is a highly sensitive test in which the body temperature is raised by 1°C by exposure of the subject to radiant heat from a heat cradle. The presence and distribution of sweating is determined by a change in color of an indicator powder placed on the skin. Abnormalities may reflect preganglionic or postganglionic lesions.

Plasma Catecholamine Levels and Norepinephrine Infusion

The resting plasma norepinephrine level provides an index of sympathetic activity. It is diminished in disorders with postganglionic (as opposed to preganglionic) pathology. The plasma norepinephrine level normally increases with change in posture from recumbent to standing, and this postural response may be markedly attenuated or absent with preganglionic or postganglionic lesions.

Another approach is to measure the blood pressure changes that occur with intravenous infusion of norepinephrine at different dosage rates up to 20 μg per minute. To increase the systolic pressure to 40 mm Hg above baseline, healthy subjects require infusion at a rate of 15 to 20 μg per minute, whereas patients with the Shy-Drager syndrome require 5 to 10 μg per minute and those with primary autonomic failure less than 2.5 μg per minute.

Bladder and Gastrointestinal Function

Bladder function is assessed by several methods. These include determination of the volume of residual urine after attempted voiding, which is normally less than 100 mL. Cystometry is an important investigative approach in which the bladder is filled with water or radiologic contrast medium or inflated with carbon dioxide by means of a self-retaining catheter, and the relationship between intravesical pressure and volume is determined. Bladder filling is normally first appreciated at volumes of 100 to 200 mL, and the desire to void increases as filling continues. Impaired sensation may reflect peripheral or central pathology. In healthy subjects, vesical pressure does not increase by more than 10 cm H_2O until maximum capacity (generally between 300 and 600 mL) is reached.

Radiologic studies are important in evaluating gastrointestinal function and especially in excluding mechanical causes of symptoms, such as dysphagia, delayed gastric emptying, intestinal pseudoobstruction, or intractable constipation. Fluoroscopy and studies of colonic transit time may also be helpful. Catheter probes connected to a pressure transducer can be used to measure anorectal pressure. In the region of the anal sphincter, resting pressure reflects internal anal sphincter tone. Activity in the external anal sphincter muscle is reflected by the pressure recorded on voluntary contraction.

Sphincter Electromyography and Evoked Potential Studies

Denervation of the urethral and anal sphincters may result from cell loss in Onuf's nucleus in the sacral cord or from pathology situated more distally. Needle electromyography permits recognition of abnormal motor unit action potentials in the sphincteric muscles in patients with dysautonomia and thereby helps to localize the lesion. With pyramidal lesions, volitional control of sphincter function is impaired, and the urethral striated sphincter muscle fails to relax during detrusor contraction (detrusor-sphincter dyssynergia).

Pharmacologic Evaluation of Pupillary Reactivity

Anticholinergic (parasympatholytic) agents applied topically cause pupillary dilation, as do sympathomimetic agents. Conversely, parasympathomimetic or sympatholytic agents cause pupillary constriction.

Pilocarpine and methylcholine are cholinergic (parasympathomimetic) agents. In patients with parasympathetic denervation, very weak solutions of pilocarpine (e.g., 0.125%) cause pupillary constriction because of denervation supersensitivity. A dilated pupil that is unresponsive to instilled cholinergic agents probably reflects sympathetic overactivity.

Sympathomimetic agents cause dilation of denervated pupils at concentrations that are ineffective on normal pupils. For example, 1% phenylephrine hydrochloride produces dilation of the pupil in Horner's syndrome because of denervation supersensitivity. Cocaine hydrochloride (4%) can also be instilled into the conjunctival sac and normally causes pupillary dilation by sympathetic activation. In Horner's syndrome resulting from peripheral sympathetic denervation, this effect of cocaine is lost.

Clinical Features of Dysautonomia

Impotence and loss of libido are common presenting features in men with dysautonomia and are often mistakenly attributed to psychogenic factors until other autonomic disturbances develop. Ejaculation may also be impaired. When retrograde ejaculation occurs, patients feel as if they have ejaculated but little is produced; the subsequently voided urine is found to be discolored and to contain spermatozoa. Abnormalities of micturition are another common early symptom and, in men, may be attributed to prostatic hypertrophy.

The most disabling symptom of autonomic dysfunction is usually postural hypotension. A feeling of impending loss of consciousness occurs on standing or walking, and the patient may ultimately fall to the ground unconscious, unless symptoms are aborted by sitting or lying down. Unlike a typical syncopal episode, there is no preceding sweating. Symptomatic postural hypotension is more likely to occur in the early part of the day, postprandially, with activity, or in hot weather because of the cardiovascular and hemodynamic alterations that occur in these circumstances.

Disturbances of bladder or bowel regulation are especially distressing. Urinary urgency, frequency, and nocturia may occur, and there may be urgency incontinence. In other instances, an atonic bladder leads to overflow incontinence. Cystometry clarifies the nature of the bladder disturbance and thereby indicates the appropriate therapeutic approach. Intractable constipation, intermittent diarrhea, or an alternation between constipation and diarrhea also occurs. Rectal incontinence during episodes of diarrhea is particularly troublesome. Nausea, postprandial fullness, and severe vomiting may reflect gastroparesis.

Visual blurring and dryness of the eyes are sometimes troublesome, and there may be a variety of pupillary abnormalities in patients with dysautonomia. The effect of lesions affecting the parasympathetic innervation (i.e., the third cranial nerve) or the sympathetic system (causing Horner's syndrome) has already been described.

Disturbances of breathing may consist of inspiratory gasps or cluster breathing in patients with the Shy-Drager syndrome of autonomic failure. Occasional patients experience central sleep apnea. Laryngeal stridor may also occur.

Selected Causes of Dysautonomia

Acute Peripheral Pathology

There are many causes of autonomic dysfunction. Neuropathies are among the most frequent causes. Acute involvement of peripheral autonomic fibers may occur in several contexts.

Autonomic dysfunction is common in patients with a peripheral neuropathy but is usually overshadowed by the coexisting motor and sensory deficits. In Guillain-Barré syndrome, however, autonomic disturbances may be life-threatening. There may be severe hypotension, paroxysmal hypertension, or extreme fluctuations in blood pressure; disturbances of cardiac rhythm, regional anhidrosis or hyperhidrosis, pupillary abnormalities, gastroparesis, constipation, urinary retention, and urinary or fecal incontinence also occur. The cause of the paroxysmal hypertension is unclear, but denervation supersensitivity to circulating catecholamines or denervation of baroreceptors may be responsible.

Autonomic dysfunction, resulting from a disturbance of cholinergic mechanisms, is a feature of botulism. Constipation, urinary retention, dryness of the eyes, anhidrosis, and xerostomia present the most common evidence of such dysfunction. Disorders of autonomic, especially cholinergic, function are well described in Lambert-Eaton myasthenic syndrome.

A number of iatrogenic neuropathies may be associated with autonomic involvement, such as those related to amiodarone, perhexiline maleate, cisplatin, or vincristine. Dysautonomia also follows exposure to acrylamide, organic solvents, and Vacor (a rodenticide). For reviews of these topics, see references 6 and 7.

Chronic Peripheral Pathology

The most common cause of dysautonomia in developed countries is probably diabetes. Diabetic autonomic neuropathy may occur in isolation or in association with any of the neuromuscular complications of diabetes, especially a symmetric sensory or sensorimotor polyneuropathy, entrapment neuropathy, mononeuropathy multiplex, polyradiculopathy, or plexopathy. Postural hypotension, abnormal cardiovagal function, and impaired distal thermoregulatory sweating are typically found; impotence, gastroparesis, constipation, fecal incontinence (often in association with diarrhea), and bladder disturbances are common.

Autonomic disturbances also occur in patients with the neuropathy of chronic renal failure, leprosy, vitamin B_{12} deficiency, and various connective tissue diseases.

A variety of hereditary disorders are associated with autonomic failure. Familial dysautonomia (Riley-Day syndrome) is a recessively inherited neuropathy that begins in infancy with disturbances of thermoregulation, lacrimation, blood pressure regulation, and gastrointestinal function. Repeated episodes of pneumonia are typical. Somatic involvement is manifest by impaired pain and temperature appreciation, weakness, poor sucking, depressed tendon reflexes, and arthropathy. Several other types of hereditary sensory and autonomic neuropathy have been described.

Autonomic dysfunction sometimes occurs in hereditary motor and sensory neuropathy type I or II and is manifest especially by pupillary changes, impaired vasomotor regulation of the distal blood vessels, and occasionally impaired cardiovascular reflexes.

Central Dysautonomias

Autonomic manifestations are a feature of many different disorders of the CNS. Seizures, for example, may have autonomic accompaniments, especially when they arise from limbic and paralimbic structures. The most common clinical evidence of autonomic involvement during seizures is a sinus tachycardia, but a variety of other cardiac manifestations may occur ictally, including atypical anginal pain, bradycardias, conduction defects, sinus arrest, tachyarrhythmias, and atrial fibrillation. Another common, possibly dysautonomic, feature of seizures—especially complex partial attacks arising from the temporal lobe—is a curious feeling of discomfort that ascends from the epigastrium, accompanied sometimes by nausea.

Lesions of the *mesial frontal lobe*, especially the cingulate gyrus, may produce disturbances of sphincter function, with urinary incontinence, fecal incontinence, or both. *Hydrocephalus* sometimes leads to a similar effect, presumably because of distortion and stretch of corticobulbar fibers.

Hypothalamic pathology may affect temperature regulation, mainly leading to hypothermia. This may be the basis of the hypothermia that is sometimes associated with episodic hyperhidrosis and agenesis of the corpus callosum. In other instances, episodic hyperthermia occurs after head injury or with hypothalamic pathology or acute hydrocephalus and may be associated with tachycardia, hyperpnea, transient hypertension, cutaneous vasoconstriction or vasodilation, pupillary changes, and increased muscle tone. Hypothalamic pathology, such as infarction, occasionally leads to an ipsilateral Horner's syndrome. Pathologic involvement of the anterior hypothalamus may affect circadian rhythms.

Diseases of the *basal ganglia* are also accompanied by autonomic dysfunction. In patients with *Parkinson's disease*, many symptoms (such as disturbances of sweating, postural dizziness, and sphincter dysfunction) suggest autonomic involvement. Autonomic symptoms also occur in other extrapyramidal disorders. Some patients with *progressive supranuclear palsy* have postural hypotension, but this does not usually exceed the postural change sometimes encountered in healthy subjects. In *Huntington's disease*, there may be sphincter disturbances, abnormalities of swallowing and respiration, hyperhidrosis, sialorrhea, polyuria, polydipsia, and hypogenitalism. Pathologic changes in this disorder involve the caudate nuclei, and these structures have been implicated in the regulation of blood pressure during change in posture.

Autonomic hyperactivity (with tachycardia, hypertension, hyperhidrosis, and hyperthermia) is a major feature of *fatal familial insomnia*, where it is conjoined with intractable insomnia and motor abnormalities, including myoclonus, ataxia, and pyramidal deficits. The disorder is associated with pathology involving the anterior ventral and dorsomedial nuclei of the thalamus.

Tumors, ischemia, or degenerative disorders affecting the *brain stem* or *cerebellum* may lead to postural hypotension or, less commonly, to hypertension. In some cases, the blood pressure abnormality precedes development of other neurologic deficits, and its cause may pass unrecognized unless the CNS is imaged.

Shy-Drager syndrome (or multisystem atrophy) is a condition in which marked autonomic dysfunction occurs in association with a somatic neurologic deficit characterized primarily by parkinsonian features but also by cerebellar signs, pyramidal deficits, and sometimes lower motor neuron involvement. It can therefore be distinguished by its neurologic accompaniments from the syndrome of primary (or pure) autonomic failure, in which the autonomic dysfunction occurs in isolation. Shy-Drager syndrome may simulate classic Parkinson's disease, but the existence of more widespread neurologic deficits and an impairment of baroreceptor reflexes distinguish it from the latter disorder. The disorder tends to pursue a progressive course, and many patients are die within 5 years of diagnosis. Dysautonomia may occur in patients with olivopontocerebellar atrophy or striatonigral degeneration; these disorders probably reflect different manifestations of multisystem atrophy.

Complete lesions of the *spinal cord*, as by trauma, have major effects on autonomic function. After transection of the cervical cord, reflex function returns to the isolated spinal segment after a variable period (usually a few weeks), but cerebral regulation of autonomic activity is lost. With lesions above T6, the resting blood pressure is reduced, and there is marked orthostatic hypotension, with an overshoot of the blood pressure on resumption of a recumbent posture. There may be disturbances of temperature regulations because of an inability to sweat or alter vasomotor function below the level of the lesion. Bladder, bowel, and sexual function are markedly impaired. Visceral, muscle, or cutaneous stimulation below the level of the lesion leads to reflex sympathetic and parasympathetic excitation, with consequent activity in a number of organs supplied by the autonomic nervous system. A marked and rapid elevation of the blood pressure may lead to intracranial hemorrhage. Stroke volume and cardiac output also increase. Cutaneous vasodilation and sweating sometimes occur above the level of the lesion, but the mechanism of this is unclear.

Cardiac arrhythmias, neurogenic hypertension, and acute pulmonary edema may occur in patients with acute intracranial pathology, such as subarachnoid hemorrhage or increased intracranial pressure, and sometimes lead to sudden death. They probably relate to excessive sympathetic activity (8).

Aging

Many subjects over the age of 70 years have a postural drop in systolic pressure of 20 mm Hg or more on standing. There may be alterations in baroreceptor sensitivity with age, and reduction in the elasticity of blood vessels and of adrenoreceptor sensitivity may also account for the postural drop in blood pressure. Syncope is a common problem in the elderly and often occurs for unclear reasons.

THE LIMBIC SYSTEM

OVERVIEW

The limbic system concept was coined by MacLean near the midpoint of this century to reintroduce and reemphasize the seminal thinking of Papez on the neuroanatomic correlates of emotion and to integrate his deductions with advances from both the clinic and the laboratory. This led to a larger "circuit for emotion" in neuroanatomic terms and a somewhat more multifaceted one in functional terms. A scant 5 years later, memory function was added to the functional correlates because acquiring new information and learning rely critically on an intact Papez's circuit.

The central theme of Papez's circuit deals with the rather simple anatomic notion of how the cerebral cortex (and, by definition, all sensory systems) influences the hypothalamus (the building up of the "emotive process") and how the latter influences the cerebral cortex ("psychic coloring"). Papez was successful in elaborating one facet of these relationships, but the full panorama was not appreciated until the past four decades with the advent of newer experimental neuroanatomic methodology and a major effort by neuroanatomists. These have led to an expanded conception of the limbic system in both neuroanatomic and functional terms. Although confusing because the term *system* implies unity of function, the core of the concept is the limbic system as the mediator of two-way communication between the cerebral cortex and the external world, and the hypothalamus and the internal world of the organism. Behavior in general, whether it be the consequence of autonomic, endocrine, or somatic effects, is governed by this interplay of the external and internal worlds.

The prefrontal cortex refers to that cortex anterior to the electrically excitable motor cortex. This is a large area of cortex in the human and nonhuman primate brain and includes a large dorsolateral sector, orbital sector, and medial sector. The latter typically includes the anterior cortices of the cingulate gyrus, which wrap around the genu of the corpus callosum and follow its rostrum ventrally and posteriorly.

Recent studies have linked this area to working memory and decision making. The former refers to the ordering of behavior, or the manner in which time is bridged neurally to enable completion of a sequence of intended acts or tasks. The latter function, decision-making behavior, complements working memory in critical ways and involves an assessment of somatic markers that impart reason and reality to choices of behavior. Some patients with frontal lobe damage have preserved intellect, analytic abilities, social consciousness, and sensory awareness but lack the ability to make accurate predictions about the outcome of decisions on their own well-being.

Our goal in this chapter is to examine recent neuroanatomic findings regarding the major parts of the limbic system and the prefrontal cortices and new concepts that emerge from them. We close with a consideration of degenerative diseases that affect these parts of the brain, evolutionary issues, and the implications that these and functional neuroanatomy have for neuropsychiatric disorders. For classic reviews, see references 9 to 11.

ANATOMY

Anatomic structures included under the term *limbic system* have diverse locations in the cerebral hemisphere and occupy parts of the telencephalon, diencephalon, and mesencephalon. In general, they are the so-called conservative parts of the brain—those found in a wide range of mammals and to some extent vertebrates in general. The cortical parts compose what are frequently termed the *older* parts of the cerebral cortex, those parts of the cerebral cortex common to many species that form the edge, or limbus, of the cerebral cortex.

A useful way of dealing with the complex anatomy of the limbic system is shown in Fig. 6-3, which divides the term into four conceptual units:

- Landmarks
- Cortical components
- Subcortical structures
- Interconnecting pathways

The cortical structures of the limbic system are well demarcated on the medial surface of the hemisphere by the cingulate sulcus dorsally and the collateral sulcus ventrally (see Fig. 6-3*A*). Bridging areas, such as the subcallosal gyrus and the posterior orbital, anterior insular, temporal polar, and perirhinal cortices, connect the cingulate and parahippocampal gyri rostrally, whereas the retrosplenal and retrocalcarine cortices provide a bridge caudally. The areas differ widely in cytoarchitecture and include Brodmann's areas

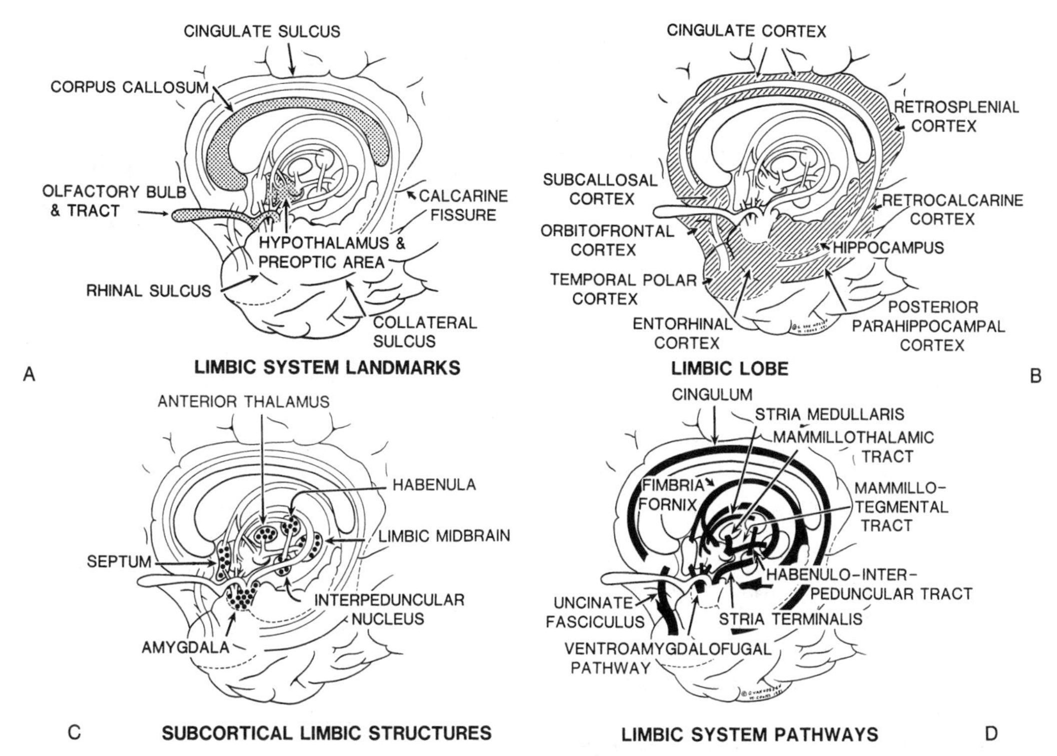

FIG. 6-3. Four schematic representations of the medial surface of the cerebral hemisphere depicting the anatomic components of the limbic system. **A:** The relevant landmarks. **B:** The limbic lobe or cortical components. **C:** Subcortical limbic structures (the nucleus basalis of Meynert and the diagonal bands of Broca, both of which contain cholinergic neurons, are not shown). **D:** Interconnecting limbic system pathways. (From Ref. 32, with permission.)

23 through 29, 35, 36, and 38. None is a true isocortical area. Rather, all fall under the categories of periallocortex and proisocortex.

The subcortical structures included in the limbic system (see Fig. 6-3*C*) vary widely among authors and are scattered throughout many parts of the hemisphere. However, it seems appropriate to include the amygdala, septum, nucleus basalis of Meynert, anterior thalamus, habenula complex, interpeduncular nucleus, and some additional limbic midbrain areas. A structural criterion common to this list relates to the fact that all are connected among themselves as well as with the hypothalamus.

The final units of the limbic system are the interconnecting pathways (see Fig. 6-3*D*), those within the limbic lobe, those that connect limbic lobe areas with subcortical limbic structures, those that connect subcortical limbic structures to each other, and finally, those that connect elements of the limbic system to the hypothalamus. These include such pathways as the cingulum, uncinate fasciculus, fimbria-fornix, stria terminalis, ventroamygdalofugal pathway, mammillothalamic tract, mammillotegmental tract, stria medullaris, and habenulointerpeduncular pathway.

Cingulate Gyrus

The cingulate gyrus is the major dorsal part of the limbic lobe on the medial wall of the cerebral hemisphere and forms the upper half of the cortical ring of gray matter. Its main portion is located dorsal to the corpus callosum and callosal sulcus and ventral to the cingulate sulcus, although it extends into the depths of the latter.

The various subsectors of the cingulate cortex and, in particular, areas 24 and 23 are coupled anatomically by a vast and organized set of intracingulate connections. The cingulate gyrus also is connected to the frontal lobe, parietal lobe, occipital lobe, temporal lobe, other parts of the limbic lobe, and the insula. In general, both the anterior and posterior divisions of the cingulate gyrus are connected to the prefrontal, orbitofrontal, posterior parietal, posterior parahippocampal, perirhinal, entorhinal, and lateral temporal cortices, as well as the presubicular and subicular CA1 parts of the hippocampal formation.

Based on the anatomic observations just summarized, several conclusions can be drawn:

1. The vast network of intracingulate connections and, in particular, those linking area 24 with area 23 provide multifaceted avenues for information exchange between anterior and posterior parts of the cingulate gyrus.
2. Widespread parts of the cingulate gyrus are linked to multimodal and limbic cortices. Multimodal sources would include prefrontal, rostral orbitofrontal, posterior parietal, and lateral temporal cortices. Limbic sources include posterior orbitofrontal, temporopolar, posterior parahippocampal, perirhinal, and entorhinal cortices. Thus, highly processed and abstract information from neocortical association areas and, perhaps, emotionally and motivationally relevant information from limbic sources can influence a wide variety of cingulate subsectors.
3. Cingulocortical connections appear to be made throughout much of the cingulate gyrus. Direct motor cortex interactions appear to occur through cingulate cortex lining the depths of the cingulate sulcus (areas 24c and 23c), and direct somatosensory cortex interactions occur through area 23c.

Summary and Functional Considerations

Available data suggest that the anterior cingulate gyrus may be implicated in the expression of affectively triggered movements related to painful stimuli. For example, units here respond selectively to a variety of noxious stimuli. In line with this observation, it has been demonstrated that the anterior cingulate cortex may mediate vocal expressions that reflect the internal state of the animal. It is well known that the periaqueductal gray matter has a role in brain-stem pain mechanisms and vocalization.

The anterior cingulate cortex has been implicated in the production of other forms of vocal expression that are linked to emotional expression. The separation cry, produced by primates for maintaining contact with a distant group of individuals and induced by separating a mother from her offspring, is adversely affected after ablation of the anterior subcallosal region of the cingulate gyrus. A different form of emotion-related vocalization, namely that of laughter, may also be mediated by the anterior cingulate gyrus. It appears that regardless of the emotional phenomena expressed, the anterior cingulate gyrus plays an important role in developing the associated motor response. It would also appear from the common engagement of the head and neck region in these responses that brain-stem centers mediating the operation of structures such as the larynx, tongue, and muscles of facial expression are heavily influenced by ongoing activity in the anterior cingulate gyrus.

The anterior cingulate gyrus may also regulate autonomic responses because its stimulation evokes pupillary dilation, piloerection, altered heart rate, and changes in blood pressure. These physiologic findings, coupled with the underlying neuroanatomic circuitry described for this part of the cingulate gyrus, has led to the suggestion that this part of the cingulate gyrus serves as a "visceral motor cortex" (12).

Hippocampal Formation and Parahippocampal Gyrus

The hippocampal formation comprises three allocortical areas:

- The pyramids that form the CA zones of the hippocampus (CA1, CA2, CA3)
- The dentate gyrus, including the CA4 polymorph neurons that are found in its hilum
- The various subicular cortices

The latter includes the subiculum proper, a true allocortical zone, and two periallocortical zones, the presubiculum and parasubiculum. The latter are multilayered and are associated closely with the hippocampal formation. They have continuity with the subiculum in their deep layers.

Input to the hippocampus arrives by means of the fimbria-fornix and perforant pathway and activates the pyramidal neurons directly, or indirectly, by means of the dentate gyrus and intrinsic pathways. The neurons that form the subiculum are responsible for a large amount of hippocampal formation output and nearly all of its diversity with regard to influencing other brain areas (Fig. 6-4). The neurons that form the subicular cortices, however, and to a lesser extent those of the CA1 zone, have extensive extrinsic projections that divide hippocampal output into major components: one to a variety of cortical areas and another to a variety of subcortical structures, such as the basal forebrain, amygdala, thalamus, and hypothalamus. Thus, hippocampal output is disseminated much more widely than previously thought, and importantly, projects not only to subcortical areas but also to cortical areas. The latter are thought to be the neural basis of whether information is stored or remembered.

Summary and Functional Considerations

The hippocampal formation is the focal point for major forebrain neural systems that are interconnected with the sensory-specific association cortices and the multimodal association cortices. These are widespread systems that involve much of the cortical mantle. As mentioned, the cortices that form the limbic lobe in general, and the amygdala and posterior parahippocampal area in particular, receive input from the various association cortices and either project directly to the hippocampal formation or first to the entorhinal cortex, which then projects to the hippocampal formation. The most compact part of this latter system, the perforant pathway, is the major output system of the entorhinal cortex. It mediates a powerful excitatory input to the hippocampal formation that culminates in extrinsic output to the septum by means of the fimbria-fornix or intrinsic output to the subicular cortices. These latter areas then project to several basal forebrain areas, including the amygdala, various diencephalic nuclei, and many parts of the limbic lobe and association cortex. Thus, hippocampal output is disseminated widely by the subicular and CA1 pyramids of the hippocampal formation (13).

Amygdaloid Complex

As mentioned previously, several amygdaloid nuclei receive inputs from the cerebral cortex (Fig. 6-5) and from a host of subcortical structures of both diencephalic and mesencephalic origin. The latter includes such structures as the hypothalamus, periaqueductal gray matter, peripeduncular nucleus, ventral tegmental area, supramammillary nucleus, and midline thalamic nuclei.

Unlike the hippocampal formation, whose input is derived largely from limbic lobe areas that receive input from the association cortices, many association areas project directly to the amygdala without relays in the limbic lobe. For example, the visual association cortices of the lateral temporal neocortex send direct projections to the lateral amygdaloid nucleus and to the dorsal part of the laterobasal amygdaloid nucleus. Some investigators have shown that the auditory association cortex of the superior temporal gyrus also projects directly to the lateral amygdaloid nucleus. Input related to somatic sensation also converges on this nucleus from the insular cortex.

The central amygdaloid nucleus is unusual in the sense that it receives input derived from all types of cortex. The superficial nuclei of the amygdala, such as the medial nucleus and

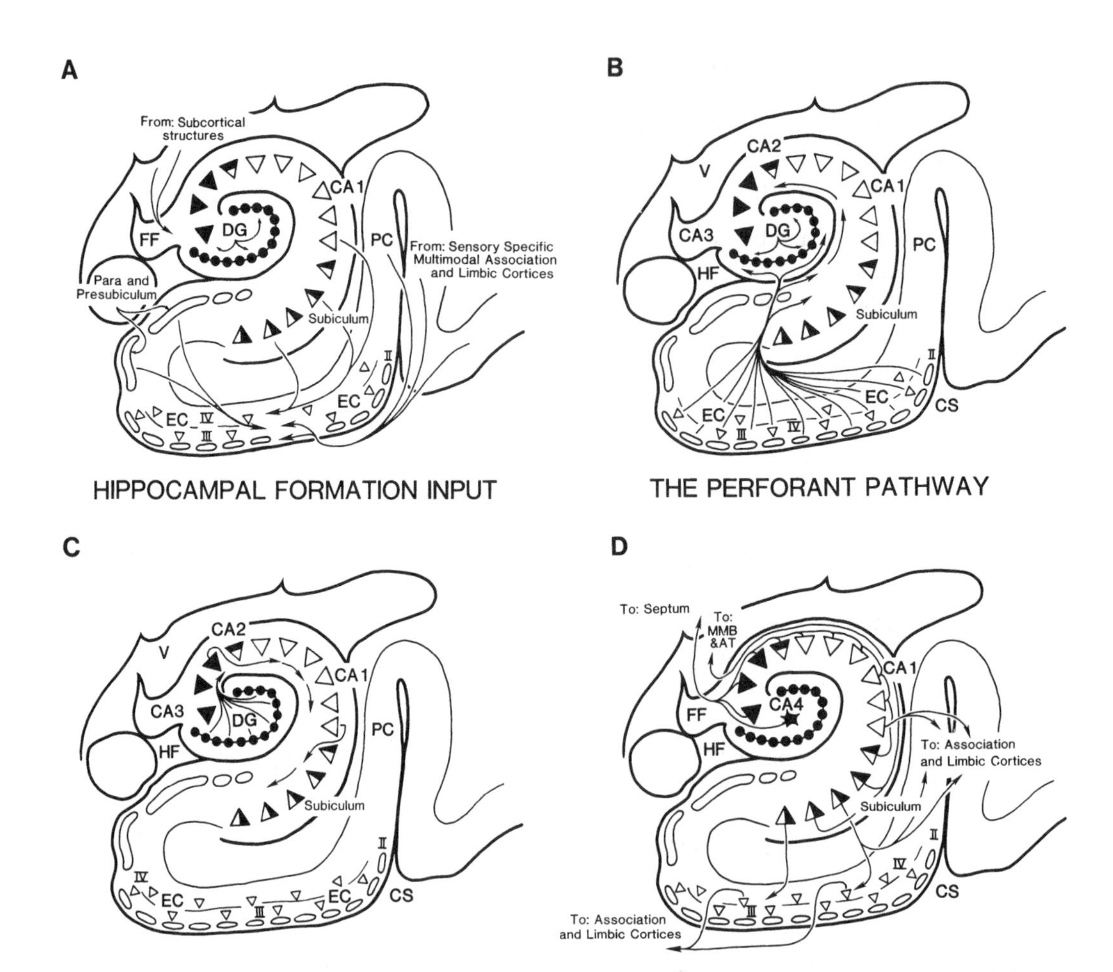

FIG. 6-4. Cross-sections of the ventromedial temporal lobe showing the major connection of the entorhinal (EC) and perirhinal cortices (PC) and the hippocampal formation. **A:** EC input from neighboring cortical areas, the subiculum, and from the sensory-specific and multimodal association areas and the limbic lobe. Subcortical input directly to the hippocampal formation arrives by way of the fimbria-fornix (FF). **B:** The origin and course of the major output pathway of the EC, the perforant pathway. Note its strong distribution to the pyramidal neurons of the subiculum, hippocampus (CA1–CA3) and dentate gyrus (DG). **C:** The major intrinsic connections of the hippocampal formation. Note in particular the DG projections to the CA3 pyramids, CA3 pyramid projections to CA1 pyramids, and CA2 pyramid projections to the subiculum. **D:** The major output projections of the hippocampal formation and the EC. Note that all hippocampal formation pyramidal neurons project to the anterior thalamus (AT) and mamillary bodies (MMB). However, these neurons give rise to output projections to the association and limbic cortices. Direct projection to the deep layers of the EC and their projections to the same areas provide the anatomic basis for a powerful hippocampal influence on other cortical areas. CS, collateral sulcus; HF, hippocampal fissure; V, inferior horn of the lateral ventricle.

the various cortical nuclei, receive input largely from allocortical regions, such as the subicular and periamygdaloid cortices and olfactory piriform cortex.

Until recently, it was believed that the major input and output relationships of the amygdala were with the hypothalamus. Such connections are strong, but the diversity of amygdaloid output is far more extensive than appreciated previously. For example, the lateral nucleus and some components of the basal complex project strongly to the entorhi-

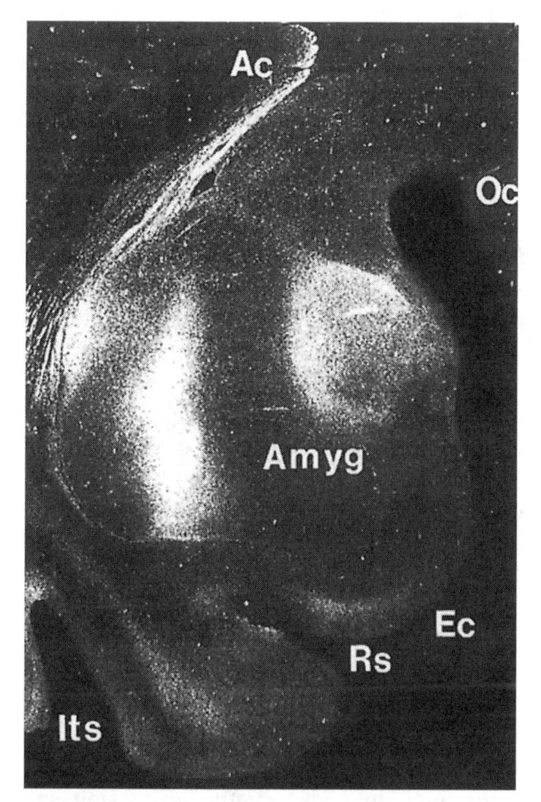

FIG. 6-5. A dark-field photomicrograph of the amygdala (Amyg) and entorhinal cortex (Ec) showing terminal labeling in a rhesus monkey experiment in which labeled amino acids were injected into the cortex of the temporal pole. The amygdaloid terminal labeling is primary over the medial part of the lateral nucleus and over the accessory basal nucleus. Ac, anterior commissure; Its, inferior temporal sulcus; Oc, optic chiasm; Rs, rhinal sulcus.

nal cortex, to several association areas, and even to the primary visual cortex. Additionally, the basal amygdaloid complex has strong reciprocal interconnections with the subiculum, the major source of hippocampal output. From these studies, it is clear that the amygdala is very much interrelated with the hippocampal formation in anatomic terms and that these two temporal neighbors undoubtedly influence each other to a great degree.

Among the more surprising aspects of amygdaloid anatomy described recently is that this structure has strong projections to many parts of the temporal association, insular, and frontal cortices.

Summary and Functional Considerations

The amygdala has powerful direct interconnections with much of the anterior cortex of the limbic lobe and with neocortical areas of the frontal, temporal, and even occipital lobes. Moreover, projections to the cingulate motor area, especially M3, provide a means for it to influence corticospinal axons. Additional projections connect the amygdala with the neostriatum and ventral striatum, involving it in basal ganglia circuitry. Certain amygdaloid nuclei also project strongly to the nucleus basalis of Meynert, whose axons provide a powerful cholinergic input to the cortex. Descending amygdaloid projections from the central amygdaloid nucleus also provide input to autonomic centers in the brain stem. Although much of the input to the amygdala, particularly from subcortical areas, cannot be characterized well in functional terms, this is not the case for amygdaloid output. Overall, it can be concluded that amygdaloid output is directed toward the origin of what may be termed *effector systems* that influence motor, endocrine, and autonomic areas along the full extent of the neuraxis. A persuasive argument could be made that the amygdala more greatly influences overt behavior, whereas the hippocampus more greatly influences more covert aspects of behavior, such as cognition and memory.

Nucleus Basalis of Meynert

The neurons that form the nucleus basalis of Meynert have attracted substantial attention because they project to the cerebral cortex and are affected frequently in Alzheimer's disease (AD). It is now clear that probably all of the nucleus basalis of Meynert projects to the cortex and all parts of it. These findings are intriguing, but they assume added significance with the demonstration that most of these neurons contain cholinergic enzymes, and, in fact, provide the major source of cholinergic input to the cortex.

The neurons forming the nucleus basalis are large, hyperchromatic, multipolar, and fusiform-shaped cells that lie among the ascending and descending limbic, hypothalamic, and brain-stem pathways that course through the basal forebrain. Part of the nucleus is found within the substantia innominata, but cholinergic neurons span the anteroposterior expanse of the ventral surface of the hemisphere all the way from the septum anteriorly to the midbrain posteriorly. They also have a lateral extension that follows the course of the anterior commissure into the temporal lobe.

It has been demonstrated recently that the nucleus basalis projects to the basal complex of the amygdala. Importantly, it is also known that this nucleus projects to the reticular nucleus of the thalamus. This places the nucleus basalis in a position to influence the cortex directly as well as indirectly because the reticular nucleus governs thalamic transmission by means of intrinsic thalamic connections.

The input to the nucleus basalis of Meynert is better understood. In terms of cortex, it has been shown that it receives projections from only a small percentage of the cortical areas to which it sends axons. These include such areas as the olfactory, orbitofrontal, anterior insular, temporal polar, entorhinal, and medial temporal cortices—all components of the limbic lobe. Subcortical projections to the nucleus basalis arise from the septum, nucleus accumbens, hypothalamus, amygdala, preoptic nucleus, and peripeduncular nucleus of the midbrain.

Summary and Functional Considerations

Many of the projections of the nucleus are not reciprocated by projections from the cortex that their axons innervate. The input to nucleus basalis neurons appears topographically organized and rather specific. At least two investigations have reported afferent input that seemingly "picks out" the clusters of nucleus basalis neurons. Some of these originate in the amygdala and may provide a highly specific, albeit indirect, manner for this structure to exert its influence on widespread parts of the cortical mantle. In this context, it should not be overlooked that the subiculum of the hippocampal formation projects both to the basal complex of the amygdala and to other basal forebrain areas that project to the nucleus basalis. Thus, a highly synthesized output from the hippocampal formation would seem plausible. It can be concluded that the major input to the nucleus basalis originates with the entire limbic system as a whole.

The nucleus basalis of Meynert receives projections from the hypothalamus. These need further study but provide a structural basis by which the internal state of the organism can indirectly influence both the motor and the sensory cortices as well as the manner by which the organism interacts with its environment (14).

Dorsomedial and Midline Thalamic Nuclei

The dorsomedial thalamic nucleus is known to play a role in many behaviors in humans, including visuospatial processing, attention, and memory. Contributing roles have also been argued for aphasia, dementia, and temporal disorientation when the nucleus is diseased or damaged.

The dorsomedial thalamic nucleus is a large midline association nucleus having powerful interconnections with the prefrontal granular association cortex.

With the exception of its prefrontal cortex connections, the neural systems of the dorsomedial nucleus are understood poorly. In fact, known input and output relationships with other structures are decidedly sparse in comparison with other nuclei of the thalamus and even other association nuclei, such as the pulvinar nuclei.

Summary and Functional Considerations

Overall, the neural systems involving the dorsomedial thalamic nucleus are not well known. The position of this nucleus ventral to two large fiber systems, the fimbria-fornix and corpus callosum; its encasement within the internal medullary lamina of the thalamus; and the fact that the mammillothalamic tract traverses its ventral parts have discouraged experimental study. Investigators' attention has focused primarily on the amygdalothalamic and temporothalamic projections. These link the dorsomedial nucleus with temporal structures known to play a role in memory (15).

Prefrontal Cortex

The frontal lobe lies anterior to the central sulcus and can be divided into two major parts: a caudal part containing the electrophysiologically "excitable" motor cortices and a rostral part containing the prefrontal association cortex. The motor cortices include the primary (M1 or area 4), supplementary (M2 or area 6m), and lateral (LPMC or areas 6D and 6V) premotor cortices.

Also included as part of the motor cortices are the frontal eye field (FEF or area 8), supplementary eye field (SEF or area F6), and presupplementary motor area (pre-SMA). The FEF and SEF are located on the lateral surface of the hemisphere. Both the FEF and SEF regulate contralateral saccadic eye movements. The pre-SMA is located on the medial wall of the hemisphere, rostral to M2, and corresponds to Walker's area 8B. Neurons in this field modulate their activity before and during movement.

The prefrontal cortex lies rostral to the motor-related cortices and extends to the frontal pole. On the medial surface, the prefrontal cortex lies anterior to the medial component of motor cortex, as well as the anterior part of cingulate gyrus. The primate prefrontal cortex is subdivided commonly on broad, anatomic grounds. Its major partitions include ventrolateral, dorsolateral, medial, and orbitofrontal regions.

Cortical Association Connections of Prefrontal Cortex

In addition to its obvious role in motor behavior, frontal lobe function has long been associated with a variety of higher-order behaviors and cognitive processes. Some of the more notable ones include working memory, motor planning, developing and implementing long-term strategies, decision making, and problem solving. When considering the higher-order functions mediated by the prefrontal cortex, the finding that prefrontal cortex is linked directly to a constellation of cortical association areas should not be surprising. Indeed, the prefrontal cortex is well known for its widespread corticocortical connections with distal parts of the cerebral cortex, specifically including primary association and multimodal association cortices (Fig. 6-6).

Primary association cortex is committed functionally to the early processing of sensory data, conveyed by the neurons of an adjacent primary sensory area. Primary association cortex operates in a more integrative fashion than primary sensory area. In contrast, multimodal association cortex is not committed to processing information related to one modality, but rather integrates highly transformed information, whose source can be traced back to multiple, sensory modalities.

The prefrontal cortex is linked to sensory association and multimodal association cortices of the parietal lobe (areas 7a, 7b, and 7m) and temporal lobe (areas V4t, MT, and MST), as well as sensory association cortices of the anterior part of the occipital lobe (area V3). Anatomic and behavioral investigations conducted during the past three decades have led to the conclusion that long association pathways, reciprocally linking posterior parietal cortex and prefrontal cortex, are particularly important for the appropriate execution of visually guided movements. Therefore, the dorsolateral part of the prefrontal cortex is thought to process information concerned with understanding *where* an object is in space. The prefrontal cortex is also influenced by other parts of the temporal lobe through a subcomponent of the ventral pathway, whose origin arises from the rostral part of the superior temporal gyrus, as well as from the temporal pole. This projec-

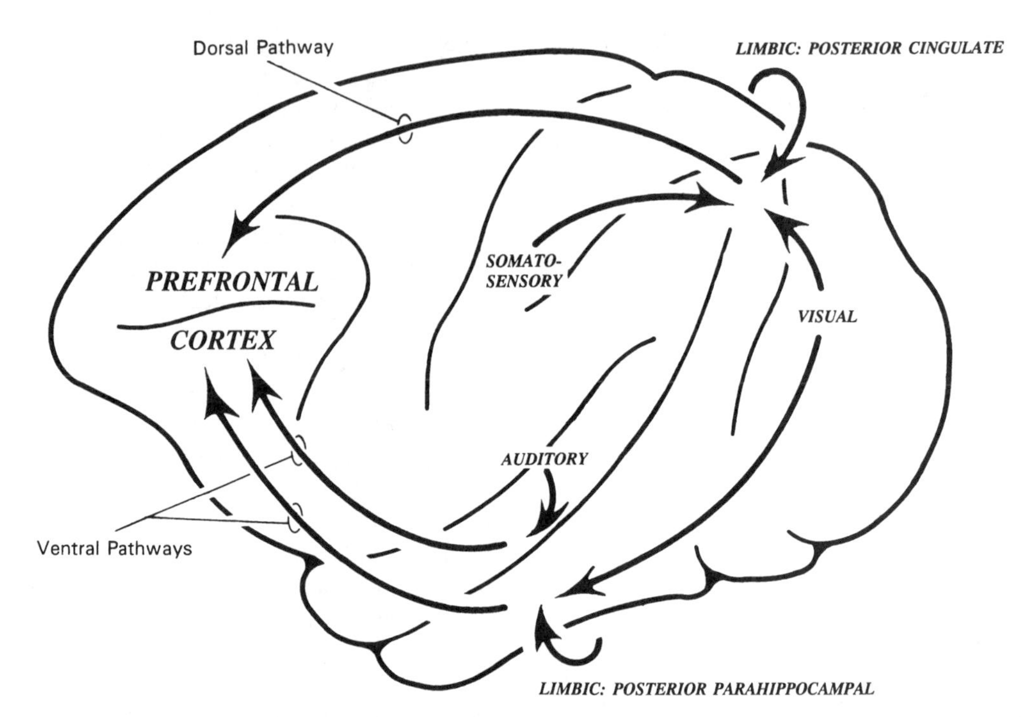

FIG. 6-6. Some of the major cortical association input to the prefrontal cortex is depicted on a lateral view of the monkey cerebral hemisphere. Note the dorsal pathway from the inferior parietal gyrus and dorsal peristriate area carrying cingulate information and the ventral pathways carrying auditory, visual, and posterior parahippocampal information. These provide cortical information from all of the other lobes, including the limbic lobe.

tion probably represents an important source of auditory input to the ventrolateral part of the prefrontal cortex.

In addition to the long association pathways linking the prefrontal cortex with the parietal, occipital, and temporal lobes, short association pathways interlock the various parts of the prefrontal cortex with one another in an organized fashion. The less differentiated (in terms of cytoarchitectonic lamination) agranular and dysgranular cortices, located posteriorly on the basal and medial surface of the prefrontal cortex, give rise to widespread *intrinsic* prefrontal connections. In contrast, the more differentiated isocortical granular areas, which are situated anteriorly and laterally, are characterized by more limited *intrinsic* connections; they account instead for a large component of the frontal lobe's widespread *extrinsic* prefrontal connections.

Cortical Limbic Connections of the Prefrontal Cortex

There are many direct connections between the limbic lobe and the prefrontal cortex. Limbic projections to the prefrontal cortex arise from diverse and widespread parts of the limbic lobe, including the cingulate, orbitofrontal, temporopolar, perirhinal, entorhinal, posterior parahippocampal, and insular cortices. Although the lateral prefrontal cortex is a target for some of these connections, the bulk of this anatomic interrelationship is established with the posterior orbitofrontal and medial prefrontal regions.

The prefrontal cortex also maintains a highly organized anatomic affiliation with the cingulate gyrus.

Early studies that relied on Marchi's technique to trace neural connections failed to recognize a strong connection between the frontal and temporal lobes; however, use of more sensitive tracing techniques enabled investigators to demonstrate that fibers forming a subcomponent of the uncinate fasciculus, as well as the extreme capsule, interconnect the frontal and temporal lobes in a strong and highly specific fashion. The strongest links with the limbic portion of temporal lobe involve the posterior orbitofrontal cortex and medial prefrontal cortex, followed by the lateral prefrontal cortex.

All parts of the prefrontal cortex are reciprocally connected with the insula, and a distinct anatomic relationship between cortex forming the orbitofrontal surface and the insula has been demonstrated. All parts of the insula appear to receive input from sensory association and multimodal association cortices. The diversity of projections to the insula implies that prefrontal input from the insula may be either as little as one synapse away from a primary sensory area or may highly processed.

Amygdala projections to prefrontal cortex arise primarily from the basolateral and accessory basal nuclei and to a lesser extent from cortical and lateral nuclei. The strongest amygdalofrontal projection ends in the posterior part (agranular and dysgranular sectors) of the orbitofrontal cortex. Another strong projection terminates in the medial prefrontal cortices. As mentioned previously, the amygdala is the recipient of a wide variety of cortical inputs from allocortical, periallocortical, proisocortical, and isocortical association areas. The latter includes converging input from both auditory and visual association areas, as well as multimodal association cortices. Therefore, there is reason to believe that amygdala output directed toward the prefrontal cortices is influenced by a variety of neural systems related to the interplay of both the internal and external environments of the organism.

Motor Cortex Connections With Prefrontal Cortex

It is well known that the prefrontal corticofugal axons are not directed to cranial nerve nuclei or the spinal cord. Since the latter part of the nineteenth century, however, it has been appreciated that prefrontal cortex plays a special and important role in guiding the outcome of voluntary motor behavior.

As summarized in Fig. 6-7, more recent efforts have shown clearly that the prefrontal cortex projects directly to parts of the motor cortices, giving rise to the corticospinal axons. In the lateral premotor cortex, only the very rostral parts of areas 6D and 6V that contain corticospinal neurons receive prefrontal input, with area 6V being the primary link. Prefrontal input to M2 converges on the rostral part of M2 that contains the face area and less so on corticospinal output zones that subserve the arm. Thus, the more rostral parts of the lateral premotor cortex and supplementary motor cortex receive prefrontal input. However, the cingulate motor cortex (M3 or area 24c) and area 23c have also been shown to receive strong prefrontal input that converges on parts of these cortices that rise to corticospinal axons. Thus, recent work implies that several anatomically distinct sources of corticospinal axons are directly influenced by prefrontal output. Although no corticospinal projections arise from the frontal eye field, it does receive strong prefrontal input.

Subcortical Connections of Prefrontal Cortex

The corticostriate projection from the prefrontal cortex is substantial. It is directed toward targets in the caudate nucleus and, to a much lesser extent, the putamen. Because a portion of the outflow of the basal ganglia is directed to the thalamus and eventually back to all parts of the frontal lobe, the corticostriate projection represents initial stages of a sequential pathway by which prefrontal cortex can influence a wide variety of neural systems.

The corticothalamic projection is one of the most studied corticofugal pathways leaving from the prefrontal cortex. Anatomic findings and functional observations have

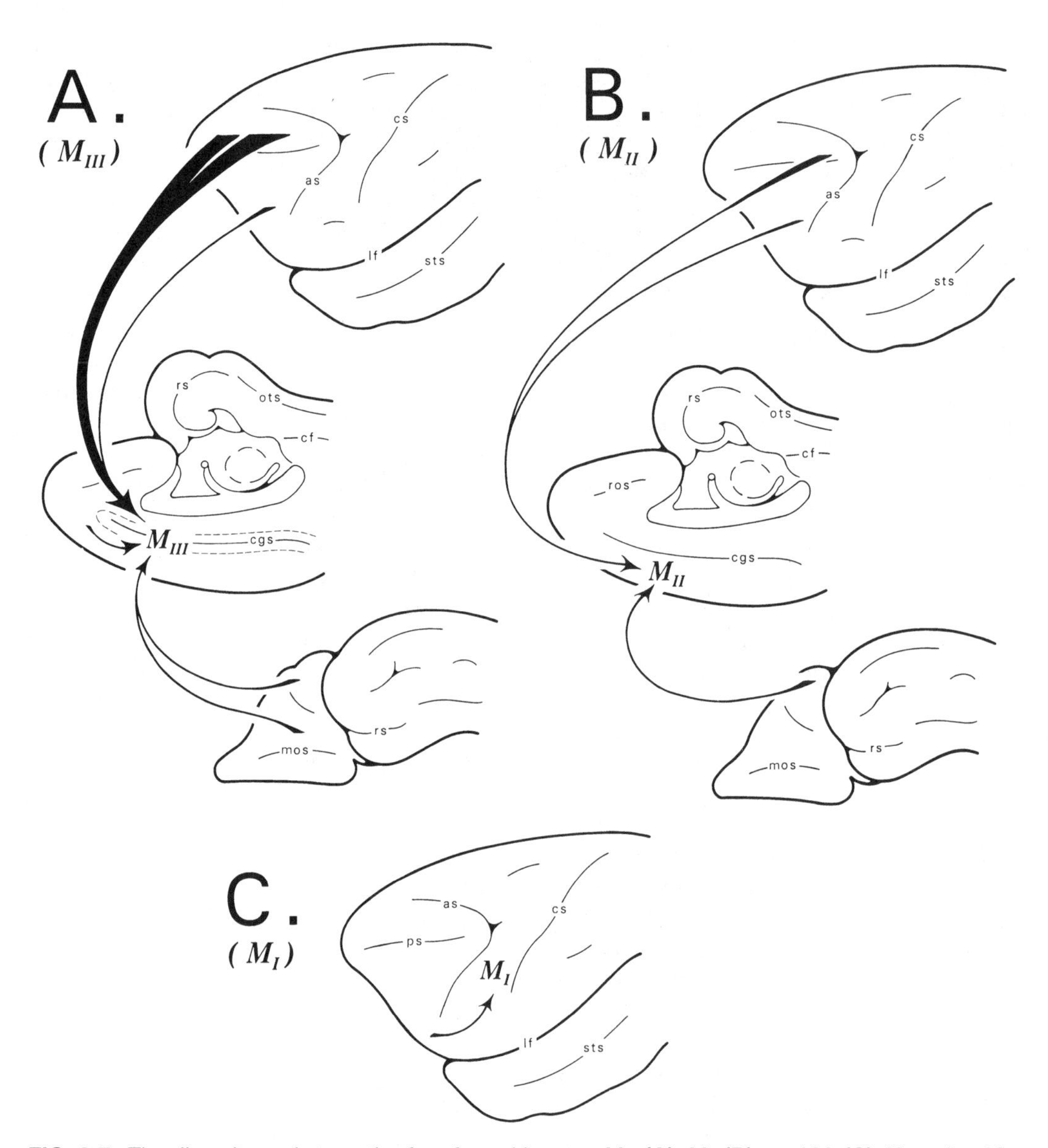

FIG. 6-7. The diversity and strength of prefrontal input to M_{III} **(A)**, M_{II} **(B)**, and M_{I} **(C)**. Note that M_{III} receives dorsolateral, medial, and orbitofrontal projections. M2, or supplementary motor cortex, receives some direct prefrontal input, but M_{I}, or primary motor cortex, receives virtually no direct prefrontal cortex input. as, arcuate sulcus; cf, calcarine fissure; cgs, cingulate sulcus; cs, central sulcus; lf, lateral fissure; mos, medial orbital sulcus; ots, occipitotemporal sulcus; ps, principal sulcus; ros, rostral sulcus; rs, rhinal sulcus; sts, superior temporal sulcus.

since combined to suggest that the prefrontal cortex can be viewed as having at least four major subdivisions. They include dorsolateral, ventrolateral, medial, and orbitofrontal divisions, with each having its own unique thalamic projection pattern.

The prefrontal cortex also gives rise to a strong corticopontine projection that ends in the medial part of the basilar pontine gray matter. This projection is a major component of the corticopontocerebellar system. Transneuronal labeling techniques have shown that cerebellar, as well as pallidal neurons are labeled after injections of herpes simplex virus type 1 into the dorsolateral prefrontal cortex. This indicates that thalamocortical input to the prefrontal cortex is influenced by both basal ganglia and cerebellar circuits. If the basal ganglia and cerebellar loops are organized in parallel, it is likely that corticopontine projection from the prefrontal cortex is involved with the cerebellothalamocortical pathway that converges back onto the prefrontal cortex (16).

Summary and Conclusions

The diverse set of associative connections converging on prefrontal cortex indicate that transformed and integrated information, associated with multiple sensory modalities, shape the outcome of prefrontal-guided behaviors. The anatomic interaction between prefrontal and limbic cortices may affect the motivational state and emotional tone or temperament of prefrontal behaviors in addition to involving memory features, such as the storage and retrieval of information.

Moreover, the prefrontal cortex gives rise to a host of descending projections that contribute to the corticostriate, corticothalamic, corticotectal, corticoreticular, and corticopontine pathways. Information directed away from the prefrontal cortex through nonreciprocating projections, such as the corticostriate and corticopontine projections, eventually converges back on the prefrontal cortex only after coursing through a sequential and parallel set of subcortical circuits or "loops." Others, such as the corticothalamic projections, being heavily reciprocated, allow for direct interaction between prefrontal and discrete subcortical diencephalic nuclei. Finally, chemically specific subcortical projections from cholinergic and monoaminergic sources to all parts of the prefrontal cortex may globally and perhaps homogeneously affect the operation of cortical states associated with arousal, attention, motivation, and learning. Likewise, projections from selected parts of the prefrontal cortex to the monoaminergic and cholinergic centers suggest that prefrontal cortex may modulate its own afferent neurochemical innervation.

CLINICAL CORRELATIONS

Degenerative Diseases Affecting the Prefrontal Cortex and Limbic System

It has been recognized for many years that the limbic system and, to some degree, the frontal lobes are the preferred targets for particular disease mechanisms. For example, this is the case for degenerative diseases such as AD and Pick's disease, as well as schizophrenia. Several of the viral encephalitides also somewhat selectively attack these brain areas. AD has been well characterized in terms of topographic neuropathology and neuroanatomy.

For many of the neural systems and connections of the limbic system and frontal lobe described previously, the projection neurons that give rise to association axons reside in layer III of the cerebral cortex and to some extent in layer V. A subset of layer III also mediates callosal connections, and various subsets of layer V give rise to corticofugal axons that course to subcortical targets. In both AD and Pick's disease, these larger projection neurons of layers III and V are targeted for pathology, whether it be neurofibrillary tangles in the former case or cell loss and cytoplasmic inclusion bodies in the latter case. Thus, although the etiology of cortical neuronal death is different in both diseases, the outcome can be somewhat similar because corticocortical connections are destroyed. The frontal-like signs of certain psychiatric diseases may be caused by subtle cortical neuroanatomic changes elsewhere yielding behavior that is more a manifestation of frontal lobe deafferentation than the direct result of diffuse and subtle lesions in the areas that provide afferent input to the frontal lobes.

Functional Observations

The idea that frontal lobe function and personality characteristics might be related had already emerged during the last century, when Harlow reported on his famous patient Phineas Gage, the so-called crowbar case. Gage underwent major personality changes after partial damage to his frontal lobes and, in Harlow's words, "was no longer Gage" after the accident.

Early in this century, it became clear that an association indeed exists between frontal lobe function and aspects of emotional behavior. At the International Neurological Congress in London in the summer of 1935, great interest was generated in the presentation of the results of frontal lobe extirpation in two chimpanzees, Becky and Lucy. After bilateral prefrontal cortex removal, Becky and Lucy's behavior changed strikingly. They did not show their usual excitement, but rather knelt quietly before the cage or ambled around. Whenever they would make a mistake when choosing between an empty or reforced cup, they showed no emotional disturbance, but quietly awaited the next trial.

Egaz Moniz, a Portuguese neurosurgeon participating in the congress, commented to Dr. Fulton and Dr. Jacobsen that, "if after frontal lobe removal, an animal no longer tends to develop an experimental neurosis and no longer has temper tantrums when frustrated, why would this not be an ideal operation for human beings suffering from persistent anxiety states?" Within 1 year of that meeting, his now famous monograph on frontal leucotomy was published, describing the neurosurgical procedure he used on psychotic patients and its outcome, which involved an improvement in the anxiety symptoms in many of his patients (17).

Modern Forms of Frontal Lobe Measurement

In the second half of this century, more serious studies emerged on the issue of the size of the frontal lobes. Blinkov and Glezer estimated the surface area of the "frontal region" (prefrontal cortex) and of the precentral region (areas 4 and 6) in relation to total surface of the hemisphere to be 32.8% in the human, 22.1% in the chimpanzee, 21.3% in the orangutan, and 21.2% in the gibbon.

Recent comparative studies of the human and ape frontal lobes have addressed the issue of size of specific cortical areas and their internal organization, as well as the issue of the volume and cortical surface of the hominoid frontal lobe and its subdivisions using modern imaging and quantitative techniques.

Summary and Functional Considerations

More modern cytoarchitectural and volumetric analyses of frontal lobe evolution are inconsistent with the assumptions and dogma that have accrued historically. Humans undoubtedly have the largest frontal lobe in absolute terms, outstripping our extant ape kin. But we have larger brains in general, and the part devoted to frontal lobe is not significantly different between apes and humans in relative terms.

In short, the human frontal lobe may be unique, not so much for quantity of frontal lobe, but instead for other reasons relating to its internal structure and convectivity. We share with our ape relatives many basic behaviors: a tendency to be social, to band together in units, to be intelligent, and to manipulate our environment. But we may be set apart by a greater balance to deal frontally with other parts of our neuraxis and the events provoked in these areas by our environment. In this regard, it is of great interest to find that white-matter volume of the prefrontal cortices is, in fact, the one anatomic variable that distinguishes humans from apes. Indeed, this is the neural substrate that governs the balance and precision of neural systems' communication. In essence, we are "better wired."

NEUROPSYCHIATRIC ASPECTS OF SEXUAL DYSFUNCTION

NEUROPHYSIOLOGY OF THE HUMAN SEXUAL RESPONSE

Reflex vasodilation of the genital vasculature in response to sexual stimuli is responsible for both male penile erection and female lubrication. Decreased vascular resistance in the penile corpora appears to be the major factor causing penile corpora vascular engorgement and penile erection. The smooth muscle of the corpora cavernosa is predominately innervated by adrenergic fibers, although cholinergic fibers are also present. α-Adrenergic impulses appear to maintain the penis in a nontumescent state. It is unclear whether a parallel innervation exists in women.

Neuroanatomic studies have demonstrated a dual innervation of the genitals in both sexes—sympathetic innervation from the T-12 to L-4 segments of the spinal cord and parasympathetic innervation from the S-2 to S-4 cord segments. Stimulation of the sacral parasympathetic fibers has been shown to elicit penile erection in many species, and ablation of these nerves interferes with reflexogenic erections. These fibers are also thought to mediate the lubrication response in women. The postganglionic neurotransmitter in these parasympathetic fibers is nitric oxide.

Stimulation and ablation experiments in laboratory animals, including mammals, have identified cerebral areas mediating erection. The major areas that elicit penile erection upon stimulation include the medial septopreoptic region and the medial part of the medial dorsal nucleus of the thalamus. Other areas involved in penile erection include septal projections of the hippocampus, the anterior cingulate gyrus, the mamillothalamic tract, and the mamillary bodies. The cerebral representation of vaginal lubrication is unknown.

Orgasm can be conceptualized as the sensory experience of a series of spinal cord reflexes. These reflexes are triggered when a series of sensory stimuli reach threshold values. In men, sensory impulses eliciting the ejaculatory reflex travel in the pudendal nerve to the sacral cord. Once a threshold value is reached, contractions of the vas deferens, seminal vesicles, and prostatic smooth muscle occur, resulting in the ejaculate being delivered into the pelvic urethra. Stimulation of the urethral bulb by the inflowing ejaculate elicits reflex closure of the bladder neck, preventing retrograde ejaculation, and rhythmic contractions of the perineal muscles and urethral bulb, resulting in expulsion of the ejaculate.

Efferent fibers mediating ejaculation arise from the thoracolumbar cord, travel in the hypogastric nerve, and synapse with short adrenergic fibers that innervate the organs involved in orgasm. These fibers appear to be mainly α-adrenergic fibers, although these organs are also innervated by cholinergic fibers. Presumably, sensory impulses from the pudendal nerve to the sacral cord are also relayed cranially to the thalamus and sensory cortex, resulting in the experience of orgasm.

Female orgasm also appears to be a genital reflex. Sensory impulses travel to the sacral cord in the pudendal nerve, and efferent fibers innervate the ovary, fallopian tubes, vaginal musculature, and uterus. Rhythmic contractions of these structures appear to be mediated by α-adrenergic fibers, although the female sexual organs also have a cholinergic innervation.

Sensory representation of genital sensations appear to be localized in the paracentral lobule.

Evidence suggests that central dopamine and serotonin pathways are involved in sexual behavior. Animal studies have demonstrated that drugs that increase brain dopaminergic activity lower thresholds for ejaculatory and erectile reflexes. Infusions of dopamine agonists into the medial preoptic region and into the lumbar cord augment male sexual behavior. Male patterns of sexual behavior are increased by drugs that lower brain serotonin levels.

From the preceding review of the neurophysiology of sexual behavior, it is clear that various neurologic lesions could interfere with sexual behavior. These include lesions in the brain, cord, and peripheral nerves. It is also clear that many commonly prescribed drugs—especially psychiatric and hypotensive drugs that affect monoamine neuromodulators—might alter both CNS function and sexual behavior (18).

Psychiatric Drugs and Sexual Behavior

Although the list of drugs with adverse sexual side effects is extensive, the worst offenders are psychiatric drugs and drugs used to treat arterial hypertensive disease.

Antipsychotic drugs have been reported to cause disturbances in libido, ejaculatory impairment, female anorgasmia, and erectile failure. Current evidence suggests that interference with orgasm is secondary to α-adrenergic blockade and that interference with libido and erectile capacity is probably secondary to central dopamine blockade. Erectile failure has been reported with chlorpromazine, pimozide, thiothixene, thioridazine, sulpiride, haloperidol, and fluphenazine. Many of these same neuroleptics have been reported to interfere with ejaculatory function. When a patient complains of sexual dysfunction caused by an antipsychotic drug, changing the patient to a different neuroleptic, such as loxapine or molindone (with lesser α-adrenergic effects), may resolve the sexual problem.

Although benzodiazepines probably do not have adverse effects on erectile function, there is evidence that these drugs cause ejaculatory delay and may be used to treat premature ejaculation.

There have been case studies showing that lithium carbonate may cause erectile impairment. This has been confirmed in a double-blind placebo-controlled trial using therapeutic doses in men with affective disorder.

Antidepressants have been reported to have a variety of sexual side effects. Heterocyclics, monoamine oxidase inhibitors, and selective serotonin reuptake inhibitors, including imipramine, desipramine, nortriptyline, amitriptyline, doxepin, protriptyline, amoxapine, trazodone, maprotiline, tranylcypromine, phenelzine, bupropion, and fluoxetine, reportedly have caused diminished libido.

Treatment approaches for antidepressant-induced low libido include the coadministration of 7.5 to 15 mg of neostigmine before coitus, the coadministration of yohimbine with fluoxetine, or the substitution of bupropion because this drug has a very low incidence of sexual side effects. Viagra (Sildenafil) has recently become available and is likely the drug of first choice. The mechanism by which antidepressant drugs influence libido is unclear. Anorgasmia and delayed ejaculation have been reported with phenelzine, amoxapine, amitriptyline, clomipramine, imipramine, fluoxetine, trazodone, and sertraline. Double-blind studies have confirmed the effects of phenelzine, imipramine, and clomipramine on orgasm. Increased serotonergic activity may be a mechanism by which antidepressant drugs inhibit orgasm. Antidepressant-induced anorgasmia has been reported to be reversed by the use of bethanechol, cyproheptadine, and yohimbine. Drug substitution with desipramine or bupropion has also been reported to be effective.

It is noteworthy that spontaneous orgasm has been reported with clomipramine and fluoxetine. Psychiatric drugs that have been reported to cause erectile and ejaculatory problems are listed in Tables 6-1 and 6-2. See reference 19 for a review.

TABLE 6-1. *Psychiatric Drugs Reported to Cause Erectile Dysfunction*

Imipramine	Amoxapine
Desipramine	Trazodone
Nortriptyline	Maprotiline
Amitriptyline	Tranylcypromine
Doxepin	Phenelzine
Protriptyline	Isocarboxazid
Clomipramine	Lithium carbonate
Chlorpromazine	Thioridazine
Pimozide	Sulpiride
Thiothixene	Haloperidol
Fluphenazine	

TABLE 6-2. *Psychiatric Drugs Reported to Cause Orgasm Disturbances*

Fluoxetine	Amitryptyline
Sertraline	Doxepin
Paroxetine	Protriptyline
Imipramine	Clomipramine
Desipramine	Amoxapine
Nortriptyline	Trazodone
Tranylcypromine	Maprotiline
Phenylzine	Alprazolam
Isocarboxazid	Chlordiazepoxide
Lorazepam	Thioridazine
Chlorpromazine	Mesoridazine
Chlorprothixine	Fluphenazine
Perphenazine	Thiothixene
Trifluoperazine	Haloperidol

Hypotensive Agents

Antihypertensive drugs are another class of drugs that have been found to interfere with sexual function. These drugs have been reported to interfere with erection, ejaculation, orgasm, and libido.

Current evidence suggests that hydrochlorothiazide, chlorthalidone, and spironolactone may decrease libido as well as cause erectile problems. Spironolactone's antilibidinal effects may be related to its antiandrogen effect. The mechanism by which other diuretics influence sexual behavior is unclear. Antihypertensive drugs with central antiadrenergic effects, such as methyldopa and reserpine, have been reported to cause diminished libido, erectile failure, and ejaculatory impairment. Current evidence also suggests that β-blockers may be associated with erectile failure. The available evidence suggests that β-blockers that are more lipophilic are more likely to cause sexual dysfunction than those that are more hydrophilic, although impotence has been reported with atenolol, a hydrophilic β-blocker. Sexual problems appear to be less frequent with pindolol, metoprolol, and nadolol. Ejaculatory problems appear to be common with guanethidine, bethanidine, labetalol, and nifedipine. Among antihypertensive drugs, the angiotensin-converting enzyme inhibitors, such as captopril, enalapril, and lisinopril, appear to lack sexual side effects.

Other Drugs

Controlled studies indicate that cimetidine can cause decreased libido and erectile failure. The mechanism for this phenomenon is unclear because cimetidine has both antiandrogenic effects and ganglion-blocking effects. Controlled studies have also found decreased libido and erectile problems associated with long-term use of digoxin. Antiepileptic drugs have been reported to decrease libido. Phenobarbital and primidone have been reported to have more adverse effects than phenytoin and carbamazepine (20).

ENDOCRINE FUNCTION AND SEXUAL BEHAVIOR

In humans, the precise relationship between sexual behavior and endocrine function has not been established, with the relationship being less well understood for women. The tremendous influence of social learning on human sexuality contributes to the difficulty in determining endocrine effects on human sexual behavior.

In men, evidence concerning the relationship between endocrine function and sexual behavior can be obtained from studies of men who have been surgically castrated, men prescribed antiandrogens, and hypogonadal men receiving androgen replacement therapy. Bilateral orchidectomy is used to treat sexual offenders in some countries and as a palliative treatment for some neoplasms. Studies of patients after bilateral orchidectomy reveal a dramatic loss of libido, usually followed by an inability to ejaculate; however, a small number of patients remain sexually active for years past castration. The effects of castration are generally reversed by the administration of exogenous androgen. Studies of the effects of the estrogenic compounds medroxyprogesterone and cyproterone acetate have shown that these drugs markedly diminish libido without interfering with erectile capacity. In men with a disease state of hypogonadism, it has been shown that androgen therapy restores libido and seminal emission. Thus, evidence from three different sources of information indicates that androgen levels are closely linked to seminal emission and sexual drive in men. Current evidence suggests that a certain minimal level of androgen is necessary for sexual function but that excess androgen above these levels has minimal or no effects.

A number of investigators have studied sexual activity across the menstrual cycle and in relationship to cyclic changes of progesterone and estradiol. To date, there are no consistent data relating cyclic changes in estrogens or progesterone during the menstrual cycle and sexual activity. There is some evidence that sexual libido and arousability may be related to serum androgen levels. Some clinicians have reported using androgen therapy to treat hypoactive sexual desire disorders; others have reported that antiandrogens lower libido in women.

Hyperprolactinemia has been reported to be associated with decreased libido in both sexes. Male patients may also complain of erectile problems. In many cases, bromocriptine therapy restores normal function (21,22).

NEUROLOGIC DISEASE AND SEXUAL DYSFUNCTION

Multiple Sclerosis

A number of different investigators have established that multiple sclerosis can cause decreased libido, decreased arousal, decreased sexual sensations, and decreased orgasmic capacity in both sexes as well as decreased erectile response in male patients. Estimates of the frequency of sexual problems in such patients have ranged widely, from 26% to 90%. Sexual life in such patients can be altered by various mechanisms, including neuropathologic lesions in the brain, as well as in the cord. In this regard, hypersexuality has been reported in a multiple sclerosis patient with frontal and temporal lesions. Psychosocial factors may also contribute to altered sexual function.

Diabetes Mellitus

Diabetes mellitus is known to be associated with erectile failure and to a lesser extent with ejaculatory disturbance. Decreased erectile function in diabetic men has been confirmed by nocturnal penile tumescence studies and by laboratory-based erotic stimulation studies. These sexual difficulties are suspected to be secondary to peripheral neuropathy of the autonomic nervous system. Decreased vaginal lubrication in women with insulin-dependent diabetes has been reported consistently.

Spinal Cord Injury

One of the most common causes of neurogenic sexual difficulty in men is spinal cord injury. Complete cervical and high thoracic cord lesions are extremely destructive to ejaculatory function. Depending on the level of the cord lesion, reflexogenic erections to local stimuli or psychogenic erections to psychic stimuli may be intact. Reflexogenic erections occur most often in patients with complete upper motor lesions, especially in men with cervical lesions. Psychogenic erections are intact most often with lower motor neuron lesions below T-9.

There is relatively little information concerning sexual behavior in female patients with cord lesions (23,24).

Damage to Autonomic Nervous System

A number of surgical procedures interfere with autonomic nervous system innervation of the pelvic organs and result in sexual impairment. These procedures include sympathectomies, retroperitoneal lymphadenectomy, abdominoperitoneal resection, anterior resection of the rectum, aortoiliac surgery, and radical retropubic and transvesical prostatectomy. Most of these same procedures have been reported to cause orgasm disorders in female patients.

Seizure Disorder

Both ictal and interictal sexual abnormalities have been noted in patients with epilepsy. Ictal sexual manifestations, such as sexual emotions, genital sensations, and sexual automatisms, are somewhat rare and appear to be more common during partial complex seizures. Sexual auras are reported to have a temporal lobe origin.

Interictal sexual abnormalities are more common than ictal sexual abnormalities. Although hypersexuality and paraphilia have been reported, hyposexuality is the more common finding. Sexual abnormalities appear to be more common in patients with partial complex seizures. Partial complex seizures of temporal lobe origin are more commonly associated with sexual abnormalities than seizures from extratemporal foci.

Some of the hyposexuality observed in some epileptics might be explained by the effects of antiepileptic drugs on endocrine function. Testosterone exists in three forms in the serum: a free form, which is biologically active; a form loosely bound to albumin; and an inactive form, which is tightly bound to sex hormone—-binding globulin. A number of antiepileptic drugs, notably carbamazepine, decrease the amount of free testosterone. Low levels of free testosterone appear to be related to decreased libido. Unfortunately, androgen replacement therapy has been only moderately successful in improving libido in men with epilepsy.

Other Neurologic Diseases

A number of clinicians have noted a high frequency of sexual problems in patients with Parkinson's disease. *Hypersexuality* has repeatedly been reported in parkinsonian patients treated with dopaminergic drugs.

Another syndrome associated with a decreased frequency of sexual activity is cerebral vascular accident. There are several potential mechanisms for the decrease in sexual activity, including hypotensive medications, partner reaction, immobility, and deformity. There is minimal evidence regarding specific brain regions infarcted and the likelihood of resulting sexual impairment.

Similarly, most studies of brain trauma patients have documented hyposexuality; however, hypersexuality has been noted after injury to the temporal lobes and to the dorsal septal region. Damage to the hypothalamopituitary region of the brain by tumor results in decreased libido, which appears to be correlated with the degree of hypogonadism. Erectile failure has been reported to be common in patients with Alzheimer's disease (25).

PSYCHIATRIC DISORDERS AND SEXUAL DYSFUNCTION

A number of investigators have documented the loss of libido and decreased sexual function in the presence of depressive disorder. Loss of libido is a common symptom in both dysthymia and major depression. After 70 years of age, loss of libido is less indicative of depression primarily because loss of libido is common among nondepressed elderly. As well as loss of libido, major depression may also be associated with decreased ability to attain and maintain penile erections. During manic episodes, an increase in sexual thoughts, conduct, and number of sexual partners may occur.

Several well-conducted studies using good methodology have reported decreased sexual interest and activity among patients with schizophrenia.

Information concerning sexual behavior in other psychiatric disorders is scarce. Women with anorexia nervosa appear to have markedly diminished sexual activity and interest. There is little evidence linking any specific personality disorder with sexual difficulties, apart from the high frequency of sexual identity issues in young adults with borderline personality disorder. Patients with sensitivity to rejection and a predisposition to anxiety and depression may be more likely to develop sexual problems (26).

NEUROPSYCHIATRIC EVALUATION OF SEXUAL DYSFUNCTION

A crucial part of the evaluation of any sexual complaint is a careful history. The clinician needs to document the precise difficulty, the associated symptoms, and the disorder's onset and course and consider if the difficulty is secondary to another psychiatric disorder, such as affective disorder. The clinician needs to ascertain gender identity and sexual preference for partners as well as assessing whether the problem appears secondary to relationship discord or life stress. The clinician also needs to delineate carefully whether the disorder is generalized or partner specific. It is unusual for biogenic sexual problems to be partner specific. Thus, a history of a partner-specific difficulty is presumptive evidence of a psychogenic difficulty.

In cases of hypoactive sexual desire disorders, the clinician needs to inquire closely regarding the frequency of romantic daydreams and sexual thoughts. A high frequency of sexual daydreams suggests that the problem is not biogenic. The clinician also needs to inquire about masturbatory frequency and the preferred masturbatory fantasy. If it appears that the decreased desire is present in all contexts, a greater suspicion that the problem might be biogenic is warranted, and evaluation of androgen and prolactin levels is the next step.

Deficient sexual arousal as evidenced by decreased vaginal lubrication as an isolated complaint is almost always caused by atrophic vulvovaginitis, related to estrogen deficiency. Psychogenic arousal disorder in women almost always occurs in conjunction with hypoactive sexual desire disorder.

Diagnostic evaluation of male erectile disorder can be complex. A careful sexual history should be a major part of the evaluation. The presence of full erections upon awakening is highly suggestive of a psychogenic erectile problem. To date, there is minimal evidence that psychometric assessment aids in the differential diagnosis of biogenic from psychogenic erectile disorder. A global screening procedure to measure erectile capacity is to monitor nocturnal erections with a nocturnal penile tumescence study. The clinician needs to be aware of a number of possibly confounding influences, including affective disorder, aging, hypoandrogenic states, and hypoactive desire states. Barring these limitations, a large amount of normative data allow the clinician to judge deviancy from these values given normal sleep parameters. Alternative nonlaboratory approaches include snap gauges and portable nocturnal erection monitors (e.g., Rigiscan DACOMED, Minnesota). The lack of sleep monitoring limits the diagnostic accuracy of these approaches. Another general screening procedure is the visual sexual stimulation method. In this procedure, the erectile response to erotica is monitored. The reliability, validity, sensitivity, and specificity of this approach have not been established.

A variety of procedures can be used to test the integrity of the vascular system. Penile blood pressure, strain gauge plethysmography, and pulse wave assessment all have been employed; however, all of these procedures only assess hemodynamics in the flaccid penis. Intracorporeal injection of papaverine-phentolamine or prostaglandin E_1 can be used to assess functional integrity of the penile vasculature. Unfortunately, it is not clear how much vascular disease must be present to cause diminution of the pharmacologically induced erection. A decreased response provides suggestive but not definitive evidence of vasculogenic impotence. More invasive procedures include cavernosometry and cavernosography. Lack of standardization currently limits the usefulness of these procedures (27).

Most of the procedures available to assess integrity of the neurologic components of erection primarily assess the pudendal nerve. Bulbocavernosus reflex latency testing consists of recording the electromyogram response of the bulbocavernosus muscle to electrical stimulation of the dorsal nerve of the penis. Absent or prolonged reflex latency indicates pathology within the reflex arc. A variation of this procedure is to stimulate the prostatic urethra, testing for autonomic neuropathy. Other procedures include measuring dorsal nerve conduction velocity, the threshold for perception of vibratory sensations in the penis, and somatosensory evoked potentials evoked by penile stimulation. All of the procedures can detect neurologic abnormalities. However, the absence of adequate normative data limits the interpretation of these findings.

Premature ejaculation rarely has an organic cause, with the possible exception of acquired premature ejaculation as an early manifestation of multiple sclerosis. If the patient complains of retarded ejaculation, one can distinguish between retrograde ejaculation and an ejaculatory orgasm by examination of the postorgasm spun urine sample. Anything that interrupts the orgasmic reflex arc can interfere with both male and female orgasmic function. Acquired orgasm disorder in the absence of marital discord mandates a search for treatable biogenic causes. Numerous drugs can cause orgasmic dysfunction. Also, numerous neuropathies (e.g., diabetes mellitus, multiple sclerosis, alcoholic neuropathy) and surgical procedures (e.g., aortoiliac surgery, lumbar sympathectomy, retroperitoneal lymphadenectomy) can cause anorgasmia. Severe hypogonadism in men can result in anorgasmia.

PSYCHOLOGICAL TREATMENT OF SEXUAL DISORDERS

The psychological treatment of sexual dysfunction is based on techniques introduced by behavioral psychologists and then elaborated by Masters and Johnson. To summarize briefly, these techniques focus on attitude change and anxiety reduction. Sexual activity is structured such that there is the progressive experience of emotional and sexual intimacy at a pace tolerable to both partners. Preferably, the couple is seen together on a once-a-week basis. Either a dual-sex therapy team or a solo therapist of either sex may be used. More detailed descriptions of this treatment approach can be found elsewhere.

PHARMACOLOGIC TREATMENT OF SEXUAL DISORDERS

There is evidence that both premature ejaculation and iatrogenic erectile disorder may respond to pharmacologic interventions. Clinicians have reported successfully using low doses of thioridazine, monoamine oxidase inhibitors, lorazepam, and clomipramine in the treatment of premature ejaculation.

Evidence suggests that yohimbine is effective in reversing certain cases of idiopathic erectile dysfunction. Studies show that yohimbine has its mechanism of action in the CNS. Viagra has recently been marketed as a treatment for erectile dysfunction of idiopathic or biogenic cause.

There is suggestive evidence that dopaminergic drugs or opioid-receptor blockers may prove useful in the treatment of human sexual disorders. Anecdotal evidence suggests that dopaminergic drugs, such as levodopa and pergolide, may increase libido. A number of independent investigators have demonstrated in controlled studies that apomorphine administered subcutaneously in the arm elicits penile erections. Unfortunately, side effects from apomorphine render it unsuitable as a treatment for erectile problems. Although it is well known that opiate abuse may diminish libido, there are relatively few studies of the effects of opioid antagonists on sexual function.

NEUROPSYCHIATRY OF GENDER IDENTITY AND SEXUAL PREFERENCE

The origins of gender identity and sexual orientation have been a subject of intense debate within the scientific community. Early sexologists, such as Kraft-Ebing, argued that homosexuality must be innate, and numerous investigations have attempted to find evidence to support this viewpoint. In the twentieth century, the work by John Money and his colleagues has been very influential in the United States. From studies of psychosexual development in children with ambiguous sexual development, it was concluded that gender identity is predominately determined by sex of rearing before age 2½ years. By contrast, Gunter Donner suggested that a neuroendocrine predisposition for homosexuality might be based on the effects of prenatal androgen deficiency on the developing hypothalamus. This theory was largely ignored by many American sexologists. A major challenge to the concept of the predominant influence of rearing on sexual identity and orientation came from the work of Imperato-McGinley and associates on the psychosexual development of children with 5-α-reductase deficiency. Genetic male children with this disorder are raised as girls and then assume masculine roles at puberty. Although others have questioned whether these children are raised unambiguously as girls early in life, the work by this group rekindled interest in the search for biologic etiologies of gender identity and sexual orientation (28–30).

In humans, gender identity may be defined as the persistent belief that one is male, female, or ambivalent. Current evidence suggests that sex assignment by parents plays a major role in determining gender identity. Gender role may be defined as behavior that society designates as masculine or feminine. Current evidence suggests that gender role behavior is influenced by gonadal hormones during development. Sexual orientation refers to the erotic responsiveness of one individual to others of the same or opposite sex.

Three basic strategies have been employed in the investigation of biologic correlates of gender identity and sexual orientation. These can be labeled as heritability, neuroendocrine, and neuroanatomic studies. The neuroendocrine and neuroanatomic studies are based on animal models of the effects of prenatal hormones on sexual behavior and sexual differentiation of the brain.

Heritability

A number of twin studies have suggested about a 50% concordance rate for homosexuality in monozygotic twins. In a study of male homosexuality, probands were solicited in a homophile publication by advertising for homosexual or bisexual men with either co-twins or adoptive brothers. Probands were asked to rate the sexual orientation of the co-twin or adoptive brother. Fifty-two percent of the 56 monozygotic co-twins were either homosexual or bisexual, as compared with 22% of the 54 dizygotic co-twins and 11% of the 57 adoptive brothers. A similar methodology was employed in the study of the heritability of sexual orientation in women. In this study, 48% of monozygotic co-twins were homosexual or bisexual as compared with 16% of dizygotic co-twins and 6% of adoptive sisters (31).

The available evidence is strongly suggestive of a genetic contribution to sexual orientation. The twin studies, despite possible sampling bias, are consistent in their findings. Pedigree studies must also be confirmed.

Neuroendocrine Studies

Most of these studies are based on the assumption that one can extrapolate from animal studies to sexual behavior in humans. The underlying model differentiates the organizational aspects (enduring effects on the developing brain) from the activating (reversible) aspects of hormones. The basic assumption is that male heterosexuality, female homosexuality, and female-to-male transsexualism all result from prenatal exposure to high levels of testicular hormones. Male homosexuality, female heterosexuality, and male-to-female transsexualism are postulated to result from lower levels of prenatal testicular hormones, thus retaining a female pattern of brain organization. Studies of gender identity and sexual orientation in syndromes involving prenatal androgen insensitivity or deficiency in male patients and studies of female patients with syndromes involving androgen excess have, by and large, not produced evidence consistent with the prenatal hormonal hypothesis.

SUMMARY

It is clear that our understanding of the neuropsychiatry of sexual dysfunction is in its infancy. One factor hindering the development of a coherent model of the biologic basis of sexual behavior is that the requisite knowledge base is spread across numerous disciplines. The adverse effects of both hypotensive and psychiatric drugs on sexual behavior, as well as the effects of specific neurologic lesions on sexual behavior, can be seen as an opportunity to deduce probable neurochemical and neuroanatomic substrates for such behavior. The shortage of literature concerning biologic treatment of sexual dysfunction attests to the need for further exploration. Fortunately, there is growing awareness that events and structures within the CNS are intimately associated with human sexual behavior.

REFERENCES

1. Loewy AD. Anatomy of the autonomic nervous system: an overview. In: Loewy AD, Spyer KM, eds. *Central regulation of autonomic functions.* New York: Oxford University Press, 1990:3–16.
2. Barron KD, Chokroverty S. Anatomy of the autonomic nervous system: brain and brainstem. In: Low PA, ed. *Clinical autonomic disorders.* Boston, MA: Little, Brown & Co, 1993:3–15.
3. Gautier-Smith PC. Sexual dysfunction and the nervous system. In: Aminoff MJ, ed. *Neurology and general medicine.* New York: Churchill Livingstone, 1989:471–486.
4. Vita G, Princi P, Calabro R, Toscano A, Manna L, Messina C. Cardiovascular reflex tests: assessment of age-adjusted normal range. *J Neurol Sci* 1986;75: 263–274.
5. Low PA. Laboratory evaluation of autonomic failure. In: Low PA, ed. *Clinical autonomic disorders.* Boston, MA: Little, Brown & Co, 1993:169–195.
6. McLeod JG. Autonomic dysfunction in peripheral nerve disease. *J Clin Neurophysiol* 1993;10:51–60.
7. Dyck PJ. Neuronal atrophy and degeneration predominantly affecting peripheral sensory and autonomic neurons. In: Dyck PJ, Thomas PK, Griffin JW, Low PA, Poduslo JF, eds. *Peripheral neuropathy.* 3rd ed. Philadelphia, PA: WB Saunders, 1993:1065–1093.
8. Sandroni P, Ahlskog JE, Fealey RD, Low PA. Autonomic involvement in extrapyramidal and cerebellar disorders. *Clin Auton Res* 1991;1:147–155.
9. MacLean P. Some psychiatric implications of physiological studies on frontotemporal portion of limbic system (visceral brain). *Electroencephalogr Clin Neurophysiol* 1952;4:407–418.
10. Papez JW. A proposed mechanism of emotion. *Arch Neurol Psychiatry* 1937; 38: 725-743.
11. Scoville WB, Milner B. Loss of recent memory after bilateral hippocampal lesions. *J Neurol Neurosurg Psychiatry* 1957;20:11–21.
12. Talairach J, Bancaud J, Geier S, et al. The cingulate gyrus and human behavior. *Electroencephaloagr Clin Neurophysiol* 1973;34:45–52.
13. Amaral DG, Insausti R. The hippocampal formation. In: Paxinos G, ed. *The human nervous system.* New York: Academic Press, 1990.
14. Whitehouse PJ, Price DL, Clark AW, Coyle JT, De Long MR. Alzheimer disease: evidence for selective loss of cholinergic neurons in the nucleus basalis. *Ann Neurol* 1981;10:122–126.
15. Goldman-Rakic PS, Porrino LJ. The primate medial dorsal (MD) nucleus and its projection to the frontal lobe. *J Comp Neurol* 1985;242:535–560.
16. Fuster JM. *The prefrontal cortex: anatomy, physiology and neuropsychiatry of the frontal lobe.* 2nd ed. New York: Raven Press, 1989.
17. Weinberger DR. A connectionist approach to the prefrontal cortex. *J Neuropsychiatry Clin Neurosci* 1993; 241–253.
18. Segraves RT, Schoenberg NW. Diagnosis and treatment of erectile problems: current status. In: Segraves RT, Schoenberg NW, eds. *Diagnosis and treatment of erectile disturbances: a guide for clinicians.* New York: Plenum, 1985:1–22.
19. Hawton K. Sexual dysfunctions and psychiatric disorders. In: Bezemer W, Cohen-Kettenis P, Slob K, van Son-Schooner N, eds. *Sex matters.* Amsterdam, Netherlands: Excerpta Medica, 1992:79–84.
20. Segraves RT, Madsen R, Carter CS, Davis JM. Erectile dysfunction with pharmacological agents. In: Segraves RT, Schoenberg HW, eds. *Diagnosis and treatment of erectile disturbances: a guide for clinicians.* New York: Plenum, 1985:23–64.
21. Davidson JM, Rosen RC. Hormonal determinants of erectile function. In: Rosen, RC, Leiblum SR, eds. *Erectile disorders: assessment and treatment.* New York: Guilford, 1992:72–95.
22. Sherwin BB, Gelfand MM. Differential symptom response to parental estrogen and/or androgen administration in the surgical menopause. *Am J Obstet Gynecol* 1985;151:153–160.
23. Yalla SV. Sexual dysfunction in the paraplegic and quadriplegic. In: Bennett AH, ed. *Management of male impotence.* Baltimore, MD: Williams & Wilkins, 1994: 181–191.
24. Berard EJJ. The sexuality of spinal cord injured women: physiology and pathophysiology. A review. *Paraplegia* 1989;27:99–112.
25. Mancall EL, Alonso RJ, Marlowe WB. Sexual dysfunction in neurological disease. In: Segraves RT, Schoenberg HW, eds. *Diagnosis and treatment of erectile disturbances.* New York: Plenum, 1985:65–86.
26. Offit AK. Psychiatric disorders and sexual functioning. In: Karasu TB, ed. *Treatments of psychiatric disorders.* Vol 3. Washington, DC: American Psychiatric Association, 1989:2253–2263.
27. Schiavi RC. Laboratory methods for evaluating erectile dysfunction. In: Rosen RC, Leiblum SR, eds. *Erectile disorders: assessment and treatment.* New York: Guilford, 1992:141–170.
28. Money J, Dalery J. Iatrogenic homosexuality: gender identity in seven 46XX chromosomal females with hyperadnenocortical hermaphrodism born with a penis, three reared as boys, four reared as girls. *J Homosex* 1976;1:357–371.
29. Donner G. Hormones and sexual differentation of the brain. In: Porter R, Whelan J, eds. *Sex, hormones, and behavior.* Ciba Foundation Symposium 62 (new series). Amsterdam, Netherlands: Excerpta Medica, 1979: 81–112.
30. Imperato-McGinley J, Peterson RE, Gautier T, Strurla E. Androgens and the evolution of male gender identity among male pseudohermaphrodites with 5 alpha-reductase deficiency. *N Engl J Med* 1979;300:1233–1237.
31. Bailey JM, Pillard RC, Neale MC, Agyei Y. Heritable factors influence sexual orientation in women. *Arch Gen Psychiatry* 1993;50:217–223.
32. Damasio AR, Van Hoesen GW. Emotional disturbances associated with focal lesions of the limbic frontal lobe. In: Heilman KM, Satz P, eds. *Neuropsychology of human emotion.* New York: Guilford Press, 1983;85–108.

7

Endocrine and Immune Systems

Beatriz M. DeMoranville, Ivor Jackson, Robert Ader, Kelley S. Madden, David L. Felten, Denise L. Bellinger, and Randolph B. Schiffer

PSYCHONEUROENDOCRINOLOGY

OVERVIEW

Psychoneuroendocrinology is the field of experimental and clinical neurosciences in which psychiatry, neurology, and endocrinology intersect. The field of psychoneuroendocrinology involves the following objectives:

1. Define the hormonal, neurotransmitter, or neuromodulatory defect in a psychiatric or neurologic disorder.
2. Use this information to elucidate the differential diagnosis of psychiatric disorders.
3. Predict the response to treatment and its prognosis by attempting to correct underlying neuroendocrine abnormalities.
4. Use this approach to characterize the mechanisms of action of the neuroactive compounds used for therapy of psychiatric conditions.

ANATOMY AND SYSTEM ORGANIZATION

Hypothalamic-Pituitary System

Hypothalamus

The hypothalamus is part of the diencephalon. Anteriorly, it is limited by the optic chiasm and the lamina terminalis, and it is continuous with the preoptic area, the substantia innominata, and the septal region. Posteriorly, it is limited by an imaginary plane defined by the posterior mammillary bodies ventrally and the posterior commissure dorsally. Caudally, the hypothalamus merges with the midbrain periaqueductal gray and tegmental reticular formation. The dorsal limit of the hypothalamus is determined by the horizontal level of the hypothalamic sulcus on the medial wall of the third ventricle, at the horizontal level of the anterior commissure. Here, the hypothalamus is continuous with the subthalamus and the zona incerta. Laterally, the hypothalamus is limited by the internal capsule and the basis of the cerebral peduncles.

The hypothalamic area involved with the regulation of the anterior pituitary has been named the *hypophysiotropic area.* Neuroendocrine cells in this area form the following nuclei: the supraoptic (SON), paraventricular (PVN), periventricular, medial preoptic, and arcuate nuclei of the hypothalamus. These nuclei project to the median eminence and secrete the hypophysiotropic factors that regulate pituitary function (1).

Median Eminence. The *median eminence* or *infundibulum* gives rise to the pituitary stalk at the base of the hypothalamus, in the floor of the third ventricle. The median eminence is the site of the following:

1. The hypothalamic neuroendocrine cells release their secretions to the primary plexus of the hypophysial portal system for regulation of the adenohypophysis (i.e., the anterior pituitary).
2. Hypothalamic neural fibers that end in the neurohypophysis and intermediate

lobe (the hypothalamic neurohypophysial tract) pass through.

3. The portal venous system, which provides the only significant blood flow to the anterior pituitary, originates.

The Pituitary Gland or Hypophysis

The *pituitary gland* or *hypophysis* lies close to the medial basal hypothalamus, to which it is connected by the pituitary stalk. In most vertebrates, it is divided into three lobes:

- Anterior lobe or adenohypophysis
- Posterior lobe or neurohypophysis
- Intermediate lobe

The Anterior Pituitary or Adenohypophysis. The *anterior pituitary* or *adenohypophysis* contains cells that secrete the following hormones: adrenocorticotropin hormone (ACTH), thyroid-stimulating hormone (TSH), luteinizing-hormone (LH), follicle-stimulating hormone (FSH), growth hormone (GH) and prolactin (PRL) (Fig. 7-1).

Five cell types have been recognized in the adenohypophysis and are responsible for the synthesis of the classic anterior pituitary hormones: somatotrophs, mammotrophs, corticotrophs, thyrotrophs, and gonadotrophs. The somatotrophs are the GH-producing cells, which account for 50% of the cells in the adenohypophysis. The mammotrophs are the PRL-producing cells, which account for 15% to 25% of the cells. The corticotrophs are ACTH-producing cells, which constitute about 20% of the anterior pituitary cells. Thyrotrophs produce TSH and constitute about 5% of the cells in the anterior pituitary. Gonadotrophs, the least numerous hormone-secreting cells in the anterior pituitary, produce LH and FSH. Additionally, there are other cell types within the anterior pituitary, including the folliculostellate cells shown to contain interleukin-6 (IL-6). Furthermore, there are numerous neural peptides produced within the anterior pituitary, including vasoactive intestinal polypeptide (VIP), which regulates prolactin and substance P. VIP may have a role in regulating the pulsatile secretion of LH and in determining circadian rhythms, along with substance P.

Neurohypophysis. The *neurohypophysis* includes the neural stalk, the neural lobe or posterior pituitary, and the specialized neurons at the base of the hypothalamus. The major nerve tracts of the neurohypophysis arise from the accessory magnocellular, the SON, the PVN, and cells scattered in the periforniсal and lateral hypothalamic areas and the bed nucleus of the stria terminalis. The supraoptic nucleus is located above the optic tract, and the PVN is located on each side of the third ventricle. Most of their unmyelinated fibers descend through the infundibulum and neural stalk within the zona interna to end in the neural lobe. Most of these fibers contain arginine-vasopressin (also known as antidiuretic hormone) and oxytocin. Some vasopressin or oxytocin fibers derived from the parvicellular division descend in the zona externa of the median eminence, where they are involved in the regulation of the anterior pituitary, specifically, ACTH release during stress. Other neuropeptides, including thyrotropin-releasing hormone (TRH), corticotropin-releasing hormone (CRH), VIP, and neurotensin, have been found to be secreted from smaller cells or parvicellular neurons. In addition, enkephalins, dynorphins, galanin, cholecystokinin, and angiotensin II have been found in the fibers of the supraopticohypophysial tract and in neuronal terminals in the neurohypophysis.

Hypothalamic Hypophysiotropic Factors

It is known that the hypothalamic neurons produce neuropeptides that regulate the function of the adenohypophysis. Two regulators are inhibitory: dopamine, a monoamine tonic inhibitory factor for PRL, and somatostatin (also known as *somatotropin-release inhibitory factor*), which inhibits the production of GH. The other known factors stimulate the release of anterior pituitary hormones and are therefore referred to as *releasing factors.*

The hypophysiotropic factors or hormones are synthesized by neurons in the hypothala-

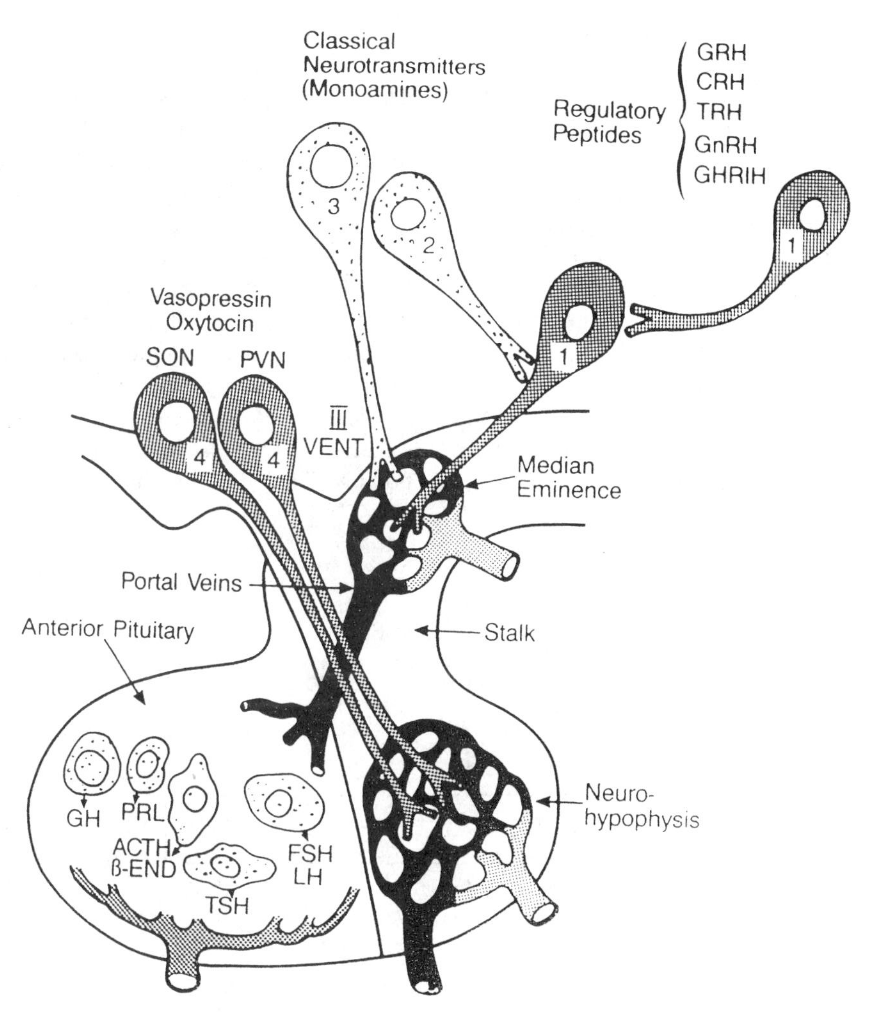

FIG. 7-1. Diagram representing the hypothalamic-pituitary axis. Neuron 4 represents the magnocellular peptidergic neurons of the hypothalamic-neurohypophysial tract with cell bodies in the supraoptic (SON) and paraventricular (PVN) nuclei and terminals in the neurohypophysis. Neurons 2 and 3 are monoaminergic neurons. Neuron 2 represents a neuron contacting the cell body of a peptidergic neuron, whereas neuron 3 represents a dopaminergic neuron projecting to the median eminence, where release of dopamine occurs. The neurons 1 are the peptidergic neurons, which secrete regulatory peptides into the pituitary portal plexus. The regulatory peptides and dopamine are involved in the control of the secretion of different anterior pituitary hormones. β-END, β-endorphin. (From ref. 1, with permission.)

mus, transported to nerve endings in the stalk-median eminence, released into the interstitial space in contiguity with the primary portal capillary plexus, and distributed to the anterior pituitary by means of the portal circulation (see Fig. 7-1).

The arcuate nucleus of the hypothalamus contains growth-hormone–releasing hormone (GHRH), the PVN contains mainly TRH and CRH, and the preoptic nucleus contains mainly gonadotropin-releasing hormone (GnRH, or luteinizing hormone–releasing hormone [LHRH]). Somatotropin-release inhibitory factor is found in the periventricular nucleus. These nuclei also contain many other peptides, including pro-opiomelanocortin

(POMC) and derived peptides; ACTH, β-endorphin, and α-melanocyte-stimulating hormone; and neuropeptide Y, galanin, substance P, enkephalins, atrial natriuretic peptide, angiotensin II, cholecystokinin, and dynorphins.

The hypophysiotropic neurons are regulated by neurotransmitters and neuropeptide modulators and by feedback effects. These are exerted both by the hormones produced by the end organs and by pituitary hormones in "short-loop" feedback.

Hypophysiotropic factors of the hypothalamus in relationship to the corresponding pituitary—end organ axis are shown schematically in Fig. 7-2.

Gonadotropin-Releasing Hormone (or Luteinizing Hormone-Releasing Hormone). This decapeptide was the first hypophysiotropic factor to be localized by immunohistochemistry. In humans, the GnRH neurons are located in the highest concentrations in the medial basal hypothalamus (infundibular and mammillary nuclei) and the preoptic area. GnRH stimulates the release of LH and FSH by the pituitary gland. In women, it controls the menstrual cycle. In men, it controls testosterone secretion and spermatogenesis.

Corticotropin-Releasing Hormone. This 41-peptide hormone is found in neuronal cell bodies of the PVN of the human hypothala-

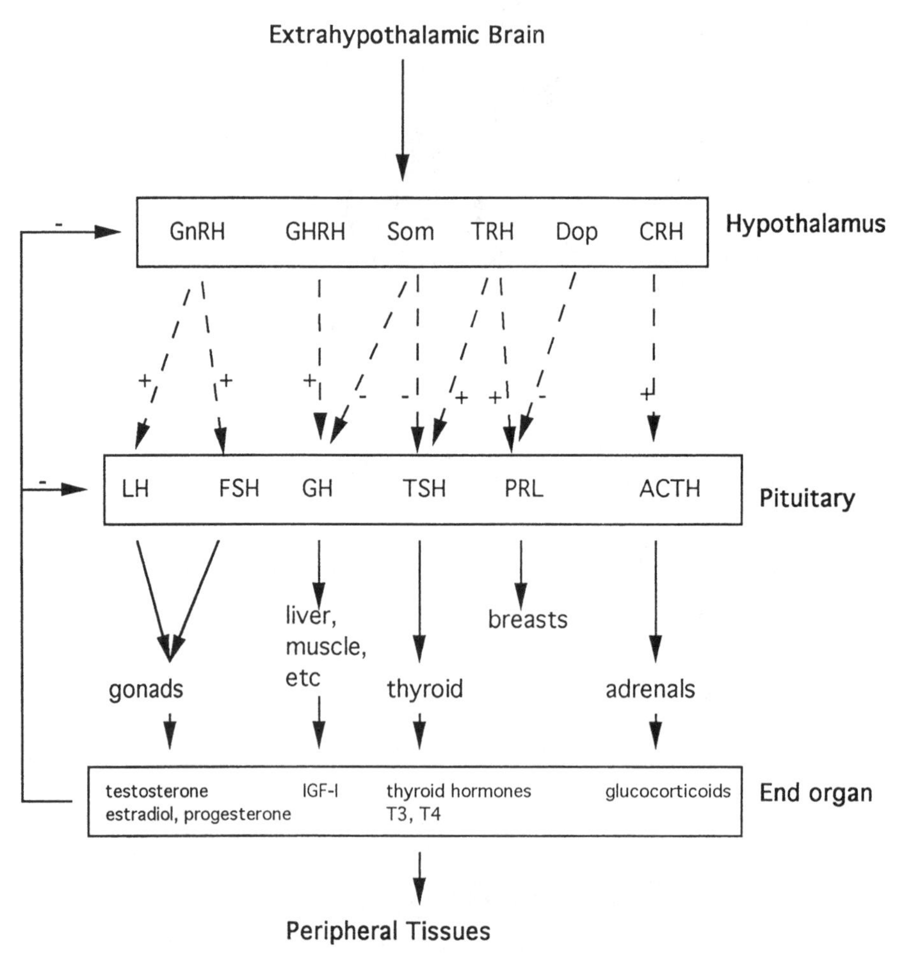

FIG. 7-2. Diagram representing the interactions of the different hypothalamic-pituitary—end organ axes.

mus. CRH is responsible for stimulating the secretion of ACTH and other POMC-related peptides from the anterior pituitary. ACTH regulates the secretion of cortisol from the adrenal cortex. The hypothalamic-pituitary-adrenal (HPA) axis is essential to the neuroendocrine response to physical or mental stress. Additionally, cerebral CRH is involved in the stress response independently of its pituitary-adrenal effects.

Thyrotropin-Releasing Hormone. TRH regulates the thyroid axis by stimulating the release of TSH from the anterior pituitary, which acts directly on receptors on the secretory cells of the thyroid gland. TRH is also a potent stimulatory factor for PRL secretion. TRH neurons are also present in extrahypothalamic tissues, particularly the raphe nucleus, where there is co-localization with serotonin (5HT). TRH may be a neurotransmitter or neuromodulator in view of its localization in nerve endings and the presence of TRH receptors in brain tissue.

Growth Hormone–Releasing Hormone. In primates, neuronal bodies containing GHRH are found in high concentrations in the arcuate nucleus. GHRH stimulates GH secretion. GH is required for normal growth and development.

Somatostatin. Bodies of the neurons containing this tetradecapeptide are mainly found in the anterior portion of the arcuate nucleus close to the infundibular recess in humans. Somatostatin inhibits GH release. It is widely distributed throughout the nervous system and in extraneuronal tissues, including the gastrointestinal tract and endocrine pancreas. Somatostatin inhibits the secretion of TSH and, under certain circumstances, PRL and ACTH. In addition, it has inhibitory effects on endocrine and exocrine secretions of the pancreas, the gallbladder, and the gut (2).

Prolactin Regulatory Factors: Dopamine. Dopamine acts on the lactotrope to inhibit the biosynthesis and release of PRL (3).

Hypothalamic-Pituitary-Adrenocortical Axis: Clinical Considerations

Regulation of the HPA Axis

At the highest level, a variety of neurotransmitter pathways (including serotonergic and cholinergic excitatory pathways and adrenergic and possibly GABA-ergic inhibitory pathways) influence the release of CRH and co-secretagogues, including arginine-vasopressin and oxytocin, into the portal circulation. In response to CRH, a large precursor molecule, POMC, is synthesized in the anterior pituitary and is cleaved into ACTH and other neuropeptides, including β-endorphin. ACTH is then secreted into the systemic circulation. At the adrenal cortex, ACTH stimulates the release of cortisol.

Regulation of the HPA axis is dependent on three major mechanisms:

1. Pulsatile CRH release linked to the endogenous circadian rhythm of the central nervous system (CNS). This mechanism involves an oscillator located in the suprachiasmatic nucleus and is mediated by serotonergic pathways.
2. Physical and psychological stresses, which affect the input from the limbic and the reticular-activating systems to CRH-secreting neurons. Vasopressin secretion, which is increased by stress, enhances the CRH-induced release of ACTH from corticotrophs.
3. Feedback loops, by which circulating glucocorticoids feed back to the pituitary, the hypothalamus, and certain extrahypothalamic regions to inhibit HPA activity

The hippocampus appears to have an important role in this inhibitory action of glucocorticoids. There are numerous glucocorticoid receptors in the hippocampus. A reduction in number of these receptors, as by prenatal dexamethasone treatment of experimental monkeys, or by hippocampal lesions such as those of Alzheimer's disease, is associated with hypersecretion of glucocorticoids, increased hypothalamic levels of CRH, and reduced suppression of ACTH by exogenous glucocorticoid administration.

Clinical Evaluation

In depression, there is both increased secretion of CRH and a neurally mediated hyperresponsivity of the adrenal gland to ACTH.

Measures of Basal HPA Activity. Several measures of HPA activity have been employed in neuroendocrine studies of depression.

Basal plasma cortisol. This increases with increased activity at any level of the HPA axis. It shows a good correlation with levels of urinary free cortisol.

Urinary free cortisol. This is an integrated measure of HPA activity over time and directly measures the unbound portion of total plasma cortisol. It reflects the cortisol secretion rate. It is less sensitive to time of collection.

Cortisol in saliva and cerebrospinal fluid (CSF). These correlate well with unbound plasma cortisol. The former provides a potential means for frequent noninvasive sampling of ambient circulating levels, whereas the latter provides a direct indication of the levels of glucocorticoids to which CNS is exposed.

Dynamic Measures of HPA Axis Activity. The Dexamethasone Suppression Test. A "psychiatric" dexamethasone suppression test (DST) is normal if there is suppression of plasma cortisol level to less than 5 μg/dL from 8 to 24 hours after an oral dose of 1 mg dexamethasone given at 11:00 PM. An abnormal DST is found in about 45% of patients with depression. The test has shown a specificity of up to 80% in some studies, but the specificity depends on the population tested. Abnormal DSTs are seen in other psychiatric disorders, including anorexia nervosa, obsessive-compulsive disorder, dementia, mania, schizophrenia, alcoholism, psychosexual dysfunction, and schizoaffective psychosis as well as gross brain diseases, including Parkinson's disease and stroke. The DST is a more sensitive test for major depression in older people and in children and adolescents. The DST has not evolved into a routine diagnostic test in clinical psychiatry because of its lack of specificity.

ACTH Stimulation Test. Adrenocortical hyperresponsiveness to exogenous ACTH is associated with the hypercortisolism of Cushing's syndrome. Depressed patients may also show an increased response to ACTH, which correlates with adrenal enlargement found by computerized tomography and in postmortem studies of adrenals of subjects who committed suicide.

CRH Stimulation Test. Patients with melancholia show blunted ACTH and cortisol response to exogenous CRH administration. Corticotrophs in the anterior pituitary are downregulated by hypercortisolism and persistent CRH release. This decreased response to exogenous CRH distinguishes depressed patients from Cushing's disease patients, who have both an ACTH and a cortisol hyperresponse to exogenous CRH. In Cushing's disease, endogenous CRH secretion is decreased; in depression, it is increased.

In depression, neuroendocrine and catecholamine dysfunctions may be linked to CRH effects on locus ceruleus (LC) neurons by means of persistent elevated levels of LC discharge and diminished responses to phasic sensory stimuli. Antidepressant drugs appear to decrease LC sensory evoked discharge after acute administration.

Stress leads to activation of both the HPA axis and the catecholaminergic systems (which include the sympathetic nervous systems and CNS catecholamines). CRH stimulates noradrenergic neurons in the LC, suggesting that brain CRH coordinates the behavioral and autonomic nervous system (ANS) responses in stress with the neuroendocrine response. There is evidence for a reciprocal effect of norepinephrine on CRH; α_1-adrenergic agonists raise the levels of CRH within the CNS.

Some forms of depression are associated with hyperactivity of the noradrenergic system, which, by direct or indirect connections, leads to enhanced CRH production in the hypothalamic PVN. This causes activation of the HPA axis, as well as increased sympathetic outflow (4).

Hypothalamic-Pituitary-Thyroid Axis

Primary thyroid disorders may present with mental and behavioral symptoms, and both physical and psychological stress can affect hypothalamic-pituitary-thyroid (HPT) axis function.

Regulation of the HPT Axis

The synthesis and secretion of the thyroid hormones, thyroxine (T_4) and triiodothyronine (T_3), by the thyroid gland is regulated primarily by TSH or thyrotropin, a glycoprotein synthesized by the thyrotropic cells of the anterior pituitary. TSH is secreted from the pituitary in a circadian rhythm determined by rhythmic TRH secretion by suprachiasmatic nucleus of the hypothalamus. The highest serum TSH concentrations occur between 9:00 PM and 5:30 AM. The TSH nadir occurs between 4:00 and 7:00 PM. Although negative feedback of T_4 and T_3 on the pituitary is the most potent factor affecting TSH secretion, the hypothalamus plays an important role through releasing and inhibiting factors. TRH is a tripeptide synthesized in the "thyrotropic" area of the hypothalamus, principally the parvicellular division of the PVN. TRH interacts with high-affinity receptors on the pituitary thyrotrophs. Two other hypothalamic factors, dopamine and somatostatin, have inhibitory effects on TSH secretion at the level of the thyrotroph and may function as physiologic thyrotropin-inhibitory factors; they reduce the degree of TSH release by TRH.

The hypothalamic neurons that secrete TRH are regulated by monoamine neurotransmitters. Serotonin, norepinephrine, and histamine can all stimulate the release of TRH.

Stress inhibits TSH release. In humans subjected to physical stress, such as in the euthyroid sick syndrome, TSH levels do not compensate for the low T_3 and T_4 levels found in these situations. Patients with this syndrome have decreased levels of TSH secretion as well as a diminished circadian periodicity of TSH release.

Thyrotropin-Releasing Hormone Test

The TRH test consists of the measurement of serum TSH before and after administration of exogenous TRH. A standardized dose of 200 to 500 μg of TRH is injected intravenously after an overnight fast, with the subject recumbent throughout the procedure. Blood samples for serum TSH are drawn at baseline and at 30-minute intervals for 3 hours. A blunted response is defined as a peak response that is less than 5 μU/mL above baseline. With this definition, about 30% of patients with primary depression have a blunted response. Hyperthyroid patients have suppressed TSH and a flat response to TRH (less than 1 μU/L). Hypothyroid patients show a high normal or elevated TSH and an exaggerated response because of lack of feedback inhibition (more than 30 μU/L). Several factors may influence the thyrotropin response; therefore, the clinical value of the TRH test is controversial. Age, especially in men older than 60 years, acute starvation, chronic renal failure, Klinefelter's syndrome, and numerous medications can also reduce the response of TSH to TRH to as little as 2 μg/mL. Many studies have evaluated the TRH test as a "marker" of depression and as a tool of prognostic value regarding outcome of treatment, but their results have been discrepant.

Loss of Nocturnal Thyroid-Stimulating Hormone Surge in Depression

T_3 can accelerate the antidepressant effect of tricyclic antidepressants in women and can induce a response to tricyclics in both male and female patients previously unresponsive to it. In addition, administration of TRH can induce an increased sense of well-being and relaxation in healthy volunteers and in some depressed patients. The mechanism of action of TRH in this context may be independent of its functions on the pituitary-thyroid axis.

Should patients with psychiatric illness be treated with thyroid hormone? At least 5% of patients with depression have an elevated TSH or hyperresponse to TRH despite normal serum T_4 levels, suggestive of latent or subclinical hypothyroidism. In such patients, a therapeutic trial of thyroid hormone, such as levothyroxine (T_4) or liothyronine (T_3, Cytomel) may be indicated, aiming to keep the serum TSH as measured by a sensitive assay within the normal range but not suppressed.

The role of thyroid supplementation is well established in rapid-cycling bipolar disorder, in which as many as 50% of patients have a subclinical hypothyroidism. High doses of levothyroxine or of T_3, given with lithium or other mood-stabilizing agents, can stop the rapid cycling, suggesting the influence of thyroid hormone in the phenotypic expression of bipolar illness. As with depression, it is not settled whether T_3 is superior to T_4 for this purpose.

REPRODUCTIVE AXIS

Sexual Dimorphism of the Central Nervous System: Sex Hormones and Central Nervous System Function

The volume and cell count of the so-called sexually dimorphic nucleus of the preoptic area (SDN-POA) shows a marked sexual dimorphism. The mean volume of the SDN-POA is 2.2 times larger in men than in women and contains about twice as many cells in men. This area might be involved in the control of male sexual behavior. Asymmetries in temporoparietal regions of the brain also differ between men and women.

Stress and Reproductive Function

Stress-related hormones (i.e., CRH, the POMC-derivatives ACTH and β-endorphin, and adrenal corticosteroids) can influence reproductive function at three different levels of the hypothalamic-pituitary-gonadal axis: in the brain, by inhibiting GnRH secretion; in the pituitary, by interfering with LH release; and in the gonads, by interfering with the stimulatory effect of gonadotropin on sex steroid secretion.

There is substantial evidence that the stress inhibition of gonadotropin secretion is mediated mainly by the direct and indirect effects of CRH on GnRH secretion. CRH infusions inhibit LH secretion in women.

Hypothalamic-Pituitary-Somatotropic System

GH release is regulated by the noncompetitive antagonism of GHRH and somatostatin. The effects of other hormones appear to be mediated by alterations in the secretion of both GHRH and somatostatin.

Somatostatin-containing neuronal cell bodies, terminals, and receptors are distributed throughout the brain, where somatostatin likely plays a role in neuronal function unrelated to regulation of pituitary GH secretion.

Clinical Issues

Suicide. Postmortem studies of brains of suicide victims support an association with altered serotonin levels and HPA axis activation. Levels of both serotonin and its metabolite 5-hydroxyindoleacetic acid are lower in brains of suicide victims than in control brains. A positive DST has been reported in 80% of patients who attempted violent suicide; elevated plasma and urine corticosteroid levels have also been reported. Brain CRH levels are increased.

Thyroid Axis and Psychiatric Disorders. About one fourth of patients with unipolar major depression have a blunted TSH response to TRH administration. Because an abnormality in brain catecholamine metabolism may underlie some depressive disorders, it is conceivable that a dopamine excess, for example, may account for the blunted TSH response to TRH seen in depression. Other hypotheses include an altered pituitary receptor sensitivity and the possibility that the blunted TRH response might reflect enhanced endogenous TRH secretion with downregulation of TRH receptors.

Thyroid Axis in Depression. Although most depressed patients appear to be euthyroid, several studies have found either high normal or borderline low thyroid hormone levels in a substantial minority of depressed samples. Reverse T_3 (rT_3), the inactive analog of T_3, is increased in patients with unipolar depression.

Lithium and Thyroid Function. Lithium at therapeutic levels inhibits thyroid hormone release by decreasing endocytosis of T_3- and T_4-laden thyroglobulin on the luminal side of the thyroid follicle. At higher levels, it can inhibit iodine uptake and organification. Clinical hypothyroidism develops in about 7% of patients taking lithium, over and above the rate expected for patients of the same age and sex not taking lithium. Women and patients with positive antithyroid antibodies are especially vulnerable.

Carbamazepine, a commonly prescribed alternative or adjunct to lithium in the treatment of bipolar disorder, can lower levels of T_4 and free T_4 without a compensatory increase in TSH. Although T_4 may fall to 70% of the basal level within 2 weeks after starting carbamazepine, patients usually remain clinically euthyroid.

Eating Disorders: Anorexia Nervosa and Bulimia.

Anorexia Nervosa. Multiple endocrine disturbances are present in anorexia nervosa that are not explained by weight loss alone. Some of them may persist despite weight recovery.

The amenorrhea of anorexia nervosa is a form of hypothalamic dysfunction, in which LH and FSH secretion reverts to a prepubertal pattern; that is, LH and FSH secretion are significantly decreased when compared with control women. LH shows a greater depression than FSH. The number of their secretory spikes is also reduced. These abnormalities may be mild or severe, ranging from a luteal phase defect to the complete picture of hypogonadotropic hypogonadism, with loss of the LH and FSH response to a clomiphene challenge. The response of LH to GnRH is blunted, whereas the FSH response may be increased. The circadian rhythm of other hormones has also been found to be altered. Growth hormone is significantly higher in many women with anorexia nervosa, and the elevation is inversely proportional to the deviation of body weight from the ideal. Basal GH levels are elevated in 29% of women with anorexia nervosa, and 22% of anorexia nervosa patients do not exhibit a significant GH nocturnal peak. An abnormal GH increase after TRH stimulation has been found in about 50% of patients.

Cortisol levels are also significantly elevated in anorexia nervosa patients and are inversely correlated with body mass index and directly correlated with the percentage decrease from ideal body weight and Hamilton Rating Scale scores for depression. Endogenous hypothalamic CRH levels are also likely elevated, and the ACTH and cortisol response to CRH is decreased, as in depression. These findings indicate a significant relationship between HPA axis abnormalities and the severity of weight loss and depressive symptoms in anorexia nervosa.

Bulimia. Studies performed in bulimic patients have shown that there is a tendency toward nonsuppression of cortisol in the DSTs of patients with bulimia. Plasma dexamethasone levels following a standard dose may be lower in bulimic patients than in controls. TRH stimulation produces blunted responses in TSH levels in 80% of patients with both bulimia and anorexia nervosa and in 22% of those with bulimia alone. Increased basal GH levels are found in 33% of bulimic patients, regardless of food restriction. A paradoxical increase of GH levels upon TRH stimulation has been found in 15% to 20% of patients with bulimia. All these results indicate that multiple neuroendocrine disturbances occur in bulimia, but to a lesser degree than in anorexia nervosa (5).

REPRODUCTIVE ENDOCRINE ASPECTS OF EPILEPSY

Epilepsy is associated with reduced fertility in women, with impotence in men, and with hyposexuality in both men and women.

Authors have observed that both carbamazepine and phenytoin increase serum levels of sex hormone–binding globulin, which, without an increase in total testosterone, reduces the free fraction present. The decreased fertility of women with epilepsy is due to impairment of ovulation. Women with epilepsy and anovulation may have either hypothalamic amenorrhea or the polycystic ovary syndrome.

Menstrual disturbances, polycystic ovaries by ultrasound, and elevated serum testosterone are all more common in women taking valproate than those taking carbamazepine or other medications. Women who start valproate before they reach 20 years of age have the highest rate of reproductive endocrine problems—80%.

Serum PRL is known to increase after generalized seizures and complex partial seizures with bilateral medial temporal involvement. PRL peaks at two to three times normal in 15 to 20 minutes and returns to baseline within 3 hours. Failure of PRL to rise after an apparent generalized seizure suggests pseudoseizures.

In summary, specific reproductive endocrine abnormalities can be demonstrated in most patients with epilepsy and altered sexual or menstrual function. Although patients with such abnormalities should be offered endocrine treatment if they are troubled by their symptoms, the efficacy of endocrine interventions in fully restoring sexual function in epileptic patients is not known.

ENDOCRINE DISEASES PRESENTING AS PSYCHIATRIC DISORDERS

Hypothalamic Lesions

Anatomic lesions of the hypothalamus can present as mental or behavioral disorders, such as explosive disorder, polyphagia and morbid obesity, hypodipsia, primary polydipsia, hypersomnolence, and dementia.

The etiologies cited are varied and include not only parasellar tumors but also granulomatous disorders, such as sarcoidosis, eosinophilic granuloma, histiocytosis X, congenital defects, pseudotumor cerebri, hemochromatosis, and whole brain irradiation.

Hypothyroidism and Hyperthyroidism

Disturbances of thyroid function can cause mental and behavioral symptoms. In addition, patients with subtle defects in thyroid function may be more susceptible to mental disorders because the prevalence of thyroid autoantibodies is higher in psychiatric patients than in the general population (Table 7-1).

The neuropsychiatric features of adult-onset hypothyroidism consist of ill-defined complaints and disturbances in cognition or mood. There is lack of attention, poor concentration, and poor memory. Symptoms may resemble those of depression, with apathy, social withdrawal, paucity of speech, and slow motor function. In severe hypothyroidism, delusions and hallucinations may occur. This may be followed by lethargy, drowsiness, and finally, stupor and coma. The pathophysiology of these symptoms and signs probably involves both direct effects of T_3 on neurons and indirect effects by way of other neurotransmitters.

Patients with *hyperthyroidism* may present with anxiety and dysphoria, emotional liability, insomnia, difficulty concentrating, irritability, restlessness, and tremulousness. This may resemble mania because motor activity is increased, but unlike mania, energy is usually decreased. In some cases, severe thyrotoxicosis may present as a psychotic illness, with delusional and paranoid thoughts. Elderly people, however, may present with apathy, lethargy, and depression *(apathetic thyrotoxicosis)*. It is probable that the interactions between the thyroid hormones and the catecholaminergic system play an important role

TABLE 7-1. *Psychiatric Manifestations of Thyroid Disease*

Hyperthyroidism	Hypothyroidism
Anxiety	Depression
Dysphoria	Inattention
Irritability	Poor memory
Insomnia	Sleepiness
Difficulty concentrating	Difficulty concentrating
Restlessness	Slow motor function
Manic symptoms	Paucity of speech
Decreased energy	Hallucinations
Delusions	Bizarre behavior
Paranoia	Lethargy
Apathy	Drowsiness
Lethargy	Stupor
Depression	Coma

in the development of symptoms and signs of hyperthyroidism. Catecholamines and thyroid hormones share the amino acid tyrosine as a precursor and have synergistic cellular actions. In hyperthyroidism, the turnover of catecholamines is decreased, but β-adrenergic receptor numbers are increased. These mechanisms may explain the role of β-adrenergic antagonists in the reduction of thyrotoxic symptoms. Successful treatment of thyrotoxicosis with antithyroid agents, thyroidectomy, or radioactive iodine usually leads to the resolution of the neuropsychiatric syndrome (6).

Hypocortisolism and Hypercortisolism

Patients with severe or chronic adrenal insufficiency may present with mild to moderate cognitive impairment, apathetic depression, or psychosis. Symptoms may include irritability, social withdrawal, poor judgment, hallucinations, paranoid delusions, bizarre thinking, and even catatonic posturing.

Hypercortisolism, such as in Cushing's syndrome or exogenous glucocorticoid excess, presents with mental or behavioral symptoms in up to 50% of cases. Symptoms and signs include emotional lability, depression, loss of energy and poor libido, irritability, anxiety, panic attacks, and paranoia. Occasional patients present with a manic psychosis.

The brain represents one of the principal targets for steroid hormones. Steroids secreted by the adrenal cortex, such as glucocorticoids, mineralocorticoids, and stress hormones, easily cross the blood-—brain barrier and exert their actions in the CNS at classic steroid receptors. Glucocorticoids have prominent actions on the hippocampus, which is the area of the brain with the major number of glucocorticoid receptors. Either a lack or an excess of glucocorticoids can produce significant functional impairment or structural change in CA3 cells in the hippocampus.

Exogenous glucocorticoids have been shown to produce impairment of higher brain functions in humans. The term *steroid psychosis* has been used to describe a psychotic syndrome in patients who are given high doses of corticosteroids for medical illnesses that have not directly affected the CNS. Administration of exogenous glucocorticoids in healthy subjects has also been associated with cognitive impairment and with poor performance on memory tasks. It has been postulated that corticosteroids may impair selective attention and may suppress the activity of the hippocampus, where the stimuli may be initially filtered (7).

PSYCHONEUROIMMUNOLOGY: INTERACTIONS BETWEEN THE BRAIN AND THE IMMUNE SYSTEM

OVERVIEW

Research in psychoneuroimmunology addresses the functional significance of the relationship among neural, endocrine, and immune processes. The brain and the immune system represent an integrated mechanism for the adaptation of the individual and the species. Within the context of neuropsychiatry, this subchapter provides a brief overview of some of the evidence that behavioral factors influence immune functions and describes some of the neural and endocrine pathways that can serve as vehicles for communication between the brain and the immune system. Refer to reference 8 for the definitive text.

BEHAVIOR: IMMUNE SYSTEM ORGANIZATION AND ANATOMIC INTERACTIONS

Conditioned Alterations of Immunity

Immune responses can be modified by classic (pavlovian) conditioning. The conditioned modulation of natural and acquired host defense mechanisms and immune responses was first investigated by Russian scientists in the 1920s, using the principles and procedures of the day. After multiple pairings of a neutral conditioned stimulus and an injection of antigen, the unconditioned stimulus, presentation of the conditioned stimulus, alone, was purported to elicit conditioned increases in nonspecific defense responses and increases in antibody titers, as well.

Current interest in conditioned alterations of immunity began with a study by Ader and colleagues (8). Using a taste aversion learning paradigm (a passive-avoidance conditioning situation), water-deprived rats were provided with a novel, distinctively flavored drinking solution (saccharin), the conditioned stimulus. Consumption of the conditioned stimulus solution was followed by an intraperitoneal injection of cyclophosphamide, an immunosuppressive unconditioned stimulus unconditioned stimulus that also causes a transient gastrointestinal malaise. Three days later, the animals were immunized with sheep red blood cells, and the conditioned population was randomly divided into three subgroups: Group CS was reexposed to the conditioned stimulus previously paired with the immunosuppressive drug and injected with saline; to control for the effects of conditioning per se, group CSo was provided with plain water and not otherwise manipulated; group UCS was also given plain water but was injected with cyclophosphamide to define the unconditioned immunosuppressive effects of the drug. A nonconditioned group initially received cyclophosphamide without the saccharin drinking solution and, following immunization, was provided with the saccharin solution whenever any subsample of group CS was reexposed to the conditioned stimulus. A placebo group was initially injected with saline following the consumption of plain water.

The pairing of saccharin consumption and an injection of cyclophosphamide resulted in an aversion to saccharin-flavored water when conditioned animals were reexposed to the conditioned stimulus, and, as hypothesized, conditioned animals reexposed to the conditioned stimulus also showed an attenuated antibody response to sheep red blood cells compared with nonconditioned animals and animals that were conditioned but not reexposed to the conditioned stimulus. These results have been taken as evidence for a conditioned suppression of immunologic reactivity.

The physiologic mechanisms underlying conditioned alterations of immunity are not known. Originally, conditioned immunosuppressive responses were considered the result of the stressful treatment because an increase in glucocorticoid elevations, equated with stress, could suppress immune responses. The data, however, do not support the hypothesis that an elevation in circulating "stress hormones," particularly adrenocortical steroids, mediates the conditioned modulation of immune responses. The fact that conditioned immunosuppression occurs when conditioned animals are reexposed to the conditioned stimulus before as well as after immunization could indicate that the mechanisms do not involve antigen-induced immune or neuroendocrine changes; it could also imply that different mechanisms are involved when conditioning is superimposed on a resting system in contrast to an antigen-activated system. As discussed later, the immune system is innervated, leukocytes and neurons share certain neuropeptide and neurotransmitter receptors, activated lymphocytes produce several neuroendocrine factors, and both cells of the immune and nervous systems produce and respond to the same cytokines. Conditioned and stressor-induced changes in the pattern of increases or decreases of neural and endocrine activity constitute potential pathways for the mediation of behaviorally induced alterations of immune function. These observations may in time have great relevance to the way in which pharmacotherapies are delivered (9).

Central Nervous System Lesions and Intracerebroventricular Infusions

A neuroanatomic approach has been used to evaluate the immunomodulatory capacity of specific regions of the CNS. Autonomic preganglionic neurons receive direct fiber projections from brain-stem nuclei (nucleus solitarius, raphe nuclei, tegmental noradrenergic nuclei); hypothalamic nuclei (paraventricular nucleus, oxytocin and vasopressin neurons, lateral hypothalamus, posterior hypothalamus, dorsal hypothalamus); limbic forebrain structures (central amygdaloid nucleus); and regions of the cerebral cortex (frontal, cingulate, and insular cortical areas, mainly zones of "limbic" cortex). In addition, indirect regulation of these systems arises from regions such as the parabrachial nuclei, central gray, and reticular formation of the brain stem; numerous hypothalamic nuclei and cell groups; limbic forebrain areas, such as the hippocampal formation and septum; and cortical association areas. These structures interconnect with the hypothalamus, the structure that lies at the crossroads of the limbic forebrain and brain-stem nuclei.

Discrete lesions in the brain, particularly in specific nuclei of the hypothalamus and limbic system, result in structural and functional changes in the immune system.

The two main pathways by which the CNS may communicate with the periphery are as follows:

- Neuroendocrine outflow by means of the hypothalamic-pituitary—-target organ axes
- ANS through direct nerve fiber connections with cells of the immune system (10)

Endocrine—-Immune System Interactions

Early studies identified the pituitary as an essential component in the regulation of immune system development and function. Surgical removal of the pituitary gland (hypophysectomy) reduced antibody responses and impaired contact sensitization.

Hypothalamic-Pituitary-Adrenal Axis: Glucocorticoids and Adrenocorticotropic Hormone

Glucocorticoids and their synthetic analogs have been used clinically at supraphysiologic levels as immunosuppressive agents in inflammatory and autoimmune diseases. Endogenous glucocorticoids regulate lymphokine production, inhibit responses to self-antigens, and limit inflammatory processes. Glucocorticoids act by binding to cytosolic glucocorticoid receptors in T and B lymphocytes and macrophages. Upon receptor binding, the glucocorticoid—-receptor complex undergoes a conformational change to allow penetration of the nuclear membrane. Glucocorticoid binds to specific regions of DNA, termed glucocorticoid-regulatory elements, to activate or silence glucocorticoid-regulatory element–containing genes. This alteration in gene expression translates into multiple mechanisms of immunoregulation in vivo.

Glucocorticoids limit immune responsiveness in vivo by the following means:

- Initiate programmed cell death (apoptosis)
- Alter lymphocyte distribution and trafficking
- Regulate lymphocyte cytokine production

Glucocorticoid-induced apoptosis is dependent on the maturational state of the lymphocyte: immature T cells in the thymic cortex and follicular B cells in secondary lymphoid organs exhibit greater susceptibility to lysis by glucocorticoids than mature (medullary) T cells and activated B cells. Glucocorticoid-induced impairment of lymphocyte emigration may inhibit immune reactivity by limiting lymphocyte encounters with antigen.

Glucocorticoids can also directly influence inflammatory processes and immune reactivity by altering cytokine production. Antiinflammatory and immunosuppressive effects of glucocorticoids are mediated in part by inhibition of macrophage production of IL-1, tumor necrosis factor, and IL-6. This pattern of cytokine production leads to inhibition of cell-mediated responses and to enhanced antibody production. By shifting from a cell-mediated response to an antibody response, the chances of acquiring cell-mediated reactivity to self is reduced, but the risk of infection by viral and bacterial pathogens that require cell-mediated immunity for elimination is increased.

Adrenocorticotropic hormone (ACTH) itself may have immunoregulatory properties. High- and low-affinity ACTH receptors have been demonstrated on normal T and B lymphocytes. ACTH and POMC, the mRNA encoding ACTH, and the opioids can be produced by endotoxin or viral-stimulated lymphocytes and macrophages, suggesting that ACTH may regulate immune reactivity locally.

When the immunosuppressive and antiinflammatory effects of the glucocorticoids were discovered, ACTH-induced release of adrenal glucocorticoids was widely considered the mechanism underlying stress-induced immunosuppression. Elevated glucocorticoids levels induced by stressful stimuli have been correlated with immunosuppression, and ACTH or cortisone treatment mimicked the diminished immune reactivity. However, not all CNS-mediated changes in immune responses, even those that are immunosuppressive, are mediated by this axis. For example, conditioned immunosuppression is not correlated with increased glucocorticoid concentrations.

Opioid Peptides

The opioid peptides consist of a series of low-molecular-weight (less than 4500 kD) peptides found in the pituitary, brain, and peripheral structures, such as the adrenal medulla; they include the endorphins, the enkephalins, and the dynorphins. In the brain, ligand interaction with opioid receptors induces profound analgesia and complex behavioral effects, many of which are blocked by naloxone and other opioid antagonists. With the finding that serum levels of β-endorphin increase in parallel with ACTH after some stressors, the search for peripheral opiate targets has focused on the cells of the immune system. Opioids may influence the immune system indirectly by way of the CNS through the sympathetics or the HPA axis, or directly through effects of opioid peptides on cells of the immune system. Functional evidence for opioid receptors on human T lymphocytes have been reported, but at present, a clear functional role has not yet emerged.

Opioid- and non–opioid-mediated effects of b-endorphin in vitro include altered T-cell proliferation and enhanced production of IL-1, IL-2, IL-4, IL-6, and interferon-g. β-Endorphin, met-enkephalin, and leu-enkephalin may themselves serve as chemotactic factors.

Growth Hormone

In several species, growth hormone (GH, somatotropin) deficiency results in small, hypocellular primary and secondary lymphoid organs that possess abnormal morphology. Reconstitution with GH restores normal morphology and the impaired immune reactivity observed in hypophysectomized animals and GH-deficient animals. In patients with GH deficiency, decreased T4-to-T8 ratios, increased circulating B cells, impaired responsiveness to alloantigen, and diminished natural killer (NK) cell activity have been reported. These findings demonstrate that GH is an important regulator of immune function in vivo and suggest that lymphocytes may provide an extrapituitary source of GH (11).

Prolactin

PRL-binding sites have been identified on macrophages, T lymphocytes, and B lymphocytes. PRL restores impaired immune reactivity in hypophysectomized animals. PRL may also be an important factor in host resistance to infectious disease. PRL may also be a cell cycle progression factor, necessary but not sufficient for lymphocyte proliferation. Lymphocytes may be an extrapituitary source of PRL. Lymphocyte secretion of a PRL-like molecule, which appears to be a structural variant of pituitary-derived PRL, has been reported, and PRL mRNA is expressed in activated murine lymphocytes.

Thyroid-Stimulating Hormone and Thyroid Hormones

TSH receptors have been reported on several B-cell lines and lipopolysaccharide-stimulated B cells. In vitro, TSH is capable of enhancing antibody responses by mouse spleen cells to both T-dependent and T-independent antigens. The TSH-induced thyroid hormones, T_3 and T_4, also have immunomodulating activities. In euthyroid animals, T_3 or T_4 administration may enhance antibody and mitogen responses.

Gonadal Steroids

The sex steroids are important regulatory hormones of the immune system and have been implicated in autoimmune processes. Females have have higher levels of serum immunoglobulin G (IgG), IgG1, immunoglobulin M (IgM), and immunoglobulin A (IgA) than males in several species. Antibody responses to T-independent and T-dependent antigens are greater in magnitude and more prolonged in females than in males. Estrogen may enhance antibody production with specificity for foreign and self-antigens. Cytokines that alter B-cell antibody differentiation are altered by estrogens. T- and B-lymphocyte expression of androgen and estrogen receptors by lymphocytes have been reported.

Women experience a higher incidence of autoimmune diseases, such as systemic lupus erythematosus (SLE) and rheumatoid arthritis. Pharmacologic manipulation of estrogen and testosterone levels can alter the course of experimental autoimmune disease in laboratory animals.

The sex steroids may influence the immune system by means of T-cell differentiation in the thymus. Androgen- and estrogen-binding sites have been reported in thymic cytosol fractions and in thymic epithelium. The mechanisms of gonadal steroid regulation of normal immunity and autoimmune processes have yet to be completely elucidated.

Neurohormones of the Posterior Pituitary

The neuropeptides derived from the posterior pituitary (neurohypophysis), vasopressin and oxytocin, exhibit immunoregulatory properties. Immunohistochemical evidence has demonstrated co-localization of oxytocin with vasopressin and IL-1 in thymic epithelial cells. Oxytocin and its protein carrier neurophysin are present in human thymic extracts and are similar to the pituitary form by molecular weight, bioactivity, and the 1:1 ratio of oxytocin to neurophysin. Concentrations of oxytocin are much higher in the thymus than circulating levels. These results constitute evidence that oxytocin is synthesized in the thymus and plays a role in T-cell differentiation.

Neural—Immune Interactions

The ANS communicates with target organs by means of a two-neuron chain. Preganglionic neuronal cell bodies are located in the thoracic and lumbar (T-1 to L-2) spinal cord intermediolateral cell column (sympathetic) or in the sacral (S-2 to S-4) intermediate gray or brain-stem autonomic nuclei (parasympathetic). These preganglionic neurons send cholinergic myelinated axons to ganglia found in the sympathetic chain or in plexuses associated with the great vessels (collateral sympathetic ganglia) or found in or near the target organ (intramural parasympathetic ganglia). Postganglionic noradrenergic sympathetic nerve fibers arborize widely in target organs, giving rise to various sites of interaction with target cells of the immune system. Norepinephrine can diffuse widely and can interact with cells immediately adjacent to, or even distant from, the sites of release.

Autonomic Innervation of Lymphoid Organs

Early studies of lymphoid tissue innervation focused on the networks of noradrenergic nerve fibers in the spleen associated with smooth muscle of the vasculature, trabeculae, and splenic capsule. Further analysis of sympathetic neural innervation of spleen and other lymphoid organs has revealed additional distribution of noradrenergic nerve fibers in the parenchyma, with compartmentalization in regions distinct from vascular or trabecular smooth muscle, and suggests that cells of the immune system may serve as target cells for neurally derived norepinephrine, particularly those cells possessing functional receptors for neurotransmitters.

Primary Lymphoid Organs

Both myelinated and nonmyelinated nerves have been demonstrated in the bone marrow. Fluorescence histochemistry for catecholamines and immunohistochemical localization of tyrosine hydroxylase, the rate-limiting enzyme for norepinephrine synthesis, have revealed extensive networks of noradrenergic nerve fibers along the vasculature and among hematopoietic and lymphopoietic cells in the substance of the bone marrow, suggesting a functional role for noradrenergic innervation of the bone marrow. The finding that the sympathetic nervous system (SNS) can be manipulated to increase hematopoietic output from the bone marrow has potential clinical applicability, especially in the area of bone marrow transplantation.

In the thymus, networks of noradrenergic nerve fibers, detected by fluorescence histochemistry and tyrosine hydroxylase immunocytochemistry, are found in plexuses associated with blood vessels and intralobular septa as well as directly in the cortical parenchyma. Nerve fibers that travel with intralobular septa branch deeply within the cortex and are found adjacent to thymocytes. The origin of these thymic sympathetic nerves has been assessed with retrograde tracing. Labeling of nuclei was found in the sympathetic chain ganglia from the superior cervical ganglia caudal to the T_3 ganglia.

Adrenergic agonists in vitro have been reported to increase the expression of markers (Thy-1, TL) associated with thymocyte differentiation. Manipulation of norepinephrine in the thymus may provide a means of boosting thymocyte proliferation in physiologic circumstances in which such cells could be beneficial, such as aging, in which naive T cells are diminished.

Secondary Lymphoid Organs

Spleen

In the spleen, tyrosine hydroxylase–positive (i.e., noradrenergic) nerve fibers have been localized to sites adjacent to T cells, B cells, and macrophages using fluorescence histochemistry and immunocytochemistry. Noradrenergic nerve fibers enter the spleen with the splenic artery and then further distribute with the central artery, with the capsular and trabecular systems, and into the parenchyma. Some noradrenergic fibers distribute into the parenchyma of the periarteriolar lymphatic sheath, where they arborize among T lymphocytes and along the macrophage zone at the marginal sinus, lined with ED3-positive macrophages and IgM-positive B cells. Some fibers travel along the margin of follicles, among OX-19–positive T cells and sIgM-positive B cells. Very fine, single tyrosine hydroxylase–positive fibers distribute into B-cell follicles, previously thought to be devoid of noradrenergic innervation by fluorescence histochemical analysis. Splenic norepinephrine is primarily neurally derived. Staining with specific antisera for tyrosine hydroxylase has allowed visualization of direct contacts between tyrosine hydroxylase–positive nerve terminals and lymphocytes in the periarteriolar lymphatic sheath and marginal zone of the rat spleen by light and electron microscopy.

Lymph Nodes and Gut-Associated Lymphoid Tissue

Cervical, mesenteric, and popliteal lymph nodes, as well as Peyer's patch and lymphoid tissue associated with the appendix, exhibit the same general pattern of innervation as that observed in the spleen. Noradrenergic nerves enter the lymph nodes at the hilus with the vasculature and distribute throughout the medullary cords among mixed populations of lymphocytes and macrophages and in the subcapsular region. Fibers from these regions then contribute to the innervation of the paracortical and cortical regions, where T lymphocytes are abundant. Fine varicosities, not associated with blood vessels, run between nodules with little distribution into the B-cell—containing follicles. The gut-associated lymphoid tissue contains noradrenergic nerve fibers that distribute through the T-dependent zones and within the lamina propria, including along the inner region of immunoglobulin-secreting plasma cells at the luminal surface.

Peptidergic Innervation and Influences on the Inflammatory and Immune Processes

Several peptides have been described that can modulate splenic vascular resistance and capsule contraction after autonomic stimulation. In addition, many neuropeptides may have direct influences on immunologic reactivity and are found in nerve fibers innervating T- and B-cell—containing compartments of primary and secondary lymphoid organs. One such peptide, NPY, is co-localized in noradrenergic nerve terminals in the rat spleen and thymus, whereas others, such as VIP and substance P, appear to be associated with non-noradrenergic nerve fibers of the arteries, veins, trabeculae, and parenchyma of the spleen.

Substance P and Somatostatin

Substance P is an 11-amino acid peptide found in the CNS, in peripheral sensory neurons, and in nerve plexuses of the gut, spleen, lymph nodes, and thymus. Substance P levels are elevated in chronically inflamed tissue, implicating substance P as a neural regulator of inflammatory processes. Substance P can induce release of histamine by mast cells; numerous mast cells can be found in contact with substance P–positive nerve fibers in the intestine and thymus. The proinflammatory cytokines IL-1, tumor necrosis factor-α, and IL-6 can be induced by substance P in the absence of endotoxin or other stimulators. These findings suggest that substance P is an early inducer of local and systemic host defense responses to inflammation and injury. Substance P also can regulate cell-mediated and humoral immune responses by means of interactions with substance P receptors on T and B lymphocytes. Substance P has also been implicated in the pathogenicity of an autoimmune and inflammatory disease, adjuvant-induced arthritis.

Somatostatin has immunoregulatory effects that are generally opposite to those of substance P. Somatostatin inhibits the proliferation of human T lymphocytes in vivo and in vitro. Somatostatin can reduce interferon-γ production stimulated by the T-cell mitogen staphylococcal enterotoxin A. High- and low-affinity somatostatin receptors have been reported on human T- and B-cell lines.

Vasoactive Intestinal Peptide

VIP is a 28-amino acid neuropeptide found in central and peripheral sites. Immunohistochemical staining of the thymus has revealed immunoreactive VIP-containing nerve fibers in the cortex, subcapsular region, and septal zones, but rarely in the medulla, suggesting a possible role for VIP in T-cell maturation.

α- and β-Adrenoceptors on Cells of the Immune System

The presence of β-adrenoceptors on T and B lymphocytes, macrophages, and neutrophils in numerous species is well established. Lymphocyte β-adrenoceptors behave similarly to β-adrenoceptors expressed on other tissues innervated by the SNS. The receptors are upregulated in the presence of β-blockers and downregulated in the presence of β-agonists. β-Receptors on lymphocytes are linked intracellularly to the adenylate cyclase system to generate cyclic AMP.

In Vivo Immune Regulation by Catecholamines

Peripheral administration of the catecholamines epinephrine and norepinephrine has been employed to demonstrate that catecholamines can modulate lymphocyte responses to antigenic challenge in vivo. This modulation of immune reactivity may occur directly through lymphocyte adrenoceptors or indirectly through interactions with other cell types, such as reticular cells, endothelial cells, or smooth muscle cells associated with the vasculature. Other studies have demonstrated that perturbations in catecholamines can alter lymphocyte migration. In humans, a single epinephrine injection can induce transient increases in the number of circulating blood lymphocytes and monocytes and can decrease proliferative responsiveness to T-cell mitogens. The differing responses between lymphoid organs indicate that SNS regulation of the immune system in vivo cannot be characterized as a simple enhancement or inhibition. Each cell type participating in the specific response may be affected differentially by catecholamines, based on environmental and temporal factors.

In Vitro Immune Regulation by Catecholamines

Catecholamines can regulate macrophage and NK cell function either directly or through regulation of cytokine production. The neuropeptide VIP, an inducer of cyclic AMP, can also potentiate norepinephrine-induced inhibition of tumor cell killing.

Immune System Communication With the Nervous System

Several investigators have reported changes in hypothalamic activity after antigen challenge. Norepinephrine concentration in the hypothalamus and brain stem has been noted to be reduced after intraperitoneal injection of supernatants from concanavalin A—-stimulated spleen cells, suggesting that secreted products of activated lymphocytes can stimulate specific regions of the CNS.

The observation that norepinephrine levels in spleen are inversely proportional to immunologic activity suggests that the immune system can influence sympathetic activity. Modulation of sympathetic activity by the immune system may be achieved by two routes:

- Control of hypothalamic autonomic centers
- At a local level, through reciprocal interactions between the cells of the immune system and nerve terminals within lymphoid organs

Investigators in the field of psychoneuroimmunology have just begun to identify the numerous modes of interactions possible between the nervous and immune systems. Perhaps the operational definition or physiologic constituents of the nervous and immune "systems" should be expanded to include each other. What lies ahead is the determination of how these systems interact physiologically under homeostatic and stressful conditions.

Clinical Issues

Effects of Stress. *Stress* refers to any natural or experimentally contrived experiential, social, or environmental circumstance that poses an actual or perceived threat to the psychobiologic integrity of the individual. In subhuman animals as well as in humans, psychosocial conditions that are perceived as a threat to the organism and to which the organism cannot adapt are accompanied by acute and chronic psychophysiologic changes that could contribute to the development of disease. An extensive literature testifies to the role of psychosocial factors in the susceptibility to or the progression of a variety of pathophysiologic processes, including allergic, infectious, autoimmune, and neoplastic diseases that, to a varying extent, involve alterations in immunologic defense mechanisms.

Experimental studies provide numerous illustrations of the fact that the outcome of stressful experiences depends on several factors, including the nature of the stressor and the pathophysiologic stimuli to which the organism is subjected. The same stressor can have different effects on different pathophysiologic processes, and different stressors can exert different effects on a single disease process.

Current research is focused on quantifying the changes in immune function induced by different stressful circumstances. The death of a family member, for example, is rated highly on scales of stressful life events and bereavement, and has been associated with depression and an increased morbidity and mortality in the case of some diseases. Also, bereavement has been associated with changes in some components of immunologic reactivity, such as reduced lymphoproliferative responses to mitogenic stimulation and impaired NK cell activity. Other reports have described similar changes in immune function associated with the affective responses to other losses, such as marital separation and divorce.

In experimental animals, losses also have immunologic consequences. Periodic interruptions of mother–litter interactions and early weaning in rodents decrease lymphoproliferative responses to mitogenic stimulation and reduce the response to subsequent antigenic challenge. In response to separation, infant monkeys and their mothers show a transient depression of in vitro mitogen responsiveness. Like the results of bereavement studies in humans, these effects are not correlated with plasma cortisol levels. The magnitude of these effects depends on the social environment in which the animals are caged after the separation. Stress is generally but not uniformly immunosuppressive. Here again, the literature indicates that the effects of stress on acquired antibody- and cell-mediated immunity and on natural immune reactions, such as NK cell activity, depend on (a) the quality as well as the quantity of the stressful stimulation and the immunogenic stimulation, (b) the temporal relationship between stressful stimulation and immunogenic stimulation, (c) the immune response (or compartment) under study, and (d) several host factors, such as strain, sex, and age (12,13).

Depression and Immunity. Associations between depressive clinical states and immune function are of importance to neuropsychiatry for two reasons. Depressive clinical states may be clinical manifestations of certain autoimmune diseases, and functional impairments of the immune system may be associated secondarily with depressive disease, rendering such patients more vulnerable to infectious or neoplastic disease.

A variety of clinical studies have described an association between certain diseases of the immune system and clinical depression. Multiple sclerosis (MS) is the best example of a T-cell—mediated autoimmune disease that has been associated with depressive syndromes. SLE is an example of a B-cell—-mediated autoimmune disease in which depressive features are frequently reported. In syndromes caused by the human immunodeficiency virus (HIV) type I, depressive features have also been described frequently.

The association of affective disorders with MS has been the subject of a recent review (14). Point prevalence estimates for diagnosable depressive disorders in MS patients are at least 27% in better-designed epidemiologic studies. Phenomenologically, these depressions tend to be moderately severe as opposed to the more psychotic or agitated states sometimes seen among psychiatric patients. Such prevalence rates of depressive illness are higher than those observed in control groups composed of healthy subjects, patients with various medical and neurologic illnesses, and patients with non-—brain-related neural disorders such as spinal cord injury. MS patients with primarily cerebral disease may have greater rates of depressive illness than MS patients with primarily spinal cord disease. Patients in exacerbation are more depressed than patients in remission. Both psychotherapy that is problem focused and antidepressant pharmacotherapy are effective in treating depression associated with MS.

In SLE, we have the opportunity to observe a primarily antibody-mediated autoimmune disease in which mood disorders also occur with some frequency. SLE is a relatively common multisystem inflammatory disease of connective tissue that, like MS, is sometimes characterized by exacerbations and remissions. The psychiatric features associated with SLE have not been as well characterized as those associated with MS but clearly include cognitive blunting, psychosis, and affective disorder. The prospective prevalence risk for SLE patients to develop such behavioral disturbances over 1 to several years approaches 50% in studies that have included longitudinal follow-up. Some of these mood disturbances may be related to a variety of circulating antiself antibodies that occur in SLE. Increased circulating oligoclonal IgG is known to occur commonly in the CSF of these patients. Certain antilymphocytic antibodies from SLE patients are known to cross-react with neurons in vitro. Serum antiribosomal antibodies have also been reported in association with psychosis and severe affective disorder, and possibly, titers of these antibodies vary with behavioral fluctuation. There is also some evidence that levels of antineuronal antibodies in CSF are elevated in SLE patients who have "organic mental disorders" (15).

Depressive and other psychiatric disturbances have also been reported in patients infected with HIV. It does not appear that there is a direct relationship between the mood disorder and severity of HIV illness, as measured by clinical staging or by CD4 cell counts. Moreover, many of the patients with HIV infection have a positive lifetime history of affective disturbance. We cannot yet conclude that there is a direct mediation by HIV in the production of these mood disorders.

Several recent reviews have described the occurrence of immunologic alterations in the setting of stress, depression, and bereavement. There is substantial evidence that in vitro measures of immune function are altered during depressive states, especially in the more severe depressive states. Persuasive evidence of alteration in immunity associated with depressive states appears when studies of immune system function are performed in depressed patients. Lymphocyte proliferation in response to mitogens and antigens, for example, appears to be impaired or decreased in some patients with clinical depression.

Immune alterations in depression are not as consistent when relatively crude measures of immune function are used, such as the enumeration of total white cells in blood, the differential count of major white blood cell types, or even lymphocyte subtype counts. On the basis of their metaanalysis of the literature on depression and immune changes, Herbert and Cohen came to the conclusion that clinical depression, particularly in older and hospitalized populations, is reliably associated with both functional and enumerative measures of immunity (15).

Depression and altered B-cell function have not been correlated extensively. There are some reports of elevated circulating antibody titers to herpesvirus and cytomegalovirus among depressed patients. Others have found no evidence of oligoclonal IgG or IgM in such patients.

The clinical significance of the in vitro immunologic changes that accompany bereavement or clinical depression remains a primary and, as yet, unresolved concern. Because the trend in most of these studies has primarily shown decreased cellular immune function, one might expect that depressed patients would be more vulnerable to certain nonpyogenic infections or malignancies. Early clinical observations suggested such a connection, but large, elegant epidemiologic studies of patients with various psychiatric diagnoses have generally not corroborated an increased mortality from medical disease among such patients. For a metaanalysis, see reference 15.

SUMMARY

The observations and research described in this chapter represent a nontraditional view of the "immune system" but also a perspective, based on interactions between behavioral and physiologic events, that is not foreign to a neuropsychiatric audience. Using stress effects, clinical depression, and conditioning phenomena as illustrations, we have described recent research showing that psychological factors are, in fact, capable of influencing immune function. We have also reviewed data indicating that the immune system can receive and respond to neural and endocrine signals.

Although recent research raises basic questions about the autonomy of the immune system, the immune system is, indeed, capable of considerable self-regulation; immune responses can be made to take place in vitro. Ultimately, however, the adaptive significance of the immune system must refer to those immunologic reactions that occur in vivo. There are now compelling data to justify the view that in vivo immunoregulatory processes influence and are influenced by the neuroendocrine environment in which they take place. In turn, that environment generates signals that resting or activated leukocytes are capable of receiving. Immune responses appear to be modulated by feedback mechanisms mediated through neural and endocrine processes as well as by "feedforward" mechanisms. The immunologic effects of conditioning, an essential feedforward mechanism, suggest that, like neural and endocrine processes, there are circumstances under which behavior can serve an in vivo immunoregulatory function.

We do not yet understand the nature of all the pathways connecting the brain and the immune system or the functional significance of those neural and endocrine interrelationships that have been identified thus far.

ACKNOWLEDGMENTS

This review of the literature is an updated and expanded version of a paper published earlier. Preparation of this chapter was supported by a Research Scientist Award (KO-5 MH06318) from the National Institute of Mental Health (RA), and was completed during the senior author's tenure as a Fellow at the Center for Advanced Study in the Behavioral Sciences with support from the John D. and Catherine T. MacArthur Foundation (#8900078).

REFERENCES

1. Pelletier G. Anatomy of the hypothalamic-pituitary axis. In: Jasmin G, Cantin M, eds. *Stress revisited. 1. Neuroendocrinology of stress.* Vol 14. Basel, Switzerland: Karger, 1991:1–22.
2. Serby M, Richardson SB, Rypma B, et al. Somatostatin regulation of the CRF-ACTH-cortisol axis. *Biol Psychiatry* 1986;21:971–974.
3. Palkovits M. Peptidergic neurotransmitters in the endocrine hypothalamus. In: *Functional anatomy of the neuroendocrine hypothalamus.* Ciba Foundation Symposium No. 168. Chichester, NY: John Wiley & Sons, 1992:3–5.
4. Lesch KP, Rupprecht R. Psychoneuroendocrine research in depression. *J Neural Transm* 1989;75:179–194.
5. Newman MM, Halmi KA. The endocrinology of anorexia nervosa and bulimia nervosa. *Endocrinol Metab Clin North Am* 1988;17:195–211.
6. Hein MD, Jackson IMD. Review: thyroid function in psychiatric illness. *Gen Hosp Psychiatry* 1990;12: 232–244.
7. Martignoni E, Costa A, Sinforani E, et al. The brain as a target for adrenocortical steroids: cognitive implications. *Psychoneuroendocrinology* 1992;17:343–354.
8. Ader R, Felten DL, Cohen N, eds. *Psychoneuroimmunology.* 2nd ed. New York: Academic Press, 1991.
9. Ader R. Conditioned immunopharmacological effects in animals: implications for a conditioning model of pharmacotherapy. In: White L, Tursky B, Schwartz GE, eds. *Placebo: theory, research, and mechanisms.* New York: Guilford Press, 1985;306–323.
10. Felten DL, Cohen N, Ader R, Felten SY, Carlson SL, Roszman TL. Central neural circuits involved in neural-immune interactions. In: Ader R, Felten DL, Cohen N, eds. *Psychoneuroimmunology, 2/e.* New York: Academic Press, 1991:3–25.
11. Kelley KW. Growth hormone in immunobiology. In: Ader R, Felten DL, Cohen N, eds. *Psychoneuroimmunology, 2/e.* San Diego, CA: Academic Press, 1991: 377-402.
12. Kiecolt-Glaser JK, Glaser R. Psychological influences on immunity. *Psychosomatics* 1986;27:621–624.
13. Glaser R, Rice J, Sheridan J, et al. Stress-related immune suppression: health implications. *Brain Behav Immun* 1987;1:7–20.
14. Minden SL, Schiffer RB. Affective disorders in multiple sclerosis: review and recommendations for clinical research. *Arch Neurol* 1990;47:98–104.
15. Herbert TB, Cohen S. Depression and immunity: a meta-analytic review. *Psychol Bull* 1993;113:472–486.

8

Pain Mechanisms

Henry J. Ralston III

OVERVIEW

Pain is a fundamental, complex sensory perception that is the product of neural processing by multiple levels of the nervous system and, finally, by the cerebral cortex. Usually, the experience or report of pain is readily associated with stimuli that are potentially or actively damaging to tissue. Often, however, there is no demonstrable cause of the pain, but the individual nonetheless reports that pain is present, and it is incumbent on the therapist to attempt to diagnose and treat this pain just as that arising from a more obvious cause. There are objective aspects of pain, such as its location on or in the body and its intensity. There are also more subjective aspects, such as the memories and emotional responses with which pain is often associated.

Given these attributes of pain, it is not surprising that research on the neural networks, functional classes of neurons, neurotransmitters and their receptors, and changes in neuronal circuitry and gene expression associated with pain constitutes a complex and rapidly expanding series of topics. In this chapter, we present a summary of current knowledge concerning the neurobiologic substrate of pain-processing mechanisms. These descriptions of the neuroanatomic organization and physiology of neurons that respond to noxious stimuli are used to suggest mechanisms by which disease or injury that damages these neurons can perturb the perception of pain, leading to anesthetic or hyperpathic states. Throughout this chapter, the basic biology of pain is related to discussions of the design of rational therapies for its amelioration. For an overall review, see reference 1.

ANATOMY AND SYSTEMS ORGANIZATION

The sensation of pain is a product of the neural networks of the brain. Noxious mechanical, thermal, or chemical stimuli activate specific peripheral nerve endings of sensory neurons (nociceptors), resulting in trains of action potentials that travel to the spinal cord or brain stem, where they are synaptically transmitted to neurons that convey the information to higher centers of the neuraxis and ultimately to the cerebral cortex. At every level of the nervous system, from peripheral tissue to the cerebral cortex, there is the potential for modulation of the signals evoked by noxious stimuli. It is at these sites of modulation that much of the therapy designed to treat pain is targeted, either by enhancing naturally occurring suppression of noxious stimuli, such as biofeedback techniques, or by decreasing the activation of neurons that carry nociceptive information. The following discussion summarizes each of these neural levels with regard to their anatomy, physiology, and neurochemistry.

Primary Afferent Neurons

Sensory Ganglion Cells

The cell bodies of these sensory neurons are located in the dorsal root ganglia of the spinal cord or in the trigeminal ganglia. Sensory ganglion cells are unipolar in that they have a single process, an axon, that arises from the cell body and then bifurcates into a peripheral branch that runs in peripheral nerve to innervate cutaneous, deep, or visceral tissue, and a central branch (primary afferent axon) that carries information through dorsal or trigeminal roots into the central nervous system (CNS), to either the spinal cord or the brain stem, respectively (Fig. 8-1). As in all neurons, the cell body of sensory neurons is the metabolic center of the cell; all peptides and proteins, such as receptors (including opiate receptors), cytoskeletal precursors, enzymes, and peptide neurotransmitters, are synthesized in the cell body and then move by axonal transport within the central and peripheral axonal branches to their appropriate places within the neuron or neuronal membrane.

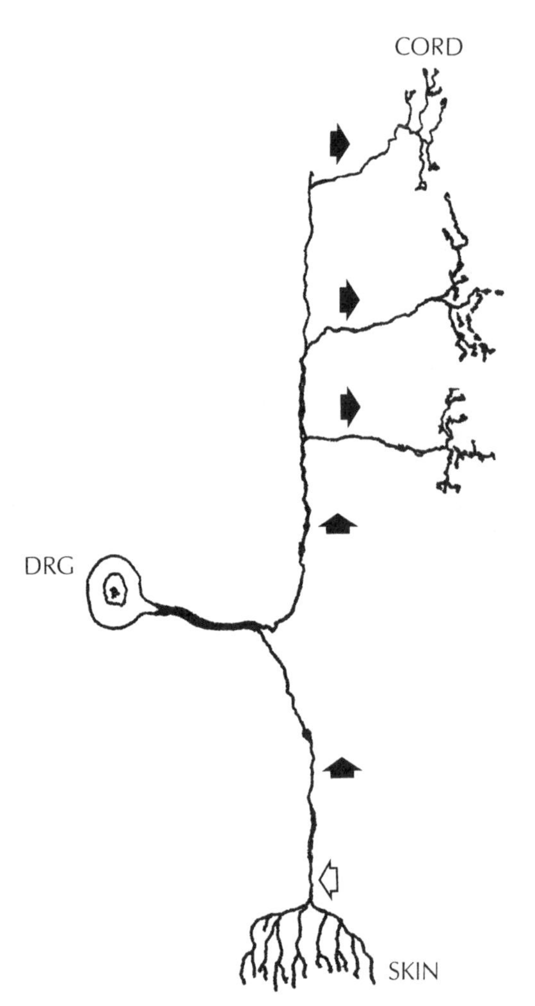

FIG. 8-1. Diagrammatic representation of a sensory neuron. The unipolar cell body is located in the dorsal root ganglion (DRG) or trigeminal ganglion. The peripheral axonal branch innervates tissue (e.g., skin) and responds to various mechanical, thermal, and chemical stimuli. The action potential is generated (*open arrow*) just proximal to the peripheral ending and travels in the direction of the closed arrows to enter the central nervous system (cord). (Modified from several drawings of Cajal SR. *Histologie du systeme nerveux de l'homme et de vertebres.* Vol 1. Paris: C.S.I.C., 1955.)

Sensory neurons are specialized cells in which the region of the axon where action potentials are initiated is immediately adjacent to the sensory endings in the tissue innervated (see Fig. 8-1), rather than at the origin of the axon from the cell body, as in most neurons. Sensory neurons that are activated by noxious stimuli are called primary afferent nociceptors (PANs). Some respond exclusively to noxious stimuli; others respond in a graded fashion as a stimulus becomes more intense (mechanical stimuli) or changes from innocuous to noxious (e.g., warm to hot). PANs usually have smaller-diameter cell bodies (10 to 15 μm) than cells activated exclusively by nonnoxious stimuli, and these cell bodies contain one or more neuropeptides, often coexisting with excitatory amino acid neurotransmitters, such as glutamate.

The best characterized neuropeptide in sensory ganglion cells (dorsal root ganglion cells [DRGs]) is the undecapeptide, substance P (SP), found in about 20% of DRGs. SP is one of the class of neuropeptides called neurokinins or tachykinins. Other peptides often coexist with SP in DRGs; for instance, calcitonin gene–related peptide is present in about 80% of SP-containing DRGs. The content of SP in DRGs can be depleted by exposure to capsaicin, a nonpeptide substance found in chili peppers that has an excitatory effect predominatey on PANs.

The axons of PANs are small in diameter: Those having a myelin sheath are usually 1 to 3 μm in size and are functionally classified as Aδ fibers; nonmyelinated axons are usually less than 1 μm in diameter and are classified as C fibers. Aδ fibers may respond to innocuous, noxious, or both classes of stimuli. Most

C fibers in primates are activated by a variety of noxious mechanical, thermal, and chemical stimuli and have thus been termed *C polymodal nociceptors* (2).

Peripheral Sensory Endings

The peripheral endings in tissue of PANs have been shown to arborize in the tissue innervated by them, leading to their being classified as free nerve endings. The peripheral terminals of PANs respond to appropriate mechanical, thermal, or chemical stimuli that evoke trains of action potentials that travel into the CNS. It is now recognized that these sensory endings are not merely the passive recipients of noxious stimuli, but rather can release neuropeptides (such as SP) into peripheral tissues that may contribute to the inflammatory response that follows injury. PANs exhibit increased excitability after repeated noxious stimulation; this excitability change is termed *sensitization* and is characterized by an increase in spontaneous activity accompanied by a lowered threshold for activation and increased firing to suprathreshold stimuli.

Sensitized PANs can contribute to hyperalgesia, such as experienced with tissue inflammation, in which a normally innocuous stimulus is perceived as painful, or in which noxious stimuli evoke heightened sensations of pain. The peripheral terminals of PANs may be directly sensitized by a variety of substances released by damaged tissue. Also, they may be indirectly sensitized by bradykinin, by mechanisms involving prostaglandin synthesis or postganglionic sympathetic axon terminals, or by the cytokine interleukin-1 produced by leukocytes after exposure to bacterial toxins or inflammatory mediators. Sensitized PANs have a lowered threshold and respond abnormally to innocuous or to noxious stimuli.

Dorsal Horn of the Spinal Cord

The spinal cord retains many of the organizational properties of the neural tube from which it developed. Surrounding the central canal (usually not patent in humans) is the aggregate of neurons that constitute the gray matter of the cord, and external to these neurons are the various pathways that make up the spinal cord white matter, owing to the myelin sheaths surrounding most CNS axons (Fig. 8-2). The gray matter is subdivided into bilateral symmetric dorsal and ventral horns, generally related to somatosensory and motor functions of the cord, respectively. The thoracic and sacral segments also contain preganglionic sympathetic and parasympathetic neurons. Dorsal horn cells can be divided into three broad classes: those that send axons to rostral targets in the brain stem or diencephalon (projection neurons), those that project to other segments of the spinal cord (propriospinal neurons), and those that serve the intrinsic excitatory and inhibitory circuitry of a given spinal segment (local circuit neurons or interneurons). Numerous local circuit neurons in the dorsal horns contain enkephalin or γ-aminobutyric acid (GABA) and are believed to participate in pain-modulatory mechanisms of the superficial dorsal horn.

Primary Afferent Axons

Primary afferent axons are the central branches of sensory neurons located in the dorsal root ganglia (or the trigeminal ganglia). They enter the spinal cord through the dorsal roots (or the brain stem through the trigeminal roots). As in peripheral nerves, PAN axons entering the CNS are either small-diameter myelinated fibers (Aδ) or nonmyelinated C fibers), reaching the dorsal horn through Lissauer's tract. The C fibers in dorsal roots terminate primarily in the superficial laminae of the dorsal horn and form a variety of axodendritic synaptic complexes with dorsal horn neurons. The larger diameter Aβ fibers, carrying nonnoxious information, terminate in the deeper laminae of the dorsal horn, as well as in the ventral horn. As would be expected from the distribution of the afferent terminals of PANs, the neurons of the dorsal horn that receive these afferents exhibit various re-

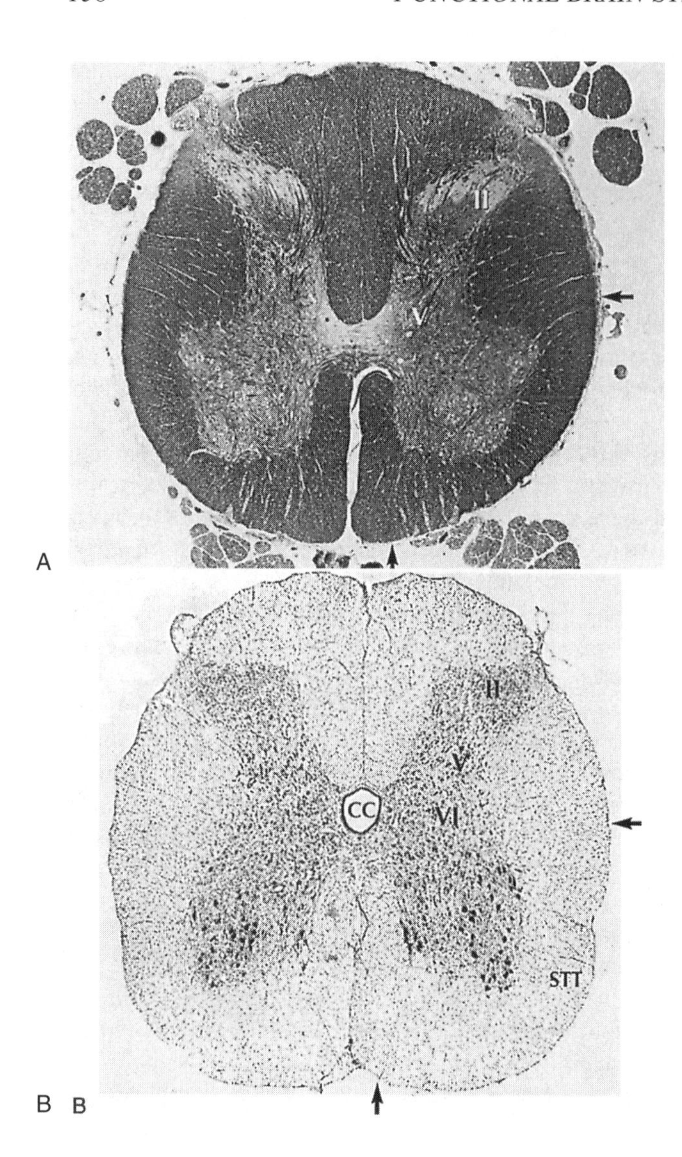

FIG. 8-2. A: Cross-section of human lumbar spinal cord, stained for myelin. The substantia gelatinosa (II) is relatively unstained. Lamina V (V) contains many spinothalamic tract cells. The arrows demarcate the approximate boundaries of the spinothalamic tract. **B:** Cross-section of macaque lumbar spinal cord, stained for neurons and glial cells (Nissl stain). Three of the dorsal horn laminae are indicated (II, V, and VI). Arrows indicate the approximate boundaries of the spinothalamic tract (STT). cc, central canal.

sponse properties after peripheral noxious and nonnoxious stimuli. Some cells in medial laminae of the dorsal horns exhibit convergent, or wide-dynamic-range (WDR) properties, responding to activation of Aβ, Aδ, and C fibers, after innocuous and noxious mechanical and thermal cutaneous stimuli. Many of these cells nearer the central canal of the spinal cord are also activated by a variety of stimuli arising from the viscera or from joints.

Neurons that project to higher centers, such as the midbrain (spinomesencephalic tract [SMT] cells) or the thalamus (spinothalamic tract [STT] cells) may be identified physiologically after antidromic activation by stimulation of the target of the axons, or anatomically by injecting the target region with a retrograde tracer that is transported back to the parent cell bodies. Using such techniques, it has been possible to identify SMT and STT projection neurons, their functional properties, and the classes of synaptic terminals that contact them.

Primary Afferent Neurotransmitters

Excitatory amino acids, particularly glutamate, numerous peptides, and adenosine triphosphate, have been identified in dorsal root axon terminals and play a role in the transmission of afferent stimuli to dorsal horn neurons. Most DRGs, and presumably most of their terminals in the dorsal horn, exhibit coexistent neurotransmitters, most often one or more neuropeptide and an excitatory amino acid. Glutamate-containing terminals have been identified synapsing on primate STT cells; glutamate evokes a fast, short-duration depolarization of STT neurons, and both glutamate and aspartate are released in the cord after activation of primary afferent axons conveying nonnoxious and noxious stimuli. Of the peptides, the neurokinins, particularly SP, have been the most studied. SP has been identified in the terminals of PANs. Initially, SP was considered to be the specific neurotransmitter associated with signaling noxious stimuli. Neurons in the dorsal horn can be activated, however, by peripheral noxious stimuli even when SP is depleted, so that it is now suggested that SP and other neuropeptides may serve some sort of neuromodulatory role, such as potentiating the spinal cord nociceptive neuronal responses to glutamate. SP may also interact with calcitonin gene–related peptide, which is found exclusively in small-diameter primary afferent axons. For additional information, see reference 3.

Opiate Mechanisms in the Dorsal Horn

The endogenous opiate enkephalin is present in neurons of the superficial dorsal horn and is liberated from axons and from dendrites (presynaptic dendrites) of these cells. The terminals of Aδ and C fibers exhibit the three known classes of opioid-binding receptors: μ, κ, and δ. Enkephalin can reduce the amount of SP released from primary afferent terminals. Morphine, acting on the μ-opioid–binding site, can also reduce SP release from PANs, and this action of morphine can be reversed by the opiate antagonist naloxone. Other opiate ligands, acting at μ- and δ-binding sites, can similarly reduce the release of neuropeptides and of excitatory amino acids from primary afferent axons. There are also opioid-binding sites on nociceptive dorsal horn neurons, including STT cells. The application of μ- and δ-opiate ligands can inhibit the responses of dorsal horn neurons to noxious stimulation. Thus, it is likely that the antinociceptive actions of opiates applied directly to the spinal cord are a result of the reduction of the release of neurotransmitters from PANs, as well as the direct inhibition of the activation of dorsal horn neurons following peripheral noxious stimulation. Both these types of analgesia are naloxone reversible.

Opioid-binding sites can exhibit plastic changes. Following lesions of primary afferent axons produced by dorsal rhizotomy, there is a substantial decrease in μ- and δ-opioid–binding sites, presumably resulting from degeneration of the axon terminals on which the binding sites are located. Within a few weeks, however, there is a gradual recovery of the binding sites, believed to be associated with sprouting of adjacent, noninjured primary afferent axons into the deafferentation zone. There are also plastic changes in dorsal horn gene expression. For instance, the mRNAs derived from the preprotachykinin gene that encode for the family of tachykinin peptides that includes SP are substantially increased in projection neurons of the rat dorsal horn after experimentally induced inflammation. These changes in neurotransmitters or their associated receptors may underlie the appearance of chronic pain states seen in many cases of peripheral nerve or dorsal root injuries (4).

Spinothalamic Tract Neurons

The dorsal horn projection neurons of the primate that have been the most extensively studied are those that project to the thalamus, termed *keep STT neurons.* As described previously, STT cells can be physiologically and anatomically characterized. Those in superficial laminae of the dorsal horns receive Aδ- or C-fiber afferents and primarily signal noxious cutaneous stimuli; STT neurons in more central laminae exhibit WDR convergent properties in that they receive input from Aβ, Aδ, and C afferents and respond to a variety of innocuous and noxious peripheral stimuli. Many of these more central STT neurons are activated in a graded fashion by peripheral mechanical or thermal stimuli in that the size of the evoked excitatory postsynaptic potentials and the frequency of firing of the STT cells increase as the intensity of the stimuli increases, moving from the nonnoxious to the noxious range.

Modulation of Pain Transmission in the Spinal Cord

It has long been known that one form of stimulus can modify the perceived quality and intensity of another. In the case of noxious stimulation, the sense of pain can be ameliorated by activation of large-diameter primary afferent fibers of the cord, a finding that led to the "gate-control" theory of pain modulation, which states that inhibition of nociceptive STT cells by impulses in large-diameter afferent axons in dorsal roots or their collaterals in the dorsal column white matter "closes the gate" by means of inhibitory interneurons, and the impulses carried by fine-diameter afferent leads to excitation of STT cells thus "open the gate" of pain transmission. Indeed, transcutaneous electrical nerve stimulation and dorsal column stimulation modulates pain transmission similarly, following the stimulation of large-diameter afferent axons. In addition, the intensity of cold-induced painful stimuli conveyed by C fibers can be suppressed by myelinated primary afferent fiber input. The gate-control hypothesis has been invaluable in generating investigations of a wide variety of modulatory phenomena in the CNS, several of which demonstrate substantial roles for descending pain modulatory systems that arise in the brain stem and that are activated by noxious stimuli and follow a specific pathway in the spinal cord to terminate in the dorsal horn. Analgesia produced by microstimulation of the periaqueductal gray (PAG) matter of the midbrain, or microinjection of morphine into PAG, has been shown to involve a descending system from PAG to medulla to the dorsolateral funiculus of the spinal cord and thence to the dorsal horn. See reference 5 for an early review of these current concepts.

Alterations of primate STT cell activity have also been shown after stimulation of cerebral cortical projections to the spinal cord dorsal horn. PAG and cortical stimulation both evoke inhibitory postsynaptic potentials in STT cells and selectively inhibit the responses of Lamina V Spinothalamic tract (VSTT) neurons activated by small-diameter afferent (noxious) input, but not those responding to large-diameter afferent (nonnoxious) fibers. These findings suggest that PAG and cortical projections to STT neurons may selectively control STT cell activation by afferent inputs, permitting switching of the coding properties of the cells according to varying behavioral states.

The appearance of the immediate early gene (IEG) has been used to assess the neuronal populations of the cord that are affected by the administration of opiates to produce behavioral analgesia. For instance, acute inflammation is associated with the appearance of fos protein in neurons of the superficial and deeper layers of the dorsal horn, and analgesia produced by the administration of systemic morphine reverses the expression of the fos protein in a dose-dependent, naloxone-reversible manner. The intracerebroventricular administration of a μ-opioid ligand sufficient to produce behavioral analgesia reduces fos protein expression to a greater degree in the deeper layers of the dorsal horn (the WDR cells) than in the superficial layers, chiefly populated by noxious-specific neurons, suggesting different roles for WDR and noxious-specific neurons in behavioral manifestations of pain. Furthermore, bilateral lesions of the spinal dorsolateral funiculus abolish the antinociceptive effects of the opiate, as well as the suppression of fos protein, providing additional evidence for the supraspinal origins of descending pain-control systems and the spinal pathway taken by the control system.

Much of what has been described also pertains to noxious mechanisms that involve the trigeminal system in that fine-diameter trigeminal afferents convey noxious information to the trigeminal nuclear complex of the brain stem, and these messages are synaptically transmitted to particular sets of neurons within the trigeminal nuclei using the same types of neurotransmitters as do dorsal root afferents. The activities of trigeminal nociceptive neurons can be modulated by processes similar to those that modulate spinal nociceptive cells.

There is thus abundant evidence that particular populations of neurons convey noxious information from the periphery to the CNS and that specialized neurons within the CNS that receive this information send it to higher centers. At every stage of this system, excitation of neuronal networks by noxious stimuli can set in motion a variety of modulatory mechanisms to facilitate or inhibit the activities of cells that transmit the information that the brain may interpret as being painful.

Ascending Pathways Conveying Noxious Stimuli

Although ascending pathways that carry noxious information are traditionally called *pain pathways,* the point once again is that the sensory experience of pain is a product of neural networks of the forebrain. Trains of stimuli ascending from the cord or trigeminal nuclei may or may not be interpreted as painful, as evident from the previous discussion of modulation of sensory processes.

This section describes four ascending systems that are concerned with pain mechanisms. The STT conveys information that is ultimately transmitted to the somatosensory cerebral cortex. Regions of the cerebral cortex other than the somatosensory cortex receive the information carried by the other three ascending systems, the SMT, the spinoparabrachial (SPB), and the spinohypothalamic (SHT) tracts. It is believed that these latter pathways are concerned with the emo–tional, memory, and autonomic mechanisms associated with pain.

Spinothalamic Tract

The best-known ascending pathway, the STT (and its homolog, the trigeminothalamic pathway) conveys somatotropically organized information arising from peripheral noxious stimuli to several thalamic nuclei. It is believed that the STT, particularly the component that terminates in the lateral thalamus (the ventrobasal and posterior nuclear complexes) is associated with the sensory discriminative aspects of pain: the peripheral location and the intensity of the noxious stimuli.

The cells of origin of the STT are found at all levels of the spinal cord. In primates, they are particularly numerous at high cervical segments of the cord. Much of the information about noxious stimuli is carried to the spinal cord by nonmyelinated primary afferents (C fibers). The axons of STT cells that carry this information in the CNS are myelinated, as indicated by electron microscopic and electrophysiologic studies.

In primates, including humans, the axons of most (greater than 90%) STT neurons cross the midline to travel in the ALQ, a zone demarcated by the denticulate ligament laterally and the anterior median fissure medially (see Fig. 8-2). The STT is intermixed with several other ascending and descending pathways of the spinal cord white matter that also lie in the ALQ of the cord. The axons of STT cells that carry noxious information are widely dispersed throughout the ALQ, but there is a segregation determined by laminar origin of the STT axons.

The terminations of the primate STT are found in several nuclei of the lateral and medial thalamus. The cortical projections from these thalamic nuclei are widespread and include parietal sensory cortex, as well as orbitofrontal cortex.

The physiologic responses of neurons in primate ventral posterior lateral (VPL) nucleus are in keeping with the anatomic distribution of STT axons, which are distributed to subsets of VPL neurons, whereas those from the dorsal column nuclei are found throughout VPL. Other VPL neurons respond only to innocuous peripheral stimuli and are driven exclusively by stimuli conveyed by large-diameter primary afferent axons, relayed in the dorsal column nuclei.

Recordings and microstimulation studies of thalamus in awake humans undergoing stereotaxic neurosurgical procedures, usually for the treatment of movement disorders, have revealed neuronal responses similar to those described in monkeys. In both cases, most neurons that respond to noxious stimuli, such as noxious heat, are found more ventrally and caudally in the thalamus, rather than in the central "cutaneous core" zone of the somatosensory thalamus. Interestingly, electrical stimulation of the human VPL does not usually elicit reports of pain, but stimulation ventral to VPL in a region analogous to the monkey posterior-SG and VPI leads the subject to complain of burning pain in a region of the body or face contralateral to the side of the thalamus being stimulated. Recordings from patients having pain of central origin, usually as a result of a stroke involving the diencephalon, have revealed abnormal burst firing of VPL neurons after innocuous and noxious peripheral stimuli.

After injury to the CNS, some patients have burning pain referred to a particular body part, a condition called *central poststroke pain,* a term replacing the previous, less accurate appellation, thalamic pain syndrome. Such patients are most likely to demonstrate a central lesion that involves the brain stem or spinal cord, the diencephalon (including the thalamus), or the cerebral hemispheres. They typically show a diminished perception of pinprick or thermal stimuli, sensory modalities carried by the spinothalamic tract that are then relayed by the thalamus to the cortex. These central pain states are probably a result of deafferentation caused by the interruption of the central pain pathways at any level of the neuraxis. The mechanisms underlying central poststroke pain are unknown but presumably involve some sort of reorganization of central somatosensory circuits or changes in the expression of neurotransmitters and their receptors that participate in the central pain pathways.

Spinomesencephalic Tract

In addition to their projections to the thalamus, spinal cord neurons that receive convergent input from cutaneous, muscle, joint, and visceral sources and that respond to noxious stimuli have been found to project to the midbrain, some sending collateral branches to both thalamic and mesencephalic targets. Studies of rat, cat, and monkey reveal that the major terminal fields in the midbrain of spinal afferents include the PAG, the nucleus cuneiformis (an important cholinergic nucleus of the brain stem), deeper layers of the superior colliculus that participate in visuomotor integrative mechanisms, and the intercollicular nuclei. Electrical stimulation of the region of PAG that receives SMT input results in inhibition of the responses of spinal nociceptive neurons, and microinjection of similar regions of the PAG with morphine results in analgesia. Thus, the SMT may contribute the ascending, afferent limb to this descending pain modulation system. The other midbrain targets of the SMT are known to participate in a variety of motor and autonomic mechanism associated with nocifensive behaviors, such as fear, flight, rage, and cardiovascular changes. Thus, the SMT contributes ascending, primarily noxious information from all types of peripheral tissues to midbrain systems involved in pain modulation and somatic motor and visceromotor responses.

Spinoparabrachial Tract

The parabrachial nucleus is found bilaterally, surrounding the superior cerebellar peduncles as they exit from the cerebellum at the dorsal junction of the pons and midbrain. Many parabrachial nucleus neurons in the rat receive afferent projections from the spinal cord; in turn, these neurons project to the hypothalamus and to the amygdaloid nuclear complex of the limbic forebrain. About two thirds of these parabrachial nucleus projection cells respond exclusively to peripheral noxious stimuli and serve to relay nociceptive information from the spinal cord and trigeminal nuclear complex to the amygdaloid nuclei.

Projections from spinal cord and trigeminal neurons of cat and monkey, presumably nociceptive in function, have also been shown to terminate in the locus ceruleus, a pigmented cell group that lies just medial to the parabrachial nucleus. The locus ceruleus is the principal source of norepinephrine for the brain and spinal cord. The noradrenergic projections from the locus ceruleus to the thalamus are believed to participate in arousal mechanisms, and locus ceruleus fibers to the spinal cord serve descending control functions that modulate the activities of spinal neurons responding to noxious stimuli. Thus, the nociceptive afferent fibers from cord to locus ceruleus may activate arousal, as well as the descending pain-modulatory circuitry of the spinal cord.

CEREBRAL CORTEX AND PAIN

The somatosensory cortex receives topographically organized projections from those thalamic regions, principally the ventrobasal complex, that receive ascending pathways conveying noxious and nonnoxious information from superficial and deep structures. Nonetheless, investigators have had great difficulty in demonstrating a particular role of the cerebral cortex in pain mechanisms in experimental animals and in humans. In their famous book reporting on the results of electrical stimulation of the exposed cerebral cortex of conscious humans, Penfield and Rasmussen did not even include the term *pain* in the index because their patients rarely reported painful sensations after stimulation of their cortices (6). During the past decade, studies in animals and humans have demonstrated conclusively that cortical neurons can be activated by peripheral noxious stimuli and that these neuronal responses can be correlated with the localization and intensity of the stimuli.

Humans are able to localize precisely the position of focal pinprick stimuli, which is almost as accurate as the ability to localize the position of innocuous tactile stimuli, indicating the presence of a highly ordered topographic representation for the neurons responding to noxious stimuli. We assume that this region must be the cerebral cortex. Cortical areas that are responsive to somatic noxious stimuli, as indicated by increased regional blood flow and by reports of the degree of pain by subjects receiving noxious thermal stimuli, are indicated in Fig. 8-3.

A recent review indicates that patients with large lesions of the cerebral cortex can still perceive painful stimuli to the contralateral body, in contrast to patients with thalamic lesions, who have a marked decrement in pain sensation. Such clinical findings have led to the view that the sensation of pain is a function of the thalamus, not the cortex. Pain is a complex sensory and emotional experience. We have much more to learn before we can understand the forebrain mechanisms that serve it. For an intersting historical review, see reference 6. For a more current review, see reference 7.

CLINICAL ISSUES

Referred Pain

Numerous neurons in the deeper spinal gray matter exhibit visceral and somatic convergent properties in that they respond to stimuli applied to both types of tissue. These cells are believed to play an essential role in referred pain. Referred pain of visceral origin is often perceived as arising from somatic tissues that are innervated by the same levels of the cord as is the visceral structure. Visceral pain is usually referred to the torso or to proximal regions of the limbs, but not often to the distal regions of the extremities. It is usually perceived as being of deep, rather than of cutaneous, origin. The phenomenon of counterirritation, in which noxious stimuli applied to one region of the body can reduce the sensation of pain arising from another somatic structure, has also been described for pain referred from visceral structures in that noxious cutaneous stimuli can inhibit neuronal and reflex responses to noxious visceral stimuli.

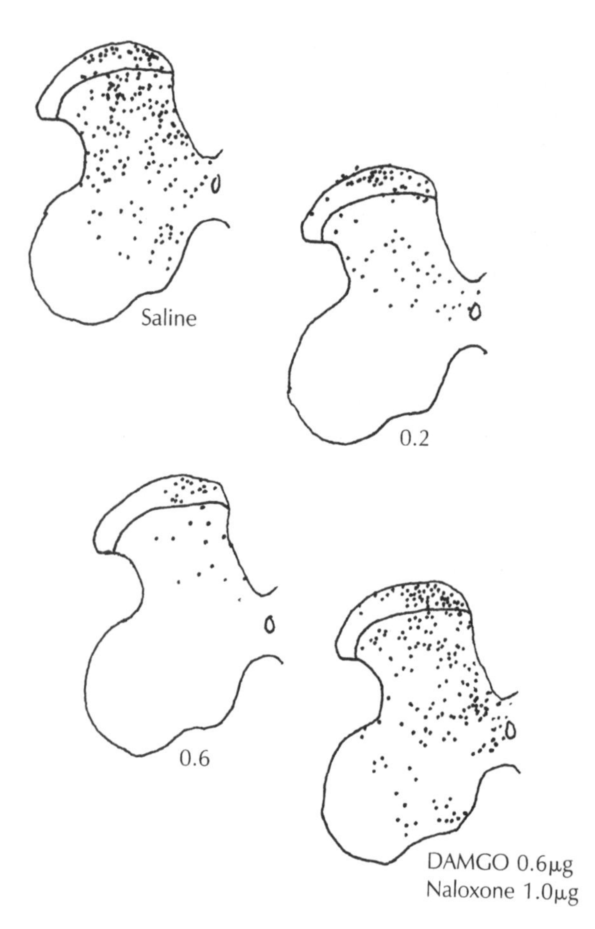

FIG. 8-3. Diagrams of the distribution of neurons of the rat spinal cord that exhibit cfos-like immunoreactivity as a result of a noxious stimulus to the paw. Saline control, in which saline was injected into the cerebral ventricles (i.c.v.), 0.2 and 0.6 show the dose-dependent distribution of cfos cells following i.c.v. injection of μg amounts of the μ opiate agonist Dala, N MePhe, Glyd enkephalin (DAMGO). The final drawing demonstrates the reversibility of the DAMGO effect by the i.c.v. injection of naloxone. (Modified from Gogas KR, Presley RW, Levine JD, Busbaum AI. The antinociceptive action of supraspinal opioids results in descending inhibitory control: correlation of nociceptive behavior and cfos expression. *Neuroscience* 1991;42:617–628.)

Surgical Procedures

A surgical procedure to treat chronic, intractable pain, usually in patients with intraabdominal or pelvic cancer, is the transection of the ALQ of the spinal cord white matter, which contains the axons of the STT that arise from dorsal horn neurons on the contralateral side of the cord (see previous discussion). The ALQ contains many other ascending and descending pathways, so that the surgically induced lesion is not only of STT axons. This procedure, spinal tractotomy, usually results in relief of pain caudal to the spinal segment of the lesion, and on the side contralateral to the lesion, because the STT axons cross the midline before they ascend in the contralateral ALQ. Patients receiving tractotomies usually survive only a few months after the surgical lesion because of the nature of their disease. In those surviving for longer periods, the pain often returns, perhaps because of the presence of other ascending systems that are not interrupted by the lesion. Tractotomy may actually cause new pain in these patients as a result of central deafferentation that is seen in central poststroke pain. Surgical lesions intended to treat pain must be used with care because damage to nerves or to central pathways that convey noxious information may actually lead to pain that is even more difficult to manage than the original pain.

CONTEMPORARY ISSUES: IMMEDIATE EARLY GENES IN THE DORSAL HORN

There are a group of IEGs whose expression and protein products have served as useful markers for dorsal horn neurons that are activated by noxious and by other stimuli. For instance, the protein product of the cfos protooncogene can be used to identify populations of dorsal horn neurons activated by peripheral noxious stimuli and to establish the relationship of the appearance of the fos protein to pain-related behavior. The changes in fos protein synthesis as a result of the administration of various opiate agonists and whether the effects of opiates on IEG protein synthesis are naloxone reversible have also been studied. The cfos protein product appears in dorsal horn neurons after visceral as well as somatic stimulation (Fig. 8-4). There is a differential time course and varying ex-

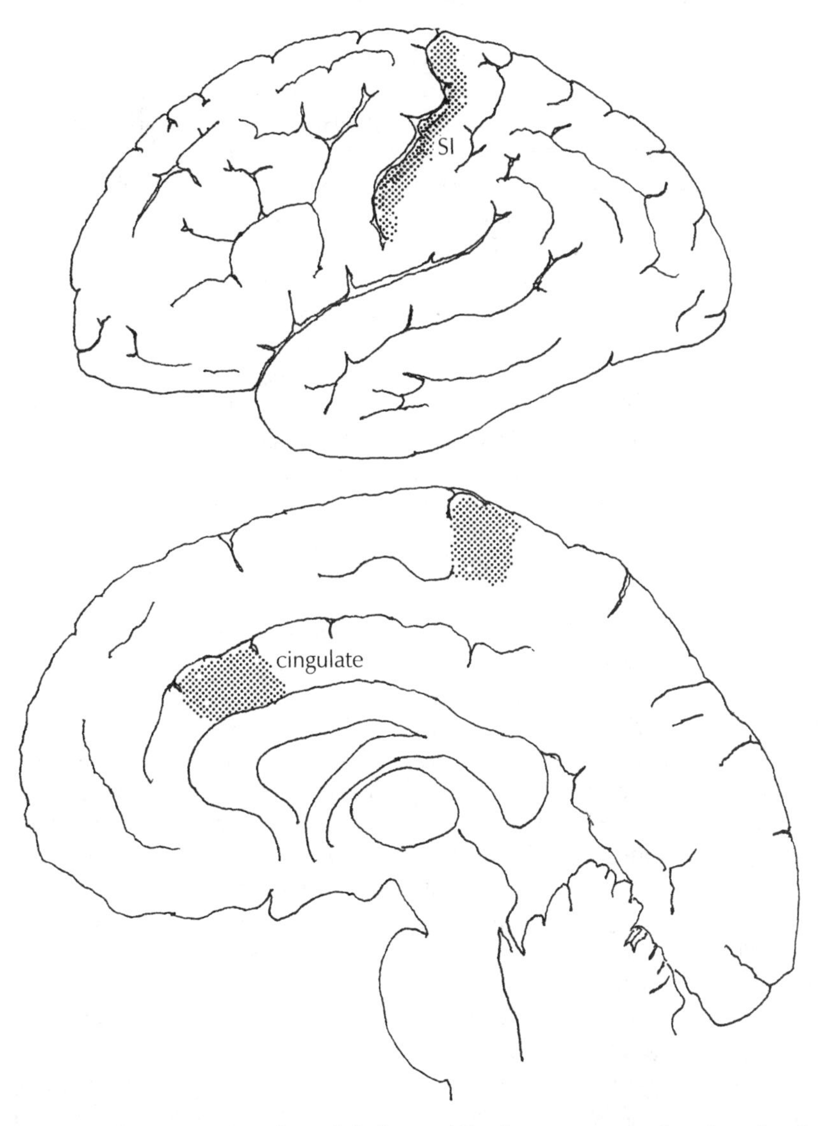

FIG. 8-4. Diagrams of the lateral and medial views of the human cerebral cortex showing regions that are activated (*stipples*) by peripheral noxious stimuli. Primary somatosensory cortex (SI); cingulate cortex. The second somatosensory cortex, SII, is located in the retroinsular area, within the sylvian fissure.

pression of a variety of IEGs as a result of subacute or chronic somatic inflammation. Although it is not known why these particular genes in certain regions of the CNS have altered expression as a result of peripheral stimuli, they have formed the basis of numerous investigations correlating structural, functional, and behavioral aspects of pain mechanisms (8,9).

PHARMACOLOGY AND SUMMARY

This chapter has described the neuronal processes and pathways that transmit information arising from noxious stimuli that ultimately are interpreted by the brain as being painful. This information can be modulated at several steps of the pathway, beginning with the peripheral sensory endings of PANs. Nonsteroidal antiinflammatory drugs, such as aspirin, inhibit the enzymatic action of cyclooxygenase, which participates in the breakdown of arachidonic acid to prostaglandins, which, in turn, leads to the inflammatory response and the sensitization of PANs. Opiates, acting principally at μ-opiate receptor sites on the sensory endings of PANs, can inhibit the hyperalgesia resulting from tissue damage and the consequent production of prostaglandin E_2.

In the spinal cord, the central synaptic terminations of PANs release SP, other neuropeptides, and excitatory amino acids. Opioids applied directly to the spinal cord (intrathecally) block the behavioral responses to noxious stimuli in experimental animals and have a powerful analgesic effect in humans, probably by inhibiting the release of SP from primary afferent terminals. The application of $GABA_A$-receptor agonists to the cord can have similar analgesic effects through the inhibitory effects of these drugs on spinal cord neurons. The activity of spinothalamic tract neurons that convey noxious information to the thalamus can be modulated by descending projections from the brain stem that use serotonin or norepinephrine as neurotransmitters to activate intrinsic neurons of the dorsal horn that contain endogenous opiate-like substances, for example, enkephalin. The descending projections have been shown to reduce the firing of dorsal horn neurons responding to noxious input, and this system is believed to be activated normally by noxious stimuli, so that animals and humans have endogenous pain control systems.

This descending control system can be accessed artificially by placing stimulating microelectrodes near the neuronal origins of the descending system, such as in the PAG of the midbrain. Such stimulus-induced analgesia was first examined in rats and has now been used therapeutically in several thousand humans, usually relatively young patients suffering from chronic intractable pain that has not responded to other forms of treatment.

Most analgesic drugs, whether opioids or nonsteroidal antiinflammatory drugs, act primarily in the periphery, cord, or brain stem. Other drugs that have a role in pain therapy are not usually considered analgesics and act on the forebrain when used for their primary purpose. Tricyclic antidepressants are examples of such agents and have been shown to be useful in a variety of painful conditions, including postherpetic neuralgia, diabetic neuropathy, and migraine headache. Animal models have shown analgesic responses to the administration of tricyclic antidepressants, and their mechanisms of action appears to be the prolongation of the actions of biogenic amine transmitters by inhibiting their reuptake into nerve terminals.

The descending modulatory systems from brain stem to cord use serotonin or norepinephrine as neurotransmitters to inhibit pain-transmitting neurons of the dorsal horn. Prolonging the activity of these amines would be expected to enhance the analgesic effects of these descending systems. Thus, the presumed sites of action in the spinal cord of tricyclic antidepressants in producing analgesia appear to be distinct from their sites of action in the forebrain associated with their use in treating depression.

The precise role of the forebrain in the perception of pain remains to be determined. Presently, drugs used to treat pain all appear to have their primary sites of action at levels caudal to the forebrain. We can expect that as the new functional imaging techniques of magnetic resonance imaging continue to improve, we will gain further insights into the forebrain regions active during pain sensation, which might then yield new approaches to pain treatment targeted on forebrain activity associated with pain.

REFERENCES

1. Willis WD, Coggeshall RE. *Sensory mechanisms of the spinal cord.* 2nd ed. New York: Plenum, 1991.
2. Ferrington DG, Sorkin LS, Willis WD. Responses of spinothalamic tract cells in the superficial dorsal horn of the primate lumber spinal cord. *J Physiol* (Lond) 1987;388:681–703.
3. Levine JD, Fields HL, Basbaum AI. Peptides and the primary afferent nociceptor. *J Neurosci* 1993;13: 2273–2286.
4. Heinricher MM, Morgan MM, Fields HL. Direct and indirect actions of morphine on medullary neurons that modulate nociception. *Neuroscience* 1992;48:533–543.
5. Wall PD. The gate control theory of pain mechanisms: a reexamination and restatement. *Brain* 1978;101:1–18.
6. Penfield W, Rasmussen T. *The cerebral cortex of man.* New York: Macmillan, 1950.
7. Roland P. Cortical representation of pain. *Trends Neurosci* 1992;15:3–5.
8. Noguchi K, Ruda MA. Gene regulation in an ascending nociceptive pathway: inflammation-induced increase in preprotachykinin mRNA in rat lamina I spinal projection neurons. *J Neurosci* 1992;12:2563–2572.
9. Gogas KR, Presley RW, Levine JD, Basbaum AI. The antinociceptive action of supraspinal opioids results from an increase in descending inhibitory control: correlation of nociceptive behavior and cfos expression. *Neuroscience* 1991;42:617–628.

9

Memory

Dean C. Delis, John A. Lucas, and Michael D. Kopelman

OVERVIEW

This chapter describes several aspects of learning and memory. We begin by reviewing the neuronal bases of learning and memory, and then we discuss some of the more common definitions of memory constructs and the neuroanatomic substrates underlying memory functioning. Next, we review a number of clinical neuropsychological techniques used to assess learning and memory and conclude with a discussion of memory disorders typically seen in clinical populations and normal aging.

NEUROBIOLOGY OF MEMORY

Most neurobiologists view memory as a special case of the general phenomenon known as *neuronal plasticity,* that is, the ability of neurons to change their structure or function in a lasting way. Early investigators hypothesized that memory reflected modification of the existing structure of the nervous system either through growth or change (1). This basic premise formed the foundation of numerous theories of memory functioning that have appeared during the past half century. Today, there is a consensus that the neuronal changes associated with learning and memory occur at the level of the synapse. Investigations of the biochemical aspects of memory focus on the identification of those biochemical processes that may underlie synaptic modification or changes in neuronal connectivity.

Memory as Change of Synaptic Efficacy

Hebb proposed that memory is reflected in structural changes of the neuronal synapses (2). Animals trained to perform specific tasks or exposed to enriched environments not only develop new synapses but also demonstrate changes in existing synapses. These changes include postsynaptic thickening and increases in the number of presynaptic vesicles, dendritic branching, and density of dendritic spines. Such changes have been observed in both young and adult animals, and they can occur subsequent to a single learning experience. This raises the question of where in the brain these changes occur. It is not likely that every neuronal synapse is modified by every experience, nor does it seem likely that each experience is represented by a change to only one corresponding neuron. Damage to any of a number of distinct brain regions can affect memory functioning; however, because memory is not a unitary construct, the type of difficulties experienced subsequent to such damage can vary, depending on the site of involvement. Consequently, before we address the question of which neuroanatomic structures underlie memory, we must first define the various constructs that comprise memory functioning.

CONCEPTUAL DIVISIONS OF MEMORY

Perhaps more than any other cognitive domain, memory has inspired researchers to hypothesize numerous conceptual divisions. This section describes several of the more widely accepted conceptual divisions of memory.

Short-Term Versus Long-Term Memory

In 1890, William James distinguished between memory that endured for a very brief time and memory that lasted after the experience had been "dropped from consciousness." The former, known as *short-term memory* (or primary memory), generally refers to recall of material immediately after it is presented or during uninterrupted rehearsal. The latter, known as *long-term memory* (or secondary memory), refers to the ability to remember information after a delay interval, during which the individual's attention is focused away from the target information. Short-term memory is thought to be of limited capacity, holding an average of seven "bits" of information at any one time. This information may be retained for up to several minutes but is lost or replaced by new information if it is not sustained by rehearsal. In contrast, long-term memory is believed to have an extraordinarily large capacity, with the potential for holding information indefinitely without the need for continued rehearsal.

The clinical significance of the distinction between short- and long-term memory is best exemplified in cases of amnesia. In perhaps the most famous case study in the neuropsychological literature, a patient, H.M., underwent surgical resection of the medial temporal lobes for treatment of intractable epilepsy. The surgery successfully treated the seizure disorder; however, H.M. was left with profound deficits in new learning and memory. One striking finding was that H.M. could repeat information immediately after presentation and hold new information by means of active rehearsal, but he would rapidly forget that information over time or with distraction. In other words, he demonstrated intact short-term memory in the presence of severely impaired long-term memory.

Encoding Versus Retrieval

Encoding refers to the process by which information is transformed into a stored, mental representation, whereas *retrieval* is the process of bringing the stored memory back into consciousness. Patients with brain damage can vary considerably in terms of whether their memory problems are at the level of encoding or retrieval. One common way to illustrate this distinction is to present new information, such as a story or word list, and to compare the patient's memory for that information using both a free recall and recognition paradigm. If information has been encoded successfully but cannot be retrieved, performance is poor on free recall, which places maximal demands on retrieval. In contrast, patients whose deficits lie in the encoding of information perform equally poorly on free recall and recognition testing.

Retroactive Versus Proactive Interference

Events or information encountered before or after the presentation of information to be remembered can interfere with recall of the target information. The two mechanisms fundamental to such memory failure are retroactive and proactive interference. *Retroactive interference* refers to the disrupting effect that later learning has on the ability to recall previously learned information. The degree of retroactive interference is a function of the similarity between the target and interference information. The more similar the interference information is to the target information, the greater the degree of interference and the poorer the recall of the target stimuli.

Proactive interference refers to the situation in which earlier learning interferes with the subject's ability to learn new information at a later time. As with retroactive interference, studies show that the degree of proactive interference in recall is dependent on the degree of similarity between the target items and the interfering information. The more similar the interference items are to the target items and the more times these items are presented, the greater the interference and the poorer the recall of the target items. Conversely, recall can be *facilitated* by making target items dissimilar to the interference information.

Anterograde Versus Retrograde Amnesia

Patients with memory disorders have been studied extensively, and two dissociable disturbances of memory have been observed. *Anterograde amnesia* refers to the inability to recall or recognize new information or events that have been encountered since the onset of the amnesic condition. In contrast, *retrograde amnesia* refers to the inability to recall or recognize information or events that were encountered before the onset of amnesia. Evidence of both anterograde and retrograde deficits are typically seen in all amnesic patients; however, the relative severity of each type of deficit may differ from patient to patient.

Anterograde amnesia typically affects a wide variety of new learning. Observing H.M., Milner wrote: "He could no longer recognize the hospital staff, apart from Dr. Scoville himself, whom he had known for many years; he did not remember and could not relearn the way to the bathroom, and he seemed to retain nothing of the day-to-day happenings in the hospital. . . " (3). On formal testing, H.M. demonstrated severely impaired memory for new verbal and nonverbal information, such as stories, faces, figures, or sequences beyond his immediate memory span. His intellectual functioning, however, remained within normal limits for his age.

Patients with retrograde amnesia are unable to remember events that occurred before the onset of the amnesic syndrome. In most cases, however, the ability to recall factual information not related to specific contexts or experiences remains fairly intact. For example, patients with true retrograde amnesia typically retain previously acquired factual knowledge, such as the name of the first president of the United States or the state capitals, as well as salient personal facts, such as their name. They may also retain basic, albeit vague, recollections of their career, marital status, and family. Patients who suffer complete loss of personal history and identity are rare and typically reflect psychological, rather than neurologic, etiologies.

Recent Versus Remote Memory

The distinction between recent and remote memory is typically applied to the temporal dimension of retrograde amnesia. *Recent memory* most often refers to the information acquired just before the onset of an amnesic syndrome, whereas *remote memory* refers to information regarding events or experiences acquired years or decades before the amnesia began. Patients with retrograde amnesia often demonstrate a temporal gradient in which memory for more recent events is disrupted to a greater extent than memory for remote events. This gradient is illustrated clearly by the case study of patient P.Z., a distinguished scientist who became amnesic secondary to alcoholic Korsakoff's syndrome (4). Several years before the onset of his amnesic syndrome, P.Z. completed an autobiography. This allowed the investigators to assess the patient's ability to recall past events that he had been able to recall premorbidly. The percentage of correct recall of autobiographic questions as a function of the decades during which the events occurred is presented in Fig. 9-1. These data illustrate a clear temporal gradient, with memory for recent experiences and events substantially worse than memory for events from his early life.

Declarative Versus Nondeclarative Memory

One of the most important insights to emerge from modern neuropsychological research is the distinction between declarative and nondeclarative memory, also known as explicit and implicit memory. *Declarative memory* refers to the acquisition of facts, experiences, and information about events. It is memory that is directly accessible to conscious awareness and can be declared. In contrast, *nondeclarative memory* refers to various forms of memory that are not directly accessible to consciousness. These include skill and habit learning, classic conditioning, the phenomenon of priming, and other situations in which memory is expressed through performance rather than recollection. Declarative memory is relatively fast and flexible. Fact-based information, for example, can usually be expressed quickly by a number of different

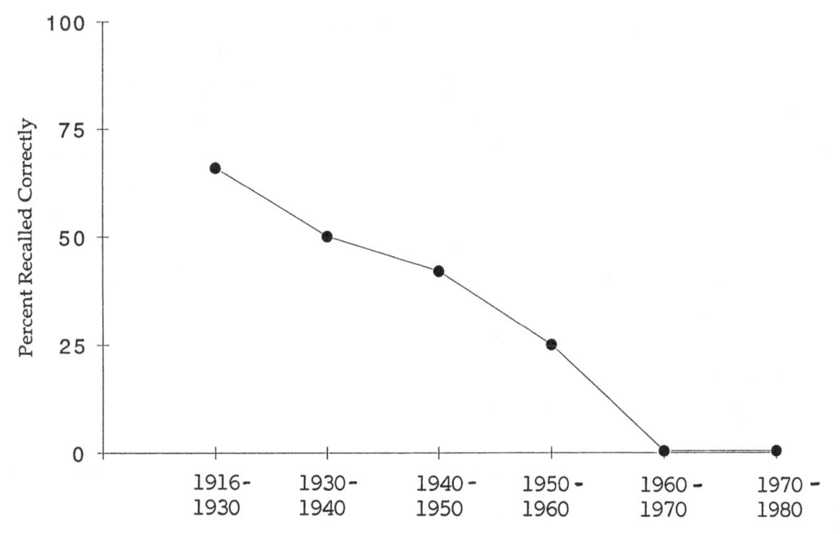

FIG. 9-1. Temporal gradient of retrograde amnesia for autobiographic information in a patient with alcoholic Korsakoff's syndrome (P.Z.). Information from earlier decades is recalled better than that from more recent decades. (From ref. 4, with permission.)

response systems. Declarative memory, however, is not always reliable, as is evident in everyday problems with retrieval of information and forgetting. In contrast, nondeclarative memory is considered quite reliable but is often slow and inflexible. The information present in a learned skill, for example, can often be expressed most readily only by the response systems that were involved in the original learning of that skill.

Episodic Versus Semantic Memory

Declarative knowledge can be divided into episodic and semantic memory. *Episodic memory* refers to information learned at a particular place and time in one's life. Asking an individual to recall what he or she ate for breakfast that morning, what he or she was doing when he or she first heard of the Space Shuttle *Challenger* disaster, or what words were presented on a list he or she heard earlier all tap into episodic memory. To recall the target information correctly, the individual must be able to access information regarding the time and place of the original event. *Semantic memory,* on the other hand, refers to general knowledge of the world that is not linked to a particular temporal or spatial context. This might include asking an individual to define the word "breakfast," to state what a space shuttle is, or to recall the alphabet. None of these tasks requires recall of where or when the information was learned. Both episodic and semantic memory are declarative, however, in that retrieval of information is carried out explicitly, on a conscious level.

Types of Nondeclarative Memory

Nondeclarative memory comprises several different forms of learning and memory abilities, including procedural memory, some forms of classic conditioning, and the phenomena of priming.

Procedural Memory

Procedures are motor, perceptual, or cognitive skills that are learned and used by an individual to operate effectively in the world. They include such behaviors as tying a shoe, riding a bicycle, or driving a car. *Procedural memory* refers to the process of retrieving information that underlies these skilled performances. Although some aspects of skills can be declared, the skill itself is most often performed automatically, without conscious retrieval of information regarding the procedure. For example, an individual skilled at riding a bicycle does not consciously retrieve information about bicycle riding in order to get on a bicycle and ride. The procedure is automatic and is performed without conscious attention to the mechanics involved. In fact, conscious attention to procedural information can often disrupt performance of the skill.

Conditioning

Classic *conditioning* is one of the most basic forms of learning and illustrates another type of nondeclarative memory. In such a learning paradigm, a stimulus that naturally produces a desired response is identified and paired with a neutral stimulus. After repeated pairings, the neutral stimulus alone elicits the desired response. For example, a dog will naturally begin to salivate when presented food, but not when presented with the sound of a bell ringing. If, however, a bell is rung immediately before the presentation of food, and this pairing is repeated over several trials, presentation of the bell alone will produce salivation. The food in this situation is called an *unconditioned stimulus* because it is a behavior that required no training. Likewise, salivation in response to food is called an *unconditioned response*. The sound of the bell ringing is called a *conditioned stimulus* because the dog becomes conditioned to salivate to the once neutral stimulus. Once this occurs, the salivation in response to the bell is called a *conditioned response*. Some researchers have argued that conditioning in humans requires conscious awareness of the conditioned–unconditioned stimulus contingency; however, studies of patients with amnesic disorders strongly suggest that associations can be conditioned without declarative knowledge.

Priming

Priming is a phenomenon in which prior experience with perceptual stimuli temporarily and unconsciously facilitates the subject's ability to later detect or identify those stimuli. Priming is the most extensively studied aspect of nondeclarative memory. In the typical priming experiment, sets of verbal or nonverbal stimuli, such as lists of words, pictures of common or novel objects, or line drawings, are presented to subjects who, at a later time, are tested using both old and new stimuli. Test procedures may involve asking subjects to identify items from fragments or to make rapid decisions concerning items. Priming is said to have occurred when task performance for previously presented stimuli is superior to that for new stimuli. Warrington and Weiskrantz demonstrated that priming is independent of the subject's ability to recall or recognize stimuli consciously (5). Normal priming effects in the presence of impaired declarative memory have been reported in amnesic patients and in nondemented patients with subcortical disease processes, such as Huntington's disease and Parkinson's disease.

NEUROANATOMY OF MEMORY

Neuroanatomic Correlates of Declarative Memory

Several regions of the brain have been implicated in declarative memory functioning. These include the temporal lobes, the medial diencephalon, and the basal forebrain. The location of these brain regions and their relationship to each other are illustrated in Fig. 9-2.

Temporal Lobes

As in the case of H.M., described earlier, bilateral surgical removal of the medial portion of the temporal lobe results in profound deficits in declarative memory functioning. Twenty-eight years after his original surgery in the 1950s, H.M. continued to demonstrate profound anterograde amnesia; he could not recall new episodic or semantic memories, such as his age, the current year, or what he had eaten during his most recent meal. Like most amnesic patients, however, H.M.

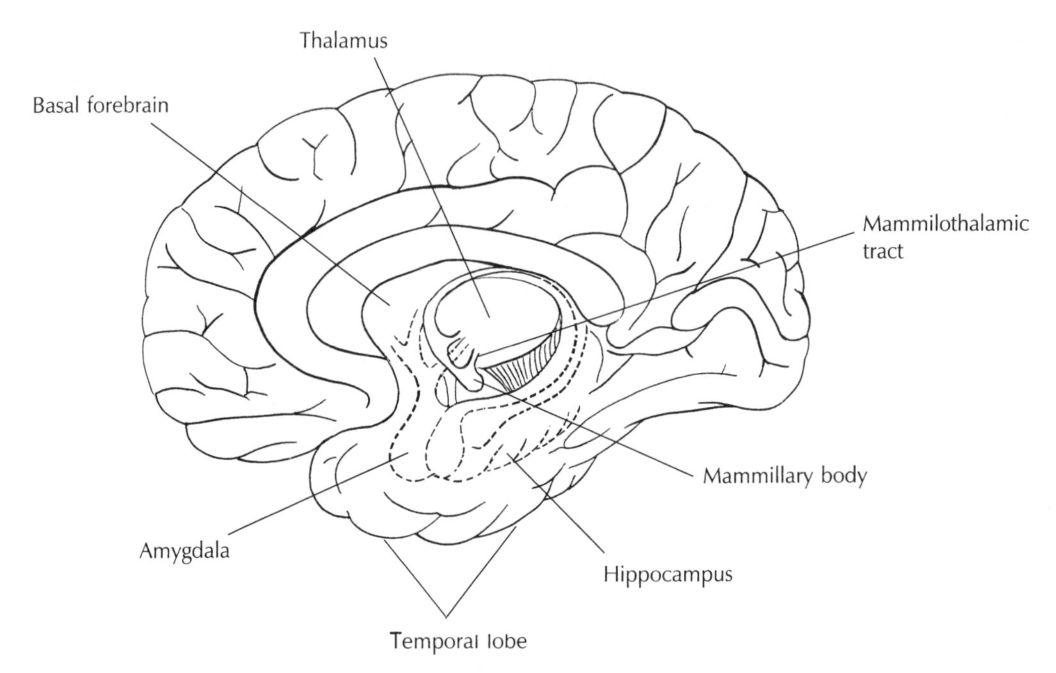

FIG. 9-2. Brain regions implicated in declarative memory functioning. (Modified from Kolb B, Whitshaw I. *Fundamentals of human neuropsychology.* New York: WH Freeman and Co.)

demonstrated evidence of some occasional new learning. He could, for example, tell the examiner what an astronaut was (a term not in use at the time of his surgery), and he was able to recall that a public person named Kennedy had been assassinated.

Although the case of H.M. and of other patients with bilateral medial temporal lobe lesions established the importance of this region in memory functioning, the investigation of patients with unilateral temporal lobe lesions and resections has provided evidence for differential, lateralized contributions of each temporal lobe to memory. Patients who have had surgical resection of the left temporal lobe have more difficulty learning and remembering verbal material, such as stories, paired verbal associations, word lists, and sequences of digits, beyond their immediate memory span than do patients with right temporal lobectomy. In contrast, patients with right temporal lobe resection are typically more impaired on tasks that require memory for nonverbal material, such as tonal patterns; visually presented, abstract geometric patterns; and faces. Despite the severity of declarative memory deficits in patients with medial temporal lobe damage or resection, the capacity for skill learning, priming, and certain kinds of conditioning remains relatively intact.

The temporal lobe is a fairly large brain region with several anatomically distinct areas. Most of the temporal lobectomy studies just described involved resections that included a number of neuroanatomic structures, many of which have been implicated in various aspects of memory functioning. Most investigators agree, however, that the hippocampus with its anatomically related cortex (i.e., entorhinal cortex, perirhinal cortex, and parahippocampal gyrus) and the amygdala hold primary importance for declarative memory functioning.

Hippocampus and Related Structures

One major observation in temporal lobectomy patients was that memory impairment developed only when the resection extended far enough posteriorly to include the hippocampal formation and associated cortical areas (6). Studies of hippocampal functioning in animals further supported the role of these structures in memory. The human hippocampal formation is located bilaterally in the cerebral hemispheres, forming a ridge that extends along the temporal horn of each lateral ventricle. The ventricular surface of the hippocampal formation is covered by a layer of myelinated axons, called the *alveus,* which arise primarily from the cells of the hippocampus. These fibers converge on the medial surface of the hippocampus to form the fimbria, which is continuous with the fornix (Fig. 9-3). As illustrated in Fig. 9-4, the hippocampal formation and its associated cortices are convoluted in structure. Proceeding from the collateral sulcus, the parahippocampal gyrus curves dorsally, transitioning into the regions of the presubiculum and subiculum. At that point, the subiculum curves medially and transitions into the prosubiculum and hippocampus, which then curves inward again, forming the hippocampal fissure. When the hippocampal fissure is opened, a narrow layer of cortex can be observed between the hippocampal fissure and the fimbria; this is known as the dentate gyrus. The hippocampus proper is composed of three substructures, which can be distinguished based on their cytoarchitecture and functional connections; these are known as areas CA1, CA2, and CA3. These areas, together with the dentate gyrus, comprise the hippocampal formation.

Amygdala

The amygdala is a complex of nuclei and specialized cortical areas situated in the dorsomedial portion of the temporal lobe rostral and dorsal to the tip of the temporal horn of the lateral ventricle. Although the earliest investigations suggested that the amygdala was

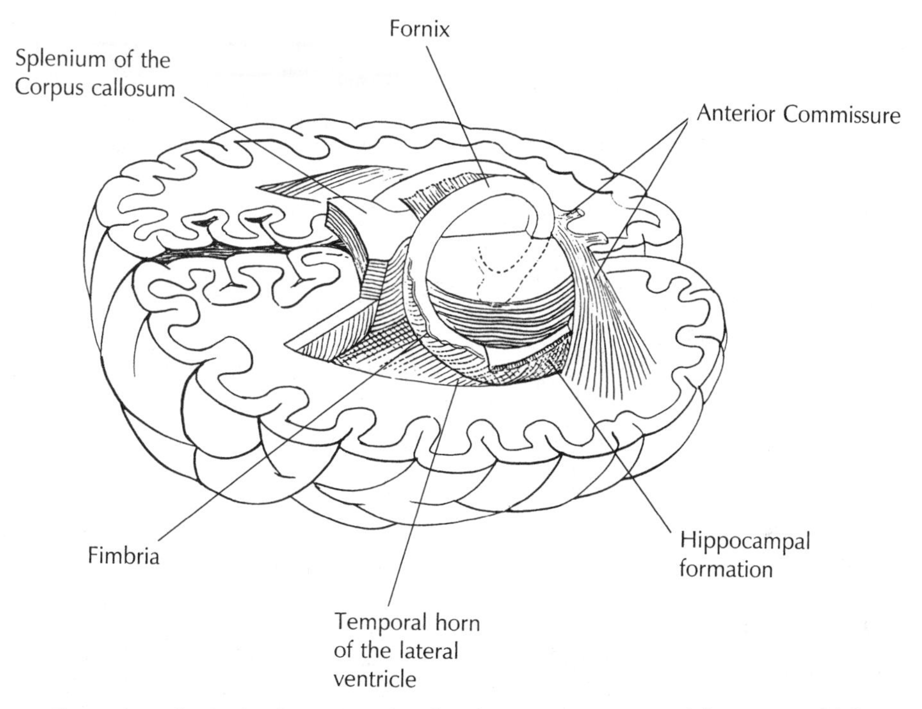

FIG. 9-3. Illustration of a brain dissection showing the gross anatomy of the temporal lobe memory system, including the hippocampal formation and fornix. (Modified from Carpenter MB. *Core text of neuroanatomy.* 3rd ed. Baltimore: Williams & Wilkins, 1985:226–264.)

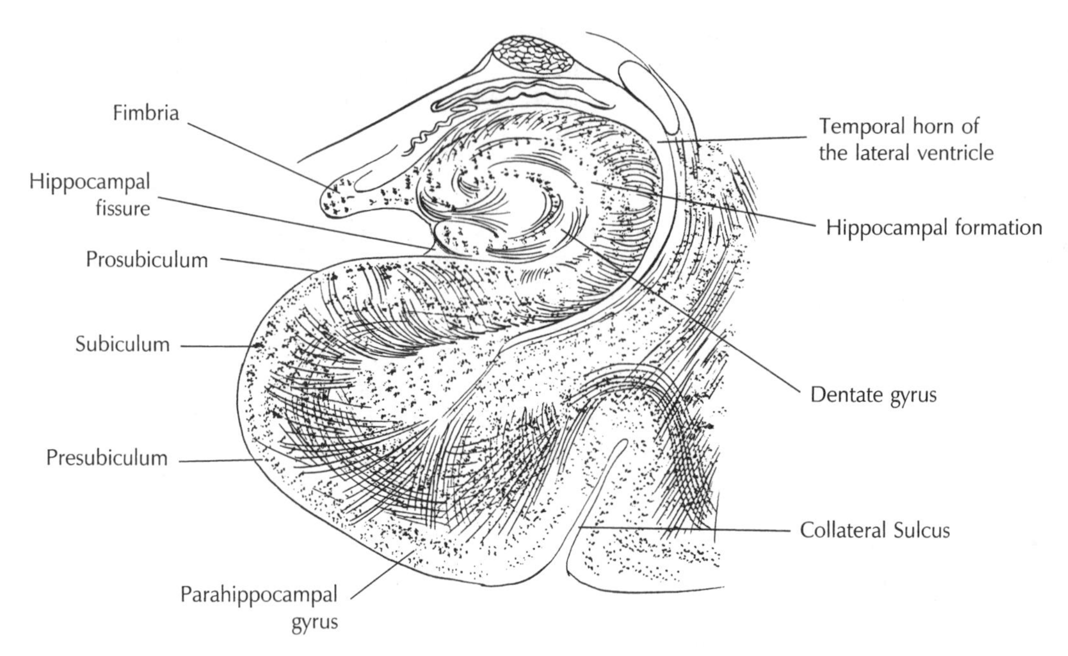

FIG. 9-4. Illustration of a transverse section through the human hippocampal formation and parahippocampal gyrus. (Modified from Carpenter MB. *Core text of neuroanatomy.* 3rd ed. Baltimore: Williams & Wilkins, 1985:226–264.)

primarily an olfactory structure, later studies revealed substantial inputs from other sensory systems as well. Reciprocal fibers connect the amygdala to the visual areas in the inferior temporal gyrus, the auditory areas in the superior temporal gyrus, and the somatosensory areas of the insula; autonomic connections to and from the brain stem are also present. In addition, the nuclei of the amygdala have axonal projections to many other parts of the brain, including the hypothalamus, thalamus, hippocampal formation, prefrontal cortex, basal forebrain, and corpus striatum. The amygdala has long been believed to play a role in the control of emotional, autonomic, reproductive, and feeding behaviors. Our understanding of the role of the amygdala in memory functioning has developed only recently.

Lesions to the amygdala do not appear to contribute to deficits in new learning and memory, such as those assessed by the delayed nonmatching to sample test. Studies of rats and monkeys, however, suggest that the amygdala is important to other types of memory. One example is the learning of conditioned fear and other types of affective memory in which the valence of a neutral stimulus is altered by experience. Such instances appear to be strongly reliant on the integrity of the amygdala.

Medial Diencephalon

Damage to the medial diencephalic region has been linked to human amnesic disorders since the turn of the century (7); however, our understanding of the specific structures involved in memory functioning and the nature of the amnesia that occurs after damage to this system has come to light only recently. The diencephalon is a region of several important nuclei located at the most rostral part of the brain stem (Figs. 9-5 and 9-6). It is bounded superiorly by the lateral ventricles, corpus callosum, and fornix, and laterally by the fibers of the posterior limb of the internal capsule, the stria terminalis, and the body and

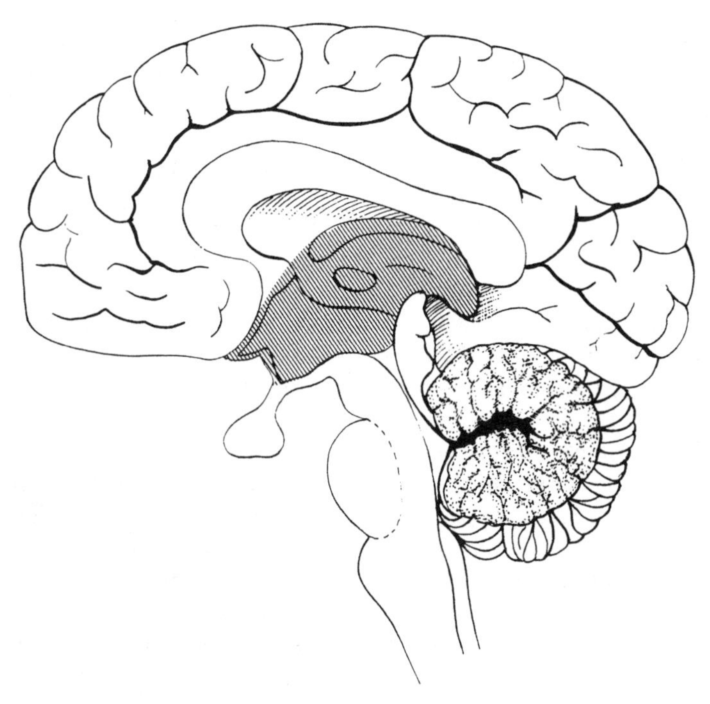

FIG. 9-5. Illustration of a midsagittal section through the human brain showing the region of the medial diencephalon *(shaded area).*

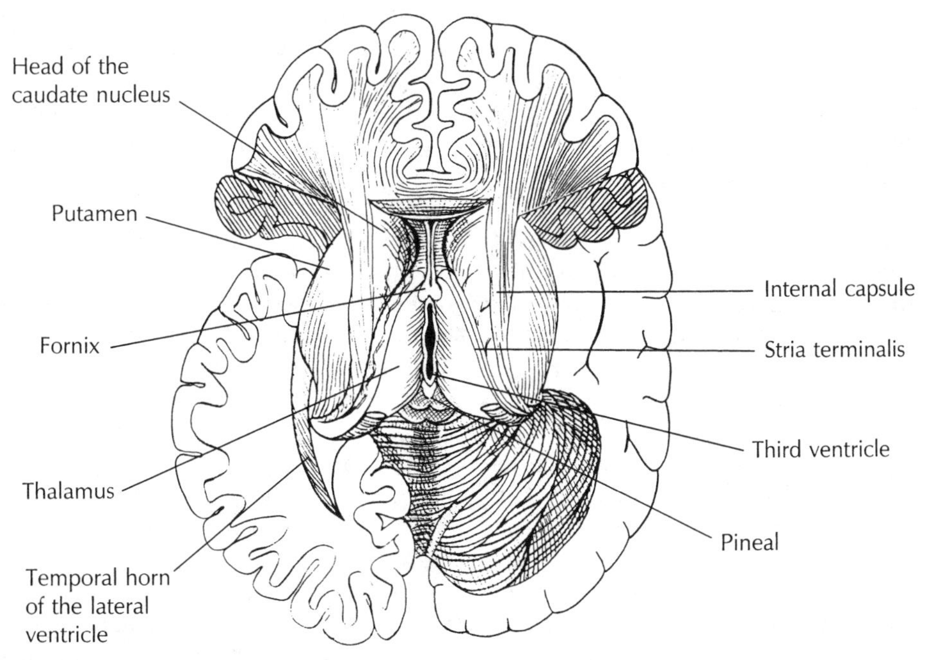

FIG. 9-6. Illustration of a brain dissection showing the gross relationship of the thalamus and related structures. (From Carpenter MB. *Core text of neuroanatomy.* 3rd ed. Baltimore: Williams & Wilkins, 1985:226–264.)

tail of the caudate nucleus. The diencephalon extends caudally to the posterior commissure and is separated into two symmetric parts by the third ventricle. Note, however, that it is common to see an interthalamic adhesion, called the *massa intermedia,* where the medial surfaces of the thalami of the diencephalon are continuous.

The diencephalon can be divided into four major parts: the epithalamus, thalamus, hypothalamus, and subthalamus (also known as the *ventral thalamus*). The epithalamus makes up the superior surface of the diencephalon and consists of the pineal body, stria terminalis, and stria medullaris. The thalamus is the largest subdivision of the diencephalon and is composed of the pulvinar, lateral, and medial geniculate bodies and numerous histopathologically distinct nuclei. The hypothalamus, as the name suggests, lies below the thalamus and extends from the optic chiasm caudally to the mammillary bodies. Finally, the subthalamus is a transition zone lying ventral to the thalamus between the hypothalamus and internal capsule. It consists primarily of the subthalamic nucleus and is traversed by many important fibers projecting to the thalamus.

With few exceptions, the phenomenon of diencephalic amnesia involves damage to regions along the midline of the thalamus and hypothalamus. There has been considerable controversy, however, concerning which of the many neuroanatomic structures and connections in this region must be damaged to cause memory dysfunction. Compared with the study of the medial temporal lobe system, identification of the neuroanatomic substrates of diencephalic amnesia has been relatively slow. To date, the structures that have been most often implicated in diencephalic amnesia are the dorsomedial nucleus of the thalamus, the mammillary bodies of the hypothalamus, and the mammillothalamic white-matter tract (see Fig. 9-2). The basis for this originated from the study of patients with alcoholic Korsakoff's syndrome. The etiology of this syndrome is believed to be a severe thiamine deficiency related to the malnutrition associated with chronic alcoholism. These patients demonstrate enduring deficits in declarative memory, including a severe anterograde and retrograde amnesia, despite maintaining relatively intact intellectual functioning.

Basal Forebrain

Significant memory dysfunction has also been associated with damage to the basal forebrain. The *basal forebrain* is a somewhat loose term used to describe the area of the brain superior to the optic chiasm. It includes the medial septal nuclei, nucleus accumbens, anterior hypothalamus, diagonal band of Broca, nucleus basalis of Meynert, and part of the prefrontal cortex (i.e., Brodmann's area 13). The structures within the basal forebrain project widely throughout the brain. For example, the septal nuclei and nucleus basalis of Meynert are known to have extensive connections to and from the hippocampal formation, amygdala, and the frontal, parietal, and temporal cortices. Most evidence suggests that these two basal forebrain structures, as well as the diagonal band of Broca, may be fundamental to memory functioning.

Basal forebrain involvement in memory functioning is implicated primarily from the study of two patient groups: patients with ruptured aneurysm of the anterior communicating artery and patients with Alzheimer's disease. The region of the basal forebrain is perfused primarily by branches of the anterior communicating artery; thus, disturbances within this flow of circulation result in infarction and necrosis of basal forebrain tissue. A number of amnesic patients have suffered damage to this region secondary to stroke or surgical repair of anterior communicating artery aneurysms; however, the damage to the basal forebrain after such events is typically too variable and extensive to identify specific structures that may be responsible for the amnesic syndrome in these patients.

The basal forebrain has also been implicated in the development of Alzheimer's disease, a disorder in which anterograde memory deficits are a prominent early symptom. Postmortem studies of patients with Alzheimer's disease reveal degenerative lesions within the basal forebrain, specifically involving the cholinergic neurons of the nucleus basalis of Meynert. These neurons produce the neurotransmitter acetylcholine and represent the major source of cholinergic input to the neocortex. This finding initially led investigators to propose that cholinergic depletion was of etiologic significance to the memory disorder seen in Alzheimer's disease. Subsequent studies, however, revealed that other neurotransmitter systems are equally depleted in Alzheimer's disease.

Prefrontal Cortex

Although often not considered a memory center per se, damage to the association area of the frontal lobes, commonly called the *prefrontal cortex,* can nevertheless result in memory disturbance. It has long been recognized that the frontal lobes play an important role in the attentional and organizational processes necessary for the registration and retrieval of information; however, recent studies provide evidence that the prefrontal cortex may be of primary importance for specific memory functions.

Patients with damage to the prefrontal cortex can typically retrieve factual information with little difficulty; however, information regarding when and where the information was obtained is frequently forgotten. This phenomenon is known as a *source memory deficit*, that is, the inability to recall the origin of learned information. Another memory function believed to be mediated by the prefrontal cortex is temporal ordering. The patient is typically asked to identify which of two stimuli was presented most recently. Patients with prefrontal damage exhibit impaired temporal ordering in the context of intact recognition memory.

Brain Systems as Memory Systems

The available data strongly suggest that medial temporal lobe and medial diencephalic structures are components of a single memory system essential to the formation of long-term declarative memory. According to Zola-Morgan and Squire, sensory information is processed by the neocortex and sent to the parahippocampal and perirhinal cortices (7). From there, the information is sent to the entorhinal cortex, hippocampal formation, and medial diencephalon, where associations are created between events or stimulus features. These associations bind together the cortical sites, where different aspects of the given information were originally processed and serve as an index of memory.

This indexing system is believed to be relatively temporary in nature. Although memory for newly acquired information is initially dependent on the integrity of the hippocampal formation, the memory gradually becomes reorganized over time and becomes less and less dependent on these structures. Patients with medial temporal lobe or diencephalic damage, for example, frequently demonstrate spared remote memories despite marked deficits in anterograde memory functioning. This suggests that the medial temporal lobes and medial diencephalon are not the storage sites for long-term memories. Rather, as time passes after learning, a more permanent memory store develops elsewhere, presumably in the neocortex.

Neuroanatomic Correlates of Nondeclarative Memory

As discussed earlier, nondeclarative memory is a performance-based behavior encompassing phenomena such as skill learning, conditioning, and priming. Unlike declarative memory, nondeclarative memory is not believed to be reliant on the medial temporal lobe–medial diencephalic system. Instead, memories underlying acquired skills or conditioned or primed responses are believed to be a function of the sensory and motor systems inherent in the involved behaviors. Consequently, no one brain system can likely account for all types of nondeclarative memory.

Procedural Memory and Skill Learning

Recent work suggests that the brain structures important for skill and procedure learning are those of the corticostriatal system. This system is composed of the corpus striatum and its projections from the neocortex. The corpus striatum represents the largest component of the basal ganglia—the subcortical nuclei located deep in the cerebral hemispheres at the upper brain stem. The basal ganglia consist of two distinct parts: the neostriatum, which includes the putamen and the caudate nucleus, and the paleostriatum, which includes the globus pallidus. The putamen and caudate nucleus receive a wide range of inputs, including projections from the thalamus, amygdala, mesencephalic structures such as the raphe nuclei and substantia nigra, and broad regions of the neocortex. The primary outputs are to the globus pallidus and back to the substantia nigra. The globus pallidus receives input from the caudate nucleus, the putamen, the nucleus accumbens, and the subthalamic nucleus, and projects primarily to the thalamus, habenula, substantia nigra, and subthalamic nucleus.

The corpus striatum is important for motor planning and for performing motor programs. Patients with Parkinson's disease and Huntington's disease (disorders caused by damage or deterioration of the substantia nigra and basal ganglia, respectively) typically demonstrate impaired performance on a wide variety of skill-learning tasks, including prism adaptation, sequence-specific procedural learning, cognitive skill learning, and motor skill learning. In contrast, patients with disorders of medial temporal lobe memory systems (e.g., Alzheimer's disease, amnesia) demonstrate normal performances on many of these same measures. In patients with Huntington's or Parkinson's disease, deficits in skill learning are believed to result from the lack of integrity of the neostriatum.

Conditioning

Considerable progress has been made in identifying the brain structures and systems that underlie the processes of conditioned learning and memory. This is especially true with regard to aversive conditioning. Stimuli that are normally neutral in emotional valence, such as lights or tones, can rapidly become conditioned to elicit fear responses when associated with aversive unconditioned stimuli, such as electric shock. A number of studies have shown that experimental manipulation of the amygdala or its connections can affect fear conditioning. Ablation of the amygdala in a variety of mammalian species results in disrupted acquisition and retention of innate and experimentally induced fear responses, such as increased blood pressure, startle response, and defensive behavioral inhibition. Electrophysiologic recordings from the central nucleus of the amygdala reveal that neuronal activity in this region changes during fear conditioning. Although the amygdala appears to be crucial to the learning of conditioned fear, it does not appear to be the site of storage for long-term fear-related memory. To date, the brain areas that have been implicated as possible storage sites for fear memories include the insular cortex and the vermis of the cerebellum.

Priming

The study of the neuroanatomic bases of priming effects is a relatively recent pursuit, and hypotheses about which brain structures and systems underlie this phenomenon continue to be formulated. Because of the numerous components that make up the physical aspects of presented words, it has been suggested that perceptual priming may occur in any of more than 30 different cortical areas known to be involved in visual information processing. Thus, the importance of a particular brain region to a specific case of priming is most likely dependent on the stimulus qualities and external demands inherent to the individual priming task.

CLINICAL ASSESSMENT OF MEMORY

The past two decades have witnessed a proliferation of memory tests.

General Memory Scales

Wechsler Memory Scale III

Perhaps more than any other clinical test of memory, the Wechsler Memory Scale (WMS) reflects the strength of the impact that neuropsychological research can have on clinical practice. The original WMS consisted of seven subtests assessing different aspects of memory, including span of attention, immediate recall of stories, learning and recall of verbal paired associates, and immediate memory of novel geometric figures. The scores for these subtests were summed and added to an age-correction factor to obtain a summary score known as a *memory quotient* (MQ). The revised versions of the WMS III represent a significant improvement over the original WMS. Delayed recall procedures are included, as are more tests of visuospatial memory. The scoring rules have been improved considerably, more extensive normative studies have been conducted, and the singular MQ index has been eliminated.

Randt Memory Test

The Randt Memory Test is composed of seven subtests, including tests of general information, immediate span of auditory-verbal attention, verbal learning and recall, picture recognition, and an incidental learning test of the names of the previous tasks (8). On some of the subtests, a Brown-Peterson distraction task is used in which the examinee must count backward by threes between presentation and recall. Twenty-four-hour delayed recall is also solicited. As the authors state, the test attempts to provide a "global survey" of patients' memory complaints; it is *not* intended to help localize brain lesions or functions, nor is it intended to tap every type of normal human memory.

Verbal Tests

Immediate Recall Span

The repetition of information immediately after its initial presentation requires attentional abilities and what many researchers call *short-term memory capacity.* Patients with classic amnesic syndromes often perform within, or close to, the normal range on such tests, whereas patients with impaired attentional skills, such as patients with severe depression, often perform poorly.

Digit Span

The examiner presents increasingly long sequences of digits, and the examinee is asked to repeat each sequence in the same order presented. Once a maximum digit span is achieved in the forward direction, a second series of increasingly long digits is presented, and the examinee is asked to repeat them backward. Some researchers have argued that repeating digits exactly as presented in the forward direction may not be the best measure of an examinee's immediate verbal recall span either. Most tests of verbal memory do not require that the target items be recalled in the exact order in which they were presented. Consequently, the length of a digit sequence accurately repeated, regardless of the order of the digits, may be the better measure of immediate recall span.

Word and Sentence Span

Tests have been developed that require the examinee to repeat increasingly long sequences of words or sentences. However, a number of variables can confound interpretation of such tests, including the number of syllables per word and the frequency, abstractness, ability to image, and meaningfulness of words, phrases, or sentences used.

Memory for Word Lists

Verbal Paired Associates

A large number of different paired-associate learning tasks can be found in the experimental literature. The examinee first reads word pairs, one at a time. The examiner then states the first word of each pair and asks the examinee to provide its associate. Errors are corrected, and the procedure is repeated for up to six learning trials to provide ample opportunity for the examinee to learn the associations. Examiners then have the option of assessing recall for the word pairs after an intervening delay period.

Selective Reminding

In this procedure, a word list is presented once, after which subjects are asked to recall as many words as possible. On the next trial, only those words the examinee failed to recall on the preceding trial are presented, after which recall for the entire list is again elicited. This procedure is repeated for several trials until the examinee recalls all the target words or until a predetermined number of trials has been administered. A number of different versions of this procedure have been developed.

Auditory Verbal Learning Test

In this test, a list of 15 unrelated words is presented over five trials, with immediate recall assessed after each presentation (9). Next, a second list of unrelated words is presented for one trial (i.e., an interference task), followed by recall and then recognition testing of the original 15 words. Many examiners also test for recall of the first list again after a delay interval.

California Verbal Learning Test

The general format of the California Verbal Learning Test (CVLT) was modeled after the Auditory Verbal Learning Test (10). One major difference between the two tests is that the 16-item word lists of the CVLT contain four words from each of four different semantic categories (e.g., fruits, tools). Words from the same category are never presented consecutively, which affords an assessment of the degree to which an examinee uses an active semantic clustering strategy in recalling the words. Another difference between the tests is that the CVLT scoring system quantifies and provides normative data for numerous learning and memory variables in addition to total levels of recall and recognition.

Hopkins Verbal Learning Test

The Hopkins Verbal Learning Test (HVLT) is similar in structure to the CVLT (11); however, the HVLT is shorter, taking only 10 minutes to administer, and has six equivalent forms. The 12-item word lists contain four words each from three semantic categories. The list is presented over three learning trials (compared with five learning trials of the Auditory Verbal Learning Test and CVLT), after which recognition memory is tested.

Memory for Stories

The most commonly used story memory test is the Logical Memory subtest of the WMS. In its standardized format, a brief paragraph is read to the examinee, who then recalls immediately as much of the story as possible. This procedure is repeated with a second, different story. Recall is assessed again after a 30-minute delay period. It is not uncommon for a patient to perform considerably better in recalling stories than word lists. These examinees often have difficulty adopting an active learning strategy, such as semantic clustering, in recalling word lists, but they are able to benefit from the thematic organization inherent in stories.

Visuospatial Tests

Visuospatial memory is often more difficult to assess than verbal memory. Finding procedures that do not confound deficits in visuospatial analysis and construction with memory, as well as stimuli that are difficult to encode verbally, has proved challenging. Often, confounds can be sorted out at least in part by assessing visuoperceptual abilities, asking patients to copy stimuli, and employing recognition techniques. The confound of verbal encoding, however, is typically more difficult to control.

Immediate Recall Span: Corsi Blocks

In this test, nine 1-inch cubes are fastened in a random but standardized pattern on a board (9). The examiner touches increasingly long sequences of blocks and the examinee is asked to re-create each sequence in the same order. In the backward condition, the examinee touches each sequence in the reverse order from the examiner. The Corsi Block Test assesses an examinee's attentional skills and span capacity for visuospatial stimuli and represents a spatial analog of the Digit Span subtest.

Memory for Visually Presented Stimuli: Benton Visual Retention Test

The Benton Visual Retention Test (BVRT) is a test of memory for geometric designs (12). The examinee is presented 10 stimulus cards, one at a time. The first two stimuli consist of one geometric shape each, and the remaining eight contain three figures each: two large "main" figures and one small "peripheral" figure. The examiner can choose between three alternate forms and four different administration conditions (e.g., immediate versus delayed recall). The BVRT was one of the first clinical instruments to use a scoring system that quantifies and provides normative data for multiple variables, including accuracy and error types.

Visual Reproduction

On this subtest of the WMS, the examinee studies each of four stimulus cards for 10 seconds and attempts to recall each one immediately after its presentation. Memory for these figures is assessed again after a 30-minute delay. The first three stimulus cards display one design each, and the last one displays two designs side by side. Like the BVRT, the figures are relatively simple, can be easily verbalized, and can be confounded by perceptual and constructional deficits.

Recognition Memory Test for Faces

This test assesses recognition memory for 50 photographs of unfamiliar faces (13). Examinees are presented each photograph one at a time and asked to rate each one as "pleasant" or "unpleasant" (in order to ensure attention to each stimulus). After all 50 stimuli have been viewed, pairs of photographs are presented side by side. One photograph is new; the other was presented among the original 50. Subjects are asked to identify which of the two faces they have seen before.

Rey-Osterrieth Complex Figure Test

In this test, the examinee is shown a complex figure and asked to copy it. When the copy is completed, both the stimulus and the copy are removed, and the examinee is asked to draw the figure a second time from memory. Recall following a delay interval is also commonly assessed (14). One advantage of this test is that the examiner can assess the relationship between the examinee's organizational strategy and subsequent recall performance. A second advantage of the Rey-Osterrieth Test is that the drawing contains both larger, configural features (e.g., the large rectangle) and smaller internal details (e.g., the dots and circle). This configural–detail stimulus parameter can help dissociate the differential processing strategies of patients with unilateral brain damage. Patients with left-hemisphere dysfunction often have difficulty remembering the internal details, whereas patients with right-hemisphere dysfunction are often impaired in their recall of the general configuration.

California Global-Local Learning Test

Unilateral brain pathology tends to disrupt analysis of wholes and parts selectively, and the California Global-Local Learning Test was developed specifically to quantify this phenomenon more rigorously (15). The test involves the presentation of visual hierarchic stimuli consisting of a larger letter or shape constructed from numerous smaller letters or shapes. These stimuli provide precise demarcation between features perceived as larger wholes (i.e., the "global" letter or shape) and smaller details (i.e., the "local" letter or shape). To control for the ease of verbalizing the stimuli, three types of stimuli are used: linguistic forms (i.e., letters), high-frequency nonlinguistic forms (i.e., shapes with established names, such as a square or trapezoid), and low-frequency nonlinguistic forms (i.e., shapes without established names). Stimuli are presented in pairs, with one stimulus in each hemispace. Each of three paired stimuli are presented for 5 seconds, followed by recall drawing of each pair immediately after its presentation. The pairs are presented for three trials to assess learning; free recall, recognition, and copy are assessed after a 20-minute delay.

Tactile Memory Tests: Tactual Performance Test

Although primarily a measure of psychomotor problem solving, the Tactual Performance Test is also used to measure memory for tactually presented information (16). The examinee sits blindfolded in front of a formboard and is asked to place cut-out shapes of high-frequency geometric figures (e.g., square, star) into their appropriate spaces as quickly as possible. This procedure is performed first with the preferred hand, then with the nonpreferred hand, and finally with both hands. Once the test is completed, the equipment is removed, the blindfold is taken off, and the examinee is asked to draw the shapes and their respective locations.

Retrograde Memory Tests

The assessment of memory for events that occurred before the onset of brain dysfunction can often be done informally by eliciting autobiographic recall. When using this technique, however, it is important to obtain verification of memories from relatives or ask the same questions a second time after a delay interval because patients with retrograde amnesia are occasionally prone to confabulation (i.e., fabricating answers). If the responses are valid, a time-line can be constructed reflecting the presence, nature, and extent of a patient's retrograde amnesia.

Boston Retrograde Amnesia Battery

A second method of testing retrograde amnesia is to ask patients about past general information, such as past public figures and events. The Boston Retrograde Amnesia Battery is one such test battery, assessing familiarity with events, politicians, celebrities, and other individuals who were in the public spotlight from the 1930s to the 1970s (17). There are three components to the battery: a famous faces test, a verbal recall questionnaire, and a multiple-choice recognition questionnaire, all of which have been updated to include stimuli from the 1980s and 1990s. Memory for information from each decade is graphed, revealing whether an existing retrograde amnesia is equally severe across all decades (i.e., flat retrograde amnesia) or less severe for more remote decades (i.e., temporally graded).

Television Test

A methodologic problem inherent in testing retrograde memory is possible differences in item difficulty across decades. Amnesic patients, for example, may show better memory for celebrities from the remote past than from the recent past because celebrities from the remote past have had a longer period of exposure and fame, thereby making them easier to recall. To circumvent this problem, Squire and Slater developed a test to maximize the potential for equivalent public exposure across years (18). They used as test items television programs that were broadcast for only one season, and they employed both recall and recognition techniques.

Information

Many clinicians use this subtest of the Wechsler intelligence scales as a rough indication of retrograde amnesia. The test consists of questions that require the examinee to recall famous people and information of the sort normally learned in school (e.g., scientific facts, geography). For middle-aged and older people, this subtest can serve as a gross measure of memory for remote semantic information. The items of this subtest, however, are neither selected systematically from different decades nor equated for ease of recall. Consequently, important qualitative features of a patient's retrograde memory cannot be evaluated. Moreover, because most of the items tap knowledge acquired during school years, older patients with retrograde amnesia extending only a few years or decades before their disease onset often perform well on this test.

MEMORY PROFILES OF SELECTED CLINICAL POPULATIONS

A brief overview of the memory profiles of selected patient populations reveals how different components of learning and memory can be selectively disrupted or spared.

Unilateral Brain Damage

For years, it was believed that patients with left-hemisphere damage (LHD) demonstrated impaired memory only for verbal material, whereas right-hemisphere damage (RHD) disrupted only memory for visuospatial material. Subsequent findings, however, have indicated that patients with RHD suffer subtle deficits in verbal memory functioning, whereas patients with LHD are impaired in remembering certain types of visual stimuli. As a group, however, patients with LHD show significantly greater impairment than patients with RHD and healthy controls on tests of story recall and memory for word lists. This is sometimes confounded by the presence of an aphasic disorder, and the examiner may conclude that language impairment, if present, precludes a valid assessment of verbal memory function. Recognition testing, however, can often be used to circumvent difficulties with expressive language and dissociate some aspects of memory from language dysfunction in patients with aphasia.

Alcoholic Korsakoff's Syndrome

Intelligence and immediate recall span of patients with alcoholic Korsakoff's syndrome ranges from average to low average, but the ability to encode new information into more permanent storage is severely impaired. This severe anterograde amnesia is restricted to declarative knowledge; nondeclarative memory, including procedural learning and semantic priming, are preserved. The declarative amnesia encompasses all stimulus categories (e.g., verbal and visuospatial material) and stimulus features (e.g., global and local forms). On delayed recall tasks, these patients frequently have no recollection of the target items or of having been presented any information at all. At times, however, they confabulate memories and present them in a matter-of-fact fashion. When recall of past autobiographic events and public information is assessed, patients typically demonstrate a severe retrograde amnesia characterized by a temporal gradient. Generally, memory of events from the 20- to 30-year period immediately preceding onset of the disorder is much worse than memory of remote events from childhood and early adulthood.

Chronic Alcoholism

Detoxified chronic alcoholic patients often display mild to moderate deficits on more challenging tests of new learning and memory. Memory for visuospatial stimuli is often worse than for verbal material, which parallels the findings that these patients tend to show greater dysfunction in visuospatial skills relative to verbal abilities. Relatively young alcoholics (in their 30s) may show little neuropsychological impairment due to alcohol. Some chronic alcoholic patients develop a global dementia that persists despite abstinence from alcohol. This alcoholic dementia is typically characterized by severe impairment of memory, conceptualization, problem solving, and visuospatial abilities. The memory disturbance is characterized by rapid forgetting of information over time, rather than retrieval or encoding difficulties. In addition, patients with alcoholic dementia are more sensitive to proactive interference. Priming abilities, however, remain relatively preserved.

Drug-Induced Memory Deficits

Alcohol is not the only substance known to affect memory functioning. Other drugs with sedating effects, such as barbiturates, are known to affect memory functioning negatively, whereas stimulants and euphoriants may augment memory, at least temporarily. Likewise, several of the most commonly used prescription medications may also affect memory functioning. Elderly people are particularly susceptible to this because of drug interactions secondary to polypharmacy and alterations in drug metabolism, distribution, binding, and excretion associated with aging. The medications most commonly associated with cognitive impairment include the sedative-hypnotics, especially long-acting benzodiazepines, and the anticholinergic agents.

Benzodiazepines

It has been known for more than a quarter century that even a single dose of benzodiazepine can impair cognitive functioning. One of the most pronounced cognitive effects is impaired ability to learn new information. The ability to recall information that has been previously learned, however, is not disrupted by benzodiazepine use. In fact, administration of benzodiazepine immediately after learning new information has been known to enhance recall of that information. This phenomenon is known as *retrograde facilitation* and is believed to be caused by the reduction of retroactive interference caused by the suppressed ability to learn new information after drug administration. The mechanism by which benzodiazepines affect memory functioning remains unclear. Part of the effect is believed to be due to the general central nervous system depressant effect of the medication; however, many believe that benzodiazepine action extends beyond the sedating effects of the medication.

Anticholinergics

Anticholinergic agents, such as atropine and scopolamine, are also known to cause significant memory impairment. In addition, a wide variety of commonly used medications can disrupt memory because of significant anticholinergic side effects. These include most heterocyclic antidepressants (e.g., imipramine, amitriptyline), over-the-counter sleep and cold preparations containing antihistamines, several antiparkinsonian medications (e.g., benztropine), and some neuroleptics (e.g., phenothiazines). It is believed that anticholinergic agents produce an imbalance in the cholinergic and adrenergic pathways of the reticular-activating system and thalamocortical projections, thereby disrupting arousal and attentional capabilities. At higher doses, anticholinergics can cause memory impairment and acute confusional episodes.

Other Medications

In addition to benzodiazepines and drugs with anticholinergic effects, a number of other prescription medications can cause memory disturbance. Anticonvulsant medications, such as phenobarbital, phenytoin, and primidone, have been found to cause significant memory impairment, whereas carbamazepine (Tegretol) has been found to have minimal cognitive sequelae. Other drugs reported to cause cognitive impairment include lithium, corticosteroids, antineoplastic agents such as interferon and azathioprine, and antihypertensive medications, including β-blockers (e.g., propranolol), antiadrenergic agents (e.g., clonidine, reserpine, methyldopa), hydralazine, and hydrochorothiazide.

Alzheimer's Disease

Anterograde amnesia for declarative knowledge is often one of the first neuropsychological findings in patients with Alzheimer's disease. Memory difficulties begin insidiously, progress gradually over time, and are typically characterized by poor learning and rapid forgetting of information. Patients with Alzheimer's disease tend to recall most items from the recency region of word lists, reflecting a highly passive learning style. They also make numerous intrusion errors. On delayed recall, they recall few target items and tend to adopt a liberal response bias on recognition testing, endorsing high numbers of both target hits and false-positive errors. On tests of retrograde memory, patients with moderate to severe Alzheimer's disease demonstrate impaired memory for past events from all periods of their lives equally (i.e., flat retrograde amnesia). Several studies have shown a dissociation of subtypes of nondeclarative memory abilities in Alzheimer's disease: Patients demonstrate deficits in priming and classic conditioning of an eye-blink response but exhibit intact procedural learning.

Huntington's Disease

Patients with Huntington's disease, a subcortical degenerative disorder, display equally impaired immediate free recall of information when compared with Alzheimer's disease patients with equivalent overall severity of cognitive dysfunction. Huntington's disease patients, however, show better retention of information over delay intervals. When presented with multiple trials of different verbal stimuli, Huntington's disease patients reveal normal sensitivity to proactive interference, and they make significantly fewer intrusion and perseveration errors than do Alzheimer's disease patients and amnesic patients. Recognition performance, although impaired, tends to be disproportionately better in Huntington's disease patients than in Alzheimer's disease or amnesic patients. Huntington's disease patients have impaired retrograde memory, but their performance differs qualitatively from that of patients with Korsakoff's syndrome. Huntington's disease patients show a flat retrograde amnesia, with equally deficient recall of events from all decades.

Frontal Lobe Dementia

The frontal lobes account for about half of the cerebrum. Given their size, the chances that the frontal lobes will be involved in any diffuse pathologic process are high. Some dementing disorders, however, preferentially affect the frontal lobes. These include one variant of Pick's disease and Jakob-Creutzfeldt disease, both of which are rare conditions. Recent evidence suggests that another more common degenerative dementia specific to the frontal lobes may exist that is histopathologically distinct from these and other known dementing illnesses. The patient with frontal lobe dementia typically presents with reports of a change in personality or adaptive behaviors that precede the onset of cognitive symptoms. When cognitive deficits appear, they typically involve disorders of planning, organization, mental flexibility, and memory. The memory deficits associated with frontal lobe dementia typically reflect poor organization, use of inefficient learning strategies, and increased susceptibility to interference. Ability to sustain attention is disturbed; however, there is no evidence of rapid forgetting of information, such as is seen in Alzheimer's disease. Errors in recall are common and include perseverations, intrusions, and source memory problems.

Affective Disorders

Patients suffering from depression represent a heterogeneous group in terms of their cognitive profiles. Depressed patients whose vegetative symptoms are not too severe but who are highly self-critical often show considerable variability in their ability to do well on tests of memory and other cognitive functioning. Effort, motivation, and subsequent performance typically wax and wane with the intensity of the depressed state. In many cases, perception of a poor performance feeds on itself, causing the patient to feel more self-critical and anxious. This, in turn, can lead to additional poor performances.

Depressed patients frequently show impaired immediate recall of span and supraspan material, presumably because their attention capacity is consumed with dysphoric preoccupation and obsessive thinking. They are deficient in adopting active encoding strategies and tend to give minimal responses, reporting only few target items and making few errors (i.e., intrusions and perseverations). The information they do report on immediate recall trials is typically retained to a normal degree after a delay interval. Their memory performance often improves with recognition testing, which probably compensates for retrieval difficulties that arise from being minimal responders. Errors on recognition testing are more likely to be missed targets than false-positive results, reflecting a negative response bias. Patients with severe vegetative symptoms of psychomotor retardation, weight loss, and insomnia appear to be a more homogeneous group cognitively. This is because they often perform close to or at the lowest levels on neuropsychological tests. A "floor effect" such as this is especially common on tests assessing high-order functions, such as memory and novel problem solving.

There is increasing evidence that the cognitive abnormalities of depressed patients may occur secondary to genuine neuropathology, often related to changes in neurotransmitter systems, such as serotonin and norepinephrine. A more sophisticated clinical examination would thus inquire whether cognitive dysfunction in a depressed patient is consistent with reversible neurochemical changes associated with the psychiatric disorder, irreversible structural changes, or some combination of the two.

Normal Aging

Decrements in memory functioning are apparent with advancing age in normal populations. The elderly commonly report that their memory is not as good as it once was, and they perform less well than younger subjects on standard tests of learning and recall. In general, the decrements in memory functioning associated with normal aging are more apparent in some aspects of memory than in others, and they appear to reflect decreased efficiency of processing rather than a true breakdown or loss of component structures. Elderly subjects do not differ significantly from younger controls on tests of immediate recall span *unless* the task requires the subject to manipulate or transform the information (such as recalling digits backward). With regard to long-term memory, several investigators have demonstrated that elderly patients normally can usually manage to perform tasks that are driven directly by the stimulus or that are strongly influenced or supported by the environment. Conversely, memory impairment is greatest when the situation requires self-initiation or the establishment of new routines or patterns. Changes in nondeclarative memory with advanced age appears to depend on the measure used. There does not appear to be a significant decline with age on measures of priming; however, skill learning, procedural memory, and the rate of developing classically conditioned responses appear to be affected negatively by age.

Transient Global Amnesia

Transient global amnesia is an amnesia occurring in clear consciousness, without focal signs or other signs of epilepsy, and resolving within 24 hours. The duration of the amnesia ranges between 15 minutes and 12 hours, and 15% of patients report multiple episodes. In the preceding 24 hours, 10% experience either headache or nausea, and 15% experience a stressful life event, whereas in others, the attack is preceded by a medical procedure or severe exercise. In 25%, there is a past history of migraine; however, the underlying cause of the attacks is unclear in 60% to 70% of patients. It has been argued that transient dysfunction in the limbic-hippocampal circuits may be the physiologic basis of the transient global amnesia syndrome. All patients show a profound anterograde amnesia on tests of both verbal and nonverbal memory. Performance on retrograde memory tests is variable. Complete recovery on neuropsychological tests of memory occurs within 1 month of the acute episode, consistent with clinical expectations.

Transient Epileptic Amnesia

Epileptic seizures give rise to amnesia, and they may be associated with epileptic automatisms or postictal confusional states. Where an automatism arises, there is always bilateral involvement of the limbic structures involved in memory formation, including the hippocampal and parahippocampal structures bilaterally, as well as the mesial diencephalon. Hence, amnesia for the period of automatic behavior is always present and is usually complete.

Head Injury

Head injury can produce a discrete or transient episode of memory loss and, if severe, can give rise to persistent memory and general cognitive impairment. In the former case, it gives rise to the familiar pattern of memory loss, consisting of a brief period of retrograde amnesia, which may last only a few seconds or minutes, a longer period of posttraumatic amnesia, and islets of preserved memory within the amnesic gap. Occasionally, posttraumatic amnesia may occur without retrograde amnesia, although this is more common in cases of penetrating lesions. Sometimes, the patient has a particularly vivid memory for images or sounds occurring immediately before the injury, on regaining consciousness, or during a lucid interval between the injury and the onset of posttraumatic amnesia.

The duration of posttraumatic amnesia is assumed to reflect the degree of underlying diffuse brain pathology. Rotational forces, such as those in a car accident, are particularly likely to produce axonal tearing and generalized cognitive impairment. The length of posttraumatic amnesia is predictive of eventual cognitive, psychiatric, and social outcomes. There appears to be a relationship with age in that older subjects tend to have a longer amnesia and more serious deficits at a given amnesia, whereas in subjects younger than 30 years of age, posttraumatic amnesia is sometimes found to be less effective as a predictor of subsequent memory impairment. In addition, contusion to the frontal and anterior temporal lobes is a common consequence of head injury.

After a mild head injury, a neurotic (posttraumatic) syndrome sometimes ensues, in which forgetfulness is a prominent complaint, in addition to anxiety, irritability, poor concentration, and various somatic complaints. The etiology of this syndrome remains controversial, but it is clear that the symptoms commonly persist long after the settlement of any compensation issues. In more severe head injury, the ability to learn new material is the slowest cognitive function to recover, and the pattern of residual memory deficit resembles, in many respects, that of a classic amnesic syndrome.

Postelectroconvulsive Therapy

Electroconvulsive therapy (ECT) is an iatrogenic form of transient amnesia. Subjects tested within a few hours of ECT show a retrograde impairment for information from the preceding 1 to 3 years and a pronounced anterograde deficit on both recall and recognition memory tests. There is general agreement that, 6 to 9 months after completion of a course of ECT, memory performance on objective tests returns to normal, apart from a persistent loss of material acquired within a few hours of the convulsions. Complaints of memory impairment may persist, however, and can still be evident 3 years or more after the course of ECT. Neuropsychological studies have found that verbal memory appears to be particularly sensitive to disruption. Unilateral ECT to the nondominant hemisphere produces considerably less memory impairment than bilateral ECT, making it important to identify the nondominant hemisphere by a valid procedure. The most effective methods of avoiding memory deficit consist of electrode placement either over the frontal rather than the temporal lobes or over the nondominant temporal lobe.

Psychogenic Fugue

A *fugue state* is an example of a psychogenic, global amnesia. It refers, in essence, to a syndrome consisting of a sudden loss of all autobiographic memories and the sense of self or personal identity, usually associated with a period of wandering, for which there is a subsequent amnesic gap upon recovery. Fugue states usually last a few hours or days only, and they appear to have occurred more commonly earlier in the century, particularly in wartime. Three main factors predispose individuals to such episodes:

- Severe, precipitating stress, such as marital or emotional discord, bereavement, financial problems, a criminal charge, or stress during wartime
- Depressed mood, which is an extremely common antecedent for a psychogenic fugue state
- History of a transient, neurologic amnesia

A confounding factor, however, is that patients are somewhat unreliable personalities with a possible legal motive for wanting to claim amnesia. The multiple personality syndrome could be perceived as an extreme instance of a fugue state, when the subject is shifting constantly between one personality and another (or many others).

Amnesia for Offenses

A type of amnesia termed *amnesia for offenses* is a situation-specific, psychogenic memory loss; this type is claimed much more commonly than are fugue states. Amnesia is claimed by between 25% and 45% of offenders convicted of homicide. A small percentage of these cases claim their amnesia in association with evidence of a frank neurologic or metabolic disorder, such as an epileptic automatism or postictal confusional state, head injury, hypoglycemia, or sleepwalking. In these instances, the presence of amnesia is essential to establish the existence of an automatism. In the vast majority of offenders claiming amnesia, however, no evidence of organic brain disease is found.

SUMMARY

The past 25 years have witnessed substantial changes in our understanding of the neurologic and cognitive processes that comprise memory. This has produced a generation of improved assessment measures incorporating constructs from the cognitive psychological literature, such as encoding strategies, recall measures, recognition discriminability, sensitivity to interference, and forgetting rates. As the investigation of cognitive processes constituting memory functioning progresses, so will our ability to measure and characterize memory disorders at both the structural and functional levels. The recognition of the existence of two dissociable memory systems, for example, has led to an explosion of interest in the neurology of memory and the search for brain structures that mediate nondeclarative memory systems.

REFERENCES

1. Squire LR. *Memory and brain.* Oxford, England: Oxford University Press, 1987.
2. Hebb DO. *The organization of behavior.* New York. John Wiley & Sons, 1949.
3. Milner B. Amnesia following operation on the temporal lobes. In: Whitty CWM, Zangwill OL, eds. *Amnesia.* London: Butterworths, 1966:113.
4. Butters N, Cermak LS. A case study of the forgetting of autobiographical knowledge: implications for the study of retrograde amnesia. In: Rubin D, ed. *Autobiographical memory.* New York: Cambridge University Press, 1986:253–272.
5. Warrington EK, Weiskrantz L. Amnesic syndrome: consolidation or retrieval? *Nature* 1970;228:628–630.
6. Squire LR, Zola-Morgan S. The medial temporal lobe memory system. *Science* 1991;253:1380–1386.
7. Zola-Morgan S, Squire LR. Neuroanatomy of memory. *Annu Rev Neurosci* 1993;16:547–563.
8. Randt CT, Brown ER. *Randt Memory Test.* Bayport: Life Science Associates, 1983.
9. Lezak MD. *Neuropsychological assessment.* 2nd ed. Oxford, England: Oxford University Press, 1983.
10. Delis DC, Kramer JH, Kaplan E, Ober BA. *The California Verbal Learning Test.* New York: The Psychological Corporation, 1987.
11. Brandt J. The Hopkins Verbal Learning Test: development of a new memory test with six equivalent forms. *Clin Neuropsychologist* 1991;5:125–142.
12. Benton AL. *The Revised Visual Retention Test.* 4th ed. New York: The Psychological Corporation, 1974.
13. Warrington EK. *Recognition Memory Test.* Windsor: Nfer-Nelson, 1984.

14. Spreen O, Strauss E. *A compendium of neuropsychological tests: administration, norms, and commentary.* New York: Oxford University Press, 1991.
15. Delis DC, Kiefner M, Fridlund AJ. Visuospatial dysfunction following unilateral brain damage: dissociations in hierarchical and hemispatial analysis. *J Clin Exp Neuropsychol* 1988;10:421–431.
16. Reitan RM, Wolfson D. *The Halstead-Reitan Neuropsychological Test Battery.* Tucson, AZ: Neuropsychology Press, 1985.
17. Albert MS, Butters N, Levin J. Temporal gradients in the retrograde amnesia of patients with alcoholic Korsakoff disease. *Arch Neurol* 1979;36:211–216.
18. Squire LR, Slater PC. Forgetting in very long-term memory as assessed by an improved questionnaire technique. *J Exp Psychol Learn Mem Cogn* 1975;104:50–54.

10

Language

David Caplan

OVERVIEW

This chapter presents an overview of language disorders in adults from a cognitive (psycholinguistic) point of view. The chapter summarizes models of normal function and descriptions of impairments, beginning with a description of the normal language system and a review of essential features of the language code and its processing in the usual tasks of language use (speech, auditory comprehension, reading, and writing). Aspects of methodology are then outlined, including psycholinguistic techniques, an approach to the diagnosis of language disorders, and neurologic observations. This is followed by a synopsis of work that describes neurogenic language disorders in terms of disturbances of linguistic representations and psycholinguistic processes. This psycholinguistic approach is compared with a more traditional clinical approach to characterizing language disorders based on "aphasic syndromes." Language disorders in psychiatric disorders are then discussed. The chapter concludes with a discussion of models of the functional neuroanatomy of language.

LANGUAGE STRUCTURE, LANGUAGE PROCESSING, AND FUNCTIONAL COMMUNICATION

Human language has many different facets. It can be vehicle for communication of information, a medium for artistic expression, a method for expression of emotions, a means for developing thought, or it can serve other functions. In its essence, however, language is a code, one whose structural properties makes it capable of serving this wide range of functions. Language is a very particular and fairly complicated code. It consists of a set of forms that are called *linguistic representations*. Different types of linguistic representations convey different aspects of meaning. These linguistic representations are organized into *levels* of the language code. The basic levels of the code are *simple words* (the *lexical level*), *words with internal structure* (the *word-formation level*), *sentences* (the *sentential level*), and *discourse* (the *discourse level*). When we use language to speak, write, listen, or read, we simultaneously make use of all the types of linguistic representations in the language code. The result is that what we produce or perceive has a complex structure that conveys a wide range of semantic information.

Structure of Language

The lexical level of language consists of simple words. The basic form of a simple word (or lexical item) consists of a phonologic representation that specifies the segmental elements (phonemes) of the word and their organization into metrical structures such as syllables. The form of a word can also be represented orthographically. Simple words are assigned to different syntactic categories, such as noun, verb, adjective, article, preposition, and so forth. The semantic values associated with the lexical level primarily consist of concepts and categories in the nonlinguistic world. Simple words tend to designate concrete objects, abstract concepts, actions, properties, and logical connectives.

The sentential level of language consists of syntactic structures—hierarchical sets of syntactic categories (e.g., noun phrase, verb phrase, sentence)—into which words are inserted. The meaning of a sentence, known as its propositional content, is determined by the way the meanings of words combine in syntactic structures. Propositions convey aspects of the structure of events and states in the world. These include who did what to whom (thematic roles), which adjectives go with which nouns (attribution of modification), which words in a set of sentences refer to the same items or actions (the reference of pronouns and other anaphoric elements), and so on. For instance, in the sentence, "The big boy told the little girl to wash herself," the agent of *told* is "the big boy," and its theme is "the little girl"; "big" is associated with "boy" and "little" with "girl"; "herself" refers to the same person as "girl." Sentences are a crucial level of the language code because the propositions they express make assertions about the world. These assertions can be entered into logical systems and can be used to add to an individual's knowledge of the world.

The propositional meanings conveyed by sentences are entered into higher-order structures that constitute the discourse level of linguistic structure. Discourse includes information about the general topic under discussion, the focus of a speaker's attention, the novelty of the information in a given sentence, the temporal order of events, and causation, among other things. The structure and processing of discourse involves many nonlinguistic elements and operations, such as a search through semantic memory, logical inferences, and a more purely linguistic structure.

Prosodic information—intonational contours—are also linguistic representations that determine aspects of discourse meaning. Phrases and sentences receive intonational contours that partially reflect syntactic structures (e.g., there tends to be a drop in pitch at the end of most sentences) and partially reflect discourse considerations. Prosody can determine whether a sentence is a statement or a question (the elocutionary force of a sentence) and whether an element conveys new or old information in a discourse. Many other linguistic devices (e.g., variation in the order of words, the use of repeated noun phrases, ellipses) determine aspects of discourse meaning. A complex interplay between linguistic and nonlinguistic domains thus characterizes discourse structure. Information conveyed by the discourse level of language serves as a basis for updating an person's knowledge of the world and for reasoning and planning action.

Language Processing

The different forms of the language code are computed by a set of processors, or components of a language processing system, each dedicated to activating particular elements of the code. Considerable research has gone into the characterization of the components of the language processing system. It is far beyond the scope of this chapter to review this literature in depth. See reference 1 and the references cited in that book for an introduction to this research. This chapter provides a brief description of some of the general properties of the language processors and their organization.

One important characteristic of the language processing system is that each processor accepts only particular types of representations as input and produces only specific types of representations as output. For instance, the processor that activates syntactic structures from auditory input may use as input many features derived from the speech signal (e.g., the syntactic categories of the words presented to it, intonational contours), but it probably does not make use of all aspects of the meanings of words in this process. Fodor uses the term *encapsulation* to refer to this property of components of the language processing system (2). Similarly, the output of this processor is a syntactic structure—it is not a representation of the logical entailments of a sentence; Fodor uses the term *domain specificity* to refer to this property (2). A major area of research is in the determination of the input and output representations that are paired by the operation of different language processors; these operations range from conversion of the acoustic signal to speech sounds to visual word recognition, spelling, the determination of sentence structure, structuring discourse, and other operations.

Information-processing models of language are expressed as flow diagrams (often called *functional architectures*) that indicate the sequence of operations of the different components that perform a language-related task. The major components of the language processing system that can be identified at this level of detail for simple words are listed in Table 10-1, and for the word formation and sentence levels in Table 10-2. Fig. 10-1 presents a model indicating the sequence of activation of components of the lexical processing system. Fig. 10-2 presents a similar model of the processing system for word formation and sentences. The

TABLE 10-1. *Summary of Components of the Language Processing System for Simple Words*

Component	Input	Operation	Output
Auditory-oral modality			
Input-side			
Acoustic-phonetic processing	Acoustic waveform	Matches acousticproperties to phonetic features	Phonologic segments (phonemes, allophones, syllables)
Auditory lexical access	Phonologic units	Activates lexical items in long-term memory on basis of sound; selects best fit to stimulus	Phonologic forms of words
Lexical semantic access	Words (represented as phonologic forms)	Activates semantic features of words	Word meanings
Output-side			
Phonologic lexical access	Word meanings ("lemmas")	Activates the phonologic forms of words	Phonologic form of words
Phonologic output planning	Phonologic forms of words (and nonwords)	Activates detailed phonetic features of words (and nonwords)	Phonetic values of phonologic segments; word stress patterns
Articulatory planning	Phonetic values	Specifies articulatory movements	Neural commands for articulation
Written modality			
Input-side			
Written lexical access	Abstract letter identities	Activates orthographic forms of words	Orthographic forms of words
Written lexical semantic access	Orthographic forms of words	Activates semantic features of words	Word meanings
Output-side			
Accessing lexical orthography from semantics	Word meanings	Activates orthographic forms of words	Orthographic forms of words
Accessing lexical orthography from lexical phonology	Phonological representations of words	Activates orthographic forms of words from their phonologic forms	Orthographic form of words
Accessing sublexical orthography from sublexical phonology	Phonologic units (phonemes, other units)	Activates orthographic units corresponding to phonologic units	Orthographic units in words and nonwords
Accessing lexical phonology from lexical orthography	Orthographic form of words	Activates phonologic forms of words from their orthographic forms	Phonologic forms of words
Accessing sublexical phonology from sublexical orthography	Orthographic units (graphemes, other units)	Activates phonologic units corresponding to orthographic units	Phonologic units in words and nonwords

TABLE 10-2. *Summary of Components of the Language-Processing System for Derived Words and Sentences (Collapsed over Auditory-Oral and Written Methods)*

Component	Input	Operation	Output
Processing affixed words			
Input-side			
Morphologic analysis	Word forms	Segments words into structural (morphologic) units; activates syntactic features of words	Morphologic structure; syntactic features
Morphologic comprehension	Word meaning; morphologic structure	Combines word roots and affixes	Meanings of morphologically complex words
Output-side			
Accessing affixed words from semantics	Word meanings; syntactic features	Activates forms of affixes and function words	Forms of affixes and function words
Sentence-Level Processing			
Input-side			
Lexicoinferential comprehension	Meanings of simple and complex words; world knowledge	Infers aspects of sentence meaning on basis of pragmatic plausibility	Aspects of propositional meaning (thematic roles; attribution of modifiers)
Parsing and syntactic	Word meanings; syntactic features	Constructs syntactic representation and combines it with word meanings	Aspects of comprehension of propositional meaning
Heuristic sentence comprehension	Syntactic categories of words	Constructs simplified syntactic structures; combines word meanings in these structures	Aspects of propositional meaning
Output-side			
Construction of functional level representation	Messages	Activates content words and assigns thematic roles and other aspects of propositional meaning	Content words; thematic roles; other aspects of meaning
Construction of positional level representation	Content words; syntactic frames; discourse features	Activates syntactic frames in conjunction with function words; inserts phonologic forms of content words into syntactic frames	Surface forms of sentences
Phonologic output planning	Surface forms of sentences	Combines lexical phonologic and sentence-level phonologic information	Phonetic values; stress and intonation

model depicted in these tables and figures outlines the way information—in this case, sets of related linguistic representations—flows through the tasks of speaking, understanding spoken language, reading, and writing.

The following points describe some of the more important aspects of these features:

1. Most processors are obligatorily activated when their inputs are presented to them.
2. Language processors generally operate unconsciously.
3. Components of the system operate remarkably quickly and accurately.
4. The operation of each component of the language processing system can be thought of as requiring "processing resources." Intuitively, one might think of each psycholinguistic operation as a machine, such as an old grist mill that demands power; the machine works effi-

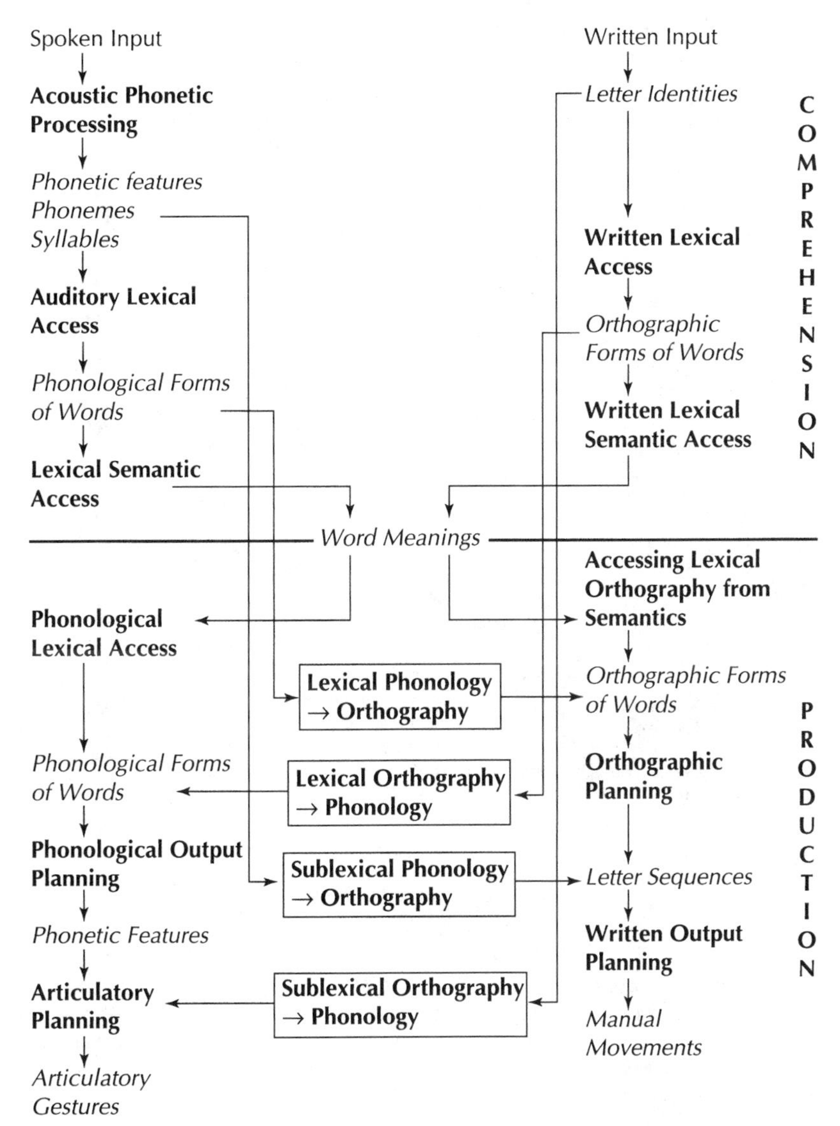

FIG. 10-1 Diagrammatic representation of the sequence of activation of components of the processing system for single words. Processing components are presented in boldface; representations are presented in italics. Arrows represent the flow of information from one processing component to another. (From Caplan D. Toward a psycholinguistic approach to acquired neurogenic language disorders. *Am J Speech Lang Pathol* 1993;2:54–83.)

ciently above a certain level of power availability and is limited by the amount of power available.

5. The operations of the language processing system are regulated by a variety of control mechanisms. These control mechanisms include both mechanisms internal to the language processor itself and those that are involved in other aspects of cognition.

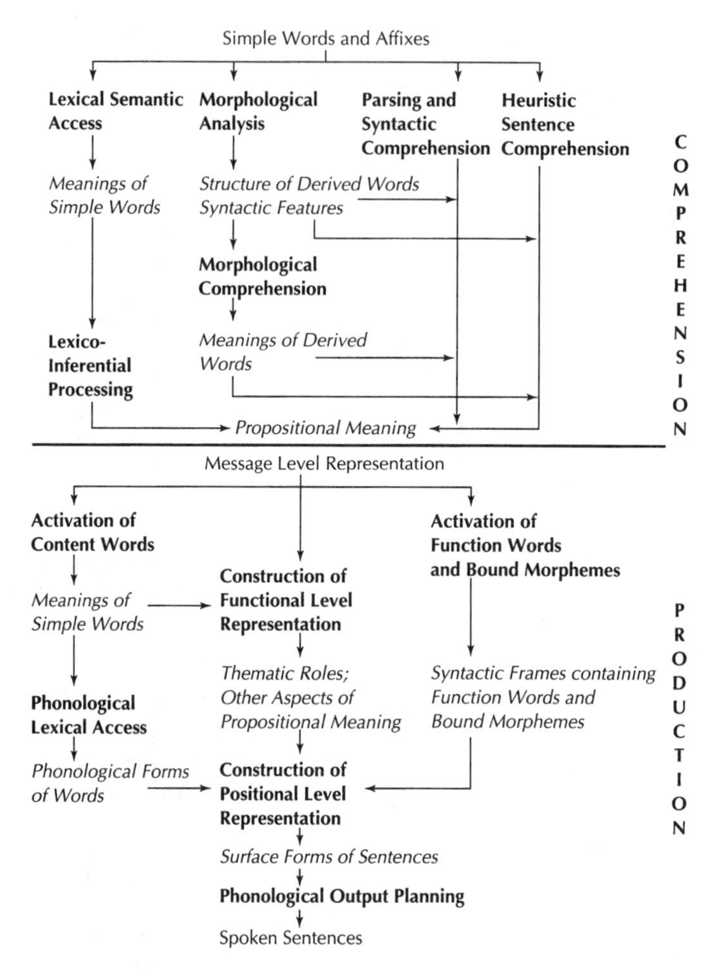

FIG. 10-2. Diagrammatic representation of the sequence of operation of components of the language processing system for morphologically complex words and sentences. Processing components are presented in boldface; representations are presented in italics. Arrows represent the flow of information from one processing component to another. (From Caplan D. Toward a psycholinguistic approach to acquired neurologic language disorders. *Am J Speech Lang Pathol* 1993;2:59–83.)

Functional Communication, Language Processing, and Communication Disorders

Most patients' complaints regarding language tend to center on impairments of the use of language for functional communication. The clinician attempting to assess language-based functional communication must realize that this function is multifactorially determined. There is no simple, one-to-one relationship between impairments of elements of the language code or of psycholinguistic processors on the one hand and abnormalities in accomplishing the goals of language use on the other. Patients adapt to their language impairments in many ways, some of which are remarkably effective at maintaining at least some aspects of functional communication; conversely, individuals with (or without) neurologic disease who have intact language processing mechanisms may fail to communicate effectively.

It is possible and valuable to identify disturbances of elements of the language code or psycholinguistic processors in many patients who have functional communicative disturbances. To do so, it is necessary to use tests that identify such disturbances. A brief description of these methods follows.

PSYCHOLINGUISTIC, CLINICAL, AND NEUROLOGIC METHODS

Experimental Studies of Language Processing and Language Disorders

Research on normal language processing has focused on the study of the obligatory, unconscious psycholinguistic processes. Researchers use methods that require a subject to make responses to ongoing language stimuli and to respond to a stimulus in a way that does not require conscious consideration of the representation under investigation. The localization of extraneous noises (clicks) in a sentence, monitoring for phonemes, and other techniques were among the earliest experimental approaches to "on-line" psycholinguistic processing. More complex tasks mark subsequent on-line methods. For instance, techniques such as self-paced reading—a task in which subjects press a key to call up subsequent words or passages—and performance of psycholinguistic tasks under dual-task conditions or when stimulus input is made more difficult to perceive—as in rapid serial visual presentation—have been used to explore the locus of increases in processing load in sentence comprehension (3).

Most of the studies of language impairments use off-line measures of language performance. In most of these studies, the test is untimed, the dependent variable is accuracy and at times error type, and the interest of the study lies in what linguistic representations a patient can and cannot deal with in what tasks. These studies provide valuable data regarding the major dissociations found in patients with respect to their abilities to process language. They are the most common type of study of psycholinguistic impairments in contemporary cognitive neuropsychological research. So-called double dissociations, in which one patient performs normally on one task and abnormally on a second and a second patient shows the opposite pattern, provide important evidence for the existence of separate processors, each involved in only one of the two tasks (4).

Clinical Assessments of Language

An achievable goal of a psycholinguistic assessment of a patient is to specify the types of linguistic representations (simple words, word formation, sentences, discourse) that are processed abnormally in each of the four major language-related tasks (speech, auditory comprehension, reading, writing). For each level of the code, the assessment specifies whether the disturbance affects linguistic forms (e.g., phonemes, syntactic structures), the semantic meanings associated with those forms, or both. It also attempts to identify selective impairments affecting each type of representation in each of these tasks and the overall level of functioning of the patient with respect to each linguistic representation in each task. This effort leads to a description of the patient's language disorder in relation to the major components of the language processing system. In most cases, more than one disorder is identified in a patient with a language impairment.

Existing aphasia batteries, such as the Boston Diagnostic Aphasia Examination, the Western Aphasia Battery, and the Porch Index of Communicative Ability (PICA), along with more specific tests, such as the Wepman Aphasia Battery, the Peabody Picture Vocabulary Test, the Boston Naming Test, and varieties of the Token Test, do not provide a systematic exploration of the levels of the language code that are disturbed in a given language processing task (5–11). New tests to screen patients psycholinguistically are beginning to be developed (12). What follows is a description of a language examination that can identify the major primary disturbances of language processing in a patient.

The test begins with an informal assessment of whether the patient understands what is said and whether he or she is able to express ideas in a comprehensible form. Problems in speech production are noted, such as limited production of nouns and verbs, errors in function words, phonemic paraphasias and neologisms, and abnormal articulation and prosody. Next, the examiner assesses the patient's ability to recognize pictures by presenting pictures of objects and asking yes-or-no questions about the object's physical and functional properties. If a patient does not demonstrate an ability to understand pictures, further testing is based only on tests that do not make use of pictures.

On the comprehension side, *phoneme discrimination* is tested by randomly presenting pairs of syllables, half identical (e.g., /ba/-/ba/) and half minimally different (e.g., /fa/-/va/); the patient must say which are the same and which different. *Word recognition* is tested by randomly presenting simple real words (e.g., *cattle, true*) and nonwords (e.g., *voot, efebant*) and asking the patient to indicate which are real and which are not. *Single-word comprehension* is tested in two ways. The first is by asking questions about words, assessing the patient's knowledge of both physical and functional properties of objects (e.g., Does a horse have hoofs or paws? Is a chisel used for cooking or making furniture?). The second is by word—picture matching (assuming the patient has done well on the test of picture recognition). Foils are from the same semantic category as the target and are visually similar to the target (e.g., the word *chisel* paired with pictures of a chisel and a screwdriver). Objects in the categories of tools, animals, and foods are tested. *Recognition of complex words* is tested by randomly presenting real complex words (e.g., *decision, runs*) and stimuli that are nonwords because of an illegal combination of a stem and an affix (e.g., *decidation, runned*) and asking the patient to say which are real and which are not. *Comprehension of morphologically complex words* is tested by asking the patient to match a complex word to one of two others, where the match depends on understanding the affix (e.g., Does *official* mean *executive* or *execution*?).

Sentence comprehension is assessed by having the patient match a sentence to one of two pictures. Both lexical foils (e.g., "The cake was eaten by the boy," matched to a correct picture and a picture showing a cake being eaten by a girl) and syntactic foils ("The man was pushed by the woman" matched to the correct picture and one of a man pushing a woman) are used. Both active sentences ("The boy ate the cake") and passive forms ("The cake was eaten by the boy") are tested in both these conditions.

On the production side, *word production* is tested by picture naming. The items to be named come from the categories of tools, animals, and foods and include words that are one syllable and three or more syllables in length. Tests of the ability to repeat long and short words and nonwords are administered. Errors of word and nonword production are classified as omissions, semantic errors, phonemic paraphasias, apraxic speech, and dysarthric speech. Tests of the ability to recognize that the word for two different pictures is a homophone (e.g., a pair of eye glasses and a set of drinking glasses) provide information about a patient's ability to access the forms of words despite a speech production impairment. *Production of morphologically complex words* is tested by having the patient change a word to fit a sentence context (e.g., Change *courage* to fit in the sentence: 'If a man is brave, we say he is ___________.'). *Sentence production* is tested by having the patient describe a picture, mentioning all the items in it and beginning with a particular item. For instance, the patient is required to describe a picture of a man pushing a bicycle, beginning with the word for bicycle; this forces him to use a complex syntactic form, such as the passive sentence, "The bicycle is being pushed by the man." The ability to produce various syntactic forms is noted.

Testing written language proceeds along the same lines. In addition to the factors just described, the regularity of the mapping from orthography to phonology and vice versa is varied in stimuli used in lexical decision, reading aloud, and spelling tests.

If tests such as these are administered, three steps must be taken to relate abnormal performance on these tasks to disturbances of components of the language processing system. First, the clinician must be able to identify abnormal performance on each task. Doing this depends on the availability of norms and Z scores for tests of this sort, which are being developed for some psycholinguistic batteries. Second, the clinician must determine that major nonlinguistic factors are not affecting performance. Neuropsychological evaluations can help rule out disturbances of nonlinguistic factors as the basis for a patient's abnormal performance. Finally, the clinician must be able to interpret a pattern of performance on a series of tests as an indication of what language processing components are affected. The examiner can attribute a *primary* deficit in a particular processing component to a patient only in the following circumstances:

- Performance on the test that requires that component is abnormal.
- Linguistic input to that component is intact (as judged by performance on other subtests).

The interpretation of the major patterns of performance on the tests described previously in terms of deficits in psycholinguistically defined language processors is listed in Tables 10-3 to 10-5.

Neurologic Observations

Until the mid-1970s, most of what was known about the neural basis for language was based on the correlation of language deficits with brain lesions described at autopsy. With the advent of computed tomography (CT) and magnetic resonance imaging (MRI), clinical and neuroradiologic correlations have greatly expanded the database relevant to localization of language processing. Positron emission tomography (PET) with radiolabeled deoxyglucose and single proton emission computed tomography (SPECT) have been used to identify regions of cerebral hypometabolism in patients, for correlation with language deficits. The regions of hypometabolism so identified are typically much larger than those in which necrosis is visible on CT or MRI.

Functional neuroimaging (PET, functional MRI) has recently emerged as a potentially powerful new way to provide data regarding the

TABLE 10-3. *Deficits in Auditory Comprehension, Defined by Performances on Psycholinguistic Tests*

Deficient Component	Pattern of Performance on Subtest[a]
Word level	
Acoustic–phonetic processing	*Abnormal performance on the phonemic discrimination test*
Auditory lexical access	Normal performance on phonemic discrimination test *Abnormal performance on the lexical decision task with words*
Lexical semantic access	Normal performance on the phonemic discrimination test Normal performance on the lexical decision task with words *Abnormal performance on any lexical comprehension test*
Affixed word level	
Morphologic analysis	Normal lexical access *Abnormal performance on lexical decision for affixed words*
Morphologic comprehension	Normal lexical decision for affixed words *Abnormal performance on any test of affixed word comprehension*
Sentence level	
Lexicoinferential comprehension	Normal lexical semantic access and morphologic comprehension *Abnormal comprehension of constrained sentences*
Parsing and syntactic comprehension	Normal lexical semantic access and morphologic comprehension Normal performance on constrained sentence comprehension *Abnormal performance on semantically unconstrained syntactically complex sentence comprehension*

[a]Subtests on which the subject must perform normally are indicated in plain type. Subtests on which abnormal performance is a prerequisite for the assignment of a particular deficit are italicized.

TABLE 10-4. *Deficits in Oral Production, Defined by Performances on Psycholinguistic Tests*

Deficient Component	Pattern of Performance on Subtest[a]
Word level	
Accessing lexical phonologic forms (from semantics)	Normal performance on picture comprehension screen *Abnormal performance on naming task* *Abnormal performance on homophone judgment task*
Phonologic output planning	Normal performance on picture comprehension screen, homophone judgment, phonemic discrimination, auditory, and written lexical decision *Phonemic paraphasias in naming, repetition, and oral reading tasks*
Affixed word level	
Accessing affixed words (from semantics)	Naming and repetition adequate for the patients' oral production of words to be recognized *Abnormal production of complex words in morphology production task*
Sentence level	
Expression of thematic roles	Normal word production in isolation Normal performance on affixed word production *Failure to produce word sequences that convey correct thematic roles on sentence production task*
Construction of syntactic structures	Normal word production in isolation Normal performance on affixed word production *Failure to produce complex structures (e.g., passives) on sentence production task*
Insertion of function words into syntactic structures	Normal word production in isolation Normal performance on affixed word production *Agrammatism and/or paragrammatism on sentence production task*
Insertion of content words into syntactic structures	Normal word production in isolation Normal performance on affixed word production *Anomia and/or phonemic paraphasias in content words on sentence production task*

[a]Subtests on which the subject must perform normally are indicated in plain type. Subtests on which abnormal performance is a prerequisite for the assignment of a particular deficit are italicized.

TABLE 10-5. *Deficits in Written SingleWord Processing, Defined by Performances on Psycholinguistic Tests*

Deficient Component	Pattern of Performance on Subtest[a]
Comprehension of written words	
Written lexical access	*Abnormal performance on the written lexical decision task*
Written lexical semantic access	Normal performance on the written lexical decision task *Abnormal performance on any written lexical comprehension task*
Production of written words	
Accessing lexical orthography from semantics	Normal performance on picture comprehension screen Normal writing of words to dictation *Abnormal performance on written picture-naming task*
Accessing lexical orthography from lexical phonology	Normal auditory lexical access Normal writing of nonwords to dictation *Abnormal (regularized) writing of irregular words to dictation*
Accessing sublexical orthography from sublexical phonology	Normal phoneme discrimination Normal writing of words to dictation *Abnormal writing of nonwords to dictation*
Reading single words	
Accessing lexica phonology from lexical orthography	Normal written lexical access Normal regular word reading and nonword reading *Abnormal (regularized) reading of irregular words*
Accessing sublexical phonology from sublexical orthography	Normal oral naming Normal word reading *Abnormal nonword reading*

[a]Subtests on which the subject must perform normally are indicated in plain type. Subtests on which abnormal performance is a criterion for the assignment of a particular deficit are italicized.

localization of language processors in healthy subjects. Measurement of electrophysiologic events associated with language processing has also increased in sophistication, with respect to both measurement and localization of event-related potentials and with respect to the design of psycholinguistic experiments. Intracortical recordings in response to language stimuli have been undertaken during neurosurgical operations for focal epilepsy. Intraoperative cortical stimulation has been used to study the localization of several aspects of language processing. The technique has been extended to the preoperative situation, through the use of grids of electrodes positioned in the subdural space. With this background, let us turn to a description of disorders of language processing.

NEUROGENIC DISORDERS OF PSYCHOLINGUISTIC PROCESSING IN ADULTS

Diseases of the central nervous system frequently result in disturbances of the ability to use the language code. These disorders are commonly referred to as *aphasia. Primary* aphasic disturbances affect the operation of one or more of the components of the language processing system directly. An example of a primary aphasic disturbance is an inability to construct aspects of the syntactic structure of a sentence after having recognized the words in the sentence. *Secondary* aphasic disturbances arise when disruptions of other cognitive functions lead to interference with the normal operation of one or more of the components of the language processing system. An example of a secondary aphasic disturbance is so-called neglect dyslexia, in which a disturbance of reading results from a disturbance of visual processing in part of the visual field. In this review, the focus is on primary psycholinguistic disturbances, beginning with those that affect single words, and then turning to word formation, sentences, and discourse.

Disturbances of Word Meanings

Most recent research on disturbances of word meanings in patients has focussed on words that refer to objects. Disturbances of word meanings cause poor performance on word—-picture matching and naming tasks. The combination of deficits in word-picture matching and naming, however, may be caused by separate input- and output-side processing disturbances that affect word recognition and production. Simultaneous deficits in naming and word—-picture matching are more likely to result from a disturbance affecting concepts when the patient does the following:

- Makes many semantic errors in providing words to pictures and definitions
- Has trouble with word–picture matching with semantic, but not phonologic, foils
- Fails on categorization tasks with pictures
- Encounters difficulty or produces errors with the same words in production and comprehension tasks

It has been argued that brain damage may affect either the storage or the retrieval of word meanings in semantic memory. There are five hallmarks of the loss of items in semantic memory:

1. Consistent production of semantic errors on particular items across different inputs (pictures, written word, spoken words)
2. Relative preservation of superordinate information as opposed to information about an item's features
3. Relative preservation of information about higher frequency items
4. Improvement of performance by priming and cueing
5. No effect of the rate at which a task is performed on performance

Disorders affecting processing of semantic representations for objects may be specific to certain types of inputs. These impairments have been taken as reflections of disturbances of verbal and visual semantic systems, although this interpretation is debated.

Semantic disturbances may also be category specific, for example, a selective semantic impairment of concepts related to living things and foods compared with man-made objects. The opposite pattern has also been found. Selective preservation and disruption of abstract versus concrete concepts and of nominal versus verbal concepts have also been reported.

Disturbances of Spoken Word Production

Disturbances affecting the oral production of single words are extremely common in patients with language impairments. There are three basic disturbances affecting word production (other than semantic deficits). They affect accessing the forms of words from concepts, planning the form of a word for articulation, and articulation itself.

A disturbance in activating word forms from concepts is a common consequence of stroke in the perisylvian region. It is manifest by an inability to produce a word from a semantic stimulus (a picture or a definition) coupled with intact processing at the semantic and phonologic levels (determined by answering questions about pictures, picture categorization tests, and repetition). The form of a patient's errors is not a good guide to whether he or she has an impairment at this level of the production process because disturbances in accessing word forms may appear in a variety of ways, ranging from pauses to neologisms (complex sequences of sounds that do not form words) and semantic paraphasias (words related to the meaning of the target item). Rarely, patients show an inability to name objects presented in one modality only, even though they demonstrate understanding of the concept associated with that object when it is presented in that modality, for example, *optic* aphasia and *auditory* aphasia. Because basic perceptual mechanisms appear to be intact in these patients, these modality-specific naming disorders are taken to reflect a failure to transmit information from modality-specific semantic systems to the processor responsible for activating the forms of words.

Disturbances of a patient's ability to convert the representation of the sound of a word into a form appropriate for articulatory production usually manifest as phonemic paraphasias (substitutions, omissions, and misorderings involving phonemes or syllables). Three features of a patient's performance suggest a disturbance in word sound planning:

1. Some phonemic paraphasias are closely related to target words (e.g., /befenit/ for /benefit/).
2. Multiple attempts are made that come closer and closer to the correct form of a word.
3. Similar phonologic errors are made in word repetition, word reading, and picture naming.

Patients often have disturbances of articulation itself, as shown by abnormalities in the acoustic waveform produced by a patient and in the movement of the articulators in speech. Investigators have identified two major disturbances of articulation: dysarthria and apraxia of speech. Dysarthria is marked by hoarseness, excessive nasality, and imprecise articulation and has been said to not be significantly influenced by the type of linguistic material that the speaker produces or by the speech task. Apraxia of speech is marked by difficulty in initiating speech, searching for a pronunciation, better articulation for automatized speech (e.g., counting) than volitional speech, abnormal prosody, omissions of syllables in multisyllabic words, and simplification of consonant clusters (often by adding a short neutral vowel sound between consonants).

Lexical phonologic access disorders and phonemic paraphasias occur after lesions in all areas of the perisylvian language zone. Apraxia of speech and dysarthria arise after lesions in the anterior portions of the language zone; dysarthria also occurs with subcortical hemispheric lesions as well as lesions in the pons. Most basal ganglia lesions affect articulation (e.g., Parkinson's disease produces a low-volume, rapid speech with articulatory distortions; Huntington's disease and Wilson's disease produce characteristic disturbances affecting the production of phonemic quality, as well as prosody). Cerebellar lesions affect the timing of speech, leading to a so-called scanning speech pattern.

Disturbances of Recognition of Auditorily Presented Simple Words

Disturbances affecting auditory comprehension of simple words have been attributed to impairments of semantic concepts, as discussed earlier, or to an inability to recognize spoken words. The latter disturbances have, in turn, been thought to have two possible origins:

- Disturbances affecting the recognition of phonemes in the acoustic signal
- Disturbances affecting the ability to recognize words despite good acoustic-phonetic processing

In most cases, single-word comprehension problems are probably multifactorial in origin and result from a complex interaction of acoustic-phonetic disturbances, disturbances in recognizing spoken words, and disturbances affecting word meanings.

Disturbances of phoneme discrimination and identification have been documented after lesions in all areas of the perisylvian cortex, and other patients with lesions in all areas of this cortex have had no impairments on tests of these functions. The lesions responsible for *pure word deafness*—an inability to understand spoken words that is much greater than any difficulty with written word comprehension—tend to be bilateral and to involve posterior perisylvian regions, especially the temporal lobes.

Repetition of Single Words

Repetition of a word can be carried out in three ways:

Nonlexically: by repeating sounds without recognizing the word (as if one were imitating a foreign language)

Lexically: by recognizing the stimulus as a word and uttering it without understanding it

Semantically: by understanding the word and reactivating its form from its meaning

Any of these routes to repetition may be disturbed. The classic localization of a repetition disorder is in the arcuate fasciculus connecting the temporal and frontal lobes, but repetition can be affected by lesions throughout the perisylvian cortex.

Disturbances of Processing Morphologically Complex Words

Disturbances of word formation can arise in either word recognition and comprehension or in word production. On the input side, researchers have observed that some patients who make derivational paralexic errors (e.g., write → wrote; fish → fishing; directing → direction) in the oral reading of complex words have particular difficulty with the recognition and analysis of written morphologically complex words compared with morphologically simple words. Disturbances affecting morphologic processing also appear in single-word production tasks. Disturbances affecting the production of morphologically complex words are most commonly seen in sentence production, where they are known as *agrammatism* and *paragrammatism.* These disorders are described in the section that follows. Too few cases with disturbances affecting word formation for isolated words have been described for any clear pattern to emerge regarding the localization of the responsible lesions.

Disorders of Sentence Production

Disturbances at the sentence production level are the inevitable result of disturbances affecting the production of simple or complex words. In addition, many patients have problems in the sentence planning process itself. Agrammatism and paragrammatism are impairments affecting the ability to produce function words and morphologic elements (see earlier discussion) and may arise only in sentence production and not when a patient produces words in isolation. The most noticeable deficit in agrammatism is the widespread omission of function words and affixes with better production of common nouns. Agrammatic patients usually produce only very simple syntactic structures.

Agrammatism is classically associated with Broca's aphasia, which classically has been related to lesions of Broca's area—the pars triangularis and pars opercularis of the third frontal convolution. However, Broca's aphasia typically is a consequence of larger lesions that include, but are not restricted to, Broca's area (13). Vanier and Caplan have documented considerable variability in the exact localization of strokes within the language zone associated with agrammatic speech (14). Paragrammatism is thought to follow more posterior lesions than agrammatism; detailed deficit—lesion correlational studies equivalent to those that have been done for agrammatism have not yet been reported for paragrammatism.

Many patients have difficulty producing prosodic aspects of speech. These disturbances may be secondary to motor output disorders or associated with other sentence production disorders. Patients with anterior and central right-hemisphere lesions have a variety of disturbances of prosody in sentence production. These patterns occurred regardless of the emotion associated with a sentence and thus are different from the aprosodías related to emotional display described after right-hemisphere disease. This reflects a primary disturbance of production of intonation in right-hemisphere—damaged patients, which differs as a function of lesion location in the hemisphere.

Disorders of Sentence Comprehension

The greatest amount of work in the area of disturbances of sentence comprehension has involved patients whose use of syntactic structures to assign meaning is not normal. Patients may have very selective disturbances affecting the use of particular syntactic structures or elements to determine the meaning of a sentence. Some patients can understand very simple syntactic forms, such as active sentences ("The man hugged the woman") but not more complex forms, such as passive sentences ("The woman was hugged by the man"). Many of these patients use strategies such as assigning the thematic role of agent to a noun immediately before a verb to understand semantically reversible sentences, leading to systematic errors in comprehension of sentences such as "The boy who pushed the girl kissed the baby." Other patients have virtually no ability to use syntactic structure. Most of these patients appear to rely on inferences based on their knowledge of the real world and their ability to understand some words in a sentence. Some patients can assign and interpret syntactic structures unconsciously but not use these structures in a conscious, controlled fashion.

Many patients who have difficulty with sentence comprehension tasks are agrammatic Broca's aphasia patients, suggesting that Broca's area is responsible for syntactic processing. No correlation between lesion site within the dominant perisylvian cortex and the severity of a syntactic comprehension deficit has been found. Sentence comprehension disorders are thus associated with lesions throughout the perisylvian cortex.

Disorders of Discourse Processing

Patients with disturbances affecting the word and sentence level of language processing would also be expected to have impairments of processing at the level of discourse. However, although aphasic patients do have some trouble with discourse structures, their ability to process discourse is often preserved to a surprising degree. This indicates that, although aphasic patients have some impairments relative to healthy subjects in tasks that require comprehending and retaining aspects of discourse structure, the aphasic patients perform well on tasks that require that they extract main ideas and topics from discourses. Contextualization may be possible in aphasic patients, even on the basis of very little lexical and propositional information, and may be able to compensate for disorders in basic linguistic abilities.

Recent studies have documented disturbances affecting the comprehension of discourse in patients without other language impairments. The most extensive studies of this type have been carried out in right-handed patients with right-hemisphere lesions (mostly strokes). These results have led to the view that the right hemisphere is responsible for aspects of the processing of discourse.

Disorders of Reading (Alexia)

The contemporary study of acquired dyslexias has focused on impairments in the ability to read single words aloud. Neuropsychologists have described disorders of three separate and partially independent routines in the brain for converting a written word into its spoken form. The first pathway (routine 1) involves recognizing a word visually, gaining access to its meaning, and then activating the sound of the word from its meaning. The second pathway (routine 2) translates the orthography of the entire word directly into a pronunciation, without first contacting the meaning. Finally, routine 3 decomposes the word into orthographic segments (graphemes and other spelling units) and derives a pronunciation by assigning each of them a spoken (phonemic) value.

The classic lesion localization in pure alexia (alexia without agraphia) is in the left occipital lobe and deep white matter and is thought to disconnect language areas in the left hemisphere from visual input to the right. In addition to this "dysconnection," other left-hemisphere lesions cause the different forms of alexia. Patients with deep dyslexia tend to have large lesions, and it has been suggested that their reading partially results from right-hemisphere mechanisms. Patients with surface dyslexia also have fairly large lesions, perhaps most often involving temporal lobe structures. Lesions in phonologic dyslexia have not been adequately described. Letter-by-letter readers (also termed *pure alexics*) have been described with lesions affecting the left parietal lobe.

Disorders of Writing (Agraphia)

The acquired dyslexias have their counterparts in the acquired agraphias. Patients with phonologic agraphia are severely impaired in their ability to spell or write nonsense words but are capable of very good performance on legitimate words, even words that are low in frequency and contain unusual spelling patterns (e.g., *leopard*). The deficit in these cases appears to lie in the ability to convert sublexical phonologic units to orthographic units (graphemes and letters). The converse impairment—an inability to access the written forms of whole words with preserved ability to convert sublexical phonologic units to orthographic units, termed *surface agraphia* or *lexical agraphia*—has also been described. A third disturbance, known as *asemantic writing* consists of the inability to write spontaneously but the retained ability to write to dictation. This suggests that the contents of the visual word-form system can be addressed from spoken input but not from the meaning of a word.

The motor scheme for a letter appears to distinguish between the movements denoting the shape of a letter and the parameters that govern scale factors like magnitude and orientation. Cases of apractic agraphia reveal a loss of the motor programs necessary for producing letters. Written characters are poorly formed and may be indecipherable, although even severely affected patients maintain the distinction between cursive and printed letters and between upper and lower case. Evidence indicates that the disturbance need not be associated with limb apraxia. Certain patients with right-hemisphere damage have no difficulty constructing written letters or words, but they exceed the correct number of strokes on letters needing repetitive movements.

NATURAL HISTORY OF LANGUAGE DISORDERS

As expected, etiology is the primary factor determining the evolution of an aphasic impairment. Most research into the natural history of language impairments has focused on the evolution of aphasic "syndromes." In the acute phase, patients tend to be severely affected. Symptomatology varies widely—for instance, speech output may range from mutism to jargon—and often suggests major disturbances in control mechanism (e.g., perseveration is frequently present). In the chronic phase of evolution of such lesions, patients tend to show stable deficits that are often highly restricted, of the sort reviewed earlier. Both clinical experience and experimental investigations of patients with long-standing aphasia indicate that considerable recovery can occur after such lesions.

The role of speech-language therapy in the management of aphasic patients is well established, but controversy exists about the effectiveness of such intervention in improving specific psycholinguistic functions (15). There are few adequate studies of this subject, and the results of these studies are contradictory.

APHASIC SYNDROMES AND THE CLASSIFICATION OF LANGUAGE DISORDERS

Physicians often describe patients with language impairments in terms of syndromes, such as Broca's aphasia, Wernicke's aphasia, conduction aphasia, and others. These syndromes are listed in Table 10-6. The major

TABLE 10-6. *Classic Aphasic Syndromes Affecting Auditory–Oral Language Functions*[a]

Syndrome	Clinical Manifestations	Postulated Functional Deficit	Usual Lesion Location
Syndromes attributed to disturbances of cortical centers			
Broca's aphasia	Major disturbance in speech production with sparse, halting speech, often misarticulated, frequently missing function words and bound morphemes	Disturbances in the speech planning and production mechanisms	Primarily posterior aspects of the 3rd frontal convolution and adjacent inferior aspects of the precentral gyrus
Wernicke's aphasia	Major disturbance in auditory comprehension; fluent speech with disturbances of the sounds and structures of words (phonemic, morphologic, and semantic paraphasias)	Disturbances of the permanent representations of the sound structures of words	Posterior half of the first temporal gyrus and possibly adjacent cortex
Anomic aphasia	Disturbance in the production of single words, most marked for common nouns with variable comprehension problems	Disturbances of the concepts and/or sound patterns of words	Inferior parietal lobe or connections between parietal and temporal lobe
Global aphasia	Major disturbance in all language functions	Disruption of all language-processing components	Large portion of the perisylvian association cortex
Syndromes attributed to disruptions of connections between centers			
Conduction aphasia	Disturbance of repetition and spontaneous speech (phonemic paraphasias)	Disconnection between the sound patterns of words and the speech production mechanism	Lesion in the arcuate fasciculus and/or corticocortical connections between temporal and frontal lobes
Transcortical motor aphasia	Disturbance of spontaneous speech similar to Broca's aphasia with relatively preserved repetition	Disconnection between conceptual representations of words and sentences and the motor speech production system	White matter tracts deep to Broca's area
Transcortical sensory aphasia	Disturbance of single word comprehension with relatively intact repetition	Disturbance in activation of word meanings despite normal recognition of auditorily presented words	White matter tracts connecting parietal lobe to temporal lobe or in portions of inferior parietal lobe
Isolation of the language zone	Disturbance of both spontaneous speech (similar to Broca's aphasia) and comprehension, with some preservation of repetition	Disconnection between concepts and both representations of word sounds and the speech production mechanism	Cortex just outside the perisylvian association cortex and/or white matter beneath perisylvian cortex

[a]Modified from Benson DF, Geschwind N. Aphasia and related cortical disturbances. In: Baker AB, Baker LH, eds. *Clinical neurology.* New York: Harper & Row, 1971.

difference between the syndrome-oriented and psycholinguistic approaches to language disorders is that the syndrome-oriented approach assigns each patient to one and only one aphasic group, whereas the psycholinguistic approach allows a patient to be assigned any number of independent language processing impairments. The syndrome-oriented approach is often said to receive support from the correlation of the classic aphasic syndromes with lesion sites in the brain. As the previous review of the neural correlates of language processing deficits reveals, however, the data relating impairments in specific language processing components to lesion localization within the perisylvian cortex are weak.

In contrast to the traditional, syndrome-oriented approach to language disorders, the psycholinguistic approach consists of identifying the disturbances in the major components of the language processing system that are present in each patient. In this approach to taxonomy, a patient is likely to have more than one deficit. Most researchers in the field now believe that little is gained by retaining the more global and less precise traditional nomenclature to describe a patient when an adequate psycholinguistic characterization of a patient's deficit is available.

PSYCHIATRIC DISEASE AND LANGUAGE DISORDERS

Language disorders have not been studied as thoroughly in psychiatric disease as in neurologic disease. The psychiatric disease whose effects on language are best documented is schizophrenia. Early descriptions of schizophrenic patients often noted language abnormalities, some of which (e.g., "clang associations") could well reflect disturbances of language processors. The older literature is difficult to interpret because some of the patients included in these studies may well have had structural neurologic disease, such as purely temporal lobe strokes, that could not be diagnosed at the time. Contemporary studies, however, in which the diagnosis of schizophrenia is well established, have also shown linguistic abnormalities in these patients. Abnormalities most clearly affect the lexical and discourse levels of language as well as certain aspects of verbal memory. They do not appear to affect the processing of syntactic form in sentence comprehension but may do so in sentence production. There are no data available regarding word formation in schizophrenia. Although language abnormalities in schizophrenia may be secondary to other cognitive impairments, some are likely to be primary disturbances of psycholinguistic processors. Schizophrenia, like many nonfocal neurologic conditions, such as Alzheimer's disease, appears to affect a variety of aspects of language processing. As with "neurologic" disease, schizophrenia does not always affect language, and some of the language processing abnormalities found in schizophrenia appear to be mild; however, many abnormalities of language are identifiable in individual schizophrenic patients.

FUNCTIONAL NEUROANATOMY OF LANGUAGE PROCESSING

The first step in understanding the neural basis for a function has traditionally been to identify the areas of the brain where that function takes place. Clinical studies indicate that the association cortex in the region of the sylvian fissure is responsible for language processing in the auditory-oral modality. This region includes the posterior half of the pars triangularis and the pars opercularis of the third frontal convolution (Broca's area), the association cortex in the opercular area of the precentral and postcentral gyri, the supramarginal and angular gyri of the parietal lobe, the first temporal gyrus from the supramarginal gyrus to a point lateral to Heschl's gyrus (Wernicke's area), and possibly a portion of the adjacent second temporal gyrus. In addition, language disorders have been described in association with electrocortical stimulation of the lingular gyrus.

The role of the perisylvian association cortex as the substrate for language in the auditory-oral modality does not vary as a function of major endogenous biologic factors, such as sex or important environmental phenomenologic factors, such as literacy. Language in the visual-gestural modality in deaf people is also based in the perisylvian cortex, although it may recruit more superior regions of cortex in frontal and parietal lobes as well. Written language is a secondary development that depends on instruction. It appears to involve areas of the brain that are more closely associated with visual processing.

The supplementary motor area is the only other cortical structure that has been suggested to play a role in language processing. Its primary function in language tasks, however, appears to be to initiate vocalization, not to activate linguistic representations through subserving a component of the language processing system per se.

Several subcortical nuclei have also been suggested to play a role in language. These include the thalamus, the caudate, and possibly parts of the striatum. White-matter tracts are thought to play important roles in transmitting the products of computations in one cortical area to another and to lower motor centers. The roles of these different portions of the hemispheres are discussed in turn.

Cortical Structures and Language

Two general classes of theories on the relationship of portions of the perisylvian association cortex to components of the language processing system have been developed, one based on "holist" or distributed views of neural function and one based on localizationist principles.

Holist Theories

The basic tenet of holist or distributed theories of the functional neuroanatomy for language is that linguistic representations are distributed widely and that specific stages of linguistic processing recruit widely scattered areas of perisylvian association cortex. The evidence supporting holist theories consists of the ubiquity of general factors in accounting for the performance of aphasic patients. Against the holist model is the finding that multiple individual language deficits arise in patients with small perisylvian lesions, often in complementary functional spheres.

Localizationist Theories

Although many localizationist models exist, the "connectionist" model of language representation and processing in the brain revived by Geschwind and colleagues in the 1960s and 1970s probably remains the best-known localizationist model of the functional neuroanatomy of language, at least in medical circles in North America (16). The basic connectionist model of auditory-oral language processing postulates three basic centers for language processing, all in cerebral cortex. The first center, located in Wernicke's area, stores the permanent representations for the sounds of words (what psycholinguists would now call a *phonologic lexicon*). The second, located in Broca's area, houses the mechanisms responsible for planning and programming speech. These localizations are thought to evolve from the relationship of these areas of the brain to primary sensory and motor regions. The third center, diffusely localized in cortex in the nineteenth-century models, stores the representations of concepts.

Geschwind proposed that the inferior parietal lobule—the supramarginal and angular gyri—is the site at which the fibers projecting from somesthetic, visual, and auditory association cortices all converge, and that as a consequence of this convergence, associations between word sounds and the sensory properties of objects can be established in this area. Geschwind argued that these associations are crucial aspects of the meanings of words and that their establishment is a prerequisite to the ability to name objects.

Language processing in this model involves the activation of linguistic representations in these cortical centers and the transfer of these representations from one center to another, largely by means of white-matter tracts. The principal evidence in favor of this model is said to be the occurrence of specific syndromes of language disorders that can be accounted for by lesions of these centers and the connections between them, shown in Table 10-3. The correlations between lesion sites and aphasic syndromes are far from perfect, however, even in vascular cases, and they become less reliable in other neurologic conditions. The variability in lesions associated with Wernicke's aphasia has been strikingly documented (17). The use of PET scanning to define lesions shows large, overlapping areas of hypometabolism in association with each of the clinical aphasic syndromes, leading to uncertainty regarding the areas of functionally abnormal brain responsible for each.

Functional neuroimaging based on PET has recently been used to study the regions of cortex that are activated during the performance of a number of language tasks by normal subjects. On the basis of these studies, a number of localizations have been suggested: visual word recognition in dominant peristriate cortex; auditory word recognition in temporal lobe; the lexicons and lexical semantic functions in frontal cortex; and rehearsal in Broca's area. The dominant inferior parietal lobe has shown little activation in language tests to date, perhaps because of some degree of insensitivity of the technique. The PET activation results are generally compatible with the results of deficit—lesion correlations. (A notable point of debate is the role of the frontal lobe in lexical semantics.)

Overall, the picture that is beginning to emerge is that different components of the language processing system are localized in different parts of the perisylvian neocortex. There is considerable evidence for some individual differences in localization, but the extent of this phenomenon is not yet clear. Several features of the effects of lesions—graceful degradation and complexity effects—are consistent with parallel distributing processing models of language; one possibility is that at least some language processors make use of this mode of representation and computation.

Subcortical Structures and Language

White-Matter Tracts

There are conflicting reports regarding the language disorders that follow white-matter lesions. According to some reports, the aphasic syndromes that follow white matter lesions do not differ from those that occur with perisylvian cortical lesions, and the classic aphasic syndromes correlate with subcortical lesion sites. The language disorders seen with subcortical CT lesions are not easily classified as any of the standard aphasic syndromes, that language disturbances of all sorts occur with lesions in all subcortical areas and that total sparing of language functions can follow lesions in identical subcortical areas. A relative sparing of language functions has been noted in multiple sclerosis; however, this may be because multiple sclerosis lesions do not affect white-matter tracts in a manner needed to interrupt language processes. Knowledge of the white-matter tracts that carry linguistic information from one processor to another and to effector motor systems remains limited.

Gray-Matter Nuclei

Aphasic disturbances can follow strokes in the thalamus, caudate, and parts of the striatum. It is likely that at least some aphasic symptoms seen after deep gray-matter lesions reflect the effects of disturbances in other cognitive functions on language. Intraoperative stimulation studies of the interference with language functions after dominant thalamic stimulation also suggest that the language impairments seen in at least some thalamic cases are due to disturbances of attentional mechanisms. However, many language impairments associated with subcortical lesions are likely to be primary.

Perhaps the most important consideration regarding language disorders after subcortical lesions is the question of whether they result from altered physiologic activity in the overlying cortex, not from disorders of the subcortical structures themselves. The availability of patients with focal strokes that are visible only subcortically on CT scans, in whom metabolic scanning is used to assess lesion site and size in both cortical and subcortical structures, provides an opportunity to investigate the role that both cortical and subcortical structures play in language. There is a perfect correspondence in published cases between the presence or absence of cortical hypometabolism or hypoperfusion and the presence or absence of aphasic impairments in patients with focal strokes visible only subcortically on CT scans. The conclusion that is suggested by this pattern is that subcortical structures play no essential role in core language processes themselves but are essential parts of complex neural systems, whose cortical regions are responsible for psycholinguistic computations.

LATERALIZATION OF LANGUAGE PROCESSES

In about 98% of familial strong right-handers, the left perisylvian association cortex accomplishes most if not all language processing functions at the word, word-formation, and sentence level. In people with anomalous dominance (handedness) profiles, these psycholinguistic functions are far more likely to involve the corresponding regions of the right hemisphere, with different likelihoods of right- and left-hemispheric involvement in language functions in different subgroups within this population. The data on differential lateralization as a function of sex are controversial. Processing of discourse appears to involve the right hemisphere in right-handed subjects. Linguistic aspects of prosody are likely to be affected by right-hemisphere disease.

Many aphasic syndromes that follow either left- or right-hemisphere lesions in subjects with anomalous dominance are often mild. To the extent that they reflect disturbances of isolated components of the language processing system, their occurrence indicates that many individual language processing components can be located in either hemisphere.

There are intriguing similarities between the phenomena of localization and lateralization of language. In both cases, the location of a particular language processing component appears to vary across the adult population as a whole. But in both situations, there are central tendencies with respect to the location of particular language processing components. There appear to be preferred sites for particular language processing functions within the perisylvian region, and there is a strong preference for language processing components to be left-hemisphere based. In the case of lateralization, these central tendencies are strongly affected by handedness profiles. It is possible that these features result from commonalities in the genetically controlled mechanisms that determine the neural substrate responsible for particular language processing operations.

SUMMARY

The view of language disorders presented here is based on linguistic theory and psycholinguistic models. The beginnings of a detailed functional neuroanatomy of language processing are emerging from deficit—-lesion correlations and functional neuroimaging studies that make use of this approach. Research using available and foreseeable psycholinguistic and neurolinguistic techniques is likely to continue to deepen our understanding of language disorders and of the way the human brain supports language functions.

REFERENCES

1. Caplan D. *Language:* structure, *processing and disorders.* Cambridge, MA: MIT Press (Bradford Books), 1992.
2. Fodor, JA. *The modularity of mind.* Cambridge, MA: MIT Press, 1983.
3. King J, Just MA. Individual differences in syntactic processing: the role of working memory. *J Mem Lang* 1991;30:580–602.
4. Shallice T. *From neuropsychology to mental structure.* Cambridge, MA: Cambridge University Press, 1988a.
5. Goodglass H, Kaplan E. *The assessment of aphasia and related disorders.* 2nd ed. Philadelphia, PA: Lea & Febiger, 1982.
6. Kertesz A. *Aphasia and associated disorders: taxon-*

omy, localization, and recovery. New York: Grune & Stratton, 1979.
7. Porch BE. *The Porch Index of Communicative Ability: administration, scoring and interpretation.* Palo Alto, CA: Consulting Psychologists 1971.
8. Wepman, JM. *Auditory Discrimination Test.* Chicago, IL: Language Research Associates, 1958.
9. Dunn LM. *Expanded manual for the Peabody Picture Vocabulary Test.* Circle Pines, MN: American Guidance Service, 1965.
10. Kaplan E, Goodglass H, Weintraub S. *The Boston Naming Test.* Boston, MA: Veterans Administration, 1976.
11. DeRenzi E, Vignolo LA. The Token Test: a sensitive test to detect receptive disturbances in aphasics. *Brain* 1962; 85:665–678.
12. Kay J, Lesser R, Coltheart M. *Psycholinguistic assessments of language processing in aphasia (PALPA).* East Sussex, England: Lawrence Erlbaum Associates, 1992.
13. Mohr JP, Pessin MS, Finkelstein S, Funkenstein H, Duncan GW, Davis KR. Broca aphasia: pathologic and clinical. *Neurology* 1978;28:311–324.
14. Vanier M, Caplan D. CT scan correlates of agrammatism. In Obler LM, ed. *Agrammatic aphasia.* Amsterdam, Netherlands: Benjamins, 1990:97–114.
15. Wertz RT, Weiss DG, Aten JL, et al. Comparison of clinic, home and deferred language treatment for aphasia: a Veterans Administration cooperative study. *Arch Neurol* 1986;43:553–568.
16. Geschwind N. Disconnection syndromes in animals and man. *Brain* 1965;88:237–394, 585–644.
17. Bogen JE, Bogen GM. Wernicke's region: where is it? *Ann N Y Acad Sci* 1976;280:834–843.

11

Emotion

Kenneth M. Heilman and Dawn Bowers

OVERVIEW

Because the brain mediates emotion, diseases that affect the brain can induce emotional changes. In this chapter, we discuss the effects of focal brain lesions on emotions. Although disorders that affect the neurotransmitter systems may also influence emotions profoundly, these are not covered in this chapter. In addition, seizures and degenerative disorders, including disorders of the basal ganglia, can affect emotions. The behavioral effects of these disorders are described in more detail in other chapters. There are two major aspects to emotions: cognitive-communicative and experiential. Although we discuss both of these aspects of emotion, we begin this chapter with a discussion of cognitive and communicative disorders.

EMOTIONAL COGNITION AND COMMUNICATION

Comprehension

Although patients with psychiatric disorders and seizures may experience emotions even in the absence of an exciting event, for most normal people, emotions are induced by environmental events or memories of these. The environmental events include the following:

- Seeing emotional displays and gestures of others, such as emotional facial displays and scenes
- Hearing emotional prosody
- Hearing words

Many other sensory and perceptual disorders in the visual and auditory system (e.g., blindness) may lead to impaired comprehension of the emotional significance of environmental events, but these are not specific for emotion and are not discussed here.

Disorders of Emotional Facial Recognition

Patients with right-hemisphere stroke are impaired in the ability to recognize emotional facial displays (1). The right hemisphere plays a dominant role in many visuospatial and visuoperceptual activities. For example, patients with right-hemisphere disease are impaired in the ability to determine whether two faces, previously unknown to the patient, are the same or different people. Therefore, the inability of patients to recognize emotional faces may be related to a visuoperceptual defect. When patients with right- or left-hemisphere brain damage are equated for visuoperceptual abilities (2), patients with right-hemisphere brain damage are still impaired in their ability to name, select, and discriminate facial emotions. The crucial factor appears to be an impairment of "categorizing" facial emotional expressions. Studies conducted in patients with callosal section and selective hemispheric anesthesia support the ablative studies. In addition, tachistoscopic studies of healthy subjects have also supported the special role of the right hemisphere in processing facial emotional gestures. Although the mechanism underlying this hemispheric asymmetry is unknown, we have proposed that the right hemisphere contains stores of prototypic facial emotional icons or facial emotional representations (3).

Patients with right-hemisphere brain damage are also impaired in their ability to derive the target emotions from verbal descriptions (4). Because the emotional displays are described verbally, the impaired performance of patients with right-hemisphere brain damage cannot be accounted for by visuospatial perceptual deficits. No differences between right- and left-hemisphere—-damaged patients are observed when patients are assessed for emotional knowledge or emotional semantics. When right- and left-hemisphere—-damaged patients are asked to perform facial emotion and object imagery tasks, the patients with right-hemisphere damage are more impaired in facial emotion imagery than object imagery, whereas the group with left-hemisphere brain damage is more impaired in object imagery than in facial emotional imagery.

Others have suggested that there are a finite number of emotional facial expressions and that these emotional faces are universally comprehended (5,6); these study results suggest that knowledge of emotional facial expressions are not learned but rather are innate. For example, infants can recognize their mother's emotional faces, suggesting that knowledge of emotional faces is inborn. Moreover, studies of patients with brain damage suggest that, in the human brain, the right hemisphere contains inborn representations of species' typical emotional facial expressions. Although the exact cerebral location of these representations remains unknown, most patients with impaired comprehension and discrimination of facial emotional expressions have temporal parietal lesions. Single cells of the monkey's temporal cortex respond selectively to emotional faces. Intraoperative stimulation of patients undergoing epilepsy surgery has also implicated the right temporal lobe in the storage of these facial emotional representations.

Clinical cases have been reported in which patients are unable to name emotional faces or point to an emotional face named by the examiner. However, these patients can determine whether two faces display the same or different emotions. These patients have a deep lesion interfering with the white-matter tracts leading to the corpus callosum. This lesion causes a disconnection, functionally disassociating the speech and language areas of the left hemisphere from the facial and emotional representations stored in the right hemisphere.

Disorders of Emotional Scene Recognition

Relatively little is known about brain-damaged patients' ability to perform emotional scene recognition. Right- and left-hemisphere disease patients are equally impaired in selecting the most humorous of a group of cartoons; however, patients with left-hemisphere disease in general perform better when viewing the cartoons without captions. Patients with right-hemisphere damage may be impaired in interpreting emotional scenes; however, the defect is more likely to be related to a visuoperceptive disorder than to a disorder in a specific emotional system.

Disorders of Emotional Prosody

Speech prosody is conveyed by changes of pitch, amplitude, tempo, and rhythm. English speakers may use prosody as a syntactic marker. For example, in declarative sentences, the pitch drops at the end of a sentence, and in interrogative sentences, it rises. However, English speakers use prosody most commonly to convey emotional content. When patients with right- or left-hemisphere brain damage are asked to listen to sentences whose literal meaning is emotionally neutral but that are intoned with four different emotional prosodies (happy, sad, angry, and indifferent), patients with predominantly right temporoparietal lesions experience particular difficulties identifying the emotional tone of the speaker. This suggests that the right hemisphere has a special role in decoding emotional prosody.

When right- and left-hemisphere—-damaged patients are compared in terms of their ability to comprehend both emotional and syntactic prosody, both groups are equally impaired on the syntactic prosody task, but the patients with right-hemisphere brain damage perform significantly worse than those with left-hemisphere brain damage on the emotional prosody task. This suggests that whereas both hemispheres may be important in comprehending propositional prosody, the right hemisphere is dominant in comprehending emotional prosody.

Further support for a special role of the right hemisphere in comprehending emotional prosody comes from studies of healthy subjects using dichotic listening. Subjects report more words broadcast to the right ear (left hemisphere) than the left ear (right hemisphere). However, when subjects are asked to detect the mood of the speaker, as determined by prosody, left ear (right hemisphere) recall is superior to right ear (left hemisphere) recall.

In our discussion of naming emotional faces, we noted that the inability to do so can be caused by at least two mechanisms, a disconnection of the facial emotional representations from the left-hemisphere speech systems and destruction of the facial emotional representations. The inability to name an emotion while listening to emotionally intoned speech may be induced by a similar mechanism. Patients with right-hemisphere temporoparietal lesions are still unable to discriminate between the same and different emotional intonations even when the patients did not have to name or classify the prosody verbally. We believe that the right hemisphere contains representations or a lexicon of emotional prosodic expressions.

Disorders of Verbal Comprehension

Does the right hemisphere plays a special role in deriving emotional meaning from verbal language? Patients with right- and left-hemisphere brain damage are not different in their ability to comprehend both the denotative and connotative meanings of emotional and unemotional words. Borod and colleagues, however, demonstrated that patients with right-hemisphere brain damage have deficits in processing emotion on lexically based tasks (7). The auditory and reading comprehension of patients with left-hemisphere brain damage may be improved by the use of emotional words. Although these findings appear to support the postulate that the right hemisphere contains a special lexicon for emotional words, emotional words may also increase arousal and interest. Increased arousal and interest may be the crucial factor in the improved comprehension.

Emotional Semantics

Certain environmental events can predictably lead to one or two specific emotions. For example, loss leads to sadness (or anger), gain to happiness, and potential physical harm to fear (or anger). The knowledge of the type of circumstance that may be associated with an emotion we term *emotional semantics.* Although patients with right-hemisphere brain damage may have problems understanding environmental stimuli that may lead to emotion (because they have an impairment of facial emotional representations, emotional prosodic representations, or even an emotional verbal lexicon), it appears that their emotional semantics may nevertheless be intact. Because unilateral hemispheric disease does not impair emotional semantics, emotional knowledge may be widely distributed and bihemispheric. Diseases that impair widely distributed networks, such as Alzheimer's disease, may impair emotional semantics. When patients are presented with complex emotional scenes, Alzheimer's patients are impaired in their ability to label scenes and match facial expressions to scenes, suggesting a defect in emotional semantics.

EMOTIONAL EXPRESSION

Emotions may be expressed through speech (e.g., words and prosody) and gesture (e.g., emotional faces).

Disorders of Facial Expression

Buck and Duffy reported that, compared with controls with left-hemisphere brain damage, patients with right-hemisphere brain damage appeared to display less emotional facial expressions when viewing emotion-inducing slides than did controls (8). Patients with right-hemisphere damage are less facially expressive than those with left-hemisphere damage. Hemispheric asymmetries of emotional facial expressions may be related to the means by which the facial affect is initiated, such that when facial emotions are produced in response to verbal commands, there may be no differences. However, when facial expressions are induced by viewing emotional scenes, other emotional faces, or emotional prosody, there are right and left asymmetries, with patients who have right-hemisphere brain impairments being less expressive. In healthy subjects, the left half of the face, controlled by the right hemisphere, is more expressive than the right half of the face.

Prosody

When asked to intone linguistically neutral sentences emotionally, patients with right-hemisphere brain damage are severely impaired (9). Instead of intoning a sentence, they often denote a target emotion by changing the words in the sentence. Ross and Mesulam described two patients who could not express emotional prosody in speech but could comprehend emotional prosody (10). Patients may have impaired comprehension but spared repetition, prompting Ross to posit that affective prosodic speech disturbances seen with right-hemisphere lesions could mirror the aphasic syndromes seen with left-hemisphere lesions (11). In contrast, patients with left-hemisphere brain damage, despite being verbally nonfluent, are able to intone stereotypic verbal expressions prosodically with a variety of appropriate emotional intonations.

Verbal Expression

In addition to being able to intone verbal expressions with emotional prosody, John Hughlings Jackson observed that nonfluent aphasic patients may become very fluent when using explicatives; he posited that the right hemisphere may be mediating this activity (12). Right-hemisphere—damaged patients use fewer emotional words in their spontaneous speech. Aphasic patients are able to write emotional words better than nonemotional words. However, the role of the right hemisphere in producing emotional words remains to be fully investigated.

Whereas the right hemisphere may contain a limited lexicon (store) for emotional words, overwhelming research clearly demonstrates left-hemisphere superiority for mediating speech output. If, as discussed, the right hemisphere is important for controlling the emotional prosodic aspects of speech, and the left is responsible for producing the phonology and the order in which words are produced (lexical syntactic aspects of speech), how are these two systems integrated? Ross proposed that the propositional and prosodic elements are integrated at the brain-stem level (11). However, it would appear that the integration of these two systems is a highly skilled activity that may require integration at the cortical level. In all likelihood, integration of the propositional and emotional messages probably occurs intrahemispherically.

Emotional Moods

Patients with hemispheric lesions appear to have a change in their emotional mood. For example, Goldstein noted that patients with left-hemisphere lesions who were aphasic often appeared depressed, anxious, and agitated (13). Goldstein called this constellation of mood changes the "catastrophic reaction." In contrast, others noted that patients with right-hemisphere disease often appear to be indifferent or even mildly euphoric despite their disability, including hemiparesis (14–16). Right-hemisphere damage is often associated with indifference, and left-hemisphere damage is associated with depression and anxious agitation. Right-hemisphere lesions are often associated with anosognosia (denial of an illness). One may hypothesize that the right-hemisphere—damaged patient's anosognosia may be related to the patient's indifference and that the catastrophic reaction is a normal response to the serious defects associated with left-hemisphere strokes. In patients undergoing barbiturate-induced hemispheric anesthesia (Wada's test), however, injection into the left carotid, which anesthetizes the left hemisphere, is also associated with a catastrophic reaction, whereas injections into the right carotid (which induces right-hemisphere anesthesia) are associated with a euphoric response.

As noted previously, patients with right-hemisphere disease have disorders of emotional expression. Those with left-hemisphere disease, however, rely more heavily on emotional expression. Therefore, patients with right-hemisphere disease who have disorders of emotional expression may appear to be indifferent, and those with left hemisphere disease who rely more heavily on emotional expression may appear to have heightened emotions. In a study that administered the Minnesota Multiphasic Personality Inventory (MMPI) to patients with right- and left-hemisphere dysfunction, patients with left-hemisphere lesions had a marked elevation of the depression scale, and patients with right-hemisphere disease did not (17). The MMPI does not require a patient to express affectively intoned speech or to make emotional faces and has been widely used as an index of affective experience and mood. These findings suggest that the emotional moods of patients with hemispheric disease cannot be solely attributed to difficulties in perceiving or expressing affective stimuli.

Benson (18) and Robinson and Szetela (19) noticed that most patients with left-hemisphere brain damage who have a depressive response had left anterior perisylvian lesions with nonfluent aphasia. About one third of stroke patients have a major and long-lasting depression. Depression is most commonly associated with cortical and subcortical lesions, especially in the region of the left caudate. The cortical lesions that cause the most severe depressions are located in the frontal pole. Many of these patients also experience anxiety associated with depression. In addition, right-hemisphere lesions, especially in the frontal lobes, are associated with indifference or even euphoria.

In contrast, depression can be seen with right-hemisphere disease and may be underdiagnosed because patients with right-hemisphere brain damage have emotional communicative disorders. When vegetative signs of depression, such as loss of appetite and sleep, are assessed, there may be no hemispheric asymmetries for depression. When patients with right- and left-hemisphere brain damage are assessed for depression with the Zung Depression Inventory and the Self-Assessment Mannequin, no differences in the rate or severity of depression are observed between patients with left- and right-hemisphere lesions.

PSEUDOBULBAR AFFECT

Patients with hemispheric lesions may laugh or cry or both without feeling sad, happy, or being exposed to something funny. Unlike normal emotional expressions, these outbursts are stereotypic and do not have different degrees of intensity. Pseudobulbar laughing and crying are usually associated with lesions that interrupt the cortical bulbar motor pathways bilaterally and thereby release reflex mechanisms for facial expression. It is unclear whether an area in the pons is responsible for emotional facial expression; when this area became disinhibited from cortical control, one may see pseudobulbar affect. Alternatively, the center for the control of emotional expressions may be located higher in the diencephalon. It is not known why some patients predominantly cry and others laugh. Although most patients have bilateral lesions, when the lesion is larger in the right hemisphere, there is a greater chance for laughter to ensue. When the lesion is larger in the left hemisphere, however, crying is more likely to occur.

TREATMENT

The first goal of treatment should always be to learn what caused the focal dysfunction and then to treat the underlying disease to prevent further damage or to reverse the damage. The second important part of treatment is education of the patient and family. Although patients typically understand that weakness and language disorders may be related to focal neurologic disease, emotional disorders are often attributed to psychodynamic factors. The patient and family need to focus on the appropriate problem. Patients with emotional communicative disorders can use alternative strategies. Therefore, when communicating an emotion to a patient with a right-hemisphere dysfunction, primarily propositional speech should be used. When communicating emotions to patients with left-hemisphere lesions, emotional faces and prosody should be used. In cases of mood changes (e.g., depression), drug therapy (e.g., antidepressants) may be helpful. Patients who have left-hemisphere brain damage and depression may be helped with antidepressants. In addition, pseudobulbar affect may also be helped with antidepressants.

SUMMARY

The right hemisphere of the human brain contains inborn representations of species' typical emotional facial expressions. The right temporal parietal region, in particular, is important in comprehending and discriminating facial emotional expressions and in decoding emotional prosody in speech. Likewise, patients with right-hemisphere damage are less facially expressive and express less emotion in speech than those with left-hemisphere damage. Patients with left-hemisphere lesions, particularly in the frontal pole, are more likely to be depressed and anxious, whereas patients with right-hemisphere lesions exhibit indifference, euphoria, and anosognosia. Treatment involves managing the cause of the underlying disease, patient and family education, and pharmacologic intervention.

REFERENCES

1. DeKosky S, Heilman KM, Bowers D, Valenstein E. Recognition and discrimination of emotional faces and pictures. *Brain Lang* 1980;9:206–214.
2. Bowers D, Bauer RM, Coslett HB, Heilman KM. Processing of faces by patients with unilateral hemispheric lesions. I. Dissociation between judgments of facial affect and facial identity. *Brain Cogn* 1985;4:258–272.
3. Bowers D, Heilman KM. Dissociation of affective and nonaffective faces: a case study. *J Clin Neuropsychol* 1984;6:367–379.
4. Blonder LX, Bowers D, Heilman KM. The role of the right hemisphere on emotional communication. *Brain* 1991;114:1115–1127.
5. Ekman P, Friesen WV. *Facial action coding system.* Palo Alto, CA: Consulting Psychologists Press, 1978.
6. Izard CE. *Human emotions.* New York: Plenum Press, 1977.
7. Borod J, Andelman F, Obler L, Tweedy JR, Welkowitz J. Right hemispheric specialization for the identification of emotional words and sentences: evidence from stroke patients. *Neuropsychologia* 1992;30:827–844.
8. Buck R, Duffy RJ. Nonverbal communication of affect in brain damaged patients. *Cortex* 1980;16:351–362.
9. Tucker DM, Watson RT, Heilman KM. Affective discrimination and evocation in patients with right parietal disease. *Neurology* 1977;17:947–950.
10. Ross ED, Mesulam MM. Dominant language functions of the right hemisphere? Prosody and emotional gesturing. *Arch Neurol* 1979;36:144–148.
11. Ross ED. The aprosodias: functional-anatomic organization of the affective components of language in the right hemisphere. *Ann Neurol* 1981;38:561–589.
12. Taylor J, ed. *Selected writings of John Hughlings Jackson.* London, England: Hodder and Stoughton, 1932.
13. Goldstein K. *Language and language disturbances.* New York: Grune & Stratton, 1948.
14. Babinski J. Contribution l'etude des troubles mentaux dans l'hemisplegie organique cerebrale (anosognosie). *Rev Neurol* (Paris) 1914;27:845–848.
15. Hàcaen H, Ajuriagurra J, de Massonet J. Les troubles visuoconstuctifs par lesion parieto-occipitale droit. *Encephale* 1951;40:122–179.
16. Denny-Brown D, Meyer JS, Horenstein S. The significance of perceptual rivalry resulting from parietal lesions. *Brain* 1952;75:434–471.
17. Gasparrini WG, Spatz P, Heilman KM, Coolidge FL. Hemispheric asymmetries of affective processing as determined by the Minnesota Multiphasic Personality Inventory. *J Neurol Neurosurg Psychiatry* 1978;41: 470–473.
18. Benson DF. Psychiatric aspects of aphasia. In: Benson DF, ed. *Aphasia, alexia, and agraphia.* New York: Churchill Livingstone, 1979.
19. Robinson RG, Szetela B. Mood change following left hemisphere brain injury. *Ann Neurol* 1981;9:447–453.

12
Perception

Lynn C. Robertson

OVERVIEW

Knowledge of how damage to different neural systems affects perception in humans has grown substantially during the past decade. This can be attributed in no small part to the advent of new imaging techniques, such as magnetic resonance imaging (MRI) and positron emission tomography (PET). This chapter addresses some of the important advances in understanding components of visuospatial functioning and how they relate to neural systems. The focus is on anatomic correlates of visuoperceptual function, but this does not imply that psychopharmacologic or electrophysiologic correlates are not equally important. Some evidence using the latter methods are mentioned in passing, but the emphasis in this chapter is on deficits that are caused by structural damage in the human brain and their application in understanding both normal brain function and dysfunction.

VISUAL PATHWAYS TO THE CORTEX

Two visual pathways that transmit visual information from the retina to the cortex have been studied extensively:

1. The retinotectal pathway sends information from the retina through the superior colliculus (SC).
2. The geniculostriate pathway sends information through the lateral geniculate nucleus (LGN) of the thalamus to striate visual cortex (V1) or Brodmann's area 17 in the occipital lobe. There are also massive back-projections to the reciprocal subcortical areas.

These two systems contribute to vision in different ways. The retinotectal pathway is considered more primitive. Cells within this system respond most vigorously to light occurring within the peripheral visual field. As a result, the system appears to be effective at detecting and localizing luminance changes. Functionally, it seems well suited to act as a monitor for potentially relevant changes in the visual field, to attract attention, and then to align the eyes with the attended location for further identification. Consistently, the retinotectal system is directly involved in the initiation of saccadic eye movements. The geniculostriate pathway responds to features that are crucial to pattern perception, such as shape, color, and orientation. Nearly 90% of the projections from the retina are within this pathway, and this system is considered the major source of information for higher-level visual analysis.

Although the SC, frontal eye field (FEF), and visual striate cortex or V1 all contain a topographic map of the contralateral visual field, cells in V1 have not been found that respond when saccades occur, and V1 is not considered an area involved in eye movements. Conversely, cells in the deep layers of the SC increase their response just before saccade initiation, and cortical cells in the FEF act in concert with SC responses. Consistently, bilateral ablation of SC and FEF in primates causes severe and permanent impairment of saccadic activity (1).

Retinotectal System and Progressive Supranuclear Palsy

Acute insults to the tectal area generally cause severe morbidity or death. However, there are several investigations of patients with a degenerative disease that affects tectal areas, known as *progressive supranuclear palsy* (PSP). Patients with PSP present with symptoms similar to those of patients with Parkinson's disease, but with a distinctive saccade deficit. This deficit is more pronounced along the vertical axis than the horizontal axis. As the disease progresses, the ability to initiate a saccade in any direction is lost. Autopsies show massive bilateral damage in the midbrain (superior colliculus and peritectal region).

In tests of covert orienting, these patients are slow to orient attention to a bright light. They also lose a normal inhibitory response to the light after long delays. This abnormality occurs along the same axis as their saccade deficit, namely vertical. PSP patients also fail to demonstrate "inhibition of return" (2), which is believed to facilitate visual search by keeping track of where attention has been. Inhibiting those locations that have been attended would encourage search of other, perhaps more relevant, locations. PSP patients show virtually no inhibition of return along the same axis as their saccadic deficits. These findings demonstrate that at least one elementary operation of normal spatial orienting, the inhibition of return, is eliminated in patients who also have a saccade deficit owing to damage in a primary structure of the retinotectal pathway.

Retinotectal System and Blindsight

Total ablation of primary visual cortex in one hemisphere causes blindness in the contralateral field. *Blindsight* refers to residual visual capacity in a hemifield "blind" owing to damage to primary visual cortex. In a thorough and intensive case study of one patient with blindsight, the patient could detect and locate light sources and motion well above chance within his blind field. Additionally, he denied any conscious awareness of stimulation when it occurred within this field (3).

Evidence for blindsight is important because it suggests that the retinotectal system is capable of transmitting sufficient visual information to affect performance (although not sufficient to culminate in full awareness of the stimulation). The major question among investigators of blindsight has been whether residual visual capacity can be attributed solely to the intact tectal system in hemianopic cases or to undetected visual capacities that remain functional in the geniculostriate system.

Effects of Frontal and Parietal Lesions

In acute stages after insult, patients with frontal or parietal damage often deviate their eyes in an ipsilesional direction. Saccades in the contralesional direction are difficult and sometimes impossible to make. Studies of patients with frontal or parietal damage have begun to shed light on the cortical visual centers involved in different aspects of this deficit.

Although saccade latency to the onset of light is typically normal in patients with dorsolateral frontal damage who do not show clinical signs of oculomotor deficits, inhibiting a saccade to a light is very poor in such cases. When subjects are told to look in the opposite direction of light onset (i.e., make an antisaccade), patients with frontal lobe damage have extreme difficulty. Conversely, there is good correspondence between eye movement deficits associated with parietal lesions and attentional mechanisms associated with unilateral visual neglect. Patients with moderate to severe unilateral visual neglect (typically associated with the inferior parietal lobe in humans) often have eye deviations in the ipsilesional direction, especially during the initial stages after insult. In most of these patients, eye deviations resolve over time.

Clinical observation suggests that patients who have recently suffered insult cannot overcome this problem. If told to "look left," they may move their eyes to the center and then move their heads toward the left, but they often cannot follow the instructions at all. It is as if the endogenous command cannot overcome involuntary eye movements into the ipsilesional field. With time, these patients are able to look left upon instruction, but their eyes drift back toward the right side. As with the syndrome of unilateral neglect, these deficits are more pronounced in patients with right-hemisphere damage than in patients with left-hemisphere damage. One question raised by these findings is whether eye movements and the covert orienting of visuospatial attention to locations in space are related to the same neural function. This question is addressed more fully in a later section.

VISUAL PATHWAYS THROUGH THE CORTEX

Studies in primates have revealed two main processing streams in cortical vision. One stream is associated with inferior occipitotemporal pathways and is involved in object identification, or determining "what" is there. The other is associated with occipitoparietal pathways and is involved in analyzing spatial relationships, or "where" objects are (4). The separation of these streams appears to begin in the retina, continues divided in geniculate and primary visual cortex, and is relatively separate (although by no means completely) as information flows anteriorly. The two major streams correspond roughly to the parvicellular and magnocellular systems respectively (Fig. 12-1). The receptive field size of both the magnocellular and parvicellular systems increases as information is processed through progressively higher levels, but this increase in size is more pronounced in the magnocellular system that projects to the parietal lobe (5). Cells in the parietal lobe often have receptive fields that include the ipsilateral and contralateral visual fields. Large receptive fields are well suited to large-scale analysis, such as for motion, low spatial frequency, figure or ground, and relative location of objects. The parvicellular system has smaller receptive fields that respond to fea-

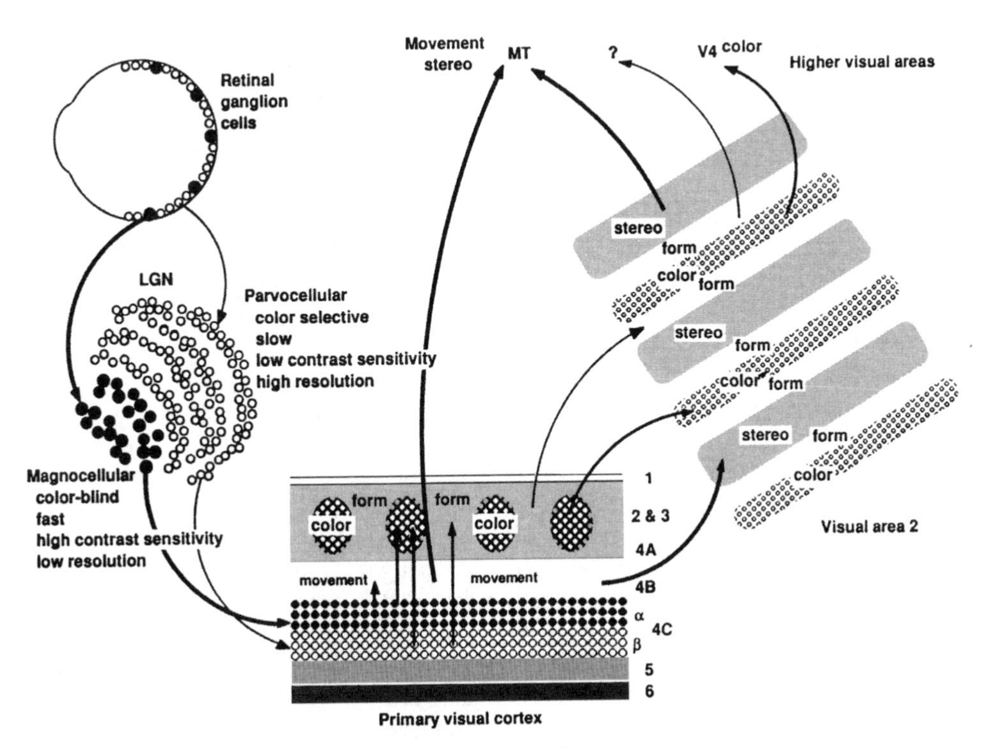

FIG. 12-1. Schematic of visual pathways. (From ref. 5, with permission.)

tures, such as color, brightness, shape, and higher spatial frequencies. These features likely provide some of the building blocks of object perception.

Primate studies have also revealed several functionally separate areas within each of the cortical visual pathways that act together to analyze objects and their spatial relationships. The number of visual areas increases in invertebrates as the ratio of brain to body weight increases across species. Between 15 and 30 visual areas in new- and old-world monkeys have been reported (6), and additional areas have been identified since then. Most of the posterior cortex in monkeys contains neurons that respond to visual stimuli. It is likely that more areas exist within the human brain.

Although the animal work has been crucial to mapping visual areas, the existence of the two pathways that correspond to identification of an object (knowing what) versus the spatial location of objects (knowing where) was first reported in humans. Extensive neuropsychological testing of English war veterans with gunshot wounds to the head revealed that penetrating wounds in the parietal lobe produced difficulty in locating objects but not in identifying them, whereas occipitotemporal damage produced difficulty in identifying objects but not in locating them (7). Consistently, posterior ventral lesions produce object agnosias, whereas dorsal lesions produce spatial deficits, as seen in unilateral visual neglect, visual extinction, or Balint's syndrome.

Despite the evidence that two processing streams are differentially involved in determining "what" and "where," this can be but a first approximation. Studies of patients with Balint's syndrome associated with bilateral parietal lobe damage have shown that the perception of spatial relationships between objects can be altered without affecting spatial relationships within objects. That is, these patients may know where a part of an object is in relation to other parts but not where the object is in relation to other objects. Patients with Balint's syndrome are often mistaken as blind. They typically report their vision as hazy or blurred. In-depth testing, however, reveals that these patients are able to recognize individual objects but have great difficulty recognizing two objects placed next to each other, either side by side or behind one another along the same line of sight. In addition to deficits in perceiving spatial relations between objects, eye movements in patients with Balint's syndrome appear fixed, and there is great difficulty tracking an object through space. These patients also suffer from a severe spatial reaching problem with no accompanying motor deficits.

DEFICITS IN PERCEIVING OBJECTS

The perception of objects can break down in highly specific ways. The investigation of deficits in recognizing objects (visual agnosias) has had an important influence on recent theories of normal and abnormal object perception. Studies of visual agnosia have also focused attention on certain brain regions as more important in object perception than others. Generally, visual agnosia occurs with ventral damage along occipital and temporal lobes. "Pure" agnosias are seldom as specific as one would like for scientific investigations, and pure cases are difficult to find. Nevertheless, the purest cases demonstrate some interesting problems that limit the way in which investigators should conceptualize normal object perception.

Associative Agnosia

Individuals with no perception deficits *cannot* look at a picture of a dog *without perceiving* a dog. Perceiving objects appears to happen effortlessly, and the concept of an object appears impossible to inhibit. When looking at or attending to a dog, one recognizes a dog. It appears that part of seeing a dog is knowing that what one sees is a dog (i.e., having a "form concept"). Neuropsychological evidence does not support this simple-minded view. There are patients who can see patterns that individuals with no perception deficits perceive as a dog, draw what they see, discriminate one dog from another, match pictures of dogs, yet not be able to report that the object is a dog or even that it is generally kept as a pet. In other words, these patients lose both the name and the meaning of the object upon visual presentation. This problem has been historically called *associative agnosia.* One striking aspect of this syndrome is that the meaning and name of an object is readily available to the patient through other modalities. He or she easily identifies a dog by touch or when it barks. This is a case in which damage to the ventral processing stream that presumably identifies objects does not disrupt the percept of the object. Rather, it disrupts the connection between the object's function and its meaning.

Integrative Agnosia and Simultanagnosia

Some patients have trouble perceiving an object as a whole but have little difficulty seeing its parts. They may perceive the parts such as the four feet, the body, and the trunk of an elephant, but they either cannot integrate the parts to form the internal representation of the elephant or are very slow to do so. If these patients do perceive a part that uniquely defines the object, such as the trunk, they are likely to guess correctly that the object is an elephant, but objects with few distinctive features are difficult for such patients to recognize. Thus, unlike patients with associative agnosia, patients with *integrative agnosia* do have form concept access, but they do not have the ability to perceive the integration of parts in the same way as people who do not have agnosia (8).

Wolpert introduced the term *simultaneous agnosia* and referred to it as an "integration apperception" (9). His patient could not comprehend the meaning of a scene but probably had a different underlying deficit. Although he could identify objects in a scene, he could not report the meaning of the scene.

Several lesion locations produce a deficit in seeing two objects simultaneously or, more accurately, in seeing two objects in the same amount of time that it normally takes to see two objects. Reaction time measures demonstrate that neurologically normal subjects recognize objects serially, but in a much shorter time window than patients. Piecemeal or serial processing is often a consequence of brain insult. Visual agnosias may lead to feature-by-feature or line-by-line processing to overcome deficits in perceiving a pattern as a whole. It is not clear what constitutes perceiving a whole, and this question poses one of the challenges for investigators of object perception. One thread that does appear throughout the history of psychology is the observation that wholes are often perceived before the parts that constitute them. Until there is better understanding about how objects are perceived as objects and how an object can be perceived before seeing its parts, the underlying cause of integrative agnosia or simultanagnosia as it occurs in vision will remain unresolved.

Prosopagnosia

Prosopagnosia refers to a recognition deficit limited to faces, although thorough testing has revealed that other categories can also be affected, such as automobiles, breeds of dogs, and species of birds. Many investigators have argued that the deficit is one of accessing the meaning of the face, as with pure associative agnosia for objects. Unlike patients with associative agnosia for objects, however, patients with prosopagnosia can identify a face as a face; they simply cannot classify it as a particular person nor accurately report whether they have seen the face before. Lesions that produce this deficit are most commonly linked to inferior right temporooccipital damage. However, there is much debate about whether a right-hemisphere lesion alone is sufficient to cause prosopagnosia or whether bilateral lesions are required; it has been argued that the fusiform gyrus of the mesial occipitotemporal region bilaterally must be affected (10).

Prosopagnosia has been considered a deficit in accessing the memory trace or structural description of a particular face. Farah has noted that visual discrimination of *unfamiliar* faces has accompanied prosopagnosia in every case in which it has been tested consistent with a primary deficit in facial discrimination (11). This investigator argues that there are higher-order perceptual deficits that may disproportionately affect face recognition. Identification of any one individual requires expertise in making subtle visual distinctions along a number of dimensions. Farah's argument can also account for cases like the prosopagnosic farmer who lost the ability to discriminate his cows or another prosopagnosic patient who lost the ability to discriminate race horses.

Other evidence collected in prosopagnosic patients demonstrates that there is intact visual analysis of faces at a preconscious level. An increased galvanic skin response to familiar faces was demonstrated in two prosopagnosic patients, even though these patients could not accurately report which faces were familiar (12). Visual discrimination must be intact at some level for such responses to occur. Thus, the issue of whether there are visual areas that respond uniquely to faces in the human brain is far from resolved.

Agnosia for Features

Some patients with neurologic damage in posterior regions involved in extrastriate occipital, temporal, and parietal lobes have difficulty with more fundamental visual analysis, including the perception of color (achromatopsia), orientation, or motion. These types of agnosia are considered primary deficits, or "bottom-up" deficits, and are generally classified as *apperceptive agnosia.* These disorders represent basic visual deficits produced by cortical lesions. Each has been observed in isolation in patients with lesions and can be evident when visual sensory function is normal. These dissociations are consistent with PET studies showing that identifying shape, color, and velocity have different profiles in extrastriate cortex (13).

Color agnosia over the full visual field is associated with lesions similar to those that produce prosopagnosia. In fact, prosopagnosia typically occurs simultaneously with achromatopsia. Achromatopsia, however, can exist in isolation and can affect only one quadrant or one hemifield. Patients may describe what they see as "dirty" or "dulled," a "washing out" of colors. Some patients do not lose the ability to discriminate between monochromatic colors but cannot accurately report the color that is presented. Achromatopsia has been associated with inferior occipitotemporal damage.

A rare case of motion agnosia has been reported. A patient had difficulty detecting motion and reported that objects in motion looked as if they jumped from place to place. Continuous motion, as in running water, looked frozen (14). There is some controversy regarding whether motion perception deficits can occur in isolation. Animal work in areas MT and MTS in the posterior superior temporal sulcus of monkeys has shown convincingly that cells in this region are sensitive to direction and velocity of movement. However, cells in parietal regions are also sensitive to movement, which may explain why movement deficits in isolation are extremely rare in humans.

DEFICITS ASSOCIATED WITH VISUAL SPACE

A great deal of evidence has shown that parietal lobes are extensively involved in spatial analysis. Animal data have demonstrated that cells in the parietal lobe are involved in the computation of abstract space. Parietal regions appear to be crucial to maintaining stable spatial structure to allow accurate movement through the world. Cells in this area attend to locations in space independent of eye movements. They appear to be involved in the covert movement of attention over the visual field, but only in particular ways, as described later. It is well established that parietal lobes are involved in location analysis and the analysis of spatial relationships. However, parietal lobes in humans are involved in covert attentional processes that attend selectively to spatial locations but also to object features.

Unilateral Visual Neglect

Unilateral neglect is considered the quintessential example of a spatial deficit. Patients with left-sided neglect often act as if they are attempting to escape from their left side, sometimes to the point of denying left paresis or other problems on the left side of their body. Upon neuropsychological examination, patients with unilateral visual neglect respond to the right side of drawings and bisect a horizontal line well to the right side of the line, as if the length of the line toward the left side was misperceived. The magnitude of neglect varies tremendously among subjects and varies even more so in acute stages. Eye deviations to the right with left-sided neglect are often observed but need not be present for neglect to occur. Patients may interact normally with a person standing on their right side and be completely unaware of a person standing on their left side.

The bulk of the evidence from studies of unilateral visual neglect suggests that the right inferior parietal lobe produces neglect more often than other areas of the brain (15); however, neglect has been found in patients with thalamic, basal ganglia, frontal, and cingulate damage. Left or right parietal lesions can also produce an extinction-like deficit that is equal in frequency and severity, but the full-blown neglect syndrome is clearly more severe and more frequent with right-hemisphere damage. This asymmetry suggests that left-sided neglect due to right hemisphere damage is a combination of an attentional deficit (that produces an extinction-like phenomenon) and some other cognitive deficit that is more associated with right-hemisphere function.

Unilateral neglect typically resolves into extinction. Patients with extinction respond to an object in their contralesional visual field if it is presented alone but do not respond to the object if presented at the same time as another object in their ipsilesional visual field. Current neuropsychological theories of neglect disagree about whether neglect is a reflection of a direct deficit in spatial attention or a deficit in the visual representation of one side of space with secondary effects on attention.

Attention, Objects, and the Parietal Lobes

One question about attentional orienting is whether attention is directed to a spatial location per se or to objects, which obviously cannot exist without a spatial location. These two possibilities are difficult to separate experimentally, but the answer appears to be that both locations in space and objects are attended. Hemispheric differences for attention to space and attention to objects have been observed in groups of patients with inferior parietal damage. A right parietal group was affected equally by movements over space whether within or between objects, whereas a left inferior parietal group was also affected by whether the movement was within objects (16).

FUNCTIONAL HEMISPHERIC ASYMMETRIES

Although few contemporary scientists would accept the oversimplification of hemispheric specialization that proposes that the left hemisphere is dedicated to language and the right hemisphere to spatial analysis, it is obvious from clinical observation that in humans, the hemispheres differ in the pattern of deficits that occur when lesioned. Generally, it is assumed that theories of hemispheric specialization refer to differences in function between homologous structures of the two hemispheres and not to the hemispheres as a whole.

Evidence from human subjects has increasingly supported hemispheric laterality in visual processing in relatively early stages of analysis. Consistent with the historically accepted notion that the right hemisphere analyzes gestalts or wholes, whereas the left hemisphere analyzes details or parts, damage extending into regions of the left superior temporal lobe with adjacent parietal involvement disrupts identification of the local form in hierarchical patterns (Fig. 12-2), whereas

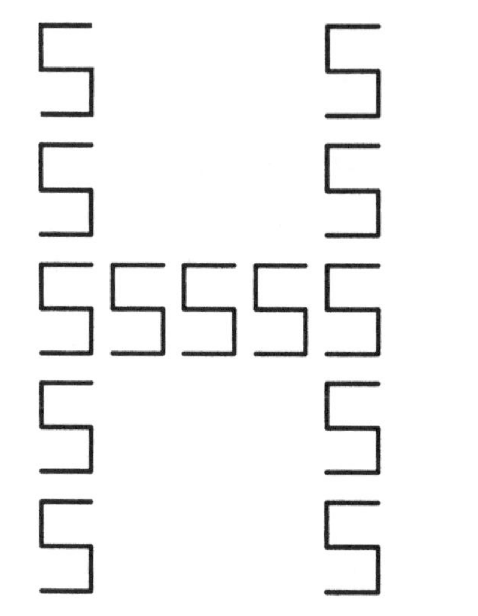

FIG. 12-2. Example of a global H created from local S's.

right hemisphere damage in similar regions disrupts identification of the global form.

REFERENCES

1. Schiller PH, True SD, Conway JL. Deficits in eye movements following frontal eye field and superior colliculus ablations. *J Neurophysiol* 1980;44:1175–1189.
2. Rafal RD, Posner MI, Friedman JH, Inhoff AW, Bernstein E. Orienting of visual attention in progressive supranuclear palsy. *Brain* 1988;111:267–280.
3. Weiskrantz L. *Blindsight: a case study and implications.* New York: Oxford University Press, 1986.
4. Ungerleider LG, Mishkin M. Two cortical visual systems. In: Ingle DJ, Goodale MA, Mansfield RJW, eds. *Analysis of visual behavior.* Cambridge, MA: MIT Press, 1982.
5. Livingstone M, Hubel D. Segregation of form, color, movement, and depth: anatomy, physiology and perception. *Science* 1988;240:740–749.
6. Kaas JH. Why does the brain have so many visual areas? *J Cogn Neurosci* 1989;1:121–135.
7. Newcombe F, Russell WR. Dissociated visual perceptual and spatial deficits in focal lesions of the right hemisphere. *J Neurol Neurosurg Psychiatry* 1969;32:73–81.
8. Riddoch MJ, Humphreys GW. A case of integrative agnosia. *Brain* 1987;110:1431–1462.
9. Wolpert I. Die Simultanagnosie: Storung der Gesamtauffassung. *Z. f. d. gesamte Neurol. u. Psychiatr.,* 93, 397–425 (as referenced by A. Benton, [1985]). Visuoperceptual, visuospatial and visuoconstructive disorders. In: Heilman KM, Valenstein E, eds. *Clinical neuropsychology.* New York: Oxford University Press, 1924.
10. Damasio AR. Disorders of complex visual processing: agnosias, achromatopsia, Balint's syndrome, and related difficulties of orientation and construction. In: Mesulum M-M, ed. *Principles of behavioral neurology.* Philadelphia: FA Davis, 1985:259–288.
11. Farah MJ. *Visual agnosia: disorders of object recognition and what they tell us about normal vision.* New York: Academic Press, 1990.
12. Tranel D, Damasio AR. Knowledge without awareness: an autonomic index of facial recognition by prosopagnosics. *Science* 1985;228:1453–1454.
13. LaBerge D. Thalamic and cortical mechanisms of attention suggested by recent positron emission tomographic experiments. *J Cogn Neurosci* 1990;4:358–372.
14. Zihl J, Von Cramon D, Mai N. Selective disturbance of movement vision after bilateral brain damage. *Brain* 1983;106:313–340.
15. Heilman KM, Watson RT, Valenstein E. Neglect and related disorders. In: Heilman KM, Valenstein E, eds. *Clinical neuropsychology.* 2nd ed. New York: Oxford University Press, 1985.
16. Egly R, Driver J, Rafal RD. Deficits following parietal damage: space based or object based? Paper presented at the annual meeting of Psychonomic Society, St. Louis, MO: November, 1992.

13

Executive Systems

Joaquin M. Fuster, Gary W. Van Hoesen, Robert J. Morecraft, and Katerina Semendeferi

OVERVIEW

The term *prefrontal cortex* is attributed to Sir Richard Owen (1). It refers to that cortex anterior to the electrically excitable motor cortex. This is a large area of cortex in the human and nonhuman primate brain and includes a large dorsolateral sector, orbital sector, and medial sector. The latter typically includes the anterior cortices of the cingulate gyrus, which wrap around the genu of the corpus callosum and follow its rostrum ventrally and posteriorly. The functions of the dorsolateral convexity, notably areas 9, 10, 45, and 46 of Brodmann (Fig. 13-1), are primarily cognitive (executive functions), in counterdistinction to medial and orbital areas, which have primarily attentional, affective, and emotional functions. Because the orbital and medial prefrontal regions cannot yet be clearly distinguished from each other, either physiologically or neuropsychologically, they are discussed together in this chapter. We first provide a summary of the functional neuroanatomy of the prefrontal lobes, which is followed by a description of the two major syndromes. The chapter ends with a discussion of treatment issues.

PREFRONTAL CORTEX

Anatomy

More modern cytoarchitectural and volumetric analyses of frontal lobe evolution demonstrate that humans undoubtedly have the largest frontal lobe in absolute terms, outstripping our extant ape kin. But we have larger brains in general, and the part devoted to the frontal lobe is not significantly different between apes and humans in relative terms. Some cell fields in humans, however, may be more elaborate than those in other species. Humans, for example, have a highly differentiated frontal pole (area 10), whereas a comparable area in the gorilla is difficult to find. The largely solitary orangutan appears to have a less elaborate orbitofrontal area, which is thought to be the neural substrate for social behavior.

The human frontal lobe may be unique, not so much for quantity of frontal lobe, but rather for other reasons relating to its internal structure and connectivity. We share with our ape relatives many basic behaviors, including the tendency to be social, to band together in units, to be intelligent, and to manipulate our environment. But we may be set apart by a greater balance to deal frontally with other parts of our neuraxis and the events provoked in these areas by our environment. In this regard, it is of great interest to find that white-matter volume of the prefrontal cortices is, in fact, the one anatomic variable that distinguishes humans from apes. Indeed, this is the neural substrate that governs the balance and precision of neural systems' communication.

The frontal lobe lies anterior to the central sulcus and can be divided into two major parts: a caudal part containing the electrophysiologically "excitable" motor cortices and a rostral part containing the prefrontal association cortex (Fig. 13-2). The motor cortices include the primary (M1, or area 4); supplementary (M2, or area 6m); and lateral

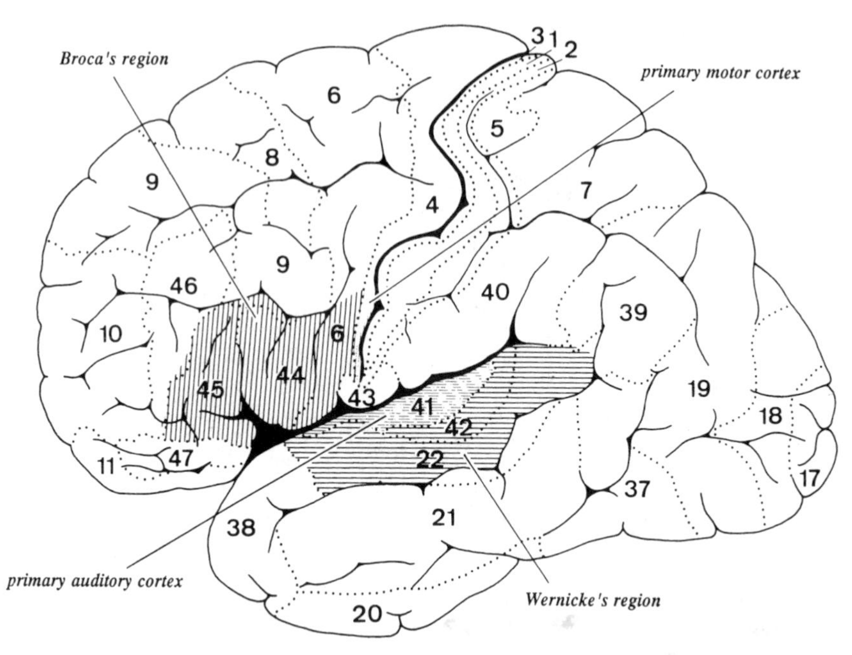

FIG. 13-1. Brodmann's cytoarchitectonic map of the human brain in lateral view. The designated four regions are of critical functional importance for language.

premotor cortices (LPMC, or areas 6D and 6V). All are characterized as agranular cortex, attesting to the fact that their internal granular layer, or layer IV, is not conspicuous. It is well known that M1 plays a crucial role in activating and facilitating independent body movements. On the other hand, M2 seems more involved with whole body movements and internally generated movements that are integrated in orderly fashion. Significant modulation of neuronal activity in the lateral premotor cortex and, in particular, the ventral part of lateral area 6, has been shown to be coupled with stimulus-triggered (visual and somatosensory cued) motor responses and with the execution of purposeful movements, such as grasping, or bringing the hand to the mouth.

Also included as part of the motor cortices are the frontal eye field (FEF, or area 8); supplementary eye field (SEF, or area F6); and presupplementary motor area (pre-SMA) (see Fig. 13-2). The FEF and SEF are located on the lateral surface of the hemisphere. The FEF is located anterior to the midportion of lateral premotor cortex and has a dysgranular cytoarchitecture. This refers to the fact that FEF is characterized by a poorly defined or incipient, internal granular layer IV. The SEF appears to be a subfield within area 6D. The rostral part of area 6D of the lateral premotor cortex has also been shown to be dysgranular. Both the FEF and SEF regulate contralateral saccadic eye movements. The pre-SMA is located on the medial wall of the hemisphere, rostral to M2, and corresponds to Walker's area 8B. Neurons in this field modulate their activity before and during movement.

The prefrontal cortex lies rostral to the motor-related cortices and extends to the frontal pole (see Fig. 13-2). On the medial surface, the prefrontal cortex lies anterior to the medial component of motor cortex and to the anterior part of cingulate gyrus. The primate prefrontal cortex is subdivided commonly on broad anatomic grounds. Its major partitions include ventrolateral, dorsolateral, medial, and orbitofrontal regions. In the monkey, the principal sulcus is located on the lateral surface of the hemisphere, and its depths form

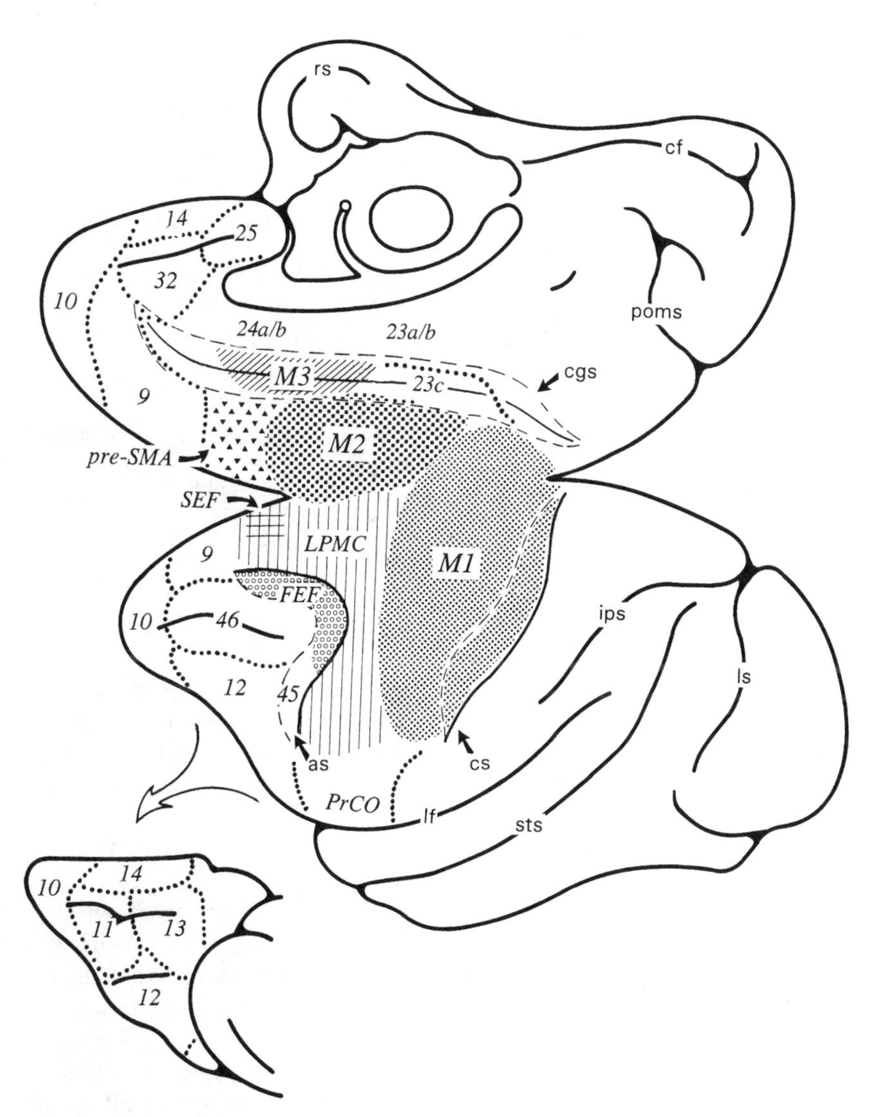

FIG. 13-2. Medial, lateral, and orbital views of the rhesus monkey cerebral hemisphere depicting the functional parts of the frontal lobe, Brodmann's and Walker's cytoarchitectural fields, and the cingulate gyrus. Note the three motor representations (M1, M2, and M3), the lateral premotor cortex (LPMC) and the frontal and supplementary eye fields (FEF and SEF). M1 is synonymous with the term *primary motor cortex* and corresponds to Brodmann's area 4. M2 is synonymous with the term *supplementary motor cortex* and corresponds to medial area 6. It, along with the LPMC, corresponds to Brodmann's lateral area 6. The supplementary motor cortex is separated from prefrontal area 9 by a presupplementary motor cortex (pre-SMA). M3 is synonymous with the term *cingulate motor area* and lies in the fundus and lower bank of cingulate sulcus. It corresponds to Brodmann's area 24, particularly area 24c. Area 23c is also thought to be a component of the cingulate motor area. All other *(unhatched)* fields correspond to prefrontal association areas or to anterior and medial parts of the limbic lobe adjacent to or within the frontal lobe. as, arcuate sulcus; cf, calcarine fissure; cgs, cingulate sulcus; cs, central sulcus; ips, intraparietal sulcus; lf, lateral fissure; ls, lunate sulcus; poms, parietooccipital medial sulcus; rs, rhinal sulcus.

the boundary between the ventrolateral and dorsolateral regions of prefrontal cortex. As named, the medial region of the prefrontal cortex is located on the medial surface of the hemisphere. Finally, the orbitofrontal region of the prefrontal cortex lies in the anterior cranial fossa above the bony orbit and forms the basal, or ventral, surface of the frontal lobe.

A large portion of the prefrontal cortex is classified cytoarchitecturally as granular cortex and has six well-differentiated layers, including a prominent external granular layer II and internal granular layer IV. However, differences in layers II and IV and in the other cortical laminae serve as a basis for partitioning the frontal granular cortex into several subfields. They are designated numerically, according to Brodmann and others. The prefrontal cortex includes areas 45, 12, 46, 10, and 9 laterally; 9, 10, and 32 medially; and 12, 13, 11, 10, and 14 ventrally (see Fig. 13-2). Many of these areas have recently been redefined and further subdivided.

Cortex on the orbitofrontal surface lobe can also be subdivided on cytoarchitectural criteria into a caudal agranular sector, an anterior granular sector, and a transitional dysgranular sector between them. From this perspective, general trends and distinguishing features of major orbitofrontal organization are clearly recognizable. For example, agranular and dysgranular components are located caudally on the orbitofrontal surface and are strongly connected to the limbic cortices, amygdala, and midline nuclei of the thalamus. In contrast, the granular component is situated rostrally on the orbitofrontal surface. This cortex is linked strongly to isocortical association areas and association nuclei of the thalamus. As expected, these unique patterns of neural interconnections would differentially influence activity and events processed in rostral versus caudal parts of the orbitofrontal cortex. Lesions rarely affect small parts of the orbitofrontal surface and thus are correlated with more global behavioral changes and not specific impairments. For example, posterior lesions disrupt limbic and medial temporal connections and lead to changes in emotional and social behavior as well as in autonomic regulation. More anterior lesions disrupt association inputs and affect more complex behaviors.

Cortical Association Connections

In addition to its obvious role in motor behavior, frontal lobe function has long been associated with a variety of higher-order behaviors and cognitive processes. Some of the more notable ones include working memory, motor planning, developing and implementing long-term strategies, decision making, and problem solving. When considering the higher-order functions mediated by the prefrontal cortex, the finding that prefrontal cortex is linked directly to a constellation of cortical association areas should not be surprising. Indeed, the prefrontal cortex is well known for its widespread corticocortical connections with distal parts of the cerebral cortex, specifically including primary association and multimodal association cortices (Fig. 13-3).

Primary association cortex is committed functionally to the early processing of sensory data, conveyed by the neurons of an adjacent primary sensory area. Primary association cortex operates in a more integrative fashion than primary sensory area. In contrast, multimodal association cortex is not committed to processing information related to one modality, but rather integrates highly transformed information, whose source can be traced back to multiple, sensory modalities. Although the traditional dogma suggests that multimodal association cortex represents the end stages of cortical processing, it is becoming clearer that information flowing in the reverse direction, that is, directed from multimodal association back to the primary association areas, may initiate and synchronize neural elements that are responsible for forming selective perceptions (2).

The prefrontal cortex is linked to sensory association and multimodal association cortices of the parietal lobe (areas 7a, 7b, and 7m) and temporal lobe (areas V4t, MT, and

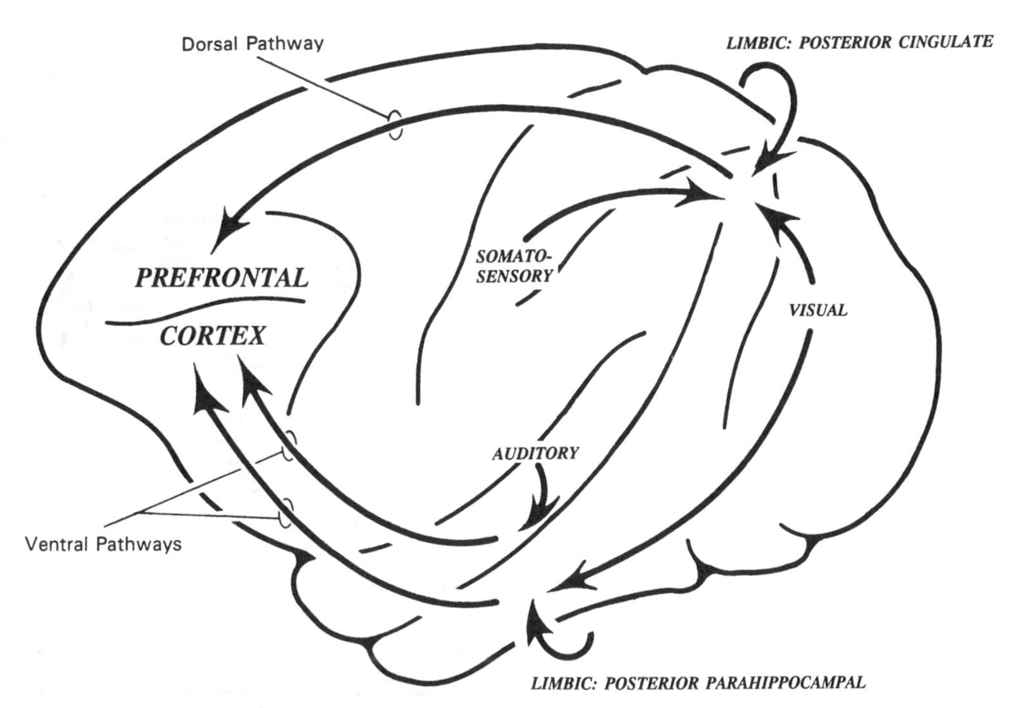

FIG. 13-3. Some of the major cortical association input to the prefrontal cortex is depicted on a lateral view of the monkey cerebral hemisphere. Note the dorsal pathway from the inferior parietal gyrus and dorsal peristriate area carrying cingulate information and the ventral pathways carrying auditory, visual, and posterior parahippocampal information. These provide cortical information from all of the other lobes, including the limbic lobe.

MST), as well as sensory association cortices of the anterior part of the occipital lobe (area V3). It has been shown in the monkey that the posterior part of the inferior parietal lobule (area 7a), anterior part of the occipital lobe, and the medial parietal lobule (area 7m) are reciprocally connected with the dorsolateral and sulcal principalis regions of the prefrontal cortex.

Anatomic and behavioral investigations conducted during the past three decades have led to the conclusion that long association pathways, reciprocally linking posterior parietal cortex and prefrontal cortex, are particularly important for the appropriate execution of visually guided movements. Presumably, somatosensory and visual inputs converge on the posterior part of the inferior parietal lobule, and information related to spatial orientation and motion analysis is conveyed to the dorsolateral prefrontal cortex (see Fig. 13-3). Therefore, the dorsolateral part of the prefrontal cortex is thought to process information concerned with understanding *where* an object is in space. It is also known that the more rostral part of the inferior parietal lobule (area 7b) and ventrolateral part of the temporal lobe project to the ventrolateral part of the prefrontal cortex. Specifically, the projection from the ventrolateral part of temporal lobe is thought to carry information dealing with form and object recognition. Therefore, it has been suggested that the ventral pathway may constitute a processing stream that addresses *what* an object represents in the extrapersonal environment.

The prefrontal cortex is also influenced by other parts of the temporal lobe through a subcomponent of the ventral pathway, whose origin arises from the rostral part of the superior temporal gyrus, as well from the temporal pole. This projection probably represents an important source of auditory input to the ventrolateral part of the prefrontal cortex.

In addition to the long association pathways linking the prefrontal cortex with the parietal, occipital, and temporal lobes, short association pathways interlock the various parts of the prefrontal cortex with one another in an organized fashion. The less-differentiated (in terms of cytoarchitectonic lamination) agranular and dysgranular cortices, located posteriorly on the basal and medial surface of the prefrontal cortex, give rise to widespread *intrinsic* prefrontal connections. In contrast, the more differentiated isocortical granular areas, which are situated anteriorly and laterally, are characterized by more limited *intrinsic* connections; they account instead for a large component of the frontal lobe's widespread *extrinsic* prefrontal connections.

Cortical Limbic Connections

The strong structural relation between prefrontal cortex and association cortex has played a dominant role in shaping our views on prefrontal organization and function. However, it is important not to neglect the structural interaction between the prefrontal cortex and limbic lobe, which is also very strong. There are many direct connections between the limbic lobe and the prefrontal cortex. Limbic projections to the prefrontal cortex arise from diverse and widespread parts of the limbic lobe, including the cingulate, orbitofrontal, temporopolar, perirhinal, entorhinal, posterior parahippocampal, and insular cortices. Although the lateral prefrontal cortex is a target for some of these connections, the bulk of this anatomic interrelationship is established with the posterior orbitofrontal and medial prefrontal regions.

The prefrontal cortex maintains a highly organized anatomic affiliation with the cingulate gyrus. The more anterior dorsolateral prefrontal cortex is connected to the posterior cingulate cortex (area 23), whereas the posterior dorsolateral prefrontal cortex is more strongly linked to the anterior cingulate gyrus. Similarly, the anterior orbitofrontal cortex is connected to area 23, and the posterior part is connected to area 24.

Fibers forming a subcomponent of the uncinate fasciculus, as well as the extreme capsule, interconnect the frontal and temporal lobes in a strong and highly specific fashion. The strongest links with the limbic portion of temporal lobe involve the posterior orbitofrontal cortex and medial prefrontal cortex, followed by the lateral prefrontal cortex.

Although a precise topography has yet to be determined, a number of other areas in the temporal portion of the limbic lobe are connected to the prefrontal cortex. They include the temporal pole (area 38), perirhinal (area 35), entorhinal (area 28), posterior parahippocampal (areas TH and TF), presubicular, and subicular cortices. All but the subicular connection have been shown to be reciprocal. The subiculum of the hippocampal formation projects to the posterior part of the orbitofrontal cortex and send some afferents to the dorsolateral part of the prefrontal cortex. A particularly heavy component of this projection terminates in the posterior part of the gyrus rectus, on the orbitofrontal surface. Hippocampal output is known to be mediated heavily by the subiculum and to some extent the CA1 sector of the hippocampal formation. Thus, subicular-CA1 output represents a direct hippocampal influence on the prefrontal cortex.

All parts of the prefrontal cortex are reciprocally connected to the insula, and a distinct anatomic relationship between cortex forming the orbitofrontal surface and the insula has been demonstrated. The agranular part of the orbitofrontal cortex is preferentially linked to the agranular part of the insula. In terms of topography, this translates into distinct connections between the posterior orbitofrontal cortex and the anterior insula. Likewise, granular orbitofrontal cortex is preferentially linked to granular insula. Anterior parts of the insula receive direct input from the gustatory and olfactory cortex, and the posterior parts receive input from primary somatosensory and auditory areas. This suggests that the insula may be a common site of *direct* convergence of all nonvisual sensory afferents. This is remarkable because integration of multimodal sensory information occurs elsewhere in the brain *after* a polysynaptic relay through sensory association cortex to multimodal cortex. All parts of the insula appear to receive input from sensory association and multimodal association cortices. The diversity of projections to the insula implies that prefrontal input from the insula may either be as little as one synapse away from a primary sensory area, or may be highly processed.

Amygdala projections to prefrontal cortex arise primarily from the basolateral and accessory basal nuclei and to a lesser extent from cortical and lateral nuclei. The strongest amygdalofrontal projection ends in the posterior part (agranular and dysgranular sectors) of the orbitofrontal cortex. Another strong projection terminates in the medial prefrontal cortices (areas 14, 25, and 32). Projections to isocortical areas on the lateral convexity (areas 9, 10, 45, and 46) are less strong. As mentioned previously, the amygdala is the recipient of a wide variety of cortical inputs from allocortical, periallocortical, proisocortical, and isocortical association areas. The latter includes converging input from both auditory and visual association areas as well as multimodal association cortices. Therefore, there is reason to believe that amygdala output directed toward the prefrontal cortices is influenced by a variety of neural systems related to the interplay of both the internal and external environments of the organism.

Motor Cortex Connections

It is well known that the prefrontal corticofugal axons are not directed to cranial nerve nuclei or the spinal cord. Since the latter part of the nineteenth century, however, it has been appreciated that prefrontal cortex plays a special and important role in guiding the outcome of voluntary motor behavior. Corticospinal neurons in M1 are influenced by prefrontal input through an indirect series of connections passing through lateral area 6, or premotor cortex. The discovery that cortex located outside the primary motor cortex, including the lateral premotor cortex, supplementary motor cortex, and cingulate motor cortex, contained corticospinal neurons, however, suggested otherwise.

The prefrontal cortex projects directly to parts of the motor cortices, giving rise to the corticospinal axons. In the lateral premotor cortex, only the very rostral parts of areas 6D and 6V that contain corticospinal neurons receive prefrontal input, with area 6V being the primary link. Prefrontal input to M2 converges on the rostral part of M2 that contains the face area and less so on corticospinal output zones that subserve the arm. Thus, the more rostral parts of the lateral premotor cortex and supplementary motor cortex receive prefrontal input. The cingulate motor cortex (M3 or area 24c) and area 23c, however, have also been shown to receive strong prefrontal input that converges on parts of these cortices that rise to corticospinal axons. Thus, several anatomically distinct sources of corticospinal axons are directly influenced by prefrontal output. Moreover, the ventral part of M1 receives input from the caudal part of the ventrolateral operculum of the prefrontal cortex. The target of this projection resides outside the M1-corticospinal projection zone and may correspond to the face representation. In many nonhuman primate models, the FEF has been shown to give rise to projections that innervate midbrain centers regulating ocular motor behavior. Although no corticospinal projections arise from the FEF, it does receive strong prefrontal input.

Subcortical Connections

The corticostriate projection from the prefrontal cortex is substantial. It is directed toward targets in the caudate nucleus and, to a much lesser extent, the putamen. Because a portion of the outflow of the basal ganglia is directed to the thalamus and eventually back to all parts of the frontal lobe, the corticostriate projection represents initial stages of a sequential pathway by which prefrontal cortex can influence a wide variety of neural systems.

The corticothalamic projection is one of the most studied corticofugal pathways leaving from the prefrontal cortex. The work of Akert and Nauta demonstrated that the mediodorsal thalamic nucleus is connected to all parts of the prefrontal cortex in the primate, including the granular sectors located anteriorly and laterally as well as dysgranular and agranular sectors located medially and ventrally (the posterior orbitofrontal cortex) (3,4). The medial part of the magnocellular division of the mediodorsal nucleus is connected primarily with the orbital surface of the prefrontal cortex and the lateral part (parvocellular division) with the lateral surface of the prefrontal cortex. The prefrontal cortex can be viewed as having at least four major subdivisions. They include dorsolateral, ventrolateral, medial, and orbitofrontal divisions, with each having its own unique thalamic projection pattern. Although the strong reciprocal anatomic relationship between the mediodorsal nucleus and prefrontal cortex is often emphasized, like all other cortical areas, numerous thalamic nuclei are linked to this brain region. They include the midline, ventral anterior, intralaminar, anterior medial, and pulvinar nuclei.

Several important brain-stem projections from prefrontal cortex have been identified that may play a role in motor control. Corticotectal projections have been demonstrated to arise from the dorsolateral principalis region in the monkey and terminate in the intermediate and deep layers of the superior colliculus. Such projections are likely candidates for influencing behaviors linked to eye movement, particularly those ocular motor tasks requiring on-line, or working, memory to order and issue an appropriate set of behavioral commands. Projections from prefrontal cortex to the pontine reticular formation have also been reported and appear to be distributed over the paramedian portion reticular formation, an area corresponding in location to the central superior nucleus.

The prefrontal cortex also gives rise to a strong corticopontine projection that ends in the medial part of the basilar pontine gray matter. This projection is a major component of the corticopontocerebellar system. Transneuronal labeling techniques have shown that cerebellar as well as pallidal neurons are labeled after injections of herpes simplex virus type 1 into the dorsolateral prefrontal cortex. This indicates that thalamocortical input to the prefrontal cortex is influenced by both basal ganglia and cerebellar circuits. If the basal ganglia and cerebellar loops are organized in parallel, it is likely that corticopontine projection from the prefrontal cortex is involved with the cerebellothalamocortical pathway that converges back onto the prefrontal cortex.

The prefrontal cortex is connected to several small but important brain-stem nuclei that synthesize and transmit selective neurotransmitters to widespread parts of the prefrontal cortex. It has been suggested that these projections may play an important role in regulating global as well as discrete behavioral states. The nucleus basalis of Meynert gives rise to cholinergic projections that innervate all parts of the prefrontal cortex. The ventral tegmental area, dorsal raphe nucleus, and locus ceruleus also belong to pharmacologically distinct classes of subcortical nuclei and, like the nucleus basalis, also project to all parts of the prefrontal cortex. The ventral tegmental area is a dopaminergic mesencephalic nucleus located ventral and caudal to the red nucleus. The dorsal raphe nucleus is situated in the midbrain and pons, immediately ventral to the periaqueductal gray matter, and consists of serotonergic neurons. Finally, neurons of the locus ceruleus give rise to norepinephrine projections.

There are several notable and interesting exceptions regarding the issue of reciprocity when considering the pharmacologically specific subcortical nuclei. For example, the nucleus basalis appears to project to all the cerebral cortex, including the prefrontal cortex, but receives input from only the limbic lobe. From the standpoint of the frontal lobe, this would include the posterior parts of the orbitofrontal cortex and medial prefrontal cortex. Also, unlike the rest of the cerebral cortex, which probably does not send projections back to the ventral tegmental area, dorsal raphe, and locus ceruleus, the dorsolateral and medial parts of the prefrontal cortex have been found to send a reciprocal subcortical efferent projection to these brain-stem centers (5). Therefore, the selective reciprocity of these connections gives distinct parts of the prefrontal cortex feedback control over their own monoaminergic and cholinergic innervation, as well as influencing innervation that distributes to widespread parts of the cerebral cortex.

NEUROLOGIC EXAMINATION

The results of the clinical neurologic examination of patients with prefrontal lobe syndrome may be entirely unremarkable, regardless of the site of the lesion. Usually, these patients show neurologic signs of malfunction under two sets of circumstances:

1. When the nosologic agent not only affects prefrontal cortex but also extends to other regions of the frontal lobe, such as the primary motor cortex or the premotor cortex, possibly encroaching on the supplementary motor area (SMA); in such cases, the patient commonly shows signs of pyramidal involvement or motor apraxia.
2. When the nosologic agent (e.g., tumor, cerebrovascular accident, or trauma) causes, in addition to prefrontal pathology, a more generalized encephalopathy; in these cases, the patient shows signs consistent with lesion in nonfrontal (cortical or subcortical) locations or with general cerebral disorder (e.g., edema with intracranial hypertension).

In any case, magnetic resonance imaging and other imaging methods are most helpful in revealing noninvasively the nature and extent of the prefrontal lesion.

DORSOLATERAL PREFRONTAL SYNDROME

To a large extent, the clinical consequences of dorsolateral prefrontal dysfunction can be predicted from what we know about the cognitive functions of dorsolateral prefrontal cortex in the primate. Thus, it is appropriate to preface our clinical considerations with a summary of that knowledge.

It has now been well established that, as a whole, the dorsolateral prefrontal cortex plays a crucial role in the *temporal organization of behavior* (6). This role extends to all domains of voluntary *action*—to skeletal movement, to ocular motility, to speech, and even to the internal domain of logical reasoning. Wherever and whenever the organism is required to form temporally extended sequences of deliberate action, the functional integrity of the dorsolateral prefrontal cortex is needed. This is especially the case when the action is new, and there are uncertainties or ambiguities in the information that determines it.

The reason that the dorsolateral prefrontal cortex is essential for temporal integration is that it—in cooperation with other cortical and subcortical structures—supports two basic and temporally complementary cognitive functions:

- A temporally *retrospective* function of *short-term memory*
- A temporally *prospective* function of *preparatory set* for action

With these two functions, the prefrontal cortex can continuously reconcile the past with the future in the course of behavior and keep the actions of the organism in logical order and on target.

Although these two temporally integrative functions are widely distributed over the dorsolateral prefrontal cortex, the behavioral domains to which they apply appear to have a somewhat discrete distribution, at least with regard to overt action. Speech appears to be based mainly in Broca's area (Brodmann's areas 44 and 45) in the inferior frontal gyrus of the left or dominant hemisphere, with some extended representation also in more rostral prefrontal cortex (see Fig. 13-1). The cognitive substrate for skeletal motility is less clearly defined but presumed to include areas 9, 10, and 46, in the anterior cortical convexity of the frontal lobe.

Table 13-1 presents a schematic outline of the most common clinical manifestations of

TABLE 13-1. *Prefrontal Disorders in Humans*

Affected Region	Functional Domain[a]	Symptom	Test Deficits
Dorsolateral	Attention[1,2]	Short attention span	Order and delay tasks, Wisconsin Card Sorting Test
	Memory[1]	Recent memory deficit	
	Planning[2]	Defective planning	
	Speech[1,2]	Central dynamic aphasia	
	Behavior[1,2]	Temporal concreteness, creative and executive deficit	
Orbitomedial	Attention[3]	Distractibility	Go–no go
	Behavior[3]	Disinhibition	

[a]Prefrontal function affected: [1]Active memory; [2]set; [3]inhibitory control.

prefrontal dysfunction in humans. In the table, those manifestations are subdivided by prefrontal region affected (dorsolateral versus orbitomedial), cognitive or behavioral domain affected, resulting symptoms, and deficits on formal psychological testing. The numbers in superscript refer to the prefrontal functions primarily impaired.

Disorders of Attention

The disorders of attention from dorsolateral damage can best be characterized as deficits in the intensive and selective aspects of attention. Aside from general disinterest in the world, the patient commonly shows difficulty in attending to particularly relevant information in the external environment, or in the thinking sphere, for either prospective behavior or mental operations. This difficulty most typically manifests as a shortened attention span, that is, trouble concentrating and maintaining attention on any given item of information. In other words, what most often fails in the dorsolateral prefrontal syndrome is the temporal continuity of the focus of attention. This failed temporal continuity results from, and feeds into, the troubles that the patient has in short-term memory and preparatory set. Interestingly, the triad of poor concentration, poor recent memory, and poor planning may be the only symptomatology of an incipient dementia from dorsolateral prefrontal degeneration, as in some cases of Pick's disease or Alzheimer's disease.

The patient with dorsolateral prefrontal damage usually has no trouble retrieving items from long-term memory. Recall and recognition are intact. In fact, established memories and routines may be a refuge from a deteriorating ability to form, retain, and use new memories. Unlike the patient with hippocampal abnormality or Korsakoff's syndrome, the patient with prefrontal pathology is not totally incapacitated for those mental operations. Rather, the prefrontal patient is beset with a weakened capacity to attend to new information and to retain it for prospective use. It is a weakening rather than failure of the ability to form new memory. Consequently, the memory for recent events is spotty and fragmentary—seemingly the result of lack of interest and attention. It has been said that the patient "forgets to remember."

Faced with the necessity to perform a task that requires the short-term retention of items for prospective action, the patient often fails. This can usually be made clear by testing the patient on the broad category of so-called delayed performance tasks. The patient usually also has difficulty with the Wisconsin Card Sorting Task (WCST), a test requiring from the subject the short-term memorization of a changing principle of classification of visual items—by color, shape, or number. Because some patients with nonfrontal (e.g., caudate, hippocampal) pathology may also fail delayed performance tests or the WCST, these tests in themselves cannot be considered specific for prefrontal dysfunction. Nonetheless, these tests, or simplified versions of them, should be part of the standard neuropsychological test battery whenever prefrontal pathology is suspected.

In conclusion, the patient with dorsolateral prefrontal pathology commonly manifests a deficit in active short-term memory. It is a deficit in a form of memory that, in the operant terms of cognitive psychology, has been called *working memory*. That deficit is partly, but not completely, attributable to the attentional deficit.

Difficulty in Planning

Perhaps the most consistent and specific symptom of dorsolateral prefrontal disorder is the patient's difficulty in planning his or her own actions. Many observers agree that this is the most characteristic manifestation of frontal lobe pathology (7). The patient is unable to plan in organized fashion any novel structures of action, to keep them in mind, and to lead them to execution. This difficulty permeates all action domains but is most apparent at the most complex levels of speech, behavior, and rational thinking. The patient cannot plan new discourse, let alone execute it; cannot plan substantial moves or departures from daily routine; and cannot elaborate novel trends of thought and lead them to their logical conclusion.

Clearly, this problem in planning is temporally the mirror image of the patient's problem in memory. It is also a memory deficit of sorts: a faulty "memory of the future." Whereas the recent-memory deficit of the patient with frontal pathology chiefly pertains to the retention of sensory material or its internal representations, the planning deficit is a deficit in representation of, and preparation for, future action. The first reflects a problem with active perceptual memory, broadly conceived; the second is a problem with active "motor memory," also broadly conceived.

The deficient motor memory of the dorsolateral prefrontal syndrome is most likely a derivative of an impairment of that second basic cognitive function of dorsolateral prefrontal cortex for which clear physiologic correlates are demonstrable in the monkey: the motor set. In these patients, the preparation for action is as disturbed as the retention of recent sensory or episodic information.

Disorders of Speech

All disorders of speech from dorsolateral prefrontal damage are attributable to malfunction of temporal integration and thus the result of memory deficit, set deficit, or both. Of course, the best known of these disorders in the spoken language is Broca's aphasia, caused by pathology in Broca's area, in the third or inferior frontal gyrus (Brodmann's areas 44, 45) of the left side. Broca's area is part of the prefrontal cortex by definition because it is directly connected to the nucleus medialis dorsalis of the thalamus. Broca's aphasia is the most drastic deficit observable in the articulation of speech, the most elementary form of motor aphasia. The patient's speech is characterized by a peculiar telegraphic style, with a general dearth of articles and liaison words, prepositions, and conjunctions. It has been designated *agrammatism*.

Lesions in more anterior prefrontal areas also produce speech deficits, but these are more subtle than Broca's aphasia. The anterior prefrontal speech defect is characterized by a general impoverishment of verbal expression. Spontaneous speech is diminished; so is verbal fluency. Sentences are short and mostly trite or commonplace. There is a general diminution in the number and quality of dependent clauses. This disorder of verbal expression from anterior dorsolateral prefrontal injury has been called "central motor aphasia." Like Broca's aphasia, anterior prefrontal aphasia is lateralized. It is more prevalent after lesions in the left or dominant hemisphere than in the right or nondominant one.

In summary, all speech disorders from dorsolateral prefrontal lesion can be ascribed to an impairment of the *syntagmatic* property of language, that is, the capability of synthesizing verbal expression in the temporal dimension. As in any other action domain, that property of the spoken language depends on the interplay of the two functions, memory and set, which we postulate are supported by dorsolateral prefrontal cortex. The speech disorders are paradigmatic of temporal integration disorders in all forms of sequential action.

The problems resulting from failure of the temporal integrative functions of dorsolateral prefrontal cortex affect not only speech but also behavior in general. The behavior of the patient with dorsolateral damage suffers from what can be characterized as *temporal concreteness*: his or her behavior is anchored to the "here and now," deprived of perspective either back or forward in time, driven by routine, and devoid of creativity. In cases of large lesions of dorsolateral cortex, that temporal concreteness is accompanied by apathy, lack of drive, and lack of spontaneity in all aspects of behavioral action. As a consequence, there is a global impairment of executive function in all action domains.

Schizophrenia

Because the executive and temporal integration functions of the dorsolateral prefrontal cortex extend to the *internal action domain*, that is, to the domain of logical thinking, this cortex has been implicated in schizophrenia. Indeed, one of the cardinal symptoms of the illness is the failure of proper temporal integration of the thought process. As manifested in speech production, the thinking of the schizophrenic is commonly characterized by blocking and looseness of associations. Is schizophrenia a disease of the dorsolateral prefrontal cortex? Clearly, the etiology of the illness cannot be reduced to such simple terms. No prefrontal lesion of any cause, locus, or extent, is known to induce a schizophrenic syndrome. Quite the contrary; decades ago, frontal lobe resection was used for the *treatment* of the psychosis, a treatment that was effective presumably insofar as it alleviated anxiety. However, lobotomy did not abolish primary symptoms of thought disorder.

Nonetheless, there is increasing evidence that a dorsolateral prefrontal disorder plays a role in the pathogenesis of schizophrenia. First, there is the indirect evidence that the dorsolateral prefrontal cortex is exceptionally rich in dopaminergic receptors and that some of the most effective antipsychotic neuroleptics are dopamine antagonists. There are morphologic and functional abnormalities in the prefrontal cortex of schizophrenics. Some of the morphologic abnormalities are revealed by neuropathologic study at the cellular level. Others are revealed by imaging techniques in the form of frontal atrophy. The most telling evidence comes from functional imaging studies; several such studies have indicated an abnormally low level of frontal metabolism ("hypofrontality") in schizophrenic patients. A striking manifestation of prefrontal dysfunction is the schizophrenic patient's failure to show the normal metabolic activation of dorsolateral prefrontal cortex when challenged by performance of the WCST (8).

In summary, it is a plausible hypothesis that schizophrenia results from a disorder of certain neurotransmitter systems, notably but not exclusively dopaminergic, and that such a disorder impairs the normal function of the dorsolateral prefrontal cortex.

ORBITOMEDIAL PREFRONTAL SYNDROME

The functions of the orbital and medial aspects of the prefrontal cortex are still largely unknown. In general terms, the orbital and medial areas, in addition to engaging in certain cognitive operations related to attention, play a more important role in emotional and instinctive behavior than do the areas of the dorsolateral prefrontal convexity. This may be related to the limbic nature and connectivity of orbitomedial cortex.

A most characteristic function of orbitomedial prefrontal cortex is the *inhibitory control of interference*, that is, the suppression of extraneous sensory or internal influences on current behavior. This kind of control, which has been well documented in animal and human neuropsychological studies, normally facilitates the attention of the organism to relevant inputs and to the pursuit of goals (9,10). When the control of interference fails, as it does in orbitomedial lesions, the subject becomes excessively *distractible*, incapable of suppressing the interference from trivial, irrelevant, or inappropriate influences from the internal or external milieu that tend to divert behavior away from its goal. Because this distractibility is commonly accompanied by motor disinhibition (see later), the resulting syndrome in the adult human resembles that of the hyperkinetic child. For this reason, the childhood disorder has been hypothetically attributed to delayed maturation of orbitomedial prefrontal cortex.

Obsessive-Compulsive Disorder

There are certain conditions in which the inhibitory control of cognition and behavior can be excessive, and this may be caused by hyperfunction of orbitomedial cortex. Attention in such cases is abnormally riveted to a given content of thought or form of behavior, which becomes overpowering and ends up by implacably controlling the patient's mind and behavior. This may be accompanied by extreme anxiety. We are referring, of course, to obsessive-compulsive disorder (OCD). Brain-imaging studies have indicated hypermetabolism of orbitofrontal cortex in OCD patients. It is therefore an attractive conjecture that the OCD is a consequence of that hypermetabolic condition and the related abnormal hyperfunction of inhibitory control.

The converse disorder, that is, the failure of interference control from orbitomedial lesion, can be demonstrated by formal psychological testing. Particularly sensitive to that failure is a category of behavioral tasks, named *go-no go tasks,* that require the inhibition of motor responses to certain stimuli in the course of sensory discriminations (11).

The deficit in the inhibitory control of interference is not the only kind of disinhibitory deficit observable in the orbitomedial syndrome. Indeed, quite commonly, the disinhibition affects motor activity in general, the thought process, and a variety of instinctual impulses. Orbitofrontal hypermotility gives the appearance that the patient has *excessive drive*, that he or she cannot stop and concentrate on anything and is driven by an incessant necessity to act, albeit in haphazard and purposeless fashion. This apparently boundless energy and impulsivity may interfere with sleep and with the orderly thought process. Also disinhibited may be the basic drives—the patient may display hyperphagia and hypersexuality in utter disregard of common good judgment and mores. Aggressivity may also be uninhibited.

The mood of orbitomedial prefrontal patients varies greatly, depending on the case and for unknown reasons. Some patients manifest a tendency to apparent depression, which may not reflect so much a sad mood as it does the apathy ("pseudodepression") that also characterizes some cases with extensive dorsolateral lesions. More commonly, however, the orbitomedial syndrome is characterized by labile mood with a general tendency to euphoria. The latter may be accompanied by a form of silly, compulsive, and childish humor that has been called *moria* (or *Witzelsucht*).

Not surprisingly, then, because of the euphoria, the uncontrolled thinking and behavior, and the insomnia that the orbitomedial prefrontal patient often exhibits, his or her syndrome may be mistaken for mania. Only a careful history and a thorough neurologic examination can exclude a primary psychiatric syndrome and confirm the "pseudomania" of a patient with orbitomedial prefrontal lesion.

The previous description can serve only as a general guideline because cases with lesions strictly circumscribed to the orbitomedial region, whether unilateral or bilateral, are uncommon. More common are cases with orbitomedial pathology in addition to pathology elsewhere in the frontal lobe. In such cases, the syndrome may be mixed, with cognitive and affective manifestations associated with both dorsolateral and orbitomedial lesions. The picture is even more complicated when the pathology extends beyond the frontal lobe.

TREATMENT

The two previous sections describe in schematic form the main features of the two principal regional syndromes from prefrontal trauma or disease. Issues of etiology have been purposefully omitted in the previous discussion because they are largely irrelevant to the particular clinical picture of any given case of frontal lobe involvement. What matters for this picture, in terms of our discussion, is the extent and location of prefrontal tissue affected; however, the etiology is extremely relevant, indeed critical, for the treatment and management of the patient with prefrontal pathology. The pathologic process may be acute or chronic, traumatic, infectious, neoplastic, vascular, or degenerative. The treatment, as in the case of any other brain lesion, has to be directly addressed to the cause when this is known.

The judicious use of pharmacologic agents can be extremely useful, albeit only symptomatically, in the treatment of frontal lobe pathology. When the cognitive syndrome from a dorsolateral lesion is accompanied by severe anxiety, a benzodiazepine may be useful. Depression may be successfully treated with an antidepressant. Neuroleptics may be useful in cases (usually orbitomedial) in which psychotic or manic manifestations are present. Medication has to be used with awareness of the patient's cognitive status and its functional relevance because some drugs (especially benzodiazepines and neuroleptics) can worsen the prefrontal cognitive deficit. The application of cognitive enhancers (e.g., cholinergics, nootropics) to prefrontal syndromes is an appealing idea, but the efficacy of such drugs for this purpose has not been demonstrated. As an adjunct, psychotherapeutic grief work is important for patients accustomed to creative and executive achievements that are no longer possible as a result of frontal damage. Supportive relationships and structured environments are needed by virtually all patients with significant frontal deficits.

SUMMARY

This chapter provides an overview of the functional neuroanatomy of the prefrontal lobes. These structures are associated with a wide range of executive functions, including working memory, attention, reasoning, and temporal information processing. Two types of prefrontal lobe syndromes have been associated with dorsolateral, as opposed to orbitomedial, lesions.

REFERENCES

1. Finger S. *Origins of neuroscience: a history of explorations into brain function.* New York: Oxford University Press, 1994.
2. Damasio AR. The time-locked multiregional retroactivation: a systems level proposal for the neural substrates of recall and recognition. *Cognition* 1989;33:25.
3. Akert K. Comparative anatomy of the frontal cortex and thalamocortical connections. In: Warren JM, Akert K, eds. *The frontal granular cortex and behavior.* New York: McGraw-Hill, 1964:372–396.
4. Nauta WJH. Neural associations of the frontal cortex. *Acta Neurobiol Exp* 1972;32:125–140.
5. Arnsten AFT, Goldman-Rakic PS. Selective prefrontal cortical projections to the region of the locus coeruleus and raphe nuclei in the rhesus monkey. *Brain Res* 1984; 306:9–18.
6. Fuster JM. *The prefrontal cortex: anatomy, physiology, and neuropsychology of the frontal lobe.* 2nd ed. New York: Raven Press, 1989.
7. Shallice T. Specific impairments of planning. *Philos Trans R Soc Lond Biol Sci* 1982;298:199–209.
8. Weinberger DR, Berman KF, Suddath R, Torrey EF. Evidence of dysfunction of a prefrontal-limbit network in schizophrenia: a magnetic resonance imaging and regional cerebral blood flow study of discordant monozygotic twins. *Am J Psychiatry* 1992;149:890–897.
9. Mishkin M. Perservation of central sets after frontal lesions in monkeys. In: Warren JM, Akert K, eds. *The frontal granular cortex and behavior.* New York: McGraw-Hill, 1964:219–241.
10. Stuss DT, Kaplan EF, Benson DF, Weir WS, Chiulli S, Sarazin FF. Evidence for the involvement of orbitofrontal cortex in memory functions: an interference effect. *J Comp Physiol* 1982;96:913–925.
11. Drewe EA. Go-no go learning after frontal lobe lesions in humans. *Cortex* 1975;11:8–16.

SECTION III

Syndromes and Disorders

14
Neurodevelopmental Disorders

John J. Ratey, Maureen P. Dymek, Deborah Fein, Stephen Joy, Lee Anne Green, and Lynn Waterhouse

OVERVIEW

This chapter summarizes the current state of knowledge regarding the etiology, diagnosis, and treatment of three neurodevelopmental disorders: mental retardation, cerebral palsy, and autism and pervasive developmental disorders (PDD).

MENTAL RETARDATION

The American Association on Mental Deficiency (AAMD) defines mental retardation as significantly subaverage intellectual functioning coupled with deficits in adaptive behavior and manifested during the developmental period (1). The prevalence of mental retardation is estimated at 3% of our population. *Mental retardation* is a term that encompasses a large and variable body of cognitive aptitude, emotional adjustment, and social development. Because of the complex classification scheme of mental retardation, behavioral variables range considerably, and it is difficult to typify an individual with mental retardation.

Levels of Retardation

Mental retardation is broken down into four categories: mild, moderate, severe, and profound. Although the diagnosis of mental retardation must be based on both intellectual ability and adaptive functioning, the salient factor this classification emphasizes is intelligence.

Mild Mental Retardation

According to standards set by the American Psychiatric Association (APA), those individuals with mild forms of mental retardation generally have intelligence quotient (IQ) levels ranging from 55 to 70 (2). This group consists of about 2% of our general population and roughly 80% of the mentally retarded. Individuals with mild degrees of intellectual impairment often appear relatively normal during the childhood years. Their predicament usually becomes apparent in early adolescence, as the social and scholastic demands placed on them increase. The mildly retarded typically are able to master essential school skills and by adulthood can achieve a satisfactory level of socially adaptive behavior. Some cases of mild retardation are attributed to structural brain disease (discussed in next section); however, many cases appear to stem from normal intellectual variation or adverse sociocultural conditions.

Moderate Mental Retardation

IQ levels in those diagnosed with moderate mental retardation stand in the range of 40 to 55. This group is much smaller than that of the mildly retarded, comprising about 10% to 15% of retarded individuals. Moderate mental retardation is usually recognized early in life, when developmental milestones such as language are delayed. By adulthood, many of these individuals have fair verbal abilities, although their rate of learning remains slow. As a whole, those with moderate intellectual impairments acquire daily living and simple vocational skills and are able to function in society with guidance. Causal factors pertaining to moderate mental retardation are less likely to be attributed to psychosocial or sociocultural factors than mild retardation. Many times poor coordination and mild body deformities are present.

Severe Mental Retardation

IQ levels of the severely retarded individual range from 20 to 39. The deficits present in these individuals are apparent at very young ages and range from severely defective speech quality to somatosensory deficits to motor handicaps. Severely retarded patients can sometimes develop limited levels of hygiene and self-help, but invariably remain dependent on the care of others.

Profound Mental Retardation

The IQ level of profoundly retarded individuals is measured at or under 20. With the exception of those individuals plagued with inborn errors of metabolism, the problems present in these individuals are often noticeable at birth, owing to the presence of severe physical malformations and other obvious symptoms of abnormality. As with the severely retarded, gross central nervous system (CNS) pathology is almost always present. The profoundly retarded show multiple handicaps and severe deficits in adaptive skills and are very resistant to learning.

Behavioral and Psychiatric Manifestations

Mild and Moderate Mental Retardation

Individuals with mild and moderate mental retardation are more susceptible to psychological problems than the general population. Roughly 50% to 60% of mentally retarded individuals have mild psychological disturbances, and 10% to 15% are plagued by severe psychiatric disorders (3). The degree of social and intellectual functioning greatly affects both frequency and type of mental illness in the mentally retarded. Individuals in the mild to moderate range of mental retardation are characterized by their simplistic cognitive schemes and heightened sensitivity to internal and external stimuli. Their thinking often uses a concrete approach and can be filled with perseverations and inflexibility. Because of their perceived capabilities, society often places unrealistically high expectations on these individuals, which can leave them prone to failure. Their reduced capacity to withstand stress, paired with limited social and peer supports, and society's emphasis on intellect and competition leaves this population extremely vulnerable to psychological and behavioral disorders.

Serious psychiatric illness, such as schizophrenia and the bipolar and unipolar affective disorders, are also diagnosed in the mildly retarded. These disorders are difficult to gauge in this population because of the limited ability of these individuals to understand and verbalize thoughts and feelings. In the verbally competent mild to moderate retardate, symptoms of schizophrenia are similar to the traditional symptoms found in normal populations, and they are likely to have paranoid and catatonic features. Characteristic features of affective disorders are also found readily among the retarded, although depression is commonly underdiagnosed.

Other behavior problems commonly found in this subgroup are aggression, hyperactivity, impulsivity, emotional lability, selective isolation, and temporary regression to primitive self-stimulatory activities. The source of these problems can be linked to psychological, environmental, and biologic factors. Poor ego development and individuation, mixed with poor social skills and reduced stress tolerance, target the mildly or moderately retarded individual for behavior disturbances. Internal catalysts, such as physical illness or premenstrual syndrome, often manifest themselves in forms of behavioral dyscontrol. Organic conditions, such as epilepsy, can also lead to a high incidence of behavior problems. The increased levels of complex partial seizures in retarded populations leads to elevated occurrences of interictal behavior syndrome, a result of frequent seizures symptomized by emotional intensity, lability, and aggression.

Severe and Profound Mental Retardation

Behaviors commonly associated with this level of retardation are characterized by stereotypy, hyperorality, aggression, and primitivism. Rudimentary use of sense modalities, such as touch and position, are common, as are autistic behaviors. Self-injurious behavior (SIB) is a widespread manifestation of the extreme forms of mental retardation and often takes form as head-banging, eye-gouging, biting, scratching, self-pinching, and rectal digging. Severe psychiatric disorders, such as schizophrenia, are extremely difficult to characterize in this population. The behavioral disturbance may arise from the specific brain dysfunction that causes the intellectual subnormality. The high incidence of psychotic behavior in individuals plagued with phenylketonuria, and the centrality of aggression and SIB in Lesch-Nyhan syndrome are examples of psychiatric disorders that are secondary manifestations of an underlying genetic or metabolic disorder.

Genetic and Neuroscience Issues

There are more than 350 known disorders and conditions that generate mental retardation, varying among genetic and acquired factors. As medical technology becomes more advanced, many forms of mild intellectual impairments are being found to have a neurobiologic basis. Subtle but significant biologic events, such as chromosomal anomalies, genetic syndromes, prenatal and postnatal infections, exposure to perinatal risks, and maternal teratogens, increasingly are implicated in mental retardation.

Prenatal Causes: Genetic Factors

Genetic factors are the single most common cause of mental retardation and can be partially implicated in up to 90% of cases (4). Syndromes of differing genetic origin are associated with distinct profiles of cognitive, emotional, and linguistic performance, as well as unique patterns of individual development. Chromosomal abnormalities, including Down's syndrome and fragile X syndrome [fra(X)], and single gene disorders, such as Cornelia de Lange's syndrome, are correlated with high levels of mental retardation. Hereditary precipitants of mental retardation are not always associated with a specific genetic syndrome and can be classified as normal intellectual variation.

Chromosomal Abnormalities

Fragile X Syndrome. Fragile X syndrome [fra(X)](Martin-Bell syndrome) is the single most common chromosomal form of mental retardation. As the name implies, this syndrome is an X-linked disorder, wherein associated phenotypic traits are usually expressed in the male and carried by the female. The disorder is induced by a fragile site or gap on the long arm of the X chromosome. About 30% of females who carry numerous active, mutated X chromosomes actually show symptoms of mental retardation. Heterozygous females may express mental retardation but do not show the characteristic physical traits found in males with fra(X). Fragile X males usually are mentally retarded and have learning disabilities that may be characterized by defective speech and language patterns. Hyperactive behavior is common, as are autistic behaviors and major affective or schizoaffective disorders. Physical characteristics of the disorder common to males are macroorchidism, abnormally structured large ears, large nose, prominent forehead, prognathism, and abnormal dermatoglyphics.

Down's Syndrome. Down's syndrome (trisomy 21) is a genetic aberration produced by extra material in the 21st chromosome, and its estimated presence is 1 in every 1000 live births. Down's syndrome appears to have a familial link (high concordance in twins and families), and the occurrences show a linear increase with the mother's age. With few exceptions, individuals with Down's syndrome are mentally retarded, usually to a mild or moderate degree. Historically, the literature has described these individuals as very sociable, affectionate, and amiable; and although this may be true, it is important to realize that they are susceptible to aggression, behavioral dyscontrol, and any other psychopathologic conditions associated with mental retardation. Specific physical features characterize the disorder, such as decreased muscle tone, thin and dry hair, small head, slanted and almond-shaped eyes, and small hands and feet. Neuroanatomic conditions, such as decreased intracranial volume exhibiting smaller gyri and simple patterns, and an underdeveloped cerebellum have also been noted. By the age of 40 years, Down's syndrome subjects exhibit neuropathologies that are strikingly similar to those found in the brains of patients with Alzheimer's disease. Many Down's syndrome patients develop the clinical manifestations of Alzheimer's disease, showing similar cognitive decline and dementia.

Single-Gene Disorders

Single-gene disorders, which are passed on through simple mendelian inheritance, account for about 15% of mental retardation.

Autosomal Dominant Syndromes. Autosomal dominant syndromes can be passed down when either parent carries the disturbed genotype. All offspring of a parent with the mutant gene possess a 50% chance of inheriting the disorder. Extent of gene expressivity in these conditions is extremely variable, and the severity and quality of the clinical manifestations vary greatly. Many autosomally dominant conditions are associated with increased risks of mental retardation, the most common of these being tuberous sclerosis, neurofibromatosis, Sturge-Weber syndrome, and Cornelia de Lange's syndrome.

Autosomal Recessive Syndromes. Autosomal recessive inheritance necessitates that both parents carry the mutant gene, and the offspring must be homozygous positive for the associated phenotype to be expressed. If the individual has only one mutant allele, he or she will not display the associated phenotype but will carry the mutant gene and has a 50% chance of passing it on to his or her offspring. Most autosomal recessive syndromes associated with mental retardation are the relatively rare inborn errors of metabolism. In these disorders, many different metabolic pathways may be involved, such as those of the amino acids, carbohydrates, lipids, plasma proteins, and vitamins. The degree of mental retardation with such disorders is rather severe. Detrimental effects of these disorders can sometimes be avoided by dietary modification or medication early in life. Examples of autosomal recessive inborn errors of metabolism implicated in mental retardation are phenylketonuria, congenital hypothyroidism, Tay-Sachs disease, galactosemia, Neiman-Pick disease, Gaucher's disease, and maple syrup urine disease.

Prenatal Causes: Environmental Factors

Numerous environmental components can alter fetal development, and because of the fragility of the CNS in utero, teratogenic effects are devastating to the fetus. Nongenetic prenatal causes of mental subnormality can be attributed to diverse maternal factors caused by such various denominators as age, improper diet, physical injury, disease or infection, toxic states, and blood incompatibility. Many maternal infections play a sizable role in the genesis of mental retardation and neurologic impairments in the offspring. Syphilis, rubella, toxoplasmosis, and cytomegalovirus are examples of infections that have the ability to diffuse through the placenta and enter the fetus. These bacterial and viral infections generally have a proclivity to settling in the CNS, eventually resulting in microcephaly, epilepsy, hydrocephalus, or cerebral calcification. Toxemias are a group of diseases common in the last 3 months of pregnancy that may result in prematurity and anoxia to the fetus, with the end result being mental retardation. In addition to acute infections, chronic diseases in the mother leave the fetus susceptible to developing mental handicaps. Conditions such as kidney disease, diabetes mellitus, thyroid disease, and hypertension all predispose the baby to diverse neurologic problems.

Maternal Intoxication

Many drugs ingested in pregnancy have teratogenic effects on the child, causing deviations in the CNS and other body systems, which often lead to mental retardation in the child. The teratogenic effects of chemical agents are caused by an interruption in the developmental stages of the body organs, particularly the CNS. Specific drugs that are especially harmful to the offspring are endocrine substances, alkylating agents, antibiotics, many psychotropic medications (hallucinogens, azodyes, thalidomide), and any drug that interferes with metabolism (antipurines, antipyrimidines, antiglutamines). Some widely used drugs, such as alcohol and tobacco, have been associated with specific syndromes of fetal malformation. Fetal alcohol syndrome often results in growth retardation, physical malformations, and mental subnormality.

The clinical manifestations found in the progeny of maternal drug users depend as much on the time in development when the drug was introduced as the specific agent itself. The CNS is susceptible to external factors during myelinization (7 months of pregnancy to the first few months after birth). Defective myelinization is associated with impaired intellectual development in the newborn; these effects may have long-lasting results. In general, the earlier a toxic drug affects the fetus, the more devastating the consequences are likely to be.

Perinatal Factors

The perinatal stage begins at birth and occupies the first several days of life. Obstetric complications, such as prolonged and difficult delivery, birth injury, anoxia, or maternal anesthesia, have a considerable effect on the neonate and are major causes of neurologic and mental handicaps. Up to 18% of cases of mental retardation are attributed to difficulties encountered in the perinatal period. Prematurity is a perinatal factor associated with mental retardation, although prematurity is an effect rather than a primary cause. The premature infant is more likely to be born with problems and is more likely to develop them over time. Postmaturity, as with prematurity, does not lead to mental retardation directly, but those factors causing postmaturity may.

Postnatal Factors

Factors such as head trauma and ingestion of toxic substances are often implicated in postnatal brain damage, but postnatal injury due to infection is by far the most serious of these postnatal factors. Encephalitis and meningitis are two of the major forms of cerebral infection. These can have primary causes from specific infections such as bacterial meningitis or viral encephalitis, where brain damage is caused by direct infection and inflammation of brain tissue. When an infection directly invades and destroys brain cells, and inflammation of the brain is present, mental retardation is a common outcome. The degree of impairment imposed by such difficulty depends on a variety of factors, including the causative agent, the age of the affected individual, and the length and severity of the disease.

Neuropathologies in Syndromes of Retardation

Autopsy studies have shown that more than 80% of moderate and severely mentally retarded individuals have structural brain damage. In recent years, many in vivo techniques have been developed to assess CNS function. Techniques that measure the chemistry of cerebrospinal fluid (CSF) offer hints about the neurochemical makeup of the individual. Neuroimaging techniques, such as computed tomography (CT), magnetic resonance imaging (MRI), and positron emission tomography (PET), are able to identify deep-seated pathologies in the CNS, thereby aiding our understanding of the organic processes that underlie cognition and behavior.

Genetic syndromes of mental handicap, such as Down's syndrome, phenylketonuria, and Lesch-Nyhan syndrome, can be correlated to specific neuroanatomic abnormalities. Individuals with Down's syndrome generally display a small cerebellum and occipital lobes, a narrow superior temporal gyrus, and a decreased total intracranial and gray-matter plus white-matter volume. The presence of senile plaques and neurofibrillary tangles in the cerebral cortex and limbic system are found in virtually all Down's syndrome individuals more than 40 years old.

Lesch-Nyhan syndrome is caused by an inborn error of metabolism that leads to mental retardation and severe compulsive self-mutilation. A low serotonin turnover is observed in these individuals, although dopamine and norepinephrine disturbances have also been noted. CSF levels of homovanillic acid (HVA; dopamine metabolite) are also reduced, in addition to low levels of 3-methoxy-4-hydroxyphenylethylene glycol (MHPG; norepinephrine metabolite) in the CSF of these individuals. It has been suggested that striatal dopamine D_1-receptor supersensitivity may be a mechanism involved in the self-mutilation of Lesch-Nyhan individuals.

Other syndromes of mental retardation have been associated with specific cerebral abnormalities. Cerebral edema and diffuse neuronal degeneration have been noted in individuals in whom lead poisoning has induced mental retardation. Multiple small patchy granulomas and cortical necrosis are common effects of mental retardation secondary to maternal toxoplasmosis. Retardation secondary to rubella often ends in cerebrovascular damage, microcephaly, hydrocephalus, and spotty necrosis in the white matter and basal ganglia. In individuals in whom hypothyroidism has led to mental retardation, cortical atrophy and incomplete cerebral and endocrine system development are often found. Lowered levels of tyrosine and serotonin are common findings in individuals with phenylketonuria, whereas dopamine, opiate, and serotonin system dysfunction have been implicated in Prader-Willi syndrome.

Epilepsy

Thirty percent of mildly retarded individuals and 50% of severely retarded individuals have epilepsy. The condition of epilepsy coincides with a variety of neuropsychiatric complications. Deviant behaviors may be ictal, as a result of actual electrical discharge, or interictal, as a product of an interictal kindling syndrome. Interictal personality traits may develop in the chronic epileptic, including depression and a schizophreniform psychosis. Some problems occurring in the epileptic patient may also be caused by side effects of anticonvulsants, especially barbiturates, that dull cognition, cause aggression, and induce depression or hyperactivity.

DIAGNOSTIC EVALUATION

For a positive diagnosis to occur, the individual must have a significantly subaverage IQ and show deficits in adaptive functioning by the age of 18 years. IQ is regarded as significantly subaverage if it is calibrated to be at least two standard deviations below the mean (70 or below) on a standardized IQ test. The most valid measures of IQ have traditionally been the Stanford-Binet and the Weschler Intelligence Scale for children or adults. By categorizing the measure of intelligence into distinct subcomponents, these tests are sensitized to individual intellectual variation, thereby giving hints to the differential diagnosis.

The significance of adaptive functioning is often overshadowed by intelligence in the diagnosis of retardation, but according to the AAMD, it is of equal diagnostic importance (1). For behavior to be considered adaptive, the individual must be able to meet the standards of personal independence and social responsibility expected by his or her age and social group. The Vineland Social Maturity Scale (VSMS) and the Adaptive Behavior Scale (ABS) are considered two of the most effective measures of adaptive level. The ABS gauges levels of daily living skills and the frequency of inappropriate behaviors, whereas the VSMS concentrates on personal maturity and independence.

Laboratory tests allow recognition of blood and chemical abnormalities that are associated with cognitive impairment. If a genetic disorder is suspected, a karyotype is useful in identifying chromosomal abnormalities. If a seizure disorder is indicated, the electroencephalogram (EEG) can be useful in the identification of epileptic patterns and seizure foci. Radiographs are constructive in identifying cranial abnormalities, such as skull calcification. Neuroimaging techniques also assist in effective diagnosis. See Table 14-1 for a complete list of syndromes.

The high rates of psychopathology in the mentally retarded attest to the importance of diagnosis and treatment of the dually diagnosed. Many diagnostic techniques that are used with nonretarded individuals generalize within the retarded population with great difficulty. The clinical interview and the *Diagnostic and Statistical Manual, 4th edition* (DSM-4), widely recognized as the cornerstones of psychiatric diagnosis, are of limited utility with the mentally retarded. Four factors (intellectual distortion, psychosocial masking, cognitive disintegration, and baseline exaggeration) contribute to the difficulty in distinguishing mental illness among retarded individuals. Intellectual distortion refers to diminished cognitive and communicative ability, which results in the individual's inability to understand and report subjective experiences and feelings. Psychosocial masking refers to the scant social skills and limited repertoire that may lead the clinician to apocryphal judgments with respect to the client's difficulties. Cognitive disintegration alludes to the inability of the retardate to cope with stressful situations and to the behavioral outbursts that may follow. Baseline exaggeration pertains to the tendency for behavior to be particularly uncontrolled during stressful times, which may lead to a misinterpretation of the actual problem during assessment. Because of the unique diagnostic challenges in assessing the mentally retarded, clinicians must learn to rely more on signs (disturbed behavior) and less on symptoms (verbally reported distress and dysfunction).

Several standardized assessments have been constructed to simplify psychiatric diagnosis in the mentally retarded. The most widely used tests are the Psychopathology Instrument for Mentally Retarded Adults and the Reiss Screen for Maladaptive Behavior. These scales cover most major forms of psychopathology and have been proved effective in diagnosing mental illness in mentally retarded populations (5,6). There has been some evidence that standardized psychopathology inventories, modified to include simplified language, may be effective in substantiating a diagnosis in mildly and moderately retarded individuals. One report reveals that modified versions of Beck's Depression Inventory and the Zung Self Rating Depression Scale were potent measures of depression in the mildly and moderately retarded (7).

TABLE 14-1. *Important Syndromes Associated with Multiple Handicaps*

Syndrome	Diagnostic Manifestations					
	Craniofacial	Skeletal	Other	Mental Retardation	Short Stature	Genetic Transmission
Aarskog syndrome	Hypertelorism, broad nasal bridge, anteverted nostrils, long philtrum	Small hands and feet, mild interdigital webbing, short stature	Scrotal "shawl" above penis		+	X-linked semidominant
Apert's syndrome (acrocephalosyndactyly)	Craniosynostosis, irregular midfacial hypoplasia, hypertelorism	Syndactyly, broad distal thumb and toe		+/–		A, D
Cerebral gigantism (Sotos' syndrome)	Large head, prominent forehead, narrow anterior mandible	Large hands and feet	Large size in early life, poor coordination	+/–		?
Cockayne's syndrome	Pinched facies, sunken eyes, thin nose, prognathism, retinal degeneration	Long limbs with large hands and feet, flexion deformities	Hypotrichosis, photo-sensitivity, t hin skin, diminished subcutaneous fat, impaired hearing	+	+	A, R
Cohen's syndrome	Maxillary hypoplasia with prominent central incisors	Narrow hands and feet	Hypotonia, obesity	+	+/–	A, R
Cornelia de Lange's syndrome	Synophrys (continuous eyebrows), thin down-turning upper lip; long philtrum, anteverted nostrils, microcephaly	Small or malformed hands and feet, proximal thumb	Hirsutism	+	+	?
Cri du chat syndrome	Epicanthic folds and/or slanting palpebral fissures, round facial contour, hypertelorism, microcephaly	Short metacarpals or metatarsals; four-finger line in palm	Catlike cry in infancy	+	+	#5p–
Crouzon's syndrome (craniofacial dysostosis)	Proptosis with shallow orbits, maxillary hypoplasia, craniosynostosis					A, D
Down's syndrome	Upward slant to palpebral fissures, midface depression, epicanthic folds, Brushfield's spots, brachycephaly	Short hands, clinodactyly of 5th finger, four-finger line in palm	Hypotonia, loose skin on back of neck	+	+	21 Trisomy
Dubovitz's syndrome	Small facies, lateral displacement of inner canthi, ptosis, broad nasal bridge, sparse hair, microcephaly		Infantile eczema, high-pitched hoarse voice	+/–	+	? A, R
Fetal alcohol syndrome	Short palpebral fissures, midfacial hypoplasia, microcephaly		+/– Cardiac defect, fine motor dysfunction	+	+	
Fetal hydantoin syndrome (Dilantin)	Hypertelorism, short nose, occasional cleft lip	Hypoplastic nails, especially 5th	Cardiac defect	+/–	+/–	
Goldenhar's syndrome	Malar hypoplasia, macrostomia, micrognathia, epibulbar dermoid and/or lipodermoid, malformed ear with preauricular tags	+/– Vertebral anomalies				?
Incontinentia pigmenti	+/– Dental defect, deformities of ears, +/– patchy alopecia		Irregular skin pigmentation in fleck, whorl, or spidery form	+/–		? D, X-linked ?, lethal in males
Laurence-Moon-Bardet-Biedl syndrome	Retinal pigmentation	Polydactyly, syndactyly	Obesity, seizures, hypogenitalism	+	+/–	A, R
Linear nevus sebaceus syndrome	Nevus sebaceus, face or neck		+/– Seizures	+	+/–	?
Lowe's syndrome (oculocerebrorenal syndrome)	Cataract	Renal tubular dysfunction	Hypotonia	+	+	X-linked recessive

Möbius' syndrome (congenital facial diplegia)	Expressionless facies, ocular palsy	+/–Clubfoot, syndactyly		+/–	+/–	?
Neurofibromatosis	+/– Optic gliomas, acoustic neuromas	+/– Bone lesions, pseudarthroses	Neurofibromas, café-au-lait spots, seizures	+/–		A, D
Noonan's syndrome	Webbing of posterior neck, malformed ears, hypertelorism	Pectus excavatum, cubitus valgus	Cryptorchidism, pulmonic stenosis	+/–	+	?
Prader-Willi syndrome	+/– Upward slant to palpebral fissures	Small hands and feet	Hypotonia, especially in early infancy, then p olyphagia and obesity; hypogenitalism	+	+	?
Robin's complex	Micrognathia, glossoptosis, cleft palate		+/– Cardiac anomalies			?
Rubella syndrome	Cataract, retinal pigmentation, ocular malformations		Sensorineural deafness; patent ductus arteriosus	+/–	+/–	
Rubinstein-Taybi syndrome	Slanting palpebral fissures, maxillary hypoplasia, microcephaly	Broad thumbs and toes	Abnormal gait	+	+	?
Seckel's syndrome	Facial hypoplasia, prominent nose, microcephaly	Multiple minor joint and skeletal abnormalities		+	+	A, R
Sjögren-Larsson syndrome		Spasticity, especially of legs	Ichthyosis	+	+	A, R
Smith-Lemli-Opitz syndrome	Anteverted nostrils and/or ptosis of eyelid	Syndactyly of 2nd and 3rd toes	Hypospadias; cryptorchidism	+	+	A, R
Sturge-Weber syndrome	Flat hemangioma of face, most commonly trigeminal in distribution		Hemangiomas of meninges with seizures	+/–		?
Treacher Collins syndrome (mandibulofacial dysostosis)	Malar and mandibular hypoplasia, downslanting palpebral fissures, defect of lower eyelid, malformed ears					A, D
Trisomy 18 syndrome	Microstomia, short palpebral fissures, malformed ears, elongated skull	Clenched hand, 2nd finger over 3rd; low arches on fingertips; short sternum	Cryptorchidism, congenital heart disease	+	+	18 Trisomy
Trisomy 13 syndrome	Defects of eye, nose, lip, ears, and forebrain of holoprosencephaly type	Polydactyly, narrow hyperconvex fingernails	Skin defects, posterior scalp	+	+	13 Trisomy
Tuberous sclerosis	Hamartomatous pink to brownish facial skin nodules	+/– Bone lesions	Seizures, intracranial calcification	+/–		A, D
Waardenburg's syndrome	Lateral displacement of inner canthi and puncta		Partial albinism, white forelock, heterochromia or iris, vitiligo, +/– deafness			A, D
Williams' syndrome	Full lips, small nose with anteverted nostrils, iris dysplasia	Mild hypoplasia of nails	+/– Hypercalcemia in infancy, supravalvular aortic stenosis	+	+	?
Zellweger cerebrohepatorenal syndrome	High forehead, flat facies		Hypotonia, hepatomegaly, death in early infancy	+	+	A, R

From Smith DW. Patterns of malformation. In: Vaughan VC III, McKay RJ Jr, Behrman RE, eds. *Nelson textbook of pediatrics.* 11th ed. Philadelphia: WB Saunders, 1979:2035.

A, autosomal; D, dominant; R, recessive.

Treatment

The most effective treatment for mentally retarded individuals is multimodal and takes into consideration the individual's strengths while minimizing weaknesses. Treatment planning that incorporates vocational programming, daily living and social skills training, and appropriate schooling is very beneficial to these individuals.

Social Matrix

The institutionalized retarded have higher rates of psychiatric disturbances than mentally retarded individuals living in community residences. The frequent lack of social skill training and independence programming found in large institutions precludes the cognitively impaired individuals from acquiring the coping strategies they so desperately need. Community residential programs, which operate on the foundation of normalization, are succeeding in many of the areas where institutions failed. Stimulating living experiences may actually heighten intellectual proclivity, whereas a lack thereof can lower IQ.

Biopsychosocial Model

A combination of behavior therapy and pharmacotherapy has been shown to be more effective than either modality alone in eliciting behavior change.

Behavior Modification

For the past 3 decades, behavioral approaches have been the most thoroughly researched and most widely used therapeutic intervention with the mentally retarded (8). Behavioral therapies stem from the belief that maladaptive behaviors result from environmental contingencies at work to promote these behaviors. The behavioral approach works to break down and objectively define specific behaviors, dissecting the contextual basis in which they occur. This functional analysis is often effective in unearthing the contingencies that are maintaining and reinforcing undesirable behaviors. With this knowledge, the clinician can devise new environments to reinforce desirable behaviors and extinguish those that are maladaptive. Reinforcement and feedback contingencies appear to be the technique of choice in replacing maladaptive behaviors with adaptive ones, although punishment, when implemented cautiously, is also effective in curbing serious problematic behaviors.

Pharmacotherapy

Neuroleptics have traditionally been the most common medication prescribed to the mentally retarded, and studies have consistently reported that between 45% to 55% of mentally retarded individuals receive antipsychotics. Low to moderate doses of neuroleptics are at least equally effective and may even be superior to high doses in most patients (9). Lower doses are also implicated in fewer unwanted side effects than high doses. Biologically and cognitively, the mentally retarded tend to be especially sensitive to drugs and vulnerable to developing pronounced side effects. Tardive dyskinesia is present in 34% of neuroleptically treated retarded individuals. Tardive dyskinesia and akathisia may agitate and frustrate the patient to a point of dyscontrol, thus indirectly exacerbating the behaviors the drugs were intended to control. Neuroleptics are commonly implicated in extreme sedation, cognitive dulling, and withdrawal. In addition to neuroleptics, benzodiazepines have been especially deleterious with mentally retarded individuals. The disinhibiting side effects of benzodiazepines actually exacerbate many of the behaviors they are intended to control.

The most effective forms of pharmacologic management employ different agents that concentrate on different effects. Pharmacotherapy must arise from the differential diagnosis and should employ specific agents that ameliorate the biochemical underpinnings of disturbed functioning respective to each patient. The optimal treatment program may require several agents, may employ doses considered homeopathic, may be used in conjunction with clinical observations of behavior, and is perfected over time in accord with research and experimentation. When employed carefully and cautiously, pharmacotherapy can be most beneficial in many mentally retarded individuals, minimizing side effects while maximizing desired effects.

Behavioral Dyscontrol

Behavioral dyscontrol is typically characterized by aggression, self-abuse, stereotypy, and agitation. Often, these problems have organic etiologies and can be ameliorated with the proper medical interventions. Behavioral dyscontrol has been cited as the single most common indication for pharmacotherapy in the mentally retarded. Overarousal and serotonergic dysfunction are thought to be the most common biologic determinants of these behaviors. The most effective treatments for these conditions have focused on replacing maladaptive behaviors with adaptive behaviors, using a combination of pharmacotherapy and behavioral interventions.

Noise and Overarousal. The developmentally disabled are especially vulnerable to overarousal because of their inability to cognitively filter internal and external stimuli. This lack of filtering results in a noisy state of stimulus overload, which can actuate internal chaos, personal distortions, impulsivity, hypervigilance, physiologic stress, and aggression. If overarousal is a precipitating force in behavioral dyscontrol, it follows that decreased arousal may increase tolerance and reduce behavioral outbursts. β-Blockers have been found to be a successful prescription for the related symptoms of behavioral dyscontrol. These agents have been efficacious in reducing anxiety and ameliorating outbursts in a variety of disadvantaged populations. Propranolol is effective in downregulating hyperarousal, subsequently reducing aggressive outbursts in the severely and profoundly retarded. Nadolol also appears to initiate a gradual disappearance of inner restlessness and tension. β-Blockers have shown little tendency to cause adverse cognitive side effects, promote the emergence of habilitative behaviors, such as speech and socialization, and may actually mediate behavioral teaching contingencies.

Serotonin. Abnormalities in serotonergic functioning have also been implicated in behavioral dyscontrol. Many studies have found correlations between decreased CSF levels of the serotonin metabolite 5-HIAA and impulsive aggression. These findings have given rise to a "low-serotonin syndrome" hypothesis of aggression and impulsive behaviors, which asserts that decreased central serotonin may be involved in precipitating impulsive aggression. A diet rich in serotonin is effective in reducing behavioral dyscontrol in the mentally retarded. Buspirone is a psychopharmaceutical that has demonstrated efficacy in lowering occurrences of destructive behaviors through serotonergic agonist mechanisms. In mentally retarded patients, low-dose buspirone is efficacious in regulating behavioral dyscontrol and aggression. Buspirone appears to reduce episodic dyscontrol with no known sedative or addictive effects. Additional serotonergic agents have shown promise in decreasing behavior and aggression outbursts. Serenics (e.g., eltoprazine) produce specific antiaggressive effects through serotonergic mechanisms.

Other Agents. Lithium carbonate has demonstrated efficacy in controlling antiaggressive behaviors in mentally retarded individuals. Overall, lithium appears to be efficacious in rapidly curbing episodic dyscontrol in some retarded patients; however, the potential toxicity of this agent and its partial sedative effects prevent it from being the primary treatment of choice for behavioral dyscontrol.

Self-Injurious Behavior

Self-abuse has been recognized as an exceedingly common problem in this population, manifesting itself in at least 10% of severely and profoundly retarded. When environmental regulation fails to control self-injurious behavior (SIB) completely, the severity of the problem warrants pharmacotherapy. As with behavioral dyscontrol, self-injury may be mediated by hyperaroused states and serotonergic dysfunction. In circumstances of SIB precipitated by hyperarousal, β-blockers have been effective in the amelioration of self-abuse. Buspirone has been reported to control self-injury in mentally retarded populations, whereas tryptophan and trazodone, serotonin enhancers, have also proved successful in treating self-injury.

Dopaminergic activity also has been repeatedly implicated in self-injury. The antagonism of D_1 and D_2 receptors in clozapine treatment stops self-mutilatory behavior. Additionally, the inhibitory responses on self-mutilatory behavior of fluphenazine, a D_1, D_2 blocker has been reported. SIB has also been associated with opioid dysfunction. Enhanced opioid activity resulting in a generalized analgesic state may predispose certain mentally retarded individuals to self-injury. Opioid antagonists, such as naltrexone, have been shown to reduce SIB in certain mentally retarded individuals.

Psychiatric Disorders

Schizophrenia. When a mentally retarded individual has been definitively diagnosed with psychotic symptoms, antipsychotic agents are clearly the treatment of choice. When prescribing an antipsychotic agent for a mentally retarded client, the clinician must be especially vigilant of cognitive side effects that dampen already low levels of intellectual functioning. Antipsychotics should not be used as a standing regimen for agitated or aggressive behavior. Low-potency neuroleptics, such as chlorpromazine and thioridazine, need to be taken in high doses to achieve a beneficial antipsychotic effect. These higher doses often lead to increased sedation. Conversely, high-potency drugs, such as thiothixene and haloperidol, may be used in smaller doses and lead to less sedation, but their potency may be more likely to induce unwanted motor reactions. Thiothixene appears to have a more rapid onset and produces fewer undesirable cognitive effects than thioridazine in the schizophrenic mentally retarded. However, higher rates of extrapyramidal side effects may ensue with thiothixene administration. Haloperidol has also demonstrated efficacy over the low-potency neuroleptics in mentally retarded populations. Clozapine or risperidone should also be given consideration for the treatment of schizophrenic symptoms in this population because of their specificity compared with most other neuroleptics and lower rate of extrapyramidal effects.

Affective Disorders. Mood disorders in the mentally retarded are rarely reported but are suspected to be relatively common. Specific depressive mechanisms have been attributed to the serotonergic and catecholinergic systems. Pharmacotherapy of endogenous affective disorders in the mentally retarded has produced positive results using lithium, carbamazepine, and traditional antidepressants. Lithium appears to be effective as a prophylactic therapy for both bipolar and unipolar affective disorders in mentally retarded populations. Carbamazepine also appears to be as effective as lithium in regulating mood, and a combination of carbamazepine and lithium is effective in the control of cycling when noncombination therapy is ineffective. Antidepressants are generally regarded as the treatment of choice for cases of unipolar depression. Heterocyclic antidepressants appear to be effective in curbing depressive symptoms in the nonverbal severely and profoundly retarded, while displaying few serious untoward effects. Fluoxetine has also demonstrated specific antidepressant effects in mentally retarded adults and is reported to have a more favorable side effect profile than traditional antidepressants.

Epilepsy

Carbamazepine and valproic acid are the safest and most effective anticonvulsants for people with compromised cognitive functioning. Carbamazepine is the best choice for complex partial seizures, whereas valproic acid is also effective for absence seizures. For the most part, these two anticonvulsants have benign cognitive side effects, although valproic acid has been associated with some toxic effects. Anticonvulsants, such as phenobarbital, phenytoin, and clonazepam, should be used with caution in this population. Phenytoin has shown negative influences on attention, memory, and overall test performance, whereas clonazepam, a benzodiazepine, should also be avoided because of its disinhibiting effects on behavior.

CEREBRAL PALSY

Cerebral palsy is a chronic and variable neurologic disorder of movement and posture induced by a nonprogressive defect of the CNS. *Cerebral palsy* is an umbrella term used to cover a medley of conditions, including movement and balance problems; delays in physical development; floppy or tight muscles; abnormalities of vision, hearing, and speech; and seizure disorders. Mental retardation is found in 50% to 75% of cases of cerebral palsy. Learning disabilities and frequent social, emotional, and interfamilial problems may also be present. Recent figures have shown cerebral palsy to have an incidence of 2.5 per 1000 live births (10), although the incidence varies among classification, with spasticity constituting at least 50% of the cases. Generally, patients have been considered to be equally distributed among the mild, moderate, and severely afflicted categories.

Classification

Cerebral palsy is classified according to the location of the neurologic lesion, the severity of the damage, the number of limbs involved, and the kind of movement affected by the disorder. The four main classifications are spasticity, athetosis or dystonia, ataxia, and mixed types. The clinical presentation of cerebral palsy varies in symptomatology and etiology according to each classification.

Spasticity

Spasticity results from damage to the motor portion of the cerebral cortex and pyramidal tract, which is responsible for the initiation of voluntary movement. Spasticity is topographically subgrouped into diplegia and hemiplegia. *Spastic diplegia* is the most common form of cerebral palsy in premature infants, resulting in 10% to 33% of the cases. This term refers to disabling of the limbs on both sides of the body, with the legs more severely affected than the arms. The result is abnormally stiff and resistant muscles, with locomotor development more impaired than manipulative skills. Mental subnormality is often concurrent and may cause retardation in social, language, and other areas of development. Spastic diplegia is almost always congenital, with a large percentage of affected children born prematurely, with low birthweight.

Hemiplegia refers to the unilateral disturbance of the extremities, independent of whether the upper or the lower portions are affected. This disorder results in increased tendon reflexes, weakness, and difficulty with discrete finger movements. When the unaffected limbs are active, associated movements, particularly in the arm, occur. Mental retardation is less common in hemiplegic cerebral palsy than in other forms, although the child's motor problems create educational difficulties even among the most intelligent. Most cases of hemiplegic cerebral palsy are congenital in origin, resulting from prenatal factors in about two thirds of cases and from perinatal brain damage in others.

Athetosis (Dyskinetic Cerebral Palsy)

Athetosis results from damage to the extrapyramidal tract, causing a form of cerebral palsy in which fine-motor movements are affected and involuntary movements of the limbs are prominent. These involuntary movements often take the form of unwanted, sympathetic movements by several limbs of the individual during an attempt to make one controlled, purposeful movement. The full-blown effects of this disorder are often not apparent until the age of 2 years, when ability to control the use of muscles is fully developed. The most common cause of athetosis is hypoxia; however, contributions are made by predisposing harmful prenatal influences. Three fourths of children with dyskinetic cerebral palsy are born at term.

Ataxia

Ataxic cerebral palsy accounts for up to 10% of cases and may result from a cerebellar lesion. Ataxia is a condition in which coordination of movement and balance are disturbed. Delays are seen in developmental milestones, such as sitting, reaching, and walking. Ataxic diplegia can be mistaken for mental retardation or neuromuscular disease owing to similar delays in development and associated hyptonia. This form of cerebral palsy differs from others in that its etiology is unknown in about 41% of cases.

Causes

Although many cases of cerebral palsy are of unknown origin, it is believed that 85% are prenatally induced. Ultrasound studies of premature infants provide evidence that many cases of cerebral palsy result from brain damage that occurred at least 2 weeks before birth. No consistent differences are observed in social and environmental factors, history of pregnancy, labor, or delivery, confirming the hypothesis that most cases are not associated with adverse obstetric factors. Prenatal influences, such as CNS malformations, chromosomal aberrations, and congenital infections, account for up to 75% of all cerebral palsy cases. Untreated medical conditions, such as infections, rubella, toxoplasmosis, cytomegalovirus, encephalitis, meningitis, herpes, and acquired immunodeficiency syndrome (AIDS) have been linked to the pathology of cerebral palsy; however, these account for only 6% of children with cerebral palsy.

A potential hereditary basis for cerebral palsy has been given substantial consideration in the past, although it has been linked to only a small percentage of the cases. The likelihood of genetic determination in spastic diplegia and quadriplegia is about 10%; the risk is greater if the spasticity is symmetric. The genetic risk for athetoid and dystonic cerebral palsy is unknown but is probably similar to that of spastic cerebral palsy. The highest likelihood of a genetically determined disorder is in the category of ataxia and ataxic diplegia. Although there are examples of multiple cases of cerebral palsy within a family, one must not assume genetic origin. Neonatal asphyxia was once believed to be a direct link to cerebral palsy, but such difficulties rarely result in neurologic deficit. In most cases, evidence for asphyxial damage is lacking, and it is now believed that asphyxia alone plays little or no role in most cerebral palsy cases.

Diagnosis

The milestones of maturity are extremely delayed in both retardation and cerebral palsy and must be scrutinized to differentiate between the two. Early reflexes, such as the Moro reflex and palmar grasp reflex, extend beyond their normal time frame in the beginning stages of cerebral palsy and can offer an important diagnostic cue to both parents and physicians. Effective diagnostic comparisons must be made not only between the patient's development and that of a normal child but also between the function of the right and left limbs and of the arms and legs. Because the motor deficit in cerebral palsy is one that evolves with maturity, assessments must be continually repeated to detect deterioration in function.

Treatment

Because cerebral palsy is a constantly evolving disorder, the most efficient treatment plan includes repeated assessments. A multidisciplinary approach has shown great benefit because it encompasses many of the problems experienced by the individual with cerebral palsy.

Although there is debate about the value of physiotherapy in treating cerebral palsy, most pediatricians believe there is some benefit. They recommend early rather than late onset of such treatment in hopes of preventing severe contractures and deformity and promoting normal motor development. Other forms of therapy include treatment of vision and hearing deficits, feeding and speech therapy, and orthopaedic therapy.

Medications are also being considered as treatment options for the motor deficits of cerebral palsy, especially in spastic forms. Baclofen has proved to be very effective in some hemiplegic and diplegic children in reducing spasticity and facilitating physiotherapy.

The assessment and treatment of psychiatric disorders in patients with cerebral palsy follows the same guidelines as for mental retardation.

AUTISM AND PERVASIVE DEVELOPMENTAL DISORDERS

PDDs generally and infantile autism in particular have been the focus of a great deal of research and clinical speculation.

Historical Development of the Concept of Autism

Pervasive developmental disorder (PDD) is the current term for what is probably a group of related neurodevelopmental disorders characterized by a similar behavioral profile; others have labeled this cluster of conditions *autistic spectrum disorders* (11). The relationship between autism and PDD is confusing. Technically, autism is a diagnostic category under the broader heading of PDD. In clinical practice, PDD is often used to describe children who have mild autism, high-functioning autism, or an incomplete set of autistic features. Autistic children can best be discriminated from PDD children by degree of social impairment.

Autism was first described by Leo Kanner and became known as *infantile autism* or *autistic disorder* (12). In the 1960s, neurodevelopmental theories outlined a variety of models of autism, positing impairment in the following areas:

- Allocation of attention to coordinate new stimuli with memories
- Vestibular mediation of perceptual processes
- Left-hemisphere–mediated linguistic abilities

Today, those advocating a primary cognitive deficit and those favoring a primary social and affective deficit agree that the basic PDD deficit stems from a neurologic dysfunction.

Differential diagnosis was a conceptual problem for early autism research. Some clinicians believed that autism was a variant of or precursor to schizophrenia. However, it was observed that autistic-like disorders virtually always begin before 3 years of age, whereas schizophrenic-like disorders almost never begin before 7 years of age (13). This realization revitalized interest in infantile autism as a distinct nosologic entity, leading to the development of more operationally precise diagnostic criteria and a reconceptualization of the syndrome as a pervasive developmental disorder, under which label it was incorporated in the 3rd edition of the *Diagnostic and Statistical Manual of Mental Disorders* (DSM-3).

Diagnostic Criteria

In 1978, Rutter proposed four essential criteria (14):

1. Onset before the age of 30 months
2. Impaired social development out of keeping with the child's intellectual level
3. Delayed and deviant language development out of keeping with the child's intellectual level
4. Insistence on sameness, as shown by stereotyped behavioral patterns, abnormal preoccupations, or resistance to change

The diagnosis was limited to cases with onset before age 30 months to differentiate maximally between autism and other symptomatically similar disorders, such as Heller's syndrome and Asperger's syndrome. Rutter recognized that some cases of autism arise at slightly older ages, but unfortunately, the 30-month guideline became a rigid rule in DSM-3. Rutter also emphasized that social impairment in autism often lessens after early childhood, although more subtle social ineptitude persists.

Ritvo and Freeman outlined slightly different criteria for autism, which were adopted by the National Society for Autistic Children (15):

- Disturbances of development rate or sequences
- Disturbances of response to sensory stimuli
- Disturbances of speech, language-cognition, and nonverbal communication
- Disturbances of the capacity to relate appropriately to people, events, and objects

Age of onset was stated to be typically before 30 months of age, but the system is descriptive and has no fixed, requisite criteria. Associated features were said to include mood lability without identifiable cause; failure to appreciate real danger; inappropriate fears; self-injurious behaviors; mental retardation; and abnormal EEGs with or without seizures. The Ritvo and Freeman criteria identify a somewhat different set of children as autistic from those that Rutter lists.

Wing and Gould took an empirical approach to the diagnosis, investigating a population of mentally handicapped children with deficits in social relatedness, communicative language, and repertoire of interests and behaviors (16). They identified three subgroups based on social behavior:

- Aloof
- Passive
- Active but odd

Two aspects of Wing and Gould's work are relevant to diagnosis. First, they broadened the concept of PDD. These investigators started with the fundamental, universally accepted symptom areas of autism and used them as the basis of classification. Second, they described social behaviors associated with different types of functioning at different developmental stages. Subgroup membership can change with development in the direction of greater sociability. Some "aloof" young children grow into "passive" or "active-but-odd" older children. However, "active-but-odd" young children are unlikely to become either "passive" or "aloof."

The revised DSM-3 established only three diagnostic classifications for severe childhood psychopathology:

- Autistic disorder
- Pervasive developmental disorder not otherwise specified (PDDNOS)
- Schizophrenia with onset in childhood

The DSM-4 followed the revised DSM-3 categories for autistic disorder and PDDNOS, but added three diagnostic categories to parallel the World Health Organization's *International Classification of Diseases,* 10th edition (ICD-10) PDD diagnostic subgroups: Rett's syndrome, other childhood disintegrative disorder, and Asperger's syndrome. The DSM-4 pervasive developmental disorders includes the following subgroups:

299.00: Autistic disorder
299.80: Rett's disorder
299.10: Other childhood disintegrative disorder
299.80: Asperger's disorder
299.80: PDDNOS (including atypical autism)

It is important to note that Rett's disorder, PDDNOS, and atypical autism (identified within PDDNOS), all carry the same diagnostic code (299.80). Diagnosticians thus actually code the PDDNOS remainder category (299.80) when they diagnose Rett's disorder, Asperger's disorder, or atypical autism.

DSM-4 and ICD-10 provide 12 parallel diagnostic criteria for autistic disorder and childhood autism. Six symptoms must be present, four of which are specified. At least two symptoms must come from a subset of four impaired social skills (nonverbal interactional behavior, friendship, joint attention, and reciprocity). One must come from a subset of four impaired communication skills (delayed or absent language, given speech—abnormal conversation, perseverative speech, and abnormal play), and at least one of the six must come from a subset of four abnormal activities (obsessive interests, rigid rituals, stereotypies, and preoccupation with parts of objects).

Behavioral, Cognitive, and Emotional Characteristics

As indicated by DSM criteria, autism is marked by a general inability to form relationships, failure to use nonverbal communicative behaviors such as eye contact, lack of reciprocity, lack of awareness of others, and failure to share experiences with others. Other well-recognized impairments in the social domain are difficulties with affection, giving and seeking comfort, awareness of social rules, social imitation, attachment and symbolic play. The most fundamental of social deficits in autism is that of impairment in basic face-to-face interaction skills. Autistic children cannot orchestrate the give and take of social interactions, including conversation and play. The autistic child is unlikely to initiate interactions but is more likely to respond in adult-initiated situations. A small subset of autistic individuals display islands of exceptional ability against a background of widespread cognitive impairment (so-called savant syndrome). Autistic savant abilities may include music and musical memory, spatial memory, decoding written material, and calculations.

Deficits in Attention

Autistic individuals are found to be overselective in their attention to stimuli, such that their focus of attention is abnormally narrow. They also have difficulties in shifting attention between modalities and in shifting spatial attention. Autistic children are able to sustain their attention at a level appropriate for their mental age, if given tangible reinforcements. Hypersensitivity to novelty and overactivation of brain-stem mechanisms of arousal causes narrowed attentional focus and stimulus overselectivity in autistic children.

Language Deficits

Verbal autistic children generally are able to acquire normal grammatical morphology and syntax, although onset is delayed. Some autistic children learn grapheme-phoneme correspondence, enabling them to write and to decode words for reading; in fact, early acquisition of written (but not spoken) language often marks high-functioning autism. However, language comprehension is significantly impaired relative to expression, and deficits in the semantic and pragmatic aspects of language are common. Even when language is relatively spared, there are usually deficits in comprehending complex language and in formulating complex output. In these functions, even high-functioning autistic children are more deficient than are children with language disorders.

Other Social and Cognition Deficits

Recent research into social and affective communication in PDD has suggested that autistic children are more impaired in emotional recognition than their cognitive deficit would warrant. Autistic subjects also appear to be impaired in the expression of emotion. Autistic children display less facial affect, especially positive affect, than other groups, but more incongruous, difficult-to-interpret expressions. High-functioning autistic subjects use communicative intonation less than the other groups but make greater use of noncommunicative patterns of intonation.

Memory Deficits

Other theoretical models of autistic cognition have been proposed. Boucher and Warrington and also DeLong have noted similarities between the behavior of autistic children and that of animals with hippocampal lesions (17,18):

- Increased general activity
- Motor stereotypies
- Reduced exploration of the environment
- Reduced responsiveness to novel stimuli when familiar stimuli are present

- Perseveration and impaired active error reduction in learning and memory tasks

These authors hypothesized that autism might be a kind of developmental amnesic syndrome. Visual and rote auditory memory tend to be spared, whereas verbal memory for semantically organized material such as stories is quite impaired. Anecdotally, many high-functioning autistic individuals show hyperamnesia for specific kinds of material, including spatial arrays, routes, music, and events, a fact that "amnesia" theories of autism must be able to explain.

Executive Function Deficits

Two tests of executive function, the Wisconsin Card Sort Test and the Trail-Making Test, reveal significant impairment in the autistic subjects. These tests require subjects to formulate problem-solving strategies and to respond flexibly to changing task demands; successful performance is thought to be mediated by frontal cortical systems.

Developmental Course and Prognosis

Autism is a developmental disorder and cannot be described from a purely cross-sectional perspective. The typical autistic child lacks interest in human relationships, lacks communicative language, expresses a strong need for environmental constancy, and engages in stereotyped behaviors, yet this presentation is most characteristic of the preschool years. During middle childhood, autistic children often master daily living skills and make some accommodation to demands made by other people, whereas their ritualized behaviors and idiosyncratic preoccupations tend to diminish. School behavior may come to resemble that of hyperactive or retarded children rather than continuing to conform to classic autistic patterns. Adolescence can be a difficult time. Besides the frequent onset of seizures during this period, a number of PDD children regress behaviorally as adolescents. Higher-functioning individuals (e.g., those with Asperger's syndrome) are especially prone to psychological problems as they realize the extent of their ineptitude in the realm of social interaction. Their feelings of helplessness may engender depressed moods or clinical depression. On a more positive note, both social and language skills usually do continue to develop during adolescence, and increased interest in relating to other people can make psychosocial intervention easier.

About half of all autistic adults remain in residential care. Many of those not in residential treatment continue to be dependent on caregivers. No more than one in five achieves truly independent living, including gainful employment; of these, many no longer meet diagnostic criteria for PDD, although social peculiarities often persist, and neither close friendships nor sexual relationships are common. The two most important predictors of good outcome in autism are IQ and language function. The presence of associated neurologic disorders is a strong negative prognostic indicator. Language development obviously is related to IQ but apparently makes an independent contribution to the prediction of long-term outcome. In particular, the emergence of communicative speech by the age of 5 years is thought to be a crucial indicator.

Epidemiology

Prevalence

Long-standing convention places the prevalence of Kanner-defined autism between 2 and 4 per 10,000 children. Based on the concept of an autistic spectrum on which Kanner's autism accounts for half of all cases of PDD, the estimated prevalence for all PDD is in the range or 4 to 8 per 10,000 children. In recent years, however, prevalence estimates have increased.

Sex Ratio

PDDs are more common among boys than among girls. Gillberg has reported that the male-to-female ratio in broadly defined PDD is between 2.0:1 and 2.9:1 (19). The preponderance of boys is greater in narrowly defined Kanner's autism, with sex ratios ranging from 2.6:1 to 5.7:1, and greater still in Asperger's syndrome, with sex ratios ranging from 7.1:1 to 10:1 across studies.

Genetic Contributions

The proportion of families with two autistic children is greater than would be expected based on chance, and reports of twins concordant for autism have appeared frequently. Attempts have been made to find morphologic, biochemical, and DNA markers in autism. There is no morphologic profile characteristic of autism, but elevated rates of minor physical anomalies are found in some samples. No DNA marker for autism has been identified. The mode of transmission of the gene or genes responsible for vulnerability to autism remains unknown. The principal hypothesis at present is that the diathesis is a polygenic, multifactorial trait, with a number of different genes contributing additively to a dimension of vulnerability.

Boundary Conditions and Co-Morbidity

In our view, autism or PDD is a clinical diagnosis that should be made whenever a patient's symptoms meet diagnostic criteria, regardless of any associated conditions. Autism is a syndrome, like mental retardation: It is applied to people with a wide range of etiologies or co-morbid disorders. If, for example, a patient displays autistic symptoms, moderate mental retardation, and the fra(X) abnormality, all three diagnosis should be made; failure to do so would impoverish the clinical portrayal of that patient. Diagnosis of associated conditions will also contribute to knowledge concerning the relationship of the disorders.

Disorders Widely Regarded as Belonging to the PDD Spectrum

Heller's Syndrome (Disintegrative Disorder)

One variant form of autism is the condition known as *infantile dementia, disintegrative psychosis,* or *Heller's syndrome.* The disorder is characterized by 2 or more years of essentially normal development followed by marked regression into a stare seemingly indistinguishable from autism, including the loss of already-acquired social abilities and linguistic competence. Heller's syndrome is poorly understood. The one point of agreement is that it is very rare, with fewer than 100 cases documented in the literature. Heller's syndrome is indistinguishable from autism on cross-sectional inspection. The syndrome could represent the extreme tail of the distribution of age of onset of PDD. The diagnostic criteria are as follows:

a) Apparently normal development for at least the first 2 years after birth
b) Loss of previously acquired skills (before age 10 years) in at least two of the following areas:
   - Expressive or receptive language
   - Social skills or adaptive behavior
   - Bowel or bladder control
   - Play
   - Motor skills
c) Impairment in at least two of the following:
   - Social interaction
   - Communication
   - Restricted, repetitive, and stereotyped patterns of behavior, interests, and activities (DSM-4).

Rett's Syndrome

Another autistic spectrum condition is Rett's syndrome, which is unusual among developmental disorders because it apparently affects only girls. Like disintegrative disorder, Rett's syndrome is characterized by a period of apparently normal development (6 to 18 months) followed by profound deterioration of social and psychomotor skills accompanying a progressive cerebral atrophy. The classic symptom involves loss of purposeful use of the hands, accompanied by the development of complex stereotypic hand movements. Rett's syndrome follows a clear longitudinal course with four well-defined stages:

1. ***Stage one***, referred to as "early-onset stagnation," typically lasts several months. The child shows reduced interest in his or her playing, may begin showing odd hand-waving behaviors at times, and head growth decelerates. There may be reductions in his or her communicative abilities and eye contact.
2. ***Stage two***, the "rapid destructive stage," can take place in a few weeks or several months. During this phase, the classic Rett's syndrome presentations (hand wringing, hand clapping, and hand washing) appear, and purposeful hand use is lost. Other motor abilities are better preserved, but the child is clumsy, with ataxia and apraxia. Breathing is often irregular, and hyperventilation may occur. Cognitively, the child displays severe dementia. Seizures sometimes develop. Behaviorally, classic autistic symptomatology is present.
3. ***Stage three***, the "pseudostationary" period, usually lasts several years. There is some cognitive recovery, and autistic symptoms diminish markedly. With the increased emotional contact of this period, the child's presentation is better described as showing severe mental retardation than as autistic. However, gross motor skills often deteriorate further, with gait apraxia and jerky movements of the trunk common. Seizures are present in about 70% of cases, and abnormal EEGs are apparently universal.
4. ***Stage four***, termed "late motor deterioration," begins anywhere from age 5 years through late adolescence. Epileptic symptoms diminish, and emotional contact with caregivers improves. Growth is retarded, but puberty occurs at the normal age. There is progressive wasting away and weakness of muscle, combined with spasticity; severe scoliosis and trophic foot disturbances are common, and most Rett's syndrome sufferers become wheelchair bound.

The estimated prevalence of Rett's syndrome is 1 in 12,000 to 15,000 live female births. Rett's syndrome may account for one fourth or more of all cases of PDD in girls. The etiology of Rett's syndrome remains unknown. Rett's syndrome has been included in DSM-4 under the PDD heading.

Asperger's Syndrome

Another condition that may belong on the autistic spectrum is Asperger's syndrome. Whereas the syndromes described by Heller and Rett are severe variants of PDD, Asperger's syndrome is a milder variant, with normal or near-normal IQ and relatively good language skills. Asperger's syndrome also appears to be more common than Rett's or Heller's syndrome, with an estimated prevalence between 10 and 26 cases per 10,000 population. Children with Asperger's syndrome typically are brought to clinical attention soon after they enter school. In early childhood, their behavior is odd but often not so unusual as to impel the parents to seek psychiatric advice. Anecdotal evidence suggests that these children often are slow to walk. Speech, too, may be delayed, but improves rapidly once begun. There may be language problems, such as pronoun reversal, but these are more transient than in cases of Kanner's autism. Imaginative play is apt to be impoverished, as in autism, but stereotypies are less common. Instead, the insistence on sameness takes the form of special interests. Children with Asperger's syndrome usually have one or two topics (such as trains or weather) to which they are passionately devoted, largely to the exclusion of all else.

When children with Asperger's syndrome enter school, their deficits soon become apparent. Single-minded pursuit of their own interests, on their own schedule, conflicts with the demands of the structured classroom. Children with Asperger's syndrome tend to excel in subjects requiring rote memorization but tend to fail when problem solving or higher-order conceptualization are needed. In elementary grades, this may entail high grades in reading and spelling but poor performance in arithmetic. Individuals with Asperger's syndrome are not socially withdrawn, but their attempts at social interaction are awkward and odd. Their speech tends to be stilted and pedantic, with flat or exaggerated prosody and gesture. These characteristics are conducive to social isolation and peer rejection in school.

Children with Asperger's syndrome crave and pursue social contact and are keenly aware of the consequences of their social deficits. Sexual frustration is a major clinical issue in adolescence and adulthood, as is their frequent failure to achieve occupational goals in keeping with their academic work. They often seek clearly stated rules of conduct to guide them in their interactions, but their clumsy, rigid enactment of these "rules" seldom succeeds in winning them friends. Depression is a frequent complication, and there is evidence of a genetic link between Asperger's syndrome and affective disorders.

The DSM-4 criteria for the diagnosis of Asperger's syndrome include the following:

a) Qualitative impairment in social interaction, as manifested by at least two of the following:
   - Marked impairment in the use of multiple nonverbal behavior
   - Failure to develop peer relationships appropriate to developmental level
   - Lack of spontaneous seeking to share enjoyment, interests, or achievements with other people
   - Lack of social or emotional reciprocity
b) Restricted repetitive and stereotyped patterns of behavior, interests, and activities, as manifested by at least one of the following:
   - Encompassing preoccupation with one or more stereotyped and restricted patterns of interest that is abnormal either in intensity or focus
   - Apparently inflexible adherence to specific, nonfunctional routines or rituals, stereotyped and repetitive motor mannerisms
   - Persistent preoccupation with parts of objects
c) Clinically significant impairment in social, occupational, or other important areas of functioning
d) No clinically significant general delay in language
e) No clinically significant delay in cognitive development or in the development of age-appropriate self-help skills, adaptive behavior (other than in social interaction), and curiosity about the environment in childhood

Nonverbal (Social-Emotional) Learning Disabilities

Discussion of dyslexia is beyond the scope of this chapter, but the less familiar social or nonverbal developmental disorders deserve mention because of their sometimes striking resemblance to the core disturbances of autism. Rourke used a psychometric approach to identify a subgroup of learning disabled children with a syndrome he labeled *nonverbal learning disability* (NLD) (20). Children with NLD were reported to be strong in phoneme-grapheme matching, rote verbal learning, and related psycholinguistic skills but were impaired on visuospatial and nonverbal problem-solving tasks, including arithmetic. Based on the awkward social behavior of NLD subjects, which includes disturbances in affect comprehension, gestural communication, and prosody, Rourke suggested that the NLD syndrome represents the mild end of the autistic spectrum (20). The justification for including Asperger's syndrome and NLD in the PDD spectrum is their autistic-like social impairment.

Disorders Not Usually Regarded as Part of the PDD Spectrum

Tourette's Syndrome and Obsessive-Compulsive Disorder

Tourette's syndrome is not regarded as a PDD spectrum disorder. Individuals with tic disorders are not impaired in social relatedness. Onset of Tourette's syndrome is later than autism (age 7 years is the mean). Because tics, unlike autistic symptoms, frequently respond to drug treatment, it is important not to allow the diagnosis of PDD to exclude the diagnosis of Tourette's syndrome.

Childhood Schizophrenia

The different age of onset and different symptomatology of the disorders differentiate them. Delusions, hallucinations, blunted or incongruent affect, and loose associations are more common in cases with onset after age 5 years (i.e., childhood schizophrenia). Gaze avoidance, odd preoccupations, disinterest in people, impoverished play, stereotypies, echolalia, and overactivity are more common in cases with onset before 3 years of age (i.e., infantile autism).

Developmental Receptive Language Disorder

Developmental receptive language disorders constitute another class of developmental dysphasias (or childhood aphasias). The most useful distinguishing characteristic is that children with receptive language disorders usually possess normal interpersonal relatedness and an interest in communicating with others by means of gestures, facial expressions, and intonation.

ASSOCIATED BIOMEDICAL CONDITIONS

No single known biomedical condition plays a casual role in all cases of PDD, but a number of conditions have been connected with some cases of the disorder. Sixteen medical conditions have been linked with autism: fra(X), marker chromosome syndrome, tuberous sclerosis, neurofibromatosis, hypomelanosis of Ito, Goldenhar's syndrome, Rett's syndrome, Möbius' syndrome, phenylketonuria, lactic acidosis, hypothyroidism, rubella embryopathy, herpes encephalitis, cytomegalovirus infection, Williams' syndrome, and Duchenne's muscular dystrophy. Two broad classes of medical conditions have been identified in autism or PDD:

- Those with a genetic basis (such as tuberous sclerosis or fragile X)
- Those traceable to prenatal, perinatal, or postnatal insults or infections (such as rubella embryopathy or herpes encephalitis)

Genetic conditions may represent familial traits or spontaneous mutations. A full discussion of all such conditions is beyond the scope of this chapter, but examples of each group are discussed in the sections that follow.

Genetic Syndromes

Of all the chromosomal abnormalities associated with PDD, fra(X) has attracted the most interest. The classic fra(X) phenotype involves large, prominent ears, a long, narrow face, and (in males, especially postpubertally) macroorchidism; at least 80% of all cases show one or more of these features. Other common characteristics include strabismus, hypotonia, tactile defensiveness, and recurrent otitis media. At least half of all fra(X) boys are hyperactive when young, and attentional deficits are universal. One in six male patients with fra(X) develops seizures. Autism and autistic features are common in this population.

Compared with other autistic patients, fra(X) patients are less likely to show echolalia but more likely to display perseverative speech. Hand flapping, hand biting, and hyperactivity are more common in fra(X) than in other forms of PDD. The most interesting distinguishing characteristic of fra(X) "autism" is that fra(X) patients generally are not aloof; rather, they are interested in social interactions, but experience intense social anxiety. Among male patients, IQ is most often in the 35 to 50 range; there is some evidence of a decline in IQ at puberty. Findings with respect to visuomotor skills are mixed. Receptive language skills are superior to language production. Spontaneous speech is often repetitious, incoherent, and littered with interjections. Prosody tends to be deviant; "jocular" and "litany-like" patterns are described.

Female patients are much more variable in functioning. Female carriers are relatively unaffected. Although female patients with fra(X) have a mean IQ of about 85, most of those with IQs above 85 still evince cognitive impairments. Typically, verbal skills are superior to visuospatial skills, and arithmetic abilities are particularly poor. The cognitive profiles of fra(X)-positive individuals resemble those of autistic spectrum disorders and nonverbal learning disabilities.

Teratogenic Factors

Some cases of autism might be traceable to neurologic damage acquired during gestation, delivery, or the first hours of life. There is a high prevalence of autism and autistic-like features among children who suffered congenital rubella. Autistic children, but not their siblings, have a high incidence of various obstetric problems: breech delivery, low birthweight, low Apgar scores, elevated bilirubin, and respiratory distress syndrome. The preponderance of evidence converges to support the hypothesis that certain pregnancy and birth complications may cause some cases of autism.

NEUROANATOMIC FINDINGS

Studies of neuroanatomic abnormalities in autism have relied on postmortem neuropathologic examinations and imaging techniques, such as CT and MRI. These methods have produced inconsistent findings and have uncovered great variability within the autistic population. Individual malformations have been reported, however, including cerebral lipidosis, microgyria and schizencephaly, slight enlargement of parietooccipital regions in the right hemisphere, and slightly lower counts of glia in the primary auditory cortex of the left hemisphere and of auditory association cortex pyramidal neurons in the right hemisphere. Neuroanatomic studies with autistic patients have provided some evidence for involvement of the hippocampus and amygdala. Many studies of the cerebellum have produced findings of neuroanatomic differences between the autistic and normal brain, including Purkinje's and granule cell loss within cerebellar tissue and reductions in total cerebellum size. Autistic individuals have been found to show hypoplasia of neocerebellar vermal lobules VI and VII.

NEUROPHYSIOLOGIC INVESTIGATIONS

Cerebral Activity: Glucose Metabolism and Cerebral Blood Flow

The function of the cerebellum has been assessed through PET measurement of regional glucose metabolic rate (rGMR). Autistic subjects display glucose utilization rates greater than or equal to rates of normal subjects. Autistic adults fail to show the hemispheric asymmetry in both cortical and specific rGMR found for normal subjects during testing state and attention tasks, as well as abnormally high GMR in varying regions. Regional cerebral blood flow analyses using single photon emission computed tomography in autistic children have shown the frontal hypoperfusion characteristic of much younger children that normalizes with maturation. ^{31}P nuclear magnetic resonance spectroscopy in dorsal prefrontal cortex of high-functioning male autistic subjects revealed low levels of high-energy phosphate and membrane phospholipid and high levels of membrane metabolites, correlating with low scores on neuropsychological measures and language tests. These findings are taken as an indication of membrane hypermetabolism and hyposynthesis in autism.

Cerebral Electrical Recording

Studies have found general abnormalities in basic EEG readings in autistic subjects, including irregular activity and desynchronous rhythms, but have failed to establish specific patterns of pathology. High-functioning autistic subjects of varying ages have abnormally small amplitudes for the long-latency, auditory-evoked wave P3b, which is thought to reflect the cognitive processes of detection and classification of target stimuli. Investigations of orientation and processing of novel stimuli in autism have revealed diminished amplitude for components A/Pcz/300 and A/Pcz/800, which are thought to reflect the physiologic process of detection of unexpected and novel "biologically significant" stimuli.

NEUROCHEMICAL CORRELATES OF AUTISM

Serotonin

Existing research has failed to characterize the precise relationship between 5HT and autism, and both central and peripheral 5HT disruptions have been implicated. The occurrence of higher levels of whole blood, plasma, and platelet 5HT in autistic children relative to normal subjects, a phenomenon referred to as *hyperserotonemia,* is the most solidly documented neurochemical finding in autism research. The precise cause of hyperserotonemia in autism is unknown.

Dopamine

Damasio and Maurer hypothesized that abnormalities in dopaminergic input from the mesolimbic and nigrostriatal pathways might account for some autistic symptoms, specifically the unusual motoric features such as stereotypies (21). Observations of hypothalamic dysregulation and hyposensitive hypothalamic dopamine receptors in individuals with autistic features may also implicate a dopaminergic system, which arises in the hypothalamus and projects to the pituitary gland.

Attempts to assess plasma and platelet dopamine and urinary homovanillic acid (HVA) have yielded conflicting results that generate evidence for both augmented and depressed levels. Elevated CSF levels of HVA have been correlated with degree of autistic symptoms, but this finding failed to replicate. Clinical pharmacologic trials have provided more consistent evidence for the role of dopamine in autism. Neuroleptics, which act as dopamine antagonists, have been found to inhibit the symptoms of autism in children, whereas the dopamine agonists, which have stimulant properties, may exacerbate the symptoms. Conversely, experimental trials with amphetamines, such as dextroamphetamine, resulted in the exacerbation of the same symptoms.

Norepinephrine and Epinephrine

The catecholamines norepinephrine and epinephrine function as hormones and neurotransmitters, and both have been found to be involved in the regulation or arousal, attention, activity level, anxiety, responses to stress, memory, and learning. Assessment of norepinephrine and epinephrine activity has been through measurement of CSF levels of norepinephrine's primary metabolite, 3-methoxy-4-hydroxy-phenylglycol (MHPG) and plasma, platelet, and urine levels of norepinephrine, epinephrine, and MHPG. Findings have generally been inconsistent.

Peptides

The endogenous opioids appear to be strong candidates as modulators of socioemotional mechanisms. This is because they are distributed in areas of the brain known to integrate sensation and emotion. Clinical pharmacologic trials with the opiate antagonists naloxone and, more reliably, naltrexone have proved somewhat successful in improving autistic symptoms. Naloxone and naltrexone administration have demonstrated decreases in the frequency of SIB; decreases in stereotypic motor behavior, decreases in hyperactivity, and increases in social and communicative behaviors.

CLINICAL ASSESSMENT

Medical Assessment

The developmental history and behavioral and mental status examinations are the basis for the diagnosis of autism or PDD. Once a diagnosis or tentative diagnosis of PDD or autism is made, assessments in specific areas should be done. The child's hearing must be assessed thoroughly. In any case where behavioral assessment of hearing is not considered reliable, the child should be referred for brain-stem evoked-response audiology. Individuals with autism are subject to a wide variety of motor impairments, especially stereotypies, hypotonia, and apraxia, as well as frank movement disorders. A complete motor examination may reveal remediable conditions.

Medical and family history and physical examination may suggest specific etiologies, such as hydrocephalus or a genetic or neurocutaneous syndrome. The genetic syndrome most often reported in association with autism is fra(X). Examination of the child's physiognomy can suggest the diagnoses of Williams' syndrome, Prader-Willi syndrome, or other neurobehavioral entities and can dictate referral to a clinical geneticist. Screening for abnormalities in amino acids, organic acids, or other metabolites in blood and urine, in the absence of a specific indication for such testing, has a very low yield.

A high proportion of autistic individuals have associated conditions that can be uncovered by extensive medical investigation. This includes blood work (chromosomes, phenylalanine, uric acid, lactic acid, pyruvic acid, and herpes titer), urine studies (metabolic screen, uric acid, calcium), CSF examination for protein, EEG, and CT or MRI to look for evidence of tuberous sclerosis, infection, neurofibromatosis, and hypomelanosis of Ito.

Epilepsy is one neurologic disorder that is often found in association with the autistic behavioral syndrome. Rates of seizures are from 11% to 42%. Adolescent onset of seizures is more common in autism than in other developmental syndromes. EEGs may be useful in determining whether unusual stereotyped behaviors represent seizure activity. When EEGs are performed, they should include sleep EEG. The use of routine CT or MRI remains controversial.

Psychiatric evaluation includes assessment for co-morbid psychiatric disorders, including attention deficit, hyperactivity disorder, depression, psychosis, anxiety, and obsessive-compulsive disorder. Family issues and interpersonal conflicts should also be assessed, including sufficient neurocognitive assessment to determine whether retardation is present.

Neuropsychological Assessment

Whenever possible, a neuropsychological evaluation should thoroughly assess cognition in the areas of language (vocabulary, syntax, and pragmatics), visuospatial skills, abstract thinking and problem solving, memory, attention, and social cognition. Language, in particular, must be thoroughly assessed. Usually, a neuropsychologist can administer a complete language assessment. If not, a referral to a speech and language pathologist is advisable.

Behavioral Assessment

A standardized instrument for the assessment of adaptive skills is the revised Vineland Adaptive Behavior Scales; it has been shown to be a powerful, highly descriptive, well-normed instrument that works well with autistic children (22). It provides age equivalents and standard scores for communication, daily living, and socialization skills as well as motor scores for younger children.

A clear description of problem behaviors is important. Behaviors central to the syndrome (such as social incapacity and resistance to change), those associated with the syndrome (such as self-injury and abnormal motor behaviors), and those ancillary to it (such as hyperactivity, aggressiveness, and passivity) should be noted. The Autism Diagnostic Interview and the Autism Diagnostic Observation Schedule are a pair of companion instruments for interviewing an informant and for direct interview of the autistic individual (23,24).

Family Assessment

Assessments of strengths, resources, and needs of families coping with autism should be described, including an assessment of personal financial resources, educational attainment, physical and emotional health, psychological characteristics, family cohesion and adaptability, and social support from the extended family, neighbors, and community.

TREATMENT

Pharmacologic Treatments

Although attention difficulties and hyperactivity are often prominent in the clinical picture of autism, the autistic child's behavior usually does not improve with stimulant medications. Exacerbation or initiation of stereotypies and psychosis has in fact been reported in autistic children placed on stimulants. Research on the use of anxiolytics with anxious or sleep-disturbed autistic individuals is sparse. Agents such as buspirone reduce agitated behavior in some autistic children. The use of neuroleptics, especially haloperidol, is controversial. Although some researchers have reported positive effects, others dispute its efficacy. Clozapine, an atypical neuroleptic, has received favorable review for its effectiveness. β-Blockers have been reported to be useful in managing impulsive, aggressive and self-abusive behavior in autistic adults. Other drugs considered to be effective with self-injury include fluoxetine, domipramine, buspirone, and opiate antagonists, especially when combined with behavioral therapy. The opiate blockers naltrexone and nalaxone have been found to have beneficial effects on self-injury and also on social withdrawal and stereotypies. Antiepileptic drugs generally are given for control of documented seizure disorders in autistic children, not for the control of behavior.

Behavioral and Educational Treatments

The major treatment modalities for autism are special education and behavioral programming. Early and extensive work on language, social interaction, preacademic and academic skills, and self-help skills are the autistic child's best hope for a positive outcome. At school age, there is a strong recent trend toward "inclusion" or "integration" in which the autistic child is placed, with or without an aide, into a "normal" public school class, sometimes with "pull-out" time for special services. An intermediate solution also on the ascent is the integrated classroom, especially in preschool, in which typical and atypical children are mixed in roughly equal proportions.

As part of the child's special education, and perhaps in addition to it, the child should receive aggressive language therapy. Except for children with severe articulatory dysfunction, this should focus more on the semantic and pragmatic use of language than on articulation and syntax. A recent development in the treatment of autism is facilitated communication. In this approach, a facilitator assists the autistic individual to communicate by means of a letter board or keyboard, by pulling the individual's hand away from the board, and sometimes by helping the individual to form a pointing gesture.

Prescription with therapies and services for the autistic individual must always include sensitivity to the often devastating effect of the disability on the family. Social support from other affected families and keeping abreast of the latest developments in treatment and other research can help families manage their affected children and their own emotional reactions. The Autism Society of America (8601 Georgia Ave., Suite 503, Silver Spring, MD 20910) publishes a regular newsletter with much information useful to parents; another good resource for parents and professionals on recent developments in autism is Rimland's newsletter *Autism Research Review International*.

SUMMARY

This chapter has summarized the current state of knowledge regarding the etiology, clinical manifestations, neurobiology, and psychosocial components of three neurodevelopmental disorders: mental retardation, cerebral palsy, and autism and pervasive developmental disorders.

REFERENCES

1. Grossman HJ. *Manual of terminology and classification on mental retardation.* Washington DC: American Association on Mental Deficiency, 1983.
2. American Psychiatric Association Task Force on Nomenclature and Statistics. *Diagnostic and statistical manual on mental disorders.* 3rd ed. (DSM III). Washington DC: American Psychiatric Association, 1980.

3. Parsons JA, May JG, Menolascino FJ. The nature and incidence of mental illness in mentally retarded individuals. In: Menolascino FJ, Stark JA, eds. *Mental illness in the mentally retarded.* New York: Plenum Press, 1984.
4. Steele MW. Genetics of mental retardation. In: Jacob I, ed. *Mental retardation.* Basel, Switzerland: Karger, 1982:27–37.
5. Matson JL. Emotional problems in the mentally retarded: the need for assessment and treatment. *Psychopharmacol Bull* 1985;21:258–261.
6. Reiss S. Assessment of a man with dual diagnosis. *Ment Retard* 1992;30:1–6.
7. Kazdin AE, Matson JF, Senatore V. Assessment of depression in mentally retarded adults. *Am J Psychiatry* 1983;140:1040–1043.
8. Rever M. Mental retardation. *Psychiatr Clin North Am* 1992:15:511–522.
9. Baldessarini RJ, Cohen BM, Teicher MH. Significance of neuroleptic dose and plasma level in the pharmacological treatment of psychoses. *Arch Gen Psychiatry* 1988; 45:79–91.
10. Aman MG, White AJ, Vaithianathan C, Teehan CJ. Preliminary study of imipramine in profoundly retarded residents. *J Autism Dev Disord* 1986;16:263–273.
11. Szatmari P. The validity of autistic spectrum disorders: a literature review. *J Autism Dev Disord* 1992;22:583–600.
12. Kanner L. Autistic disturbances of affective contact. *Nervous Child* 1943;2:217–250.
13. Kolvin I. Studies in the childhood psychoses. I. Diagnostic criteria and classification. *Br J Psychiatry* 1971;118:318–384.
14. Rutter M. Diagnosis and definition of childhood autism. *J Autism Childhood Schizophrenia* 1978;8:139–384.
15. Ritvo E, Freeman B. National Society for Autistic children definition of the syndrome of autism. *J Autism Childhood Schizophrenia* 1978;8:162–169.
16. Wing L, Gould J. Severe impairments of social interaction and associated abnormalities in children: epidemiology and classification. *J Autism Dev Disord* 1979;9: 11–29.
17. Boucher J, Warrington E. Memory deficits in early infantile autism: some similarities to the amnesic syndrome. *Br J Psychol* 1976;67:73–87.
18. DeLong GR. Autism, amnesia, hippocampus and learning. *Neurosci Biobehav Rev* 1992;16:63–70.
19. Gillberg C. Autism and autistic-like conditions: subclasses among disorders of empathy. *J Child Psychol Psychiatry* 1992;33:813–842.
20. Rourke B. *Nonverbal learning disabilities: the syndrome and the model.* New York: Guilford Press, 1989.
21. Damasio A, Maurer R. A neurological model for childhood autism. *Arch Neurol* 1978;35:777–786.
22. Sparrow S, Balla D, Cicchetti D. *Vineland Adaptive Behavior Scales.* Circle Pines, MN: American Guidance Service, 1984.
23. LeCouteur A, Rutter M, Lord C, et al. Autism diagnostic interview: a standard investigator-based instrument. *J Autism Dev Disord* 1989;19:363–387.
24. Lord C, Rutter M, Good S, Heemsbergen J. Autism diagnostic observation schedule: a standardized observation of communication and social behavior. *J Autism Dev Disord* 1989;19:185–212.

15

Neurobiologic Mechanisms of Human Anxiety

Dennis S. Charney, Linda M. Nagy, J. Douglas Bremner, Andrew W. Goddard, Rachel Yehuda, and Steven M. Southwick

OVERVIEW

The anxiety disorders discussed in this chapter are those included in the *Diagnostic and Statistical Manual of Mental Disorders*, 4th edition (DSM-4: panic disorder and agoraphobia, social phobia, specific phobia, posttraumatic stress disorder [PTSD], acute stress disorder, and generalized anxiety disorder). Initially, the neurobiologic foundation needed to understand the etiology of the major anxiety disorders is described. Subsequently, for each disorder, epidemiologic data, a basic description of the syndrome, comparisons to other psychiatric and nonpsychiatric disorders, and current etiologic theories are reviewed.

NEUROANATOMY AND NEUROBIOLOGY

The sensory information contained in a fear- or anxiety-inducing stimulus is transmitted from peripheral receptor cells to the dorsal thalamus. An exception is the olfactory system, which does not relay information through the thalamus, and whose principal targets in the brain are the amygdala and entorhinal cortex. Visceral afferent pathways alter the function of the locus ceruleus (LC) and the amygdala, either through direct connections or through the nucleus paragigantocellularis and the nucleus tractus solitarius. The thalamus relays sensory information to the primary sensory receptive areas of the cortex. In turn, these primary sensory regions project to adjacent unimodal and polymodal cortical association areas. The cortical association areas of visual, auditory, and somatosensory systems send projections to other brain structures, including the amygdala, entorhinal cortex, orbitofrontal cortex, and cingulate gyrus. The hippocampus receives convergent, integrated inputs from all sensory systems by way of projections from entorhinal cortex.

There is a pivotal role for the amygdala in the transmission and interpretation of fear- and anxiety-inducing sensory information because it receives afferents from thalamic and cortical exteroceptive systems, as well as subcortical visceral afferent pathways. The neuronal interactions between the amygdala and cortical regions, such as the orbitofrontal cortex, enable the individual to initiate adaptive behaviors to threat based on the nature of the threat and prior experience.

The efferent pathways of the anxiety–fear circuit mediate autonomic, neuroendocrine, and skeletal-motor responses. The structures involved in these responses include the amygdala, LC, hypothalamus, periaqueductal gray, and striatum. Many of the autonomic changes produced by anxiety- and fear-inducing stimuli are produced by the sympathetic and parasympathetic neural systems. The sympathetic activation and hormonal release associated with anxiety and fear are probably mediated in part by stimulation of the hypothalamus through projections from the amygdala and LC.

The vagus and splanchnic nerves are major projections of the parasympathetic nervous system. This innervation of the parasympathetic nervous system may relate to visceral symptoms associated with anxiety, such as gastrointestinal and genitourinary disturbances.

The regulatory control of skeletal muscle by the brain in response to emotions is complex. Adaptive mobilization of the skeletal motor system to respond to threat probably involves pathways between the cortical association areas and motor cortex, cortical association areas and striatum, and amygdala and striatum (1).

Effects of Prior Experience

Memories and previously learned behaviors influence the responses to anxiety- and fear-inducing stimuli through such neural mechanisms as fear conditioning, extinction, and sensitization. There is considerable interaction between storage and recall of memory and affect. This is exemplified by the crucial role of the amygdala in conditioned fear acquisition, sensitization, extinction, and the attachment of affective significance to neutral stimuli.

The hippocampus and amygdala are sites of convergent reciprocal projections from widespread unimodal and polymodal cortical association areas. It is probably through these interactions, as well as cortical–cortical connections, that memories stored in the cortex are intensified and develop greater coherence. These interconnections can explain how a single sensory stimulus such as a sight or sound can elicit a specific memory. Moreover, if the sight or sound was associated with a particular traumatic event, a cascade of anxiety- and fear-related symptoms can be activated (2).

NEURAL MECHANISMS OF ANXIETY AND FEAR

Fear Conditioning

In many patients with anxiety disorders such as panic disorder with agoraphobia, simple phobias, and PTSD, vivid memories of a traumatic event, autonomic arousal, and even flashbacks can be elicited by diverse sensory and cognitive stimuli that have been associated with the original trauma. Modality-specific and contextual fear conditioning may explain some of these observations.

Contextual fear conditioning, which involves more complex stimuli from multiple sensory modalities, may require projections to the amygdala from higher-order cortical areas that integrate inputs from many sources and the hippocampus. The hippocampus may have a time-limited role in associative fear memories evoked by contextual sensory (polymodal), but not unimodal sensory, stimuli.

Several behavioral paradigms indicate an important role for noradrenergic neuronal systems in the processes involved in fear conditioning. Neutral stimuli paired with shock produce increases in brain norepinephrine metabolism and behavioral deficits similar to those elicited by the shock. There is also a body of evidence indicating that an intact noradrenergic system may be necessary for the acquisition of fear-conditioned responses (3).

Extinction

Experimental *extinction* is defined as a loss of a previously learned conditioned emotional response after repeated presentations of a conditioned fear stimulus in the absence of a contiguous traumatic event. Results in several studies suggest that the original associations are intact after extinction. Re-presentation of the unconditioned stimulus, even up to 1 year after extinction, is sufficient for reinstating extinguished response to a preextinction level. These data indicate the essentially permanent nature of conditioned fear and the apparent fragility of extinction. This phenomenon may help to explain the common clinical observation that traumatic memories may remain dormant for many years, only to be elicited by a subsequent stressor or unexpectedly by a stimulus long ago associated with the original trauma. Extinction of conditioned fear responses may represent an active suppression by the cortex of subcortical neural circuits (thalamus, amygdala) that maintain learned associations over long time periods.

Behavioral Sensitization

Sensitization generally refers to the increase in behavioral or physiologic responsiveness that occurs after repeated exposure to a stimulus. Behavioral sensitization can be generally context dependent or conditioned, such that animals will not demonstrate sensitization if the stimulus is presented in a different environment. If the intensity of the stimulus or drug dose is high enough, however, behavioral sensitization occurs even if the environments change. Moreover, *cross-sensitization* (augmented response to a different stimulus than the original evoking stimulus) may occur.

The conditioned components of sensitization are related to increased dopamine release in the nucleus accumbens, according to animal experiments. Behavioral sensitization to stress may also involve alterations in noradrenergic function. Dopamine D_2 receptor antagonists block the development but not the maintenance of sensitization. Conversely, α_2-receptor agonists and benzodiazepine agonists block the maintenance but not the development of sensitization. In contrast, lesions of the hippocampus and frontal cortex have no effect (4).

NEUROTRANSMITTERS IMPLICATED IN ANXIETY AND FEAR

Noradrenergic System

Stressful stimuli of many types produce marked increases in brain noradrenergic function. Stress produces regional selective increases in noradrenergic turnover in brain regions identified as part of the neural circuitry of anxiety, including the LC, hypothalamus, hippocampus, amygdala, and cerebral cortex. Anxiolytic agents reverse the effects of stress on noradrenergic metabolism. The heightened responsiveness of the noradrenergic system to stress is consistent with the notion that the elevated sense of fear or anxiety connected to stress may be a crucial factor in the neurochemical effects observed.

Locus Ceruleus Activity and Behavioral States Associated With Stress and Fear

In laboratory rats, chronic stress results in an increased firing of the LC. Animals exposed to chronic inescapable shock, which is associated with learned helplessness, have an increase in responsiveness of the LC to an excitatory stimulus compared with animals exposed to escapable shock. These increases in LC function are accompanied by sympathetic activation.

A parallel activation of LC neurons and splanchnic sympathetic nerves is produced by noxious stimuli. The LC-like sympathetic splanchnic activity is highly responsive to various peripheral cardiovascular events, such as alterations in blood volume or blood pressure. Internal events that must be responded to for survival, such as thermoregulatory disturbance, hypoglycemia, blood loss, an increase in PCO_2, or a marked reduction in blood pressure, cause robust and long-lasting increases in LC activity (5,6).

Behavioral Effects of Locus Ceruleus Stimulation

Electrical stimulation of the LC produces a series of behavioral responses similar to those observed in naturally occurring or experimentally induced fear. These behaviors are also elicited by administration of drugs, such as yohimbine and piperoxone, which activate the LC by blocking α_2-adrenergic autoreceptors. Drugs that decrease the function of the LC by interacting with inhibitory opiate (morphine), benzodiazepine (diazepam), and α_2- (clonidine) receptors on the LC decrease fearful behavior and partially antagonize the effects of electrical stimulation of the LC in the monkey. These studies suggest that abnormally high levels of LC activity, producing increased release of norephinephrine at postsynaptic projection sites throughout the brain, may augment some forms of fear or pathologic anxiety.

Dopaminergic System

Acute stress increases dopamine release and metabolism in a number of specific brain areas. However, the dopamine innervation of the medial prefrontal cortex (mPFC) appears to be particularly vulnerable to stress; sufficiently low-intensity stress (such as that associated with conditioned fear) or brief exposure to stress increases dopamine release and metabolism in the prefrontal cortex in the absence of overt changes in other mesotelencephalic dopamine regions. Benzodiazepine anxiolytics prevent selective increases in dopamine use in mPFC after mild stress (7).

Serotonergic Function

Animals exposed to a variety of stressors, including foot shock, tail shock, tail pinch, and restraint stress, have all been shown to produce an increase in serotonin turnover in the mPFC. On the other hand, inescapable stress paradigms producing "learned helplessness" behavioral deficits have been associated with reduced in vivo release of serotonin in cerebral cortex. Serotonin antagonists produce behavioral deficits resembling those seen after inescapable shock. Drugs that enhance serotonin neurotransmission (serotonin reuptake inhibitors) are effective in reversing learned helplessness.

A primary distinction in the qualitative effects of serotonin may be between the dorsal and median raphe nuclei—the two midbrain nuclei that produce most of the forebrain serotonin. The serotonergic innervation of the amygdala and the hippocampus by the dorsal raphe is believed to mediate anxiogenic effects by means of 5-HT_2 receptors. In contrast, the median raphe innervation of hippocampal 5-HT_{1A} receptors has been hypothesized to facilitate the disconnection of previously learned associations with aversive events (8).

Benzodiazepine Receptor Systems

Benzodiazepine receptors are present throughout the brain, with the highest concentration in cortical gray matter. Benzodiazepines potentiate and prolong the synaptic actions of the inhibitory neurotransmitter γ-aminobutyric acid. Administration of inverse agonists of benzodiazepine receptors, such as B-carboline-3-carboxylic acid ethyl ester (B-CCE), result in behavioral and biologic effects similar to those seen in anxiety, including increases in heart rate, blood pressure, and plasma cortisol and catecholamine levels.

Neuropeptides Implicated in Anxiety and Fear

Corticotropin-Releasing Factor

Considerable data indicate that corticotropin-releasing factor (CRF) has anxiogenic-like properties when injected centrally. Furthermore, CRF appears to play an important role in the neuroendocrine, autonomic, and behavioral responses to stress. Severe stressors produce increases in CRF concentrations in the amygdala, hippocampus, and LC. The brain sites mediating the CRF responses to stress have not been established; however, there is increasing evidence that these effects of CRF may be produced by interactions with LC noradrenergic neurons. CRF in a dose-dependent fashion increases the firing rate of LC-norepinephrine neurons, and stressors that activate norepinephrine neurons increase CRF concentrations in the LC.

Opioid Peptides

A primary behavioral effect of uncontrollable stress is analgesia, which results from the release of endogenous opiates. These effects are mediated, in part, by a stress-induced release of endogenous opiates in the brain stem because the analgesia is blocked by naltrexone and shows cross-tolerance to morphine analgesia.

Cholecystokinin

Cholecystokinin (CCK), an octopeptide originally discovered in the gastrointestinal tract, has been found in high concentrations in the cerebral cortex, amygdala, and hippocampus in mammalian brain. CCK has anxiogenic effects in laboratory animals and appears to act as a neurotransmitter or neuromodulator in the brain. Benzodiazepine anxiolytics antagonize the excitatory neuronal and anxiogenic effects of CCK, and CCK_B antagonists have anxiolytic actions in animal models. Studies in healthy human subjects have demonstrated that CCK-4 induces severe anxiety or short-lived panic attacks. This effect is reduced by lorazepam.

Neuropeptide Y

Low doses of neuropeptide Y administered intraventricularly have anxiolytic effects in several animal models of anxiety. These actions may be mediated by neuropeptide Y in the amygdala.

ANXIETY DISORDERS: DESCRIPTION AND PATHOPHYSIOLOGY

Panic Disorder and Agoraphobia

Panic disorder is characterized by recurrent discrete attacks of anxiety accompanied by several somatic symptoms, such as palpitations, paresthesias, hyperventilation, diaphoresis, chest pain, dizziness, trembling, and dyspnea. Usually, the condition is accompanied by agoraphobia, which consists of excessive fear (and often avoidance) of situations, such as driving, crowded places, stores, or being alone, in which escape or obtaining help would be difficult. Current DSM-4 classifications include panic disorder without agoraphobia, panic disorder with agoraphobia, and agoraphobia without history of panic disorder (9).

Epidemiology

The Epidemiologic Catchment Area (ECA) Study reports prevalence estimates based on DSM-3 diagnoses, which were separated into agoraphobia (analogous to DSM-4 diagnoses of agoraphobia without history of panic attacks and panic disorder with agoraphobia) and panic disorder (analogous to DSM-4 diagnosis of panic disorder without agoraphobia). Lifetime prevalence rates at the three sites varied between 7.8% and 23.3% for all phobias (including social and simple) and between 1.4% and 1.5% for panic disorder. Six-month prevalence rates were 2.7% to 5.8% for agoraphobia and 0.6% to 1.0% for panic disorder. The lifetime rate for female patients was 2.4 to 4.3 times greater than that for male patients diagnosed with agoraphobia and 1.3 to 3.5 times greater for panic disorder.

Age of onset of panic disorder is typically in the late teens to early 30s and is unusual after the age of 40 years. Most patients (78%) describe the initial attack as spontaneous (occurring without an environmental trigger). In the remainder of patients, the first attack is precipitated by confrontation with a phobic stimulus or use of a psychoactive drug. Onset of the disorder often follows within 6 months of a major stressful life event, such as marital separation, occupational change, or pregnancy.

Both 6-month and lifetime rates are lower in patients older than 65 years of age. Panic disorder is increased among family members of those with the disorder. A history of childhood separation anxiety disorder is reported by 20% to 50% of patients. Preliminary findings of high rates of behavioral inhibition in the offspring of patients with panic disorder are consistent with the hypothesis that the disorder may have developmental antecedents (10,11).

Description and Differential Diagnosis

Patients may feel constantly fearful and anxious after the first attack, wondering what is wrong and fearing it will happen again. Usually, patients gradually become fearful of situations which they associate with the attacks and in situations in which they would be unable to flee if the attack occurred, could not find readily available help, or would be embarrassed if others should notice they are experiencing an attack (although attacks are not usually evident to others).

The differential diagnosis of panic disorder and agoraphobia includes secondary anxiety disorder due to medical conditions; substances such as caffeine, cocaine, or amphetamines; withdrawal from alcohol, sedative-hypnotics, or benzodiazepines; and other phobic conditions, generalized anxiety disorder, and psychosis. Endocrine disturbances, such as pheochromocytoma, hyperthyroidism, or hypoglycemia, can produce similar symptoms and can be excluded with appropriate clinical history and laboratory evaluations.

Panic disorder differs from generalized anxiety disorder in that the former is distinguished by recurrent discrete, intense episodes of panic symptoms, although in both disorders, anticipatory anxiety and generalized feelings of anxiety may be present. Although some of the same situations may be feared, agoraphobia differs from social and simple phobias in that the fear is related to feeling trapped or being unable to escape and that the fears often become generalized.

Panic disorder is frequently associated with major depression, other anxiety disorders, and alcohol and substance abuse. In clinical samples, as many as two thirds of patients with panic attacks report experiencing a major depressive episode at some time in their lives. Similarly, studies of patients seeking treatment for major depression report high rates of panic in these patients and their relatives.

Genetic Epidemiology

Genetic epidemiologic studies have consistently demonstrated increased rates of panic disorder among first- and second-degree relatives of panic disorder probands. The reported recurrence risk of illness in first-degree relatives of panic disorder probands is 15% to 18% by patient report, as compared with 0% to 5% in controls, and is 20% to 50% by direct interview of relatives of panic probands, as compared with 2% to 8% in relatives of controls. Segregation analysis indicates the pattern of familial transmission is consistent with single-locus, autosomal-dominant transmission with incomplete penetrance, although a multifactorial mode of inheritance (additive effects of more than one gene) or genetic heterogeneity (different gene defects producing similar clinical syndromes) have not been excluded as possibilities. Twin studies indicate concordance for anxiety disorder, with panic attacks much higher in monozygotic twin pairs than in dizygotic twin pairs.

Pathophysiology

The evidence for an abnormality in noradrenergic function is most compelling for panic disorder and PTSD. Panic disorder and PTSD patients frequently report cardiovascular, gastrointestinal, and respiratory symptoms. Because the LC is responsive to peripheral alterations in the function of these systems, minor physiologic changes in these patients may result in abnormal activation of LC neurons and, consequently, panic attacks and flashbacks. These functional interactions may explain the association of anxiety symptoms with tachycardia, tachypnea, hypoglycemia, and visceral and organ distention, as well as the marked sensitivity of panic disorder and PTSD patients to interoceptive stimuli. The important role of the noradrenergic system in fear conditioning may account for the development of phobic symptoms in these patients. Finally, the involvement of noradrenergic neurons in learning and memory may relate to the persistence of traumatic memories in both disorders.

Regulation of Noradrenergic Function in Panic Disorder

Yohimbine is a presynaptic α_1-antagonist. About 60% to 70% of panic disorder patients experience yohimbine-induced panic attacks. These patients have larger yohimbine-induced increases in plasma 3-methoxy-4-hydroxyphenylglycol (MHPG), blood pressure, and heart rate than do healthy subjects and patients with other psychiatric disorders. In panic disorder patients, yohimbine significantly reduced frontal regional cerebral blood flow rates in patients compared with healthy subjects (5).

β-Adrenergic Receptor Function

Infusion of isoproterenol, a peripherally acting compound that is selective for the β-adrenoceptor, has been reported to trigger anxiety responses in panic patients compared with controls. These studies are consistent with the hypothesis of increased β_1-adrenoceptor sensitivity in panic disorder.

Noradrenergic Function and Treatment of Panic Disorder

Evidence is emerging that the efficacy of some tricyclic and monoamine oxidase inhibitor drugs against panic may be related to their regulatory effects on noradrenergic activity. The effects of chronic treatment with these agents on the regulation of noradrenergic activity are complex. Some of these effects, such as reduced tyrosine hydroxylase activity, LC firing rate, noradrenergic turnover, and postsynaptic β-adrenergic receptor sensitivity diminish noradrenergic function.

Benzodiazepines are highly effective treatments for panic disorder, generally at higher doses than those needed for generalized anxiety disorder. It has been hypothesized that the antipanic properties of benzodiazepines may also relate to inhibitory effects on noradrenergic function because these drugs reduce LC neuronal activity.

The antipanic efficacy of the potent serotonin reuptake inhibitors, such as clomipramine, fluvoxamine, zimeldine, and fluoxetine, is well documented. The mechanism of action of these agents in panic disorder has not been established; however, 5-HT_2, 5-HT_{1C},, and 5-HT_{1A} receptors are unlikely to be directly involved because ritanserin, a 5-HT_2 and 5-HT_{1C} antagonist, and buspirone, a 5-HT_{1A} agonist, lack antipanic efficacy. If the noradrenergic system is dysregulated in panic disorder, the mechanism of action of antipanic therapy may be its ability to decrease the wide and unpredictable fluctuations in noradrenergic activity.

Serotonin Function

Benzodiazepine Receptor Function

Evidence from clinical studies suggests a possible role for alterations in benzodiazepine receptor function in the anxiety disorders. Administration of the benzodiazepine receptor antagonist flumazenil to patients with panic disorder results in an increase in panic attacks and subjective anxiety in comparison with controls. Both oral and intravenous flumazenil have been shown to produce panic in a subgroup of panic disorder patients, but not in healthy subjects.

The benzodiazepine receptors are highly concentrated in gray matter structures, including cortical areas, such as temporal, parietal, and frontal cortex, and subcortical areas, such as the hippocampus. Patients with panic disorder may have abnormalities of the right temporal lobe, with some studies suggesting atrophy and areas of increased signal activity detectable by magnetic resonance imaging (12).

Hypothalamic-Pituitary-Adrenal Axis Function

Hypothalamic-pituitary-adrenal (HPA) axis function does not appear to be markedly disturbed in panic disorder. Most dexamethasone suppression test studies in panic disorder have shown normal suppression. The 24-hour urinary free cortisol levels are normal in panic disorder patients. The cortisol response to clonidine is also normal in panic disorder.

Respiratory System Dysfunction

Intravenous infusions of lactate produce panic anxiety in susceptible patients but not in healthy subjects. Lactate response appears to be specific for panic disorder compared with other anxiety disorders and psychiatric conditions. Moreover, treatment of panic with imipramine blocks the effects of lactate. The panicogenic mechanism of lactate has not been established.

Panic can also be provoked by increases in PCO_2 (hypercapnia). This can be done slowly, such as by rebreathing air or by breathing 5% to 7% CO_2 in air. This mechanism may be related to the hypothesis that panic disorder patients suffer from a physiologic misinterpretation of a suffocation monitor, which evokes a suffocation alarm system.

Brain Imaging Studies

Single Photon Emission Computed Tomography Studies of Cerebral Blood Flow

Patients with panic disorder, compared with healthy controls, have been shown to have a relative decrease from baseline in the ratio of frontal cortex to cerebellar blood flow measured with single photon emission computed tomography (SPECT) and ^{99m}Tc-labeled hexamethyl-propylenamine oxime (HMPAO) after administration of yohimbine. In addition, panicking patients had a slightly lower lactate-induced occipital cortex increase in blood flow compared with nonpanicking patients.

Positron Emission Tomography Studies

Using positron emission tomography (PET) with $H_2{}^{15}O$ to measure cerebral blood flow, lactate-sensitive panic disorder patients have been shown to have a decreased left-to-right parahippocampal ratio of blood flow, in comparison with controls. Investigations of cerebral metabolism using ^{18}F-deoxyglucose have found decreases in left inferior parietal lobule metabolism and decreased left hippocampal–to–right hippocampal ratio of metabolism in panic disorder patients, in comparison with controls.

Posttraumatic Stress Disorder

Posttraumatic stress disorder (PTSD) can be an immediate or delayed response to a catastrophic event.

Epidemiology

PSTD was not recognized as an independent diagnosis until the publication of DSM-3 in 1980. Lifetime and current prevalence estimates of PTSD range from 13% to 3%, respectively. Victims of sexual assault are at especially high risk for subsequent mental health problems and suicide. One site of the ECA Study estimated a population prevalence of 1% to 2%, which may be an underestimate. A survey of a population aged 20- to 30-years in a large health maintenance organization found that 39.1% had been exposed to a traumatic event, and the lifetime rate of PTSD in those exposed was 23.6% (i.e., lifetime population prevalence, 9.2%). Risk factors for PTSD after exposure to trauma included separation from parents during childhood, family history of anxiety, preexisting anxiety or depression, family history of antisocial behavior, female sex, and neuroticism. A latent period of months or years may intervene between the trauma and the onset of symptoms, or an exacerbation or relapse can occur after a period of remission. The disorder can occur in childhood.

Description and Differential Diagnosis

The symptoms are clustered into three categories: reexperiencing the trauma, psychic numbing, or avoidance of stimuli associated with the trauma and increased arousal. Reexperiencing phenomena include intrusive memories, flashbacks, nightmares, and psychological or physiologic distress in response to trauma reminders. Intrusive memories are spontaneous, unwanted, distressing recollections of the traumatic event. Repeated nightmares contain themes of the trauma or a highly accurate and detailed recreation of the actual event. Flashbacks are dissociative states in which components of the event are relived, and the person feels as if he or she is experiencing the event for a few seconds to as long as days. The level of distress, impairment, and duration of the symptoms can be difficult but are important measures. Recent research suggests that a 3-month duration may be a threshold between "acute" and "chronic" PTSD. In adjustment disorder, the stressor is usually less severe, and the characteristic symptoms of PSTD, such as reexperiencing and avoiding, are not present. Many of the symptoms of PSTD resemble those of major depression. If a full depressive syndrome also exists, both diagnoses should be made (13).

Pathophysiology

Noradrenergic Function

Effects of Stress on Noradrenergic Function in Healthy Subjects. Accumulated evidence suggests that brain noradrenergic systems play a role in mediating normal-state anxiety and the response to stress in healthy human subjects. Increases in heart rate, blood pressure, and alerting behaviors, essential for the response to life-threatening situations, may be mediated by brain noradrenergic systems, although the evidence of this activation is largely peripheral.

States of anxiety or fear appear to be associated with an increase in norepinephrine release in healthy subjects. During public speaking, epinephrine levels increase twofold, whereas during physical exercise, norepinephrine levels increase threefold. Levels of the norepinephrine metabolite MHPG have also been found to increase in healthy subjects during emotional stress.

Studies of Noradrenergic Function. Two studies have found significantly elevated 24-hour urine norepinephrine excretion in combat veterans with PTSD, as compared with healthy subjects or patients with schizophrenia or major depression. Since the early 1980s, there have been a series of well-designed psychophysiologic studies that have further documented heightened autonomic or sympathetic nervous system arousal in combat veterans with PTSD. Combat veterans with PTSD have been shown to have higher resting mean heart rate and systolic blood pressure, as well as greater increases in heart rate, when exposed to visual and auditory combat-related stimuli, as compared with combat veterans without PTSD, patients with generalized anxiety disorder, and healthy subjects. These data are consistent with the hypothesis that noradrenergic hyperreactivity in patients with PTSD may relate to the conditioned or sensitized responses to specific traumatic stimuli.

Regulation of Noradrenergic Function. About two thirds of PTSD patients experience yohimbine-induced panic attacks. In addition, 40% of PTSD patients report flashbacks after yohimbine. As a group, PTSD patients also have greater yohimbine-induced increases in plasma MHPG, sitting systolic blood pressure, and heart rate than do healthy subjects.

Possible Sites of Noradrenergic Dysfunction. The data just reviewed suggest that some panic disorder and PTSD patients may share a common abnormality in noradrenergic system function. The cause of the disturbance in noradrenergic function may differ in the two disorders. Panic disorder is generally believed to be familial, suggesting that the noradrenergic dysfunction may represent an inherited trait. On the other hand, the frequency of panic disorder in the family members of PTSD patients is no higher than in the general population.

The abnormal noradrenergic function observed in panic disorder and PTSD may reflect abnormalities in the LC-noradrenergic system at the level of the LC, the LC projection areas, and the interaction of the peripheral sympathetic with the central noradrenergic system. Dysfunction at the LC, producing noradrenergic hyperactivity, could be caused by decreased functional sensitivity of the α_2-adrenergic autoreceptor or at one or more of the neuronal systems that have regulatory actions on LC activity.

There is abundant evidence implicating the opiate, endogenous benzodiazepine, and serotonin neuronal systems in the development of anxiety and fear. Because each of these systems has inhibitory effects on LC activity, it is possible that decreased activity of any one of these systems could also account for the increased response to drugs, such as yohimbine, which activate noradrenergic neurons.

Projection areas of the LC are likely to be involved in the enhanced anxiogenic effects of yohimbine in panic disorder and PTSD. The LC projects to many brain regions implicated in the pathophysiology of anxiety and fear, including the amygdala, hypothalamus, thalamus, hippocampus, and cerebral cortex. An enhancement of norepinephrine metabolism occurs in these brain areas after a variety of stressful stimuli.

Traumatic Memories and Noradrenergic Function. A striking effect of yohimbine is its ability to increase the severity of the core symptoms associated with PTSD, such as intrusive traumatic thoughts, emotional numbing, and grief. This may be due to the involvement of noradrenergic systems in the mechanisms by which memories of traumatic experiences are reawakened by a variety of stimuli and stressors. Experimental and clinical investigations have demonstrated that memory processes are susceptible to modulating influences after the information has been acquired. Activating the LC-norepinephrine system that projects to the amygdala by frightening and traumatic experiences may facilitate the encoding of memories associated with the experiences.

Hypothalamic-Pituitary-Adrenal Axis Function

The finding of low urinary cortisol excretion in PTSD patients, as compared with other psychiatric groups and normal controls, has been replicated in several studies. Results from three studies have demonstrated an increased lymphocyte glucocorticoid receptor number in combat veterans with PTSD, as compared with nonpsychiatric and psychiatric comparison groups. These findings are consistent with observations of low cortisol levels in patients with chronic PTSD.

Dexamethasone Suppression Test

The dexamethasone suppression test (DST) involves the administration of 1 mg dexamethasone at 11:00 PM when normal cortisol secretion is at its nadir in the diurnal cycle. The inhibition of CRF and adrenocorticotropic hormone results in a decrease in the amount of cortisol released from the adrenal gland. In healthy subjects, a dose of 1 mg usually suppresses plasma cortisol to a level below 5 μg/dL at 8:00 AM; it remains below that level at 4:00 PM.

The cortisol response to 1 mg of dexamethasone has been investigated in five studies of PTSD. These studies have all reported that PTSD patients without major depression show a "normal" suppression to dexamethasone. Closer examination, however, reveals that PTSD patients as a group show an exaggerated response to dexamethasone. A recent metaanalysis of the psychiatric literature on the dexamethasone suppression test, averaging the mean cortisol data across all published studies, revealed a cortisol value in nondepressed PTSD subjects of 1.74 μg/dL, a value well below the established cutoff of 5 μg/dL.

Serotonin Function

Although only two studies have directly examined the 5HT system in PTSD, there is a large body of indirect evidence suggesting that this neurotransmitter may be important in the pathophysiology of trauma-related symptomatology. In humans, low 5HT functioning has been associated with aggression, impulsiveness, and suicidal behavior. Patients with PTSD are frequently described as aggressive or impulsive and often suffer from depression, suicidal tendencies, and intrusive thoughts that have been likened to obsessions. The observation has also been made that serotonin reuptake inhibitors are effective in treating PTSD symptoms, such as intrusive memories and avoidance symptoms.

Opioid Peptide Function

There is some evidence of altered baseline opioid peptide metabolism in traumatized humans. Significantly lower morning and afternoon plasma β-endorphin levels in 21 PTSD patients compared with 20 controls has been reported (14). The results were viewed as support for Van der Kolk's hypothesis that patients with PTSD have a chronic depletion of endogenous opioids and that hyperresponsiveness is related to endogenous opiate withdrawal (15). According to this hypothesis, patients with PTSD seek out or provoke recurrent stressors and trauma in order to increase opiate release and hence decrease endogenous opiate withdrawal.

It has been hypothesized that symptoms of avoidance and numbing are related to a dysregulation of opioid systems in PTSD. Further, it has been suggested that the use of opiates in chronic PTSD may represent a form of self-medication. Animal studies have shown that opiates are powerful suppressants of central and peripheral noradrenergic activity. If, as suggested earlier in this chapter, some PTSD symptoms are mediated by noradrenergic hyperactivity, opiates may serve to "treat" or weaken that hypersensitivity and accompanying symptoms.

PHOBIAS

Social Phobia

Social fears are commonly experienced by healthy people, especially in initial public-speaking experiences. For some people, this fear becomes persistent and overwhelming, limiting their social or occupational functioning because of intense anxiety and, often, avoidance.

Epidemiology

Preliminary estimates of 6-month prevalence rates of social phobia ascertained by screening questionnaires in the ECA Study are 1.2% to 2.2%: 0.9% to 1.7% in men; and 1.5% to 2.6% in women. The distribution is fairly even across the age span, although rates may be lower in the those older than 65 years of age. Onset is usually between 15 and 20 years of age, and the course tends to be chronic and unremitting. Complications include interference with work or school, social isolation, and abuse of alcohol or drugs. In inpatient alcoholism treatment programs, 20% to 25% report social phobia beginning before the onset of alcoholism or persisting after 1 year of abstinence. Significant depressive symptoms may also occur in social phobia; in one study, one third of social phobia patients reported a history of major depression.

Description and Differential Diagnosis

Social phobia is characterized by a persistent and exaggerated fear of humiliation or embarrassment in social situations, leading to high levels of distress and avoidance of those situations. The fear may be of speaking, meeting people, or eating in public and relates to the fear of appearing nervous or foolish, making mistakes, being criticized, or being laughed at. Often, physical symptoms of anxiety, such as blushing, trembling, sweating, and tachycardia, are triggered when the patient feels he or she is being evaluated or scrutinized. Social phobics may experience situational panic attacks resulting from anticipation or exposure to the feared social situation. Sometimes, social phobia and panic disorder coexist. Social phobia can be differentiated from simple phobias in that the latter do not involve social situations involving scrutiny, humiliation, or embarrassment.

Genetic Epidemiology

Animal studies demonstrate heritability of various fear, anxiety, and exploratory, escape, or avoidance behaviors, often mediated by combinations of genes; these studies may be relevant to social phobia and other anxiety disorders (16). Human studies of general population samples have suggested some genetic heritability for traits, such as fear of strangers, shyness, social introversion, and fear of social criticism. There is greater monozygotic than dizygotic twin concordance for social phobic features such as discomfort or trembling when eating with strangers or when being watched eating, writing, or working (17).

Specific Phobia

Specific phobia shares many of the basic features of the general phobias, but the fear is limited to a specific object or situation, such as dogs or heights, so that the extent of interference in a patient's life tends to be mild.

Epidemiology

Six-month prevalence rates of specific phobia reported in the ECA Study are between 4.5% and 11.8%; rates are higher for female subjects than for male subjects. The onset of animal phobia is usually in childhood. Blood-injury phobia usually begins in adolescence or early adulthood and may be associated with vasovagal fainting on exposure to the phobic stimulus. Age of onset may be more variable for other simple phobias. Many childhood-onset phobias may remit spontaneously.

Description and Differential Diagnosis

As with the other phobias, the fear is excessive and unrealistic, exposure to the phobic stimulus produces an anxiety response, expectation of exposure may produce anticipatory anxiety, and the object or situation is either avoided or endured with considerable discomfort. Unlike social phobia, however, the fear does not involve scrutiny or embarrassment, and, unlike agoraphobia, the fear is not of being trapped or of having a panic attack. The nature of the fear is specific to the phobia, such as a fear of falling or loss of visual support in height phobia, or fear of crashing in a flying phobia. Isolated fears are common in the general population; a diagnosis of simple phobia is reserved for situations in which the phobia results in marked distress or some degree of impairment in activities or relationships.

Pathophysiologic Studies of Social and Specific Phobias

In patients with specific phobias, increases in subjective anxiety and increased heart rate, blood pressure, plasma norepinephrine, and epinephrine have been associated with exposure to the phobic stimulus. Patients with social phobia have been found to have greater increases in plasma norepinephrine on orthostatic challenge, as compared with healthy controls and patients with panic disorder. In contrast to panic disorder patients, the density of lymphocyte β-adrenoceptors is normal in social phobic patients.

GENERALIZED ANXIETY DISORDER

Generalized anxiety disorder (GAD) is characterized by excessive and uncontrollable worry about multiple life circumstances. It is accompanied by symptoms of muscle tension, restlessness, fatigue, concentration problems, difficulty falling or staying asleep, and irritability. The anxiety is unrelated to panic attacks, phobic stimuli, obsessions, having an illness, or traumatic events (as in PTSD).

Epidemiology

An epidemiologic study estimated a 2.5% 1-month prevalence rate for GAD using research diagnostic criteria. Of the patients with GAD, 80% had at least one other anxiety disorder in their lifetime, and 7% had major depression. In this study, GAD was slightly more common in young to middle-aged women, nonwhites, those not currently married, and those of lower socioeconomic status. Rates for GAD were not reported in the ECA Study. Age at onset is variable but is usually in the 20s to 30s. GAD may begin as childhood overanxious disorder. In clinical samples, the prevalence in male and female patients appears to be equal, and the course tends to be chronic (18).

Description and Differential Diagnosis

GAD is characterized by chronic excessive anxiety about life circumstances accompanied by symptoms of motor tension, autonomic hyperactivity, vigilance, and scanning. The individual often "awakens with" apprehension and unrealistic concern about future misfortune. The current diagnostic criteria require 6-month duration of symptoms to differentiate the disorder from more transient forms of anxiety, such as adjustment disorder with anxious mood. DSM-4 criteria emphasize that the worry is out of proportion to the likelihood or impact of the feared events, is pervasive (focused on many life circumstances), is difficult to control, and is not related to hypochondriacal concerns or part of PTSD and that anxiety secondary to substance-induced or nonpsychiatric medical etiologies is excluded. This is accompanied by tension or nervousness as manifested by at least three of the following symptoms:

- Restlessness
- Easy fatigueability
- Feeling "keyed up" or "on edge"
- Difficulty concentrating; mind going "blank"
- Irritability

In addition, significant functional impairment or marked distress is required for the diagnosis.

Genetic Epidemiology

Twin studies have found no evidence for genetic transmission of GAD. Diagnostic heterogeneity of GAD is suggested by the high frequency of nonanxiety psychiatric disorders in co-twins. However, a family study of GAD probands has reported an increased rate of GAD but not other anxiety disorders in first-degree relatives, suggesting some degree of familial transmission and separation of GAD from panic disorder and agoraphobia.

Pathophysiology

Benzodiazepine Function

Patients with generalized anxiety disorder have been found to have decreases in peripheral-type benzodiazepine receptor binding as assessed by [3_H]PK11195 binding to lymphocyte membranes, although the relationship to central benzodiazepine receptor function is unclear. In addition, reduced binding to [3_H]PK11195 has been reversed with benzodiazepine therapy in patients with anxiety disorders.

Brain Imaging Studies

Patients with GAD have been found to have a decrease in metabolism at baseline measured with PET in basal ganglia and white matter, and an increase in normalized left occipital cortex, right posterior temporal lobe, and right precentral frontal gyrus metabolism, in comparison with healthy controls. Administration of benzodiazepine therapy in patients with GAD results in a decrease in glucose metabolism in the occipital cortex, a brain region high in benzodiazepine receptors.

TREATMENT

The neural circuits and neural mechanisms of anxiety and fear described in this chapter may have implications for the psychotherapy and pharmacotherapy of anxiety disorders. As noted, fear conditioning may contribute to the symptoms of panic disorder, PTSD, and a variety of phobias. Cognitive and behavior therapies have been designed to reverse the impact of fear conditioning. Panic control treatment, which focuses on reversing the effects of fear conditioning on somatic symptoms associated with the efferent arm of the circuit, has been shown to be particularly effective. Some patients may not respond to such therapies because of an inability to extinguish intrusive memories and maladaptive behaviors. For such patients, therapies need to be developed that are specifically designed to facilitate extinction through the use of conditioned inhibitors and the learning of "new memories."

For information about current drug therapies of the anxiety disorders, see the pharmacology chapter in this volume. There is considerable overlap with regard to efficacy of psychoactive drugs on both generalized anxiety and panic disorder. Buspirone is a recently available $5HT_{1A}$ agonist that has antianxiety properties. Benzodiazepines and tricyclic antidepressants are about equally effective for initial treatment of panic disorder. These should be considered the first-line therapies, with serotonin reuptake inhibitors and monoamine oxidase inhibitors as second-line agents.

CONTEMPORARY ISSUES

Panic Disorder

Recent advances in 5HT receptor pharmacology have identified at least seven distinct families of 5HT receptor and multiple subtypes within those families. The challenge studies just reviewed have manipulated "global" 5HT functioning in "panic disorder" or have used probes that lack selectivity (e.g., mCPP) without taking into account the full complexity of the 5HT system and subsystems. Future studies will use selective $5HT_{1A}$ receptor subtype agonists and antagonists as probes. There may exist $5HT_{1A}$ receptor subsensitivity in the pathophysiology of panic disorder. In conclusion, the 5HT system or one of its subsystems may have a role in the pathophysiology of panic disorder, the precise nature of which needs to be delineated by further investigation.

Despite the modest findings of the available investigations regarding abnormalities in benzodiazepine receptor function in panic disorder patients, clearly, more direct assessments of benzodiazepine receptor systems in anxiety disorders are indicated. These include measuring the density of benzodiazepine receptors and the behavioral and cerebral metabolic effects of benzodiazepine agonist and inverse-agonist drugs, using PET and SPECT imaging techniques.

The Role of Cholecystokinin in Panic Disorder

Preclinical studies have shown that CCK agonists are anxiogenic in laboratory animals. In healthy volunteers, intravenous administration of CCK-4 (a tetrapeptide that crosses the blood-—brain barrier more readily than CCK-8) can induce severe anxiety or short-lived panic attacks. The anxiogenic effect of CCK can be blocked by the benzodiazepine lorazepam, although this may merely be pharmacologic opposition and not true antagonism.

Panic disorder patients may be more sensitive to the anxiogenic effects of CCK-4 and a closely related peptide, pentagastrin. The mechanism responsible for the enhanced sensitivity to CCK-4 has not been elucidated. Because CCK has important functional interactions with other systems implicated in anxiety and fear (noradrenergic, dopaminergic, benzodiazepine), these interactions need to be evaluated in panic disorder patients. CCK β-antagonists are being tested as antipanic drugs.

Identification of the relevant brain structures mediating anxiety and fear symptoms and knowledge of the functions of neurotransmitters and neuropeptides located within these structures may permit the discovery and testing of drugs that act on neurochemicals heretofore not associated with anxiety or fear.

SUMMARY

The proposed brain structures, neural mechanisms, and neural circuit related to anxiety and fear provide a basis to increase the understanding of the pathophysiology of anxiety disorders, such as panic disorder, PTSD, phobic disorders, and GAD. To date, the primary focus of neurobiologic research of these conditions has focused more on single neurotransmitter or neuropeptide theories, such as the role of noradrenergic, benzodiazepine, serotonergic, and CCK systems, than on interactions among neurotransmitter and neuropeptides and specific brain structures, or of the neural mechanisms involved in the genesis of the symptoms of anxiety and fear.

Brain imaging techniques, such as PET, SPECT, and functional magnetic resonance imaging, provide a means to further the pathophysiologic investigations of anxiety along these lines of inquiry. Although definitive patterns have not yet emerged, alterations in cerebral blood flow or metabolism have been identified in cortical association areas, orbitofrontal cortex, cingulate cortex, and the hippocampus during anxiety states.

There are several levels within the neural circuitry of anxiety and fear that may be dysfunctional in anxiety disorders. There may be abnormalities in peripheral sensory receptor systems, the relay of sensory information through the thalamus, the processing of sensory data in cortical and subcortical structures, the attachment of affect based upon prior experience by the amygdala, and the autonomic, neuroendocrine, neurochemical, and neuromotor efferent responses.

Panic disorder, which is characterized initially by spontaneous panic attacks, may be due to dysfunction in the efferent arm of the circuit. For example, increased responsiveness of the LC-norepinephrine system has been proposed. Based on the neural circuit outlined earlier in this chapter, additional functional neuroanatomic hypotheses can be developed to account for spontaneous panic attacks. Situational panic attacks and agoraphobia more likely result from modality-specific and contextual fear conditioning and the associated brain structures and neurotransmitters.

PSTD is characterized by intrusive traumatic memories manifested by recurring dreams, flashbacks, and psychological distress following exposure to events that symbolize or resemble the original trauma, persistent avoidance of stimuli associated with the trauma or a numbing of general responsiveness, and persistent symptoms of increased arousal. The neurobiologic basis of some of the symptoms of PTSD can be understood in the context of neural circuits and neural mechanisms of anxiety and fear. Persistent intrusive memories may be due to the strength of neuronal interactions between cortical regions, where many such memories are stored, and subcortical regions, such as the amygdala, which serve to attach affect to the memories. The psychological distress and physiologic responses to trauma reminders involve the mechanisms of fear conditioning and extinction. Contextual fear conditioning may be particularly relevant to severe cases of PTSD, in which stimulus generalization is a cardinal feature. The autonomic hyperarousal may be mediated by brain structures of the efferent arm of the anxiety circuit.

Simple phobias are a persistent fear of a circumscribed object or stimulus. Exposure to the phobic stimulus invariably provokes an anxiety response that includes somatic symptoms, such as sweating, tachycardia, and dyspnea. This disorder is likely to be mediated, in part, through the neural mechanisms underlying fear conditioning.

The essential feature of GAD is unrealistic or exaggerated anxiety about several life circumstances. The anxiety is frequently accompanied by symptoms of motor tension, autonomic hyperactivity, and excessive vigilance. It is frequently questioned whether GAD represents a homogeneous anxiety disorder. Some GAD symptoms overlap considerably with the hyperarousal category in the PTSD criteria. Further, many GAD patients suffer from limited symptom panic attacks or infrequent full-criteria panic attacks. The symptoms of many GAD patients may relate to early childhood or adult traumatic experiences. It is possible that the learning associated with psychotherapy will result in changes in the neuronal systems that mediate anxiety and fear responses.

REFERENCES

1. LeDeux JE. Nervous system V. Emotion. In: Blum F, ed. *Handbook of physiology.* Washington, DC: American Physiological Society, 1987:419–459.
2. Swanson LW. The hippocampus and the concept of the limbic system. In: Seifert W, ed. *Neurobiology of the hippocampus.* London: Academic Press, 1983:3–19.
3. Davis M. Animal models of anxiety based upon classical conditioning: the conditioned emotional response and fear potentiated startle effect. *Pharmacol Ther* 1990;47:147–165.
4. Kalivas PW, Striplin CD, Steketee JD, Klitenick MA, Duffy P. Cellular mechanisms of behavioral sensitization to drugs of abuse. *Ann N Y Acad Sci* 1992;654: 128–135.
5. Charney DS, Woods SW, Price LH, Goodman WK,

Glazer WM, Heninger GR. Noradrenergic dysregulation in panic disorder. In: Ballenger JC, ed. *Neurobiology of panic disorders.* New York: Alan R Liss, 1990: 91–105.
6. Roth RH, Tam S-Y, Ida Y, Yang J-XX, Deutch AY. Stress and the mesocorticolimbic dopamine systems. *Ann N Y Acad Sci* 1988;537:138–147.
7. Dunn AJ. Stress-related activation of cerebral dopaminergic systems. *Ann N Y Acad Sci* 1988;537:188–205.
8. Graeff F. Role of 5-HT in defensive behavior and anxiety. *Rev Neurosci* 1993;4:181–211.
9. Bremner JD, Charney DS. Anxiety disorders. In: Rakel RE, ed. *Conn's current therapy.* Philadelphia: WB Saunders, 1994:1103–1108.
10. Robins LN, Helzer JE, Weissman MM, et al. Lifetime prevalence of specific psychiatric disorders in three sites. *Arch Gen Psychiatry* 1984;41:949–958.
11. Myers JK, Weissman MM, Tischler GL, et al. Six-month prevalence of psychiatric disorders in three communities. *Arch Gen Psychiatry* 1984;41:959–967.
12. Roy-Byrne PP, Cowley DS, Greenblatt DJ, Shader RL, Hommer D. Reduced benzodiazepine sensitivity in panic disorder. *Arch Gen Psychiatry* 1990:47:534–538.
13. Orr SP. Psychophysiologic studies of post traumatic stress disorder. In: Giller EL, ed. *Biological assessment and treatment of post traumatic stress disorder.* Washington, DC: American Psychiatric Press, 1990:114–120.
14. Hoffman L, Burges Watson P, Wilson G, Montgomery J. Low plasma beta-endorphin in post-traumatic stress disorder. *Aust N Z J Psychiatry* 1989;23:269–273.
15. Van der Kolk BA, Greenberg MS, Orr SP, et al. Endogenous opioids, stress induced analgesia and post-traumatic stress disorder. *Psychopharmacol Bull* 1981; 25:417–421.
16. Marks IM. Genetics of fear and anxiety disorders. *Br J Psychiatry* 1986;149:406–418.
17. Fyer AJ, Mannuzza S, Gallops MS, et al. Familial transmission of simple phobias and fears: a preliminary report. *Arch Gen Psychiatry* 1990;47:252–256.
18. Merikangas KM, Weissman MM. Epidemiology of anxiety disorders in adulthood. In: Cavenar JO, ed. *Psychiatry,* vol. 3, Philadelphia: JB Lippincott, 1986:86–102.

16

Mood Disorders and Suicide

Robert G. Robinson, Javier I. Travella, Yeates Conwell, and Robin E. Henderson

OVERVIEW

The neuropsychiatry of mood disorders is based primarily on the study of patients with structural brain lesions such as stroke, traumatic brain injury (TBI), and Parkinson's disease. Thus, much of the early information about emotional disorders associated with brain disease has drawn on data obtained from a heterogeneous group of patients, some with clearly defined and localized brain disease and others without it. In this chapter, we present the disturbances (depressive, mania, and bipolar disorders) associated with the specific neuropsychiatric disorders of stroke, TBI, and Parkinson's disease, along with a review of what is presently known concerning the neuropsychiatry of suicide.

CLINICAL DISORDERS

DEPRESSIVE DISORDERS

Depression Associated With Stroke

The incidence of stroke increases steadily with age, rising from 10 per 100,000 population under age 35 years to 5970 per 100,000 over age 75 years. The American Heart Association estimates there will be 400,000 new victims of stroke each year. About 75% of stroke survivors are left with physical or intellectual impairments of sufficient severity to limit their vocational capacity.

Prevalence

Depression is the most common emotional disorder associated with cerebrovascular disorder. Investigators who have used structured psychiatric interviews and established diagnostic criteria have usually identified two forms of depressive disorder associated with brain disease. One type is major depression, which has been defined by the *Diagnostic and Statistical Manual of Mental Disorders*, 3rd edition (DSM-3) or revised 3rd edition (DSM-3-R) criteria (excluding the criteria that precludes an organic factor). The second type of depression is dysthymic depression as defined by DSM-3 or DSM-3-R criteria (excluding the 2-year duration criterion and the exclusionary organic factor).

The prevalence of these depressions varies, depending on the duration of the underlying illness, as well as the nature and location of the brain disorder. In a study of 103 consecutive patients admitted to the hospital with acute cerebrovascular lesions, Robinson and colleagues reported that 27% met symptom criteria for major depression, whereas 20% met symptom criteria for dysthymic (minor) depression (termed *minor depression* because the 2-year duration criterion was not met) (1). Most other studies of patients with cerebrovascular lesions have reported similar frequencies of depression, ranging from 25% to 50% of the population studied.

Longitudinal Course of Depression

Investigators have found that about two thirds of poststroke depression (PSD) patients continue to be depressed by 8 to 9 months after the initial evaluation. By the 1-year follow-up, the major depressive episodes have mostly resolved. Those patients with minor depression have a less favorable prognosis, with only 30% having recovered by 2 years after the stroke.

Biologic Markers

The dexamethasone suppression test has been investigated as a possible biologic marker for functional melancholic depression. The sensitivity of this test for major depression appears to be about 67%, but the specificity is only in the range of 70%. False-positive tests are found in 30% of patients (2).

Relationship to Impairment in Activities of Daily Living

Empirical studies have consistently failed to find a strong relationship between severity of depression and severity of physical impairment. Studies have demonstrated that severity of physical impairment is one of several factors that contributes to depression and, in some subpopulations, may be an important contributing factor to depression. Patients with in-hospital PSD tend to have a significantly poorer recovery than nondepressed stroke patients, even after their depression had subsided (Fig. 16-1).

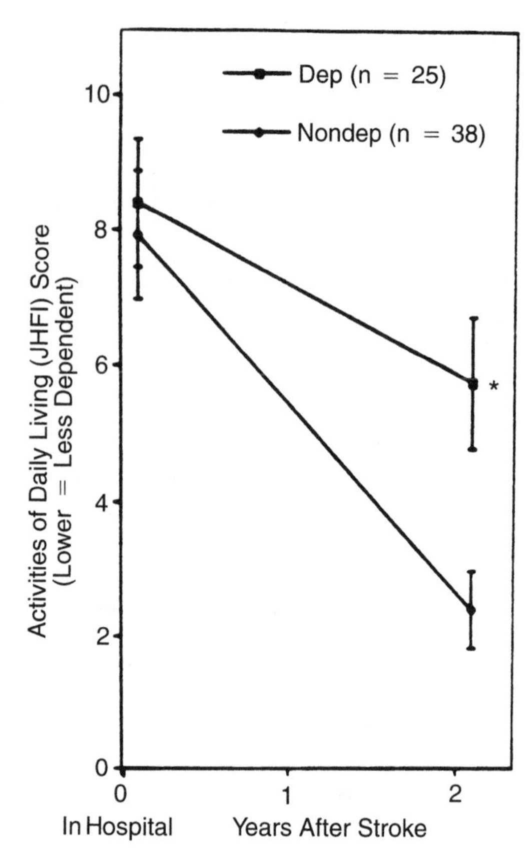

FIG. 16-1. Change in activities of daily living scores for depressed (major or minor) patients and nondepressed patients at the time of the in-hospital evaluation and 2 years later. Depressed patients show less recovery than nondepressed patients. (From Parikh RM, Robinson RG, Lipsey JR. The impact of poststroke depression on recovery in activities of daily living over a 2 year follow-up. *Arch Neurol* 1990;47:787.)

Relationship to Cognitive Impairment

Patients with major PSD after left-hemisphere lesions have significantly more cognitive deficits than nondepressed patients with a similar size and location of brain lesion. In contrast, among patients with right-hemisphere lesions, patients with major depression do not differ from nondepressed patients on measures of cognitive impairment (3). These findings suggest that left-hemisphere lesions (particularly left frontal and left basal ganglia lesions) that lead to major depression may produce a different kind of depression from depression caused by comparable right-hemisphere lesions.

Relationship to Lesion Location

Robinson and Starkstein have systematically examined the association between lesion location and PSD. They have found in different series of patients with acute stroke that major or minor PSD showed a significantly higher frequency of lesions in anterior areas of the left hemisphere, namely the left frontal dorsolateral cortex. They also found a relationship between PSD and specific subcortical lesions. In summary, the frequency of depression appears to be higher among patients with lesions of the left anterior hemisphere (left frontal dorsolateral cortex) and left basal ganglia than among those patients with any other lesion locations (4).

Perhaps the most consistent finding in PSD has been the association of depressive symptoms with intrahemispheric lesion location. There is an inverse correlation between severity of depression and distance of the anterior border of the lesion from the frontal pole (Fig. 16-2). This phenomenon is also found in other groups of patients with purely cortical lesions of the left hemisphere or purely left subcortical lesions and in left-handed patients with single left-hemisphere lesions (5). Longitudinal studies have found that proximity of the lesion to the frontal pole is significantly associated with severity of depression most strongly during the first 6 months after the stroke, suggesting that this phenomenon is a dynamic one that changes over time.

Relationship to Vascular Territory of Stroke

It has been reported that patients with middle cerebral artery (MCA) lesions show a higher incidence and longer course of PSD when compared to those with posterior circulation lesions (i.e., temporooccipital and cerebellar–brain-stem regions). Major or minor depression occurs in 48% of the patients in the MCA group and in 35% of patients in the posterior circulation group. At follow-up 1 to 2 years after the stroke, frequencies of depression are in the range of 68% and 0%, respectively (6). This finding suggests that the mechanism of depression after MCA infarcts may differ from that after posterior lesions.

Risk Factors for Depression Following Stroke

Patients with major PSD may have significantly more subcortical atrophy as measured by the ratio of third ventricle to brain (i.e., the area of the third ventricle divided by the area of the brain at the same level) and the ratio of lateral ventricle to brain (i.e., the area of the lateral ventricle contralateral to the brain lesion divided by the brain area at the same level). Thus, a mild degree of subcortical atrophy may be a premorbid risk factor that in-

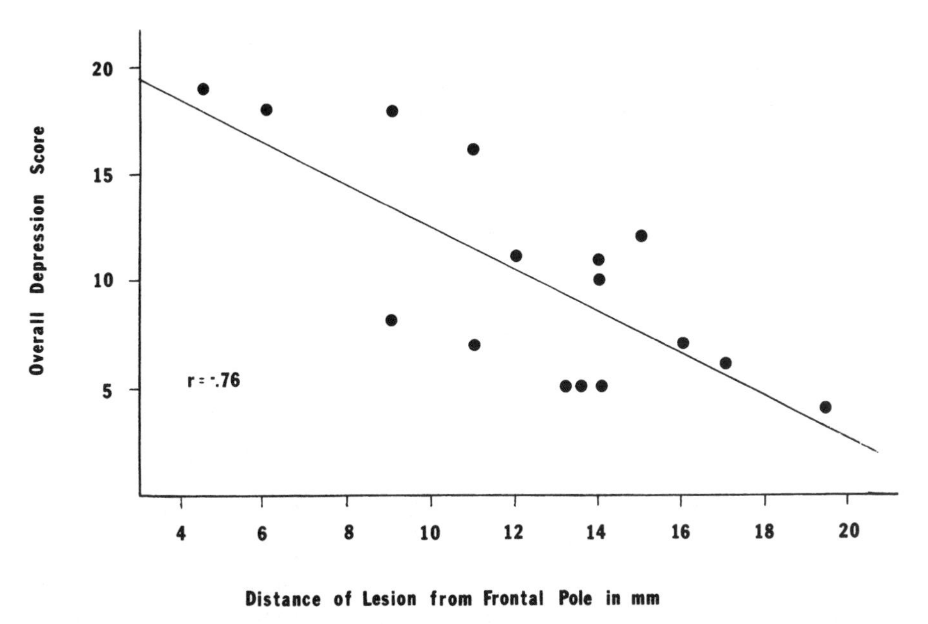

FIG. 16-2. Relationship between severity of depression and proximity of the computed tomography (CT) scan visualized lesion to the frontal pole for patients with stroke or traumatic brain injury (TBI) involving the left hemisphere. Lesions whose anterior border was closer to the frontal pole were associated with more depressive symptoms. (From Robinson RG, Szetela B. Mood change following left hemisphere brain injury. *Ann Neurol* 1981;9:450.)

creases the risk for developing major depression after stroke.

Investigators have also found that patients who develop PSD after a right-hemisphere lesion have a significantly higher frequency of family history of psychiatric disorders than nondepressed patients with right-hemisphere or left-hemisphere lesions. This suggests that a genetic predisposition for depression may play a role in the development of major depression after right-hemisphere lesions.

Mechanism of Depression Following Stroke

Although the cause of PSD is not known, it has been hypothesized that disruption of the amine pathways by the stroke lesion may play an etiologic role. The noradrenergic and serotonergic cell bodies are located in the brain stem and send ascending projections through the median forebrain bundle to the frontal cortex. Lesions that disrupt these pathways in the frontal cortex or the basal ganglia may affect many downstream fibers. Based on these neuroanatomic facts and the clinical findings that the severity of depression correlates with the proximity of the lesion to the frontal pole, it has been suggested that PSD may be the consequence of depletion of norepinephrine or serotonin produced by frontal or basal ganglia lesions.

Another possibility is that both the frontal dorsal lateral cortex and the dorsal caudate play an important role in mediating motor, intellectual, and instinctive behavior through their connection with the supplementary motor area, temporoparietal association cortex, and limbic system. A lesion of these anterior brain areas may result in low activation of motor, sensory, or limbic areas and produce the autonomic and affective symptoms of depression.

Treatment of Depression Following Stroke

Two randomized double-blind studies have confirmed the efficacy of antidepressant drug treatment in PSD; the superiority of nortriptyline over placebo at doses of 75 to 100 mg/d has been reported (Fig. 16-3) (7). Trazodone has also been reported to be effective (8).

Depression Associated With Parkinson's Disease

Although the frequent association of emotional disorders with Parkinson's disease was recognized more than 50 years ago, it has only

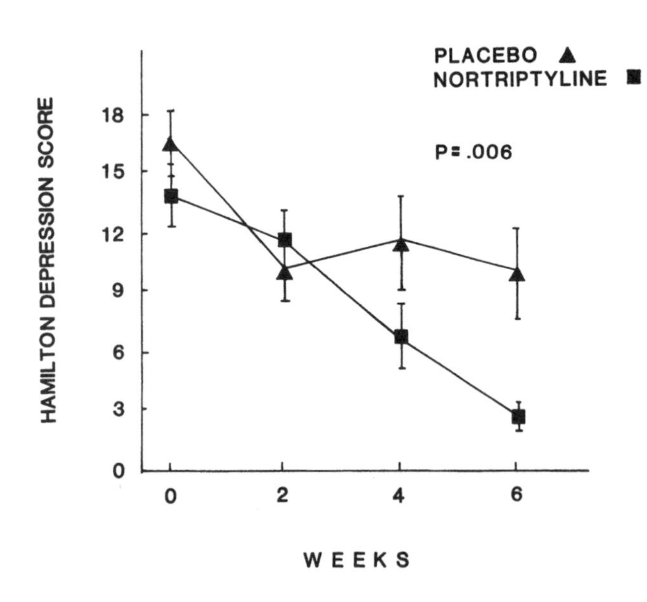

FIG. 16-3. Hamilton depression scores during a 6-week double-blind treatment trial of nortriptyline versus placebo for patients with depression (major or minor) following stroke. Patients receiving active treatment showed significantly greater improvement than those receiving placebo. (From Lipsey JR, Robinson RG, Pearlson GD. Nortriptyline treatment for poststroke depression: a double blind trial. *Lancet* 1984;1:299.)

been within the past several years that investigators have begun to examine the empirical nature of the relationship. Most investigators have found no significant correlation between the severity of depression and the severity of the physical impairment and have suggested that depression may be a consequence of neurobiologic changes in specific brain areas (9).

Prevalence

The frequency of depression in patients with Parkinson's disease has been reported to be about 40%. The highest frequency of depression is found in the early and the late stages of Parkinson's disease (Fig. 16-4) (10).

Relationship to Cognitive Impairment

Mayeux and associates, using a modified Mini-Mental State Examination, reported a significant correlation between cognitive deficits and severity of depression (i.e., severe depression was associated with severe cognitive impairments). These findings suggest that cognitive deficits may primarily be a result of motor impairments; but when depression also occurs, the cognitive deficits are greater, and they increase in severity as the Parkinson's disease progresses (11).

Mechanism

Several studies have demonstrated that patients with Parkinson's disease and dementia may show senile plaques and neurofibrillary tangles compatible with the diagnosis of Alzheimer's disease, as well as severe depletion of cholinergic neurons in the nucleus basalis of Meynert or Lewy bodies in cortical regions. Few neuropathologic studies, however, have been carried out in patients with Parkinson's disease and depression. A recent report has implicated marked loss of pig-

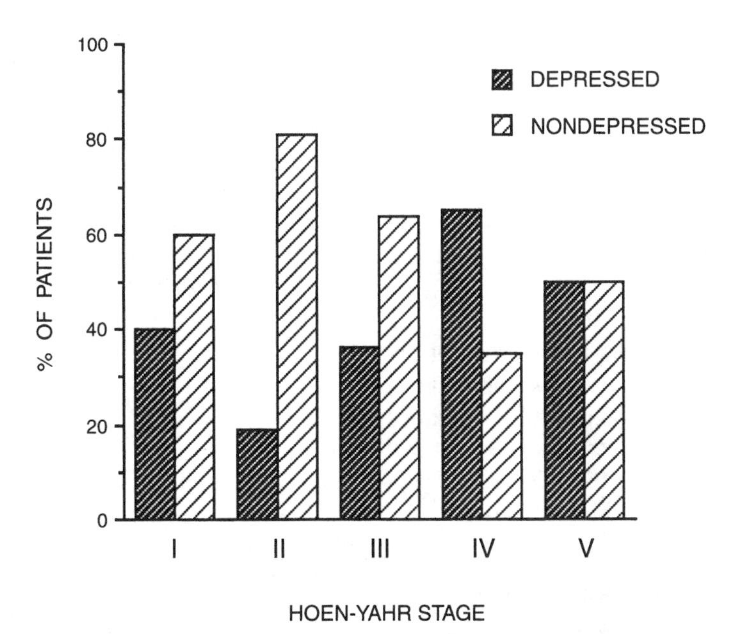

FIG. 16-4. The percentage of patients (total of 105 patients) at each stage of Parkinson's disease who were depressed. All patients were attending an outpatient care clinic and ranged in duration of disease from a few months to more than 15 years. The relative frequency of depression was higher in both the early and late stages of the illness compared with the middle stages.

mented neurons in the ventral tegmental area in a small group of patients with Parkinson's disease, dementia, and depression (12).

Depression in Parkinson's disease may also be related to changes in other biogenic amines. Patients with Parkinson's disease and depression were shown to have lower 5-hydroxyindoleacetic acid (5-HIAA) (a metabolite of serotonin) concentrations in the cerebrospinal fluid (CSF) than patients with Parkinson's disease without depression (11). Patients with Parkinson's disease and both dementia and depression have the lowest 5-HIAA CFS values.

Depression Associated With Traumatic Brain Injury

Traumatic brain injury (TBI) has an annual incidence of 2 million cases and represents the most common cause of death in people younger than 45 years of age in the United States. Although there is extensive literature on the neurobehavioral consequences of head injury, few studies have examined the frequency or course of mood disorders that occur after TBI (13).

Prevalence

Most of the studies of emotional or depressive symptoms among patients with TBI have relied on rating scales or relatives' reports rather than on structured interviews and established diagnostic criteria (e.g., DSM-3-R). Resulting reports have indicated variable levels of depression in the populations studied. Recent studies have reported risks for major depressive disorder among patients admitted to a head trauma unit in the range of 40% during a 1-year follow-up (14). These data suggest that major depression constitutes a significant psychiatric complication in this population.

Longitudinal Course of Depression

Mood disorders after TBI may be transient syndromes lasting for a few weeks, or they may be persistent disorders lasting for many months.

Risk Factors for Depression Following Traumatic Brain Injury

Several premorbid factors may influence patients' emotional responses to acute TBI and may therefore be relevant to the etiology of depressive disorders after TBI. There is a significantly greater frequency of previous personal history of psychiatric disorder in the depressed TBI group compared with nondepressed patients.

Relationship to Lesion Location

TBI is characterized by the presence of diffuse and focal lesions that may be the direct result of traumatic shear injury or secondary to ischemic complications. Still, there is some evidence that depression is more frequent in TBI patients who have more anterior left-hemisphere focal damage (13).

In conclusion, some acute-onset depressions appear to be related to lesion characteristics and may be caused by biologic responses such as neurochemical changes. Left dorsolateral frontal and left basal ganglia lesions are strongly associated with major depression during the initial in-hospital evaluation and may represent a strategic lesion location that elicits biochemical responses that finally lead to depression. In addition, however, other major correlates of depression are previous history of psychiatric disorder and impaired social functioning, suggesting psychosocial factors in the mechanism of some post-TBI depressions.

Treatment of Traumatic Brain Injury Depression

No double-blind, placebo-controlled studies have been performed on the efficacy of pharmacologic treatments of depression in TBI patients. The selection of antidepressant drugs for the treatment of post-TBI depression is usually guided by their side-effect profiles. Mild anticholinergic activity, minimal lowering of seizure threshold, and low sedative effect are the most important factors to be considered in the choice of an antidepressant drug in this population.

There are case reports of successful treatments of post-TBI depression with psychostimulants. These include dextroamphetamine (8 to 60 mg/d), methylphenidate (10 to 60 mg/d), and pemoline (56 to 75 mg/d). Patients taking stimulants need close medical monitoring to prevent abuse or toxic effects. The most common side effects are anxiety, dysphoria, headaches, irritability, anorexia, insomnia, cardiovascular symptoms, dyskinesias, or even psychotic symptoms.

Electroconvulsive therapy is not contraindicated in TBI patients and may be considered if other methods of treatment prove to be unsuccessful. Finally, the role of social interventions and adequate psychotherapeutic support should be included in the treatment of depression.

MANIA

Mania Associated With Stroke

Mania is a relatively rare consequence of acute stroke lesions. Less than 1% of patients with acute stroke have mania. Most of the patients included in studies of mania following stroke present with manic symptoms and are only secondarily found to have brain injury. These patients, however, have typical manic syndromes, and their symptoms are not significantly different from those of patients with mania who do not have brain injury (15).

Relationship to Lesion Location

An increased frequency of right-hemisphere lesions has been found in patients with mania after stroke. These lesions involve the basal and polar areas of the right temporal lobe, as well as subcortical areas of the right hemisphere, such as the head of the caudate and right thalamus. In a recent study using positron emission tomography with ^{18}F-deoxyglucose, metabolic abnormalities were examined in three patients with mania after right basal ganglia stroke. These patients were found to have focal hypometabolic deficits in the right basotemporal cortex. This finding suggests that lesions that lead to secondary mania may do so through their distant effects on the right basotemporal cortex (16).

Risk Factors for Mania Following Stroke

Patients with secondary mania have a significantly higher frequency of familial history of psychiatric disorders, as well as significantly more subcortical brain atrophy (as determined by increased ventricle-to-brain ratios) than patients with similar brain lesions but without mania.

Three risk factors for mania after stroke have been identified:

1. Genetic burden for psychiatric disorders
2. Increased subcortical atrophy
3. Seizure disorder

Most patients with secondary mania have right-hemisphere lesions, which involve the orbitofrontal or basotemporal cortex, or subcortical structures, such as the thalamus or head of the caudate. Secondary mania may result from disinhibition of dorsal cortical and limbic areas or dysfunction of asymmetric biogenic amine pathways.

Treatment

Clonidine, 600 mg/d, has been reported to reverse secondary manic symptoms rapidly. Other treatment modalities, such as antiepileptic drugs (valproate and carbamazepine), neuroleptics, and lithium, have been reported to be useful in treating secondary mania. None of these treatments has been evaluated in double-blind, placebo-controlled studies.

Mania Associated With Traumatic Brain Injury

Prevalence

Mania is more common among patients with TBI than among patients with stroke lesions. As many as 9% of patients with acute TBI develop manic-like symptoms. These manic episodes, however, are short-lived, with a mean duration of 2 months. Some secondary manic patients develop brief episodes of violent behavior at some point during a 1-year follow-up.

Relationship to Impairment Variables

The severity of mania after TBI is not known to be associated with severity of brain injury, degree of physical or cognitive impairment, personal or family history of psychiatric disorder, or availability of social support, or quality of social functioning. The present data suggest that mania is not a response to the associated impairments.

Relationship to Lesion Location

The cortical areas most frequently affected by closed head injury are the ventral aspects of frontal and temporal lobes. There is also evidence that damage to subcortical, diencephalic, and brain-stem structures are involved. The major neuroanatomic correlate of mania after brain injury is the presence of anterior temporal lesions. Factors such as personal history of mood disorders or posttraumatic epilepsy do not appear to significantly influence the frequency of secondary mania in this group of patients.

NEUROPSYCHIATRY OF SUICIDE

OVERVIEW

Great advances in the neurosciences in the past two decades have altered our understanding of suicide and self-injury. Evidence links suicide and self-injurious behaviors to central nervous system function. In this section, we review the known associations between suicide and specific psychiatric, medical, and neuropsychiatric syndromes and their implications for the diagnosis and treatment of patients at risk for self-injury (17).

Nosology

Striking demographic differences between suicide attempters and completers suggest that they are distinct but overlapping populations. Whereas the risk for completed suicide is highest in elderly men, the risk for attempted suicide is highest in young women. The margins of overlap for those populations are indistinct. Suicidal ideation is common to both groups but is common in the general population as well. Estimates of nonclinical populations found that from 50% to as high as 80% of individuals have considered suicide at some point in their lives. Much of the most compelling evidence of a neurobiologic basis for suicide is derived from postmortem studies of suicide victims (18).

CLINICAL BACKGROUND

Epidemiology of Suicide

Attempted Suicide

In the Epidemiological Catchment Area (ECA) Study, 2.9% of the population reported that they had attempted suicide at some time in their lives. By far, those at highest risk for suicide attempt were individuals with a lifetime diagnosis of psychiatric disorder. Additional risk factors included female gender, being separated or divorced, and a lower socioeconomic status. Whites had significantly higher rates of suicide attempts than blacks, and with increasing age, a smaller proportion had a past history of suicidal behavior. Other studies have estimated a prevalence of 120 to 730 attempts per 100,000 population per year, as compared with rates of about 12 per 100,000 per year for completed suicide (19). Of those who attempt suicide, about two thirds make no further attempt; 10% to 20% make an additional attempt within 1 year; 1% to 2% complete suicide within the year of an attempt, and about 10% to 15% of suicide attempters eventually take their own lives.

Completed Suicide

As Fig. 16-5 illustrates, suicide rates in the United States have fluctuated greatly over the course of the twentieth century. In 1992, at a rate of 12.0 per 100,000, there were more than 30,000 suicide deaths in the United States. For the general population, suicide was the ninth leading cause of death. It was the third leading cause of death for young adults, and the thirteenth leading cause of death in the elderly (20).

In the general population, rates rise through young adulthood to an initial peak of about 15 per 100,000 in the 35- to 44-year-old group, plateau through midlife, and rise again to a high of 22.8 per 100,000 in those age 75 to 84 years.

In 1992, the suicide rate for white Americans was two times that for blacks (13.0 versus 6.8 per 100,000). This differential is present at all ages across the life course, although the relative proportion changes. In young adulthood, rates for blacks approximate those for whites, whereas in late life, whites are at about six times greater risk.

At all ages, men are at greater risk than women within each racial subgroup (Fig. 16-6). Although women make about three times as many suicide attempts as men, the overall ratio of women to men in completed suicides is 1 to 4. This difference has been ascribed in large part to men choosing more violent, and hence potentially lethal, means.

As depicted in Fig. 16-7, rates within the United States are lowest in the Northeast. They increase as one moves south and west, peaking in the Mountain region, and decreasing again in the Pacific states. Satisfactory explanations for these sometimes substantial differences have not been established.

Table 16-1 lists rates per 100,000 population in a wide range of industrialized nations. Stillion and colleagues reviewed suicide rates by age and sex for the 12 countries that reported suicide statistics between 1983 and 1985, demonstrating remarkable consistency in age- and sex-specific patterns. Eleven of the 12 reporting countries showed suicide rates rising progressively with age in men. Rates in women, however, showed more variability. In the United States, Australia, Denmark, Poland, and Sweden, rates for women peaked for in women 45 to 54 years of age. In Canada and the United Kingdom, rates for women peaked somewhat later, at 55 to 64 years of age, whereas in the remaining five countries (Austria, France, Italy, Japan, and the Netherlands), the suicide rate for women, like men, peaked at age 65 years or older (21).

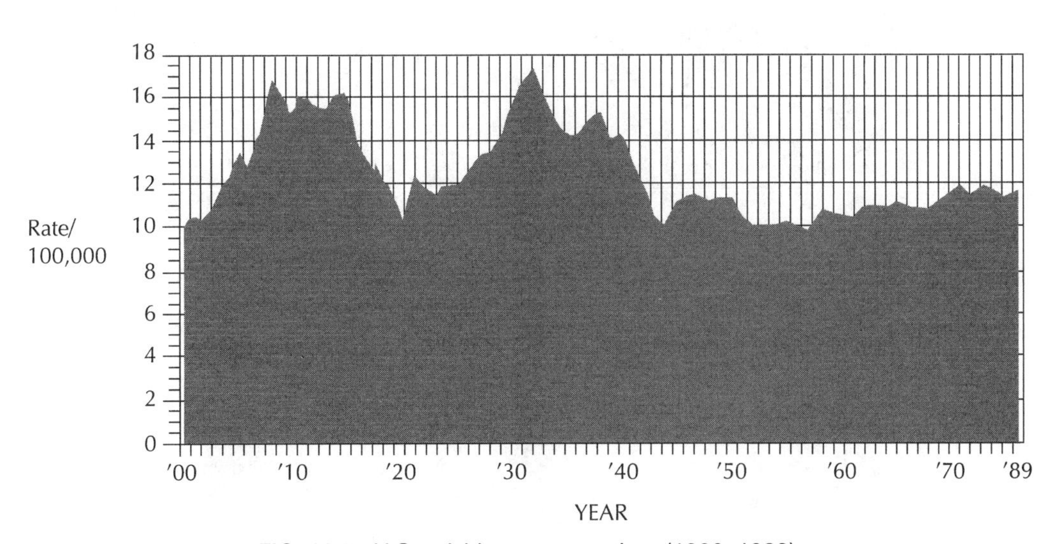

FIG. 16-5. U.S. suicide rates over time (1900–1989).

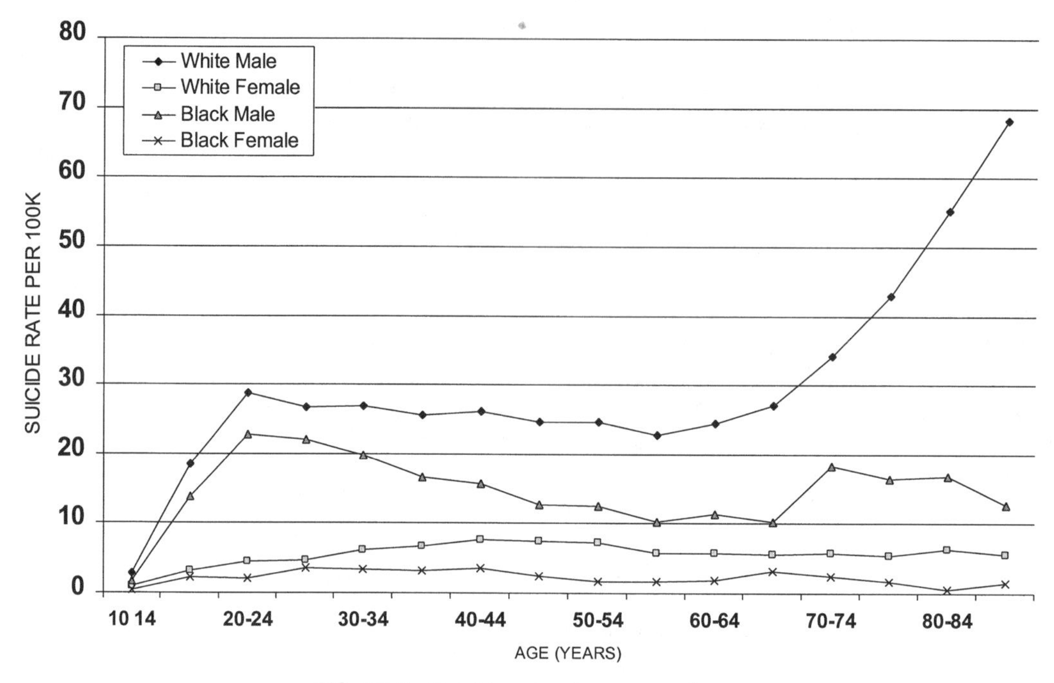

FIG. 16-6. Suicide rates by race and sex.

Stressful Life Events

Studies of suicide attempters have found that an increased number of stressful life events occur in the months preceding the self-destructive act. High rates of stressful events have been noted in the lives of completed suicides as well. More specifically, the nature of the stressful life events immediately preceding suicide differs as a function of both age

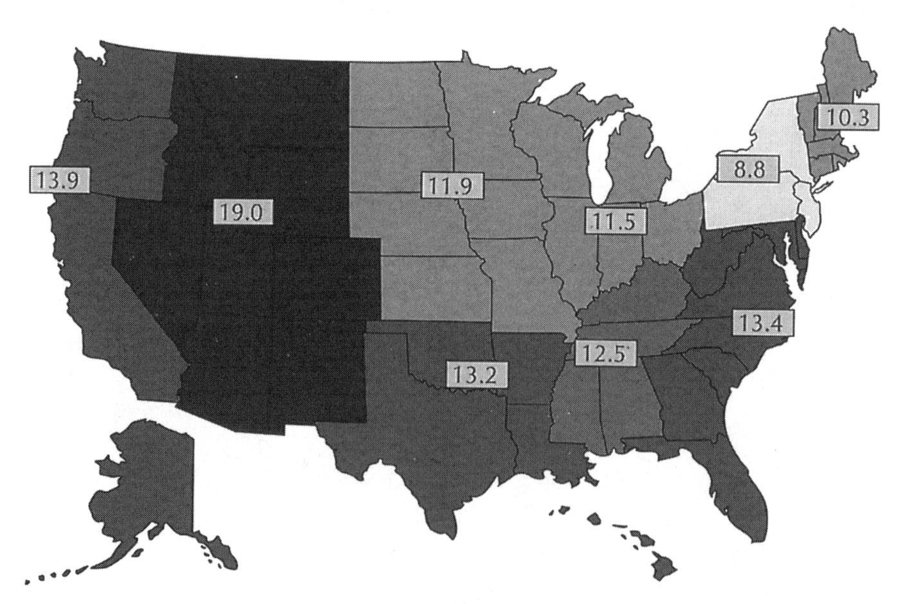

FIG. 16-7. U.S. regional suicide rates; 1988 rate/100,000 population.

TABLE 16-1. *Suicide rates per 100,000 population in industrialized nations*

Hungary	39.9
Denmark	26.8
Finland	26.7
Austria	22.6
Belgium	21.0
France	20.9
Switzerland	20.7
Czechoslovakia	18.1
Japan	18.1
Sweden	17.3
Bulgaria	16.0
W. Germany	15.5
Norway	15.4
Australia	14.0
Canada	13.9
New Zealand	13.1
Poland	13.0
United States	12.5
Scotland	11.4
Netherlands	10.8
England	8.0
Portugal	7.9
Italy	7.6
Spain	6.6

and diagnosis. Job, financial, and interpersonal difficulties are the events most commonly associated with suicide in young adulthood and middle age. The elderly most frequently commit suicide in the context of physical ill health and loss.

Physical Illness

Physical illness is a stressor associated far more frequently with suicide in the elderly than in younger populations. The precise quantitation of this relationship has proved difficult because concerns about physical ill health expressed by people in the days and weeks before they take their own lives are for some a reflection of distorted or delusional ideation rather than objectively evident organic disease.

Family History of Suicide

In one report, 6% of suicide victims had a parent who had died by suicide (22). Additional evidence for a genetic component to suicide has been provided by twin studies that demonstrate a higher concordance for suicide among monozygotic than dizygotic twins. Of 10 twin pairs concordant for suicide, however, all were monozygotic; in five of these twin pairs, the twins were also concordant for either depression or schizophrenia (23). The available data cannot at this point distinguish a genetic factor for suicide that is distinct from a genetic predisposition to major mental disorders.

NEUROBIOLOGY OF SUICIDE

Animal Models

Stress-related behaviors in animals have the greatest inherent similarity to self-destructive behaviors in humans. Frequently cited examples include self-mutilation in animals confined to small cages and removed from social stimuli. Among laboratory models, learned helplessness in rats and social separation in primates elicit behaviors reminiscent of depression in humans, which respond in turn to known human antidepressant therapies. Each has an incidence of self-destructive behavior associated as well (24).

Neurochemical Systems and Suicide in Humans

Serotonin: Postmortem Studies

Tissue Levels of 5HT and 5HIAA

The first evidence linking neurochemical changes in the brain and suicidal behavior was provided by studies of 5HT and 5HIAA in tissue obtained postmortem from suicide victims. Table 16-2 lists 12 studies in which indolamine and metabolite levels from a variety of brain regions were compared in suicide victims and controls. In five of these studies, investigators found no differences in 5HT levels between suicide victims and controls, and in four, no differences in 5HIAA levels could be demonstrated. 5HT levels were significantly decreased in specific brain regions of suicide victims in five studies, including brain

TABLE 16-2. *Studies comparing levels of 5HT and 5HIAA in brains of suicide victims and controls*

Study	Subjects	Tissue	Findings
Crow et al., 1984 (48)	10 suicides (7 depressed) 9 depressives with natural deaths 20 Alzheimer's disease 19 controls	Frontal cortex	Nonsignificant trend for ↓ 5HIAA in suicides
Bourne et al., 1968 (49)	23 suicides 28 controls	Brain stem No difference in 5HT	↓ 5HIAA in suicides
Cochran et al., 1976 (50)	10 depressed suicides 9 alcoholic suicides 12 controls	33 brain regions	No difference in 5HT
Beskow et al., 1976 (51)	23 suicides 62 controls	7 brain regions	No difference in 5HT ↓ 5HIAA in suicides (no difference after controlling for PM interval)
Shaw et al., 1967 (52)	28 suicides 17 controls	Hindbrain	↓ 5HT in depressed suicides
Cheetham et al., 1989 (53)	19 suicides 19 controls Brain stem	Cortex Hippocampus	↑ 5HIAA in amygdala of suicides ↓ 5HT in putamen
Pare et al., 1969 (54)	24 suicides 15 controls Caudate	Brain stem Hypothalamus	No difference in 5HIAA ↓ 5HT in brain stem of suicides
Owen et al., 1986 (55)	19 suicides 19 controls Hippocampus	Occipital and frontal cortex	↑ 5HIAA in hippocampus of suicides
Owen et al., 1983 (56)	17 suicides 20 controls	Frontal cortex	No difference in 5HIAA
Lloyd et al., 1974 (57)	5 suicides 5 controls	Raphe nuclei	↓ 5 HT in nucleus dorsalis and centralis inferior
Korpi et al., 1986 (58)	30 schizophrenics 14 nonschizophrenic suicides 29 normal controls	14 brain regions	↓ 5HT in hypothalamus of nonschizophrenic suicides
Arató et al., 1987 (59)	13 suicides 14 controls	Frontal cortex Hippocampus	No difference in 5HT or 5HIAA

stem, raphe nuclei, the putamen, and the hypothalamus of victims with diagnoses other than schizophrenia. Only two studies showed significant reduction in 5HIAA levels in brain of suicide victims relative to controls. At most, these studies suggest a trend for decreased indolamine and metabolite levels in brain-stem nuclei of suicide victims.

Presynaptic Receptor Studies

Table 16-3 lists results of studies comparing measures of binding at the serotonin reuptake site in the brains of suicide victims and controls. Although the greatest number of studies found no difference in frontal cortex or other brain regions, others have found significant decreases in imipramine binding in the brains of suicide victims. One study that showed greater imipramine binding in frontal cortex of suicide victims than controls ascribed this inconsistent finding to premorbid drug exposure in the suicide group.

Postsynaptic 5HT Receptor–Binding Studies

The $5HT_2$ receptor is located on the postsynaptic neuron, with its greatest density in frontal cortex projections of the ascending

TABLE 16-3. *Studies comparing binding at the serotonin reuptake site in brains of suicide victims and controls*

Study	Subjects	Tissue	Findings
Crow et al., 1984 (48)	10 suicides (7 depressed) 9 depressives—natural deaths 20 Alzheimer's disease 19 controls	Frontal cortex	[^{3}H]imipramine binding ↓ only in depressed suicides
Owen et al., 1986 (55)	19 suicides 19 controls	Frontal cortex Occipital cortex Hippocampus	[^{3}H]imirapamine binding; no difference between groups
Gross-Isseroff et al., 1989 (60)	12 suicides 12 controls	Multiple cortical and subcortical regions	↑ [^{3}H]imipramine binding in specific cell layers of hippocampus fields ↓ imipramine binding in postcentral and insular cortex and claustrum; no difference in prefrontal cortex or brain stem
Arató et al., 1987 (59)	13 suicides 14 controls	Frontal cortex Hippocampus	[^{3}H]imipramine binding in frontal cortex of suicides L > R hemisphere; in controls R > L hemisphere; no difference in hippocampus
Lawrence et al., 1990a (61)	22 suicides 20 controls	10 brain regions	No difference in [^{3}H]paroxetine binding in any region
Lawrence et al., 1990b (62)	8 suicides 8 controls	Frontal cortex Putamen Substantia nigra	[^{3}H]paroxetine binding—no interhemispheric differences; no differences in either hemisphere between suicides and controls
Meyerson et al., 1982 (63)	8 suicides 10 controls	Frontal cortex	↑ [^{3}H]imipramine binding in suicides
Stanley et al., 1982 (64)	9 suicides 9 controls	Frontal cortex	↓ [^{3}H]imipramine binding in suicides
Arora and Meltzer 1991 (65)	6 suicides 10 controls	Frontal cortex	No interhemispheric differences; no differences between suicides and controls
Arora and Meltzer 1989 (66)	28 suicides 28 controls	Frontal cortex	[^{3}H]imipramine binding—no differences between groups
Arató et al., 1991 (67)	23 suicides 23 controls	Frontal cortex	↑ [^{3}H]imipramine binding in left hemisphere of suicides vs. controls; ↓ binding in right hemisphere of suicides vs. controls; differences more pronounced in violent suicides

dorsal raphe nucleus. As shown in Table 16-4, some investigators have found increased $5HT_2$ receptor density in the frontal lobes of completed suicide victims. Four studies investigated this question using ^{3}H-ketanserin as the ligand, all of which failed to show a significant difference between suicide victims and controls in frontal cortex.

$5HT_1$ receptors have both presynaptic and postsynaptic subtypes, with highest density in areas of the hippocampus and basal ganglia receiving projections from the median raphe nucleus. Table 16-5 lists studies examining $5HT_1$ receptor binding in brains of suicide victims and controls assayed by a variety of methods. Using ^{3}[H]-5HT, a number of studies have found no difference between suicide victims and controls in frontal, occipital, or temporal cortex.

In summary, evidence from studies in both suicide attempters and victims of completed suicide supports a potential role for abnormalities in central serotonin function in the expression of suicidal behavior. In general, these findings are consistent with a decrease in presynaptic serotonin function, reflected in decreased serotonin levels, serotonin turnover, and presynaptic receptor binding, coupled with postsynaptic receptor upregulation as a compensatory response. Frontal cortex is the brain region most intensively studied, with few significant abnormalities reported in temporal cortex tissue.

Cerebrospinal Fluid Studies

Cerebrospinal fluid (CSF) serotonin studies exist from suicide attempters. In general,

TABLE 16-4. *Studies comparing $5HT_2$ receptor binding in brains of suicide victims and controls*

Study	Subjects	Tissue	Findings
Owen et al., 1983 (56)	17 suicides 20 controls	Frontal cortex	[^{3}H]ketanserin binding; trend for ↓ in suicides
Owen et al., 1986 (55)	19 suicides 19 controls	Frontal cortex Occipital cortex Hippocampus	[^{3}H]ketanserin binding; no difference at any site
Mann et al., 1986 (68)	21 suicides 21 controls	Frontal cortex	Spiroperidol binding—↑ in suicides.
Arora and Meltzer 1989 (69)	32 suicides 37 controls	Frontal cortex	[^{3}H]spiperone binding—↑ in suicides (males > females; violent > nonviolent)
Gross-Isseroff et al., 1990 (70)	12 suicides 12 controls	Frontal cortex Hippocampus Other	[^{3}H]ketanserin binding—↓ in frontal cortex of young suicides vs. controls; no difference for suicides > age 50 years vs. controls
Stanley and Mann 1983 (71)	11 suicides 11 controls	Frontal cortex	[^{3}H]spiroperidol binding—in suicides
Arango et al., 1992 (72)	13 suicides 13 controls	Prefrontal cortex Temporal	^{125}I-LSD binding ↑ in suicides in frontal cortex only
Cheetham et al., 1988 (76)	19 suicides 19 controls	Frontal, temporal, and occipital cortex Amygdala Hippocampus	[^{3}H]ketanserin binding—↓ in hippocampus; no difference in other regions
Arango et al., 1990 (73)	11 suicides 11 controls	Prefrontal cortex Temporal cortex	^{125}I-LSD binding—↑ in suicides in frontal cortex only
Crow et al., 1984 (48)	10 suicides 9 depressives, natural deaths 20 Alzheimer's disease 19 controls	Frontal cortex	[^{3}H]ketanserin binding; no difference between groups

TABLE 16-5. *Studies comparing $5HT_1$ and $5HT_{1A}$ receptor binding in brains of suicide victims and controls*

Study	Subjects	Tissue	Findings
Owen et al., 1983 (56)	17 suicides 20 controls	Frontal cortex	[^{3}H]5HT receptor binding; no difference between groups
Owen et al., 1986 (55)	19 suicides 19 controls	Frontal cortex Occipital cortex Hippocampus	[^{3}H]5HT receptor binding; no difference between groups
Dillon et al., 1991 (74)	14 suicides 14 controls	Numerous cortical and subcortical brain regions	OH-DPAT binding; no difference between groups; negative correlation with age in males
Matsubara et al., 1991 (75)	23 suicides 40 controls	Frontal cortex	OH-DPAT receptor binding; negative correlation with age in males controls, but not females or suicides of either sex; ↑ binding in nonviolent suicides
Mann et al., 1986 (68)	21 suicides 21 controls	Frontal cortex	[^{3}H]5HT receptor binding; no difference between groups
Cheetham et al., 1990 (76)	19 suicides 19 controls	Frontal cortex Temporal cortex Hippocampus	[^{3}H]5HT receptor binding; ↓ in hippocampus of suicides; no difference in cortical regions
Crow et al., 1984 (48)	10 suicides 9 depressives 20 Alzheimer's disease 19 controls	Frontal cortex	[^{3}H]5HT receptor binding; no differences between groups

these studies offer support to the postmortem studies suggesting altered central serotonin function in suicide. Åsberg and colleagues noted that depressives with low CSF levels of 5HIAA had a higher incidence of suicide attempts (25). Of 68 patients with depressive illness in their study, 15 had made a suicide attempt during the index episode. Concentrations of CSF 5HIAA were bimodally distributed in the sample, with 40% of the low 5HIAA group having attempted suicide, as compared with 15% of patients with normal CSF 5HIAA levels. Furthermore, all patients who attempted suicide by violent means clustered in the low 5HIAA group, two of whom subsequently killed themselves during the study period. A number of subsequent studies have confirmed Åsberg's original finding, whereas others have failed to do so (26).

Diagnoses Other Than Depression

Schizophrenic patients with a history of suicide attempts have lower levels of CSF 5HIAA than either controls or schizophrenic patients without history of suicidal behavior. Analogous studies exist for suicide attempters with diagnoses of alcoholism and adjustment disorders (27).

Linnoila and colleagues studied individuals with personality disorders incarcerated after violent acts (28). Dividing the group into "impulsive" and "nonimpulsive" offenders, they found significantly lower CSF 5HIAA concentrations in the former group. Some subjects, however, had a past history of suicide attempt; these individuals had the lowest metabolite levels. The results were interpreted to support the hypothesis that poor impulse control is associated with low levels of monoamine metabolites.

Summary

Data from studies using a broad range of technologies and tissues support the existence of an association between suicidal behavior and functioning of the central serotonin system. It seems most likely that serotonin plays its role in suicidal behavior through the mediation of impulse control and aggression. Deficits in this system may predispose individuals to violent behavior, as has been demonstrated in animal models, under specific circumstances, such as the development of a co-morbid depressive disorder or individually significant stressful life circumstances.

Dopamine

Many authors have noted an association in the CSF between levels of 5HIAA and the dopamine metabolite homovanillic acid (HVA). Several authors have reported low levels of CSF HVA paralleling decreases in 5HIAA in suicidal depressives (29). In combination with CSF HVA levels in suicidal subjects, these data indicate that although decreased dopamine function may have little bearing on the expression of suicidal behavior in other diagnostic groups, it may signal vulnerability in patients with affective illness. Other lines of evidence suggest a role for dopamine in suicidal behavior. In the epidemiology of suicide, we noted the striking rise in rate with age in men. There are well-documented reductions in dopaminergic neurons, the function of receptors and uptake sites, and synthetic enzyme activity with increasing age. Furthermore, as discussed later in this chapter, suicide risk is elevated in both Huntington's disease and Parkinson's disease, disorders in which degenerative changes of the central dopaminergic systems have been implicated.

Norepinephrine and Epinephrine

Positive findings concerning the noradrenergic system and suicide include the observation of an inverse correlation between CSF levels of the norepinephrine metabolite 3-methoxy-4-hydroxyphenylglycol (MHPG) and ratings of suicidal intent in unipolar and bipolar depressives, and arsonists and violent offenders. MHPG in the CSF is negatively correlated with a lifetime history of aggression in children and adolescents with disruptive behavior disorders as well (30).

Only a small number of studies have examined α_1-adrenergic receptor function in suicide-prone patients. Whereas one study using in vitro quantitative autoradiography with tritiated prazosin found significant decreased binding in the prefrontal cortex of suicide victims compared with controls, a second study with similar methodology found an increased binding specific to layers IV to V of the prefrontal cortex in suicide victims (31).

Neuroendocrine Systems and Suicide

Thyroid Axis

People with depressive disorders have a blunted thyrotropin (thyroid-stimulating hormone) response to the infusion of thyrotropin-releasing hormone (TRH). Retrospective analyses in two studies have demonstrated an association between blunted thyrotropin response and attempted and completed suicide. As with other neurobiologic measures, the distinction between violent and nonviolent suicidal behavior may be important because a number of studies have found associations between thyroid axis abnormalities and suicide in those who used violent methods only (32).

Hypothalamic-Pituitary-Adrenal Axis and Suicide

Observations have been made of elevated urinary excretion of cortisol and 17-hydroxy corticosteroids in patients who attempted or completed suicide. Elevated plasma cortisol has been associated with completed suicide at follow-up as well. In postmortem studies, CSF corticotropin-releasing factor (CRF) levels have been reported to be significantly higher and CRF-binding sites in the frontal cortex significantly reduced in suicide victims, as compared with sudden death controls. Not all investigators have corroborated these findings (33,34).

A number of groups have found that escape from dexamethasone suppression predicts suicide risk, whereas others have failed to find that association. Furthermore, failure of the dexamethasone suppression test to normalize has also been associated with subsequent completed suicide (35).

SUICIDE IN NEUROPSYCHIATRIC DISORDERS

Psychiatric Illness

Studies of Suicide in the General Population

Since the mid-1950s, more than a dozen retrospective studies of suicide in the general population have been conducted worldwide. Psychiatric diagnoses established for victims in six of the most comprehensive of these studies are listed in Table 16-6. They differ in sample size, the years that they were conducted, and the criteria by which diagnoses were made. Nonetheless, all these investigations show a remarkable and important consistency in finding diagnosable psychopathology in 90% or more of cases. In general, affective disorders were most common, present in 30% to 80% of cases, closely followed by substance use disorders in 19% to 60% of cases. Schizophrenia was diagnosed in 2% to 14% of completed suicides in these studies.

To explore the marked variation in suicide rates as a function of age and sex, a number of authors have examined differences between men and women who completed suicide. A comparison of 143 men and 61 women in a suicide study sample concluded that women had a significantly greater mean age at death. There was no difference in the proportion in each group with a substance use disorder diagnosis; however, women were significantly more likely to have had a major depressive disorder. Men used firearms significantly more often and drugs or poison significantly less often as a means of death than did women (36).

Affective Disorder

Suicide rates in patients with affective illness are far higher than those in the general population. Patients from 30 studies who had primary depressive disorders were followed longitudinally; analysis outcome was that about 15% of affected individuals ultimately die by suicide (37).

In an attempt to define the clinical characteristics unique to those depressed patients at highest risk for suicide, a number of studies have compared depressed suicide completers with living patients. Barraclough and Pallis found that depressed suicide victims studied by the psychological autopsy method were more likely to be male, single, and living alone than the comparison group of ambulatory patients with affective disorder diagnoses (38). Symptoms of insomnia, self-neglect, impaired memory, and a history of suicide attempts also were distinguishing features.

The posthospitalization period is the time of greatest risk for suicide in affective illness, decreasing progressively after the first several years after discharge. Hospital discharge, therefore, represents a time for special vigilance on the part of patients and their caretakers.

Schizophrenia

Epidemiologic studies in the United States and other countries estimate that the preva-

TABLE 16-6. *Psychological autopsy studies: percent of completed suicides with selected diagnoses*

Study	Affective disorder	Substance use disorder	Schizophrenia	Axis II	No. diagnosis
Robins et al., 1959 [N = 134]	45	25	2	—	6
Dorpat et al., 1960 [N = 108]	30	27	12	9	0
Chynoweth et al., 1980 [N = 135]	55	34	4	3	11
Barraclough et al., 1974 [N = 100]	80	19	3	—	7
Rich et al., 1986 [N = 283]	44	60	14	5	6
Conwell et al., 1991 [N = 85]	55	42	8	18	11

lence of schizophrenia in the general population is about 1%, yet about 5% to 10% of completed suicides carry that diagnosis. Based on suicides in follow-up intervals from 4 to 40 years, suicide rates for schizophrenics have been estimated to be as high as 615 per 100,000, more than 50 times greater than that of the general population (39).

Like patients with other psychiatric disorders, the schizophrenic patient is at highest risk for suicide in the first weeks and months after admission and discharge from the hospital. In studies of suicide committed while in the hospital, schizophrenia is the most common diagnosis.

Substance Use Disorders

Far fewer data are available on the suicidal behavior and risk of other substance users. There are some reports of suicide among patients with sedative-hypnotic dependence in 4- to 6-year follow-up. Suicide rates for heroin addicts have been estimated to range from 82 to 350 per 100,000, with 2.5% to 7% dying by their own hand. These age-adjusted rates are 5 to 20 times greater than those for the general population.

The mechanism of an association between suicide and substance abuse is obscure. Comorbidity is common. Data from the ECA Study show that individuals with either an alcohol or drug use disorder are more than seven times more likely than the rest of the population to have a second addictive disorder, and that 37% of those with alcoholism and 53% of those with other drug addictions have a co-morbid mental illness (40).

Neurologic Illness

The neurologic conditions in which suicide has been studied in greatest detail include Huntington's disease, multiple sclerosis, epilepsy, and traumatic spinal cord injury. For the most part, these investigations have been retrospective, using archival databases such as death certificates, hospital records, and federal or state disease rosters. Less frequently, the information gathered has been supplemented by seeking interviews with identified patients or with family members when the patient is deceased.

Huntington's Disease

Huntington's disease is a neurodegenerative, autosomally dominant genetic disorder of adult onset. The major neuropathologic changes include neuronal degeneration in the caudate nucleus, putamen, and associated pathways. Using the National Huntington's Disease Research Roster, Farrer examined the proportion of deaths in 831 patients that were due to suicide, through a detailed questionnaire sent to families (41). Of 452 individuals identified as deceased, 25 had died by suicide, four times more than would be expected in a general population sample. Of the suicide victims, 27.6% had attempted suicide at least once (18.5% of men, 22.2% of women). Interestingly, age at onset of Huntington's disease was not correlated with completed suicide; however, suicide did tend to occur in early to middle stages of the illness (41).

The advent of preclinical genetic testing has fostered examination of psychiatric sequelae, including suicidal behavior, in symptom-free people at risk for expression of the Huntington's disease gene. On questionnaire studies of at-risk subjects, 11% reported they would consider suicide if they tested positive (42).

Multiple Sclerosis

Multiple sclerosis (MS) is a chronic demyelinating disease of the central nervous system that is more common in women than in men (female-to-male ratio, 1.7:1). The course is progressive but unpredictable, and patients may have multiple periods of illness exacerbation and remission. Of four studies that attempted to establish the prevalence of suicide in patients with MS, all but one found patients to be at increased risk. In an epidemiologic investigation of 6088 Danish patients with MS, adjusted for age, gender, and effects of early mortality in MS from causes other than suicide, the expected suicide rate in this population was significantly lower than observed (2.0% versus 3.7%). The risk for suicide in MS was greatest in men, in individuals with disease onset before 30 years of age, and in those diagnosed before 40 years of age (43).

Epilepsy

Comorbid psychological variables, including suicide risk, have been a topic of considerable interest in the clinical epilepsy literature. Estimates of suicide risk in patient with epilepsy are three to four times higher than in the general population (44).

The most comprehensive and methodologically sound study of completed suicide in epilepsy was conducted by White and associates (45). Evaluating the records of 2099 epilepsy patients admitted to a British epilepsy center, they concluded that suicide was among the chief causes of excess death in this population. Of 636 deaths that had occurred, 21 were by suicide, a rate 5.4 times greater than expected. Attempted suicide rates among epileptics are also high; patients with epilepsy are estimated to be at four to five times greater risk for attempted suicide than the general population.

Acute Spinal Cord Trauma

Increased mortality patients with spinal cord trauma has been well established, as has increased frequency of death by suicide. Risk estimates ranging from two to six times that of the general population have been reported. Younger age and quadriplegia place patients with spinal cord injury at increased risk, particularly in the first few years after injury (46). Preinjury variables associated with suicide have included a history of alcohol abuse, drug abuse, or depression. Postinjury variables of import included despondency, shame, apathy and helplessness, anger, alcohol abuse, destructive behaviors, and attempted suicide.

Summary

Increased risk for suicide has been well established in MS, epilepsy, and traumatic spinal cord injury, with probable increased risk associated with Huntington's disease as well. The relative contributions of a high incidence of co-morbid psychiatric sequelae, particularly depressive illness, has also been well documented. The degree to which depressive illness or other psychiatric phenomena, such as psychosis, serve to mediate suicidal behavior in neurologic illness, as compared with suicidal behavior existing as an independent disease-related behavior, is unclear.

PREVENTION AND TREATMENT

Despite decades of research and progress in the recognition of risk factors for suicide, efforts to construct a useful and sufficiently sensitive and specific predictive scale have thus far failed. Although many hope that the development of biologic markers will greatly enhance our ability to identify individuals at risk for suicide, the body of knowledge we now possess can serve only in a nonspecific manner to enhance our clinical judgment (Table 16-7).

The multiple causality of suicide suggests the possibility of intervention at numerous

TABLE 16-7. *Summary of risk factors for suicide*

Older
Male
White
Living alone
History of suicide attempt(s)
Psychiatric illness
Affective disorder
Substance abuse or dependence
Schizophrenia
Medical illness
Neurological illness
Multiple sclerosis
Epilepsy
Traumatic spinal cord injury
Huntington's disease
Life stressors
Recent or threatened loss
Lack of available support
Inability to accept help

sites, any of which may be effective in resolving at least the immediate crisis. As a general rule, any intervention that decreases the subject's "intolerable psychological pain" will diminish suicide risk. We have structured a summary discussion using the terms of preventive medicine:

Primary prevention: those strategies concerned with preventing the occurrence of suicidal behavior or the development of the suicidal state

Secondary prevention: early recognition of the suicidal crisis, intervention to reduce its progression to complete suicide, and treatments that lead to its full resolution

Tertiary prevention: interventions targeted at those who survive the death of a loved one by suicide, decreasing the psychiatric and social morbidity of this special at-risk group

Primary Prevention

The definition of demographic and psychosocial risk factors for suicide suggest that interventions made at the level of social policy may have a beneficial effect on suicide rates. For example, if social isolation were diminished for depressed elders through more vigorous outreach and socialization programs, fewer might choose to end their lives. In young adult cohorts, drug abuse prevention programs should be expected to diminish rates of death by all violent means, including suicide.

A great many myths about suicide are commonly held by the public. One such misconception is that, if a person truly wants to commit suicide, there is little that can be done to prevent it. Other myths are that people who threaten suicide will not follow through or that discussing suicide with a depressed person may lead them to the act. Broadly aimed educational programs may favorably affect some of these belief systems. In addition to education of the general public about suicide, suicidal elderly may particularly benefit from education of healthcare professionals.

Secondary Prevention

Secondary prevention efforts can be divided into those concerning assessment of individuals at risk and treatment interventions. Generally, assessment of an individual's suicide risk must begin with a direct, yet sensitive, discussion with the patient about the nature and extent of the suicidal ideation, including the potential lethality of any plans the individual has considered. One should carefully define the nature and extent of the patient's social support network and the elements in the patient's environment that may be mobilized to ameliorate a suicidal crisis.

In the presence of more highly developed suicidal ideation and planning, the absence of helpful supports in the patient's social network, and resistance by the patient to available interventions, hospitalization may be life-saving. In most cases, suicidal thoughts are transient, a symptom of illness that emerges at times of episodic crisis and that resolves with treatment.

Psychiatric consultation should be obtained for all suicidal patients to help assess the extent of risk and establish a psychiatric diagnosis and formulation.

Pharmacotherapeutic interventions are primarily directed at the psychiatric condition of which suicidal ideation and behavior may be symptomatic manifestations. In addition to somatic therapies, the suicidal patient may respond to a range of psychological, cognitive, and behavioral treatment modalities, all of which have been thoroughly reviewed elsewhere (47).

Tertiary Prevention

There is a growing consensus that the grief following a death by suicide involves a complex combination of depression, guilt, and anger that is distinctive from "normal" bereavement. It is a grief that may place the survivor at increased risk for psychiatric morbidity and suicide mortality. Therefore, treatment should not end with the death of a patient by suicide. The important tasks that remain include attention to the impact of the suicide on the patient's survivors—family, friends, and healthcare providers, including the physician.

SUMMARY AND CONCLUSIONS

The neuropsychiatry of mood disorders involves the investigation of similarities and differences in mood disorders associated with several neurologic disorders. This chapter has focused on mood disorders associated with stroke, Parkinson's disease, and TBI.

Depression occurs in about 40% of patients with acute stroke lesions, and its natural evolution is from 1 to 2 years, although patients with subcortical or brain-stem lesions may show depression of shorter duration. Both intrahemispheric and interhemispheric lesion location appear to contribute to the development of depression. Major depression is significantly more common among patients with left-hemisphere lesions involving anterior cortical (frontal) or subcortical (basal ganglia) regions than any other lesion location. On the other hand, depression following right-hemisphere lesions is sometimes associated with a genetic vulnerability and frontal or parietal lobe damage. Finally, an important risk factor for the development of PSD is the presence of subcortical atrophy before the stroke lesion.

Mania that develops after a stroke has a phenomenology similar to that of mania without known neuropathology. Secondary mania, a rare complication in stroke patients, is almost always the consequence of lesions involving right cortical (orbitofrontal or basotemporal) or right subcortical (head of the caudate or thalamus) limbic-related regions. Among these areas, dysfunction of the basotemporal cortex seems to be particularly important to the development of secondary mania, and basotemporal dysfunction may be produced by direct or indirect (diaschisis) damage. Risk factors for secondary mania include a familial history of psychiatric disorders or prior subcortical atrophy.

Among patients with TBI, the frequency of major depression is about 26%. Although some patients remain depressed for more than a year, the average duration of depression is 4 to 5 months. Risk factors for depression include previous history of psychiatric disorder and poor social functioning. Intellectual or physical impairment is not associated with the degree of depression. In addition, transiently depressed patients are more likely to have sustained injury to left dorsal lateral frontal or left basal ganglia structures than are patients with more prolonged depressions. Similarly, manic patients are significantly more likely to have temporal basopolar lesions than are depressed or nondepressed patients.

Many areas can benefit from future research. The most important elements of social functioning that contribute to depression need to be explored, as does the effect of social intervention. The role of antidepressants in treating these depressive disorders has not been systematically explored and deserves study.

Finally, the mechanism of these depressive disorders, both those associated with psychosocial factors and those associated with neurobiologic variables (e.g., strategic lesion locations), need to be investigated. It is only through the discovery of their mechanisms that specific and rational treatment strategies for these disorders will be developed.

ACKNOWLEDGMENTS

The authors are indebted to Drs. Sergio E. Starkstein, Thomas R. Price, John R. Lipsey, Rajesh M. Parikh, J. Paul Fedoroff, Helen S. Mayberg, and Karen Bolla, who participated in many of these studies. This work was supported by the following NIMH grants: Research Scientist Award MH00163 and MH40355.

REFERENCES

1. Robinson RG, Starr LB, Kubos KL. A 2 year longitudinal study of post-stroke mood disorders: findings during the initial evaluation. *Stroke* 1983;14:736–741.
2. Lipsey JR, Robinson RG, Pearlson GD. Dexamethasone suppression test and mood following strokes. *Am J Psychiatry* 1985;14:318–323.

3. Downhill J, Robinson RG. Longitudinal assessment of depression and cognitive impairment following stroke. *J Nerv Ment Dis* 1994;182:425–431.
4. Robinson RG, Kubos KL, Starr LB, Rao K, Price TR. Mood changes in stroke patients: relationship to lesion location. *Brain* 1984;107:81–93.
5. Robinson RG, Szetela B. Mood change following left hemisphere brain injury. *Ann Neurol* 1981;9:447–453.
6. Starkstein SE, Robinson RG, Berthier ML, Price TR. Depressive disorders following posterior circulation as compared with middle cerebral artery infarcts. *Brain* 1988;11:375–387.
7. Lipsey JR, Robinson RG, Pearlson GD. Nortriptyline treatment for post stroke depression: a double blind trial. *Lancet* 1984;1:297–300.
8. Reding MJ, Orto LA, Winter SW. Antidepressant therapy after stroke: a double blind trial. *Arch Neurol* 1986; 43:763–765.
9. Mayeux R, Williams JBW, Stern Y: Depression in Parkinson disease. *Adv Neurol* 1984;40:242–250.
10. Starkstein SE, Preziosi TJ, Berthier ML. Depression and cognitive impairment in Parkinson's disease. *Brain* 1989;112:1141–1153.
11. Mayeux R, Stern Y, Rosen J. Depression, intellectual impairment and Parkinson disease. *Neurology* 1981;32: 645–650.
12. Torack RM, Morris JC. The association of ventral tegmental area histopathology with adult dementia. *Arch Neurol* 1988;45: 211–218.
13. Silver JM, Hales RE, Yudofsky SC. Depression in traumatic brain injury. *Neuropsychiatry, Neuropsychology and Behavioral Neurology* 1991;4:12–23.
14. Fedoroff JP, Starkstein SE, Forrester AW, Geisler F, Jorge RE, Robinson RG. Depression in patients with acute brain injury. *Am J Psychiatry* 1992;149:7.
15. Robinson RG, Boston JD, Starkstein SE, Price TR. Comparison of mania with depression following brain injury. *Am J Psychiatry* 1988;145:172–178.
16. Starkstein SE, Bolduc PE, Preziosi TJ. Depression in Parkinson's disease. *J Nerv Ment Dis* 1990;178:27–31.
17. Farberow NL. Cultural history of suicide. In: Farberow NL, ed. *Suicide in different cultures.* Baltimore, MD: University Park Press, 1975:1–15.
18. Berman AL. Self-destructive behavior in suicide: epidemiology and taxonomy. In: Roberts AR, ed. *Self-destructive behavior.* Springfield, IL: Charles C Thomas, 1975:5–20.
19. Moscicki EK, O'Carroll P, Rae DS, Locke BZ, Roy A, Regier DA. Suicide attempts in the epidemiologic catchment area study. *Yale J Biol Med* 1988;61:259–268.
20. Murphy GE, Wetzel RD. Suicide risk by birth cohort in the United States, 1949 to 1974. *Arch Gen Psychiatry* 1980;37:519–523.
21. Stillion JM, McDowell EE, May JH. *Suicide across the lifespan: premature exits.* New York: Hemisphere Publishing, 1989.
22. Farberow NL, Simon MD. Suicide in Los Angeles and Vienna: an intercultural study of two cities. *Public Health Rep* 1969;84:389–403.
23. Roy A. Genetics of suicide. *Ann N Y Acad Sci* 1986;487: 97–105.
24. Crawley JN, Sutton ME, Pickar D. Animal models of self-destructive behavior and suicide. *Psychiatr Clin North Am* 1985;8:299–310.
25. Åsberg M, Thoren P, Traskman L, et al. "Serotonin depression": a biochemical subgroup within the affective disorders? *Science* 1976;191:478–480.
26. Åsberg M, Nordstrom P, Traskman-Bendz L. Biological factors in suicide. In: Roy A, ed. *Suicide.* Baltimore: Williams & Wilkins, 1986:47–71.
27. Banki CM, Arato M, Papp Z, et al. Biochemical markers in suicidal patients: investigations with cerebrospinal fluid amine metabolites and neuroendocrine tests. *J Affect Disord* 1984;6:341–350.
28. Linnoila M, Virkkunen M, Scheinin M, Nuutilla A, Rimon R, Goodwin FK. Low cerebrospinal fluid 5-hydroxyindoleacetic acid concentration differentiates impulsive from non-impulsive violent behavior. *Life Sci* 1983;33:2609–2614.
29. Roy A, Karoum F, Pollack S. Marked reduction in indexes of dopamine metabolism among patients with depression who attempt suicide. *Arch Gen Psychiatry* 1992;49;447–450
30. Roy A, Pickar D, DeJong J, Karoum F, Linnoila M. Suicidal behavior in depression: relationship to noradrenergic function. *Biol Psychiatry* 1989;25:341–350.
31. Gross-Isseroff R, Dillon KA, Fieldust SJ, Biegon A. Autoradiographic analysis of α_1-noradrenergic receptors in the human brain postmortem: effect of suicide. *Arch Gen Psychiatry* 1990;47:1049–1053.
32. Linkowski P, VanWettere JP, Kerkhofs M, Gregoire F, Brauman H, Mendlewicz J. Violent suicidal behavior and the thyrotropin-releasing hormone-thyroid-stimulating hormone test: a clinical outcome study. *Neuropsychobiology* 1984;12:19–22.
33. Bunney WE, Fawcett JA, Davis JM, Gifford S. Further evaluation of urinary 17-hydroxy-corticosteroid in suicidal patients. *Arch Gen Psychiatry* 1969;21:138–150.
34. Krieger G. The plasma level of cortisol as a predictor of suicide. *Dis Nerv Syst* 1979;35:237–240.
35. Coryell W, Schlesser MA. Suicide and the dexamethasone suppression test in unipolar depression. *Am J Psychiatry* 1981;138:1120–1121.
36. Rich CL, Young D, Fowler RC. San Diego suicide study. I. Young vs. old subjects. *Arch Gen Psychiatry* 1986;43: 577–582.
37. Miles CP. Conditions predisposing to suicide: a review. *J Nerv Ment Dis* 1977;164:231–246.
38. Barraclough BM, Pallis DJ. Depression followed by suicide: a comparison of depressed suicides with living depressives. *Psychol Med* 1975;5:55–61.
39. Caldwell CB, Gottesman II. Schizophrenia—a high risk factor for suicide: clues to risk reduction. *Suicide Life Threat Behav* 1992;22:479–493.
40. Regier DA, Farmer ME, Rae DS, et al. Co-morbidity of mental disorders with alcohol and other drug abuse. *JAMA* 1990;264:2511–2518.
41. Farrer LA. Suicide and attempted suicide in Huntington disease: implications for preclinical testing of persons at risk. *Am J Med Genet* 1986;24:305–311.
42. Kessler S, Field T, Worth L, Mosbarger H. Attitudes of persons at risk for Huntington disease toward predictive testing. *Am J Med Genet* 1987;26:259–270.
43. Stenager EN, Stenager E. Suicide and patients with neurologic diseases. *Arch Neurol* 1992;49:1296–1303.
44. Mendez MF, Lanska DJ, Manon-Espaillat R, Burnstine TH. Causative factors for suicide attempts by overdose in epileptics. *Arch Neurol* 1989;46:1065–1068.

45. White SJ, McLean AEM, Howland C. Anticonvulsant drugs and cancer: a cohort study in patients with severe epilepsy. *Lancet* 1979;2:458–461.
46. DeVivo MJ, Black KJ, Richards JS, Stover SL. Suicide following spinal cord injury. *Paraplegia* 1991;29: 620–627.
47. Blumenthal SJ, Kupfer DJ, eds. *Suicide over the life cycle: risk factors, assessment, and treatment of suicidal patients,* 1st ed. Washington, DC: American Psychiatric Press, 1990.
48. Crow TJ, Cross AJ, Cooper SJ, et al. Neurotransmitter receptors and monoamine metabolites in the brains of patients with Alzheimer-type dementia and depression, and suicides. *Neuropharmacology* 1984;23(12B):1561–1569.
49. Bourne HR, Bunney WE Jr, Colburn RW, et al. Noradrenaline, 5-hydroxytryptamine, and 5-hydroxyindoleacetic acid in hindbrains of suicidal patients. *Lancet* 1968;2 (7572):805–808.
50. Cochran E, Robins E, Grote S. Regional serotonin levels in brain: a comparison of depressive suicides and alcoholic suicides with controls. *Biol Psychiatry* 1976;11(3): 283–294.
51. Beskow J, Gottfries CG, Roos BE, Winblad B. Determination of monoamine and monoamine metabolites in the human brain: post mortem studies in a group of suicides and in a control group. *Acta Psychiatr Scand* 1976; 53(1):7–20.
52. Shaw DM, Camps FE, Eccleston EG. 5-Hydroxytryptamine in the hind-brain of depressive suicides. *Br J Psychiatry* 1967;113(505):1407–1411.
53. Cheetham SC, Crompton MR, Czudek C, Horton RW, Katona CL, Reynolds GP. Serotonin concentrations and turnover in brains of depressed suicides. *Brain Res* 1989; 502(2):332–340.
54. Pare CM, Yeung DP, Price K, Stacey RS. 5-hydroxytryptamine, noradrenaline, and dopamine in brainstem, hypothalamus, and caudate nucleus of controls and of patients committing suicide by coal-gas poisoning. *Lancet* 1969; 2(7612):133–135.
55. Owen F, Chambers DR, Cooper SJ, et al. Serotonergic mechanisms in brains of suicide victims. *Brain Res* 1986; 362(1):185–188.
56. Owen F, Cross AJ, Crow TJ, et al. Brain 5-HT-2 receptors and suicide. *Lancet* 1983;2(8361):1256.
57. Lloyd KG, Farley IJ, Deck JH, Hornykiewicz O. Serotonin and 5-hydroxyindleacetic acid in discrete areas of the brainstem of suicide victims and control patients. *Adv Biochem Psychopharmacol* 1974;11(0):387–397.
58. Korpi ER, Kleinman JE, Goodman SI, et al. Serotonin and 5-hydroxyindleacetic acid in brains of suicide victims. Comparison in chronic schizophrenic patients with suicide as cause of death. *Arch Gen Psychiatry* 1986;43 (6):594–600.
59. Arato M, Tekes K, Tothfalusi L, et al. Serotonergic split brain and suicide. *Psychiatry Res* 1987;21(4):355–356.
60. Gross-Isseroff R, Israeli M, Biegon A. Autoradiographic analysis of tritiated imipramine binding in the human brain post mortem: effects of suicide. *Arch Gen Psychiatry* 1989;46(3):237–241.
61. Lawrence KM, De Paermentier F, Cheetham SC, Crompton MR, Katona CL, Horton RW. Brain 5-HT uptake sites, labelled with [3H]paroxetine, in antidepressant-free depressed suicides. *Brain Res* 1990;526(1):17–22.
62. Lawrence KM, De Paermentier F, Cheetham SC, Crompton MR, Katona, CL Horton RW. Symmetrical hemispheric distribution of 3H-paroxetine binding sites in postmortem human brain from controls and suicides. *Biol Psychiatry* 1990;28(6):544–546.
63. Meyerson LR, Wennogle LP, Abel MS, et al. Human brain receptor alterations in suicide victims. *Pharmacol Biochem Behav* 1982;17(1):159–163.
64. Stanley M, Virgilio J, Gershon S. Tritiated imipramine binding sites are decreased in the frontal cortex of suicides. *Science* 1982;216(4552):1337–1339.
65. Arora RC, Meltzer HY. Laterality and 3H-imipramine binding: studies in the frontal cortex of normal controls and suicide victims. *Biol Psychiatry* 1991;29(10): 1016–1022.
66. Arora RC, Meltzer HY. 3H-imipramine binding in the frontal cortex of suicides. *Psychiatry Res* 1989;30(2): 125–135.
67. Arato M, Tekes K, Tothfalusi L, et al. Reversed hemispheric asymmetry of imipramine binding in suicide victims. *Biol Psychiatry* 1991;29(7):699–702.
68. Mann JJ, Stanley M, McBride PA, McEwen BS. Increased serotonin2 and beta-adrenergic receptor binding in the frontal cortices of suicide victims. *Arch Gen Psychiatry* 1986;43(10):954–959.
69. Arora RC, Meltzer HY. Serotonergic measures in the brains of suicide victims: 5-HT2 binding sites in the frontal cortex of suicide victims and control subjects. *Am J Psychiatry* 1989;146(6):730–736.
70. Gross-Isseroff R, Salama D, Israeli M, Biegon A. Autoradiographic analysis of [3H]ketanserin binding in the human brain postmortem: effect of suicide. *Brain Res* 1990;507(2):208–215.
71. Stanley M, Mann JJ. Increased serotonin-2 binding sites in frontal cortex of suicide victims. *Lancet* 1983;1(8318): 214–216.
72. Arango V, Underwood MD, Mann JJ. Alterations in monoamine receptors in the brain of suicide victims. *J Clin Psychopharmacol* 1992;12(2 Suppl):8S–12S.
73. Arango V, Ernsberger P, Marzuk PM, et al. Autoradiographic demonstration of increased serotonin 5-HT2 and beta-adrenergic receptor binding sites in the brain of suicide victims. *Arch Gen Psychieatry* 1990;47(11): 1038–1047.
74. Dillon KA, Gross-Isseroff R, Israeli M, Biegon A. Autoradiographic analysis of serotonin 5-HT1A receptor binding in the human brain postmortem: effects of age and alcohol. *Brain Res* 1991;554(1-2):56–64.
75. Matsubara S, Arora RC, Meltzer HY. Serotonergic measures in suicide brain: 5-HT1A binding sites in frontal cortex of suicide victims. *J Neural Transm Gen Sect* 1991;85(3):181–194.
76. Cheetham SC, Crompton MR, Katona CL, Horton RW. Brain 5-HT1 binding sites in depressed suicides. *Psychopharmacology (Berl)* 1990;102(4):544–548.

17
Psychosis

D. Frank Benson, D. Gregory Gorman, Dilip V. Jeste, Douglas Galasko, Jody Corey-Bloom, Stanley Walens, and Eric Granholm

OVERVIEW

Premier among the clinical findings indicative of psychiatric disorder, delusions and hallucinations are not only the most dramatic but also the most visible. For most psychiatrists, a delusion establishes the presence of psychosis, and for almost all physicians, the presence of either delusions or hallucinations implies psychiatric disorder. Interpretations of hallucinatory experiences and delusional beliefs have varied with the wide shifts in accepted psychiatric reasoning over the years but have almost always been accepted as prime examples of "functional" (nonorganic) mental disorder (1). Schizophrenia remains the architypal psychotic behavioral illness and is discussed in a separate section.

DELUSIONS VERSUS HALLUCINATIONS

In both delusions and hallucinations, the fundamental component concerns incorrect interpretation. Both can vary in the tenacity with which the incorrect impression is held. Variations in content and degree have led to numerous categorizations and subclassifications. A sensory experience without external stimulation and firmly accepted as real is considered pathologic; a similar mental experience recognized as unreal is normal. A hallucinatory experience may appear real at the time and only in retrospect be recognized as unreal. Determining whether the experience is actually hallucinatory can be difficult, at times almost impossible.

Hallucinations are distinguished from *illusions*, which are actual external stimuli, perceived but misinterpreted. In contrast, the hallucinatory experience occurs in the absence of relevant external stimulus. *Pseudohallucinations* are partial hallucinatory states in which clear, vivid images are perceived and misinterpreted but are recognized as unreal. *Hallucinosis*, an ongoing series of hallucinatory experiences, is most often linked to toxic brain disorders, has an acute onset, and tends to persist or recur. Although often considered distinct, hallucinations and hallucinosis differ mostly in degree. Although distinct by definition, hallucinatory experiences and delusional thinking are often intertwined.

Almost all definitions of *delusion* emphasize two points: the presence of a false belief and the persistent, unshakable acceptance of the false belief. One basic definition of delusion appears in The American Psychiatric Diagnostic and Statistical Manual of Mental Disorders, 4th ed. (Washington: American Psychiatric Association, 1994):

> . . . a false personal belief based on incorrect inference about external reality and firmly sustained in spite of what almost everyone else believes and in spite of what constitutes incontrovertible and obvious proof or evidence to the contrary. The belief is not one ordinarily accepted by other members of the person's culture or subculture (i.e., it is not an article of religious faith).

A basic distinction between hallucinations and delusions concerns content. With hallucinatory experiences, sensory input is improperly processed; a delusion is an idea or belief—a complex thought content—that is abnormal. Delusions are almost universally categorized in terms of thought content, whereas the most consistently used classifications of hallucinations are based on the sensory modality involved.

CLINICAL SYNDROMES

HALLUCINATIONS

The most commonly used means to classify hallucinations is based on the misperceived sensory element. Division into visual, auditory, olfactory-gustatory, somesthetic (haptic), and vestibular categories has proved most practical for identifying types of hallucinatory experiences. In this chapter, discussion of hallucinatory experiences is based on the sensory modality, and when possible, the described entities start from external and move to central involvement.

Visual Hallucinations

The most readily demarcated and, therefore, most clearly categorized hallucinatory experiences are those involving the visual system. Disorder may occur at the end organ (retina) during sensory transmission or be based on abnormal cortical reception, perception, or interpretation.

Several purely ophthalmologic entities meet the definition of hallucination. Thus, *Moore's lightning streaks*, vertical bands of light that occur in the temporal visual fields during eye movement, represent true hallucinations; most often, they indicate vitreous detachment. *Ocular phosphenes*, the phenomenon produced by vigorously rubbing the eyes, are similar to Moore's lightning streaks; they can occur spontaneously in the aging eye or in cases of optic neuritis. *Auditory-visual synesthesias* are photisms induced by sound and can occur in some patients with ocular pathology.

After enucleation of one eye, hallucinations that are analogous to the phantom limb syndrome and appropriately called *phantom vision* can occur. A related phenomenon, *palinopsia* (the persistence of a visual image in a disordered or blind visual field) is well known. A single image (e.g., flower, face) can appear wherever the patient fixes a gaze and may persist from minutes to days. Palinopsia has been described in such diverse entities as nonketotic hyperglycemia, cocaine abuse, and occipital lobe seizures.

Charles Bonnet's syndrome has been used to describe certain hallucinations that are not frightening and occur without delusions in the elderly. Charles Bonnet was a Swiss naturalist who reported (in 1760) that his 89-year-old grandfather, blind because of cataracts, was entertained by complex visions but was fully cogent of their unreality and could still lecture in a clear, coherent fashion, an indication that he was not demented. Although the phenomenon is always related to ocular pathology, its occurrence in old age suggests that brain alterations plus ocular disorder may be essential to the phenomenon (2).

A better-known variety of visual hallucination, *hypnagogic hallucinations*, are vivid, often terrifying hallucinatory experiences that accompany or precede sleep paralysis and usually occur in people with narcolepsy. The hallucination may be auditory, vestibular, or somesthetic but most often features a strong visual component.

Peduncular hallucinations are dreamlike states in which small (lilliputian) people, animals, or objects carry out activities. They occur in people who have both vision and insight preserved. The patient's relationship with the hallucination is characteristic, almost diagnostic. At first, the hallucination may be troubling, but it then becomes interesting and pleasantly entertaining. Eventually, however, the patient becomes irritated by the persistence of the hallucinatory experience. Lhermitte originally described the syndrome and suggested that a structural change in the midbrain peduncular region was the source. Subsequent reports indicate that a variety of disorders affecting the upper midbrain can be associated with these characteristic hallucinatory experiences. Others suggest that decreased blood flow in the vessels feeding the occipital and medial temporal regions is the true source of these hallucinations. None of the localization theories has been substantiated (3).

Visual release hallucinations have traditionally been associated with visual field defect. Visual release hallucinations tend to last minutes or hours, are variable in content, and can be modified by altering visual input (e.g., opening or closing the eye). In a broad sense, all of the visual hallucinatory phenomena described thus far can be considered examples of release hallucinations.

Visual ictal hallucinations tend to be stereotyped in content and are often geometric in design or produce a recollection of previously experienced visual images. They are not confined to any single portion of the visual field. They may be experienced as auras or distortions (e.g., micropsia, macropsia, or metamorphopsia) and can be artificially produced by direct brain stimulation. Although these experiences are most commonly related to epilepsy, similar hallucinatory experiences have been described with ecstasy and migraine.

Disorders Associated With Visual Hallucinations

Disorders that decrease or alter the level of conscious awareness have been associated with visual hallucinations. About one third of patients who suffer acute confusion (delirium) report hallucinatory experiences; most of the hallucinations seen in acute confusion are visual and increase with diminished visual stimulation ("sundowning"). Sleep deprivation and narcolepsy are often associated with visual hallucinations, and a vast array of drugs can produce visual hallucinatory experiences. Three groups of drugs are particularly likely to produce visual hallucinations (4):

- Hallucinogens
- Sedative-hypnotic drugs
- Toxic quantities of certain other drugs

Hallucinogenic drugs, such as lysergic acid diethylamide (LSD), mescaline, or psilocybin, are defined by their ability to produce hallucinations. Drug-induced hallucinations, particularly from LSD, mescaline, and the like, can recur at a later, drug-free time; these experiences, called "flashbacks," may occur up to several years after LSD use. Many other substances can produce hallucinations, including drugs such as the bromides, most of the narcotic substances, levodopa, barbiturates, benzodiazepines, sedative-hypnotic drugs, and substances of abuse such as alcohol. Most of these tend to produce hallucinations during periods of withdrawal. Any type of hallucination may be produced, but visual experiences predominate.

Visual hallucinations occur in many traditional psychiatric disorders. The visual hallucinations associated with psychiatric disorders are often complex, may be enhanced by auditory misperceptions, and often lead to delusional belief. Although auditory hallucinations are far more common in schizophrenic subjects, visual hallucinations are reported in a sizable number, particularly by those most severely afflicted.

It should be recognized that hallucinatory-like experiences also occur in healthy people. Thus, dreams, the transient mental (mostly visual) experiences that occur in the sleep state, are, by any definition, perceptions without immediate external stimulus. Daydreams and fantasies are similar hallucinatory experiences occurring during the waking state. In some societies, "visions" of this type are valued as a source of guidance and inspiration. Eidetic imagery, daydreams, the imaginary friends of childhood, sleep deprivation, narcolepsy, hypnosis, sensory deprivation, and intense emotional experiences (such as hostage situations) are all capable of producing visual hallucinations in the absence of either primary brain disorder or significant psychiatric disease.

Auditory Hallucinations

Tinnitus, the perception of ringing, whistling, buzzing, or even a drumbeat, represents an auditory experience without external stimulus. In some instances, tinnitus can be experienced as a complex series of sounds, a state resembling release hallucinations. When auditory input is decreased (total or partial deafness), this form of auditory hallucination becomes more notable and may resemble articulate speech; hallucinations of threatening or disparaging voices may be an important factor leading to the paranoid state of acquired deafness.

Palinacousis, the continued recurrence (echo) of a prior auditory perception, resembles palinopsia but is less common. Some cases have been related to temporal lobe seizures.

Musical hallucinations, a variation of release hallucinations featuring the perception of music without an external source, are associated with many disorders ranging from simple deafness to focal temporal lobe pathology. When based on brain pathology, musical hallucinations are more often correlated with right-hemisphere disorder.

Auditory hallucinations are commonly reported as *auras of epilepsy*. At times, focal pathology in the temporal lobe may be associated with sounds or voices that are independent of reality but occur without overt seizure activity.

Withdrawal states, particularly from alcohol, are well recognized as causes of auditory hallucinations and have been called *alcoholic hallucinosis*. The hallucinations most often feature voices that are self-deprecating or threatening. They tend to affect older people with a long record of heavy alcohol consumption.

A variety of psychiatric disorders are known to cause auditory hallucinations. Auditory hallucinations due to primary mental illness are often difficult for the patient to localize in space (often sensed as occurring inside the head), and the gender and age of the voice are often vague. The voice most often comments on the patient's behavior or echoes the patient's thoughts. Although auditory hallucinations are often considered an indication of primary mental illness, the possibility of a secondary psychosis always deserves consideration. A significant number of patients with ictal psychosis present schneiderian first-rank symptoms, suggesting schizophrenia. Psychiatric conditions other than schizophrenia may also produce auditory hallucinations. Psychotically depressed or manic patients often have periods in which voices speak to them. Patients with multiple personality disorder report auditory hallucinations at times, leading to misinterpretation of the disorder as schizophrenia (5).

Olfactory and Gustatory Hallucinations

Disorders of smell and taste are difficult to distinguish from hallucinatory experiences. Thus, apparent hallucinations of smell or taste can be the result of damage to the olfactory bulb, and at times, the symptom can be "cured" by surgical resection of the olfactory bulb. Tumors at the base of the brain that extend into the olfactory apparatus can produce apparent olfactory hallucinations. Migraine patients occasionally describe auras of olfactory or gustatory type.

Patients with complex partial seizures commonly report indescribable odors or tastes, traditionally called *uncinate fits*. A seizure focus, if present, is most often localized in the insula or opercular area, but olfactory hallucinations have been reported in patients with parietal and temporal area lesions and even with a colloid cyst of the third ventricle. Olfactory hallucinations have been described by some demented patients, particularly those with Alzheimer's disease. Olfactory hallucinations may occur in drug abuse situations. Among patients with primary mental illness, olfactory or gustatory hallucinations are most commonly seen in those who present a delusion of being poisoned.

The complaint that one's body is malodorous and socially offensive, the *olfactory reference syndrome*, can be seen in severe depressive states; although this represents a symptom based on involvement of a single sensory system, it is usually accompanied by other incorrect perceptions, often systematized, and is thus classed as a delusion.

Haptic and Pain Hallucinations

Somesthetic (haptic) hallucinations include both unusual body experiences and pain. After limb amputation, a hallucination of *phantom limb* is almost universal, and pain in the nonexisting limb is present in as many as 85% of amputation patients. Similar phantom hallucinatory experiences are seen after amputation of breasts, genitals, or other body parts and are commonly reported to affect the body in areas below spinal cord transection. Over time, the phantom hallucination changes, most often perceived as a shrinking in size of the affected body area. Even children born without limbs have phantom experiences, and it has been suggested that the brain contains a genetically inherited template of the body image that is the source of these haptic hallucinations (6).

Hallucinatory duplication of a limb or other body segment or a sensation of distorted body shape may occur with migraine, as an epileptic aura, with toxic encephalopathy, or after use of hallucinogenic drugs.

Formication hallucinations, the feeling that bugs are crawling on the skin, occur in a variety of neurologic and psychiatric conditions. They are common in alcohol and drug withdrawal states and occur in a number of toxic and metabolic conditions. They are also common in schizophrenia and affective disorder. If the sensation is unilateral, formication hallucinations may indicate thalamic or parietal lesions.

Vestibular Hallucinations

Vestibular hallucinations, producing a feeling of giddiness or vertigo, are almost impossible to distinguish from primary vestibular disease. Better-defined experiences, such as sinking into a hard object, flying off through the air, walking on waves or cork, or perceiving the ground as rising, lowering, or turning, suggest that the experience is hallucinatory. Nonetheless, a careful search for primary vestibular pathology as the source of these experiences is indicated.

Haptic-vestibular hallucinations, in which the patient describes abnormal heaviness, lightness to the point of levitation, and similar feelings, have been reported in schizophrenia, drug withdrawal, and seizure disorders.

Synesthesia

Synesthesia is a term used to represent crossed modality hallucinations, for example, the experience of hearing a vision or seeing a sound. Taste may be geometrically structured, and smells may provoke a color identification. *Auditory-visual synesthesia* (sound-induced photisms) may occur as release hallucinations, particularly if there is a defect in the visual field. Synesthesia may occur in the presence of impaired primary sensory pathways, with brain-stem impingement, or with multimodal association area involvement.

DELUSIONS

Although no single classification of the varieties of delusion is satisfactory, most emphasize alterations of thought content. Nine of the most common and dramatically distinct types of delusion are presented here.

Delusions of Persecution

Commonly referred to as paranoid delusions, delusions of persecution are common, occur in a wide variety of neuropsychiatric disorders, and can take many different forms.

Delusions of Reference

One common presentation of the delusion of persecution features belief by patients that people are talking about, slandering, or spying on them. When associated with depression, this belief can become a delusion of guilt in which the patients insist that they are being persecuted, justifiably, on the basis of some previous or ongoing action. Delusions of reference are frequent in patients with paranoid schizophrenia, and similar symptoms also occur in a variety of organic conditions.

Delusions of Loss of Property

Some individuals are convinced that "they" are attempting to rob them, deprive them of an inheritance, or cheat them out of earnings from an invention, written material, or a business arrangement.

Delusions of Poison or Infection

Some patients, particularly those suffering morbid jealousy, develop a firm belief that they are being poisoned. These people tend to blame their current mental or physical problems on sinister alterations to their environment. Delusions of poisoning are often accompanied by hallucinations of smell or taste.

Delusions of Influence

Delusions of influence include feelings of passivity under the influence of hypnotism or control of the person's movements or thoughts through radio waves, atomic rays, radar, or other mysterious manipulations. Delusions of influence are most commonly seen in schizophrenic patients and have long been considered a significant diagnostic criterion for this diagnosis.

Delusions of Innocence

Opposite in nature but apparently related to the delusion of influence is the delusion of innocence, the firmly held belief in self-justification or acquittal noted in some people accused—or who believe themselves accused—of crime, cowardice, or other unacceptable behavior. The delusion of innocence appears to exist only in combination with a more primary delusion, almost always that of persecution.

Delusions of persecution occur in a wide and disparate variety of disorders. They make up the major, almost the defining, manifestation of most paranoid states, including paranoid schizophrenia, and are often present in patients suffering severe depression or other overwhelming breakdown of personal integrity. In addition, delusions of persecution occur in both acute and chronic organic mental states. Of these, the best known are the psychotic states produced by certain intoxications; amphetamine and cocaine, both of which tend to produce bizarre, complex delusions of persecution and fear, are the prime examples.

Hypochondriasis

Concern about bodily health can become sufficiently severe to be considered delusional. *Briquet's syndrome* is a true *hypochondriasis fantastica*; by definition, the patient with this syndrome presents, over a period of years, a series of physical complaints that involve several different physical systems (e.g., reproductive, nervous, cardiac, pulmonary). The fourth edition of the *Diagnostic and Statistical Manual of Mental Disorders* (DSM-4) simplified the clinical description and renamed the problem *somatization disorder*. To fulfill this definition, the patient, predominantly (but not exclusively) female, must have presented at least eight complaints before 30 years of age, for which no medical justification could be found.

Monosymptomatic Hypochondriasis

In some patients, only a single symptom, organ system, or body part becomes the source of a fixed delusion. The patient may complain that the stomach is being eaten away, that the brain is dissolving, that the body is infested by insects or worms, or that a blemish represents a severe, disfiguring anatomic defect (dysmorphophobia or body dysmorphic disorder). One of the most common variations concerns infestations with bugs, parasites, or other unseeable objects. Parasitosis, also called Ekbom's syndrome, is common in dermatologic practice. Monosymptomatic hypochondriacal delusions are relatively common and can be extremely difficult to treat. Body dysmorphic disorder, however, may respond to serotonin reuptake inhibitors.

Somatic Delusions

The most severe and bizarre examples of delusions of ill health, termed somatic delusions, are usually linked with serious mental disorders. The health problem is often fantastic, bizarre, and unbelievable and is tenaciously maintained against all efforts to reassure or disprove. Treatment of the delusion (rather than the somatic complaint) is actively resisted. Antipsychotic medications can be helpful in some cases, and if depression is the cause, electroshock therapy (ECT) may be of considerable help.

Delusions of Grandeur

Also known as *grandiose delusions, expansive delusions,* or *happiness psychosis,* delusions of grandeur are known to occur in a variety of disorders. The delusion is often supported by fantastic confabulations, detailed accounts of business or political accomplishments, great athletic feats, international honors, and so forth. At times, the degree of confabulation may be sufficient to warrant the term *fantastic hallucinosis.*

In the nineteenth century, delusions of grandeur were often described in the general paresis variant of tertiary neurosyphilis. Delusions of grandeur have been reported in primitive groups (e.g., cargo cults), but mania is by far the most common underlying cause of delusions of grandeur.

If treated, the delusion of grandeur is usually short-lived and without residuals. In some instances, however, the involved person may influence others and develop a following, with the potential of producing a political or cult organization; this scenario is especially likely to occur among primitive or oppressed people.

Treatment for delusions of grandeur is closely tied to the etiology. Penicillin and related antispirochetal agents are the treatment of choice for neurosyphilis; adequate treatment of the luetic infection is usually effective in controlling the delusions. Many treatments are suggested for mania and are reviewed elsewhere in this volume. In most instances, the delusional quality of the manic episode is either self-limited or comes under control with appropriate treatment.

Delusions of Poverty

The opposite of the grandiose delusion is an unreal belief of extreme poverty. The involved person feels depleted, impoverished, and, despite adequate finances or even considerable wealth, expresses a feeling of being destitute or in immediate danger of becoming so. Delusions of poverty tend to be associated with a broader, more pervasive nihilistic delusion, a general belief by the patient that he or she is of no value.

Delusions of poverty are most often seen in psychotic depression, and the course usually mirrors the response to the treatment of depression. Both medication and ECT have been successful. Although the delusion may disappear with appropriate therapy, an underlying feeling of monetary concern often persists as part of an ongoing depressive personality.

Delusions of Possession

The belief that one's body is possessed by God, by a mystic power figure, by the devil, by a lesser demon, or by an animal dates to antiquity. Serious mental illnesses, such as schizophrenia, severe depression, delirium, or dementia, can produce possession states. Organic factors are seldom suggested as the source of possession states.

Delusions of possession are difficult to treat. Understanding an individual case of demonic possession requires comprehension of the underlying cultural construct, as well as the patient's personal problems. Successful management involves culturally sensitive treatment of the patient's mental illness.

Delusions of Love

Erotic delusions are relatively uncommon. Pertinent literature is sparse, and categorization within the disorder remains difficult. Two documented variations—the phantom lover syndrome and erotomania (belief that one is loved by a powerful or prominent person)—have been detailed. Whether the object of the delusion is imagined (phantom lover) or a real person (as in de Clérambault's syndrome) is the distinguishing feature. In both, the predominant finding is a *feeling* by the patient, most often a female patient, that she is loved, admired, and adored by another person, often of a high social rank (e.g., royalty, movie star).

Management of an erotic delusion, either the phantom lover syndrome or de Clérambault's syndrome, is difficult and may actually be dangerous. Psychotropic medications or ECT can reduce the power of the delusion, occasionally bringing about a total cure but more often producing a state of loss. Careful, skillful psychotherapy is essential. A delusion of love is powerful, and attempts to remove the delusion demand extreme caution on the part of the treating physician.

Delusions of Jealousy

Delusions of jealousy, also known as *morbid jealousy* or *delusions of infidelity,* are fairly common and occur in a variety of organic and functional states. The patient is often said to have had a sensitive, suspicious, and mildly jealous nature before onset of the psychosis. Metabolic intoxications, particularly alcohol addiction, are common in people who describe morbid jealousy. During the course of the illness, a suspicious, insecure person becomes convinced of infidelity by his or her spouse or significant other. The severity of the delusion fluctuates; during psychotic episodes, the patient may question the spouse incessantly, keep him or her awake for hours at night, search his or her clothes for stains or body hairs, and proffer physical stigmata on the part of the spouse (e.g., insomnia, poor appetite, tired look, bags under the eyes) as "evidence" of infidelity. Except for the morbid jealousy, the patient often displays no symptoms suggestive of serious mental disorder.

Delusions of Reduplication

Reduplicative phenomena have been found to occur when frontal damage or dysfunction is present. They also suggested that, although frontal disorder alone was not sufficient to produce the phenomenon, it is an essential element for psychotically held reduplications, and the presence of frontal dysfunction in other psychotic syndromes is a currently popular concept (7).

Although not common in either psychiatric or neurologic practice, delusions of reduplication, including the delusion that a person has been replaced by an imposter or that a building or possession has been duplicated elsewhere, deserve consideration among the recognized delusions. Phenomenologically, reduplication syndromes are related, differing only in the object being duplicated. Four variations are described here.

Reduplicative Paramnesia

Originally described in brain-injured subjects, reduplicative paramnesia has always been considered a brain abnormality. The patient involved almost invariably recognizes and remembers the name of the hospital or institution but insists that it is located in a different community or area, usually closer to his or her own home. The patient may be normally competent in other cognitive functions (e.g., memory and visuospatial skills). The disorder is most commonly described as a stage of recovery from severe brain trauma. The cases reported since the availability of brain imaging have almost always had demonstrable frontal pathology.

Reduplication of Person

Reduplication of person has come to be known as *Capgras' syndrome* and is usually discussed as an unusual psychiatric problem. The patient with this disorder holds a strong, fixed belief that a person or persons (usually a family member or someone closely associated with the patient) has been replaced by an imposter of nearly identical physical appearance and possessing similar identifying features (e.g., same name, age, number of children, location of home). Despite constant assurances by the individual who is the object of the delusion and by multiple friends, acquaintances, and therapists, the patient remains convinced that the person is an imposter.

The bizarre nature and strongly fixed quality of the delusion has suggested schizophrenia, and most cases of Capgras' syndrome are seen and reported by psychiatrists. More recently, Capgras' syndrome has been described in patients with obvious organic brain disease.

A related but phenomenologically distinct disorder, the *delusion of de Fregoli* identifies a situation in which a specific person, usually a suspected tormentor, is believed by the patient to change into different people encountered by the patient. Many dissimilar individuals are misidentified by the patient as the specified person (the opposite of Capgras' phenomenon).

Misidentification of Objects

Some people, particularly those suffering persecutory feelings, insist that an object (e.g., their coat, their car, an object on their dressing table) has been moved or has been replaced with a similar item. Although relatively common as a symptom, particularly in paranoid populations, only rarely does this phenomenon reach sufficient severity to disable the patient and warrant classification as an individual delusion.

Treatment of Delusions

Treatment of delusions of persecution depends on the underlying etiology. Those produced by chemical intoxicants sometimes respond to simple withdrawal, but intractable persecutory delusions often remain after successful withdrawal from cocaine addiction. The long-lasting fixed delusions seen with structural brain disorder (e.g., frontal brain tumor, traumatic brain injury) and those associated with major psychiatric disorders, such as schizophrenia or depression, demand disease-specific treatment. High-potency neuroleptic medication or ECT may be indicated. Skillful management, including a secure environment (often in the hospital), carefully molded behavioral modification measures, and supportive psychotherapy are often necessary. Although intense treatment tends to be successful, the results are often transient.

Along with the altered ideas concerning etiology, treatment of the delusions of reduplication has changed over the years. Psychodynamically directed behavioral conditioning therapy was attempted, particularly in cases with Capgras' syndrome, but reduplicative phenomena have not, in general, been aided by psychotherapy. If an acute organic psychosyndrome is the cause (e.g., after ECT or recent brain trauma), the prognosis for eventual recovery is relatively good. If, however, a stable, noncorrectable organic problem (such as an old stroke or chronic residua of brain trauma) is the source of the reduplicative syndrome, the prognosis for recovery must be considered guarded. Indirect therapy aimed at symptom control, such as behavioral conditioning or psychotropic medication, becomes necessary.

THE SCHIZOPHRENIAS

The schizophrenias represent the dominant clinical problem among the psychotic disorders. Schizophrenia is a serious psychiatric illness that can involve massive disruptions of thinking, perception, emotions, and behavior. With a worldwide lifetime prevalence rate of 1%, schizophrenia is a major cause of long-term psychiatric disability.

The startling cognitive, affective, and conative dysfunctions in schizophrenia present clinicians and theoreticians with a conundrum whose understanding is essential to the understanding of many other psychopathologic disorders. Yet despite the amount of research that has been devoted to schizophrenia, it remains an enigmatic condition, and its etiology and pathophysiology remain obscure. Although schizophrenia has often been conceptualized as a monadic entity, current theories have moved toward viewing schizophrenia as a collection of etiologically disparate disorders with similar clinical presentations—as a syndrome with heterogeneous symptoms.

Current approaches to schizophrenia acknowledge it to be essentially a neurobiologic disorder and have focused on neurophysiologic approaches to its pathogenesis; however, schizophrenia presents no consistent or gross neuropathology, unlike Alzheimer's disease, and the localization and nature of the lesions that "produce" schizophrenia remain unknown. It is likely that schizophrenia is not a single disease but a syndrome with heterogeneous manifestations, implicating a diffuse and diverse neuropathology (8).

Neuropsychology

Several extensive reviews have described numerous neuropsychological deficits in patients with schizophrenia. The most consistently replicated deficits in schizophrenia patients have been observed on tasks measuring attention and information processing, learning and memory, and executive (i.e., frontal systems) functions. Findings of impairment in basic language and visuospatial functions have been inconsistent, and deficits are typically reported on complex tasks that make significant demands on attention (e.g., distraction) or executive (e.g., planning) functions. This pattern of neuropsychological impairments in patients with schizophrenia has led to the hypothesis that specific frontal-subcortical brain systems are involved in the pathophysiology of schizophrenia (9).

Attention and Information Processing

The extensive literature on the performance of schizophrenia patients on these tasks has shown deficits, which are most apparent when the processing demands (workload) of the tasks are increased by increasing the number or complexity of cognitive operations that must be completed at a given moment. One of the most consistently replicated impairments in this domain in schizophrenia patients has been found on reaction time tasks. Schizophrenia patients show slower reaction times relative to healthy subjects on these tasks that require a rapid response to a single imperative stimulus, for example, a tone.

Studies using dichotic listening tasks have also shown that schizophrenia patients are abnormally susceptible to the effects of distraction. On these tasks, subjects are asked to listen to a message and repeat (shadow) it out loud, while ignoring an irrelevant message played simultaneously either in a different ear or in a different voice.

Impairments on some of these attentional and information-processing tasks have been observed in schizophrenic subjects in a remitted state, in nonpsychotic biologic mothers of schizophrenia patients, in foster children whose biologic mothers have schizophrenia, and in individuals in the general population with schizotypal or "psychosis-prone" characteristics. These findings of impairment in individuals across the schizophrenia spectrum suggest that some of these tasks are sensitive, not only to the psychotic state, but also to more subtle "core" deficits of schizophrenia that may mark a genetic vulnerability and possibly a genetically transmitted pathophysiology of schizophrenia (10).

The brain systems that govern the availability of processing resources are thought to involve frontal-basal ganglia-thalamic-reticular circuits. The frontal lobes modulate sensory input to cortical resource pools through connections with the nucleus reticularis thalami and modulate activation of cortical resource pools in accordance with task demands through connections (directly or through the basal ganglia) with the midbrain reticular formation. Reduced cortical resource availability in patients with schizophrenia may result from a failure to mobilize appropriate cortical resources by midbrain reticular centers, or from a failure to gate information flow through the thalamus to the cortex.

Learning and Memory

A common finding is that recall, but not recognition, performance is deficient in patients with schizophrenia. Intact recognition is, however, not universally found in schizophrenia patients. Unlike normal controls, schizophrenia patients do not make normal use of semantic categorical clustering of word lists. Memory for visual stimuli has received less attention in neuropsychological studies of schizophrenia, but these patients typically show comparable levels of impairment on verbal and nonverbal memory tasks (11).

The learning and memory deficits in schizophrenia appear to be related to a failure to use spontaneously contextual cues (e.g., semantic categories) and strategic processes to organize their encoding and retrieval of information. Because patients with frontal lobe dysfunction show similar impairments in the spontaneous use of strategic mnemonic processes and patients with temporal-hippocampal dysfunction show similar impairment in the encoding of semantic information, the memory impairments in schizophrenia have been interpreted as being consistent with frontal or temporal-limbic abnormalities.

Executive Functions

Executive functions are those neuropsychological processes important for adapting to the environment, such as preparation, initiation, and modulation of action; maintenance of arousal; cognitive set maintenance and set shifting; abstract reasoning; hypothesis testing; and monitoring of ongoing purposeful behavior. The executive function deficits shown by schizophrenia patients include sorting fewer categories and making more perseverative errors than healthy controls on the Wisconsin Card Sorting Task (WCST); increased errors on the Halstead Category Task; reduced verbal fluency on Benton's Controlled Oral Word Association Task; psychomotor slowing on the Trail Making Test—Part B; and impaired memory for the temporal ordering of information. These executive impairments are consistent with frontal system dysfunction in schizophrenia patients.

Structural Neuroimaging

One of the most consistent findings in the myriad of computed tomography (CT) and magnetic resonance imaging (MRI) studies of schizophrenia has been that of enlarged ventricular spaces. It remains unclear, however, what the significance of the enlarged ventricles is to the pathophysiology of the disease and whether the ventricular enlargement reflects tissue loss. Most CT and many early MRI studies of schizophrenia subjects have relied on clinical ratings, essentially visual inspection, or linear and area measurements of selected sections. A new generation of MRI techniques has recently made it possible to quantify the volume of both whole brain and specific small brain structures.

In MRI studies of schizophrenia, the most frequently described morphologic abnormalities have involved the temporal lobes, frontal lobes, and basal ganglia; however, the pattern and magnitude of findings varies from one study to another. A number of recent MRI studies have reported reductions in temporal lobe volumes in chronic schizophrenia patients as compared with controls, especially on the left side. One study described decreased gray matter volume in the left hippocampus, left parahippocampal gyrus, and left superior temporal gyrus without a decrease in overall left temporal lobe volume (12,13).

The basal ganglia have also been implicated in the pathophysiology of schizophrenia. Larger left caudate volumes have been observed by a number of investigators in patients with chronic schizophrenia. Abnormalities of the corpus callosum plus adjacent septum pellucidum and decreased thalamic volumes have also been described in MRI studies of schizophrenia patients (14).

Neuropathology and Neurochemistry of Schizophrenia

Three sets of studies that controlled for variables such as age, gender, weight, and height found that the overall brain weight was decreased in schizophrenia patients. The extent of the decrease was modest but consistent, about 5% to 8% (15).

In agreement with neuroimaging studies on living patients, autopsy studies on the brains of schizophrenia patients have generally shown ventricular enlargement. This selective enlargement suggests that structures adjacent to the temporal horn bear the brunt of the pathology, including the temporal cortex, hippocampus, and amygdala. Planimetric studies have shown decreased volumes or areas on cross-section of temporal lobe structures, notably the hippocampus, parahippocampal gyrus, amygdala, and inner pallidal segment.

Recent histologic studies have paid special attention to matching cases with controls, studying brains "blind" to clinical diagnoses, and using computer-assisted cell-counting methods of serial thin sections. Several investigators have reported decreased numbers and disordered arrangement of neurons in temporal lobe structures—especially the parahippocampal gyrus, where neuron loss as high as 20% has been reported. Other studies have found a more modest degree of neuron loss in several layers of the prefrontal and cingulate cortex and, to a lesser extent, in other brain areas. The balance of evidence from these controlled studies supports the presence of nonspecific pathology in temporolimbic areas in the brains of schizophrenia patients (16). The severity of neuron loss or other changes does not appear to increase in proportion to the duration of illness. This implies that the pathology is established at the time of initial clinical presentation and does not necessarily progress. The concept of schizophrenia as a neurodevelopmental lesion that manifests clinically in adolescence is thus supported.

Many studies have consistently reported that inflammation or gliosis does not accompany neuronal abnormalities or atrophy in the brains of schizophrenia patients. This further supports the notion that the pathology likely reflects a developmental event, occurring in utero, because acquired lesions generally elicit a glial response.

The neurochemical basis of schizophrenic symptoms has been a separate focus of inquiry. Until recently, the most accepted explanation was the dopamine hypothesis, which attributed the illness to an excess of dopaminergic neurotransmission. The dopamine hypothesis derived from two major findings. First, neuroleptics, which block D_2 receptors, improve the positive symptoms of schizophrenia. Their clinical efficacy is roughly proportional to their affinity for D_2 receptors. Second, amphetamine and other aminergic agents, which increase the activity of transmission of dopamine and other catecholamines, can provoke a psychotic state or worsen the symptoms of some schizophrenia patients, especially those who are more resistant to treatment.

The technique of radioligand receptor-binding studies was used to examine the distribution and levels of dopamine receptors in schizophrenia. Findings of increased D_2 receptors in the caudate and nucleus accumbens, targets of mesolimbic catecholaminergic projections, have been attributed to the effect of treatment with neuroleptics rather than to the disease itself. Nonetheless, neuroleptic-treated patients with Alzheimer's or Huntington's disease show lesser degrees of increased D_2-receptor binding, implying indirectly that medications do not fully account for the increased receptor binding found in schizophrenia.

Damage to the ventral tegmental area system may potentially explain hypofrontality, thought to be an important basis of the deficit syndrome of schizophrenia. Thus, frontal hypodopaminergic function accompanied by episodic hyperdopaminergic function in temporolimbic areas may account more closely for the clinical features of schizophrenia. The atypical antipsychotics, such as risperidone and clozapine, may shed light on other important neurochemical pathways in schizophrenia. These drugs block serotonin *and* dopamine receptors, control both positive and negative symptoms of schizophrenia, and are less likely to induce extrapyramidal side effects. Clozapine binds strongly to D_4 receptors but poorly to other subtypes, whereas typical neuroleptics show relatively greater binding to D_2 than to D_4 receptors (17).

CONTEMPORARY ISSUES

Functional Neuroimaging

Numerous studies have demonstrated abnormalities of cerebral perfusion and glucose metabolism in patients with schizophrenia. Initial efforts focused on the measurement of either perfusion or glucose use during the "resting state." There is, however, substantial variance in the resting state among individual patients. Therefore, more recent studies have emphasized the use of paired examinations of control conditions and cognitive challenges in the same patient.

Thus far, the data that have emerged from functional brain imaging studies have generated a number of important hypotheses about the neural basis of schizophrenia. Frontal, temporal, and basal ganglia dysfunction have all been demonstrated.

Beginning with the seminal studies of Ingvar and Franzen using intracarotid ^{133}Xe regional cerebral blood flow, numerous functional imaging studies have reported diminished cerebral activity in the frontal lobes of schizophrenia patients. A substantial literature now exists that has addressed the hypofrontality issue. A variety of potential confounders, including use of neuroleptics, chronicity of illness, clinical presentation, and nature of the cognitive task during functional imaging, appear to be important factors. The effects of neuroleptics and other medications on frontal lobe metabolism, for example, are not yet clear. Evaluation of metabolic activity in the frontal lobes and other brain structures is best ascertained in patients who have never previously been treated with neuroleptics. Large sample studies of neuroleptic-naïve patients are rare, however. A recent study indicated that clozapine tended to normalize asymmetry but exaggerated hypofrontality (18). Despite all of the uncertainties and variability among studies, results of many functional evaluations to date suggest that there may be a disturbance in frontal lobe function in schizophrenia. A number of two-dimensional regional cerebral blood flow studies have reported hypofrontality in both medicated and drug-free schizophrenia patients.

Since the time of Kraepelin and Bleuler, many researchers have viewed schizophrenia as resulting from frontal lobe pathophysiology. The modern version of this hypothesis considers the frontal cortex in the context of its subcortical connections with basal ganglia, thalamus, and reticular structures. Some investigators have suggested that dysfunction in the flow of information through frontostriatal cognitive and motor "loops" is involved in schizophrenia. Others have proposed a diffuse lesion in schizophrenia patients involving the periventricular limbic and diencephalic nuclei and connections between these structures and the dorsolateral prefrontal cortex, which occurs early in development but results in symptoms later when the brain areas involved reach maturity. There may be a negative feedback loop involving frontal, striatal, thalamic, and reticular structures, which together modulate cortical sensory input (through thalamic nucleus reticularis sensory gating) and cortical tone (through arousal centers in the midbrain reticular formation). Disruption in the balance of glutamate and dopamine in this system may lead to inability to modulate the influx and processing of information, which may lead to psychosis. Andreasen and colleagues also hypothesized that deficient gating of sensory input due to an abnormality in the thalamus and related circuitry could explain the symptoms of schizophrenia. For a discussion of the background issues, see references 19 to 21.

It is now possible to examine neuroreceptors in vivo in patients with schizophrenia, using radioligands for the dopamine receptor with positron emission tomography and single photon emission computed tomography technology. This is an exciting and important application of functional brain imaging because alterations in dopamine receptors have been suggested in this illness. The nature and pattern of dopamine receptor abnormalities in the striatum of schizophrenic patients is an area of active investigation.

Summary

Neuroimaging techniques have provided unique opportunities for exploring brain anatomy and function in schizophrenia; however, the precise morphologic and physiologic substrates of this disabling disorder remain ambiguous. Evidence suggests that the frontal lobes, temporal lobe structures, basal ganglia, and thalamus, either alone or in concert, may be involved in the pathophysiology of schizophrenia. As the technology improves and measurements are standardized and refined, these techniques will likely become an even more powerful aid in understanding the primary brain lesions responsible for producing the symptoms and signs of schizophrenia. Future studies should involve combined structural and functional modalities in large cohorts of well-characterized and carefully selected subjects and appropriate controls.

Treatment

Neuroleptic medications have been shown to be the most effective treatment modality for schizophrenia in general. In a significant proportion of schizophrenia patients, many of the positive symptoms, such as hallucinations and delusions, and some negative symptoms, such as apathy and social withdrawal, of the disorder can be brought under control by the use of neuroleptic drugs. In about two thirds of patients with acute schizophrenia, neuroleptic medication can reduce the psychotic symptoms, sometimes within a matter of weeks. In addition, a maintenance program of continuous neuroleptic medication can aid in preventing psychotic relapses in remitted patients. Some cases are treatment resistant; there is little evidence that intensive treatment with massive prolonged doses of neuroleptics has any beneficial effect in these patients. Atypical antipsychotics, such as clozapine, should be considered in these patients.

Neuroleptic drugs carry a significant risk for side effects, including but not limited to sedation, hypotension, dryness of the mouth, blurred vision, tachycardia, cardiac effects, amenorrhea, galactorrhea, hyperpyrexia, pigmentary retinopathy, allergic reactions, and seizures. Some patients experience extrapyramidal side effects, including dystonia, akathisia, and parkinsonism, which may respond to anticholinergic medication. Prolonged use of neuroleptics can produce Tardine dyskinesia (TD), and a small number of patients may develop neuroleptic malignant syndrome, a potentially fatal condition.

Given the potential side effects of neuroleptic drugs, the current conventional clinical strategy is to minimize the total lifetime amount of neuroleptic medication administered to a patient by prescribing the lowest effective dosage and avoiding unnecessary use of neuroleptics. Because commonly prescribed typical or conventional neuroleptic medications appear to be equally efficacious in equivalent doses in controlling the active symptoms of schizophrenia, selection of a specific neuroleptic drug in a given instance is often dictated by careful consideration of the drug's side effects and the patient's particular medical condition. Atypical antipsychotics, such as clozapine, risperidone, olanzapine, and quetiapine, are useful in controlling the symptoms of schizophrenia and may carry a reduced risk for TD. A unique feature of these drugs, which block serotonin and dopamine receptors, is their efficacy against negative symptoms of schizophrenia. A trial of an atypical neuroleptic is indicated in patients who are unresponsive to treatment with typical neuroleptics.

Pharmacotherapy must be supplemented by psychosocial support. Successful treatment requires that each patient be engaged in a therapeutic process fashioned to his or her personal situation and directed toward providing the patient with an expanded repertoire of interpersonal skills. The therapy should involve behavior modification designed to help the patient to develop necessary social or other skills. A trusting therapist—patient relationship can facilitate the success of other modes of treatment, providing a secure environment in which the patient can learn and practice strategies of interpersonal relationship and communication.

SECONDARY PSYCHOSES AND GROSS BRAIN DISEASES ASSOCIATED WITH SCHIZOPHRENIC SYMPTOMS

Typically, in secondary psychoses (psychoses secondary to general medical conditions or focal brain lesions), fragments of the schizophrenic syndrome, especially delusions, occur rather than the full repertoire of positive and negative symptoms. The relationship between psychotic symptoms and brain lesions is discussed later, with conditions being grouped as neurodegenerative diseases, infections, focal lesions, and metabolic or toxic encephalopathies. For more complete lists of disorders reported in combination with delusions or hallucinations, see references 22 and 23.

Neurodegenerative Diseases

In dementing conditions, delusions and hallucinations may be prominent. In Alzheimer's disease, the prototypical cortical dementia, about 30% to 40% of patients have delusions at some point during the illness. Usually, the delusions in Alzheimer's disease are simpler than those in schizophrenia and are not systematized. They often have a paranoid quality and may be related to memory impairment. Psychotic symptoms occur less often in Pick's disease and Creutzfeldt-Jakob disease.

Idiopathic Parkinson's disease is associated with psychotic symptoms in two situations: in patients who develop dementia, usually late in the course of the illness, and as a result of overstimulation with L-dopa. Clozapine has been reported as being beneficial for psychosis in Parkinson's disease.

About 50% of Huntington's disease patients develop psychosis, sometimes as the presenting feature. Rarer conditions involving the basal ganglia, such as Wilson's disease and idiopathic calcification of the basal ganglia, and cerebellar or multisystem degenerations may also produce psychosis.

Infections and Focal Lesions

Psychotic symptoms occur in a small proportion of patients with human immunodeficiency virus encephalopathy; these symptoms respond to relatively low doses of neuroleptics.

The relationship between epilepsy and psychosis has been extensively studied and comprehensively reviewed (24). In brief, complex partial seizures are more likely than other types of seizures to manifest with psychosis, which may be ictal, interictal, or postictal and may bear a poor relationship to the adequacy of seizure control. Interictal psychosis can be a major management problem in patients with complex partial seizures.

Metabolic (and Toxic) Encephalopathies

Psychosis may appear in the setting of delirium or fluctuation of consciousness. Renal, hepatic, or pulmonary failure; hypoxemia; disorders of the thyroid, parathyroid, or adrenal glands; collagen-vascular diseases, such as systemic lupus erythematosus and temporal arteritis; and vitamin deficiencies (B_{12}, folate, thiamine) are examples of such conditions. The psychoses seen in intensive care units or postoperatively are usually multifactorial.

An extensive list of medications reported as being potentially associated with delusions or hallucinations has been compiled by Cummings (4). Therapeutic drugs affecting neurotransmitter function, such as anticholinergics, dopaminergic agents, antidepressants, anticonvulsants, antihistamines, and antihypertensive agents, may provoke psychotic symptoms, as may many other drugs such as cimetidine, benzodiazepines, corticosteroids, and digoxin. Many drugs of abuse may produce psychiatric symptoms that include hallucinations or delusions. LSD, phencyclidine (PCP), psilocybin, and cocaine may cause acute excited states with hallucinations. More sustained psychotic symptoms that sometimes persist after withdrawal of the drug occur with amphetamines, LSD, PCP, and mescaline, and with inhalation of glue or other organic solvents. Withdrawal from alcohol, barbiturates, opioids, and, more rarely, other drugs may precipitate psychosis.

From the plethora of conditions associated with secondary psychotic symptoms, some patterns emerge linking brain sites of lesions and specific symptoms. Delusions associated with disease processes affecting the cortex tend to be less elaborate than those related to subcortical or limbic lesions. Left-hemisphere lesions are overrepresented in case series and case reports, as are temporal lobe lesions.

Diagnosis and Treatment of Secondary Psychotic Symptoms

Apart from the physical examination findings, patients with secondary psychosis are generally distinguished from those with schizophrenia by a later age of onset. Many toxic or metabolic disturbances produce psychotic symptoms in association with delirium. These patients may need a period of observation and relevant blood and urine tests to establish the diagnosis. Neuroimaging studies are needed in patients in whom the underlying disease process is not evident on clinical examination and laboratory tests.

SUMMARY

The information presented in this chapter proposes that hallucinatory experiences and delusional thinking are products of disturbed neural function that is almost certainly multifactorial and that may be substantially affected by individual psychological processes.

As the previous sections have illustrated, multiple variations of hallucinations and delusions are documented and recognized. Phenomenologically, hallucinations and delusions are discrete entities, easily distinguished from each other by definition; however, the similarities are many and the boundaries insecure. The defining difficulty underlying hallucinatory experiences is an *abnormality of percept*, whereas the difficulty underlying delusions is an *abnormality of thought*. In both disorders, the basic problem can be considered misinterpretation. The boundary between hallucinations and delusions, although distinct, is also artificial, because they share the essential feature of disordered interpretation.

For purposes of treatment, the most approachable of the multiple factors active in the production of hallucinations and delusions must be selected.

ACKNOWLEDGMENTS

This work was supported in part by: Scottish Rite Benevolent Foundation's Schizophrenia Research Program, N.M.J., U.S.A.; NIMH grants MH 43693, MH 45132, MH 49671-01; and the Department of Veterans Affairs.

REFERENCES

1. Maltbie AA. Psychosis. In: Cavenar JO, Brodie HKH, eds. *Signs and symptoms in psychiatry.* Philadelphia: JB Lippincott, 1983:413–432.
2. Berrios GE, Brook P. The Charles Bonnet syndrome and the problem of visual perceptual disorders in the elderly. *Age Ageing* 1982;11:17–23.
3. Dunn DW, Weisberg LA, Nadell J. Peduncular hallucinations caused by brainstem compression. *Neurology* 1983;33:1360–1361.
4. Cummings JL. *Clinical neuropsychiatry.* Orlando, FL: Grune & Stratton, 1985.
5. Perez MM, Trimble MR, Murray NMF, Reider I. Epileptic psychosis: an evaluation of PSE profiles. *Br J Psychiatry* 1985;146:155–163.
6. Melzack R. Phantom limbs and the concept of a neuromatrix. *Trends Neurosci* 1990;1:88–92.
7. Benson DF, Stuss DT. Frontal lobe influences on delusions: a clinical perspective. *Schizophr Bull* 1990;16: 403–411.
8. Nasrallah HA, Weinberger DR:. *Handbook of schizophrenia:* the neurology of schizophrenia. Amsterdam, Netherlands: Elsevier, 1986.
9. Goldstein G: The neuropsychology of schizophrenia.

In: Grant I, Adams KM, eds. *Neuropsychological assessment of neuropsychiatric disorders.* New York: Oxford University Press, 1986.
10. Granholm E: Processing resource limitations in schizophrenia: implications for predicting medication response and planning attentional training. In: Margolin DI, ed. *Cognitive neuropsychology in clinical practice.* New York: Oxford University Press, 1992:43–69.
11. Gold JM, Randolph C, Carpenter CJ, Goldberg TE, Weinberger DR: Forms of memory failure in schizophrenia. *J Abnorm Psychol* 1992;101(3):487–494.
12. Shenton ME, Kikinis R, Jolesz FA, et al. Abnormalities of the left temporal lobe and thought disorder in schizophrenia: a quantitative magnetic resonance imaging study. *N Engl J Med* 1992;327:604–612.
13. Gur R, Mozley P, Resnick S: Magnetic resonance imaging in schizophrenia. I. Volumetric analysis of brain and cerebrospinal fluid. *Arch Gen Psychiatry* 1991;48: 407–412.
14. Andreasen NC, Arndt S, Swayze VI, et al. Thalamic abnormalities in schizophrenia visualized through magnetic resonance image averaging. *Science* 266:294–298, 1994b.
15. Bruton C, Crow T, Frith C, et al. Schizophrenia and the brain: a prospective clinico-neuropathological study. *Psychol Med* 1990;20:285–304.
16. Benes F. Neurobiological investigations in cingulate cortex of schizophrenic brain. *Schizophr Bull* 1993;19 (3):537–549.
17. Davis KL, Kahn RS, Ko G, Davidson M. Dopamine in schizophrenia: A review and reconceptualization. *Am J Psychiatry* 1991;148:1474–1486.
18. Potkin SG, Buchsbaum MS, Jin Y, Tang C. Clozapine effects of glucose metabolic rate in striatum and frontal cortex. Long Island Jewish Medical Center and Case Western Reserve University School of Medicine Symposium on Clozapine. *J Clin Psychiatry* 1994;55:63–66.
19. Andreasen NC, Flashman L, Flaum M, et al. Regional brain abnormalities in schizophrenia measured with magnetic resonance imaging. *JAMA* 1994;272:1763–1769.
20. Carlsson M, Carlsson A. Schizophrenia: a subcortical neurotransmitter imbalance syndrome? *Schizophr Bull* 1990;16:425–432.
21. Weinberger DR, Berman KF, Zec RF. Physiological dysfunction of dorsolateral prefrontal cortex in schizophrenia. I. Regional cerebral blood flow evidence. *Arch Gen Psychiatry* 1986;43:114–124.
22. Davison K, Bagley C. Schizophrenia-like psychoses associated with organic disorders of the central nervous system: a review of the literature. In: Hetherington R, ed. *Current problem in neuropsychiatry.* British Journal of Psychiatry Special Publication #4. Kent, R.N. Headley Bros, 1969:113–184.
23. Cummings JL. Organic delusions: phenomenology, anatomical correlations, and review. *Br J Psychiatry* 1985:146:184–197.
24. Trimble M. Interictal psychosis. In: Trimble M, ed. *The psychoses of epilepsy.* New York: Raven, 1991:109–149.

18

Aggression

Jeffrey L. Saver, Stephen P. Salloway, Orrin Devinsky, and David M. Bear

OVERVIEW

Human aggression is an urgent social and clinical problem. In the United States, homicide is the twelfth leading cause of death and the second most common cause of mortality among young, healthy people (1).

Aggression is an inescapable clinical challenge in diverse neuropsychiatric patient populations. Relatives of people with traumatic brain injury identify temper and irritability as major behavioral difficulties in 70% of patients. About 10% to 20% of psychiatric inpatients commit acts of violence or assault during the 2 weeks before their admission, and 3% to 37% assault staff or patients during their hospitalization. Costs of in-patient care for assaultive patients are more than 40% greater than for their nonassaultive counterparts (2).

Epidemiologic studies demonstrate that key environmental variables contributing to the development of repeatedly violent people include rearing in disordered households, physical or sexual abuse in childhood, and social deprivation. Every violent behavior, however, whether motivated by disturbed rearing or the highest political and religious ideals, requires a neurobiologic substrate to orchestrate the complex array of perceptual, motor, and autonomic components of acts that constitute aggressive conduct. In humans, acquired brain lesions may disrupt the neural systems that ordinarily regulate hostile behavior. In some instances, patients with focal brain injury may exhibit aggressive behavior that has no relevant developmental or environmental precipitant, or only minimal social provocation. More often, damage to neural circuits controlling aggression leads not to random acts of overt aggression but to alterations in temperament and inappropriate choices of targets and settings for aggressive behavior. The recognition, diagnostic evaluation, and treatment of such patients are challenges facing the neuropsychiatrist and behavioral neurologist.

This chapter outlines a multiregional, hierarchical model of the neural regulation of aggression that draws on converging sources of evidence from evolutionary studies, ethology, neurophysiology, pharmacology, and clinical neuroscience. Recognizing multiple, hierarchical controls over hostile behavior affords the neuropsychiatrist a framework for identifying distinctive clinical syndromes of aggression due to brain injury, considering the differential diagnosis and diagnostic workup of patients, and implementing rational, pathophysiologically directed treatment.

NEUROSCIENCE OF AGGRESSION

Biologic Origins of Human Aggression

Recent formulations of evolutionary theory suggest that competitive selection favors the development of closely regulated and intertwined aggressive and peacemaking behaviors. The need for a neural system to regulate aggression is greater in social animals than in species leading a solitary existence. In social primates, its expression is precisely controlled at multiple anatomical levels (3).

Concerning the outward display of emotion, natural selection has fostered principles

TABLE 18-1. *Behavioral classification of aggression*

Type	Eliciting stimulus	Form
Predatory	Natural prey	Efficient, little affective display
Territorial	Boundary crossing	—
Intermale	Conspecific male	Ritualized responses
Fear-induced	Threat	Autonomic reactions, defensive behaviors
Maternal	Distress calls, threat to offspring	Initial attempts to avoid conflict
Irritable	Frustration, deprivation, pain	Hyperactivity, affective display
Instrumental	—	—

Modified from Moyer KE. *The psychobiology of aggression.* New York: Harper & Row, 1976.

of signaling among conspecifics that recur in invertebrates, dogs, cats, primates, and humans. Emotional displays are highly stereotyped, and opposing emotions are frequently conveyed by "antagonistic" postures, assuring accurate communication of an organism's emotional state. An efficient neurobiologic mechanism for regulating emotional expressions and modifying the probability of aggressive versus submissive responses would employ opposing neuromodulators to excite or inhibit groups of motoneurons that enact

TABLE 18-2. *Neuroanatomic correlates of aggression in experimental animals*

Triggers	Suppressors
Predatory offensive aggression	
Anterior hypothalamus	Prefrontal cortex
Lateral hypothalamus	Ventromedial hypothalamus
Lateral preoptic nuclei	Basolateral amygdala
Ventral midbrain tegmentum	Mammillary bodies
Ventral midbrain	
Ventromedial periaqueductal gray matter	
Inter-male (competitive) aggression	
Laterobasal septal nuclei	Dorsolateral frontal lobe
Centromedial amygdala	Olfactory bulbs
Ventrolateral posterior thalamus	Dorsomedial septal nuclei
Stria terminalis	Head of caudate
Fear-induced aggression	
Centromedial amygdala	Ventromedial hypothalamus
Fimbria fornix	Septal nuclei
Stria terminalis	Basolateral amygdala
Ventrobasal thalamus	Ventral hippocampus
Maternal-protective aggression	
Hypothalamus	Septal nuclei
Ventral hippocampus	Basolateral amygdala
Anterior hypothalamus	Frontal lobes
Ventromedial hypothalamus	Prefrontal cortex
Dorsomedial hypothalamus	Medial prepiriform cortex
Posterior hypothalamus	Ventromedial hypothalamus
Anterior cingulate gyrus	Head of caudate
Thalamic center median	Dorsomedian nucleus of thalamus
Ventrobasal thalamus	Stria terminalis
Ventral hippocampus	Dorsal hippocampus
Ventral midbrain tegmentum	Posterior cingulate gyrus
Ventromedial periaqueductal gray matter	Periamygdaloid cortex
Cerebellar fastigium	
Sex-related aggression	
Medial hypothalamus	Septal nuclei
Fimbria fornix (male)	Fimbria fornix (female)
Ventral hippocampus	Cingulate gyrus
	Dorsolateral amygdala

From Treiman DM. Psychobiology of ictal aggression. *Adv Neurol* 1991;55:343.

aggressive postures. The lobster provides an example of such a hormonal system that has been characterized in detail. Serotonin biases the organism to assume an aggressive or fighting stance (extension of major muscle groups), whereas octopamine promotes a submissive posture (flexion of major muscle groups) (4).

The substantial role of subcortical neuromodulators, such as norepinephrine, acetylcholine, and serotonin, in modifying aggressive propensities in humans likely developed from their phylogenetically ancient functions in promoting relevant peripheral skeletal postures and autonomic responses. The complex array of controls over aggression in social primates, however, clearly could not be achieved by simple changes in levels of subcortical neuromodulators or peripheral neurohormones.

One source of the complexity of neural regulation of aggression in mammals is the existence of several distinct subtypes of aggressive behavior. In general, aggressive acts are triggered, targeted, promptly terminated, and specifically inhibited by various classes of environmental stimuli. Ethologists have identified several classes of aggressive behavior, each with a specific outward display and set of determining stimuli. Moyer's widely recognized classification scheme divides hostile behavior into predatory, territorial, inter-male, maternal, defensive, fear-induced, irritable, and instrumental subtypes (Table 18-1). Each subtype is probably controlled by distinct, albeit somewhat overlapping, neuranatomic and neurochemical substrates. In various animal models, neuronal recording and lesion studies have identified important loci participating in neural networks controlling these discrete assertive behaviors (Table 18-2) (5).

Brain-Stem Regulation of Aggression

Pontine and mesencephalic nuclei mediate fragments of aggressive displays. Inputs to the system include spinoreticular proprioceptive and nociceptive sensory circuits. Outputs incorporate pontine (facial) and descending motor centers, leading to stereotypic movements. Medullary sympathetic and parasympathetic nuclei exert direct autonomic effects on cardiovascular, respiratory, and gastrointestinal peripheral organ systems. In humans, however, full-fledged aggression-related behavior patterns are not produced at the brain-stem level. Response coordination and decision making are carried out at higher processing stations. Consequently, brain-stem lesions, although sometimes disturbing fragments of aggressive behavioral output, generally do not produce a syndrome of altered aggressivity (6).

One important exception is aggressive behavior arising from disruption of brain-stem regulation of sleep—-awake states, most notably rapid eye movement (REM) sleep behavior disorders. Normally, during dreaming REM sleep, neurons in the vicinity of the locus ceruleus actively inhibit spinal motoneurons, suppressing motor activity. As Jouvet and Delorme first demonstrated in the cat, bilateral pontine tegmental lesions may compromise REM sleep muscle atonia and permit the enactment of oneiric behaviors, frequently including biting and other attack conduct (7). Such a parasomnia has recently been delineated in humans. Most commonly, middle-aged men experience a violent dream in which they are attacked by animals or unfamiliar people. In response, the dreamer engages in vigorous, coordinated motor acts that are often violent in nature. Patients may jump off the bed, smash furniture, or attack their bed partner. Physical injuries are common to both the patient and spouse. This disorder is now called the *REM sleep behavior disorder*.

Hypothalamic Regulation of Aggression

The primate hypothalamus receives inputs conveying information regarding the internal state of the organism through chemoceptors, osmoceptors, and viscerosensory cranial nerves. In contrast to these rich sources of interoceptive data, the hypothalamus does not directly receive sensory input regarding the external world from primary sensory or association neocortex. Important outputs of the hypothalamus are to the pituitary gland through releasing factors and directly transported peptides, to the autonomic nervous system, for which it serves as "head ganglion," and to midbrain and spinal motor centers that coordinate stereotypic movements.

Studies in animals suggest that the hypothalamus contributes important regulatory control to aggressive behavior. Bard demonstrated in cats that when all neural structures rostral to the hypothalamus were destroyed, the decorticate animals periodically entered a state of "sham rage" (8). With little or no provocation, they exhibited a combination of hissing, piloerection, pupil dilation, and extension of the claws. Later studies pinpointed the posterior lateral hypothalamus as the responsible site; stimulation reliably elicited sham rage in animals who had undergone cortical ablation. In the intact feline or rodent brain, stimulation of the posterior lateral hypothalamus shortens the latency for species-specific predatory attack. Attack behavior can similarly be facilitated by instilling acetylcholine and cholinomimetics into the lateral hypothalamus, promoting biting attacks of a cat on a mouse or rat, or of a rat on a mouse or frog, even by previously docile animals. Injection of cholinergic antagonists eliminates a biting attack, even in usually aggressive cats or rats.

Conversely, ventromedial hypothalamic stimulation may inhibit rather than facilitate aggression or can lead to a defensive posture. Ablation studies also suggest that the ventromedial area restrains hostile behavior, and its removal may permit the activity of regions that promote aggression to proceed unchecked.

Several case reports and small case series suggest a broadly similar role of the hypothalamus in human aggression. Neoplasms that destroy the ventromedial area bilaterally are associated with attacks on attendants reminiscent of animal aggression following ventromedial lesions (9).

Amygdala and Temporolimbic Cortex Regulation of Aggression

In contrast to the hypothalamus, the amygdaloid complex is reciprocally connected to multiple cortical sensory systems capable of conveying highly processed information regarding the external world. Rich connections are established with a variety of both unimodal and polymodal sensory regions, such as the perirhinal cortex and the superior temporal sulcus, allowing convergence of information from visual, auditory, tactile, and gustatory cortices. Of particular note, the basolateral amygdala receives extensive projections from unimodal visual cortices in the inferotemporal cortex (such as area TE in primates) that are specialized for recognizing objects, such as faces, in central vision. Extensive intrinsic connections within the amygdala promote further coordination of sensory information.

Important outputs from the amygdala in primates are to the hypothalamus through the stria terminalis and ventral amygdalofugal pathway, to brain-stem centers controlling heart rate and respiration through the central nucleus projection pathway, and to the extrapyramidal motor system, especially the ventral striatum, also through the stria terminalis and ventral amygdalofugal pathway (Fig. 18-1).

The amygdala appears to provide a crucial link between sensory input processed in the cortical mantle to produce a model of external reality and hypothalamic and somatomotor centers evoking pain, fear, and other basic drive-related emotions. Many observations of animals and humans suggest that a fundamental function performed by the amygdaloid complex and related temporolimbic structures is linking perceived objects with appropriate emotional valences. The result is a qualitative steering of behavior rather than quantitative regulation of threshold.

The importance of the amygdaloid complex in the recall of the affective significance of stimuli is demonstrated by the drive-object

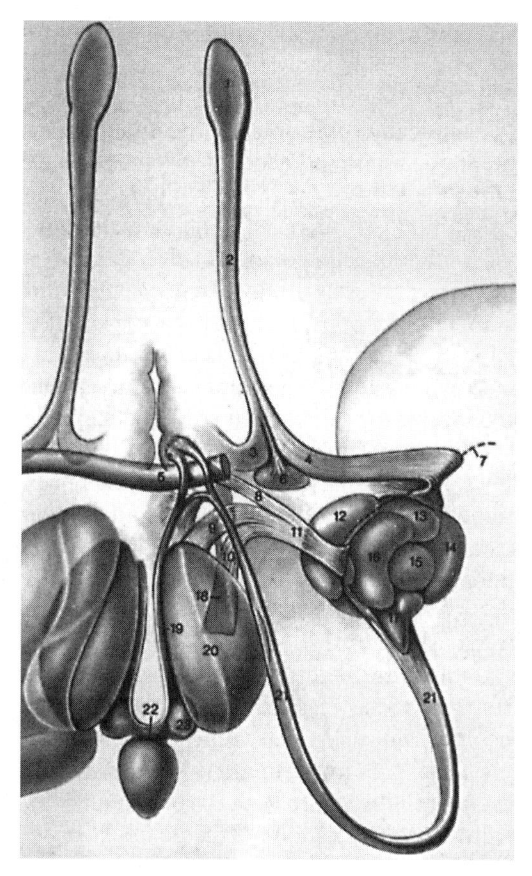

FIG. 18-1. The human amygdala and selected subcortical efferents. 11, Ventral amygdalofugal pathway; 12-17, The amygdala; 12, Cortical nucleus; 13, Anterior nucleus; 14, Lateral nucleus; 15, Central nucleus; 16, Medial nucleus; 17, Basal nucleus; 18, Lateral hypothalamic area; 20, Medial thalamic nucleus; 21, Stria terminalis. (Modified with permission from Nieuwenhuys R, Voogd J, Van Huijzen C. *The human central nervous system: a synopsis and atlas.* 3rd ed. Berlin: Springer-Verlag, 1988:301.)

dysregulation of the Kluver-Bucy syndrome, observed in animals when the amygdala (and often overlying temporal neocortex) are removed bilaterally. Monkeys with such lesions move about in a continuous olfactory and oral exploration of their environment in which each object evokes the same response of tasting and sniffing as though the monkey had never encountered it previously. The animals fail to distinguish food from inedible objects and eat metal bolts and feces as readily as normal dietary items. Animals have difficulty distinguishing appropriate from inappropriate sexual partners; similarly, lesioned cats attempt copulation with chickens or other animals. These results suggest that lesioned animals cannot identify particular objects as being appropriate or inappropriate to satisfy hypothalamic drives (10).

The effects of bilateral amygdalectomy on aggressive behavior are consistent with such a hypothesis. Amygdala removal results in taming and placidity in most animals. Objects that previously evoked signs of fear or provoked attack seem to lose their past associations. Monkeys no longer behave aggressively toward experimenters, becoming tame and easy to handle. Unilateral amygdalectomy with lesions of all commissural pathways produces taming when stimuli are presented to the operated hemisphere but appropriate hostile responses when the stimuli are shown to the unoperated hemisphere. Amygdalectomy in submissive monkeys, however, has led to a maintained or increased level of aggression, consonant with the view that the fundamental effect of amygdalectomy on aggression is not a change in aggressive threshold, but rather a modification of previously acquired patterns of linking stimuli with aggressive responses. Fundamentally, appetitive drives, such as feeding and reproduction, are released onto inappropriate targets. An instrumental drive such as aggression is no longer elicited or suppressed according to past, learned responses of the animal.

In humans, extensive bilateral temporolimbic damage produces behavior that is similar to that of lesioned monkeys, frequently accompanied by amnesia, aphasia, and visual agnosia. Patients may engage in indiscriminate oral and tactile exploration of their environment (hyperorality and hypermetamorphosis) and change their sexual preferences. Such patients exhibit a flattened affect and report diminished subjective emotional responses to stimuli. Aggressive behaviors become uncommon, and apathy with lack of either strongly positive or negative responses becomes the rule.

An intriguing contrast to the behavioral alterations that result from removal of the temporal lobes is provided by far a more common clinical condition, temporal lobe epilepsy, in which abnormal neuronal excitability develops within temporolimbic cell populations. Although patients are rarely aggressive during ictal discharges, recent interest has focused on interictal behavioral alterations among patients with a long-standing temporal lobe focus.

Within the temporal lobe, the amygdaloid complex is particularly sensitive to the phenomenon of kindling, in which repeated stimulation of neurons leads to a progressive lowering of the threshold for discharge. Because many processing pathways converge on the amygdala, activity of epileptic foci throughout (and in fact beyond) the temporal lobe can affect amygdalic excitability. The resulting enhancement of amygdaloid activity may, in a general sense, be the converse of the decreased activity underlying Klüver-Bucy syndrome. Kindling may lead to long-term changes in limbic physiology that alter and enhance aggressive and other emotional responses to both drive-related and neutral stimuli. Rather than losing previously acquired associations between sensory stimuli and drives, some temporal lobe epilepsy patients appear to forge new, fortuitous associations. Rather than a lack of emotional response to stimuli, they exhibit deepened and generalized affective associations.

Prefrontal Cortical Regulation of Aggression

The dorsolateral prefrontal cortex receives extensive afferents from multiple posterior neocortical association areas. The orbitofrontal cortex is reciprocally connected to the rest of the neocortex principally through the dorsolateral convexity of the frontal lobe. Projections from the hypothalamus inform the frontal lobes of both internal (hypothalamus) and external (neocortical association to temporal lobe) stimuli of affective significance. The prefrontal cortex has direct outputs to the pyramidal motor system, the neostriatum, temporal neocortex, and hypothalamus. Schematically, prefrontal cortices appear to integrate a current account of the outside world, the state of the internal milieu, and the recognition of drive-relevant objects with knowledge of learned social rules and previous experiences relating to reward and punishment. The prefrontal cortices construct a behavioral plan that is consistent with experience and especially the rules of socialization, to optimize the satisfaction of biologic drives. The simplest summary of these complex functions in humans is judgment. It has been proposed that, in selecting among alternative response options, prefrontal cortices are guided by internal, somatic state markers, physiologic cues that allow rapid choice of previously rewarded, effective options.

Damage to the orbital undersurface of the frontal lobe has classically been described as resulting in superficial, reflexive emotional responses to stimuli in the immediate environment. Patients are impulsive without foresight or consideration of the remote consequences of their actions. Orbital frontal lesions thus lead to episodes of transient irritability. Often a patient strikes out quickly after a trivial provocation, with little consideration of social prohibitions limiting aggressive behavior or untoward future consequences. The targets of aggression are categorically appropriate, but patients are unable to apply abstract rules that would override the immediate environmental provocation. Case series of murderers have demonstrated a high incidence of frontal structural abnormalities on computed tomography (CT) and magnetic resonance imaging (MRI), frontal hypofunction on positron emission tomography (PET), and abnormal neuropsychological performance on frontal systems tasks (11).

Combined Lesions of Temporal and Frontal Lobes

Some pathologic processes may simultaneously damage the multiple levels of neural circuitry crucial to the regulation of aggression. The orbitofrontal surface and rostral temporal poles are particularly susceptible in closed head injuries, and conjoint lesions in the same patient are not uncommon.

Aggressive behavioral syndromes may result that have features associated with dysfunction of several brain regions. For example, a patient may, as a result of brain trauma, develop a temporolimbic epileptic focus, as well as a contusion of the orbital frontal cortex. Such a patient can display the deepened emotions and anger associated with the interictal behavior syndrome, as well as a failure to inhibit or modulate hostile responses typical of a frontal lesion. Evidence for frontal and temporolimbic dysfunction in violent criminals was recently renewed by Raine and colleagues (11). They demonstrated frontal lobe hypometabolism by PET scans in murderers found not guilty by reason of insanity.

Hemispheric Asymmetries in the Regulation of Aggression

Several lines of neuropsychological research suggest differences in left- and right-hemisphere specialization for the processing of emotion, including anger and aggression. The left hemisphere plays a greater role in decoding linguistically conveyed emotional information, and the right hemisphere is more important in processing nonverbal emotional cues, such as prosody and facial expression of emotion. Moreover, the right hemisphere may be more highly specialized for mediating emotional responses in general, and negative emotional responses such as fear and anger in particular (12).

Neuropsychological studies in unselected populations of violent criminal offenders also support an important role of hemispheric specialization in the genesis of hostile behaviors. When neuropsychological deficits are observed in studies of violent groups, they tend to involve not only frontal-executive functions but also verbal comprehension, expressive speech, and other left-hemisphere language functions. These findings are consonant with a large number of studies in conduct-disordered and delinquent juveniles, indicating that the lowered intelligence quotient (IQ) found in these groups frequently reflects disproportionately lowered verbal IQ (language—left hemisphere) compared with performance IQ (visuospatial—right hemisphere).

Psychophysiologic studies in violent offenders also show a trend toward left-hemispheric dysfunction. Although plagued by methodologic defects and not adequately addressing laterality, early studies of large samples of violent offenders suggested that a substantial proportion, about 50%, exhibit electroencephalogram (EEG) abnormalities. More recent studies employing computerized EEG spectral analysis have suggested that persistently violent behavior among psychiatric inpatients is linked to increased δ-band slow-wave activity in left frontotemporal derivations. Divided visual field, dichotic listening, skin conductance asymmetry, and lateral preference studies additionally suggest subtle abnormalities of left-hemispheric function in sociopathic individuals without overt neurologic lesions.

NEUROCHEMISTRY OF AGGRESSION

Most experimental and clinical reports on "the neurochemistry of aggression" likely describe the effects of neuromodulators at peripheral, brain-stem, and hypothalamic sites, which can reduce or raise the person's overall predisposition to aggression. Still lacking are studies that evaluate the multisynaptic integration of parallel-processed streams of complex sensory and limbic information that link amygdala, orbitofrontal, and other higher cortical centers. Quite likely, these networks are subserved by diverse messenger systems and not exclusively controlled by a single neurotransmitter. Because neurochemical studies provide the basis for pharmacologic interventions in aggressive patients, however, data from such studies are clinically invaluable.

Serotonin

Animal and human studies suggest that serotonin is a crucial modulator of aggressive behavior. Studies in humans have focused on serotonergic markers in suicidal patients and violent criminals. See Chapter 16 on mood disorders and suicide for a review.

Violent juvenile delinquents have been reported to have *decreased* platelet $5HT_2$ binding in one recent study (13). Cerebrospinal fluid (CSF) levels of 5-hydroxyindoleacetic acid (5-HIAA), a metabolite of serotonin, are decreased in patients attempting suicide and in violent criminal offenders. Violent and impulsive self-injurious acts tend to have the strongest negative correlation with CSF 5-HIAA values, suggesting that low CSF 5-HIAA levels are a marker not simply of depression and suicidal risk but also of a tendency to aggressive and impulsive behavior. Several studies in criminal and interpersonally violent psychiatric populations support this view. Lowered CSF 5-HIAA levels have been found in samples of impulsive fire setters and in individuals committing impulsive manslaughter. In a group of soldiers with behavior problems, there was a negative correlation between CSF 5-HIAA and aggressive behavior, and CSF 5-HIAA was reduced in a group of borderline patients with aggressive and suicidal behavior.

Acetylcholine

Some of the earliest work on the neurochemistry of aggression focused on acetylcholine. Electrical stimulation of the lateral hypothalamus in rats leads to predatory attack on mice in animals that previously had tolerated mice in their cage without attacking them. Applying carbachol, a cholinergic agonist, to the lateral hypothalamus provokes the stereotypic aggressive response, which can be blocked by atropine and facilitated by acetylcholinesterases. This cholinergic-induced predatory response is target specific—directed only at the animal's usual prey—and without affective display. Applying carbachol to the amygdala also induces a predatory response. Aggressive behavior after human exposure to cholinesterase inhibitors has been observed in several clinical case reports (14).

Norepinephrine and Dopamine

Catecholamine systems are associated with aggressive behavior in several animal models and clinical populations. Peripherally administered norepinephrine enhances shock-induced fighting in rats. α_2-Receptor agonists increase rat aggressive behavior, whereas clonidine decreases rodent aggressive behavior acutely. Long-term treatment with β-adrenergic blocking agents such as propranolol can decrease aggressive behavior in laboratory animals and in diverse neuropsychiatric patient groups with violent behaviors (15).

L-Dopa can induce aggressive behavior in rodents and humans. Apomorphine, a potent dopamine agonist, can induce fighting in rats. Dopamine antagonists tend to reduce aggression but usually at doses that also slow motor and cognitive performance.

γ-Aminobutyric Acid

Several lines of evidence suggest that γ-aminobutyric acid (GABA) inhibits aggression in animals and humans. GABA injected into the olfactory bulbs in rats inhibits mouse killing. GABA antagonists can induce mouse killing behavior. Benzodiazepines and other agents that facilitate GABA can decrease isolation-induced fighting in mice and attenuate aggression caused by limbic lesions. Benzodiazepines most likely exert their antiaggressive effects by binding to sites on the GABA receptor. In humans, despite their tranquilizing and antiaggressive effect in most patients, benzodiazepines rarely can lead to a transient increase in aggressive behavior ("paradoxical rage").

Testosterone and Other Androgens

Testosterone is an important mediator of aggressive responding in diverse mammalian species. In rats, dominant males have higher levels of testosterone than submissive males. In vervet monkeys, increases in serum and salivary testosterone levels correlated with the number of aggressive encounters. Moyer suggested that androgens increased intermale, irritable (but not predatory), sexual, fear-induced, and maternal forms of aggression. An interaction between androgens and other neuromodulators, such as the monoamine neurotransmitters, appears to govern aggressive responding. Testosterone-induced dominance in rats is reduced after treatment with $5HT_{1A}$, $5HT_{1B}$, and $5HT_{2A/2C}$ receptor agonists.

Human studies support an important link between circulating androgens and aggressive behavior. Elevated testosterone levels in adolescent boys correlate with low frustration tolerance and impatience. Boys with increased testosterone are more likely to respond aggressively when provoked. Increased levels of free testosterone have been measured in the saliva of incarcerated violent criminals. Victorious collegiate wrestlers show a greater rise in their serum testosterone than their defeated counterparts. Violent behaviors have been reported in people taking anabolic steroids for body-building programs. Archer's metaanalysis of reported studies demonstrated a strong positive correlation between testosterone levels and observer-rated aggressiveness (16). More recently, Bergman and Brismor showed that male alcoholics who abused other people had higher levels of testosterone and lower levels of cortisol than those who did not (17). Loosen and colleagues showed that inhibiting gonadal function with a gonadotropin-releasing hormone antagonist reduced outward-directed aggression (18).

Caution must be employed when generalizing findings across species, particularly when comparing responses between humans, other primates, and other mammalian orders. Many aggression-related neurotransmitters are conserved across species, but there are probably important variations in receptor subsystems.

Precision in defining and measuring aggression, impulsivity, and irritability in many animal models and humans is difficult to achieve. Also, many studies fail to recognize and fully clarify state-trait distinctions. The neurochemical substrates for responses related to rapidly shifting behavioral states and those related to enduring dispositional traits may differ.

A CLINICAL NEUROPSYCHIATRIC APPROACH TO THE AGGRESSIVE PATIENT

General Approach

The recognition that specific neurologic lesions may lead to violent behavior in humans and that abnormalities at different levels of the neuraxis result in distinctive types of aggressive behavior provides a guiding schema for the evaluation and treatment of inappropriately aggressive (and inappropriately hypoaggressive) people (Tables 18-3 to 18-5). In addition to emphasizing the need for careful neuropsychiatric evaluation of every violent patient, the hierarchical model for the regulation of aggression suggests important parameters that may help to characterize any aggressive act. The neuropsychiatrist should integrate information regarding the clinical manifestations of aggressive behavior, additional aspects of the history (particularly related to other drive-related behaviors), the neurologic and psychiatric examinations, and structural and functional laboratory studies to classify patients among major syndromes of dysregulation and aggression.

Hypothalamic and Brain-Stem Syndromes

In patients with hypothalamic lesions, outbursts of violent behavior may be precipitated by internal or visceroceptive states, such as hunger, fatigue, light deprivation, or hormonal stimulation; alternatively, patients may exhibit a heightened general level of aggression. Most frequently, attacks are on individuals who happen to be near the patient, without the formation of complex plans. Patients often have diminished insight into the reasons

TABLE 18-3. *Features that distinguish clinical syndromes of hyperaggression*

Syndrome	Eliciting stimulus	Provocation	Outbursts	Complex plans	Memory for acts	Remorse	Mode
Hypothalamic	Individuals who happen to be present	Basic drive (e.g., hunger); unprovoked	Yes	No	Yes	Yes	Simple kicking, biting, scratching; infrequently, throwing of readily available objects
Ictal	Any nearby individual or inanimate object	None	Yes	No	No	Yes	Random, undirected; wild swinging, kicking movements
Postictal	Caretakers	Attempts to restrain or protect patient	Yes	No	No	Yes	Disoriented, irritable; simple kicking, biting, scratching
Interictal	Individuals	Threat, including misinterpretation of trivial stimulus; perception of moral injustice	Occasional	Yes	Yes	Yes, may be intense	1) Morally driven; complex plans; may be enacted over long time period; may involve weapons 2) Quick-tempered lashing out
Orbitofrontal	Individuals	Minor provocations	Yes	No	No	No	1) Irritable lashing out 2) Rarely may carry out plan of minor or intermediate complexity

TABLE 18-4. *Selected differential diagnosis of syndromes of hyperaggressivity*

Syndrome	Diseases
Hypothalamic	Hamartoma
	Craniopharyngioma, astrocytoma
	Parasomnias
Temporal lobe epilepsy	Mesial temporal sclerosis
	Glioma
	Vascular malformation
	Traumatic brain injury
Orbitofrontal systems	Traumatic brain injury
	Herpes simplex encephalitis
	Anterior communicating artery aneurysm
	Orbital meningioma
	Pick's disease
	Frontal dementia without specific histologic features
	Huntington's disease
Combined frontotemporal	Traumatic brain injury
	Pick's disease
	Herpes simplex encephalitis
Multifocal or poorly localized	Attention deficit disorder
	Toxic-metabolic encephalopathies
	Vitamin B_{12} deficiency
	Multiple sclerosis
	Subcortical vascular dementia
	Alcohol, cocaine
Delusional cognition	Paranoid schizophrenia
	Endogenous depression (unipolar or bipolar)
	Late-life paraphrenia
	Mania
	Alzheimer's disease
	Vascular dementia

TABLE 18-5. *Differential diagnosis of syndromes of hypoaggressivity*

Syndrome	Diseases
Bilateral amygdalotemporal	Pick's disease
	Herpes simplex encephalitis
	Bilateral posterior cerebral artery infarction
	Traumatic brain injury
	Urbach-Wiethe disease
	Temporal lobectomy
Dorsolateral frontal systems	Subdural hematoma
	Glioma
	Progressive supranuclear palsy
	Anterior cerebral artery infarction

for their actions, although they recall and may demonstrate remorse for their actions. Subjects with hypothalamic lesions may demonstrate altered patterns of sleeping or eating, polydipsia, and polyuria, or deficient regulation of sex hormones, thyroid, or adrenocortical function. Heteronymous visual field impairments may be evident if lesions extend to involve the optic nerves, chiasm, or tracts. The workup of patients with suspected hypothalamic lesions should include MRI or other structural imaging of the region of the third ventricle, endocrine studies, and formal visual fields. The differential diagnosis includes benign and malignant tumors such as craniopharyngiomas and astrocytomas, which present with subacute alterations in behavior. Another etiology is hypothalamic hamartoma, which usually presents with a distinctive clinical profile. Patients in childhood have the onset of gelastic epilepsy (ictal laughter), sometimes accompanied by precocious puberty, along with interictal bouts of uncontrolled rage.

In patients with brain-stem and hypothalamic-mediated sleep-related disorders of violence, aggressive actions are generally nocturnal, associated with incomplete maintenance of REM or non-REM sleep states. The REM sleep behavior disorder predominantly occurs in middle-aged men. Violence most commonly occurs during a vivid and frightening dream, often is directed at a bed partner mistaken for a dream figure, and is unplanned, without the use of weapons. Affected patients are difficult to arouse from their dreaming state. Afterward, they generally recall the dream material that provoked aggression but report believing they were attacking animal or human oneiric figures rather than their furniture or spouse. They exhibit great remorse about their actions and, before a diagnosis is made, frequently self-treat their disorder by tying themselves in restraining devices at night.

The REM sleep disorder must be distinguished from other parasomnias that may be injurious to patient and spouse, such as somnambulism, sleep drunkenness, and sleep terrors, which arise out of non-REM sleep. Nocturnal seizures must also be excluded. The diagnostic workup in suspected cases includes a thorough history of sleep complaints from patient and bed partner, neurologic and psychiatric examination, overnight polysomnographic study, and MRI. REM sleep behavior disorder has been associated with a variety of neurologic conditions, including dementia, stroke, multiple sclerosis, and alcohol withdrawal. However, more than 50% of cases are idiopathic. Pontine tegmental lesions, which might be expected from animal studies, are rare, possibly because pontine injury frequently produces devastating motor and arousal deficits that preclude expression of the disorder. About 90% of patients exhibit sustained improvement when treated with clonazepam.

Temporolimbic Epilepsy Syndromes

Failure to distinguish adequately between aggressive actions in the ictal, postictal, and interictal periods has contributed greatly to the confusion regarding the relationship between temporolimbic epileptic foci and violent behavior. Ictal aggression does occur, but with extreme rarity. An international consensus panel found that only 7 of 5400 patients studied on video-EEG monitoring exhibited aggressive behavior during recorded seizures. Hostile behaviors ranged from physical violence directed toward inanimate objects to mild verbal or physical aggression directed toward a person. When aggressive acts during complex seizures occur, they may appear without provocation or in response to an environmental stimulus and are characterized by spontaneous, undirected aggression toward nearby objects or people. The patient is amnestic for actions and often expresses remorse. See Chapter 25 on epilepsy for greater detail.

A much more common form of aggressive behavior in epilepsy is resistive violence during the postictal period. After a complex partial seizure or, more frequently, a generalized convulsion, patients may be disoriented and confused. During this epoch, well-intended attempts at physical restraint can provoke aggression, which almost always ceases when restraint is withdrawn. The attacks generally involve striking out without the use of a weapon or sometimes with objects that happen to be close at hand. Patients have no memory of their actions upon clearing of consciousness and typically express dismay if they have injured others. Aggression immediately after the end of a seizure activity can also occur in the context of postictal psychosis, especially in patients with paranoid delusions and threatening hallucinations.

Overt aggression related to the interictal behavior syndrome of temporolimbic epilepsy is unusual because the heightened moral and religious values that are features of the syndrome preclude violent actions. In rare circumstances, however, intense emotional reactions to perceived injustice or threat can lead subjects to formulate and carry out complex plans of violent response. Attacks may be directed against a specific person and could involve the use of a weapon. Not all hostile actions by these patients involve long-term planning; rarely, the intensity of feelings evoked in a particular situation leads to an immediate response. Patients fully recall their actions and often exhibit extreme remorse. Some, however, continue to feel their acts had ample moral justification.

In patients with epilepsy, sedative antiepileptic medications such as barbiturates may contribute to hostile behaviors by impairing impulse control. Irritability, a common complaint among patients with poorly controlled partial and primary generalized seizures, may result from environmental factors or medications or may be related to the underlying cerebral pathology or epileptogenic process.

In patients in whom ictal, postictal, or interictal aggression is suspected, history taking should first be directed at eliciting symptoms of simple or complex partial seizures. Although tonic-clonic convulsions and loss of consciousness are easily recognized, focal manifestations of temporolimbic seizures may escape notice or be dismissed as psychological in origin. Noxious odors, epigastric pain, forced feelings of sadness or fear, sensations of familiarity and unfamiliarity, and other partial seizure phenomena should be directly inquired after.

These subjects should also be questioned for elements of the interictal behavior syndrome: hypergraphia, heightened philosophic concerns, change in sexual interest, and deepened emotionality. Routine scalp EEG may show interictal spikes or slow-wave abnormalities in the temporal region but may miss physiologic discharges occurring in deep, mesial temporal regions. Sleep-deprived and sleep recordings and the use of nasopharyngeal or sphenoidal leads may increase the yield of scalp recording. In selected cases, ambulatory EEG, long-term in-patient video-EEG monitoring with scalp and sphenoidal electrodes, or invasive subdural grids or depth electrodes may be necessary to establish the seizure focus.

CT and especially MRI are used to exclude slowly growing gliomas and other mass lesions. Volumetric MRI or careful visual analysis of the hippocampus and amygdala (seen best on T1-weighted coronal cuts) may aid in the diagnosis of mesial temporal sclerosis by demonstrating unilateral or bilateral atrophy. Metabolic imaging with PET or SPECT may show increased blood flow and hypermetabolism in mesial temporal structures during ictal discharges and decreased blood flow and hypometabolism interictally. PET is the more sensitive technique interictally; SPECT is more practical for capturing ictal events. Common causes of temporolimbic epilepsy include mesial temporal sclerosis, hamartomas, dysplasia, low-grade astrocytomas, oligodendrogliomas, vascular malformations, and traumatic brain injury.

Orbitofrontal Systems Syndromes

Patients with lesions in orbitofrontal cortices or associated subcortical structures, such as the caudate nucleus, may engage in directed acts of aggression (Fig. 18-2); however, they

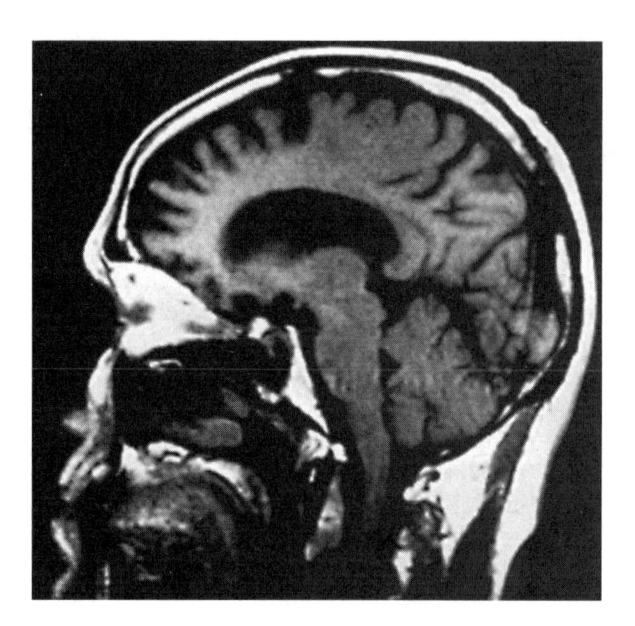

FIG. 18-2. Aggression with bifrontal lesions. A 56-year-old man who, over 3 years, developed ritualistic behaviors and became disinhibited and jocular. He exhibited inappropriate violent outbursts and persistent foul language. Magnetic resonance imaging (MRI) demonstrates frontal lobar atrophy in an advanced "picket fence" pattern consistent with Pick's disease. Aggressive behavior eventually diminished with high doses of propranolol. However, the patient required being confined to a chair with a locking top and occasional treatment with neuroleptics.

are often incapable of planning or executing a complex response that requires an extended sequence of actions. Failure to consider long-term and especially social consequences of violent outbursts is a salient feature. Frequently, the patient engages in impulsive, unreflective responses to identifiable but trivial environmental provocations. Patients remember their actions but often lack remorse, and they fail to link aggressive actions with punishment or other long-term adverse outcomes, contributing to repeated offenses.

The neurologic examination may reveal anosmia due to damage to the olfactory nerves or tracts on the undersurface of the frontal lobes and release phenomena, such as the grasp reflex. If the lesion is confined strictly to orbitofrontal cortices, subjects may show few deficits on conventional IQ tests or even on neuropsychological tests explicitly concerned with frontal-executive function. When lesions trespass on dorsolateral frontal territories, deficits in go—no-go testing, verbal and nonverbal fluency, and set-shifting may be evident. CT and especially MRI studies are helpful in screening for structural lesions. Common causes include traumatic brain injury, anterior communicating artery aneurysm rupture, anterior cerebral artery infarction, orbital meningioma, the frontal lobe dementias, and Huntington's disease.

Syndromes of Diffuse or Multifocal Brain Injury

Several medical conditions that produce aggression have effects on the brain that are diffuse or multifocal. A large body of neuropsychiatric literature has demonstrated the frequency of "minimal brain dysfunction" and poorly localized neurologic "soft signs" in violent patients. Neurologic findings are common in violent individuals in juvenile reform school and on death row. Attention deficit disorder was significantly correlated with criminal and violent offenses in a prospective study. Not surprisingly, developmental or acute medical conditions producing scattered minor neurologic impairments places an individual at higher risk of expressing aggressive impulses. Diffuse or multifocal brain insults are likely to affect one or several circuits within the multiregional, hierarchical aggression regulatory system. In addition, a history of being abused or reared in an unstable household is likely to interact synergistically with multifocal brain injuries. An individual who has learned a model of acting on impulse and possesses a limited repertoire of other response options is all the more likely to demonstrate diminished flexibility and inhibition of aggression after diffuse, especially frontal, injuries. Etiologies of diffuse brain dysfunction with episodic violence include, in addition to attention deficit disorder, toxic-metabolic encephalopathies (such as hyperglycemia and hypoglycemia, vitamin B_{12} deficiency, and thiamine deficiency), multiple sclerosis, and subcortical vascular dementia.

Delusional Syndromes

People with delusions are prone to violent behavior. In these patients, aggression-related neural systems may, like the intellect, be placed in service of the psychosis. In a variety of neuropsychiatric disorders, including schizophrenia, endogenous depression, mania, Alzheimer's disease, and other dementias, the presence of thought disorder and hallucinations, especially the persecutory type, increases the risk for violent outbursts. Dementia patients with paranoia and aggressive behavior have an increased rate of early institutionalization. Delusions are more likely to lead to aggression when frontal systems are also impaired.

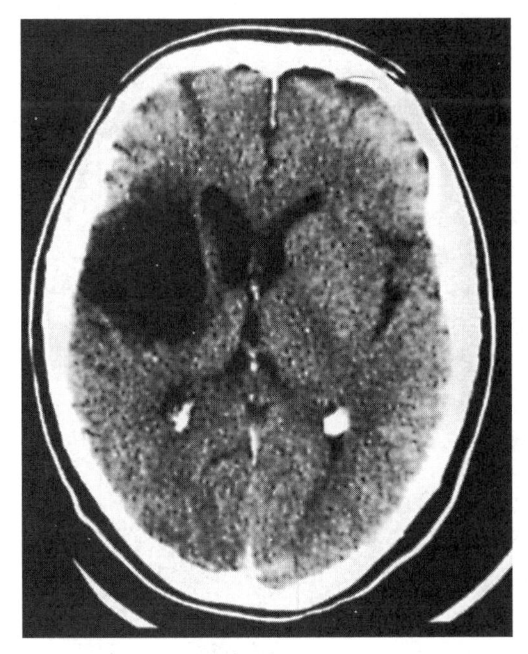

FIG. 18-3. Aggression related to organic delusional syndrome. A 70-year-old right-handed man sustained a large right middle cerebral artery infarction, with residual left hemiparesis. After the stroke, he believed that his wife was having an affair with a number of men in the neighborhood, aged 16 to 84 years. He attacked her in bed with his cane because he was convinced that she was sleeping with a covert sexual partner. The pathologic jealousy (Othello) syndrome persisted for several years. Psychotherapy was not helpful. The intensity of jealous complaints and physical outbursts decreased on a combination of pimozide and sertraline.

The clinical approach to these patients is focused on the diagnosis and treatment of the underlying psychotic disorder. Etiologies of delusional disorders include paranoid schizophrenia, affective illness, late-life paraphrenia, Alzheimer's disease, multiinfarct dementia, and subcortical dementias (Fig. 18-3).

TREATMENT

Consideration of multiregional neural processing may guide selection of appropriate environmental and biologic interventions to control aggression (Table 18-6). Much of our current pharmacologic armamentarium for the treatment of violence is targeted on neuromodulators within the brain stem and hypothalamus, relying on a pharmacologically induced bias against all types of aggressive responses. Brain-stem structures, for example, are the likely sites of action for drugs that block β-adrenergic receptors, activate $5HT_1$ receptors, or enhance GABA activity. By raising the threshold for aggressive responding, such agents may have an ameliorative effect on aggression resulting from lesions at all levels of the neuraxis. Agents modulating muscarinic cholinergic receptors in the lateral hypothalamus might be systematically investigated for antiaggressive effects, based on observations that cholinomimetics directly instilled in the lateral hypothalamus elicit aggression.

Intervention at the temporolimbic level of control introduces additional considerations. Patients in whom a focal seizure disorder is

TABLE 18-6. *Therapeutic approaches to aggressive behavior*

Clinical setting of aggression	Strategies	Specific agents	Comment
Impulsive aggressive acts in setting of congenital brain abnormality (mental retardation) or diencephalic injury	1. Control of appetite, sleep, diurnal cues 2. β-adrenergic blocker 3. Selective serotonin reuptake inhibitors 4. Cholinergic (muscarinic) antagonists (?) 5. Avoidance of barbiturates, benzodiazepines, or sedatives	Propranolol Fluoxetine, sertraline, paroxetine Trihexyphenidyl	Hydrophilic β-blockers may have delayed (up to 4–6 weeks) onset of action. 5HT agonists (e.g., eltoprazine) may have special utility. Anticholinergic agents are of theoretic value but have not been systematically evaluated.
Aggression related to deepened affect or ideation of the interictal behavior syndrome of temporolimbic epilepsy (moralistic conviction)	1. Antiepileptic medications 2. Selective serotonin reuptake inhibitors 3. Reality-oriented psychotherapy 4. Avoidance of lithium carbonate (can compromise seizure control and worsen behavior)	Carbamazepine, valproic acid Fluoxetine, sertraline, paroxetine	Aggressive acts occur in clear consciousness as interictal, not ictal, events. Reality-oriented therapy may improve patient's understanding of deepened emotions and the need to control them.
Superficial (reflexive) aggression in response to transient environmental stimuli—signs of social disinhibition (prefrontal, dysexecutive personality syndrome)	1. Explicit and concrete social structure 2. Selective serotonin reuptake inhibitors 3. β-adrenergic blockers 4. Avoidance of barbiturates, benzodiazepines, and sedatives	Fluoxetine, sertraline, paroxetine Propranolol	The environment must supply cues that the patient cannot retain in "working memory." Avoid disinhibiting medications and alcohol.
Aggression prompted by hallucinations, delusions from underlying psychotic illness	1. Antipsychotic agents	Haloperidol, perphenazine, risperidone, clozapine	High-potency agents may be preferable because sedation can lead to disinhibition, increased paranoia. When used to treat chronic aggression, tardive dyskinesia risk may be increased. Akathesia may exacerbate paranoia.
Irritability related to manic or hypomanic states in bipolar illness or secondary mania	1. Mood stabilizers	Lithium carbonate Carbamazepine, valproic acid	Antipsychotic agents may also be necessary are generally preferable to benzodiazepines, which may lead to behavioral disinhibition.
Acute violence or agitated aggression	1. Neuroleptics 2. Benzodiazepines	Haloperidol, thioridazine Lorazepam, diazepam	Exploit both sedative/hypnotic and antiaggressive properties of agents. Rare "paradoxical rage" may be seen with benzodiazepines.

diagnosed should be treated with agents effective for complex partial seizures, especially those such as carbamazepine and valproate, which have mood-stabilizing effects. Neuroleptics, which lower the seizure threshold, should be employed with caution. Avoiding restraint and providing gentle reassurance and reorientation in the postseizure period reduces postictal resistive violence. Patients whose clinical seizures remain refractory to antiepileptic therapy and who have well-defined unilateral foci are candidates for surgical resection. Temporal lobectomies or amygdala-hippocampectomies produce excellent seizure control in most of such patients and may have a beneficial effect on aggression in patients whose epilepsy is associated with violent behaviors.

Processing within the orbitofrontal cortices represents an advanced stage of synaptic interaction occurring at the convergence of multiple streams of prior sensory evaluations. It is unlikely that a single neurotransmitter could be modulated to duplicate or restore prefrontal functions. Conversely, drugs that nonselectively inhibit neuronal function through inhibition of chloride channels or other mechanisms, such as ethanol, benzodiazepines, and barbiturates, exert a disproportionate effect n prefrontal functioning, which is particularly dependent on polysynaptic inputs. A paradox in the current psychopharmacology of aggression may be illuminated by considering the prefrontal effects of agents that simultaneously act at other levels of the neuraxis. Benzodiazepines, for example, have antiaggressive properties in many species and tranquilizing effects in humans, likely mediated by their potentiating effects at GABA-ergic inhibitory receptors. Yet these compounds are known to precipitate paradoxical rage, the release of previously inhibited hostile responses. A possible explanation is that benzodiazepines impair prefrontal processing through nonselective neuronal inhibition in a manner similar to ethanol intoxication. In some patients, the resulting loss of social insight and judgment more than offsets the general tranquilizing effects of these agents.

Serotonergic agents may also exert substantial effects on prefrontal function. Serotonergic efferents from prefrontal cortices appear to serve an important role in the inhibition of impulsive responses. Lowered levels of 5-HIAA acid and blunted responses to fenfluramine in impulsive aggressive individuals are consonant with this formulation, suggesting that serotonergic agents, especially those acting at $5HT_1$ receptors, may be beneficial to patients with orbitofrontal dysfunction. α_2-Adrenergic agonists, such as clonidine and guafacine, have been shown to ameliorate frontal focused attention deficits and may thus be helpful to patients with orbitofrontal derangements.

Supportive psychotherapy is helpful to many patients with temporolimbic and orbitofrontal aggression syndromes. Insight-oriented therapy is unlikely to be beneficial to the patient with prefrontal injury who may verbally comprehend a behavioral problem and propose solutions but cannot reliably call on this knowledge to control behavior. Some patients with interictal behavior changes associated with temporal lobe epilepsy are amenable to insight psychotherapy, but many accept neither criticism nor advice. In our experience, however, by alerting patients to their intensified emotional responses and capitalizing on their heightened moral and religious sensitivities, a therapist may reduce their likelihood of aggressive actions.

CONTEMPORARY ISSUES

One view concerning the neural regulation of aggression is that each hemisphere performs complementary processing related to hostile behavior, and functional abnormalities of either hemisphere may produce disturbed aggressive responding through distinctive mechanisms. Left-hemisphere dysfunction, implicated in a preponderance of studies, may lead to overt expression of negative affects mediated by the right hemisphere, diminished linguistic regulation over behavior, and adverse social encounters due to impaired verbal communication. Right-hemisphere dysfunction may lead to improper intrahemispheric decoding and encoding of prosody, facial expressions, or other nonverbal emotional responses; overreliance on semantic processing; and a different pattern of altered aggression.

Recent investigations of the relationship between serotonin and aggression have shifted to exploring agents that act at specific serotonergic receptor subtypes. Buspirone, a $5HT_{1A}$ agonist, produced a normal prolactin release when given intravenously to healthy male volunteers. This effect was blocked by the nonselective 5HT-receptor antagonist metergoline and by pindolol, a β-adrenergic and $5HT_1$ antagonist, in a dose-related fashion. Prolactin response to buspirone was inversely correlated with levels of "irritability" in patients with personality disorders, suggesting that decreased sensitivity of the $5HT_{1A}$ receptor may be responsible for components of impulsive-aggressive behavior in patients with personality disorders. Future research should deepen our understanding of receptor subtype specificity in the control of aggression, leading in time to exploration of links to receptor genes.

Interactions between neurotransmitter systems and humoral systems need to be addressed in future clinical research concerning aggression. It is possible that serotonin interacts with other neurotransmitter and neurohumoral systems in modulating impulsivity and aggression. For example, one group of investigators examined the effects of testosterone and serotonin administration on dominance and aggression in rats. In their model, male rats given testosterone became dominant. Quipazine, a serotonin agonist, blocked aggression in both naturally dominant and testosterone-induced dominant rats. Nonspecific serotonin antagonists blocked aggression only in testosterone-induced dominant males. This study begins to address the pharmacoselectivity of different forms of aggression and foreshadows an important future avenue for pharmacologic research (19).

Progress in basic investigations of the neurochemical and neuroendocrine mediators of aggression may set the stage for advances in the pharmacotherapeutics of violent disorders. As receptor subtypes involved in modulating specific aggressive behaviors are isolated, it becomes feasible to tailor pharmacologic diagnostic probes and treatments to specific clinical settings. When a better understanding of the individual neurochemical and neuroendocrine factors contributing to aggression is attained, interactions among the multiple agents operating convergently and divergently at hierarchical sites in the neuraxis to regulate hostile behavior may be more fully explored.

Recent studies have suggested that excitatory amino acid receptors, such as the class activated by *N*-methyl-d-aspartate (NMDA), may be critically involved in the associative process of long-term potentiation within the hippocampus and amygdala. Agents that modulate this response, such as selective NMDA-receptor blockers or nitric oxide synthetase inhibitors, merit investigation for antiaggressive and serenic properties in temporolimbic epileptic patients.

SUMMARY

We recognize that many aggressive behaviors are neither maladaptive nor the result of neurologic disease. Even when violent behavior and neurologic lesions coexist, they may not be causally related. A violent lifestyle may lead to head trauma and neurologic abnormalities that are the consequence, rather than the cause, of aggression.

Although fully accepting these qualifications, we believe that neuropsychiatric care is enriched by recognizing both that a diverse array of neurologic lesions may contribute to violent behavior in human beings and that abnormalities at different levels of the neuraxis produce distinctive subtypes of aggression. Future basic and clinical studies that consolidate and extend our understanding of the multiregional, hierarchical neural networks regulating aggression are urgently needed to refine diagnostic and therapeutic approaches to violent individuals.

REFERENCES

1. Violence Prevention Panel. Prevention of violence and injuries due to violence. *MMWR* 1992;41:5–7.
2. Davis S. Violence by psychiatric inpatients: a review. *Hosp Community Psychiatry* 1991;42:585–590.
3. Maynard Smith J. The theory of games and the evolution of animal conflicts. *J Theor Biol* 1974;47:209–221.
4. Kravitz EA. Hormonal control of behavior: amines and the biasing of behavioral output in lobsters. *Science* 1988;241:1775–1781.
5. Moyer KE. Kinds of aggression and their physiological basis. *Commun Behav Biol* 1968;2A:65–87.
6. Delgado JMR. Social rank and radiostimulated aggressiveness in monkeys. *J Nerv Ment Dis* 1967;144:383–390.
7. Jouvet M, Delorme JF. Locus coeruleus et sommeil paradoxal. *C R Seances Soc Biol Fil* 1965;159:895–899.
8. Bard P. A diencephalic mechanism for the expression of rage with special reference to the sympathetic nervous system. *Am J Physiol* 1928;84:490–515.
9. Reeves AG, Plum F. Hyperphagia, rage and dementia accompanying a ventromedial hypothalamic neoplasm. *Arch Neurol* 1969;20: 616–624.
10. Kluver H, Bucy PC. Preliminary analysis of functions of the temporal lobe in monkeys. *Arch Neurol Psychiatry* 1939;42:979–1000.
11. Raine A, Leucz T, Scerbo A. Antisocial behavior: neuroimaging, neuropsychology, neurochemistry, and psychophysiology. In: Ratey JJ, ed. *Neuropsychiatry of personality disorders.* Cambridge, MA: Blackwell Scientific, 1995;50–78.
12. Ross ED. Nonverbal aspects of language. *Neurol Clin* 1993;11:9–23.
13. Blumensohn R, Ratzoni G, Weizman A, et al. Reduction in serotonin $5HT_2$ receptor binding on platelets of delinquent adolescents. *Psychopharmacology* 1995;188(3): 354–356.
14. Grossman SP. Chemically induced epileptiform seizures in the cat. *Science* 1963;142:409–411.
15. Mattes J. Comparative effectiveness of carbamazepine and propanolol for rage outbursts. *J Neuropsychiatry* 1990;2:159–164.
16. Archer J. The influence of testosterone on human aggression. *Br J Psychiatry* 1991;82:1–28.
17. Bergman B, Brismor B. Hormone levels and personality traits in abusive and suicidal male alcoholics. *Alcohol Clin Exp Res* 1994;18(2):311–316.
18. Loosen PT, Purdon SE, Pavlou JN. Effects on behavior of modulation of gonadal function in men with gonadotropin-releasing hormone antagonists. *Am J Psychiatry* 1994;151(2):271–273.
19. Bonson K, Winter J. Reversal of testosterone-induced dominance by the serotonergic agent quipazine. *Pharmacol Biochem Behav* 1992;42:809–813.

19

Clinical Pain Syndromes

Howard L. Fields and David A. Fishbain

OVERVIEW

Persistent pain presents a major clinical challenge. This is because pain is subjective, and the knowledge required for optimal evaluation and treatment crosses traditional disciplinary lines. Severe persistent pain can have a strikingly destructive impact on the psychological state of the patient. Furthermore, psychological problems that either predate or are induced by persistent pain can add to the intractability of the problem. This chapter reviews the clinical features of patients with neuropathic pain, the current thinking about the pathophysiology of different syndromes, and a strategy for the treatment of patients with this condition.

CLINICAL FEATURES OF PAIN SYNDROMES

Epidemiology of Pain and Chronic Pain in the General Population

A recent New Zealand study indicated that 81.7% of the general population has had a pain experience severe enough to have led to a consultation with a doctor or other health professional or to the use of medications for the pain or that interfered with life or activities "a lot." In general, the prevalence of pain increased with age, except for headache and abdominal pain. In a suburban Australian community, 4.3% of the population was in constant pain. Finally, it has been reported that 14.4% of the U.S. population between the ages of 25 and 74 years may suffer from chronic pain related to the joints and musculoskeletal system (1–3). Based on these reports, the following observations can be drawn:

1. Pain is extremely common in the general population and is reported in 4.3% to 40% of patients.
2. The lifetime prevalence of persistent pain may be as high as 80%.

The following is a list of the common features of neuropathic pain:

Association of the pain with evidence of neural damage, particularly sensory deficits. Although not conclusive, this feature is by far the most reliable indicator of neuropathic pain. Except in specific, easily identifiable syndromes, such as trigeminal neuralgia or pain due to epilepsy, the absence of this feature should raise doubt about the diagnosis. Commonly, the reported location of the neuropathic pain is at least partially coextensive with a sensory disturbance; usually, this is a sensory loss, but sometimes hyperresponsiveness to sensory stimuli is present without a deficit.

The sensory quality of the pain is unique to the patient's experience. Burning and tingling are frequently used descriptors. Shooting, shocking, and electrical feelings are also common terms that such patients use. Crawling, tightness, cramping, and tearing are also used. The point is that the pain is unusual and difficult to ignore. Neurologists use words like *dysethesias* (unpleasant sensation) or *paresthesias* (abnormal sensations) to describe these phenomena.

There is often a significant delay between the causative insult and the onset of pain. For example, in patients with pain due to central nervous system (CNS) lesions, the pain often begins after a delay of months, usually after the patient has achieved partial recovery of motor or sensory function.

Hypersensitivity phenomena are present. It is not unusual for patients with painful injuries of any type to complain of tenderness of the affected part and to avoid threatened contact with it. With neuropathic pain, such hypersensitivity phenomena are exaggerated, often to the point of contributing in a major way to the patient's disability. For example, many patients with posttraumatic neuralgias complain that any movements involving the affected nerve trigger severe pain. Many neuropathic pain patients report that light, moving stimuli, such as a gust of wind or the brushing of their skin by clothing, induces rapid bursts of pain. This phenomenon, in which severe pain is evoked by very light, moving tactile stimuli, is termed *allodynia* and is suggestive of neuropathic pain. Other hypersensitivity phenomena include reduced threshold for heat pain, a striking buildup of reported pain with repeated stimuli that are near threshold (summation), the spatial spread of perceived pain from the site of the noxious stimulus, and a prolonged paroxysm of pain following a brief stimulus (after discharge). These hypersensitivity phenomena are often thought of as defining the hyperpathic state (4).

NEUROBIOLOGIC MECHANISMS OF NEUROPATHIC PAIN

Until better diagnostic tools are developed to determine the proximate cause of the pain, the optimal strategy for treatment of an individual patient will remain a sequence of therapeutic trials targeted on specific pain-generating mechanisms.

Deafferentation Hyperactivity

Deafferentation is associated with pain. It seems paradoxical that interruption of the pathways involved in pain transmission should result in pain. Usually, this does not occur. Loss of small-diameter primary afferents, such as occurs in some polyneuropathies, usually leads to loss of pain sensation. Furthermore, in most cases, interruption of central pain transmission pathways results in impaired pain sensation without spontaneous pain. This is the basis for the use of cordotomy for the treatment of certain pain patients. On the other hand, CNS lesions can cause pain in some patients. Such central pain syndromes are almost always accompanied by impaired pain and temperature sensation.

The most persuasive clinical example of deafferentation pain results from avulsion of the brachial plexus. In this condition, backward hyperextension of the arm places severe traction on the brachial plexus, and some dorsal roots are anatomically separated from the spinal cord. Most of these patients have spontaneous pain referred to the anesthetic extremity. In animal studies, cutting the dorsal roots results in the development of high-frequency spontaneous activity in dorsal horn neurons, some of which may be pain transmission neurons. In support of the concept that spontaneous activity in dorsal horn neurons contributes to the pain of brachial plexus avulsion, investigators have shown that destructive lesions of the dorsal horn can give significant relief to patients with pain due to avulsion of the brachial plexus (5).

It is possible that some pain syndromes associated with lesions of the CNS are caused by deafferentation hyperactivity of central pain transmission neurons. For example, thalamic pain could be caused by deafferentation of the cortical neurons to which they project.

Loss of the Inhibition Produced by Myelinated Primary Afferents

Primary afferents are classified by their response to peripheral stimuli, their axonal diameter, and their conduction velocity. Primary afferent nociceptors have axons that are of small diameter, mostly unmyelinated, and therefore conduct at slow velocities. Early psychophysical work had shown that when large-diameter axons in a peripheral nerve are blocked selectively by pressure or ischemia, pain threshold is unaffected, whereas discriminative aspects of sensation (joint position sense, two-point discrimination, vibration sense) are lost. More important, experimental blocking of myelinated axons produces an exaggerated response to noxious stimuli. In fact, stimuli that are usually not painful can produce significant pain when only unmyelinated primary afferents (C fibers) are functioning in a peripheral nerve. In parallel animal studies, dorsal horn neurons that respond to noxious stimuli show an exaggerated response to nociceptor activation when their predominantly inhibitory input from myelinated axons is blocked.

These observations lead to the conclusion that some central pain transmission neurons have inhibitory input from myelinated primary afferents. It is, thus, possible that pain can result from damage to peripheral nerves when it is relatively selective for myelinated fibers. Traumatic mononeuropathies caused by compression would be most likely to produce damage of this type because larger-diameter, myelinated axons are more susceptible to compression and ischemic damage. Consistent with this idea is the clinical evidence that selective electrical stimulation of large-diameter axons in a peripheral nerve can be dramatically effective in relieving pain caused by traumatic nerve injury.

Ectopic Impulse Generation

Although it is likely that deafferentation hyperactivity of central pain transmission neurons and selective damage to large-diameter primary afferents are major contributing factors in some patients with neuropathic pain, such patients are undoubtedly a minority. Other mechanisms must be postulated for most patients. For example, patients with such common painful conditions as postherpetic neuralgia and painful diabetic neuropathy often have minimal deafferentation, and small-diameter axons may show relatively greater damage than large-diameter axons.

A major breakthrough in our understanding of neuropathic pain was the discovery that many primary afferents become active spontaneously when they are damaged. In rats, when the sciatic nerve is cut, axons sprout from the proximal cut end and form a neuroma. Many of the primary afferents that grow into the neuroma, including some that are unmyelinated, develop spontaneous activity. Furthermore, when primary afferents are damaged, ectopic impulses can be generated at a site near the dorsal root ganglion, as well as at the damaged and regenerating distal axon tip. Such spontaneous activity in primary afferent nociceptors is a likely source of pain in patients with nerve injury.

In addition to providing a possible explanation for the association of pain with damage to small-diameter primary afferents, the discovery of ectopic impulse generation may help us understand why certain patients with neuropathic pain obtain relief with membrane-stabilizing drugs. For example, the antiepileptic drugs (e.g., carbamazepine) and the antiarrhythmics (e.g., lidocaine) are not known to have general analgesic efficacy, but they can be dramatically helpful for some patients with neuropathic pain (see discussion that follows). In fact, such membrane-stabilizing drugs block the ectopic impulse generation in damaged primary afferents at concentrations that spare normal axonal conduction.

Peripheral Release of Proinflammatory Neuropeptides From Primary Afferent Neurons

Many unmyelinated primary afferent axons contain neuropeptides that are released from their peripheral terminals. There are a number of such peptides, but the most extensively studied is the 11-amino acid peptide substance P. Activity of certain primary afferents causes the release of substance P from their peripheral terminals. Substance P is a potent vasodilator and a chemoattractant for white blood cells and can also elicit the release of histamine from mast cells and serotonin from platelets. By releasing biologically active peptides like substance P, activity in primary afferent nociceptors can contribute to a local inflammatory process. This effector function of neurons that have traditionally been thought of as purely afferent is undoubtedly part of their normal tissue protective function.

Sympathetic Nervous System Outflow

Although electrical stimulation of the sympathetic chain is normally painless, a small percentage of patients with peripheral nerve injury develop a severe pain syndrome that is exacerbated by sympathetic activity and reversed by blockade of the sympathetic nervous system.

Mitchell was the first to describe what is probably the most dramatic example of sympathetically maintained pain: causalgia (6). *Causalgia* is the syndrome of burning pain that occasionally follows peripheral nerve injury. In addition to the pain, which is the most prominent and disabling feature of the syndrome, patients with causalgia are often observed to have a cold, sweaty, and swollen extremity, especially distally in the limb. In addition, there may be focal arthritis and exquisite hypersensitivity to light, moving touch. In many patients, the pain is exacerbated by loud noises, movements, and cold.

Experimental models of neuropathic pain have provided a basis for understanding how the sympathetic nervous system can elicit and maintain pain. After partial injury to their spinal nerves, rats develop hypersensitivity to mechanical stimuli, which is abolished by sympathectomy. Sympathetic efferent activity sensitizes primary afferent nociceptors, but only in damaged peripheral nerve. Furthermore, damaged nociceptive afferents that have regenerated into a neuroma can be excited by activating sympathetic efferents that have grown into the same neuroma. This ability of sympathetic outflow to evoke afferent activity can persist for months in rats.

Prolonged Changes in Central Neurons Generated by Synaptic Activity

There is evidence from experimental studies in animals that tissue-damaging stimuli can elicit prolonged changes in the excitability of central pain-transmission neurons. Even brief stimuli, especially if they are intense and involve deep somatic and visceral structures, are capable of eliciting a prolonged hyperexcitable state in spinal cord pain transmission neurons. Depending on the stimulus intensity, duration, and particular nerve stimulated, this hyperexcitable state may last for hours. There is evidence that a sensory barrage occurring at the time of injury can contribute to the hyperpathic syndrome that develops later.

These long-term changes in the CNS following intense nociceptor activation make a convincing case for the theory that there is a persistent memory trace for pain. The clinical evidence supporting this idea is controversial; however, there are reports that patients having preoperative or intraoperative local anesthetic block to prevent the massive sensory input from surgical trauma have less severe postoperative pain. Similarly, preoperative anesthetic block appears to lower the incidence of postamputation phantom limb pain (7).

The presence of a prolonged central hyperexcitable state, although not unique to neuropathic pain, may help to explain the hyperpathia often observed in these patients. For example, in patients with causalgia or postherpetic neuralgia, severe pain can usually be elicited with light, moving mechanical stimuli. This phenomenon, termed *allodynia*, is mediated by activity in large-diameter myelinated fibers, whose activity normally elicits light tactile sensations. Perhaps spinal pain transmission neurons can be sufficiently excited by large-diameter myelinated afferents to elicit pain sensations only when they are sensitized by prior noxious input.

Nociceptive Nerve Pain

It is important to discuss the possibility that pain may arise from activation of nociceptive primary afferents that innervate the connective tissue sheath of nerve trunks. Although such an innervation of the nerve sheath by nociceptive primary afferents is unproved, the concept is supported by the clinical observation that inflammation of nerve trunks is painful. Peripheral nerve sheaths are known to be innervated by nervi nervorum. Furthermore, there is clear evidence that the vasa nervorum are innervated by axons containing peptides, including substance P. Because substance P is a marker for unmyelinated primary afferents, mostly nociceptors, it is likely that these axons render the nerve sensitive to noxious stimuli and inflammation.

Thus, there are two mechanistically different types of pain arising from injured nerves: one due to cutting axons or otherwise interfering with normal nerve conduction, and the other due to "physiologic" activation of the nociceptors that innervate the nerve sheath. We refer to this latter type of pain as *nociceptive nerve pain.* It may account for the deep aching pain of nerve root irritation or brachial neuritis.

MEASUREMENT OF PAIN

The clinical measurement of pain has been a difficult problem, infrequently evaluated as an end point of treatment. Currently, the following *indirect* methods are available for the measurement of clinical pain:

Rating scale methods in which patients rate pain experiences on structured scales with clearly defined limits

Psychophysical methods that attempt to define pain threshold and pain tolerance in terms of experimentally induced pain and then ask the patient to *match* the perceived experimentally induced pain to his or her current clinical pain

Measurement of drug dosage required to relieve the pain

Measurement of observed pain behaviors, such as moving in bed, bracing, rubbing an affected part

Magnitude estimation procedures in which judgments of perceived pain are translated into cross-modality matching techniques such as handgrip force

Measurement of performance ability on laboratory tasks

Human physiologic correlates, such as direct recording from peripheral nerves and evoked potentials

Some specific approaches to the use of rating scale methods, psychophysical methods, and measurement of observed pain behaviors are described in the sections that follow.

Rating Scale Methods

Visual Analog Scale

Of the rating scale methods, the Visual Analog Scale (VAS) is the most researched and accepted method, and it is the most widely used in measuring clinical pain. The VAS consists of a 10-cm line anchored by two extremes of pain, "no pain" and "pain as bad as it could be." Patients are asked to make a mark on the line that represents their level of perceived pain intensity, and the scale is scored by measuring the distance from the "no pain" to the patient's mark. There are various types of VAS scales, for example, scaled from 0 to 10, but the most practical index uses a 101-point numeric scale. An example of this scale is shown in Fig. 19-1 (8).

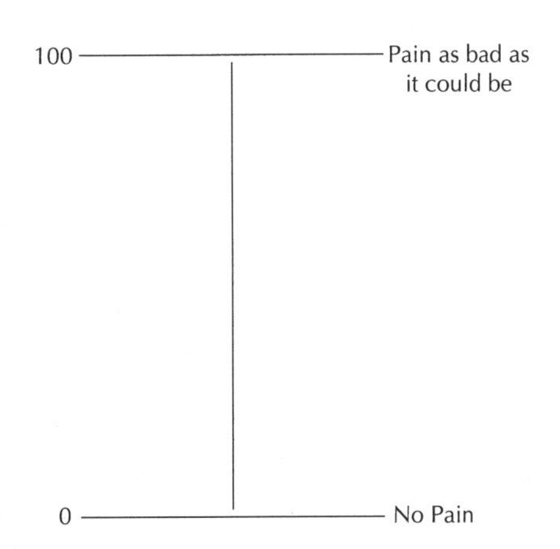

FIG. 19-1. The Visual Analog Scale (VAS) (10 cm).

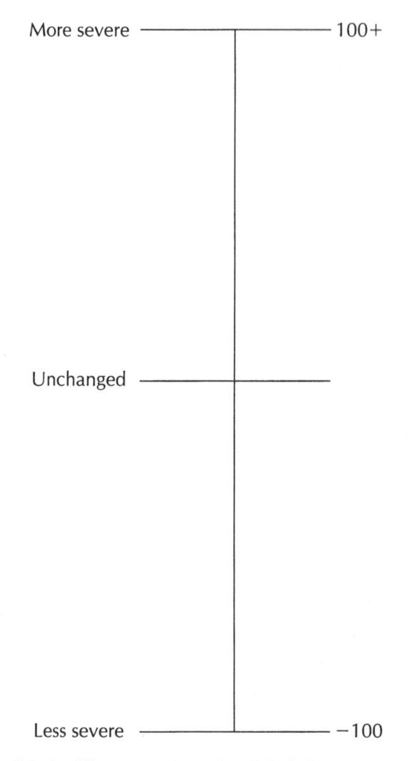

FIG. 19-2. The comparative VAS (20 cm, total).

A common problem with the VAS is that it assumes pain to be a one-dimensional experience that varies only in intensity. Of the rating scale methods, the VAS is preferable for clinical application. However, even the VAS may have poor sensitivity to treatment effects. Therefore, it has been recommended that, when measuring treatment effects, two types of VAS be employed:

- Absolute VAS (described in preceding section)
- Comparative VAS

The comparative VAS is 20 cm in length and has three points: less severe, unchanged (at the midline), and more severe (Fig. 19-2).

Description Differential Scale

The Description Differential Scale (DDS) consists of a list of adjectives describing different levels of pain intensity (Fig. 19-3). Chronic pain patients (CPPs) are asked to rate the intensity of their pain as more or less than the word on the list by placing a check mark either to the left (less) or right (more) of the word being rated. If the word describes their pain level, they place a check mark under the word. There are 10 points to the right and left of each word, giving a 21-point rating scale. Pain intensity is defined as the mean of each rating and ranges from 0 to 20. The DDS scale has high internal consistency. The major advantage of this scale is the fact that chronic pain patients (CPPs) can be checked to see how consistently they are using the scale by examining the relationship among different items (9).

Pain Measurement Inventories

McGill Pain Questionnaire

The McGill Pain Questionnaire is an inventory rather than a rating scale. It is designed to quantify three dimensions of pain: sensory, affective, and valuative. The questionnaire is made up of 20 sets of pain word descriptions, each set containing up to six words. Pain patients are asked to circle words in each set that are relevant. The investigator scores the number of words chosen on the total number of word sets that apply to the pain. Because the words within each set have been ranked, one can compute a total score (10).

Million Scale

There is now convincing evidence that chronic pain intensity levels and measures of functional impairment and disability are correlated (11–20). From these data, one may conclude that pain itself results in functional impairment or disability, that the pain is perceived by the patient as a disability, or both. This evidence has been used in developing rating scales and inventories that tap both the perceived pain and perceived functional impairment aspects of the pain experience. The Million Scale is an example of this concept. This is a 15-item inventory scored on a VAS

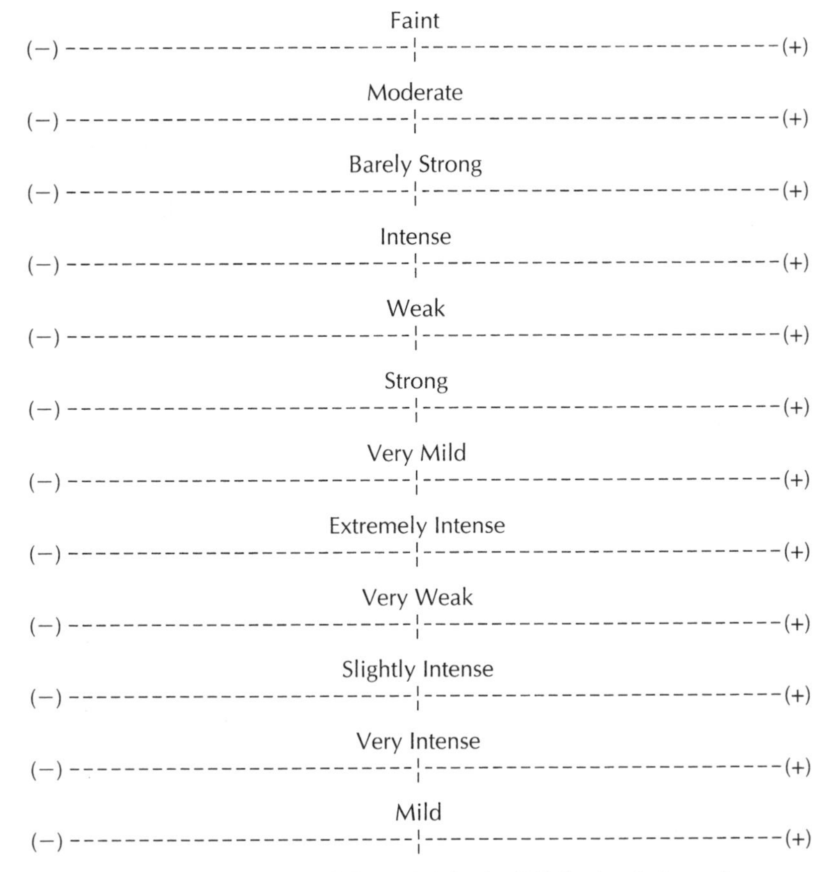

Instructions: Each word represents an amount of sensation. Rate your sensation in relation to each word with a checkmark.

Faint
(−) ------------------------|------------------------ (+)

Moderate
(−) ------------------------|------------------------ (+)

Barely Strong
(−) ------------------------|------------------------ (+)

Intense
(−) ------------------------|------------------------ (+)

Weak
(−) ------------------------|------------------------ (+)

Strong
(−) ------------------------|------------------------ (+)

Very Mild
(−) ------------------------|------------------------ (+)

Extremely Intense
(−) ------------------------|------------------------ (+)

Very Weak
(−) ------------------------|------------------------ (+)

Slightly Intense
(−) ------------------------|------------------------ (+)

Very Intense
(−) ------------------------|------------------------ (+)

Mild
(−) ------------------------|------------------------ (+)

FIG. 19-3. Descriptor Differential Scale (DDS) of pain intensity.

concerning the association of pain with several activities and self-perceived functional impairment. It appears to be valid and reliable in chronic low back pain patients. Low scores on this scale appear to predict return to work (21).

Pain Behavior Methods

Pain behavior has been defined as "any and all outputs of the patient that a reasonable observer would characterize as suggesting pain, such as (but not limited to) posture, facial expression, verbalizing, lying down, taking medications, seeking medical assistance and receiving compensation" (22). A list of identified pain behaviors is presented in Table 19-1. Pain behaviors can often be elicited during physical examination and correlate with physical examination findings, number of operations, and longer pain histories. To measure pain behavior in a systematic fashion, the University of Alabama Pain Behavior Scale was developed. This scale quantifies the observed pain behavior and has been shown to be reliable. In addition, ratings of pain behavior using this scale appear to be *significantly* related to both VAS sensory and VAS affective ratings (23).

TABLE 19-1. *Behaviors utilized in pain behavior rating scales*

1. Sits stiffly
2. Moves Slowly
3. Limps
4. Shifts position frequently
5. Stands bent forward
6. Walks bent forward
7. Distorted gait
8. Grimaces
9. Holds painful part
10. Rubs painful part
11. Moans
12. Groans
13. Writhes
14. Uses cane
15. Takes medications for pain
16. Moves in a guarded fashion
17. Uses heat or ice
18. Uses prosthetic devices
19. Rests frequently (lying down)
20. Avoids physical activity

Pain-Matching Methods

Although the VAS scale is efficient and simple to use in pain measurement, it suffers from one major problem: pain threshold (level of stimulation at which the subject begins to perceive pain in a pain stimulation experiment) and pain tolerance (level of stimulation at which the subject can no longer tolerate the induced pain in a pain stimulation experiment) vary among individuals. Although a pain stimulus may approach pain tolerance or "pain as bad as it could be" for one patient, another may perceive it as mild. Therefore, the VAS is most effective in measuring pain intensity and pain change in the same person but is less reliable when it is used for between-subject pain comparisons. In an attempt to solve this dilemma, the pain-matching method was developed. This method allows a more *direct* measurement of the actual clinical pain a patient experiences and comes closest to the true experimental situation. Various methods can be used in humans to induce clinical experimental pain, including heat, electric shock, noise, and pressure. Of these, pressure has been noted to approximate most closely chronic, gnawing pain. A pressure algometer was developed for this type of pain induction and improved to give greater reliability while producing sensations more closely related to clinical pain (24).

Dilemma of the Severity of Pain and Its Relationship to Function and Disability

The major problem in the evaluation and treatment of chronic pain is the frequent discrepancy between the patient's reported level of pain and resultant functional status and the physician's perception of what the patient should be able to do functionally. Chronic pain patients perceive their pain as a disability that limits their functional status. The great number of patients who claim disability based on pain alone has forced the U.S. Social Security Administration Commission on the Evaluation of Pain to recommend the development of a listing based on "impairment due primarily to pain" (25). Also, chronic pain patients often lack documentation of clear structural organic pathology, thereby having minimum "medical impairment" as currently rated by the American Medical Association (AMA) Guides to Permanent Impairment. At the same time, many chronic pain patients demonstrate higher disability than medical impairment ratings (26).

Medical impairment ratings vary widely from one physician to another, as do measures of disability. These differences in impairment ratings have been blamed on evaluation schedules that are not scientifically based and that do not take functional status into consideration.

In cases in which patients report severe pain, physicians have great difficulty in estimating patients' functional status. In addition, physicians have great difficulty in estimating anxiety, pain, and activity limitation from patients' self-report and the physical examination. These three dimensions are underestimated 35% of the time, and the activity limitation dimension is the one most often underestimated. Patient-generated statements about severity of pain and functional impairment from pain have been shown to correlate. However, the measures show few relationships to other measures of severity of pain: medication consumption, health care use, activity level, and frequency counts of engaging in a set of commonplace activities.

Chronic pain patients display a reduced activity level attributable to pain resulting in a consistent negative relationship between exercise and pain behavior; that is, the more exercise performance, the fewer the pain behaviors.

ASSOCIATION OF PAIN WITH PSYCHIATRIC SYNDROMES

What Is Chronic Pain?

There is no consensus definition of chronic pain. The current definitions can be summarized as follows.

The Institute of Medicine Committee Report on Pain and Disability recognized chronic pain as a syndrome and has defined it as follows: "pain lasting for long periods of time . . . more than 6 months . . . may be associated with a residual structural defect, . . . may be pain persisting past healing time without objective physical findings of residual structural defect, . . . may be pain that recurs regularly and frequently over long periods." In addition, the committee pointed out that there was no agreed-on operational definition of chronic pain in the studies reviewed (27).

The most recent issue of the *AMA Guides to the Evaluation of Permanent Impairment* recognized chronic pain as a syndrome and pointed out that the medical profession had been slow to identify chronic pain as a specific medical disorder. The AMA defined acute pain, acute recurrent pain, and chronic pain. Acute recurrent pain was defined as "episodic noxious sensations resulting from tissue damage in chronic disorders, e.g. arthritis, tic douloureux." Chronic pain was defined as "not a symptom of an underlying acute somatic injury . . . [but rather] a pathological disorder in its own right [that is] . . . chronic, . . . long-lived, [and] . . . progressive [and in which] . . . tissue damage has healed and does not serve as a generator of pain, . . . although applied to pain of greater than 6 months duration a chronic pain syndrome can be diagnosed 2 to 4 weeks after onset." In addition, the AMA awarded a 5% impairment (lumbar) for discal herniation or other soft tissue lesions with a minimum of 6 months medically documented pain associated with muscle spasm and rigidity (26).

In the classification of chronic pain by the *International Association for the Study of Pain Subcommittee on Taxonomy*, chronic pain was supposed to be coded on Axis IV as 1 month, 16 months, or greater than 6 months. In their definitions, the subcommittee advised that "many people report pain in the absence of tissue damage. . . . [I]f they regard their experience as pain, it should be accepted as pain. . . . [T]his definition would avoid tying pain to the stimulus" (28).

Based on the previous statements, a number of conclusions can be drawn:

1. Chronic pain is now recognized as a distinct syndrome.
2. Quoted authorities appear to consider pain chronic if it has lasted for more than 6 months, but they point out that chronic pain may begin earlier.
3. There is disagreement about the importance of the presence or absence of continued tissue damage and the presence or absence of objective physical findings in the definition.
4. Overall, there appears to be no universally accepted operational definition of chronic pain.

Patients demonstrate having chronic pain through pain behavior, impaired function, and disability perception. Patients define themselves as suffering from chronic pain if the periods of what they perceive as intolerable pain are frequent enough to interfere with normal function. As noted earlier, however, there is neither a totally satisfactory definition for chronic pain nor an operational diagnosis for chronic pain. This major problem has interfered with delineating the frequency of chronic pain within the general population and within the psychiatric population.

Pain in Psychiatric Disorders

Pain is a common documented symptom in psychiatric patients, but there are few studies

TABLE 19-2. *Similarities and differences in pain characteristics between psychiatric patients and chronic pain patients*

	Psychiatric patients	Chronic pain patients
Pain as a presenting complaint	Rarely	Always
Consistency of pain	Transient	Chronic
Location	Cephalic Trunkal	Low back Neck
Onset of pain with injury	Insidious	Usually associated
Pain sites	Multiple	Generally low back and/or neck
Neuroanatomic correlation	Poor	Poor to good
Exacerbating factors	Nonspecific	For low back pain, usually lifting, bending, sitting, or standing
Alleviating factors	Nonspecific	For low back pain, usually walking, lying down, changing positions, ice, or heat

of pain complaints in these patients. France and associates have described a number of characteristics of the pain in psychiatric disorders (29). To delineate the differences between psychiatric patients with pain and chronic pain patients seen in pain centers, we have placed France's pain characteristics into table form (Table 19-2). It is clear from this table that the pain seen in psychiatric patients differs from chronic pain seen within pain centers.

Few incidence and prevalence data for pain are available within these diagnostic groups. Available incidence data indicate that 75% of nonpsychotic psychiatric inpatients had been bothered with pain within the last 3 months. In this study, pain was most frequent in patients with neurosis and personality disorders and was associated with unskilled work. The prevalence rate for chronic pain in a psychiatric outpatient clinic has been reported to be 14.37% (30,31).

Relationship Between Chronic Pain and Depression

Within pain treatment facilities, most CPPs are depressed. The reported point prevalence of major depression in the chronic pain population has varied between 1.5% to 54.5%; the reported lifetime prevalence of major depression among CPPs varies between 20% to 71%. The reported prevalence of dysthymia has ranged from 0% to 43.3%. Adjustment disorder with depressed mood has been reported in 28.3% of CPPs (32).

In addition to the reported association between pain and depression, there is strong evidence that pain affects mood and the severity of depression. Pain severity has been found to be associated with negative mood. It also appears that negative mood increases and becomes fixed as pain continues and becomes persistent. Furthermore, CPPs may demonstrate a distinctive depressive syndrome. This is compatible with the observation that dysthymia and atypical depression are associated with greater severity of pain than major depression or adjustment disorder with depressed mood.

Does the Depression Seen in Chronic Pain Patients Precede or Follow the Development of Chronic Pain?

Although depression can precede pain as an independent phenomenon, there is empirical evidence that persistent pain causes depression. A recent longitudinal study of rheumatoid arthritis patients, for example, found that pain severity predicted subsequent depression. The causal relationship between pain severity and depression occurred only after the first 12 months of the study. The situation is complicated because evidence indicates that psychological events may be risk factors for the development of chronic pain but that emotional disturbance is likely the result of chronic pain (33).

Is There Evidence for Preexisting Psychiatric Pathology in Chronic Pain Patients?

This issue has not been extensively explored. An excellent study using the Diagnostic Interview Schedule (DIS) found that, in 81% of the CPPs, alcohol use disorders preceded pain onset. CPPs, when matched against age-matched controls, had significantly higher pre-pain rates of alcohol use disorders but not of depression. A second study used a self-designed questionnaire. In this study, 46% of the CPPs had pre-pain stress-related illness, 34% had a history of psychiatric illness, and 17% had been previously disabled. In a third study, the Standardized Clinical Interview for Depression (SCID-I) was administered to 98 patients with chronic low back pain who had a diagnosis of somatoform pain disorder. Thirty-nine percent of the CPPs admitted to a preexisting substance abuse disorder (41% of men; 33% of women). Twenty-nine percent had at least one episode of major depression before the onset of chronic pain (36% of females; 25% of males). Twenty-one percent had had preexisting symptoms consistent with an anxiety disorder (34–36) (Fig. 19-4).

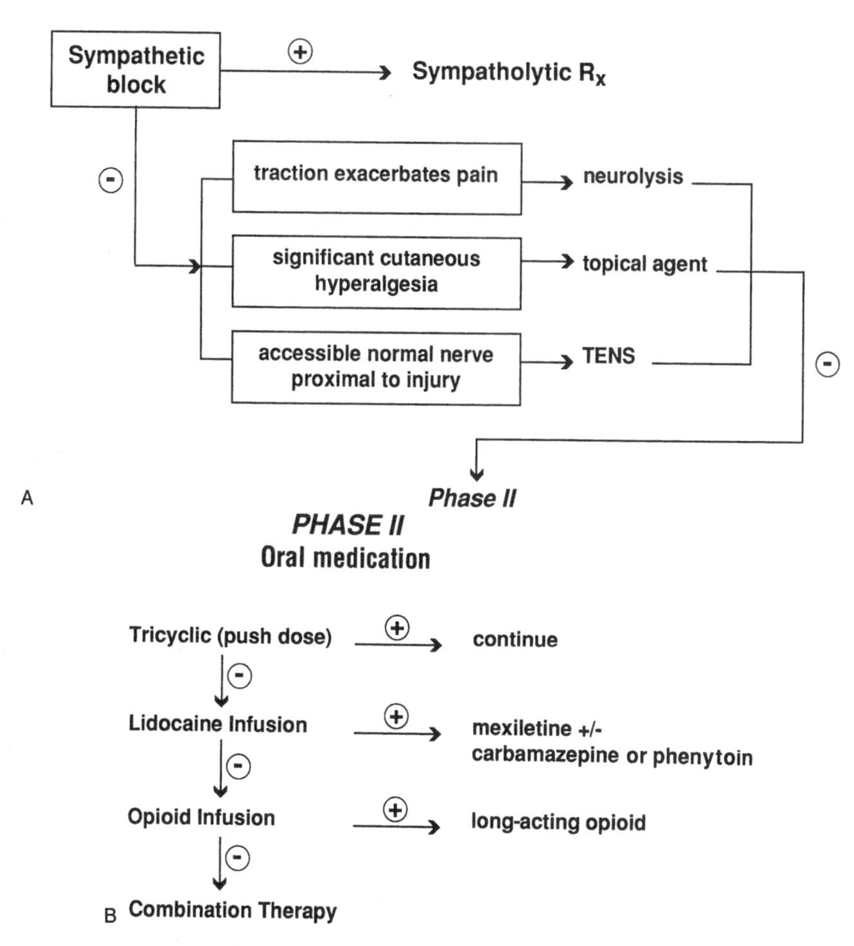

FIG. 19-4. Algorithm for treatment of pain associated with peripheral nerve damage. Phase I outlines the steps to be taken before instituting long-term medical management. Phase II illustrates one systematic approach to long-term pharmacological treatment. (Adapted from Fields HL, Rowbotham MC. Multiple mechanisms of neuropathic pain: a clinical perspective. In: Gebhart GF, Hammond DL, Jensen TS, eds. *Progress in pain research and management.* Vol. 2. Seattle, WA: IASP Press, 1994: 437–454.)

TREATMENT APPROACHES TO PATIENTS WITH NEUROPATHIC PAIN

Sympathectomy

The knowledge that sympathetic outflow can cause pain, particularly in patients with partial nerve injury, has important practical implications. First, the use of sympathetic blockade becomes a major tool in the evaluation of patients with neuropathic pain. If a patient responds to sympathetic blockade, most likely there is a major sympathetic component. Such patients should be treated with repeated sympathetic blocks, physical therapy, and antiinflammatory drugs.

There are a variety of ways to carry out a sympathetic block. The traditional method is by regional blockade of the sympathetic chain with local anesthetic. Recently, the use of intravenous phentolamine has become popular because it is less invasive and produces little discomfort. Furthermore, it does not produce a false-positive response owing to local anesthetic block of nearby somatic sensory axons. Regional infusions with adrenergic antagonists, such as bretylium and guanethidine, have also been used; however, guanethidine is not available in the United States. There is anecdotal evidence that oral sympatholytic agents, such as phenoxybenzamine (Dibenzyline), guanethidine, or prazosin, are helpful. On the other hand, dramatic responses are unusual, and there are no controlled clinical trials demonstrating their efficacy.

Capsaicin

Capsaicin, the active ingredient of the hot chili pepper, is a compound that has been shown specifically to activate unmyelinated primary afferents, primarily nociceptors. At increasing concentrations, capsaicin can activate, reversibly inactivate, peptide-deplete, or destroy primary afferent nociceptors. In human subjects, 0.1% formulations of capsaicin produce reversible cutaneous analgesia. Capsaicin has been used topically as a folk remedy for pain for many years, and there is evidence that a 0.075% preparation can provide relief for patients with postherpetic neuralgia, postmastectomy syndrome, and diabetic neuropathy.

Alternative Treatments

Carbamazepine and tricyclic antidepressants have been the mainstays for neurologists in the treatment of neuropathic pain. Carbamazepine's popularity among neurologists is, no doubt, based on its efficacy in trigeminal neuralgia and its relative safety. Tricyclics are popular because they have broad efficacy, which includes several types of neuropathic pain as well as both tension and migraine headache. Although useful for some patients, antiepileptic drugs and antidepressants are often either ineffective or have unacceptable side effects. Fortunately, recent research has both increased the number of therapeutic options and refined the use of previously available modalities. These alternative treatments for neuropathic pain are described in the sections that follow, along with an algorithm to optimize therapy quickly in an individual patient.

Tricyclic Antidepressants

Tricyclic antidepressants are effective for a broad range of problems, including migraine and muscle contraction headache, low back pain, cancer pain, and, of course, any problem accompanied by significant depression. These tricyclics are effective for a different range of conditions from those of the nonsteroidal antiinflammatory drugs (NSAIDs), which is the other major category of drug that has achieved wide acceptance for chronically painful conditions.

Amitriptyline (Elavil) is the most commonly prescribed tricyclic for the treatment of pain. It is unclear which of its manifold pharmacologic actions is responsible for its analgesic action. It not only blocks the reuptake of serotonin and norepinephrine but also blocks α-adrenergic, muscarinic, cholinergic, and histamine receptors. This broad range of actions produces a significant number of dose-related side effects from the harmless but annoying dry mouth to the more serious orthostatic hypotension, urinary retention, cardiac conduction abnormalities, and memory disturbance. It also produces sedation, which is desirable in some patients and unwanted in others. So pervasive and unpleasant are the side effects that patient compliance requires starting amitriptyline at a very low dose (10 mg every hour of sleep) and building slowly to therapeutic effect.

Other tricyclics are effective for the treatment of neuropathic pain. For example, desipramine is effective for postherpetic neuralgia and is as effective as amitriptyline for diabetic neuropathy. Desipramine is less potent as a histamine and acetylcholine antagonist and is thus less sedating and less likely to cause memory disturbance. As with amitriptyline, it is best to initiate therapy with a very low dose of this drug (10 to 25 mg/d) and raise it about every third day. In nondepressed patients, relatively low doses of a tricyclic may be sufficient for optimal pain control (e.g., 75 mg/d of amitriptyline or desipramine). If the patient's pain does not respond to lower doses of a tricyclic, the dose should be increased into the antidepressant range.

Phenothiazines

Some reports suggest that drugs of the phenothiazine class are useful for neuropathic pain, especially when used as adjuncts to tricyclics; however, controlled studies have failed to demonstrate phenothiazine efficacy for pain management. Furthermore, the side-effect profile of these drugs should discourage their use.

Membrane-Stabilizing Agents

The group of membrane-stabilizing agents includes the anticonvulsants and the local anesthetic-antiarrhythmic drugs. Compared with the tricyclics, this group of drugs is helpful for a more restricted patient population. Carbamazepine and phenytoin have been used for many years in the treatment of neuralgic pain. They are effective for trigeminal neuralgia, and carbamazepine has been reported to be helpful in some patients with painful diabetic neuropathy. Beyond these specific conditions, antiepileptic drugs are most likely to be helpful when there is a shooting, shocklike, lancinating pain. If carbamazepine and phenytoin are ineffective or are helpful but not well tolerated, alternative choices include baclofen (a second-line drug for trigeminal neuralgia), clonazepam, and valproic acid.

In clinical studies, lidocaine, given by intravenous infusion, has been shown to provide immediate and often dramatic pain relief. A variety of neuropathic pain types respond, including postherpetic neuralgia. Because of the effectiveness of lidocaine-like drugs in the treatment of cardiac arrhythmias, several oral drugs of this class are available; tocainide and mexiletine are two examples. Tocainide has considerable toxicity, making mexiletine the preferred first-line drug of this class. Mexiletine has been shown to be effective for diabetic neuropathy and other neuropathic pain.

Opioids

The use of opioids in patients with neuropathic pain of nonmalignant origin is a subject of significant controversy. Although there are open-label, uncontrolled studies indicating long-term benefit of opioids in such patients, there are also studies that have shown no benefit. Although many patients obtain significant relief with opioids, for many, the relief is incomplete and, for some, the benefit is minimal to nonexistent. Unfortunately, in the absence of a therapeutic trial, it is difficult to predict the degree of relief that will be obtained by a given patient. The major argument against a trial of opioids in nonmalignant neuropathic pain is the possibility that the patient will become addicted. Available evidence, however, indicates that the risk for addiction is extremely small in patients on short-term opioid therapy. If a decision is made to use long-term opioid therapy, our preference is to use long-acting opioids. The pharmacokinetics of such compounds (methadone, levorphanol, sustained release morphine) avoids the plasma level peaks (associated with increased side effects) and valleys (associated with breakthrough pain and mild abstinence).

Some patients treated with opioids experience a fading of their efficacy over time. This could represent a change in the underlying pain problem or the development of tolerance. Dose escalation in these patients should be undertaken cautiously and only after the patient is reevaluated both neurologically and psychologically and adjuvant drugs have been added (e.g., tricyclic antidepressants, membrane-stabilizing drugs, or α_2-adrenergic agonists such as clonidine).

A General Algorithm for the Management of Patients With Neuropathic Pain

This algorithm is useful as a rough guideline to evaluate and treat patients with peripheral neuropathic pain. It is unnecessary to use this approach if the diagnosis is obvious and there is an accepted treatment of choice.

Phase I: Initial Evaluation, Local Treatments

The initial step is a diagnostic sympathetic block by phentolamine infusion, particularly in cases with pain in a single extremity that has a burning quality, is made worse by cold, and is associated with swelling (Fig. 19-4). If phentolamine infusion is negative, but there is a high index of suspicion for sympathetically maintained pain, local anesthetic block of the paraspinal sympathetic chain should be carried out.

If the patient has pain associated with a traumatic or compressive mononeuropathy and the pain is made worse by movement, it is possible that the pain is due to traction on a mechanically sensitive neuroma at the site of injury. Such patients can often be helped by either decompression or neurotomy and moving the nerve to reduce the traction on it.

Topical agents should be considered in patients who experience cutaneous discomfort. Such an approach appears to be especially effective if the patients have allodynia or hyperalgesia to cutaneous stimulation. Capsaicin preparations are commercially available. Topical NSAIDs and local anesthetic preparations are not currently available; however, aspirin tablets can be crushed in chloroform, and topical lidocaine formulations can be made by a hospital pharmacist.

Phase II: Oral Medications

If local approaches leave the patient with significant pain, therapy can be initiated with a tricyclic antidepressant. If further relief is needed, a trial of a membrane-stabilizing drug can be given. If there is no relief with an infusion of lidocaine, we have found that oral membrane-stabilizing drugs will not work, and we can save the patient the time it would take to build up to the dose required for a full evaluation of related oral medications. If there is immediate and dramatic relief with the lidocaine infusion, we are willing to push oral mexiletine to high levels (up to 1200 mg/d, provided plasma levels are within the acceptable therapeutic range).

For many patients, tricyclics, membrane-stabilizing drugs, and oral sympatholytics either alone or in combination are inadequate. For these patients, opioids can be useful. We either begin the patient on an oral opioid or use a fentanyl infusion as a predictive test. Finally, we assess the efficacy of various combinations of the aforementioned drugs.

PSYCHOTROPIC DRUGS FOR THE TREATMENT OF NONNEUROPATHIC CHRONIC PAIN

The utility of the following psychotropic drug groups (the World Health Organization classification) in the treatment of nonneuropathic chronic pain is discussed: antidepressants (tricyclics, heterocyclics, serotonin reuptake inhibitors, monoamine oxidase [MAO] inhibitors); neuroleptics; antihistaminics; psychostimulants; and antiepileptic drugs.

Antidepressants (Tricyclics, Heterocyclics, and Serotonin Reuptake Inhibitors): Evidence for Analgesic Efficacy in Chronic Pain

Antidepressants have been used for numerous syndromes for which chronic pain is believed to be nociceptive, due to a psychological condition, or both. These include headache, facial pain, arthritic or rheumatic pain, ulcer pain, fibrositis, low back pain, neck pain, pelvic pain, cancer-associated pain, depression-associated pain, and idiopathic pain. Antidepressants have also been used in studies in which the pain conditions were of *mixed* etiology.

Low Back Pain

Of 12 placebo-controlled studies, six studies have demonstrated an analgesic effect significantly greater than placebo. The most consistent responses are seen with doxepin and desipramine in doses above 150 mg/d.

Cancer Pain

Currently, there is a lack of controlled studies for antidepressant treatment of chronic pain associated with cancer. However, it appears that antidepressants may be opiate sparing in chronic pain associated with cancer. In the reviewed reports, the onset of analgesic action occurs in most cases in less than 6 days. Withdrawal of the antidepressant is associated with breakthrough pain in 24 to 48 hours. Thus, it is likely that the antidepressants are not acting through effects on mood.

Arthritic and Rheumatic Pain

Although a large number of controlled studies have been completed for the treatment of this syndrome with antidepressants, only a few studies have demonstrated the antidepressant used to be more effective than placebo. It appears that the older antidepressants with some serotonergic activity (amitriptyline, imipramine) are overrepresented in the studies that have demonstrated an effect significantly better than placebo. Whether antidepressants boost the efficacy of analgesics in the management of arthritic and rheumatic pain has not been settled.

Fibrositis and Fibromyalgia

The preponderance of studies of antidepressants in the treatment of fibrositic and fibromyalgic pain have indicated an analgesic effect greater than that of placebo. Amitriptyline appears to be superior to naproxen, whereas clomipramine is superior to maprotiline.

Facial Pain

Only a few controlled studies have looked at the use of antidepressants to treat facial pain, and these indicate a favorable pain response. One controlled study demonstrated a reduction in pain *without* a decrease in the level of depression. Researchers in this area have concluded that antidepressants are useful therapy in the treatment of chronic facial pain.

Headache

There have been more controlled studies of antidepressants in the treatment of chronic headache than any other pain syndrome. Most of these controlled studies have found the antidepressant used to be significantly better than placebo. In the controlled studies, the following antidepressants were shown to be efficacious: doxepin, amitriptyline, femoxetine, maprotiline, and clomipramine. The response to headache treatment was often demonstrated at a dose lower than that used to treat major depression, for example, less than 150 mg of amitriptyline. Posttraumatic headache has not been shown to be responsive to antidepressants.

Pain as a Symptom of Depression

It appears that antidepressants, both heterocyclics and MAO inhibitors, are effective in alleviating depression-associated pain.

Psychogenic (Idiopathic) Pain

There have been a surprising number of placebo-controlled studies (eight) performed on the treatment with antidepressants of alleged psychogenic (idiopathic) pain. Of these eight studies, seven demonstrated a significant improvement versus placebo. It can be concluded that antidepressants are effective in the treatment of idiopathic pain. Antidepressants can relieve pain even if they have no significant effect on depressed mood. Onghena and van Houdenhove, in a metaanalysis of 39 antidepressant placebo-controlled studies on the treatment of chronic pain, concluded that the size of the analgesic effect from antidepressants is not much different for pain having an "organic" versus a "psychogenic" basis (37).

Summary

Antidepressants appear to be effective for the analgesic treatment of a wide range of nonneuropathic pain conditions. Most studies showed antidepressants to be superior to placebo. This observation has recently been supported by a metaanalysis of these data:

1. The beneficial effect from antidepressants for most pain conditions is not related to mood.
2. The beneficial effect appears to be mild to moderate.
3. There is strong evidence for the pain effect of amitriptyline, doxepin, and clomipramine and to a lesser extent for imipramine and dothiepin. The antidepressants with more selective neurochemical effects, for example, the serotonin reuptake inhibitors, may be less effective for pain in some specific pain syndromes.
4. CPPs may have an analgesic response to lower dosages of antidepressant than usual for treating depression.
5. The delay in the onset of action of the pain effect varies from a few days to several weeks.

Mechanism of Action for the Antinociceptive Effect of Antidepressants

A number of hypotheses have been advanced as a potential explanation for the mode of action of the antinociceptive effect of antidepressants. These hypotheses are summarized in Table 19-3.

Onghena and van Houdenhove now believe that a biochemical hypothesis is correct: that antidepressants have hitherto unsuspected intrinsic analgesic properties (37). Their metaanalysis did not establish the superiority of smaller antidepressant doses for analgesia, nor did they support the superiority of one antidepressant over another. Thus, one must question the hypothesis that the

TABLE 19-3. *Hypotheses for the mode of action of the antidepressant antinociceptive effect*

Antidepressant effect (manifest or masked depression)
Stabilizing aberrantly conducting neurons or inhibiting their afferent transmission at the spinal cord (migraine headaches, neuropathic pain)
Antiepileptic (suppressing epileptiform activity in deafferented neurons)
Selective serotonin inhibition
Facilitating analgesia by facilitating central monoamine transmission (inhibiting serotonin and norepinephrine re-uptake at the synapse)
Facilitating the descending pain modulation system (resultant decrease in afferent input through the spinothalamic tract)
Interaction with opioid receptors (altering binding characteristics of the receptor to morphine and enkephalins)
Affecting a "core disorder" (common psychobiologic abnormalities but multiple clinical and diagnostic presentation, e.g., chronic pain and depression)
Peripheral antiinflammatory action (inflammatory disorders)
Central skeletal muscle relaxant action (depressing polysynaptic reflexes causing muscle spasm or tension)
Sedative effect
Placebo effect

Adapted from Ref. 46.

analgesic properties of antidepressants are related to inhibition of serotonin reuptake. In summary, manifest depression, masked depression, placebo, sedation, and selective serotonin inhibition are no longer viable hypotheses for the antinociceptive effect of antidepressants. The other hypotheses presented in Table 19-3 are still viable alternatives.

Practical Guidelines for the Use of Antidepressants in Chronic Pain

Based on the concepts discussed previously, two algorithms for the antidepressant treatment of chronic pain are offered (Figs 19-5 and 19-6). When using the algorithm in Fig, 19-5, the following factors should be taken into account:

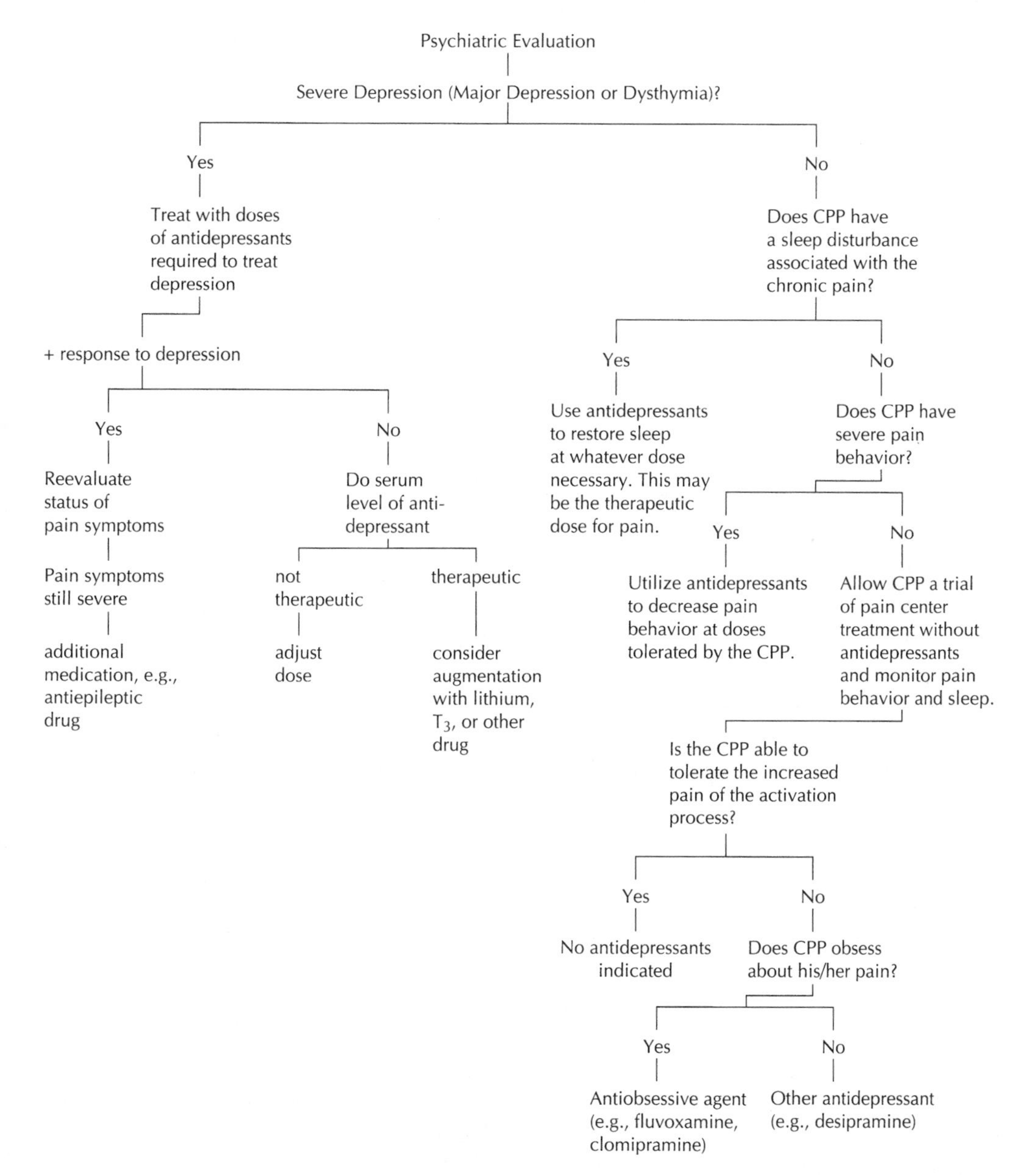

FIG. 19-5. Algorithm for the use or nonuse of antidepressants with chronic pain patients (CPP).

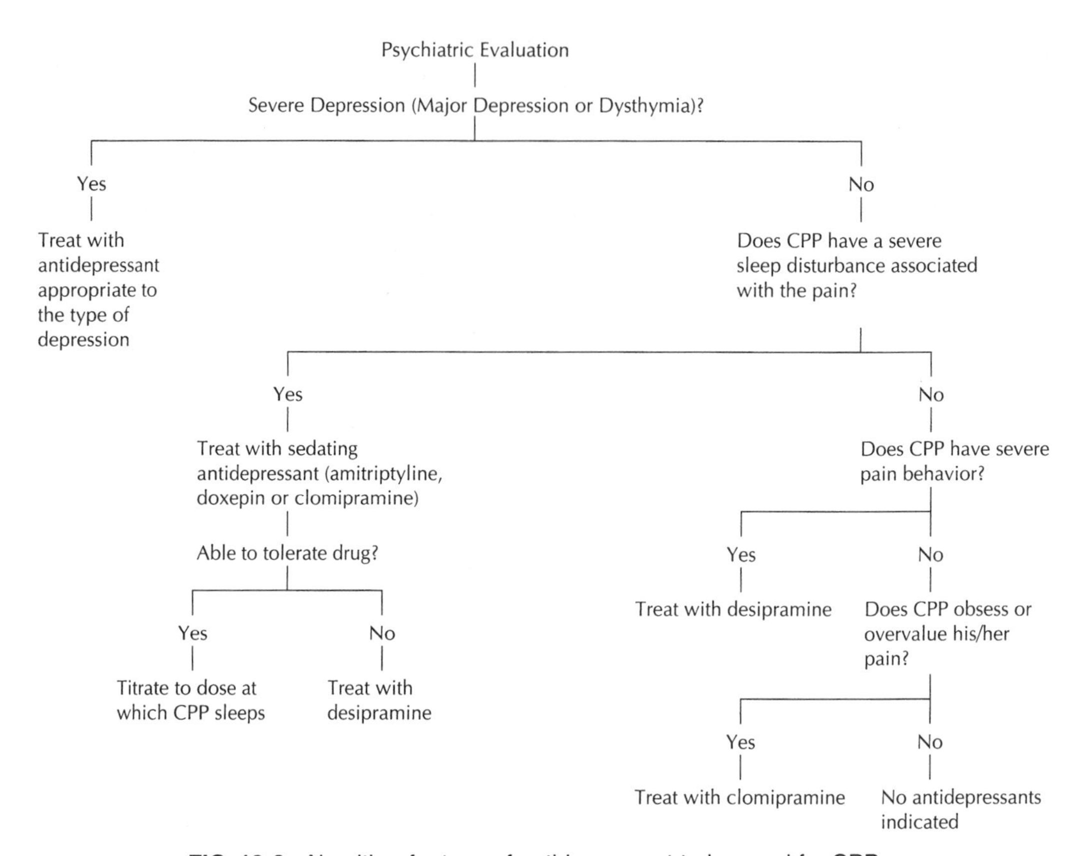

FIG. 19-6. Algorithm for type of antidepressant to be used for CPPs.

Depression. If severe depression is present, it should be treated with antidepressants at usual therapeutic doses for depression.

Sleep. If sleep is disturbed, insomnia should be treated by antidepressants at whatever dose facilitates sleep.

Pain. Antidepressants should be used at a dosage tolerated by the patient but adequate to reduce pain behavior.

A final issue is which antidepressants should be used in the treatment of chronic pain (see Fig. 19-6). The evidence appears to favor the tricyclic antidepressants: amitriptyline, doxepin, clomipramine, and desipramine.

Other Drugs Used to Treat Pain

Monoamine Oxidase Inhibitors

Table 19-4 reveals that only a few studies have evaluated monoamine oxidase inhibitors (MAO) inhibitors for the treatment of pain. Of these, two were controlled and demonstrated a significant analgesic effect. One study demonstrated an analgesic effect superior to that of amitriptyline. One study, however, used facial pain patients, whereas the other utilized patients with various types of pain. At this juncture, it is not yet established that MAO inhibitors have an analgesic effect, although it is likely (38,39).

Neuroleptics

Although neuroleptics, such as fluphenazine and perphenazine, appear to be effective in treating neuropathic pain, the situation is less clear for nonneuropathic pain. This table indicates that there is strong evidence that some neuroleptics may have a significant analgesic effect. Methotrimeprazine in a number of well-designed studies has been shown to have as

TABLE 19-4. *Monoamine oxidase inhibitors (MAO) in nonneuropathic pain*

Study	N	Type of pain	Type of drug	Dose (mg/d)	Controlled study	Percentage of patients reporting pain relief
Anthony & Lance, 1969 (47)	25	Headache	Phenelzine	45	No	80
Lascelles, 1966 (48)	40	Facial pain	Phenelzine	45	Yes	Phenelzine more effective than placebo or other comparison drug.
Raft et al. 1981 (49)	23	Various	Phenelzine vs. amitriptyline vs. placebo	1.5 mg/kg 3.5 mg/kg	Yes	Phenelzine superior to amitriptyline and placebo for pain relief
Raskin, 1982 (50)	1	Chest pain	Tranylcypromine	30	No	Marked decrease in chest pain
Rees & Harris, 1978–79 (51)	4	Facial pain	Phenelzine	?	No	100% complete pain relief or mild brief attacks

strong or stronger analgesic effect than morphine. In addition, neuroleptics have been demonstrated to have a strong analgesic effect on various types of headaches: acute migraine, tension, and cluster.

The risk for tardive dyskinesia associated with neuroleptics is a major deterrent to the use of neuroleptics for the treatment of chronic pain. We recommend that neuroleptics should not be used as first-line drugs in the treatment of nonneuropathic chronic pain. These drugs should be used only when other drugs have failed and only with the informed consent of the patient, the latter including specific mention of the risk for tardive dyskinesia.

Antihistamines

A number of studies have attempted to use antihistamines for the treatment of nonneuropathic pain. In general, antihistamines could have an analgesic effect in nonneuropathic pain, and this effect can be additive to that of other analgesics. This was also the conclusion of Rumore and Schlichting, who in a recent review of 27 controlled clinical trials of antihistamines, concluded that there is evidence for a direct analgesic effect of these drugs (40). Diphenhydramine, hydroxyzine, orphenadrine, and pyrilamine were shown to produce analgesia. This, however, was not the case for chlorpheniramine and phenyltoloxamine. Based on these reports, it has become common practice to combine an opioid with an antihistamine such as hydroxyzine.

Psychostimulants

A limited number of studies have been conducted on the treatment of nonneuropathic pain with psychostimulants. The studies to date indicate that, at least in cancer pain, methylphenidate may have an analgesic effect. Other psychostimulants may also have an analgesic effect. Dextroamphetamine has been found to potentiate the analgesic effect of morphine in acute clinical pain. Fenfluramine was also demonstrated to increase significantly the analgesic potency of morphine and had a mild analgesic effect alone.

As a psychostimulant, cocaine presents a special case. Cocaine has been included within the Brompton cocktail for the treatment of cancer pain. Recent evidence indicates that cocaine has an analgesic effect on experimental pain and potentiates opiate analgesia.

The reviewed studies indicate that there may be a place for the use of psychostimulants as adjuncts to the treatment of cancer-associated chronic pain, especially when sedation limits opiate dosage.

Antiepileptic Drugs

It is now generally accepted that antiepileptic drugs have some efficacy in neuropathic pain syndromes. The issue is less clear, however, for nonneuropathic pain because there have only been a limited number of studies for the treatment of nonneuropathic pain with these drugs. Overall, the available studies indicate that antiepileptic drugs could have an analgesic effect in some pain conditions. Experimentally, phenytoin has been demonstrated to have an analgesic effect against suma-methonium—induced myalgia; however, its mechanism of action is still unclear. In addition, phenytoin, carbamazepine, valproate, and clonazepam have different modes of action. Therefore, it is difficult to postulate one mechanism of action for the analgesic effect of these drugs in nonneuropathic pain.

Finally, as pointed out previously, it is difficult to make a distinction between neuropathic pain and nonneuropathic pain based on diagnosis. It is unclear whether all lancinating, shooting, stabbing, and burning pains have a neuropathic component. Some authors consider these pains neuropathic in nature, but these pains may not fit the diagnostic nomenclature for neuropathic pain, for example, as do diabetic neuropathy and herpetic pain. In our experience, most CPPs have these kinds of pain and yet do not have a neuropathic diagnosis. These patients are usually diagnosed as suffering from failed back surgery syndrome or myofascial pain syndrome, among others. It is therefore possible that some or many CPPs have more than one type of pain, that is, both neuropathic and nonneuropathic. The pain clinician needs to keep this in mind in treating CPPs. Drug choice should be based not only on pain diagnostic category (e.g., peripheral neuropathy) but also on the pain description.

Lithium

Lithium has been used extensively in the treatment of migraine and cluster headache. An earlier reviewer concluded that lithium aggravates or exacerbates the symptoms of migraine except in cyclic migraine, in which it possibly has a positive prophylactic effect (52). As far as cluster headaches are concerned, the reviewer concluded that lithium is effective for the prophylaxis of chronic cluster headache. Patients with chronic cluster headache may respond to lithium in the first 2 weeks of therapy. They usually respond to typical mood-stabilizing serum levels (e.g., 0.8 to 1.5 mE/L).

MULTIDISCIPLINARY PAIN CENTER TREATMENT OF CHRONIC PAIN

Recent evidence from well-designed outcome studies indicates that multidisciplinary pain centers (MPCs) do fulfill some of their goals of treatment. Our group has recently completed two review studies of this outcome literature, including a metaanalysis for return to work. Both of these studies concluded that MPCs return CPPs to work, that the increased rates of return to work are the result of treatment, and that the benefits of treatment are not temporary (41,42).

In a landmark study, Fordyce and co-workers demonstrated that operant conditioning treatment could modify pain behavior (43). In this study, nurses withheld social reinforcement when CPPs displayed pain behaviors and provided attention when CPPs displayed "well behaviors," such as exercise. In addition, the reinforcing effects of pain medications were removed, using instead "pain cocktail" detoxification. The results demonstrated a dramatic increase in exercise tolerance and activity and a decrease in pain ratings and medication intake. Since that study, Fordyce's assertion that operant conditioning is the cause of pain behavior has been criticized and modified somewhat. These techniques are still used in pain center treatment programs, but it is now considered that behavior treatments such as exercise quota systems are effective not because of reinforcement but because of a deconditioning process. At this juncture, one can conclude that pain behaviors are probably not a consequence of operant conditioning but can be modified by operant conditioning. Thus, it has been proposed that all forms of behavioral interventions, including operant conditioning, may exert an influence on chronic pain by changing the way in which the CPP thinks about the pain.

Table 19-5 summarizes the pain management techniques that are currently being used within MPCs. Cognitive and behavioral methods are made up of a variety of these techniques and generally encompass A through E in the table. Recent reviewers have concluded that cognitive behavioral methods *do* show that various cognitive strategies can increase pain tolerance levels. In addition, cognitive coping strategies, a subcategory of cognitive and behavioral methods, have been shown to be more effective in alleviating pain than either no treatment or expected treatment.

TABLE 19-5. *Pain management techniques*

A. Suggestion
   - Information
   - Direct verbal suggestion
   - Programmed suggestion
   - Hypnosis
   - Self-hypnosis

B. Distraction
   - Internal (imagination)
   - External (ceiling TV)

C. Cognitive awareness
   - Detailed information
   - Mental preparation
   - Rehearsal of strain to reduce pain
   - Stimulus control (identification of stimulus and rearrangement to minimize exposure)

D. Anxiety-reduction techniques
   - Relaxation
   - Desensitization

E. Behavior skills training
   - Social skills
   - Stress management
   - Self-management
   - Self-improvement training
   - Self-control
   - Self-efficacy
   - Anxiety management
   - Decision making
   - Social intervention
   - Psychoeducation

F. Self-control
   - Biofeedback
   - Autogenic training
   - Progressive muscle relaxation
   - Imagery relaxation
   - Breathing exercises
   - Tension control

G. Cognitive therapy
   - Transactional analysis
   - Specific therapy to reevaluate meaning of pain, i.e., subjective ideas about possible aversive consequences of pain

H. Operant techniques
   - Decrease attention in environment to pain behavior

I. Conditioning

J. Total-push programs
   - All of the above techniques

Indications for MPC treatment have not yet been delineated in the literature. In general, reasonable criteria for such referrals include the presence of chronic pain refractory to available or standard medical and surgical therapies. Unrealistic and overoptimistic expectations on the part of the patient or referring physician may be relative exclusionary criteria for such referrals.

CONTEMPORARY ISSUES

As neuroscientific research provides increasingly specific agents targeted on neurochemical systems in the brain, new therapies for pain are likely to emerge. Animal studies indicate that this state of hyperexcitability can be partially and selectively blocked by antagonists to the NMDA-type glutamate receptor. For example, ketamine and dextrorphan have *N*-methyl-d-aspartate (NMDA)-receptor–blocking action and have been shown to reduce hyperalgesia in a rat model of neuropathic pain. There is also evidence that the neuromodulator nitric oxide contributes to this long-term change. This knowledge has provided encouraging leads (44,45).

The newer generation of nontricyclic antidepressant agents have not yet been shown to affect neuropathic pain. Fluoxetine is a good example; it is almost a pure 5-HT uptake inhibitor and is an excellent antidepressant. Furthermore, because it has no cholinergic, adrenergic, or histamine-blocking action, it is remarkably free of the troubling side effects that are common with the tricyclics. Unfortunately, it has little or no effect in patients with neuropathic pain who are not depressed. These clinical observations suggest that 5-HT uptake blockade is insufficient to produce pain relief. Other new antidepressants are becoming available, which have more complex neurochemical actions than fluoxetine, and it is possible that some of these will be helpful in treating neuropathic pain syndromes.

SUMMARY

Chronic pain is a common complaint in the general population. A wide range of psychiatric disorders are present in the chronic pain population, the most frequent of which are the mood disorders. There is evidence both that psychiatric disorders usually follow the development of chronic pain and that CPPs may have preexistent psychiatric pathology, particularly alcohol-related problems, episodes of major depression, and personality disorders. It is likely that operant factors are not the reason for pain behaviors but that operant conditioning can aid in the treatment of chronic pain. The most effective chronic pain treatment includes both behavioral and psychopharmacologic interventions, individualized according to the specific neuropsychiatry of each patient.

This chapter has outlined a rational mechanism-based approach to the evaluation and treatment of patients with neuropathic and nonneuropathic pain. It is important to point out that this approach is still evolving and that as we learn more about the mechanisms of neuropathic pain and new treatment options become available, the algorithm will change. Furthermore, the therapeutic outcome for a given patient will be suboptimal unless the assessment and treatment of the psychosocial aspects of his or her problem are carried out concurrently with the somatically based algorithm outlined in this chapter.

REFERENCES

1. James FR, Large RG, Bushnell JA, Wells JE. Epidemiology of pain in New Zealand. *Pain* 1991;44:279–283.
2. Baum FE, Cooke RD, Kaluncy E. The prevalence of pain in a suburban Australian community. *Pain* 1990;S5:S335 (A642).
3. Magni G, Caldieron C, Regatti-Luchini S, et al. Chronic musculoskeletal pain and depressive symptoms in the general population: an analysis of the First National Health and Nutrition Examination Survey Data. *Pain* 1990;43:299–307.
4. Fields HL. *Pain.* New York: McGraw-Hill, 1987: 133–170.
5. Nashold BS Jr, Ostdahl RH. Dorsal root entry zone lesions for pain relief. *J Neurosurg* 1979; 51:59–69.
6. Mitchell SW. *Injuries of nerves and their consequences.* New York: Dover Publications, 1965.
7. Bach S, Noreng MF, Tjellden NV. Phantom limb pain in amputees during the first 12 months following limb amputation, after reoperative epidural blockade. *Pain* 1988; 33:297–301.
8. Price DDR, McGrath PA, Rafii A, Buckingham B. The validity of visual analogue scales as ration scale measures for chronic and experimental pain. *Pain* 1983;17: 45–56.
9. Gracely RH, Kwilosz DM: The descriptor differential scale: applying psychophysical principles to clinical pain assessment. *Pain* 1988;35:279–288.
10. Melzac R. The McGill Pain Questionnaire: major properties and scoring methods. *Pain* 1975;7:275–299.
11. Gronbold M, Lukinmaa A, Konttinen YT. Chronic low back pain: intercorrelation of repeated measures for pain and disability. *Scand J Rehabil Med* 1990;22:73–77.
12. Tooney TC, Mann JD, Abashian S, et al. Description of a brief scale to measure functional impairment in chronic pain patients. *Pain* 1990;S5:A578.
13. Millard RW, Palt RB. A comparison of measures for low back pain disability. *Pain* 1990;S5:A5743.
14. Waddell G, Newton M, Henderson WI. Pain and disability. *Pain* 1990;S5:S966.
15. Sonty N, Tart RC, Chibnall JT. Use of the psychosomatic symptom checklist (PSC) as a screening instrument in patients with chronic pain. *Proceedings of the 10th Annual Meeting of the American Pain Society,* New Orleans, 1991:A91404, 124.
16. Put CL, Sitkower A. Pain and impairment, beliefs in patients treated in an interdisciplinary inpatient pain program. *Proceedings of the 10th Annual Meeting of the American Pain Society,* New Orleans, 1991:A91408, 126.
17. Turk DC. Associations among impairment, pain perception, and disability: results of a national survey. *Proceedings of the 10th Annual Meeting of the American Pain Society,* New Orleans, 1991:A91230, 51.
18. VonKorff M, LeResche L, Whitney CW, et al. Prediction of temperomandibular disorder (TMD) pain disability. *Pain* 1990;S5:A631, S329.
19. Riley JF, Ahern DK, Follick MJ. Chronic low back pain and functional improvement: assessing beliefs about their relationship. *Arch Phys Med Rehabil* 1988;69:579–582.
20. VonKorff M, Ormel J, Keefe FJ, Dworkin SF. Grading the severity of chronic pain. *Pain* 1992;50:133–149.
21. Deyo RA. Measuring the functional status of patients with low back pain. *Arch Phys Med Rehabil* 1988;69: 1044–1053.
22. Turk DC, Matyas TA. Pain-related behaviors: communication of pain. *Am Pain Soc J* 1992;1(2):109–111.
23. Keefe FJ, Crisson JE, Maltbie A, et al. Illness behavior as a predictor of pain and overt behavior patterns in chronic low back pain patients. *J Psychosom Res* 1986; 30:543–551.
24. Forgione AG, Barber TX. A strain gauge pain stimulator. *Psychophysiology* 1971;8:102–106.

25. Turk DC, Rudy TE, Stieg RL. The disability determination dilemma: towards a multi-axial solution. *Pain* 1988; 34:217–229.
26. American Medical Association. *Guides to the evaluation of permanent impairment.* 3rd ed (revised). Chicago: American Medical Association, 1990.
27. Osterweiss M, Kleinman A, Mechanic D, eds. *Pain and disability: clinical, behavioral, and public policy perspectives.* Washington, DC: National Academy Press, Institute of Medicine Committee on Pain, Disability and Chronic Illness Behavior, 1987.
28. International Association for the Study of Pain Subcommittee on Taxonomy. Classification of chronic pain. *Pain* 1986;(Suppl S3).
29. France RD, Rama Krishnan KR, eds. *Chronic pain.* Washington, DC: American Psychiatric Press, 1988.
30. Jensen J. Pain in non-psychotic psychiatric patients: life events, symptomatology and personality traits. *Acta Psychiatr Scand* 1988;78:201–207.
31. Chaturvedi SK, Michael A. Chronic pain in a psychiatric clinic. *J Psychosom Res* 1986;30:347–354.
32. Dworkin RH, Gitlin MJ. Clinical aspects of depression in chronic pain patients. *Clin J Pain* 1991;7:79–94.
33. Brown GK. A causal analysis of chronic pain and depression. *J Abnorm Psychol* 1990;99:127–137.
34. Atkinson JH, Slater MA, Patterson TL, et al. Prevalence, onset, and risk of psychiatric disorders in men with chronic low back pain: a controlled study. *Pain* 1991;45:111–121.
35. Ciccone DS, Grzesiak RC. Psychological vulnerability to chronic back and neck pain. *Proceedings of the 9th Annual Meeting of the American Pain Society,* St Louis, 1990:A248, 60.
36. Polatin PB, Kinney RK, Gatchel RJ. Premorbid psychopathology in somatoform pain syndrome. *Proceedings of the 14th Annual Meeting of the American Psychiatric Association,* New Orleans, 1991:NR553, 181.
37. Onghena P, Van Houdenhove B. Antidepressant-induced analgesia in chronic non-malignant pain: a meta-analysis of 39 placebo-controlled studies. *Pain* 1992;49: 205–219.
38. Lascelles RG. Atypical facial pain and depression. *Br J Psychiatry* 1966;112:651–659.
39. Raft D, Davidson J, Wasik J, et al. Relationship between response to phenelzine and MAO inhibition in a clinical trial of phenelzine, amitriptyline and placebo. *Neuropsychobiology* 1981;7:122–126.
40. Rumore MM, Schlichting DA. Clinical efficacy of antihistaminics as analgesics. *Pain* 1986;25:7–22.
41. Fishbain DA, Rosomoff HL, Goldberg M, et al. The prediction of return to work after pain center treatment: a review. *Clin J Pain* 1993;9:3–15.
42. Cutler RB, Fishbain DA, Rosomoff HL, et al. Does nonsurgical pain center treatment of chronic pain return patients to work? A review and meta-analysis of the literature. *Spine* 1994;19:643–652.
43. Fordyce WE, Fowler RS, Lehmann JF, DeLateur BJ, Sand PL, Trieschmann RB. Operant conditioning in the treatment of chronic pain. *Arch Phys Med Rehabil* 1973; 54:399–408.
44. Mao J, Price DD, Hayes RL, Lu J, Mayer DJ, Frenk H. Intrathecal treatment with dextrorphan or ketamine potently reduces pain-related behaviors in a rat model of peripheral mononeuropathy. *Brain Res* 1993;605: 164–168.
45. Tal M, Bennett GJ. Dextrorphan relieves neuropathic heat-evoked hyperalgesia in the rat. *Neurosci Lett* 1993; 151:107–110.
46. Satterthwaite JR, Tollison CD, Kriegel ML. The use of tricyclic antidepressants for the treatment of intractable pain. *Compr Ther* 1990;16(4):10–15.
47. Anthony M, Lance JW. Monoamine oxidase inhibition in the treatment of migraine. *Arch Neurol* 1969;21(3): 263–268.
48. Lascelles RG. Atypical facial pain and depression. *Br J Psychiatry* 1966;112(488):651–659.
49. Raft D, Davidson J, Wasik J, Mattox A. Relationship between response to phenelzine and MAO inhibition in a clinical trial of phenelzine, amitriptyline and placebo. *Neuropsychobiology* 1981;7(3):122–126.
50. Raskin DE. MAO inhibitors in chronic pain and depression. *J Clin Psychiatry* 1982;43(3):122.
51. Rees RT, Harris M. Atypical odontalgia. *Br J Oral Surg* 1979;16(3):212–218.
52. Yung CY. A review of clinical trials of lithium in neurology. *Pharmacol Biochem Behav* 1984;21 Suppl 1:57–64.

20
Substance Abuse

Edythe D. London, Steven J. Grant, Michael J. Morgan, Stephen R. Zukin, Walter Ling, Peggy Compton, Richard Rawson, and Donald R. Wesson

OVERVIEW

Drug abuse has important social and legal dimensions; however, from a biologic standpoint, abuse potential is an attribute of a drug, and it derives from the reinforcing properties of the drug. Drugs of abuse compose a chemically heterogeneous group that represents a very small percentage of all drugs. This chapter reviews biologic aspects of the acute and chronic effects of drug abuse. Information about neurochemical and anatomic substrates of these effects is presented with a discussion of the techniques used to elucidate them. We review basic diagnostic procedures for the major drug abuse syndromes, as well as approaches to pharmacologic and behavioral management.

Drug abuse: Broadly defined, this term refers to the use of a drug for nonmedicinal purposes, with intent to affect consciousness, distinct from *misuse*, which implies inappropriate medicinal use (e.g., for the wrong indication, at the wrong dose, or for the wrong amount of time). Contemporary American psychiatry views abuse as a maladaptive pattern of use, leading to recurrent, significant adverse physical, legal, occupational, or social consequences. The diagnosis of substance dependence preempts the diagnosis of abuse in the same drug class.

Dependence: This term includes both psychic dependence, characterized by compulsive drug-seeking behavior, and physical dependence, characterized by the presence of withdrawal symptoms upon abrupt discontinuation of the drug.

Addiction: This term refers to a state of physical and psychic dependence, although it can refer to psychic dependence alone when physical dependence is not apparent.

Tolerance: Tolerance is the body's compensatory response to the effect of the drug and refers to the decreased response to drug effects such that an increasingly larger dose is required to achieve the same effect.

NEUROBIOLOGY OF DRUG ABUSE

Receptors: Initial Targets for Actions of Abused Drugs

Drugs of abuse interact with specific neuronal receptors (Table 20-1). A considerable and rapidly expanding body of data concerns these central target sites at which drugs of abuse initiate their effects.

Dopamine Receptors and the Dopamine Transporter

Although drugs of abuse generally do not interact directly with dopamine (DA) receptors, as shown later, activation of the mesolimbic DA system is crucial to the rewarding effects and perhaps to other behavioral actions of these drugs.

Dopamine Receptors

Both D_1 and D_2 receptors are coupled to adenylate cyclase by guanine nucleotide regulatory proteins (G proteins). In addition, both synergistic and opposing interactions between D_1 and D_2 receptors occur. A proposal that the

TABLE 20-1. *Receptors for drugs of abuse*

Prototype compounds	Chemical or pharmacologic classification	Receptor	Receptor family	Drug effect	Endogenous ligands of drug-binding sites
Morphine	Opiate alkaloid	μ-Opiate receptor	G protein–coupled receptors	Inhibits adenylate cyclase	Enkephalins Endorphins Morphine (?)
Δ^9-Tetrahydro-cannabinol	Cannabinoid	Cannabinoid receptor	G protein–coupled receptors	Inhibits adenylate cyclase	Anandamid (?)
Diazepam	Benzodiazepine	Benzodiazepine site of $GABA_A$ receptor complex	Ligand-gated anion channels	Enhances GABA effect; increases Cl^- influx	Diazepam-binding inhibitor?
Phencyclidine (PCP)	Arylcyclohexylamine (dissociative anesthetic)	PCP site within channel of NMDA receptor complex	Ligand-gated cation channels	Blocks channel; noncompetitive NMDA antagonist	?
Nicotine	Alkaloid	Nicotinic cholinergic receptor	Ligand-gated cation channels	Activates nicotinic receptor	Acetylcholine
Cocaine	Alkaloid (psychomotor stimulant)	DA transporter	Na^+/Cl^--dependent neurotransmitter transporters	Blocks DA uptake	Dopamine?
Amphetamine	Psychomotor stimulant	DA terminal (DA transporter monoamine oxidase)	Na^+/Cl^--dependent neurotransmitter transporters	Promotes DA release	?
Lysergic acid diethylamide	Hallucinogen	$5HT_2$-serotonin receptor	Second-messenger–coupled receptor	Partial agonist; increases phosphoinositol hydrolysis	Serotonin

GABA, γ-aminobutyric acid; NMDA, *N*-methyl-D-aspartate; DA, dopamine.

D_2 receptor gene is a marker for vulnerability to substance abuse has created considerable controversy (1,2).

Dopamine Transporter

It has been known for more than three decades that cocaine and other drugs can block the reuptake of DA and other amine and indoleamine transmitters into their synaptic terminals. The development of new ligands that are highly selective for the DA transporter has permitted detailed study of the binding properties of the transporter. The DA transporter has been sequenced and cloned. It is a membrane-bound protein consisting of 12 membrane-spanning regions and a large extracellular loop with the C- and N-terminal strands located in the intracellular domain. The structure of the DA transporter is similar to that of transporters for norepinephrine and serotonin (5-hydroxytryptamine [5HT]), suggesting that these molecules form a superfamily of proteins analogous to the G-protein—linked receptor superfamily.

Opiate Receptors

About 20 years ago, biochemical evidence of the existence of stereo-specific–binding sites for opiates in the brain emerged from several laboratories. These sites bound opiate agonists and antagonists in a rank order of affinities consistent with the analgesic potencies of the drugs. Of the distinct families of endogenous opioid neuropeptides, the dynorphins interact preferentially with M receptors. Because of the preferential interaction of the enkephalins with a receptor type having properties distinct from those of μ or M receptors, the existence of a δ-receptor was postulated. In addition, subtypes of μ and M receptors have been identified (3). The validity of classifying opioid receptors into μ, δ, and M classes with distinct patterns of ligand selectivity, stereo-specificity, and neuroanatomic distribution has now been confirmed by molecular cloning of brain μ, δ, and M receptors. Opiates acting preferentially at μ or δ receptors display reinforcing properties in animal models and human subjects, whereas those acting preferentially at M receptors tend to lack reinforcing properties and to induce dysphoria.

The anatomic distribution of opioid receptors has been the subject of numerous autoradiographic studies of rodent brain. High levels of μ receptors, labeled with the μ-selective radioligand \[D-Ala^2, methyl-Phe^4, Gly^5-ol]-enkephalin (DAMGO), occur (in approximate rank order) in striatal "patches" and "streaks," accessory olfactory bulb, nucleus accumbens, ventral subiculum and dentate gyrus of the hippocampal formation, amygdala, central gray, superior and inferior colliculi, geniculate bodies, thalamic nuclei, and substantia nigra. By contrast, high levels of δ receptors, labeled with the selective δ radioligand \[^{3}H][D-Pen^2, D-Pen^5]-enkephalin ([^{3}H]DPDPE), are found in nucleus accumbens, external plexiform layer of olfactory bulb, olfactory tubercle, striatum, amygdala, and layers I to II and V to VI of cortex. The distribution pattern of M receptors, labeled with [^{125}I]dynorphin, in the presence of specific blockers of μ and δ receptors, reveals a narrower range of densities than those of μ and δ receptors, with high densities in the tail of the striatum, hypothalamus (medial preoptic area, suprachiasmatic nucleus), globus pallidus, and nucleus accumbens. Opiate receptors in sensory and accessory sensory brain areas presumably mediate the analgesic effects of opiate drugs, whereas limbic, extrapyramidal, and hypothalamic-neuropituitary opiate receptors may be associated with reinforcing, motor, and endocrine effects of opiates, respectively.

Nicotinic Acetylcholine Receptor

The actions of nicotine are mediated by receptors for the natural transmitter acetylcholine in the central and peripheral nervous systems. Cholinergic receptors are classified as either muscarinic or nicotinic, depending on their sensitivities to the natural alkaloids muscarine and nicotine. Nicotinic cholinergic receptors (nAChRs) represent a class of heterogeneous receptors, including those found in the electric organ (electroplax) of *Torpedo* skeletal muscle, autonomic ganglia, and the brain. Unlike muscarinic receptors, which are coupled to G proteins, nAChRs (which give a faster response to agonists) are rapidly responding ligand-gated sodium ion (Na^+) channels. The four subunits of the nAChR belong to a ligand-gated channel superfamily that also includes the A subtype of the γ-aminobutyric acid ($GABA_A$) receptor, and the glycine receptor.

The availability of radioligands that can bind with high affinity to central nicotinic receptors has allowed the mapping of the distribution of nAChR in the brain. Quantitative in vitro autoradiographic studies of [^{3}H]nicotine to slices of rat brain have shown densest labeling in the interpeduncular nucleus and medial habenula; dense labeling in thalamic nuclei, brain areas related to sensory function, and the cerebral cortex; and moderate labeling in the molecular layer of the dentate gyrus and subiculum.

N-Methyl-D-Aspartate Receptor

A stereo-specific brain receptor site for phencyclidine palmitate (PCP) and related drugs was demonstrated biochemically in 1979. Considerable biochemical, neuroanatomic, functional, and clinical evidence indicates that this receptor represents the central target site at which PCP-type drugs initiate their unique effects. The PCP receptor is located within the ion channel gated by the *N*-methyl-D-aspartate (NMDA) class of glutamate receptors. The major excitatory neurotransmitter of the central nervous system (CNS), l-glutamate, acts at a minimum of four types of receptors, named for their selective agonists: NMDA, kainate, α-amino-3-hydroxy-5-methyl-4-isoxazole propionic acid (AMPA), and *trans*-1-amino-cyclopentyl-1,3-decarboxylate (ACPD). Receptor activation is regulated at a number of sites and requires the binding of two molecules of l-glutamate to the agonist sites, as well as the presence of glycine (Fig. 20-1).

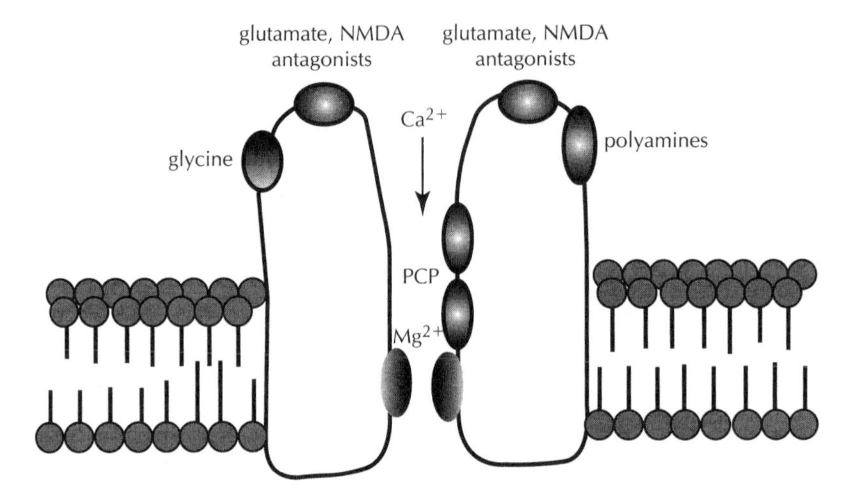

FIG. 20-1. Schematic diagram of the *N*-methyl-D-aspartate (NMDA) receptor. The receptor is a polymeric complex that gates a nonselective cation channel. It bears binding sites for agonists (glutamate and NMDA). Competitive antagonists can also can bind to these sites. A strychnine-insensitive binding site for the coagonist glycine (Gly) has been characterized, as have sites for various modulators, including H^+ and polyamines. Phencyclidine palmatitate (PCP) and Mg^{2+} bind within the channel to block to influx of Ca^{2+}. (Modified from Kameyama T, Nebeshima T, Domino EF, eds. *NMDA receptor related agents: biochemistry, pharmacology, and behavior.* Ann Arbor, MI: NPP Books, 1991.)

γ-Aminobutyric Acid Receptor

A molecular target of benzodiazepines, barbiturates, and ethanol, the $GABA_A$ receptor is an inhibitory ligand-gated receptor-channel complex, consisting of distinct α, β, and δ subunits surrounding a central pore. Each subunit exists in multiple isoforms, indicating several thousand possible distinct structures for the $GABA_A$ receptor.

The $GABA_A$ complex comprises multiple functional domains (Fig. 20-2). Ligands for

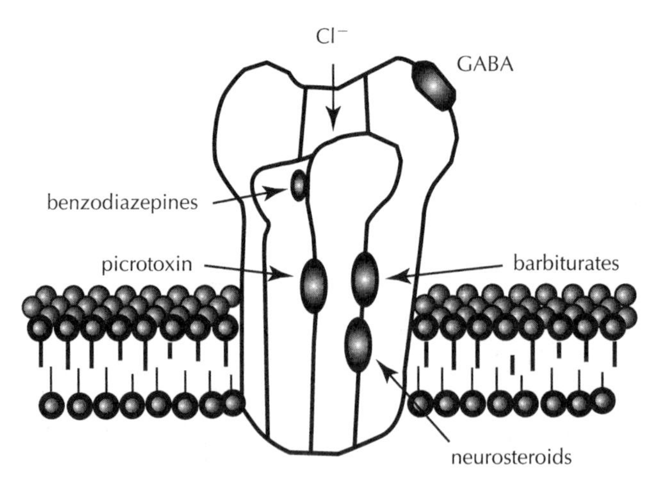

FIG. 20-2. Schematic diagram of the γ-aminobutyric acid (GABA)-benzodiazepine receptor. The receptor is a polymeric complex that gates a Cl^- channel. It bears binding sites for GABA and agonistic drugs, such as muscimol, and for a variety of drugs (benzodiazepines, barbiturates, and convulsants, such as picrotoxin). In addition, the activity of the receptor is modulated through the actions of endogenous neurosteroids, such as pregnenolone sulfate, dehydroepiandrosterone sulfate, and desoxycorticosterone sulfate.

the benzodiazepine site can be divided into three groups on the basis of their effects. Clinically used anxiolytic, sedative, and anticonvulsant benzodiazepines (such as diazepam), classified as agonists of the benzodiazepine-binding site, increase the sensitivity of the complex to GABA. Inverse agonists, such as the β-carbolines, decrease the sensitivity of the complex to GABA, thus exerting anxiogenic effects. Antagonists, such as flumazenil, lack significant intrinsic effect on agonist sensitivity of the receptor complex but can block the effects of benzodiazepine agonists or of inverse agonists.

Barbiturates also enhance $GABA_A$ receptor function, acting at a domain of the complex independent of the benzodiazepine-binding site.

Ethanol represents the third type of drug that has abuse liability and that interacts with the $GABA_A$ receptor (4).

Cannabinoid Receptor

The active principle of marijuana, Δ^9-tetrahydrocannabinol (Δ^9-THC), proved a challenging probe for a CNS recognition site because of its extreme hydrophobicity. Identification of the cannabinoid receptor was facilitated by demonstration of cannabinoid enantioselectivity and by synthesis of a series of novel cannabinoids.

THEORIES OF ADDICTION AND THE SELF-ADMINISTRATION PARADIGM

During the 1950s, researchers began to find evidence that laboratory animals would learn to perform behavior that resulted in drug injections. It soon became apparent that they were responding to the drug infusion as if it were a positive reinforcer. In subsequent studies, animals self-administered doses of morphine that were too low to produce physical dependence. They also self-administered stimulants, such as cocaine, and no obvious withdrawal syndrome on cessation of stimulant use had been described. Further studies demonstrated that animals even self-administered certain abused substances without any previous experience of having received the drug. Studies such as these had an immense impact on the concept of motivation underlying drug abuse in humans and have resulted in a shift in emphasis away from the physical dependence model toward the view that drug abuse is largely driven by the motivation to seek and experience the incentive (rewarding or reinforcing) properties of abused drugs (5).

In addition to the roles played by negative and positive primary reinforcement in drug addiction, secondary reinforcement has recently been found to play a significant role. Secondary reinforcers or incentives are otherwise neutral stimuli present in the environment that become associated with drug-taking after repeated administration through a process of incentive learning. Once acquired, secondary reinforcement can become an extremely powerful source of motivation for drug-seeking behavior (6).

Brain Reward Systems

One of the most basic principles of behavior is the "law of effect," which states that behaviors that lead to a desired outcome are repeated. These outcomes are referred to as *reinforcers*. A precise definition of the term reinforcer has been difficult to formulate and is beyond the scope of this review. In general, however, a reinforcer is an event that facilitates learning. The terms *reward* and *hedonic* are not synonymous with reinforcement. For example, when animals are trained on a conditioned emotional response, in which they acquire an association between a tone and a withdrawal response produced by mild foot shock, the posttraining, noncontingent consumption of sucrose, or a brief application of intense foot shock have identical memory improving effects. Thus, both rewarding and aversive events can produce reinforcement.

The notion that the brain contains circuits that mediate reward or pleasure is central to our current understanding of the neural basis of drug abuse. The identification of the specific neural systems involved in reward originated with the work of Olds and Milner, who discovered that laboratory animals avidly self-administer electrical stimulation delivered by electrodes to specific deep regions of the brain (7). This observation suggested that there are anatomically specific circuits in the brain that mediate reward or pleasure. It was *later* found that drugs that are abused by humans facilitated brain stimulation reward when administered to animals. This finding suggested that drugs of abuse influence the same brain reward system that is activated by electrical stimulation or by natural rewards.

Self-Administration Paradigm

The most striking feature of drug self-administration is how closely the classes of drugs that are self-administered by animals correspond with those that are abused by humans. With ratio schedules of operant reinforcement, laboratory animals have been observed to respond for amphetamine, cocaine, methamphetamine, methylphenidate, nicotine, caffeine, opiates, ethanol, nicotine, barbiturates, benzodiazepines, PCP, and THC. Many drugs, including opioid antagonists, neuroleptics, and tricyclic antidepressants, either do not maintain self-administration in animals or stimulate voluntary termination of drug infusion. Drugs that have not been found to support self-administration in animals are, almost without exception, not abused by humans. Lysergic acid diethylamide (LSD) is one of the few exceptions.

Neural Substrates of Drug Self-Administration Reward

Systemic Pharmacologic Challenge

The systemic pharmacologic challenge technique involves the noncontingent co-administration of an agonist or an antagonist of a particular neurotransmitter and an examination of the effect of this manipulation on drug self-administration responding. The rationale for challenge with an agonist of a particular neurotransmitter is that if the neurotransmitter facilitates reward, the agonist should substitute for the self-administered drug and should temporarily decrease self-administration of the drug. Conversely, challenge with an antagonist should produce a selective increase in responding to compensate for the reduced effectiveness of the self-administered drug. Higher doses, however, would further reduce rewarding efficacy of the self-administered drug and would therefore reduce responding.

Lesion Studies

Lesion studies have provided considerable information about the localization of central sites that are important to drug-induced reward. Such studies have demonstrated that the mesolimbic DA system plays a crucial role (Fig. 20-3). The cell bodies of this system originate in the ventral tegmental area (VTA). They project rostrally to the nucleus accumbens and to more anterior regions, including the amygdala, olfactory tubercle, and prefrontal cortex. Selective lesions of the nucleus accumbens, produced by the catecholamine-specific neurotoxin 6-OHDA, disrupt self-administration of cocaine and amphetamine. Similarly, selective 6-OHDA lesions of the VTA disrupt self-administration of cocaine.

Lesion studies have also demonstrated that the mesolimbic DA system is crucial to the self-administration of other classes of abusable drugs. Selective 6-OHDA lesions of this system have been found to disrupt self-administration of heroin and morphine. Therefore, the mesolimbic DA system appears to include a common neuronal substrate that mediates both opioid and psychomotor stimulant reward.

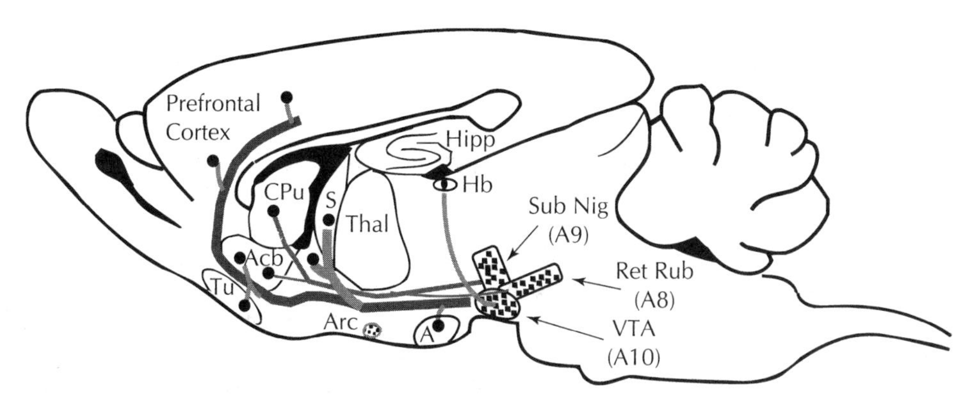

FIG. 20-3. Chemical neuroanatomy of (rat) dopamine (DA) systems relevant to drug abuse. Cell bodies of DA-containing neurons are indicated by dotted regions. The mesocorticolimbic dopaminergic pathway (heavy stippling) has been specifically implicated in the rewarding properties of drugs. The mesocorticolimbic DA system consists of projections from dopaminergic neurons found mainly in the ventral tegmental area (A10) to the nucleus accumbens (Acb), olfactory tubercle (Tu), bed nucleus of the stria terminalis and septum (S), amygdala (A), lateral habenula (Hb), and portions of the prefrontal cortex. The nigrostriatal system *(light stippling)* consists of dopaminergic neurons in the substantia nigra (A9) that project primarily to the dorsal portions of the caudate and putamen (CPu). Dopamine (DA) neurons in the retrorubral (Ret Rub) region (A8) contribute to both systems. The tuberoinfundibular DA neurons in the arcuate nucleus (Arc) are of interest because of their regulation of neuroendocrine function, such as inhibition of prolactin release. (Modified from Cooper JR, Bloom FE, Roth RH, eds. *The biochemical basis of neuropharmacology.* 6th ed. New York: Oxford University Press, 1991.)

Intracranial Self-Administration of Drugs

Studies using the intracranial self-administration procedure have shown that animals will voluntarily self-administer microinjections of amphetamine into the nucleus accumbens and the prefrontal cortex (orbitofrontal cortex of the rhesus monkey) but not into other brain regions. Cocaine is also voluntarily self-administered into the prefrontal cortex. Morphine is self-administered into the VTA, lateral hypothalamus, and nucleus accumbens, all of which are either nuclei or terminal projection regions of the mesolimbic DA system. Other synthetic and endogenous opioids are also self-administered intracranially into the mesolimbic system. The first evidence that opiate receptors in the vicinity of the periaqueductal gray may play a role in intravenous self-administration of opioid drugs derives from the observation that intracranial microinjections of methylnaltrexone into this brain region produced dose-related increases in responding for heroin, but not for cocaine, by rats on a continuous reinforcement schedule (8).

Self-Administration of Direct and Indirect Dopaminergic Agonists

There is now a considerable body of evidence that implicates DA as a crucial neurochemical substrate of drug self-administration reward. The most obvious is that the psychomotor stimulants amphetamine and cocaine, indirect DA agonists that prolong the actions of DA in the synaptic cleft, are avidly self-administered by laboratory animals. It has been known for two decades that cocaine blocks the reuptake of DA in the brain. This observation indicated that the rewarding properties of cocaine were due to interaction of the drug with the DA transporter (9).

Self-Administration of Nicotine

In addition to the evidence implicating dopaminergic neurotransmission in the self-administration of direct and indirect dopaminergic agonists, studies of nicotine self-administration provide support for the view that nicotine-induced reward requires intact functioning of the mesolimbic DA system. Neurochemical evidence supports the feasibility of a role for mesolimbic DA in reinforcement due to nicotine. Nicotinic receptors are present in the midbrain area containing the substantia nigra and the VTA and in the terminal fields (nucleus accumbens, olfactory tubercle) of the mesolimbic DA system. Systemic nicotine increases the firing rate of neurons in the VTA, and intracellular recording from presumed DA-containing cells of the VTA in vitro reveals that nicotine directly depolarizes the neurons.

Electrical Brain Stimulation Reward Paradigm

Studies on the effects of electrical stimulation of the human brain have confirmed that stimulation of certain brain regions can produce intense feelings of euphoria. Furthermore, a notable common feature of drugs, such as opiates, stimulants, barbiturates, benzodiazepines, PCP, alcohol, and marijuana, which are self-administered systemically by laboratory animals, is that they enhance brain stimulation reward. These findings provide support for the hypothesis that abused drugs activate the same reward system that is activated by electrical stimulation of specific anatomic sites in the brain (10).

Neuroanatomic and Neurochemical Substrates

Shortly after his discovery of intracranial self-stimulation, Olds found that brain stimulation reward is strongly attenuated by drugs that block catecholaminergic transmission. He subsequently identified the medial forebrain bundle (MFB) in the region of the lateral hypothalamus as a highly effective locus for intracranial self-stimulation. By the end of the 1980s, extensive research suggested that DA is the most important catecholamine involved in reward produced by stimulation of the MFB.

There are two major dopaminergic systems in the MFB: the nigrostriatal system, which terminates in the caudate-putamen, and the mesolimbic system, which innervates several forebrain regions, including the nucleus accumbens, the olfactory tubercle, the ventral pallidum, and parts of the frontal cortex (see Fig. 20-3).

Virtually all substances that are self-administered by animals have the effect of increasing basal neuronal firing or basal neurotransmitter release in brain circuits that are relevant to reward. This common feature among all drugs that are self-administered by animals provides the basis for the theory that drug self-administration activates the same reward circuits as those activated by electrical brain stimulation reward, and that the mechanisms involve an important dopaminergic component.

Pharmacologic interventions have demonstrated an involvement of serotonergic systems as well. Treatments that would be expected to enhance serotonergic tone *reduce* self-administration of drugs of abuse. For example, pretreatment with l-tryptophan, a precursor of 5HT, or feeding with a diet enriched in tryptophan attenuated amphetamine self-administration. Fluoxetine likewise can reduce the self-administration of ethanol and cocaine in rats. These and other findings support the view that inhibition of 5HT uptake opposes reinforcing effects of amphetamines and related drugs (11).

Another system that has been implicated in drug abuse–related reward is glutamatergic transmission involving the NMDA subtype of glutamate receptor. There is evidence from microinfusion studies in animals that blocking glutamatergic transmission at NMDA receptors of the nucleus accumbens reduces the rewarding value of cocaine but not of heroin.

LONG-TERM EFFECTS OF DRUGS OF ABUSE ON THE CENTRAL NERVOUS SYSTEM

The long-term consequences of chronic drug administration are defining features of drug abuse and addiction. The most prominent and persistent of these effects include alterations in the magnitude of the response to the drug, effects that emerge upon cessation of drug taking, and responses to environmental cues that become associated with drug taking.

The term *neuroadaptation* is commonly used to describe the neural mechanisms underlying alteration in magnitude of drug effect because both decreases (tolerance) and increases (sensitization) in the response can accompany chronic exposure to a drug. The emergence of withdrawal symptoms when the drug is no longer present is thought to reflect the consequences of such enduring neuroadaptations.

Tolerance, or a diminution in the response to a given dose of the drug, is the classic consequence of chronic drug administration. Although the development of tolerance to a particular drug, by definition, requires that higher doses of drug are needed to obtain the same level of effect, the presence of tolerance to a particular drug does not necessarily imply a potential for abuse of the drug. In fact, the ability to produce tolerance is a nearly universal property of all drugs. Altered bioavailability constitutes the simplest mechanism of neuroadaptation. In this case, tolerance can result from reduced absorption, increased metabolism or excretion, or impaired distribution of the drug (e.g., through reduced blood flow). Repeated exposure to a drug can also lead to an increase in the response to the drug. This phenomenon is termed *sensitization* or *reverse tolerance.* Sensitization has been increasingly emphasized in neurobiologic hypotheses of addiction, and it is thought to contribute to the escalation of drug-taking or relapse after a period of abstinence.

Long-term changes in the specific receptors that are the targets of the drugs of abuse have proved complex. Both upregulation and downregulation of receptor function accompanying chronic drug exposure have been documented. In addition to changes at the ligand-binding site, alterations can also occur along signal transduction pathways.

The presence of a withdrawal syndrome was long considered the defining feature of addiction. It is currently thought that the cellular mechanisms underlying tolerance and withdrawal are often related.

CLINICAL ASSESSMENT

Although the general approach to assessing drug abuse disorders follows traditional lines of history taking and physical and mental status examinations supported by laboratory tests, the circumstance under which evaluation takes place merits special consideration. This is because it can greatly influence the patient's attitude toward the process of evaluation and the information to be gathered. An empathetic, nonjudgmental attitude on the part of the evaluator is necessary. Patient interviews should be conducted in private. Physicians should be aware that substance abuse treatment information is, under the law, treated with a degree of confidentiality more stringent than general medical information (12).

The history is the cornerstone of clinical evaluation. Because substance abuse and dependence disorders are clinically defined as maladaptive patterns of use, it is important to explore such issues as amount of use, presence of withdrawal on cessation of use, amount of time spent in procurement and use, past attempts to control or cut down, neglect of social responsibility, and continuous use in the face of adverse physical and social consequences. For each drug class, age of first use and age of first regular use or intoxication must be determined.

Past periods of heavy use and periods of abstinence, and the surrounding life circumstances that may give clues to events precipitating relapse or motivation for abstinence, should also be ascertained. This information can also aid in planning treatment. The pattern of use during the 3 to 4 weeks before evaluation, including the amount and time of last use, should be explored in detail. This information is crucial in deciding whether a period of hospitalization will be necessary to establish abstinence because of withdrawal from certain substances.

Screening for substance abuse disorders in a general medical or neuropsychiatric context differs from evaluating patients seeking treatment. In the former, the subject matter can sometimes be approached by means of a questionnaire, such as the Michigan Alcohol Screening Test (13,14).

The history should be supplemented by careful observation and a thorough physical examination. Common physical findings of injection drug abuse include fresh needle marks and old scars, thrombosed veins, abscesses and congested nasal mucosa, enlarged liver, and local lymph nodes. Spider nevi and gynecomastia are common in alcoholism, as are cardiac murmur or arrhythmia in cocaine addiction. Signs of intoxication or withdrawal may be present, depending on the time of the last "fix." The mental status examination should include observation for any unusual behaviors or mannerisms, speech, orientation, attention, concentration, thought content, simple calculation, and recent and remote memory. Sometimes, it is useful to employ a standardized cognitive screening test, such as the Mini-Mental State Examination. Taken together, these components of patient evaluation should lead naturally to a formal diagnosis (15).

CLINICAL DIAGNOSIS

The formal diagnosis of drug-related neuropsychiatric syndromes, covered under Substance-Related Disorders in fourth edition of the *Diagnostic and Statistical Manual of Mental Disorders* (DSM-4), contains three distinct but related diagnostic categories (16):

1. Generic diagnosis of substance dependence and substance abuse (substance use disorders)
2. Clinical syndromes directly related to intoxication and withdrawal (substance-induced disorders)
3. Drug-induced psychiatric disorders phenomenologically related to other specific psychiatric disorders (substance-induced mental disorders)

Table 20-2 lists the salient features of substance dependence and substance abuse.

TABLE 20-2. *DSM-4 Criteria for substance dependence and substance abuse*

Dependence (three or more in a 12-month period)	Abuse (one or more in a 12-month period)[a]
Tolerance (marked increase in amount; marked decrease in effect)	Recurrent use resulting in failure to fulfill major role obligations at work, home, or school
Characteristic withdrawal symptoms and substance taken to relieve withdrawal	Recurrent use in physically hazardous situations
Substance taken in larger amount and for longer period than intended	Recurrent substance-related legal problems
Persistent desire or repeated unsuccessful attempt to quit	Continued use despite persistent or recurrent social or interpersonal problems caused or exacerbated by substance
Much time and activity to obtain, use, recover	
Important social, occupational, or recreational activities given up or reduced	
Use continues despite knowledge of adverse consequences (e.g., failure to fulfill role obligation, use when physically hazardous)	

DSM-4, *Diagnostic and Statistical Manual,* 4th edition.
[a]Symptoms must never have met criteria for substance dependence for this class of substance.

Laboratory Testing

Screening for drugs of abuse is invaluable in investigating patients suspected of drug overdose, confirming the diagnosis of drug dependence for planning and initiating treatment, differentiating drug-induced psychopathology from syndromes of other causes, detecting multiple drugs of abuse, monitoring compliance and treatment progress, and detecting relapse. Drug testing can be performed on several body fluids and tissues, including blood, urine, saliva, sweat, and hair. Several test methods are available for drug screening. Thin-layer chromatography (TLC), the least expensive, is also the least sensitive. Its primary advantage besides cost is that a single test can detect multiple drugs. It is highly suitable for detecting high-dose drug use of recent origin, as in cases of suspected overdose in emergency rooms when the types of drugs may be unknown but the need to detect low levels of drugs is not an issue. Many labs confirm results of TLC by an alternative method, and alone, TLC is insufficient for forensic purposes. Several immunoassay methods are available, based on antigen–antibody interactions. These include the enzyme multiplied immunoassay test (EMIT), radioimmunoassay (RIA), and fluorescent polarization immunoassay (FPIA). These methods are highly sensitive, and the tests are quick and easy to perform. Unfortunately, however, they can test for only one drug at a time; thus, a drug that is not thought of is not detected.

The detectability of a drug depends not only on its characteristics but also on a host of other factors, including the size of the dose, the frequency of its use, the time of the last use, the time of the sample collection, the metabolites of the drug, and the type of body fluid used for testing and the sensitivity and specificity of the analytic methods.

CLINICAL SYNDROMES

Each class of drugs of abuse can produce a number of neuropsychiatric syndromes. Alcohol, for example, leads to intoxication, withdrawal, seizures, and dementia. On the other hand, each syndrome can be produced by more than one class of drugs of abuse. For example, seizures can result from acute stimulant abuse or from withdrawal from alcohol or sedative-hypnotics. Table 20-3 shows the common syndromes seen with each class of drugs of abuse.

TABLE 20-3. *Neuropsychiatric syndromes associated with drug abuse*

Syndrome	Drug-related etiology
Seizure	Alcohol intoxication
	Cocaine, CNS stimulant toxicity
	Alcohol, sedative-hypnotic withdrawal
Stroke	Cocaine, CNS stimulant intoxication, and toxicity
	Opiate toxicity
	Alcohol abuse
	Intravenous drug use
Coma	Opiate toxicity
	Alcohol, sedative-hypnotic toxicity
	Drug-induced stroke or seizure
	Life-style–related (head trauma, hypoglycemia)
Delirium	Cocaine, CNS stimulant intoxication
	Alcohol, sedative-hypnotic withdrawal
CNS infection	Intravenous drug use
Peripheral neuropathy	Alcohol abuse
	Injection drug use

CNS, central nervous system.

Neuropsychiatric Syndromes by Drug Class

Alcohol

Nearly 14% of the adults in the United States meet diagnostic criteria for alcohol abuse or dependence at some point in their lives. About 10,000,000 Americans over the age of 21 years are classified as heavy drinkers, defined as drinking five or more drinks per occasion on five or more occasions in the past 30 days.

Alcohol Intoxication

At low doses, alcohol acts primarily as a CNS intoxicant, whereas at high doses, its sedative and anesthetic actions predominate. Thus, neurologic syndromes associated with acute alcohol ingestion vary with alcohol dose (Table 20-4).

Alcohol acutely depresses the integrating and inhibitory functions of the cerebral cortex and reticular activating system. The action of alcohol in the CNS is nonspecific, which accounts in part for the high doses needed to achieve, relative to other psychoactive drugs, comparable levels of intoxication. The sensitivity with which neurotransmitter systems respond to alcohol appears to have a genetic basis (17).

The average adult metabolizes 7 to 10 g (1 oz) of alcohol per hour. Once this clearance rate is exceeded, blood-alcohol levels begin to rise, and CNS manifestations of intoxication occur. Blood-alcohol level, a commonly used measure of intoxication, accurately reflects the CNS effects of alcohol in nontolerant people. Severity of intoxication at a given blood level is typically greater when the blood-alcohol level (BAL) is rising as opposed to falling. At levels between 0.01% and 0.10%, mild declines in coordination and cognition are experienced, and individuals present with disinhibited behavior, euphoric mood, and decreased anxiety. At concentrations slightly above this level, alcohol users are notably intoxicated, demonstrating dysarthria, ataxia, poor coordination, decreased sensory function, impaired judgment and psychomotor skills, decreased attention span, and mood swings. At 0.20%, individuals are severely intoxicated, manifesting nausea and vomiting, diplopia, gross motor incoordination, incoherence, and confusion. Alcoholics have been shown to perform well on psychomotor or cognitive tasks at BALs between 0.20% and 0.30%.

During an alcohol blackout, a classic example of anterograde amnesia, the intoxicated person appears alert, is able to ambulate, carry on conversation, and perform previously learned tasks. No impairment of long-term memory or immediate recall is noted. Upon becoming sober, however, the person has total and permanent amnesia for the period. At higher BALs, the depressant and anesthetic actions of alcohol predominate. At 0.30%, individuals become stuporous, and by 0.40%, they may lose con-

TABLE 20-4. *Neurologic syndromes associated with acute alcohol ingestion*

Syndrome	Blood-alcohol concentration (g/100 mL)[a]	Neurologic findings
Intoxication	0.02–0.25	Dysarthria, vestibular and cerebellar impairment (ataxia, impaired balance, poor coordination), hyperreflexia, autonomic excess; sluggish pupillary response, nystagmus, diplopia, nonconvergence; impaired judgment, impaired cognition, decreased attention span, emotional lability, anterograde amnesia
Overdose	0.30–0.90	Stupor to coma, fluctuating level of consciousness, absence of focal neurologic signs, depressed respirations, normal to dilated pupils, sluggish pupillary response

[a]Blood levels for nontolerant individuals.

sciousness, thus protecting themselves from further increases in BAL. A BAL of 0.50% to 0.70% results in coma, respiratory depression, and eventually death.

Alcohol Withdrawal

As with intoxication, the nature and course of alcohol withdrawal depends on whether the individual is alcohol dependent (Table 20-5). Alcohol withdrawal is characterized by rebound effects in those physiologic systems that were initially modified by its ingestion. Alcohol withdrawal results in adrenergic hyperactivity at the level of the cortex, limbic system, and brain stem.

Uncomplicated alcohol withdrawal, or "hangover," is commonly experienced by the non–alcohol-dependent person after an occasion of heavy drinking. It is characterized by tremulousness, mild autonomic hyperactivity, nausea and vomiting, photophobia, malaise, irritability, and vascular headache. The syndrome typically occurs 4 to 6 hours after the last drink and subsides within 48 hours. For the individual physically dependent on alcohol, sudden abstinence or a significant decrease in alcohol intake precipitates a withdrawal syndrome of varying severity. Alcoholic tremulousness, alcoholic hallucinosis, and withdrawal seizures appear within 6 to 24 hours after the last drink. Most alcoholics experience the mildest of these syndromes, *alcoholic tremulousness*, as BAL drops to between 0.10% to 0.20%. Postural tremors of the distal upper extremities and tongue, mild tachycardia, headache, irritability, anxiety, anorexia, insomnia, moderate diaphoresis, hyperreflexia, and systolic hypertension are common. Alcoholic hallucinosis, in which the individual experiences vivid, typically auditory, and often threatening hallucinations, also occurs in early withdrawal.

Alcohol withdrawal seizures also take place during the early stage of withdrawal, often within several hours after the last drink. They are characteristically generalized tonic-clonic seizures, occurring in groups of two to six, with 90% occurring within the first 48 hours of alcohol abstinence. Treatment

TABLE 20-5. *Neurologic syndromes associated with alcohol withdrawal*

Syndrome	Alcohol dependent?	Time since last drink	Duration	Neurologic findings
Uncomplicated withdrawal	No	4–6 h	24–48 h	Tremors, vascular headache, photophobia, mild autonomic excitation, irritability
Acute withdrawal: Early stage	Yes	6–24 h	24–48 h	
Alcohol tremulousness				Tremors, headache, irritability, anxiety, mild to moderate autonomic excitation, hyperreflexia, potential transient hallucinations, clear sensorium
Alcoholic hallucinosis				Vivid, threatening, auditory hallucinations; clear sensorium; tremors; moderate autonomic excitation
Withdrawal seizures				Generalized tonic-clonic seizures, occurring in groups of two to six, without EEG focus, postictal confusion and disorientation
Acute withdrawal: Late stage	Yes	36–72 h	24–72 h	
Alcohol withdrawal delirium				Profound disorientation; fluctuating level of consciousness; poor remote and immediate memory; visual, auditory, or tactile hallucinations; gross tremor; dilated pupils; extreme autonomic hyperactivity
Protracted withdrawal	Yes	2–3 wk	6–24 mo	Autonomic dysfunction, sleep disturbance, EEG changes, impaired short-term memory, fatigue

EEG, electroencephalogram.

should consider the entirety of the alcohol withdrawal syndrome because other signs of full-blown withdrawal usually develop quickly. Thiamine, 100 mg intravenously, should be part of the initial fluid and electrolyte management. There is no convincing evidence, however, to warrant prophylactic anticonvulsants to prevent alcohol withdrawal seizures.

The late stage of alcohol withdrawal syndrome, known as *alcohol withdrawal delirium* or, more commonly, *delirium tremens* (DTs), occurs 36 to 72 hours after the last drink. It is relatively rare, occurring in about 5% of alcoholics. Its presence suggests concomitant medical problems, such as pneumonia, subdural hematoma, pancreatitis, fractures, liver disease, or malnutrition. Alcohol withdrawal delirium is characterized by extreme autonomic hyperactivity, profound disorientation, gross disturbances of recent and remote memory, fluctuating levels of consciousness, incoherence, and confusion. The individual is agitated and restless, with terrifying visual, auditory, and tactile hallucinations that cannot be differentiated from reality. Autonomic hyperactivity is evidenced by hyperventilation, tachycardia (commonly 130 to 150 beats/min), hypertension (blood pressure greater than 150/100 mm Hg), hyperthermia (38° to 39.5°C), pronounced diaphoresis, facial flushing, gross tremor, and dilated pupils. Mortality rates of alcohol withdrawal delirium range from 5% to 15%.

Syndromes Associated With Chronic Alcohol Abuse

Alcohol abuse–associated syndromes (Table 20-6) typically present after at least a decade of heavy drinking and, in many cases, are permanent conditions. Acute intoxication and withdrawal in people who continue to drink often complicate the clinical picture.

Wernicke's Encephalopathy. The onset of Wernicke's encephalopathy is typically abrupt, but it may develop over several days to weeks and often is accompanied by drowsiness, headache, nausea, and vomiting. The so-called classic triad of ataxia, encephalopathy, and ophthalmoplegia were present in only one third of people in whom Wernicke's lesions were found on autopsy. Oculomotor palsies of the third and sixth cranial nerves are common, as is horizontal or vertical nystagmus. Pupils react sluggishly and may be unequal. Ataxia is typically cerebellar, evident in the gait and lower extremities, with the upper extremities and speech being relatively spared. Untreated, Wernicke's encephalopathy progresses to stupor and coma. Fortunately, it is responsive to parenteral thiamine therapy, with improvement noted (especially ocular symptoms) within days to weeks. If memory deficits persist, a diagnosis of alcohol amnestic disorder must be considered.

Alcohol Amnestic Disorder. More commonly known as Korsakoff's psychosis, alcohol amnestic disorder appears to result from

TABLE 20-6. *Neurologic syndromes associated with chronic alcohol abuse*[a]

Syndrome	Neurologic findings
Wernicke's encephalopathy	Mental confusion, oculomotor disturbances, gait ataxia; may present with headache, nausea, and vomiting
Alcohol amnestic disorder	Pronounced anterograde and retrograde amnesia, confabulation, perseveration, placid, emotionally flat
Alcoholic dementia	Stable cognitive dysfunction of unknown origin, sensorium clear, aphasia absent, heterogeneous presentation
Cerebellar degeneration	Gait and upper extremity ataxia, bilateral cerebellar incoordination, absence of oculomotor disturbances
Polyneuropathy	Distal, bilateral sensory and lower motor neuron deficits, decreased reflexes, and atrophy of affected areas

[a]All patients report a 10-year or longer history of heavy alcohol intake.

cumulative effects of Wernicke's encephalopathy, hence the term Wernicke-Korsakoff syndrome. Alcohol amnestic disorder is primarily a gross memory disorder of both anterograde and retrograde function, although anterograde deficits predominate. Intelligence, language, and speech are unaffected, and sensorium is clear. These patients are typically unaware of and unconcerned about their amnesia.

Alcoholic Dementia. The diagnosis of alcoholic dementia is made when slowly progressive but stable cognitive dysfunction, attributed to metabolic or organic disease, is noted in alcoholics with extensive drinking histories. Current thought is that the dementia relates to the neuropathology of Wernicke's encephalopathy, in that these lesions are commonly found on autopsy of people diagnosed with alcoholic dementia. Unlike most dementias, there is no disruption of language or loss of primary motor and sensory function. Cognitive deficits typically center in the areas of nonverbal intelligence and visuospatial abilities. Memory deficits are present, but never to the extent noted in alcohol amnestic disorder. Sensorium is typically clear, and aphasia is not present.

Cerebellar Degeneration. Degeneration of Purkinje's cells in the cerebellar cortex has been noted in certain alcoholic patients. Symptoms related to these lesions tend to develop subacutely over a period of months and worsen with an uneven trajectory as drinking continues. Symptoms include gait ataxia and may be accompanied by upper extremity ataxia and dysarthria. As with alcoholic dementia, mild improvement of symptoms is reported with abstinence (18).

Neuropathy. Inadequate nutrition has been shown to play an important role in the development of polyneuropathy in alcoholics, perhaps related to general vitamin B deficiencies as opposed to thiamine deficiency specifically. Alcohol-induced neurotoxic effects on axonal transport may be a contributing factor as well. Alcoholic neuropathy tends to progress gradually and affects sensory and motor nerves more than autonomic nerves. Deficits are distal, bilateral, and asymmetric, are seen in a glove-and-stocking distribution, and affect all sensory modalities.

Cocaine

The cocaine epidemic of the 1980s, combined with the switch to the potent smoked (alkaloidal) form, or "crack," has resulted in better clinical description of the intoxication, withdrawal, and cerebrovascular syndromes associated with cocaine use (Table 20-7). In 1991, more than 24,000,000 Americans reported having at some time used cocaine, with about 1 in 10 of these reporting cocaine use in the past month. About 0.2% of the U.S. population has met dependence or abuse criteria for cocaine at some point over their lifetime. The more dramatic neurologic complications of

TABLE 20-7. *Clinical syndromes of cocaine intoxication and withdrawal*

Clinical syndromes	Clinical findings	Other associated features
Intoxication	Euphoria or affective blunting, sociability, hypervigilance, anxiety, tremor, stereotyped movement, dyskinesia and dystonia, impaired judgment	Tachycardia, papillary dilation, elevated or lowered blood pressure, sweating or chills, nausea and vomiting, weight loss, psychomotor agitation or retardation, muscle weakness, respiratory depression, cardiac arrhythmia, confusion, seizure
Withdrawal	Phase I: "crash," depression, fatigue, craving, anxiety, suspiciousness, paranoia, exhaustion, sleep Phase 2: dysphoria, anhedonia, craving (cue induced) Phase 3: intermittent craving	

cocaine abuse tend be related to acute cocaine toxicity, whereas withdrawal-related neurologic symptoms appear to be relatively mild.

Stimulants

The long-term consequences of psychostimulant abuse differ from those of other abused drugs, such as opioids and sedative-hypnotics. In addition to the psychological problems engendered by addiction, the detrimental consequences of prolonged use of these drugs include cardiovascular pathology (hypertension, cardiac infarctions, arrhythmias, and neuropathy), hepatotoxicity, neurotoxicity, seizures, and psychopathology (paranoid psychosis). On the other hand, long-term use of stimulants does not invariably lead to tolerance, nor does cessation of chronic use produce clear withdrawal signs (19).

Tolerance and Sensitization. After repeated administration of stimulants, the response to a challenge dose of the drug can either be attenuated (tolerance) or enhanced (sensitization). Because sensitization can persist long after discontinuation of drug administration, it has been proposed that sensitization is the basis not only for the pathologic consequences of long-term stimulant use but also for other essential features of addiction (e.g., craving and relapse). It is unclear whether tolerance or sensitization develops to the rewarding effects of stimulants.

Recognizing that increased dopaminergic transmission in the mesolimbic system was responsible for the behaviors that exhibit sensitization (e.g., locomotor activity), researchers hypothesized that sensitization is due to a progressive enhancement of the ability of stimulant drugs to increase the levels of synaptic DA in the nucleus accumbens. The reader is directed to recent reviews for a more comprehensive treatment (20,21).

Stimulant Withdrawal. Unlike withdrawal syndromes from opioids and sedative-hypnotic drugs, abstinence from cocaine and other psychostimulants is believed to be characterized primarily by emotional and motivational disturbances.

The description of a cocaine abstinence syndrome by Gawin (22) has provided a provocative heuristic framework for both clinical and preclinical research. Gawin proposed that cocaine withdrawal consists of three distinct temporal phases, each marked by progressive changes in mood and drug craving. The first stage ("crash") is analogous to an alcohol hangover. The "crash" phase lasts for 1 to 4 days after cessation of cocaine use and is characterized by exhaustion and intense craving for cocaine. The true withdrawal syndrome was considered to start about a week after the last dose of cocaine and to last for 1 to 10 weeks, during which there was a high probability of relapse. This phase is marked by a subtle dysphoric syndrome, consisting of decreased activity, loss of motivation, increased boredom, and anhedonia, punctuated by episodes of drug craving. If abstinence is maintained for 1 to 2 months, mood, although somewhat labile, begins to normalize. This normalization of mood represents the beginning of a protracted extinction phase, during which the craving response to cues gradually decreases if abstinence is maintained, although intermittent episodes of drug craving and relapse can occur.

Even though craving is central to descriptions of stimulant withdrawal and is a major factor in relapse, the neural basis of craving is unknown. One problem with this line of investigation is that craving does not have a widely accepted definition or physiologic measure.

Cocaine Intoxication

Oral ingestion of cocaine hydrochloride results in a delayed onset (10 to 30 minutes) and prolonged duration (45 to 90 minutes) of action, with relatively low peak plasma levels; intranasal use results in similar peak plasma levels, but with a more rapid onset (2 to 3 minutes) and shorter duration (30 to 45 minutes) of drug effects. Smoking and intravenous use result in rapid onset of cocaine effects, within 5 to 45 seconds, lasting between 5 to 20 minutes, with peak plasma levels 5 to 10 times those achieved by oral and intranasal routes. The plasma half-life of cocaine is between 40 and 60 minutes. Cocaine is detectable in the urine for up to 36 hours after use.

At low doses, cocaine produces euphoria and a sense of enhanced self-image. Users report increased energy and mental acuity, heightened sensory (including sexual) awareness and self-confidence, and diminished appetite. The euphoric "rush" following intravenous or crack cocaine use is described as intense and orgasmic, lasting from seconds to several minutes. The sympathomimetic effects of cocaine result in increased motor activity and tremor and increased heart rate, blood pressure, and temperature.

Psychotic disorders associated with cocaine abuse range from delusional disorders to schizophrenia-like symptoms and are characterized by anxiety, paranoia, agitation, and impaired judgment, with relative sparing of general cognitive function. Cocaine-induced delusions are commonly persecutory, jealous, or somatic, with the latter including parasitosis or cocaine "bugs." These psychoses correlate with high cocaine plasma levels and are believed to be the result of sustained DA release. Symptoms gradually resolve as plasma levels fall, and they clear within 24 to 48 hours in relatively inexperienced cocaine users and in several days to weeks in chronic abusers. Increasingly, a hyperarousal delirium syndrome has been noted secondary to cocaine use, characterized by sudden onset of disturbances in perception and attention, disorientation, and cognitive impairment. Untreated, the delirium can progress to include hyperthermia, mydriasis, autonomic instability, and fatal respiratory collapse.

Cocaine-induced seizures have been well described and are believed to arise from lowered seizure threshold by the local anesthetic effects of the drug. Seizures typically occur immediately after or within hours of cocaine use and are more likely to occur when cocaine is injected intravenously or smoked as crack. Typically, these seizures are single and generalized and not associated with lasting neurologic deficits. Multiple or focal seizures suggest underlying brain pathology and other substance abuse. Status epilepticus and death have occurred after ingestion of large amounts of cocaine, as happens when latex cocaine-filled condoms rupture inside "body packers." Death from cocaine-related seizures are believed to result from cardiac arrhythmias and hyperthermia.

Cerebrovascular accidents in cocaine abusers result from specific effects of cocaine on the cerebral vasculature in combination with the sympathomimetic effects of cocaine on vascular tone. It has been estimated that the relative risk for stroke among drug abusers after controlling for other stroke risk factors is 6.5, making drug abuse the most common cause of stroke in people under the age of 35 years. Hemorrhagic and ischemic strokes occur with about equal frequency in cocaine abusers. In almost half of the hemorrhagic stroke cases, underlying vascular abnormalities (saccular aneurysms and arteriovenous malformations) are present; in the remainder of cases, thalamic and basal ganglia bleeds are common. Transient ischemic attacks have also been reported in cocaine abusers, many of which escape medical attention owing to their spontaneous resolution.

Cocaine Withdrawal

Physiologic dependence on cocaine is evidenced by the development of tolerance to its euphorigenic effects and by emergence of distinct withdrawal phenomena within a few hours of last cocaine use. Withdrawal symptoms are commonly recognized after heavy and prolonged cocaine use. Symptoms characteristic of cocaine withdrawal attributed to acute DA depletion and receptor upregulation include dysphoric mood, fatigue, vivid and unpleasant dreams, insomnia or hypersomnia, increased appetite, and either psychomotor agitation or retardation. The most serious problems associated with cocaine withdrawal are suicidal ideation and suicide attempts, associated with profound withdrawal depression.

Three phases of cocaine abstinence have been described. The first phase, commonly referred to as the crash, occurs within hours of last cocaine use and lasts 3 to 4 days. Agitation, anorexia, and intense cocaine craving appear early in the crash but are quickly replaced with fatigue, hypersomnolence, and hyperphagia as craving subsides. Depression persists throughout this first phase. During the second phase, which occurs over the next 8 to 10 weeks, patients experience an early return to normalized mood and sleep pattern with little cocaine craving, although as the anhedonia, anergia, anxiety, and craving for cocaine gradually intensify, patients are at high risk for relapse. The third and final extinction phase occurs with continued cocaine abstinence. During this period, the risk for relapse decreases as mood and hedonic response normalize, although cocaine craving continues to emerge intermittently, especially in the presence of conditioned cues.

Opiates

Recent estimates indicate that more than 6% of the U.S. population has used analgesics for nonmedical reasons, whereas less than 2% report having used heroin in the past year. The lifetime prevalence of people meeting diagnostic criteria for opiate dependence has been estimated to be 0.7%, with 0.1% meeting dependence criteria within the past 6 months.

Opiate Intoxication

The CNS effects of opiates include analgesia, drowsiness, changes in mood, and mental clouding. Respiratory depression, miosis, nausea and vomiting, and cough reflex suppression are other opiate effects mediated centrally. Opiate overdose occurs when the respiratory depressant effects result in a decreased level of consciousness. Pinhole pupils, slow, shallow respirations, and coma are the classic triad of acute opiate overdose. If hypoxia persists, pulmonary edema, hypotension, and cardiovascular collapse occur, and eventually death supervenes. Establishment of an airway and intravenous administration of naloxone, an opiate antagonist, are the key elements in the treatment of opiate overdose.

Tolerance

Cellular adaptive changes to chronic opioid administration have been best characterized in the signal transduction pathways activated by μ-receptors on noradrenergic neurons in the locus ceruleus (LC). Most of the noradrenergic cell bodies in the brain are found in the LC, a nucleus adjacent to the floor of the fourth ventricle at the pontine–medullary junction. Acute administration of morphine or selective μ-opioid agonists produces dose-dependent decreases in the spontaneous activity of LC neurons. These decreases are blocked by specific μ-antagonists, such as β-chlornaltrexamine and β-funaltrexamine, as well as by prototypical opiate antagonists, such as naloxone and naltrexone.

Neurons in the LC develop tolerance to opioids in rats chronically treated with morphine. After 3 to 5 days of chronic treatment, spontaneous firing rates of LC neurons are no longer suppressed and are equivalent to those in drug-naïve animals. Larger doses of opiates are then required to suppress firing. This tolerance appears to result from the uncoupling of opiate-binding sites from their associated signal transduction pathways.

Opiate Withdrawal

CNS hyperexcitability characterizes the opiate withdrawal syndrome. Early symptoms include lacrimation, rhinorrhea, yawning, and sweating, with progression of symptoms to include dilated pupils, anorexia, gooseflesh, restlessness, irritability, and tremor. In fully developed opiate withdrawal, increased irritability, insomnia, marked anorexia, violent yawning, severe sneezing, lacrimation, nausea, vomiting, diarrhea, weakness, pronounced depressed mood, tachycardia, and hypertension are noted. The symptoms of protracted abstinence, including hypotension, hyperalgesia, and poor stress tolerance, have been identified as important reasons for relapse after opiate withdrawal. The severity of the syndrome is dependent on the dose and duration of use. The symptoms begin to appear 8 to 12 hours after the last dose of an opioid agonist, depending on the pharmacokinetics of the individual drug. The syndrome reaches a peak over the next 48 to 72 hours and is virtually absent by 7 to 10 days.

Although many brain areas, including thalamic, hypothalamic, limbic, and hindbrain areas, are activated during the physical withdrawal from opiates, it is now generally acknowledged that hyperactivity of noradrenergic innervations in the CNS contributes substantially to the expression of many of the physical symptoms seen during opiate withdrawal. The most compelling evidence for a noradrenergic role in opiate withdrawal is that pharmacologic suppression of noradrenergic activity attenuates morphine withdrawal in both animals and humans. Systemic and local (intracerebral) administration of clonidine, a specific α_2-adrenergic receptor agonist that decreases the spontaneous activity of LC neurons, attenuates both the increase in LC firing due to withdrawal and most of the accompanying physiologic signs.

Affective Aspects of Withdrawal. Dysphoria is a prominent part of opioid withdrawal in human subjects. Opioid withdrawal interferes with motivational states, as seen by reductions in the performance of operant schedules for food reward. Opioid withdrawal also acts as a negative unconditioned stimulus in conditioning paradigms. The nucleus accumbens appears to play a crucial role in the affective components of withdrawal, complementary to its role in opioid reinforcement. Clonidine is less effective in alleviating dysphoria than other signs in humans during opioid withdrawal.

As with alcohol, syndromes commonly associated with opiate use are temporally related to the last opiate use and chronicity of use (Table 20-8). The duration of opiate intoxication and onset of withdrawal vary with the pharmacologic half-life of the particular opiate. For example, heroin and meperidine have relatively short half-lives, and withdrawal symptoms from these opiates can emerge within 8 hours of drug abstinence. Tolerance and physical dependence develop rapidly after repeated opiate exposure; mild physical dependence has been demonstrated after short-term postoperative opiate analgesic administration. Tolerance, however, does not develop uniformly to all opiate effects; notable tolerance develops to the analgesia, euphoria, sedation, and respiratory depression, and little to the gastrointestinal, pupillary, and endocrine effects.

TABLE 20-8. *Syndromes of opiate intoxication and withdrawal*[a]

Syndrome (onset and duration)	Characteristics
Opiate intoxication	Conscious, sedated, "nodding"; mood normal to euphoric; pinhole pupils; history of recent opiate use
Acute overdose	Unconscious; pinhole pupils; slow, shallow respirations
Opiate withdrawal	
Anticipatory[b] (3–4 hours after last fix)	Fear of withdrawal; anxiety; drug craving; drug-seeking behavior
Early (8–10 hours after last fix)	Anxiety; restlessness; yawning; nausea; sweating; nasal stuffiness, rhinorrhea; lacrimation; dilated pupils; stomach cramps; drug-seeking behavior
Fully developed (1–3 days after last fix)	Severe anxiety; tremor; restlessness; piloerection;[c] vomiting, diarrhea; muscle spasms;[d] muscle pain; increased blood pressure; tachycardia; fever, chills; impulse-driven drug-seeking behavior
Protracted abstinence (may last up to 6 months)	Hypotension; bradycardia; insomnia; loss of energy, appetite; stimulus-driven opiate cravings

[a]The times given in the table refer to heroin. Withdrawal will develop more slowly with long-acting opiates.
[b]Anticipatory symptoms begin as the acute effects of heroin begin to subside.
[c]The piloerection has given rise to the term "cold turkey."
[d]The sudden muscle spasms in the legs have given rise to the term "kicking the habit."

Neuropsychiatric Syndrome Associated With Opiate Abuse

Opiates themselves do not directly cause either neuropathology or psychopathology; rather, it is the common routes of administration (subcutaneous and intravenous); the adulterants used to cut or prepare abused opiates (e.g., quinine, lead, chloroquine, MPTP); and the lifestyle associated with illicit opiate use that account for the majority of neurologic complications of opiate abuse. Other than those associated with human immunodeficiency virus (HIV) infection and acquired immunodeficiency syndrome (AIDS; see later discussion), most infective neurologic complications of intravenous drug use are those that accompany endocarditis, which is typically right-sided with tricuspid valve involvement. Infectious material filtered in the pulmonary tree is circulated systemically when arteriovenous shunts develop secondary to pulmonary hypertension, and can result in intraparenchymal and extraparenchymal abscess of the brain and spinal cord, meningitis, embolic cerebral infarction, diffuse vasculitis, and septic aneurysm. Other neurologic complications of infectious origin in street opiate addicts include hepatitis, encephalopathy, CNS syphilis, tetanus, and botulism.

Sedative-Hypnotics

Sedative-Hypnotic Intoxication

The acute effects of sedative-hypnotics consist of slurred speech, incoordination, ataxia, sustained nystagmus, impaired judgment, and mood lability. In large amounts, sedative-hypnotics produce progressive respiratory depression and coma. The amount of respiratory depression produced by the benzodiazepines is much less than that produced by the barbiturates and other sedative-hypnotics.

Sedative-Hypnotic Withdrawal

Drugs in this class produce withdrawal syndromes that are qualitatively similar, but not identical. Behaviorally, the predominant symptoms are the converse of the primary effects of the sedative-hypnotics. They include anxiety, increased locomotor activity, insomnia, tremors, and convulsions. As with opioid withdrawal, the severity of the syndrome is dependent on the dose, duration of use, and especially the rate of elimination of the drug.

Barbiturate and ethanol withdrawal symptoms appear soon after discontinuation of drug use, even after short-term use of moderate doses. Benzodiazepine withdrawal symptoms can take several days or weeks to emerge, although faster onsets have been noted after long durations of use. This delayed development of symptoms has been attributed to the slow decline in plasma levels of most benzodiazepines in humans, even when drug administration is discontinued abruptly.

Hyperactivity of the central noradrenergic system may also contribute to the anxiolytic-sedative abstinence syndrome, although evidence for noradrenergic involvement is less well established in this situation than it is for opioid withdrawal. The potential involvement of noradrenergic systems in withdrawal from opiates and anxiolytic-sedative drugs may explain the commonalities between the two withdrawal syndromes. These include gastrointestinal distress (decreased food intake, retching, and vomiting); weight loss; "wet-dog" shakes; increased startle responses; vocalizations; and piloerection.

If hyperactivity of the noradrenergic system is involved in withdrawal from anxiolytic-sedative drugs, treatment that reduces noradrenergic activity should attenuate this withdrawal syndrome as it does the opioid abstinence syndrome. A preliminary clinical study demonstrated that clonidine decreased the severity of withdrawal from alcohol and the time required for tapered withdrawal from alprazolam (23).

The withdrawal syndrome from sedative-hypnotics is similar to the withdrawal syndrome from alcohol. Signs and symptoms include anxiety, tremors, nightmares, insomnia, anorexia, nausea, vomiting, postural hypotension, seizures, delirium, and hyperpyrexia. The syndrome is qualitatively similar for all sedative-hypnotics; however, the time course depends on the particular drug. With short-acting sedative-hypnotics (e.g., pentobarbital, secobarbital, meprobamate, oxazepam, alprazolam, and triazolam), withdrawal symptoms typically begin 12 to 24 hours after the last dose and peak in intensity between 24 to 72 hours. With long-acting drugs (e.g., phenobarbital, diazepam, and chlordiazepoxide), withdrawal symptoms peak on the fifth to eighth day.

Tolerance

Continued use of sedative-hypnotic drugs (e.g., benzodiazepines, barbiturates, ethanol, aldehydes) produces tolerance to their sedative, anticonvulsant and, anxiolytic effects. However, tolerance develops differentially to each of these effects. Tolerance to the sedative and anticonvulsive effects develops within days. Tolerance to the anxiolytic effects has been difficult to demonstrate.

Tolerance to other sedative-hypnotic drugs may also involve changes to the GABA–benzodiazepine receptor complex. This mechanism is most likely the basis of cross-tolerance between these drug classes. Tolerance to ethanol also involves direct changes in the function of the NMDA receptor. Acutely, ethanol decreases NMDA receptor function, probably through an action within the ion channel.

Hallucinogens

Hallucinogens are a chemically diverse group of compounds that produce perceptual distortion and prominent visual hallucinations. Perceptual changes include depersonalization, derealization, illusions, and synesthesias, in which sensory stimuli in one sense produce changes in another; for example, a beating drum is seen as a flashing light. The hallucinogenic drugs considered here are those commonly used by drug abusers in the United States (Table 20-9). Hallucinogens do not produce physical dependence, and because tolerance to the psychological effects develops rapidly, they are typically used episodically.

Although 8.2% of the U.S. population report having used a hallucinogen, less than 0.3% have ever met diagnostic criteria for dependence. The true prevalence of use is difficult to determine because most users do not come to medical attention.

Hallucinogen Intoxication

The effects of hallucinogens depend on the drug, the dose, the setting, and the circumstances of use. Typically, CNS effects include dilated pupils, tachycardia, sweating, palpitations, and tremors. Perceptual alterations associated with intoxication include changes in mood, intensification of perceptions, depersonalization, and derealization. Hallucinations are usually visual, but auditory and tactile hallucinations sometimes occur.

Neuropsychiatric Syndromes Associated With Hallucinogen Abuse

Sporadic case reports suggest that LSD can cause significant vasospasm and stroke. Subarachnoid hemorrhage has been reported in association in MDMA abuse. Neuropsychological impairment has been reported in

TABLE 20-9. *Chemical classification of drugs used for hallucinogenic effects*

Chemical class	Drugs	Common or street names
Indoles	Lysergic acid diethylamide	LSD, acid
	Dimethyltryptamine	DMT
	Psilocybin and psilocin	Magic mushrooms
Phenethylamines	Mescaline	
	2,5-dimethoxy-4-methylamphetamine	DOM
	3,4-methylenedioxyamphetamine	MDA
	3,4-methylenedioxymethylamphetamine	MDMA, XTC, Ecstasy
	3,4-methylenedioxyethylamphetamine	Eve

heavy users of MDMA. *Flashbacks* are transient recurrences of perceptual alterations like those that occurred under the influence of the hallucinogen. Flashbacks can occur up to 5 years or more after abstinence. LSD and MDMA can cause flashbacks and have also been reported to induce panic disorder and chronic paranoid psychosis.

Marijuana

Marijuana intoxication has been demonstrated to impair learning and short-term memory and the performance of complex motor tasks. In an extensive review of the literature, Wert and Raulin found no significant differences between volunteer and heavy user samples and normal controls on computed tomography scan or electroencephalogram measures nor upon detailed neurologic examination (24). Furthermore, most studies reviewed provided no evidence that chronic marijuana use resulted in diminished psychological or neuropsychological performance.

Nicotine

Tolerance

Chronic administration of nicotine produces tolerance. Furthermore, it is now clear that deprivation of nicotine, in subjects that have been exposed chronically to the drug, also leads to the emergence of withdrawal signs, indicating dependence upon the drug. The mechanisms for producing these effects appear to involve cellular adaptation rather than dispositional changes.

Reversible tolerance to some of the effects of nicotine can develop in a matter of minutes. Tolerance is not explained by pharmacokinetics. Another central effect of nicotine that shows the development of tolerance is hypothermia.

Sensitization

Whereas tolerance has been the predominant effect of chronic nicotine on nicotine-induced hypothermia and suppression of locomotor activity, chronic nicotine enhances behavioral activation, which follows the early suppression of locomotor behavior by nicotine. Such stimulation has been observed 20 to 80 minutes after a nicotine challenge in rats given repeated administration of low doses of nicotine.

Dependence and Withdrawal

Acceptance of the view that nicotine produces physical dependence derives from observations of withdrawal signs and symptoms in human subjects that were previously exposed chronically and then deprived of cigarettes or smokeless tobacco products. The syndrome is characterized by anxiety, irritability, difficulty in concentration, sleep disturbances, increases in appetite, and craving for tobacco.

Studies of the effects of chronic or subchronic nicotine on the binding of radioligands to nAChRs in the brain have supported the view that nAChR is at least one of the cellular targets that contributes to the long-term changes in behavioral responses to nicotine. The findings generally indicate that chronic nicotine produces upregulation of nAChRs, particularly in the cerebral cortex.

Unlike cocaine, amphetamines, and opioids, behavioral sensitization after chronic nicotine does not appear to involve potentiation of the mesolimbic DA transmission. Despite the ability of nicotine to activate DA impulse activity and to increase extracellular DA in the nucleus accumbens, these effects are not enhanced by repeated administration of nicotine. Behavioral sensitization to nicotine may involve direct effects on neurons of the nucleus accumbens or actions on cortically projecting DA neurons, or changes in other monoaminergic neurons (25).

Co-Morbidity of Mental Health and Substance Use Disorders in the General Population

In the Epidemiologic Catchment Areas (ECA) study, the lifetime prevalence in the

United States of coexisting mental health and substance dependence disorders was 32.7%. Mental health disorders, other than those that are alcohol- or drug-related, were diagnosed in 22.5% of the sample, whereas alcohol abuse or dependence was found in 13.5%, and drug abuse and dependence was found in 6.1% of the sample. These rates are substantially higher in some special populations. In a sample of prisoners, 82.1% were found to have coexistent mental health and substance abuse and dependence disorders, with alcoholism prevalent in 56.2% and substance abuse and dependence prevalent in 53.7%. Likewise, among a homeless population in Los Angeles, California, the lifetime prevalence of mental health disorders other than substance abuse was 68.8%, and the lifetime prevalence of substance abuse disorders was 62.9% (26).

Alcoholism

The frequency of secondary mental health diagnoses among alcoholics (36.6%) almost doubles the rate for mental disorders in the general population (19.9%), and the prevalence of other drug disorders (21.5%) in the alcoholic population is nearly six times the rate (3.7%) of drug disorders in nonalcoholics. Table 20-10 illustrates the relationship between alcoholism and other drug dependence disorders and the incidence of co-morbid psychiatric disorders. The 36.6% rate of mental health disorders in alcoholics is 2.3 times the rate found in nonalcoholics. One of the most consistent epidemiologic findings is the greater prevalence of alcohol dependence and abuse disorders in men compared with women. Moreover, whereas 44% of male alcoholics have a secondary psychiatric diagnoses, 65% of female alcoholics have concurrent diagnoses.

Drug Dependence

The prevalence of mental disorders among patients with drug abuse and dependence disorders is 53.1%, or 4.5 times the rate in the non–drug-abusing or dependent population. Similarly, the prevalence of alcohol abuse or dependence among drug abusers is 47.3%, or 7.1 times the rate in the non–drug-abusing, nondependent population. In combination, the prevalence of any mental or alcohol disorder among drug abusers is 71.6%, or 6.5 times the non–drug-abusing, nondependent population. As with alcohol disorders, the rate of antisocial personality in the drug-dependent population is much higher (17.8%) than that found in non–drug-dependent groups, with an odds ratio of 13.4. Among marijuana users, the prevalence of alcohol abuse or dependence is 45.2%, or 6 times the nonuser group. Among opiate, amphetamine, and hallucinogen users, the rate of alcohol abuse or dependence ranges between 62% and 66%; however, among those diagnosed with cocaine disorders, the rate of alcohol abuse or dependence is 84.8%, or 36.3 times the rate of non–cocaine-abusing groups. This strong relationship between cocaine abuse or dependence and alcohol abuse or dependence complicates treatment because patients are typically unaware of the severity and significance of their alcohol disorder.

Substance Abusers With Human Immunodeficiency Virus

Substance abusers who administer drugs parenterally, support their drug use by sex work, or engage in high-risk sexual behavior while intoxicated are at significantly increased risk for HIV infection. The neurocognitive decline associated with HIV disease, referred to as AIDS dementia complex, can occur at any time during the course of the disease, although detection in symptom-free patients remains the subject of debate. The AIDS dementia complex involves many of the same neurologic deficits noted with chronic substance abuse, including problems with verbal memory, attention, concentration, psychomotor speed, cognitive flexibility, and, in some cases, nonverbal skills and memory (27,28).

TABLE 20-10. *Prevalence of alcohol- and drug-dependence disorders and other co-morbid psychiatric disorders*

	Any alcohol		Any other drug		Marijuana		Cocaine		Opiates		Barbiturates		Amphetamines		Hallucinogens	
Comorbid disorder	%	OR	%	OR	%	OR	%	OR	%	OR	%	OR	%	OR	%	OR
Any mental	36.6	2.3	53.1	4.5	50.1	3.8	76.1	11.3	65.2	6.7	74.7	10.8	62.9	6.2	69.2	8.0
Schizophrenia	3.8	3.3	6.8	6.2	6.0	4.8	16.7	13.2	11.4	8.8	8.0	5.9	5.5	3.9	10.0	7.4
Any affective	13.4	1.9	26.4	4.7	23.7	3.8	34.7	5.9	30.8	5.0	36.4	6.6	32.7	5.7	34.3	5.8
Any anxiety	19.4	1.5	28.3	2.5	27.5	2.3	33.3	2.9	31.6	2.8	42.9	4.5	32.7	2.9	46.0	5.0
Antisocial personality	14.3	21.0	17.8	13.4	14.7	8.3	42.7	29.2	36.7	24.3	30.3	19.0	24.5	14.3	28.5	15.6
Alcohol abuse or dependence			47.3	7.1	45.2	6.0	84.8	36.3	65.9	12.8	71.3	16.9	61.7	11.1	62.5	10.9

OR, odds ratio.
Adapted from ref. 26, with permission.

Implications for Treatment and Recovery

Neuropsychological impairment in treatment-seeking substance abusers necessitates consideration of how these impairments affect treatment and recovery. It is recommended that interventions with newly abstinent drug abusers should minimize cognitively demanding approaches. Complex skill training and vocational and educational interventions should not be introduced until several weeks of abstinence have been achieved. The degree of neuropsychological impairment may be a significant predictor of treatment outcome. Substance abusers with cognitive deficits are at greater risk for relapse than those without.

Summary of Neuropsychological Impairment

Predictable patterns of neuropsychological impairment frequently accompany chronic substance abuse, with specific deficits varying according to the drug of abuse. Although there is evidence that neuropsychological deficits may predispose individuals to substance use disorders or make them more susceptible to the neurotoxic effects of drugs of abuse, there is also evidence that certain deficits may clear slowly with continued abstinence. Neuropsychological assessment upon treatment entry can help structure treatment interventions and can predict the patient's ability to remain abstinent.

PHARMACOLOGIC TREATMENT OF DRUG ABUSE

Opiate Pharmacotherapies

Pharmacotherapy for opiate dependence includes treatment of acute overdose, detoxification, and long-term substitution (Table 20-11). Naloxone, a highly potent, virtually pure opiate antagonist, is the specific treatment for acute opiate overdose. A transient response should suggest overdose from long-acting opiates like *l*-acetyl-α-methadol (LAAM) or methadone. Coma may return in these patients after the effects of naloxone wears off.

Opiate Detoxification (Gradual Withdrawal)

Methadone and clonidine are the two most commonly employed medications for opiate detoxification. A synthetic full opiate agonist, methadone effectively suppresses the symptoms of opiate withdrawal and can be conveniently administered once daily in an oral preparation, usually diluted with juice. The starting dose for detoxification of street heroin addicts is usually 30 to 40 mg. After several days of stabilization, the dose is tapered to zero over 2 to 3 weeks. The failure rate for patients treated in out-patient methadone detoxification clinics is very high.

Clonidine, an α-adrenergic agonist marketed for treatment of hypertension, suppresses the physiologic manifestation of opiate withdrawal related to rebound sympathetic

TABLE 20-11. *Opiate addiction pharmacotherapies*

Medication	Drug class	Indication	Clinical action
Methadone	Opiate agonist	Maintenance therapy, detoxification	Suppress withdrawal; block effects of subsequently administered opiates
LAAM	Opiate agonist	Maintenance therapy	Suppress withdrawal; block effects of subsequently administered opiates
Buprenorphine[a]	Partial opiate agonist	Maintenance therapy, detoxification	Suppress withdrawal; block effects of subsequently administered opiates; produce less physical dependence than full agonists
Naltrexone	Opiate antagonist	Postdetoxification	Block effects of subsequently administered opiates
Clonidine	α_2-Adrenergic agonist	Withdrawal	Suppress withdrawal-induced noradrenergic hyperactivity

LAMM, *l*-acetyl-d-methadol.
[a]Not yet FDA-approved for the treatment of opiate dependence.

hyperactivity. It is not a scheduled medication and is therefore available to physicians in general. Several dosing strategies have been described. Its major drawbacks are orthostatic hypotension and sedation, which limit the amount of medication that can be safely administered (29).

Opiate Maintenance

Two relatively long-acting opiate agonists, methadone and LAAM, are currently available for chronic maintenance treatment for opiate dependence. Naltrexone, an orally effective, long-acting antagonist, has been available for more than a decade for the prevention of relapse but has enjoyed only limited clinical success.

Methadone

Methadone maintenance is the prototype for opiate substitution pharmacotherapy, having been administered in specially licensed clinics for more than 30 years. It is a full opiate agonist that, when administered orally, suppresses the emergence of withdrawal symptoms in opiate-dependent individuals and blocks the effects of subsequently administered opiates for up to 72 hours. Methadone has a slower onset and longer duration of action than heroin and other short-acting, abused opiates and thus does not provide a reinforcing "rush" upon administration. Most patients can be effectively maintained with once-daily dosing and, because of its blocking action at the μ-opiate receptor, addicts no longer need to use drugs to counteract or avoid withdrawal. Because it is orally administered, secondary reinforcements related to needle use are obviated. High-dose methadone maintenance (70 to 90 mg/d) has proved more effective than low-dose maintenance in reducing injection drug use, which in turn is important in reducing HIV transmission among the drug-abusing population.

LAAM

l-Acetyl-d-methadol (LAAM) is a methadone derivative. LAAM is a synthetic opiate agonist, with actions qualitatively similar to morphine, a prototype μ-agonist that affects the CNS and smooth muscles. Its slow onset of action makes it less subject to abuse, greatly decreases its street value, and, thus, minimizes its risk for street diversion. Its long duration of action provides a more stable blood level and allows for three-times-a-week dosing. Because LAAM is also a schedule II narcotic, its use is governed by regulations similar to methadone. LAAM's efficacy in reducing illicit opiate use and keeping patients in treatment, as well as its safety with chronic administration, have been demonstrated convincingly (30).

Cocaine Pharmacotherapies

No U.S. Food and Drug Administration (FDA)-approved pharmacotherapy for cocaine addiction currently exists. In a recent review, Tennant and colleagues listed more than 30 agents potentially useful in cocaine withdrawal or abstinence but concluded that little clinically significant reductions in cocaine use or improved treatment had been demonstrated (31).

Alcohol Pharmacotherapies

In theory, the intoxicating effect of alcohol can be reduced or reversed by inhibiting its absorption, antagonizing its effect at the receptor sites, altering or counteracting its physiologic effects, or enhancing its elimination from the body. In practice, the search for a sobering agent has been singularly unsuccessful. Treatment of acute withdrawal has enjoyed considerably more success. Most currently employed medications for suppression of withdrawal are benzodiazepines. Some examples are diazepam, in doses of 5 to 20 mg orally every 4 to 6 hours; chlordiazepoxide, 25 to 100 mg orally every 4 to 6 hours, and oxazepam, 15 to 60 mg orally every 4 to 6 hours. Clonidine, an α-adrenergic agonist, has been useful in reducing the neurophysiologic effects of withdrawal, as has propranolol, which counteracts the tachycardia, anxiety, and tremor of acute withdrawal.

Pharmacologic strategies for relapse prevention have involved medications that produce an adverse reaction when alcohol is consumed and medications that modify craving and consumptive behavior. Disulfiram (Antabuse) irreversibly inactivates acetaldehyde dehydrogenase, a step in alcohol metabolism. This leads to the accumulation of acetaldehyde when alcohol is consumed and the appearance of the disulfiram–alcohol reaction, characterized by flushing, throbbing headache, nausea, vomiting, thirst, sweating, palpitation, chest pain, tachycardia, confusion, agitation, and, when severe, respiratory depression and cardiovascular collapse. The dosage used ranges from 125 to 1000 mg/d, although most patients are treated with 250 to 500 mg/d.

Medications that decrease alcohol craving and consumption aim at manipulating the brain neurotransmitter systems mediating reward, mood, and appetite behavior. Naltrexone, the narcotic antagonist developed originally for relapse prevention of opiate dependence, has received FDA approval as a recommended adjunct to alcohol relapse prevention in the context of comprehensive treatment programs. Based on these clinical studies, the FDA has approved naltrexone, with the new trade name ReVia, for the treatment of alcoholism, the first medication so approved since disulfiram.

CONTEMPORARY ISSUES

The search continues for new endogenous ligands for receptor systems of relevance to the neurochemistry of drug abuse. The characterization of a specific cannabinoid receptor has raised the question of the existence of endogenous ligands. Recently, a novel derivative of arachidonic acid, arachidonylethanolamide (anandamide), isolated from porcine brain, was identified as a ligand of the cannabinoid receptor and was proved to exert cannabinoid-like behavioral effects (32).

The neurophysiology of neural-based reward systems continues to be under investigation. To explain the inconsistency between the physiologic properties of dopaminergic neurons and those activated by rewarding stimulation, it has been proposed that directly stimulated, non-DA "first-stage" neurons synapse with "second-stage" dopaminergic neurons in the midbrain tegmentum and stimulate ascending DA pathways indirectly. The neurochemistry of the first-stage, myelinated, caudally projecting fiber system remains an open question. There is some evidence that at least some of these neurons may be cholinergic. Cholinergic neurons in the pedunculopontine and laterodorsal tegmental nucleus make synaptic contact with dopaminergic neurons.

Effects of intracranial microinjections of abusable drugs on electrical brain stimulation reward suggest that these substances produce reward by direct actions on a small subset of second-stage dopaminergic, unmyelinated axons. The neural substrates for the enhancement of brain stimulation reward by amphetamine may lie within the nucleus accumbens and neostriatal forebrain terminal projections of mesotelencephalic dopaminergic neurons. Similarly, the substrate for the enhancing effect of morphine on brain stimulation reward was localized to cell body regions of the mesotelencephalic DA system, including the posterior hypothalamus, and less consistently, the VTA. Such interconnections of neurochemical reward systems may allow us to understand the DA system as a final common pathway of sorts in electrical brain stimulation reward.

The role of gene activation and suppression in all phases of the organism's response to drug administration will likely be an important horizon of future research. The inactivation of the G proteins during chronic opiate treatment probably reflects changes in gene regulation. Acute administration of opiates decreases the expression of the immediate early genes, c-fos and c-jun. Conversely, expression of these genes increases several-fold during naltrexone-induced withdrawal. Because the time course for gene regulation is much longer than the decrease in spontaneous activity of the cell in response to acute opiates, decreased expression of these immediate early genes may play a role in adaptation of the cells to chronic treatment (33).

The role of newly understood neurotransmitter systems, such as the gases, will increasingly come under investigation in the neuropsychiatry of drug abuse. Recently, inhibition of NO synthesis was shown to attenuate the behavioral signs of opioid withdrawal. This observation reflects the evidence reviewed previously proposing a crucial role for NO in the development of morphine tolerance at both the behavioral and cellular level. At this time, it is not known whether inhibition of NO prevents LC hyperactivity during opioid withdrawal, or if the actions of NO are mediated at some other brain site. Furthermore, the dense collection of NO synthase–containing neurons in the laterodorsal tegmental (LDT) nucleus, immediately adjacent to the LC, is a likely source of NO that might diffuse to act in the LC. If the LDT is the source of NO input to the LC, a cholinergic involvement in opioid withdrawal would be implicated because all NO synthase–containing neurons in the LDT are cholinergic (34).

SUMMARY

The past two decades have seen remarkable advances in our understanding of the mechanism of action and drug abuse. Receptor pharmacology has provided major insights about the initial sites of drug action in the brain. It is now recognized that drugs of abuse act on classic receptors or transporter molecules associated with specific neurotransmitter systems. Starting with the first assays of opiate-receptor binding, using radiolabeled naloxone in the 1970s, relevant receptors for every major drug of abuse have been identified in the brain. The impact of chronic administration of drugs of abuse on receptor function has become characterized in increasing detail, which has explained tolerance and other long-term drug effects at a cellular level. Furthermore, rapid advances in molecular biologic techniques have allowed the cloning of these receptive molecules, with elucidation of their amino acid sequences.

Equally important are advances in our understanding of how receptor-mediated events influence neural systems and how effects in these systems produce the behavioral manifestations of addiction. One crucial feature of drugs of abuse is their rewarding properties, which are required for the initiation and maintenance of drug use. It is now well established that the mesolimbic DA system serves as a common substrate for this phase of addiction. On the other hand, a plethora of evidence has implicated noradrenergic systems in the physical aspects of withdrawal. The neurobiologic substrates of other aspects of the addictive process, such as sensitization, conditioning, and predisposition, are not characterized as well. Current investigations are uncovering mechanisms that include involvement of serotonergic, glutamatergic, and other neurotransmitter systems in these processes.

Until now, most of the major advances in our understanding of substance abuse have been derived from investigations of laboratory animals. The challenge for the future is to verify and to extend this work in human investigations. Noninvasive techniques, such as positron emission tomography (PET), for imaging brain function with biochemical and anatomic resolution, will be paramount in this effort. The availability of PET has already facilitated investigations of the biologic correlates of drug-induced alterations of mood and feeling state. Studies of volunteers with histories of polysubstance abuse have shown that doses of either morphine or cocaine that produce positive affect, as measured by standard rating scales, also produce widespread reductions of cerebral glucose use. In fact, a wide variety of substances that are self-administered reduce cerebral glucose metabolism, as measured by PET in human volunteers. In addition to morphine and cocaine, this group of substances includes alcohol, amphetamine, barbiturates, and benzodiazepines. It has been hypothesized that this action of drugs of abuse is related to the euphorigenic properties of these drugs and may result from effects on dopaminergic neurotransmission. Ultimately, the anatomic resolution and subcortical definition provided by PET will be combined with the exquisite time resolution of electrophysiologic measures, such as cortical evoked-potential mapping, to provide a meaningful picture of the acute and chronic cerebral mechanisms of substance abuse.

ACKNOWLEDGMENT

Preparation of this manuscript was supported by NIDA grant R18 DA6082 to Friends Medical Research Center, Inc. The authors gratefully acknowledge the assistance of Mrs. Sandy Dow and of Mindy Blum, Ph.D. The authors are also grateful for the outstanding work and dedication of Mrs. Cindy Ambriz, who did extensive library and editorial work and prepared the manuscript.

REFERENCES

1. Strange PG. The structure and mechanism of neurotransmitter receptors: implications for the structure and function of the central nervous system. *Biochem J* 1988;249:309–318.
2. Schwartz J-C, Giros B, Martres M-P, Sokoloff P. The dopamine receptor family: molecular biology and pharmacology. *Semin Neurosci* 1992;4:99–108.
3. Pasternak GW. Pharmacological mechanisms of opioid analgesics. *Clin Neuropharmacol* 1993;16:1–18.
4. Harrison NL, Majewska MD, Harrington JW, Barker JL. Structure-activity relationships for steroid interaction with the γ-aminobutyric $acid_A$ receptor complex. *J Pharmacol Exp Ther* 1987;241:346–353.
5. Gardner EL. Brain reward mechanisms. In: Lowinson JH, Ruiz P, Millman RB, Langrod JG, eds. *Substance abuse: a comprehensive textbook.* Baltimore, MD: Williams & Wilkins, 1992:70–99.
6. Lamb RJ, Preston KL, Schindler CW, et al. The reinforcing and subjective effects of morphine in post-addicts: a dose-response study. *J Pharmacol Exp Ther* 1991;259:1165–1173.
7. Olds J, Milner P. Positive reinforcement produced by electrical stimulation of septal area and other regions of rat brain. *J Comp Physiol Psychol* 1954;47:419–427.
8. Corrigall WA, Vaccarino FJ. Antagonist treatment in nucleus accumbens or periaqueductal grey affects heroin self-administration. *Pharmacol Biochem Behav* 1988; 30:443–450.
9. Ritz MC, Lamb RJ, Goldberg SR, Kuhar MJ. Cocaine receptors on dopamine transporters are related to self-administration of cocaine. *Science* 1987;237:1219–1223.
10. Esposito RU, Porrino LJ, Seeger TF. Brain stimulation reward: measurement and mapping by psychophysical techniques and quantitative 2-[^{14}C]deoxyglucose autoradiography. In: Bozarth MA, ed. *Methods of assessing the reinforcing properties of abused drugs.* New York: Springer-Verlag, 1987:421–445.
11. Cone EJ, Risner ME, Neidert GL. Concentrations of phenethylamine in dog following single doses and during intravenous self administration. *Res Commun Chem Pathol Pharmacol* 1978;22:211–232.
12. *State methadone maintenance guidelines.* Center for Substance Abuse Treatment, U.S. Department of Health and Human Services, Washington, D.C., 1992:337–369.
13. Selzer ML. The Michigan Alcoholism Screening Test: the quest for a new diagnostic instrument. *Am J Psychol* 1971;127:1653–1658.
14. Ewing J. Detecting alcoholism: the CAGE questionnaire. *JAMA* 1984;252:1905–1907.
15. Folstein M, Folstein S, McHugh P. The Mini-Mental State Examination. *J Psychiatr Res* 1975;12:189–198.
16. American Psychiatric Association. *Diagnostic and statistical manual of mental disorders.* 4th ed, revised. Washington, DC, APA Press, 1994.
17. Miller NS, Gold MS. *Alcohol.* New York: Plenum, 1991.
18. Victor M, Adams RD, Collins GH, eds. *The Wernicke-Korsakoff syndrome and related neurologic disorders due to alcoholism and malnutrition.* Philadelphia, PA: FA Davis, 1989.
19. Gibb JW, Hanson GR, Johnson M. Neurochemical mechanisms of toxicity. In: Cho AK, Segal DS, eds. *Amphetamine and its analogs:* psychopharmacology, *toxicology, and abuse.* San Diego, CA: Academic Press, 1994:269–296.
20. Galloway MP. Neuropharmacology of cocaine: effects on dopamine and serotonin systems. In: Lakoski JM, Galloway MP, White FJ, eds. *Cocaine:* pharmacology, *physiology, and clinical strategies.* Boca Raton, FL: CRC Press, 1992:163–190.
21. Zahniser NR, Peris J. Neurochemical mechanisms of cocaine induced sensitization. In: Lakoski JM, Galloway MP, White FJ, eds. *Cocaine:* pharmacology, *physiology, and clinical strategies.* Boca Raton, FL: CRC Press, 1992:229–260.
22. Gawin FH. Cocaine addiction: psychology and neurophysiology. *Science* 1991;251:1580–1586.
23. Wilkins AJ, Jenkins WJ, Steiner JA. Efficacy of clonidine in treatment of alcohol withdrawal state. *Psychopharmacology* (Berlin) 1983;83:78–80.
24. Wert RC, Raulin ML. The chronic cerebral effects of cannabis use. I. Methodological issues and neurological findings. *Int J Addict* 1986;21:605–628.
25. Grenhoff J, Svensson TH. Pharmacology of nicotine. *Br J Addict* 1989;84:477–492.
26. Regier DA, Farmer ME, Rae DS, et al. Comorbidity of mental disorders with alcohol and other drug abuse: results from the Epidemiologic Catchment Area (ECA) Study. *JAMA* 1990;264:2511–2518.
27. Wellman MC. Neuropsychological impairment among intravenous drug users in pre-AIDS stages of HIV infection. *Int J Neurosci* 1992;64:183–194.
28. Silberstein CH, O'Dowd MA, Chartock P, et al. A prospective four-year follow-up of neuropsychological function in HIV seropositive and seronegative methadone-maintained patients. *Gen Hosp Psychiatry* 1993; 15:351–359.
29. Ling W, Wesson DR. Drugs of abuse-opiates. *West J Med* 1990;152:565–572.
30. Ling W, Blaine JD. The use of LAAM in treatment. In: Dupont RL, Goldstein A, O'Donnell J, eds. *Handbook on drug abuse.* Washington, DC: U.S. Government Printing Office, 1976.
31. Tennant FS, Rawson RA, Pumphrey E, Seecof R. Clinical experience with 959 opioid-dependent patients

treated with levo-alpha-acetylmethadol (LAAM). *J Subst Abuse Treat* 1986;3:195—202.
32. Devane WA, Dysarz FA III, Johnson MR, Melvin LS, Howlett AC. Determination and characterization of a cannabinoid receptor in rat brain. *Mol Pharmacol* 1988;34:605–613.
33. Nestler EJ. Molecular mechanisms of drug addiction. *J Neurosci* 1992;12:2439–2450.
34. Adams ML, Kalicki JM, Meyer ER, Cicero TJ. Inhibition of the morphine withdrawal syndrome by a nitric oxide synthase inhibitor, N^G-nitro-L-arginine methyl ester. *Life Sci* 1993;52:PL245–PL249

21

Basal Ganglia Disorders

Mary Sano, Karen S. Marder, George H. Dooneief, Mary M. Robertson, and Jessica Yakeley

OVERVIEW

The most common features of diseases of the basal ganglia are movement disorders, psychiatric syndromes, and cognitive impairment. The movement disorder usually is the most prominent, as well as the presenting and defining feature. These movement disorders, which reflect disturbances of the subcortical motor systems, range from subtle findings, such as asymptomatic rigidity, apparent only with facilitating maneuvers, to disabling impairments of posture, gait, and purposeful limb movement. The range of psychiatric manifestations in these diseases includes depression, psychosis, hallucinations, delusions, personality change, anxiety, agitation, paranoia, mania, and hyperactivity. Table 21-1

TABLE 21-1. *Summary of motor, cognitive, and psychiatric manifestations of selected basal ganglia diseases*

Disease	Age of onset	Motor disturbance	Cognitive deficits	Psychiatric manifestations
Parkinson's disease (PD)	4th–6th decade; mean age, 65 years	Rigidity, tremor, masked face, bradykinesia, shuffling gait	Executive functions, memory, fluency, visuospatial and construction abilities, dementia	Depression, anxiety, mania, psychosis
MPTP toxicity	Depends on age at exposure	Full range of PD features	Executive functions, motor sequencing, construction, may occur in asymptomatics	Depression ?
Huntington's disease	1st–8th decade	Chorea, gait impairment, abnormal eye movement, impaired alternating movements	Executive functions, memory, motor sequencing, dementia	Depression, apathy, irritability, mania
Progressive supranuclear palsy	4th–7th decade; mean age, 63 years	Supranuclear gaze palsy, axial rigidity, dysarthria, bradykinesia	Subcortical dementia, frontal lobe–like syndrome	Personality change, emotional incontinence
Hallevorden-Spatz syndrome	1st or 2nd decade; late onset after 20 years of age	Gait disturbance, rigidity, dystonia, dysarthria, pyramidal dysfunction, mutism	Slowing of thought processes, deficits of variable frequency and severity	Impulsivity, aggressiveness, mood disturbance (generally infrequent)
Wilson's disease	2nd–3rd decade	Tremor, bradykinesia, dysarthria, dysphagia, chorea, dystonic posturing, rigidity	Psychomotor deficits of variable frequency and severity	Anxiety, depression, suicide, psychosis 25%–33% have symptoms early; 50% have symptoms at some point
Tourette's syndrome	Before 20 years of age	Multiple tics, vocalization	Attention, concept formation, planning, executive functions	Attention deficit disorder, obsessive-compulsive disorder, depression
Diffuse Lewy's body disease	5th–7th decade	Rigidity, bradykinesia, gait disturbance, frequent falls, tremor (less common)	Mild memory deficit to dementia; visuospatial abilities	Confusion, hallucinations, delusions, agitation, depression

PD, Parkinson's disease.

lists many of the diseases affecting the basal ganglia.

Subcortical Dementia

The term *subcortical dementia* was initially used to describe the cognitive loss syndromes associated with Parkinson's disease (PD), Huntington's disease (HD), and progressive supranuclear palsy (PSP). Tasks that are thought to be especially sensitive to "subcortical" dysfunctions are those that require timed responses, rapid retrieval, and mental manipulation of known information. Multistaged planning tasks, often labeled as *executive functions,* have also been described as subcortical, with the understanding that these tasks also require frontal lobe function and connections to the frontal lobes from deeper structures involving the basal ganglia (1).

CLINICAL DISORDERS

Parkinson's Disease

Parkinson's disease is characterized by tremor, muscular rigidity, bradykinesia, and loss of postural reflexes. The age of onset is typically between 50 and 65 years. Genetic and environmental factors have been implicated in its etiology, although the cause remains unknown. Neuropathologically, Parkinson's disease is characterized by neuronal loss and depigmentation in the substantia nigra and locus ceruleus. Pathologic findings also include the presence of Lewy's bodies in these areas. The principal biochemical deficit, a depletion of dopamine in the nigrostriatal tract, accounts for much of the clinical picture.

Psychiatric Manifestations

The most common psychiatric disturbance in Parkinson's disease (PD) is depression. A review of the literature reveals prevalence rates ranging from 12% to 90%. Because the earliest motor manifestations of PD (loss of agility, loss of facial expression, and a sense of slowness interpreted as psychomotor retardation) may lead an observer to suspect depression, the diagnosis of an affective disturbance must be confirmed by establishing the presence of depressed mood and anhedonia (2).

There is no relationship between age, age at onset, or duration of illness and the presence of depression in PD. The course of depression varies and may differ from that of primary depressive illness. The profile of depressive features in PD has been examined. Using the Beck Depression Inventory, several authors have noted elevated levels of dysphoria as well as pessimism about the future, irritability, sadness, and suicidal ideation, but overall little guilt, self-blame, or feelings of failure or punishment.

Depletion of serotonin appears to play a role in major depression in PD ecause cerebrospinal fluid (CSF) concentrations of 5-hydroxyindoleacetic acid (5HIAA), a serotonin metabolite, are decreased in depressed patients with PD compared with both age-matched controls and PD patients without depression. However, PD patients without depression have levels of 5HIAA intermediate between those of depressed patients with PD and controls (3).

There is no association between the severity of PD and the presence or severity of depression, nor is there an association between CSF levels of the dopamine metabolite homovanillic acid and depression.

Atypical depression with anxiety has also been reported in PD. An imbalance in dopamine and norepinephrine concentration in the locus ceruleus has been implicated in anxiety in PD. Anxiety has also been reported in relation to dopamine replacement therapy. Up to 20% of patients who initiate dopaminergic drug treatment may experience anxiety, which is usually reduced when drug dosage is lowered.

Mania is rare in PD and is usually associated with excessive dopaminergic treatment. Euphoria can occur in up to 10% of patients taking levodopa, and hypomania has been reported in patients taking dopamine agonists (bromocriptine and pergolide). Studies with large numbers of patients suggest that the rate of hypomania is close to 2%. In general, these symptoms are reversible with reduction of the dose of dopamine agonists.

Psychosis can also occur in PD, and is usually characterized by paranoid delusions, hallucinations, and confusion. Historically, these symptoms were reported before the advent of significant levodopa treatment. However, with current use of both direct and indirect dopaminergic stimulation, 20% to 30% of patients with PD experience these symptoms. The presence of dementia or a preexisting history of psychiatric problems increases the risk of these side effects.

Treatment

Psychiatric Manifestations

Because psychosis can be the result of dopaminergic stimulation, when possible, reduction of antiparkinson medication is the first course of action. Traditional neuroleptics have been used, but they can further increase parkinsonian symptoms. Clozapine, a dibenzodiazepine derivative, an atypical antipsychotic, has been used with success. Some reports suggest that doses as small as 12.5 mg/d may be efficacious. The most commonly reported side effect is sedation, which can often be managed by administering clozapine in the evening, adjusting the dose or using an alternating-day treatment regimen. More serious, although infrequent, side effects, namely, agranulocytosis and leukopenia, require monitoring of the white blood cell count weekly.

Depression

Double-blind studies have demonstrated that tricyclic antidepressants, which primarily increase noradrenergic function, are effective in the treatment of depression in PD (4,5). The selective serotonergic reuptake inhibitors (SSRIs) may be reasonable treatment choices, although no clinical studies have been reported in depressed PD patients. In some cases, however, they can aggravate parkinsonian symptoms by reducing dopamine turnover. The combination of SSRIs and selegiline (a relatively selective monoamine oxidase type B inhibitor, which may slow the disability of PD) has recently been associated with a wide range of adverse events, including serotonergic reactions and extrapyramidal signs.

Electroconvulsive therapy has been used in depressed patients with PD, and benefits in both the motor and depressive symptoms have been reported. Usually, improvement in motor symptoms occurs after one or two treatments and before there is improvement in mood. The motor symptoms return soon after electroconvulsive therapy is completed, although some patients remain improved for several weeks to months.

Cognitive Impairment and Dementia

The most common cognitive deficits are in executive function, visuospatial function, and memory. The more subtle cognitive findings associated with PD do not appear to be related to the severity of the motor disturbance. Patients with longer duration of illness have a greater likelihood of cognitive impairment. Studies of dementia in PD provide estimates of the prevalence ranging from 6% to 81%.

Overall cognitive findings suggest that at least two types of cognitive deficit can be identified in PD:

- Mild or focal deficit that parallels the description of subcortical deficits
- More severe and global impairment consistent with dementia and having a pattern of both cortical and subcortical deficits

The presence of substantive memory impairment distinguishes the group with dementia.

MPTP Toxicity

Davis and colleagues reported the development of parkinsonism apparently resulting from an injection of a self-prepared meperidine analog, 1-methyl-4-phenyl-propionoxypiperidine (MPPP). The clinical syndrome was attributed to the byproduct, MPTP, which resulted from shortcuts taken in the synthesis of MPPP. The patient responded to standard antiparkinsonian therapy but died and was found at autopsy to have neuropathologic changes (depletion of the dopamine-containing pigmented neurons of the pars compacta of the substantia nigra) characteristic of PD, but without Lewy's bodies or degeneration of the locus ceruleus (6).

Subsequent reports identified additional cases in intravenous drug users exposed to MPTP. Examination of a number of exposed individuals revealed that some remained symptom free; others had subtle symptoms of PD; and still others had mild parkinsonism, suggesting that MPTP could produce a full range of disease stages comparable to PD. The similarity of the pattern of intellectual deficit seen in MPTP-exposed individuals to patients with MPTP-induced parkinsonism and to patients with PD provides evidence that the dopamine system mediates a specific set of cognitive functions and that changes in these functions can occur even in the absence of overt motor signs.

Huntington's Disease

Huntington's disease (HD) is an autosomal-dominant disorder with complete penetrance, characterized by a triad of symptoms and signs:

- Movement disorder
- Cognitive impairment
- Psychiatric features

The movement disorder of HD includes involuntary movements, such as chorea, dystonia, athetosis, and motor restlessness in the classic form of the disease, and tremor and myoclonus in the juvenile (early-onset) variant. Although there are many phenotypic presentations, the disease has been identified as an unstable nucleotide repeat (CAG)n in a gene, IT15, on chromosome 4p16.3 (7).

Psychiatric Impairment

A wide spectrum of psychiatric disorders has been described in HD, including mood disorders (major depression and bipolar illness), dysthymia, schizophrenia, and personality disorders such as obsessive-compulsive disorder (OCD).

Depression in HD is extremely common. The lifetime prevalence rate for mood disorder in this disease is at least 40%. Most reports indicate that affective symptoms precede chorea or dementia by 2 to 20 years (average, 5.1 years). This suggests that mood disorder is not solely a reaction to the illness (8).

Suicide is more common among HD patients with affective disorder. Suicide tends to occur in patients in the early stages of the disease. In a large study, more than half of the suicides occurred in patients not yet diagnosed.

Mania, although also reported, is less common than depression. Manic episodes with typical symptoms of elation, expansiveness, and self-importance may follow periods of depression. Periods of hypomania are much more common than frank mania. They may be very brief, sometimes lasting only a few days, and are present in up to 10% of patients.

Apathy and irritability are very common troublesome symptoms to patients and families. The apathy in patients with HD has been described as situational apathy, which is defined as a state of inactivity and lack of spontaneous expression that can be modified by the active participation of others (9).

Treatment

Depression

HD patients with depressive syndromes may respond to a tricyclic antidepressant, although it may be important to start at a very low dose, especially in elderly patients and in those with dementia. Other options include serotonin reuptake blockers, such as fluoxetine or sertraline.

Psychiatric Manifestations

Although lithium is usually employed in the treatment of mania, there may be an increased likelihood of lithium toxicity in HD because these patients are prone to dehydration. Therefore, carbamazepine may be a more effective first-line treatment.

There is no established pharmacologic treatment for apathy. However, the use of haloperidol for movement disorder should be minimized in apathetic patients. Atypical neuroleptics (e.g., risperidone) may be superior in treating psychosis in the patient with HD.

Neuropsychological Impairment

Neuropsychological deficits, progressing to dementia, are seen in HD. Cognitive impairment appears early and is a major factor in reducing functional capacity. Advanced HD patients show a nonfocal pattern of deficits, with severe decrease on full-scale, verbal, and performance IQ scores, memory tests, and a test of verbal fluency. Recently diagnosed HD patients demonstrate memory impairment out of proportion to performance on IQ testing.

Although in any patient, the cognitive, motor, and behavioral manifestations might contribute to disease severity, functional impairment is far more likely to be secondary to the psychiatric and cognitive problems than to the more obvious chorea.

Progressive Supranuclear Palsy

Progressive supranuclear palsy (PSP) is a progressive neurologic disorder with the following characteristics:

- Supranuclear ophthalmoplegia (especially vertical gaze palsy)
- Pseudobulbar palsy with prominent dysarthria
- Dystonic rigidity of the neck and upper trunk

There are four basic pathologic findings:

- Neurofibrillary tangles (NFT)
- Granulovacuolar degeneration of nerve cells
- Loss of nerve cells
- Gliosis

The pathologic findings occur in a characteristic distribution involving the basal ganglia (globus pallidus, substantia nigra, subthalamic nucleus), red nucleus, superior colliculi, pontine tegmentum, periaqueductal gray matter, and dentate nucleus. Senile plaques are not seen, and the NFTs possess a different ultrastructure from NFTs seen in Alzheimer's disease (AD), postencephalitic PD, and the Guamanian Parkinson's disease–dementia complex.

Prevalence has been reported to be of 1.4 per 100,000 population, with an adjusted prevalence for the population older than 55 years of age of 7 per 100,000. Mean age at onset is about 63 years (range, 44 to 75 years). Cognitive impairment is noted early but remains mild. On neurologic examination, there is a paucity of expression. Dysarthria is generally severe and invariably progressive, with dysphagia sometimes appearing late in the course of disease. The most striking feature of the disorder is the ophthalmoplegia. Typically, there is a loss of voluntary vertical gaze, affecting downgaze more than upgaze, in the presence of normal vestibuloocular and caloric responses.

Laboratory evaluation is of little value in making the diagnosis. Computed tomography (CT) and magnetic resonance imaging (MRI) usually reveal atrophy, which may be especially pronounced in the brain stem, and positron emission tomography (PET) demonstrates decreased glucose consumption in the frontal cortex.

Psychiatric Impairment

PSP is often first manifest by vague changes in personality, and the triad of personality change, gait disturbance, and dysarthria often characterizes the early stages of this disorder. The personality changes, per se, have been poorly described in the literature. They are often considered mild and are thought to reflect mental slowing, bradyphrenia, or early dementia.

Cognitive Impairment and Dementia

The dementia of PSP is characterized by the following features:

- Forgetfulness
- Slowing of thought processes

- Emotional and personality changes (apathy or depression with occasional outbursts of irritability)
- Impaired ability to manipulate acquired knowledge

Treatment

In general, dopaminergic, cholinergic, and serotonergic agents have been ineffective in attempts to improve function in PSP. Noradrenergic agents have been tried in an attempt to improve cognitive function specifically, also without success.

Hallervorden-Spatz Disease

Hallervorden-Spatz disease (HSD) is a rare disorder characterized by extrapyramidal signs, other motor system dysfunction, and progressive mental deterioration. The etiology is unknown, but the disease follows a pattern suggestive of autosomal-recessive inheritance. The neuropathologic findings consist of lesions of the globus pallidus and the pars reticulata of the substantia nigra, accumulations of pigment, primarily iron, in these regions, and widespread neuroaxonal spheroids.

Psychiatric Manifestations and Cognitive Impairment

Intellectual retardation or decline is present in most patients. The clinical course varies from rapid deterioration over 1 to 2 years to slow progression or even plateaus for many years. Impaired cognitive function, often beginning with slowness of thought and disturbances in conceptual ability, progressively worsens to dementia. Neuropsychological studies have revealed a general loss of intellectual ability. Impairment is seen in those verbal tests requiring conceptual and analytic strategies (e.g., similarities, arithmetic).

Treatment

Trihexyphenidyl and levodopa or carbidopa have been effective in treating the motor aspects, especially dystonia; however, no therapy for the intellectual deterioration has been of value.

Wilson's Disease

Wilson's disease (WD), also known as hepatolenticular degeneration, is an autosomal-recessive inherited disorder of copper metabolism. The prevalence of WD is about 30 per 1 million population. The gene for WD has been localized to chromosome 13. Although the exact biochemical abnormality remains unknown, excretion of copper by the liver is impaired.

Patients generally present in the second or third decade of life. The initial manifestations may include neurologic, psychiatric, hepatic, ocular, renal, or articular abnormalities. Hepatic disease may take the form of an asymptomatic rise in liver enzymes, hepatitis, jaundice, or cirrhosis. Kayser-Fleischer rings (golden-brown corneal rings) are generally visible by slit-lamp examination once neurologic signs are present and are pathognomonic for the disorder. The initial neurologic features are usually tremor, slowness of movement, dysarthria, dysphagia, choreic movements, and dystonic posturing.

Diagnosis

Low serum ceruloplasmin (less than 20 mg/dL), low serum copper (less than 80 μg/dL), and increased urinary copper excretion (more than 100 μg/24 hours) confirm the diagnosis. basal ganglia lesions, generalized atrophy, and ventricular dilation are commonly seen in both CT and MRI scans.

Pathology

The pathologic changes involve primarily, although not exclusively, the lenticular nuclei. Changes in the caudate, thalamus, cerebellar nuclei, subthalamic nuclei, pontine nuclei, and surrounding white matter have all been described. These changes take the form of shrinking and discoloration, which may progress to cystic degeneration or cavitary necrosis. Presumably, the excess copper or ischemia (or both) is responsible for these effects.

Psychiatric Symptoms

WD has a broad spectrum of psychiatric manifestations. Reports have included cognitive impairment, confusional states, dementia, mental retardation or poor school performance, anxiety, irritability, emotional lability, "neurosis," psychosis, schizophrenia-like states, "incongruous behavior" (described as a "dissociation between environmental cues and behavior"), anorexia nervosa, alcohol abuse, criminality, affective disorders, and suicide. WD may present with psychiatric symptoms in 10% to 65% of cases.

Cognitive Impairment

WD has been considered an example of subcortical dementia characterized by memory loss, impairment of manipulation of acquired knowledge, personality changes, and slowness of mentation without aphasia, apraxia, or agnosia. The basal ganglia, in addition to their involvement in motor control, also have a role in the flow of information during cognitive processes and interact with the supplementary motor areas as an integration center for sensorimotor information.

Treatment

Therapy in WD is based on decreasing the body's copper burden with the chelating agent D-penicillamine (1 to 2 g/d). In the patient unable to tolerate penicillamine, zinc acetate (50 mg zinc five times daily) or trientene (1 to 1.5 g/d) may be substituted. Improvement in measures of general intelligence and memory have been seen with treatment and have been more marked with longer duration of therapy. Discontinuation of therapy is usually followed by death in about 3 years. When hepatic failure is evident, therapy is ineffective, with the possible exception of liver transplantation.

Diffuse Lewy Body Disease

Diffuse Lewy body disease (DLBD) is characterized neuropathologically by the presence of Lewy bodies (the primary pathologic feature associated with PD) and clinically by dementia, parkinsonism, and psychosis. In DLBD, Lewy bodies are found in the brain stem, as they are in PD, and also in cortical areas. A transitional form has also been identified in which Lewy bodies are found in subcortical and diencephalic areas. DLBD can co-occur with the pathologic features of AD. Such cases have been labeled the *Lewy body variant* (LBV) of AD. The literature has been well summarized by Hansen and Galasko (10).

Clinical Presentation

Most clinicopathologic series suggest that DLBD patients usually present with dementia or extrapyramidal features and therefore typically carry an initial diagnosis of either AD or PD. Several reports estimate that between 12% and 22% of those who initially present with dementia or extrapyramidal symptoms have DLBD or LBV. Gait disturbance is a commonly reported feature. Other common extrapyramidal features are bradykinesia, rigidity, and masked facies.

Psychiatric Manifestation, Cognitive Impairment, and Dementia

Psychiatric disturbances are common in DLBD, with reports of hallucinations, delusions, anxiety, and depression. Psychiatric disturbances occur more often in DLBD (and LBV) than in "pure" AD (11).

A fluctuating course has been described by several authors. Changes in cognition, occurring over short intervals, have been identified as a hallmark of this disease. The fluctuations may include episodic confusion interspersed with lucid intervals.

Treatment

Case reports of levodopa therapy have demonstrated mixed results, with some reporting a benefit, others reporting no benefit, and one reporting a transient response. The overall impression from these reports is that a benefit from levodopa therapy may be apparent at the earliest stage, particularly if motor symptoms are a primary problem.

Gilles de la Tourette Syndrome and Other Tic Disorders

Diagnostic Criteria

The generally accepted diagnostic criteria for Gilles de la Tourette syndrome (GTS) are those included in the *Diagnostic and Statistical Manual of Mental Disorders*, 4th edition (DSM-4) of the American Psychiatric Association (APA) and the 10th edition of *The International Statistical Classification of Diseases and Related Health Problems* (World Health Organization). These include both multiple motor and one or more vocal tics, which do not necessarily present concurrently, but which occur many times a day (usually in bouts) and last more than 1 year (12–14).

Prevalence

The exact prevalence of GTS is unknown, but a currently accepted figure is 0.5 per 1000 population. The generally accepted figure of 0.5 per 1000 (5 per 10,000) means about 110,000 patients in the United States and 25,000 in the United Kingdom, but rates range from 0.77 per 10,000 for male patients and 0.22 per 10,000 for female patients, to 2.87 per 10,000, and finally, 105.05 per 10,000 for boys and 13.18 per 10,000 for girls. Most studies indicate that GTS occurs three to four times more commonly in males than in females. The syndrome is found in all social classes.

Neurobiology

Structural abnormalities of the brain in GTS have been sought by a variety of neuroimaging techniques, including CT and MRI. Although abnormalities have been few and only small numbers of patients have been examined to date, there are implications that abnormalities of the basal ganglia are found in GTS (15,16).

Functional Evidence

One study using PET showed abnormalities in five GTS patients when compared with controls. In the GTS patients, there was a fairly close positive association between metabolism in the basal ganglia (particularly the corpus striatum) and metabolism throughout the cerebral cortex (17). Regional glucose metabolism appeared to also have a close inverse association with the severity of vocal tics in the middle and inferior parts of the frontal lobes bilaterally, extending posteriorly from the frontal poles to the central gyrus. Coprolalia, in contrast, was inversely correlated with hypometabolism in the left parasylvian region. Single photon emission computed tomography (SPECT) has shown hypoperfusion in the basal ganglia, thalamus, and frontal and temporal cortical areas as well as elevated frontal cortex blood flow (relative to basal ganglia, inclusive of caudate nucleus and putamen) in GTS patients.

In summary, despite the small numbers, functional imaging using PET and SPECT have demonstrated metabolic and perfusion abnormalities in the basal ganglia and frontotemporal areas, with special reference to the putamen.

Neuropathologic Postmortem Evidence

Postmortem investigations have suggested decreased 5HT and glutamate (especially in the subthalamus) in many areas of the basal ganglia, as well as a reduction in the second-messenger cyclic AMP and an increased number of dopamine uptake carrier sites in the striatum.

Biochemical Evidence

Dopaminergic System

The neurochemical basis for GTS is not known. Dopamine has received the most support because haloperidol and other dopamine antagonists reduce the symptoms in a large number of patients, whereas stimulants such as pemoline and methylphenidate exacerbate the symptoms.

Genetic Evidence

At present, nevertheless, the evidence is mostly in support of a major autosomal-dominant gene. Recent studies using complex segregation analysis on independent samples and on large families suggest the presence of a major autosomal-dominant gene with varying but usually high penetrance. Some have shown that obsessive-compulsive behavior is a phenotypic variation of the putative GTS gene (18,19).

Clinical Characteristics

The age at onset of symptoms ranges from 2 to 15 years, with a mean of 7 years being most common, and with the most frequent initial symptoms involving the eyes (e.g., eye-blinking, eye-rolling). Although often referred to as a *tic disorder,* patients with GTS usually demonstrate a variety of complicated movements, including licking, hitting, jumping, smelling, squatting, abnormalities of gait, and forced touching.

The onset of vocalizations is usually later than that of the motor tics, with a mean age of onset at 11 years, with throat-clearing, sniffing, grunting, coughing, barking, snorting, humming, clicking, colloquial emotional exclamations, low- and high-pitched noises, and inarticulate sounds being the usual utterances. Vocalizations have been reported as initial symptoms in 12% to 37% of subjects, of which the most frequent was repeated throat-clearing. Many GTS patients also describe premonitory "sensory" experiences that are distinct from the actual motor or vocal tic. Coprolalia (the inappropriate and involuntary uttering of obscenities) occurs in less than one third of clinic patient populations and in very few children or mildly affected cases. Copropraxia (involuntary and inappropriate obscene gestures) is reported in 1% to 21% of clinic samples.

Echolalia (the imitation of sounds or words of others) and echopraxia (the imitation of movements or actions of others) occur in 11% to 44% of patients. Palilalia (the repetition of the last word or phrase in a sentence or the last syllable of a word uttered by the patient) occurs in 6% to 15% of patients. Characteristically, the course of GTS over the person's lifetime is punctuated by the appearance of new tics and the disappearance of older ones.

Initially characterized as a psychiatric disorder, GTS has been reclassified as a movement disorder with presumed basal ganglia involvement and is perhaps the most widely studied tic disorder. Criteria for the diagnosis of GTS require onset before 21 years of age, the presence of multiple tics, and at least one vocalization. The type of tic may fluctuate, with progression of tics following a rostral to caudal pattern. Although phenomenologically coprolalia is highly associated with this entity, large-scale studies suggest that it occurs in only one third of patients.

Clinical Course

Several studies have suggested that the tic disturbance ameliorates with age. Patients tend to find their symptoms are worse in the first decade after diagnosis and then gradually improve with each succeeding decade. Demographic studies suggest that the prevalence of GTS decreases by about one fourth in adulthood. Of note is the relative absence of disability in adults with GTS and minimal intrusion into private and professional life. These findings suggest that the psychosocial and cognitive impact of GTS may be manifest as developmental delay rather than permanent disability.

Psychiatric Manifestations

Psychopathology, Associated Features, and Behaviors

Some types of behavior, such as obsessive-compulsive behaviors, are intimately linked to GTS from both a phenomenologic and genetic standpoint and thus represent an integral part of the syndrome. Others, such as attention deficit hyperactivity disorder (ADHD), are associated with GTS, although the precise nature of the relationship is unclear.

ADHD occurs in a substantial proportion of GTS patients, ranging from 21% to 90% of GTS clinic populations, but the precise relationship between the two disorders is complex and remains unclear.

In conclusion, it appears that patients with GTS are especially prone to depression, which may be related to the duration of the GTS, partly connected to the reality that GTS sufferers have a chronic, socially disabling, and stigmatizing disease. Anxiety is also high in patients with GTS, but some of this could be accounted for by the depression. ADHD and self-injurious behavior (SIB) occur in a substantial number of GTS patients, but the precise relationship between them are unclear. There is no generally accepted association between GTS and psychosis.

Cognitive Impairment

There is little evidence of intellectual impairment or focal neuropsychological deficits in patients with pure GTS. Mild executive function deficits are reported in adults. Typically, they complain of disorganization at work and in personal affairs, such as bill paying, keeping appointments, and correspondence. They have difficulty completing tasks and trouble returning to tasks if interrupted. Deficits in formal testing of vigilance and in sustained and focused attention, as well as impulsivity, may also be evident in adulthood.

Diagnostic Evaluation

Assessment and Rating of Tic Severity

Three self-report schedules are used in the evaluation of GTS (20,21):

1. The Tourette Syndrome Questionnaire (TSQ) is a structured parental and self-report schedule that was originally developed for an epidemiologic survey and offers a systematic way of obtaining relevant information, such as personal and demographic data, developmental history, family history, general medical and treatment history, the course of tic behaviors, and the impact of GTS on the person's life.
2. Stefl and colleagues developed a parent-report and self-report instrument for a needs assessment survey conducted for the Tourette Syndrome Association of Ohio (20).
3. The Tourette Syndrome Symptom Checklist (TSSL) was devised by Cohen and co-workers to assist parents in making daily or weekly ratings of tic severity. The schedule, which takes into account the frequency and disruption of both tics and behavioral symptoms, has been used successfully to monitor the longitudinal course of GTS and to document changes during medication trials (20,21).

There are also several clinician or observer schedules available for the assessment of tics and GTS. Tanner and colleagues developed a scale that enables an observer to count the patient's tics under several different conditions that are videotaped (patient alone, with the examiner, sitting quietly, and performing a task) (22). The Tourette Syndrome Severity Scale (TSSS) was developed by Shapiro and colleagues for use in a clinical trial evaluating pimozide in the treatment of GTS (23). The Tourette Syndrome Global Scale (TSGS), which combines a variety of ratings for tic symptoms and social functioning into an overall global score for severity, was developed by Harcherik and colleagues (24). The Yale Global Tic Severity Rating Scale (GTSS) was developed to refine measurement of GTS symptoms, building on the developers' experience with the TSGS (25).

Neurologic Examination

Subtle neurologic deficits have been found in a number of GTS patients. Left-handedness has been reported, as well as minor motor asymmetry. A significant minority demonstrate chorea or choreoathetoid movements. Other reported neurologic abnormalities have included chorea, dystonia, torticollis, dysphonia, dysdiadochokinesis, postural abnormalities, reflex asymmetry, motor incoordination, nystagmus, and unilateral Babinski's reflexes.

Neuropsychological Evaluation

Early neuropsychological studies note an average IQ in GTS patient cohorts, with verbal and performance discrepancies of 15 points (performance being the lower); specific deficits in reading, writing, and arithmetic; and dysfunction on the Halstead-Reitan Neuropsychological Assessment Battery. Language skills appear to be largely unimpaired, whereas consistent deficits in visuopractice performance have been documented by a number of authors. Specific learning problems are found in 36% of 200 children with GTS.

Differential Diagnosis

Most of the conditions that should be considered in the differential diagnosis of GTS are described in reviews cited previously (14). These include the athetoid type of cerebral palsy, seen commonly among learning-disabled populations with an age at onset between birth and 3 years. Dystonia musculorum deformans, usually presenting with a torsion dystonia often in the legs, is usually progressive, with a crippling state resulting 10 to 15 years after the onset; remissions are rare. One of the most difficult differential diagnoses is that of tics of childhood, which commence between the ages of 5 to 10 years, but many remit spontaneously and usually improve with age. For a review of the diagnostic criteria for transient tic disorder, chronic motor tic disorder, and GTS, see the article by Woody and Laney (26).

Treatment

Psychosocial Management

For many adults with mild GTS, explanation and reassurance are often sufficient. For the moderate to severely afflicted patient, who may have the associated features of obsessive-compulsive behavior, ADHD, SIB, and possibly antisocial behavior, the management is more complex.

Massed practice—over rehearsal of the target tic—and other forms of behavior therapy can sometimes be helpful. Support in helping the patient and family to cope are important, as is guidance to teachers. For the severely affected patient, who may have the associated features of ADHD, SIB, and aggressive behavior, the management is complex. In these cases, management includes counseling, regular assessment of mood and danger to the individual (depression and SIB), and often the prescribing of more than one medication (e.g., an SSRI and sulpiride). When the GTS symptoms and associated behaviors are severe and life-threatening, although not common, neurosurgical treatment has been used (27).

Drug Therapy

Dopamine antagonists, such as haloperidol, pimozide, and sulpiride, have been widely and successfully used. Doses are relatively small, beginning with, for example, haloperidol, 0.25 to 0.5 mg/d and increasing by 0.5 mg per week, often with 2 to 3 mg/d being sufficient. Extrapyramidal side effects, sedation, and dysphoric states are common with haloperidol but less so with pimozide. Sulpiride causes fewer extrapyramidal problems, including tardive dyskinesia and dystonia, as well as less cognitive and sedative side effects when compared with haloperidol and pimozide.

Clonidine has also been used successfully in GTS patients. Clonidine may well be the agent of choice if a child has GTS and associated ADHD.

SSRIs (e.g., fluoxetine) can be used to treat the obsessive-compulsive aspects of GTS, and augmentation of these antiobsessional effects of SSRIs (e.g., fluvoxamine) by neuroleptics (e.g., pimozide) has been reported in patients with OCD and GTS.

Prognosis

There have been no substantial long-term follow-up studies to document the exact course of the syndrome; however, from the literature of case reports and from clinical experience, it is clear that it is usually a lifelong illness. Tics and associated GTS symptoms appear to become worse during adolescence. There is some evidence that temper tantrums, aggression, and explosiveness appear in the preadolescent period, become severe in teenage years, and gradually recede thereafter; the tic symptomatology of GTS also ameliorates with age.

New Developments and Future Research

With no one transmitter convincingly implicated to date, the importance of animal models and further postmortem studies is stressed. GTS may well reflect the interaction among genetic, neurophysiologic, behavioral, and environmental factors. There may be a complex interaction of androgenic and immunologic factors in the susceptibility to neurodevelopmental disability and, therefore, GTS. The observation of Balthasar of an increased number of neurons in the caudate and putamen, suggesting the persistence of an immature neuronal pattern should be followed up (28). Bonnet addressed the anatomic localization of GTS and concluded that the biochemical structure of the limbic forebrain structures, particularly the anterior cingulate cortex, and their interrelationships with other specific nuclei suggest they are the anatomic site for GTS (29). These hypotheses await further neuroanatomic studies.

Genetic linkage studies are underway in several centers in the United States and Europe. A range of chromosomal anomalies in GTS has been reported, which have led to investigations for linkage. Reports have tentatively assigned the gene to chromosomes 3, 11 (D_2 receptors), and 18 and to the dopamine D_3 receptor gene, but a consensus does not yet exist. Chromosomes 7 and 18, 11, and the D_1 receptor have been excluded. The findings of these reports, plus a review of the literature, suggest that no one chromosomal abnormality can be said to be absolutely characteristic of GTS. Bilineality (i.e., affected patients on both maternal and paternal sides of index cases) has been reported, which suggests that homozygosity is not uncommon in GTS and might further explain difficulties in localizing the gene defect by linkage analysis. Further studies on the mechanism of inheritance may well be assisted by more precise definition of the phenotypes involved. Future studies likely show that the mechanism is more complicated than a single gene because previous studies have not entirely ruled out the possibility of other alleles, multifactorial inheritance, or genetic heterogeneity. Genetic studies suggest that GTS is inherited by autosomal-dominant transmission with high penetrance. Although more than 50% of the genome has already been excluded, this is likely the result of imprecise definition of the clinical phenotypes or genetic heterogeneity. Future research, therefore, will emphasize genetics and more precise definition of the phenotype. Once genetic linkage is established, it will be possible to identify the actual gene and its causative mutations, followed by prenatal diagnosis and genetic counseling as well as development of new treatments based on a knowledge of the disease pathways.

Obsessive-Compulsive Disorder

The currently accepted definition of obsessive-compulsive disorder (OCD) appears in the American Psychiatric Association DSM-4. OCD is characterized by recurrent obsessions or compulsions sufficiently severe to cause marked distress that are time-consuming or interfere significantly with the person's normal routine, functioning, social activities, or relationships. Obsessions are recurrent ideas, thoughts, images, or impulses that enter the mind and are persistent, intrusive, and unwelcome. The individual recognizes them as a product of his or her own mind. Compulsions are repetitive, purposeful behaviors performed in response to an obsession and are designed to neutralize or prevent discomfort or some dreaded event or situation. The affected person recognizes that his or her behavior is unreasonable.

Clinical Characteristics

Obsessional thoughts are intrusive words, ideas, and beliefs that are often upsetting and unwelcome. Obsessional doubts by these patients involve excessive and inappropriate concern about previous actions and their consequences. Obsessional fears are feelings of uneasiness and dread about imagined events that might happen, which may be highly improbable or unreasonable. Obsessional ruminations are recurring thoughts of a complex nature, such as the ending of the world. Obsessional images are unwelcome mental pictures that often suddenly intrude upon the mind, often of an unacceptable nature, possibly involving violent or sexual scenes. Obsessional impulses are urges to carry out an act that is usually socially unacceptable.

Compulsions (also called *obsessional rituals*) are repeated and meaningless rituals performed in a purposeful and stereotyped way that may be performed to neutralize an obsessional thought. Although the compulsion may seem to reduce the anxiety resulting from the obsession and produce a tension release in its performance, the person strongly resists carrying out the ritual and does not derive pleasure from it. The compulsion may occasionally have no apparent connection with an obsession, except for an obsessional urge to carry out the act, such as ordering things in a particular way. Compulsions mostly take the form of cleaning, checking, repeating, counting, and ordering. Other compulsions involve hoarding and collecting; the need to tell, ask, or confess; and a special way of dressing.

Epidemiology

The Diagnostic Interview Survey has been suggested as a good instrument for determining the prevalence of OCD and has been used in the Epidemiological Catchment Area (ECA) studies, nine of which have been performed, including five in the United States. This survey was designed to determine the prevalence of DSM-3 Axis I psychiatric disorders in the general population. Eight of the surveys have given figures for lifetime prevalence of OCD ranging from 1.9% to 3.2%. These studies revealed that OCD was 50 to 100 times more common than previously thought and that it was the most common psychiatric illness after the phobias, substance abuse, and depression (30).

The age at onset is often early, with an estimated 60% of people with OCD reporting their first symptoms before the age of 25 years and 30% experiencing the onset of obsessions and compulsions between the ages of 5 and 15 years. The mean age at onset of first symptoms was 21.4 years for men and 19.6 years for women. The disorder is, however, being increasingly recognized in children and adolescents. The lifetime morbid risk of developing OCD within a normal life-span was calculated to be 5.4% from the data obtained from the study in Edmonton, Canada (31).

The ECA studies found that only one third of patients experience spontaneous remission, sometimes after years of illness, and in 10%, the illness runs a continuous deteriorating course.

Psychopathology and Associated Disorders

The wide range of psychiatric and neurologic disorders found in association with OCD is summarized in Table 21-2. Evidence emerging from epidemiologic, genetic and family, phenomenologic, and treatment stud-

TABLE 21-2. *Co-morbid conditions associated with obsessive-compulsive disorder*

Affective disorders
Major depressive disorder
Dysthymia
Anxiety disorders
Separation anxiety disorder
Panic disorder
Agoraphobia
Generalized anxiety disorder
Simple phobia
Social phobia
Tourette's syndrome and other tic disorders
Eating disorders
Anorexia nervosa
Bulimia nervosa
Obsessive-compulsive (anankastic) personality disorder
Habit disorder
Trichotillomania
Onychophagia
Body dysmorphic disorder
Neurologic disorders
Epilepsy
Sydenham's chorea
Postencephalitic Parkinson's disease

ies demonstrates significant overlap in the clinical characteristics of many of these disorders and suggests etiologic relationships between them.

Mood Disorders

Symptoms of depression are extremely common in patients with OCD. Up to 80% have dysphoric moods, two thirds have a lifetime history of a major depression, and one third have a major depression at the time of diagnosis. Obsessive-compulsive symptoms are seen in 23% to 28% of depressed patients, but these are usually related to depressive symptomatology in their content, such as obsessive thoughts of guilt. Some authors attempt to distinguish whether depression or OCD is the primary illness by the temporal relationship between the two, diagnosing OCD only if it occurs first. Thus, it has been found that most (85%) of these patients have an affective disorder secondary to their OCD, whereas the remainder have a coexisting primary unipolar recurrent depression.

Anxiety Disorders

OCD is classified as an anxiety disorder and has a significant overlap with other anxiety disorders. The ECA epidemiologic studies found that 22% of OCD patients had a lifetime history of simple phobia, 18% had social phobia, 12% had panic disorder, and 2% had separation anxiety disorder. Other associated anxiety states include generalized anxiety disorder and agoraphobia.

Eating Disorders

Many mental health workers have noted similarities in the clinical features of OCD and anorexia nervosa and bulimia nervosa. The extreme preoccupations with food, weight, and body image that are characteristic of eating disorders can be viewed as obsessions similar to those occurring in OCD. Likewise, ruminative calorie counting and ritualistic behavior regarding food may resemble the rituals and compulsions of OCD. Neurobiologic similarities also exist between OCD and the eating disorders. Abnormalities in serotonin metabolism and regulation are implicated in both types of these disorders, and neuroimaging techniques, such as PET, have demonstrated increased metabolic activity in the caudate nucleus in both illnesses (32). Several studies have investigated the prevalence of eating disorders among patients with OCD, and vice versa, and have demonstrated significant co-morbidity.

Habit Disorders, Dysmorphophobia, and Other Disorders

The so-called habit disorders, such as trichotillomania, characterized by chronic severe hair pulling, and onychophagia, or severe nail biting, are abnormal behaviors that bear some resemblance to the compulsions seen in OCD in that they are repetitive, purposeful, stereotyped, and anxiety relieving. These similarities and the evidence that trichotillomania and onychophagia, like OCD, respond to serotonergic uptake inhibitors such as clomipramine and fluoxetine have led to the theory that these disorders are variants of OCD. Trichotillomania, onychophagia, and dysmorphophobia are included in a group of associated disorders that resemble OCD, which have been called the *obsessive-compulsive spectrum disorders*. Other conditions included in this group are hypochondriasis, GTS, eating disorders, and impulse-control disorders, such as kleptomania, compulsive gambling, pyromania, and the paraphilias.

Neurologic Disorders

Obsessive-compulsive symptoms can occur after head trauma, either at birth or as a result of accidents. Up to one third of patients with Sydenham's chorea have obsessive-compulsive symptoms, which has also been reported in patients with Huntington's chorea and rarely in those with diabetes insipidus. Obsessive-compulsive symptoms have been noted to develop in cases of epilepsy.

Childhood Obsessive-Compulsive Disorder

Epidemiology

The only epidemiologic study to date of OCD in an unselected population of adolescents gave a lifetime prevalence rate of 1.9%, much higher than previous clinical estimates but in concordance with the rates found in adults from the ECA studies (33).

Course and Prognosis

There have been few prospective follow-up studies of childhood OCD, but they all concluded that long-term outcome was poor, with about half of the subjects reporting no improvement on follow-up 1 to 7 years after diagnosis. The availability of drug treatment appears not to have improved this poor prognosis.

Etiology

Until very recently, obsessions and compulsions were thought to be products of intrapsychic conflict, and psychodynamic theories have dominated the etiologic hypotheses of OCD for most of this century. In the past decade, however, evidence from brain imaging, psychopharmacologic studies, and advances in treatment have challenged these concepts. OCD is now viewed as a model neuropsychiatric illness.

Structural and Functional Abnormalities

Neuroradiologic Evidence

Luxenberg and associates suggested involvement of the basal ganglia in the pathogenesis of OCD because CT scanning of young male patients showed reduced caudate volumes (34). The association of OCD symptoms with some cases of neurologic disease that involve dysfunction of the basal ganglia, such as postencephalitic parkinsonism and Sydenham's chorea, is further evidence to support the involvement of this part of the brain in OCD.

MRI studies have given similar results. Garber and colleagues also found no structural abnormality unique to OCD but observed regional tissue abnormalities, particularly in the orbital frontal cortex, which correlated strongly with symptom severity in patients with OCD (35).

Neuroimaging techniques, such as PET, which allow functional rather than anatomic assessment of the brain, have extended the findings of neuroanatomic studies by indicating that orbitofrontal-basal circuitry may mediate obsessive-compulsive behavior. PET studies of cerebral glucose metabolism in patients with OCD have found increased rates in the caudate nuclei and left orbital gyri; in the left orbitofrontal, right sensorimotor, and bilateral prefrontal and anterior cingulate regions; and in both orbital gyri (36).

Neuropsychological Evidence

Neuropsychological testing of OCD patients has detected frontal lobe dysfunction, memory impairment, and reduction of visuospatial skills, performance IQ, and motor functioning, which are predominantly right-sided activities. Recent studies designed to avoid the methodologic problems encountered previously found cognitive deficits consistent with basal ganglia or right-hemisphere disturbance but no frontal lobe dysfunction.

In summary, these electrophysiologic, brain imaging, and neuropsychological studies have yielded somewhat conflicting results. Still, the overall evidence points toward the orbitofrontal and anterior cingulate-basal ganglia-thalamocortical circuits as representing the neuroanatomic substrate for OCD.

Biochemical Abnormalities

The Serotonin Hypothesis

The neurotransmitter serotonin, or 5-hydroxytryptamine (5HT), has been thought to play an important role in the pathogenesis of OCD. Convincing evidence from several different lines of research has led to a serotonergic neurochemical hypothesis for OCD. The studies to date that have investigated such a role for serotonin can be grouped under three main separate headings (37):

- Drug treatment studies
- Peripheral markers of 5HT function
- Pharmacologic challenge studies

Drug Treatment Studies

It was first noted in the 1960s that the tricyclic antidepressant clomipramine has antiobsessional properties, even when evidence of an underlying depressive illness was lacking. Since 1980, clomipramine has been found to be a superior drug in its antiobsessive effects to placebo. Its efficacy has also been proved in a large recent multicenter trial in the treatment of patients with OCD.

Additional evidence for the role of serotonin in OCD came with the development of more selective and potent 5HT uptake inhibitors. Various studies confirmed their effectiveness in OCD. Thus, double-blind, placebo-controlled clinical trials of fluvoxamine, fluoxetine, and sertraline have confirmed their antiobsessional action and their superiority to desipramine. Zimeldine and trazodone, a serotonin $5HT_2$ antagonist, may also have antiobsessional properties.

Drug response data have strengthened a dopamine–serotonin hypothesis of OCD. SSRIs have been shown to have dopamine-blocking activity, which may contribute to their antiobsessional effect. In addition, there are reports that neuroleptic dopamine antagonists have been successfully used in conjunction with a 5HT uptake inhibitor in the treatment of some cases of GTS concomitant with OCD and of OCD unresponsive to the 5HT uptake inhibitor alone. Fluoxetine has been associated with extrapyramidal symptoms similar to those seen with neuroleptic treatment, and decreased CSF levels of the principal dopamine metabolite homovanillic acid have been found after fluoxetine treatment. These findings have been interpreted as fluoxetine-facilitating serotonergic inhibition of striatal dopamine neurons.

Family and Genetic Evidence

Twin studies have revealed higher concordance rates for OCD among monozygotic twins than among dizygotic twins as well as obsessive-compulsive traits among twins. In no cases, however, did concordance approach unity, indicating that expression of OCD is probably influenced by nongenetic factors. Most family studies have shown that there is a significantly higher incidence of OCD among first degree relatives (FDRs) of patients with OCD than in the general population.

There is convincing evidence that GTS and OCD share a genetic etiology, but not all cases of OCD are associated with GTS. The findings from these genetic studies and data described earlier in this chapter suggest that OCD can be divided into at least three categories:

- Sporadic (no family history)
- Familial (positive family history)
- OCD associated with a family history of tics or GTS

This challenges traditional assumptions of etiologic homogeneity in OCD (38).

Diagnosis

Two simple screening questions have been designed to isolate symptoms of the disorder. These questions can be asked routinely in suspected cases, while taking a history, as follows: "Are you bothered by thoughts coming into your mind that make you anxious and that you are unable to get rid of?" "Are there certain behaviors that you do over and over that may seem silly to you or to others but that you feel you just have to do?" If the answers to these questions are affirmative, the clinician can proceed to inquire further about the nature, frequency, onset, duration, and severity of the obsessions and compulsions expressed by the patient.

A number of standardized inventories or rating scales have been constructed to assess OCD. The Leyton Obsessional Inventory was the first rating scale to quantify subjective reports of obsessive feelings and behavior. The original inventory consisted of 69 questions and was designed as a supervised card-sorting procedure that gives information about feelings of resistance and interference with other activities in addition to the straightforward answers to the questions. It has been found to be reliable in distinguishing OCD adolescents from normal controls and from other psychiatric patients with severe OC symptoms. It can be also used as a valid instrument in treatment studies (39).

Other rating scales that have been used to tap obsessive-compulsive symptoms are the Sandler-Hazari Scale, the Obsessive-Compulsive Check List, and the Maudsley-Obsessive Comprehensive Inventory. Recently, the Yale-Brown Obsessive Compulsive Scale (Y-BOCS) has been designed to provide a specific measure of the severity of OCD symptoms independent of the particular type of obsession or compulsion present (40–42).

The differential diagnosis of OCD is large and includes all of the associated psychiatric and neurologic conditions, such as the other anxiety disorders, depression, GTS, eating disorders, trichotillomania, body dysmorphic syndrome, and personality disorders discussed previously. Diagnostic accuracy may be paramount in distinguishing whether a GTS patient with complex tics also has comorbid OCD; recent evidence suggests that such patients who are resistant to treatment with SSRIs alone may respond to a combination of fluvoxamine and a neuroleptic. Most likely, in the future, OCD will be regarded as a heterogeneous disorder with separate phenomenologic subgroups that require different treatment.

Treatment

There is little evidence that psychoanalysis and dynamic psychotherapy are effective in managing OCD, and these approaches to the treatment of this disorder have largely been abandoned. Advances in behavioral theories and techniques, and more recently in pharmacotherapy, have led to rational and integrated psychopharmacologic treatment regimes for a condition that was until a decade ago considered mainly treatment resistant. Psychosurgery is now reserved as a final option for only the most refractory cases. The importance of education, reassurance, and self-help must not be underestimated. Family therapy and helping the OCD patient to cope at work or school may be as crucial to the outcome as pharmacobehavioral intervention.

Behavior Therapy

The most widely evaluated and useful techniques are exposure and response prevention. Exposure in vivo therapy requires the patient to maintain contact with the stimuli that provoke the obsessions or compulsions until habituation occurs. Response prevention instructs the patients to refrain from ritualistic behavior despite often overwhelming urges to the contrary (43).

Poor response to behavior therapy is most often seen in patients with concomitant depression, those who use CNS-depressing drugs, and patients who are obsessional ruminators with no obvious compulsive behaviors. Cognitive therapy has been advocated by some authors to use as an adjunct to existing behavioral techniques in the treatment of OCD, based on theories that obsessions and compulsions originate in thoughts, but to date, controlled trials are lacking.

Recent studies have shown that the combination of behavior and drug therapies may be the optimum management strategy for OCD.

Drug Therapy

Double-blind drug treatment studies have demonstrated that clomipramine is superior in its antiobsessional effect to nortriptyline, amitryptyline, clorgyline, imipramine, and desipramine. Fluvoxamine has been shown in a single-blind study and in subsequent double-blind studies to be more effective than placebo in OCD patients. Studies reporting beneficial treatment with fluoxetine have been less rigorously controlled, involving for the most part open trials. Few studies directly compare the efficacy of the different SSRIs in OCD. Comparing the response rates in independent trials of clomipramine, fluvoxamine, and fluoxetine in OCD patients, no drug emerges as significantly superior in efficacy. Clinician choice, therefore, depends more on matching the patient with the anticipated side-effect profile than on drug efficacy.

The empirical use of higher doses of antidepressant drugs in OCD than in depression is common, but there is little evidence that this is necessary. A recent 10-week study showed no advantage for the administration of a daily dose of 40- or 60-mg doses over 20 mg of fluoxetine in OCD patients. The antiobsessional effect of these drugs tends to take longer to emerge than the antidepressant effect observed in the treatment of depression, with the initial response sometimes not seen until 4 to 6 weeks after starting medication. A trial of therapy should therefore continue for at least 10 weeks. It is unclear how long maintenance therapy should be continued. Most clinicians would treat for a period of 6 to 12 months and then gradually taper the dose, but in view of high reported relapse rates, patients may need medication indefinitely. There is emerging evidence that treatment with lower doses than those used initially to produce a response may be effective in maintaining OCD patients free of symptoms.

Although SSRIs can be successful in improving the symptoms of some OCD patients, 40% to 60% of patients exhibit minimal to no change after treatment with an adequate trial of an SSRI alone. These treatment-refractory patients, therefore, represent a large proportion of the population of OCD sufferers, and a number of different biologic therapeutic strategies have been advocated for this group of patients. There may be some value in substituting one SSRI for another if the patient fails to respond to an adequate trial of the first. Combination therapies have also been recommended. Thus, other agents that also modify serotonergic function, such as the amino acid precursor of serotonin tryptophan, the serotonin releaser and reuptake blocker fenfluramine, and the serotonin type 1A agonist buspirone, have been added to ongoing SSRI therapy with variable success. The role of neuroleptics in the treatment of OCD has been reevaluated. Neuroleptics alone do not appear to be effective, but combination therapy involving treatment with SSRIs and low-dose dopamine-blocking agents may be of use, although the hazards of long-term treatment with the latter must be appreciated. Trazodone, clonidine, and antiandrogens have also been described in a few case reports as producing a positive response in patients with OCD, but these results must be considered with caution.

The mainstay of treatment of OCD is an 8- to 12-week trial of pharmacotherapy with one of the SSRIs, which have been shown in double-blind controlled trials to be superior to placebo in improving the symptoms of OCD. There is no convincing evidence to suggest that one SSRI has more efficacy than another; hence, clinician choice depends on side-effect profile, availability, and patient preference. The duration of treatment should be at least 6 months and often longer, considering the high rates of relapse after cessation of treatment. Although published controlled research support for combination treatments is scanty, augmentation of SSRI treatment with agents such as lithium, buspirone, or a neuroleptic may be useful for the patient who has failed to respond to an SSRI alone, particularly if there is concurrent depression, anxiety disorder, or tics.

Neurosurgical Treatment

Neurosurgery is reserved for the minority of patients who fail to respond to behavior therapy and SSRIs and have chronic, unremitting illness for at least 2 years with severe life disruption. Follow-up studies show that of all the disorders treated by neurosurgery, OCD consistently responds best. With a greater understanding of functional neuroanatomy and the development of more precise and accurate techniques, such as stereotactic cryosurgery, thermocoagulation, and multifocal leukocoagulation, adverse effects, such as epilepsy, cognitive deficits, or personality change are rare, and mortality approaches zero. Procedures that interrupt the pathways from the frontal cortex to the basal ganglia, such as cingulotomy, anterior capsulotomy, and orbital frontal leukotomy, are most effective in ameliorating disabling OC symptoms in such patients. Neurosurgery is reserved for the most severe refractory cases. Recent evidence suggests that behavior therapy and SSRIs are complementary, and this combination may be the optimum treatment available for the OCD patient.

New Developments and Future Research

The neuropsychiatric boundaries between OCD and other disorders should come under continuous investigation. Some studies suggest a link between OCD and eating disorders. It has even been proposed that eating disorders represent a variant of OCD, starting with the adolescent period.

We can expect greater clarification of the complexity of the neurochemistry underlying OCD as neuropharmacology becomes more sophisticated. Results of earlier neuroendocrine and behavioral challenge studies should be reinterpreted in light of more advanced understanding of the complexity of brain 5HT pathways and the multiple subsystems and receptors. Dysfunction of 5HT receptor subtypes may be important in the pathogenesis of OCD. Future double-blind placebo-controlled studies using more selective test compounds are required to delineate the precise serotonin deficit underlying this disorder.

Although genetic linkage studies are being conducted in GTS to try to establish a relationship between a marker and the hypothetical gene for the disorder, this methodology has not been used to date in OCD. It is hoped that these techniques will be applied to future research, but until biologic markers are identified, the family history of OCD is the strongest evidence suggesting a genetic foundation underlying at least some forms of OCD.

SUMMARY

The diseases described in this chapter demonstrate the broad range of cognitive and psychiatric disorders that are associated with basal ganglia pathology. Some of these deficits result from the direct impact of damage to the basal ganglia and to the associated neurochemical systems. Other findings are related to projection areas innervated by the basal ganglia, especially the frontal cortex. It is important to characterize the full range of clinical findings both for diagnostic reasons and for management. There is a wide range of efficacy of treatments for the movement disorders described here. In some diseases (PSP, HSD, DLBD), limited efficacy is expected at best; in others (PD, HD, and GTS), relatively good efficacy is expected; and in one (WD), complete reversal can be achieved.

In general, the efficacy of treatment for mood disorders is similar among patients with and without basal ganglia diseases. However, treatment of psychosis is problematic because neuroleptics usually exacerbate motor disturbances in patients with basal ganglia diseases and impair voluntary purposeful motor function, even when they suppress involuntary movements. Atypical neuroleptics, such as clozapine and risperidone, offer the hope of better treatment, but these have not been tested apart from the use of clozapine in PD. There are few treatments for cognitive impairment in any condition and, to complicate matters further, in basal ganglia disease, treatment for either behavioral or motor symptoms can worsen cognitive function. Clearly, the treatment of one problem can lead to exacerbation of another. For each symptom, it is necessary to weigh the degree of disability it produces, the likelihood of benefit with a given treatment, and the probability of exacerbating other symptoms with such treatment in order to choose the best therapeutic approach.

ACKNOWLEDGMENTS

This work was supported by the Charles S. Robertson Memorial Gift for Alzheimer's Disease Research and by federal grants AG02802, AG07232, AG0737 and RROO654.

REFERENCES

1. Albert ML. Subcortical dementia in Alzheimer's disease. In: Katzman R, Terry RD, Bick KL, eds. *Senile dementia and related disorders.* New York: Raven, 1979:173–180.
2. Cummings JL. Depression and Parkinson's disease: a review. *Am J Psychiatry* 1992;149:443–454.
3. Mayeux R, Stern Y, Sano M, Williams JBW, Cote L. The

relationship of serotonin to depression in Parkinson's disease. *Mov Disord* 1988;3:237–244.
4. Andersen J, Aabro E, Gulmann N, Hjelmsted A, Pedersen HE. Antidepressant treatment in Parkinson's disease: a controlled trial of the effect of nortriptyline in patients with Parkinson's disease treated with L-dopa. *Acta Neurol Scand* 1980;62:210–219.
5. Goetz CG, Tanner CM, Klawans HL. Bupropion in Parkinson's disease. *Neurology* 1984;34:1092–1094.
6. Davis GC, Williams AC, Markey SP, et al. Chronic parkinsonism secondary to injection of meperidine analogues. *Psychiatr Res* 1979;1:249–254.
7. Kremer B, Goldberg P, Andrew S, et al. A worldwide study of the Huntington's disease mutation: the sensitivity and specificity of measuring CAG repeats. *N Engl J Med* 1994;330:1401–1406.
8. Folstein ES, Abbott HM, Chase AG, Jensen AB, Folstein FM. The association of affective disorder with Huntington's disease in a case series and in families. *Psychol Med* 1983;13:537–542.
9. Caine DE, Shoulson I. Psychiatric syndrome in Huntington's disease. *Am J Psychiatry* 1983;140:6.
10. Hansen LA, Galasko D. Lewy body disease. *Curr Opin Neurol Neurosurg* 1992;5:889–894.
11. McKeith IG, Perry RH, Fairbairn SJ, Perry EK. Operational criteria for senile dementia of Lewy body type (SDLT). *Psychol Med* 1992;22:911–922.
12. American Psychiatric Association. *Diagnostic and statistical manual of mental disorders.* 4th ed. Washington, DC: American Psychiatric Association, 1994.
13. World Health Organization. *International classification of diseases and health related problems.* 10th revision. Geneva, Switzerland: World Health Organization, 1992.
14. Robertson MM. The Gilles de la Tourette syndrome: the current status. *Br J Psychiatry* 1989;154:147–169.
15. Peterson B, Riddle MA, Cohen DJ, et al. Reduced basal ganglia volumes in Tourette's syndrome using three-dimensional reconstruction techniques from magnetic resonance images. *Neurology* 1993;43:941–949.
16. Singer HS, Reiss AL, Brown JE, et al. Volumetric MRI changes in basal ganglia of children with Tourette's syndrome. *Neurology* 1993;43:950–956.
17. Chase TN, Foster NL, Fedio P, et al. Gilles de la Tourette syndrome: studies with the fluorine-18-labelled fluorodeoxyglucose positron emission tomographic method. *Ann Neurol* 1984;15(Suppl):S175.
18. Pauls DL, Leckman JF. The inheritance of Gilles de la Tourette syndrome and associated behaviors: evidence for autosomal dominant transmission. *N Engl J Med* 1986;315:993–997.
19. Robertson MM, Trimble MR. Normal chromosomal findings in Gilles de la Tourette syndrome. *Psychiatr Genet* 1993;3: 95–99.
20. Kurlan R, McDermott M. Rating tic severity. In: Kurlan R, ed. *Handbook of Tourette's and related disorders.* New York: Marcel Dekker, 1993:199–220.
21. Cohen DJ, Leckman JF. Tourette syndrome: advances in treatment and research. *J Am Acad Child Adolesc Psychiatry* 1984;23:123–125.
22. Tanner CM, Goetz CG, Klawans HL. Cholinergic mechanisms in Tourette syndrome. *Neurology* (NY) 1982;32:1315–1317.
23. Shapiro AK, Shapiro ES. Controlled study of pimozide versus placebo in Tourette's syndrome. *J Am Acad Child Adolesc Psychiatry* 1984;23:161–173.
24. Harcherik D, Leckman J, Detlor J, Cohen DF. A new instrument for clinical studies of Tourette's syndrome. *J Am Acad Child Adolesc Psychiatry* 1984;23(2):153–160.
25. Leckman JF, Riddle MA, Pradin MT, et al. The Yale Global Tic Severity Scale: initial testing of clinician rated scale of tic severity. *J Am Acad Child Adolesc Psychiatry* 1989;28:566–573.
26. Woody RC, Laney M. Tics and Tourette's syndrome: a review. *J Arkansas Med Soc* 1986;83:53–55.
27. Kurlan R, Kersun J, Ballantine HT Jr, Caine ED. Neurosurgical treatment of severe obsessive compulsive disorder associated with Tourette's syndrome. *Mov Disord* 1990;5:152–155.
28. Balthasar K. Uber das anatomische substrat der generalisierten Tic-Krankheit (maladie des tics, Gilles de la Tourette): Entwicklungshemmung des corpus striatum. *Arch Psychiatr Nervenkrankheiten* (Berlin) 1957;195: 531–549.
29. Bonnet KA. Neurobiological dissection of Tourette syndrome: a neurochemical focus on a human neuroanatomical model. In: *Advances of neurology.* Vol 35. New York: Raven Press, 1982:77–82.
30. Bebbington P. The prevalence of OCD in the community: current approaches to "obsessive compulsive disorder." *Duphar Medical Relations* 1990:7–19.
31. Bland RC, Newman SC, Orn H. Epidemiology of psychiatric disorders in Edmonton. *Acta Psychiatr Scand* 1988;77 (Suppl):338.
32. Murphy DL, Pigott TA. A comparative examination of a role for serotonin in obsessive-compulsive disorder, panic disorder, and anxiety. *J Clin Psychiatry* 1990;51 (Suppl):53–58.
33. Flament M, Whitaker A, Rapoport JL, et al. Obsessive compulsive disorder in adolescence: an epidemiological study. *J Am Acad Child Adolesc Psychiatry* 1988;27: 764–771.
34. Luxenberg JS, Swedo SE, Flament MF, et al. Neuroanatomical abnormalities in obsessive-compulsive disorder detected with quantitative x-ray computed tomography. *Am J Psychiatry* 1988;145:1089–1093.
35. Garber HJ, Ananth JV, Chiu LC, et al. Nuclear magnetic resonance study of obsessive-compulsive disorder. *Am J Psychiatry* 1989;146:1001–1005.
36. Nordahal T, Benkelfat C, Semple W. Cerebral glucose metabolic rates in obsessive-compulsive disorder. *Neuropsychopharmacology* 1989;2:23–28.
37. Barr LC, Goodman WK, Price LH, McDougle CJ, Charney DS. The serotonin hypothesis of obsessive compulsive disorder: implications of pharmacologic challenge studies. *J Clin Psychiatry* 1992;53(Suppl 4):17–28.
38. Comings DE, Comings BG. Hereditary agoraphobia and obsessive-compulsive behavior in relatives of patients with Gilles de la Tourette's syndrome. *Br J Psychiatry* 1987;151:195–199.
39. Cooper J. The Leyton Obsessional Inventory. *Psychol Med* 1970;1:46–64.
40. Mears R. Obsessionality: the Sandler-Hazari scale and spasmodic torticollis. *Br J Med Psychol* 1971;44: 181–182.
41. Rackman C, Hodgson R. *Obsessions and compulsions.* Engelwood Cliffs, NJ: Prentice-Hall, 1980.
42. Goodman WK, Price LH, Rasmussen SA, et al. The Yale-Brown Obsessive-Compulsive Scale. II. Validity. *Arch Gen Psychiatry* 1989;46:1012–1016.
43. Meyer V. Modification of expectations in cases with obsessive rituals. *Behav Res Ther* 1966;4:270–280.

22

Dementia

Charles E. Wells and Peter J. Whitehouse

OVERVIEW

Alzheimer's disease (AD) is the most common degenerative dementia and is characterized clinically by progressive loss of abilities in a number of cognitive and behavioral domains. The term *degenerative* implies progressive decline in the number of neurons in the central nervous system. AD remains a clinical impression that needs to be confirmed by analysis of tissue obtained at autopsy. Even with the examination of necropsy material, however, diagnostic dilemmas remain. For example, it is still not clear whether there are qualitative, as well as quantitative differences in the neuronal loss and cellular changes found in AD when compared with normal aging. Moreover, the clinical heterogeneity has a biologic basis. For example, the relationships between Lewy body dementia and AD are controversial. The term AD itself may well encompass several processes or diseases. Perhaps some of the biologic changes we review represent the final common pathway of a number of etiologic events.

SYSTEMS NEUROSCIENCE

Neocortex

Neuronal loss occurs in a number of brain areas in patients with AD, although the pathologic changes were originally observed by Alois Alzheimer in the neocortex. Gross cortical atrophy is accompanied by loss of neurons and synapses, with associated astrogliosis. Neurofibrillary tangles (NFT) and senile plaques are observed in the cortex, primarily in association areas. White matter may also be involved, perhaps partly due to the loss of neurons. Alterations in blood vessels also occur, including deposition of amyloid (amyloid angiopathy), as well as neurotransmitter receptor and other neurochemical changes.

The cortical neurons that are affected in AD use a number of neurotransmitters, including excitatory amino acids, such as glutamate, as well as neuropeptides, corticotropin-releasing factor, and somatostatin. Variable changes occur in other neuromodulators, such as substance P and neuropeptide Y.

Hippocampus and Hippocampus-Associated Circuits

Prominent pathology occurs in the hippocampus and may underlie some of the significant memory problems that occur in the disorder. Loss of pyramidal cells is greatest in CA1, CA2, and the entorhinal cortex. Granulovacular degeneration is often commonly seen in the hippocampus. Pathology also occurs in associated structures such as the amygdala.

Basal Forebrain Cholinergic Systems

Loss of cells in the cholinergic basal forebrain, which includes the nucleus basalis of Meynert, nucleus of diagonal band, and medial septum, is also a consistent feature of AD. This abnormality has attracted particular attention because the reductions in choline acetyltransferase associated with the loss of neurons in the basal forebrain have been most strongly associated with cognitive impairment. In association with the loss of cholinergic neurons, there is also loss of specific neurotransmitters, including muscarinic receptors (particularly M_2), and nicotinic receptors (1).

Brain-Stem Monoaminergic Systems

Loss of cells in the noradrenergic locus ceruleus and serotonergic raphe nucleus also occur in AD. Dysfunction in cells in these populations may be associated with some of the noncognitive symptoms found in the disorder, such as depression and psychosis.

CELLULAR NEUROSCIENCE

Neurofibrillary Tangles

NFTs are perhaps most strongly associated with specific neuronal loss because they occur in populations of neurons in which cell dysfunction and eventual death occur. NFTs are composed of 15-nm straight filaments and 10-nm paired helical filaments. Immunocytochemically, the NFTs have been found to have several associated cellular constituents, including microtubular-associated proteins, such as tau and ubiquitin. Many believe that tau or a modification of tau forms the core of the NFT. Abnormalities in phosphorylation may also occur (Fig. 22-1).

Senile Plaques

Senile plaques occur with high frequency in amygdala, hippocampus, and neocortex. They are spherical structures that have three principal components: abnormal neurites and glia cells surrounding a central core of amyloid. The main constituent of the core of amyloid is a protein referred to as the β-A4 peptide. This approximately 42-amino acid peptide is part of a much larger precursor protein, the amyloid precursor protein (APP). Various forms of APP of different molecular weights exist. Whether the amyloid protein itself is neurotoxic is controversial, but great efforts are underway to understand the processing of amyloid, that is, how β-A4 forms from APP in the brain and how the process is altered in AD. The origin of the amyloid is unclear, some believing that it is bloodborne and others believing it is derived from neurons or glia. A major unsolved part of the amyloid process is how the β-A4 protein, a small peptide, forms the β-pleated sheet, a configuration that is amyloidogenic. These senile plaques and NFTs have been claimed to be as-

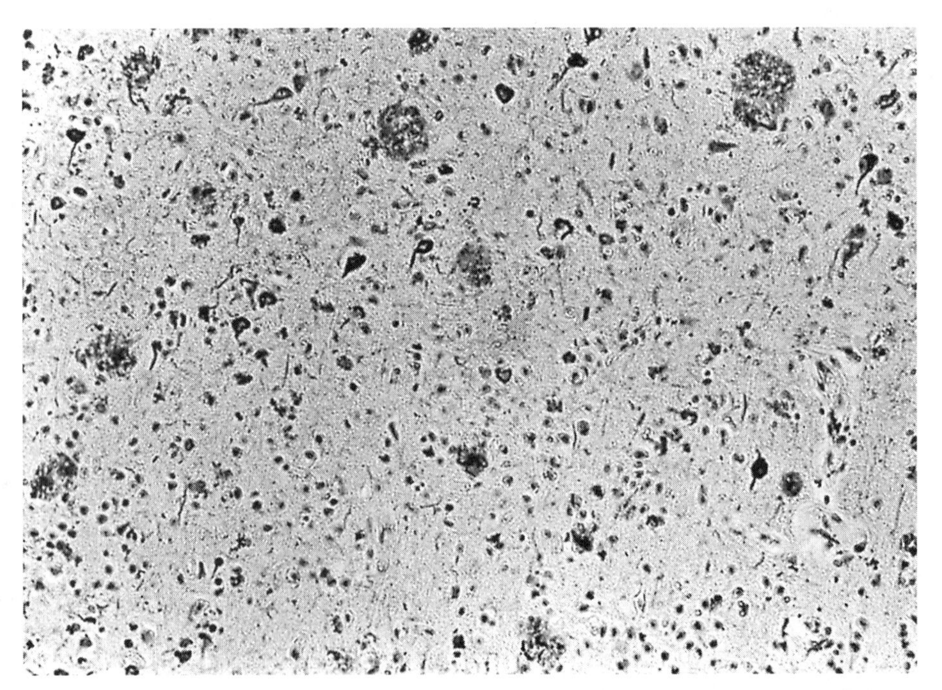

FIG. 22-1. Low-power view showing numerous neurofibrillary tangles and extracellular neuritic plaques.

sociated with the severity of clinically apparent dementia. It is more likely that the loss of neurons, synapses, and neurotransmitter elements is a more direct cause of the cognitive impairment (2).

MOLECULAR NEUROSCIENCE

It has been known since the 1930s that some cases of AD are familial and that some pedigrees demonstrate autosomal-dominant inheritance. Individuals with trisomy 21 invariably develop senile plaques and NFTs after about the age of 35 years, and some develop dementia superimposed on the preexisting mental retardation. An explosion of knowledge has occurred recently in which linkage studies have identified other possible genetic abnormalities on chromosomes 14 and 19 as well as 21. Specific genetic mutations have been described in the APP, which is known to be located on the 21st chromosome. On chromosome 21, these mutations occur at two locations, involving several different possible amino acid substitutions. Their existence strongly suggests that at least one cause of AD is an abnormality in the APP. How these mutations affect the processing of the protein and result in the accumulation in senile plaques is not known (3).

A linkage exists between chromosome 19 and late-onset disease; specifically, apolipoprotein E, a lipid-processing protein, has been associated with both late-onset and sporadic disease. This finding has particularly important implications because this risk factor (specifically homozygosity for E4 allele) may be associated with a large number of cases of AD (4).

Other factors that may contribute to the development of AD include metals and other environmental toxins. Claims have been made, for example, that aluminum is found in NFTs and senile plaques. Although some studies have observed this association, it is not clear that the aluminum is more than a secondary effect after a primary pathogenetic event has occurred. Epidemiologic studies have contributed a number of interesting observations, including the possibility that head injury and cardiovascular disease are risk factors for AD. Both insults might affect the blood-brain barrier and the deposition of amyloid in brain. Other studies have also shown that higher education can lower the risk of AD, possibly through a mechanism of increased functional brain reserve. Also, some epidemiologic studies have suggested that smoking may be protective in AD; this observation is particularly interesting given the consistently reported reductions of nicotinic cholinergic receptors in cortical specimens from patients with AD and the fact that smoking increases nicotinic receptors.

CARE OF THE DEMENTED PATIENT

Diagnosis

Studies began appearing in the late 1960s and early 1970s suggesting that perhaps as many as 15% of patients with dementia (or appearing to be demented) were suffering from treatable and thus potentially reversible disorders. Normal-pressure hydrocephalus, vitamin B_{12} deficiency, hypothyroidism, depression, and a host of other medical, neurologic, and psychiatric disorders were sought, and appropriately so, in every patient thought to be demented. In this area of medicine, diagnosis is always a two-step process in which the physician seeks to answer two questions sequentially:

1. Does the patient have dementia?
2. If so, what is the disease causing the dementia?

Syndrome Diagnosis

The criteria for the diagnosis of dementia are set out clearly and explicitly in the *Diagnostic and Statistical Manual of Mental Disorders,* 4th edition, published by the American Psychiatric Association (5). To summarize, the diagnosis of dementia can be made whenever a patient demonstrates impaired memory plus defects in at least one other area of cognition or changes in personality; when this impairment interferes with work, social activities, or relationships with others; when this is not due to delirium; and when "the disturbance cannot be accounted for by any nonorganic mental disorder."

When all of these criteria are met, except for the last, what features should cause the clinician to doubt the diagnosis of dementia? Several factors come immediately to mind: a history of recent onset and rapid progression of symptoms; *marked* inconsistency in performance on cognitive testing from one examination to another; persistent clinical manifestations of either hypomania or depression; a lack of fit or congruence between various elements in the history, clinical examination, and results of ancillary diagnostic testing; or a sense on clinical examination that the patient is dissimilar, in some precise or imprecise way, from most other demented patients for whom one has cared. Patients may have significant dementia in the absence of any demonstrable abnormality on the electroencephalogram (EEG), computed tomography (CT), or magnetic resonance imaging (MRI) examination of the brain or on neurologic examination. Among the readily available ancillary diagnostic procedures, only neuropsychological testing possesses the sensitivity to rule out dementia.

Patients with depression alone, especially among the elderly, often present defects in memory and in other cognitive functions. These are manifested most prominently by a failure to keep up with the details of everyday life (perhaps due to impaired attention and concentration), problems with names, slowness in thinking, and usually a striking awareness of the defects and of the effort required to answer questions and to commit new facts to memory. At the same time, these patients often demonstrate no difficulty in carrying out activities that require memory; usually show no circumstantiality, perseveration, or tangentiality in their speech; and when not prodded by questions, often demonstrate a clear recollection of events both recent and remote. It is vitally important for the patient's well-being that the neuropsychiatrist determine, during the first visit if possible, that the depressed patient with cognitive impairment is *not* demented. It is well established that a sizable percentage of patients who are demented also satisfy the diagnostic criteria for a coexisting major depression. When a patient presents with clinical features of depression as well as dementia (even if the evidence for dementia is modest), effective treatment of the depression is not likely to result in clinically significant improvement in the dementia and most certainly will not slow the progressive course of the dementing illness. This does not mean, of course, that depression in the demented patient should not be treated vigorously.

Disease Diagnosis

The single most important requirement for disease diagnosis is to be sure that one does not overlook a treatable cause for dementia. Neuropsychiatrists now seldom identify a potentially treatable cause for dementia, much less actually bring about improvement in the cognitive dysfunction per se. It is almost certainly a tribute to the improved skills of internists and general physicians that few demented patients now arrive at neuropsychiatrists' offices for consultation requiring any diagnostic procedures to establish a diagnosis.

Whether any background information about the patient is available, the patient should almost invariably be interviewed alone and in private before the physician talks with family members or other caretakers. One almost never encounters a patient who is so disorganized or so agitated that this practice must be foregone. To do otherwise is potentially degrading and demeaning to the patient. Most of the examiner's initial comments should be nondirective ("Tell me about," rather than "What is?"), with requests for precise data evolving in a natural way out of the patient's verbalizations. Such an interview must be slow-paced, nonhurried, and nonthreatening. The mental status examination as it is often performed may come across as an assault on the patient.

At this initial encounter, one may or may not carry out a neurologic examination. One does so, of course, if there are features that suggest a focal brain lesion or lesions, or the examination can be done later if additional information strengthens the suspicion. In patients seen primarily because of symptoms of dementia, the neurologic examination per se seldom yields new information that is decisive either in diagnosis or in therapy.

The interview with family members or other caretakers usually follows this initial patient encounter. This "family interview" (usually without the patient) has two basic objectives:

- To obtain a fuller history than the patient may be able to provide
- To determine exactly what it is that the patient and family may be seeking

The examiner is attentive to details beyond the usual history of dementia that might indicate a diagnosis other than AD or another disorder superimposed on AD.

When there are unusual features, either by history or on clinical examination, and when the diagnosis remains uncertain, the neuropsychiatrist usually has little problem setting forth a schedule of ancillary diagnostic procedures aimed at achieving an accurate disease diagnosis.

Ancillary Diagnostic Studies

If the patient has not had a recent medical evaluation, the physician should usually order a chemistry profile, vitamin B_{12} level and folate levels, and thyroid function studies (usually thyroxine and thyroid-stimulating hormone) and then refer the patient for full medical evaluation. If the patient has had a recent medical evaluation, but vitamin B_{12} and folate levels and thyroid function studies were not done, these should be ordered. The demented patient's interests are not best served by the use of a standard battery of diagnostic procedures for each patient (although this may be necessary in the research setting). The physician should be cautious and sparing in the use of diagnostic procedures both because of their cost and because of the often fragile nature of these patients' adjustment (e.g., neuropsychological testing can result in catastrophic reactions, and MRI may be terrifying for an uncomprehending subject).

Thus, an EEG may serve this patient population when there is uncertainty about the existence of a physical brain disease, when one suspects a delirium may be superimposed on a dementing illness (when sequential EEGs may help in following its course), or when some clinical feature suggests the possibility of a subclinical seizure disorder. Unless there is a specific reason to suspect a disorder other than AD or multiinfarct dementia, a CT scan or MRI of the brain may be of little value except for cases in which the family needs these for reassurance that the patient has no intracranial tumor.

There is likewise no justification for routine examination of the spinal fluid. When dementia is the salient clinical feature, cerebrospinal fluid examination seldom contributes significantly to either diagnosis or plans for patient care.

Neuropsychological testing is most helpful in two groups of patients:

- Those who fear they are demented but clearly are not (in whom negative findings can be used to bolster the diagnosis of "no dementia")
- Those suspected to have an early dementia but whose performance in interview does not demonstrate definite defects (in whom testing will usually demonstrate performance still within the normal range but at a lower level than would have been predicted from education and occupation)

More sensitive tests continue to be developed to aid in recognition of the second group.

Positron emission tomography and single photon emission computed tomography are the most recent arrivals on the diagnostic scene. Although each of these tests may yield information of clinical interest and value in the patients with dementia, to date there is no convincing evidence that they are of great value in patients whose diagnosis is uncertain on clinical grounds.

The most commonly used criteria for the diagnosis of AD are those proposed by the National Institute of Neurological and Communicative Diseases-Alzheimer's Disease and Related Dementias Association (NINCDS-ADRDA) Work Group (6), which can be seen in Table 22-1. The American Psychiatric Association's criteria include all the criteria for the

TABLE 22-1. *Criteria for clinical diagnosis of probable and possible Alzheimer's disease*

Criteria for clinical diagnosis of *probable* Alzheimer's disease
- Dementia demonstrated by clinical examination
- Deficits in two or more cognitive functions
- Slowly progressive worsening of cognitive functions
- Consciousness unimpaired
- Onset between ages 40 and 90 years
- Absence of systemic or other brain disorders that could account for the clinical dysfunction

Criteria for clinical diagnosis of *possible* Alzheimer's disease
- Presence of dementia in the absence of systemic or other brain disorders that might account for the dementia, even when certain features in onset, presentation, or clinical course are unusual
- Other systemic or brain disorders may be present but are not considered to be the cause of the dementia.

Modified from McKhann G, Drachman D, Folstein M, Katzman R, Price D, Stadlan EM. Clinical diagnosis of Alzheimer's disease: report of the NINCDS-ADRDA work group under the auspices of Department of Health and Human Services Task Force on Alzheimer's disease. *Neurology* 1984;34:939–944.

diagnosis of dementia (see earlier discussion) plus an "insidious onset with a generally progressive deteriorating course" and "exclusion of all other specific causes of Dementia by history, physical examination, and laboratory tests."

Treatment

Cognitive Problems

Since the discovery that acetylcholine is reduced in the brains of patients with AD, efforts have been made to use pharmacologic agents to increase the brain's supply of acetylcholine. Efforts have centered chiefly on increasing the supply of acetylcholine precursors available to the brain (by increasing blood levels of choline, lecithin, or both) and the manipulation of enzyme functions to increase the levels of acetylcholine at neurotransmitter sites. There is general agreement that augmenting supplies of acetylcholine precursors has been of no value in AD.

Tacrine (1,2,3,4-tetrahydro-9-acridinamine monohydrochloride monohydrate) was the first cholinesterase inhibitor proved to have efficacy with regard to the cognitive deficits of AD. The clinical effect proved modest, however, and the use of this agent was associated with impaired liver function in some patients. This drug has largely been superceded by donepezil, a piperazine-based cholinesterase inhibitor. Donepezil, in doses of 5 or 10 mg given once daily, has been shown to be superior to placebo in improving cognitive function in mild to moderate AD and to maintain its efficacy for up to 38 weeks. Currently, donepezil is a standard treatment and should be considered for AD patients who do not exhibit some contraindication (7).

Depression

The rates of depressive disorders in AD have been reported between 0% and 86%, although most well-researched series report rates between 17% and 29%. With a frequency this high, depression in AD offers the neuropsychiatrist a promising window for therapeutic intervention because it can often be treated with remarkable effectiveness. Demented patients with depression do not always complain of depression or even perceive themselves to be depressed. Thus, the physician often must depend on the statements of caretakers and on behavioral observations rather than on what the patient says. Even when the demented patient denies feelings of depression, chronic irritability and irascibility or a plethora of ill-defined somatic complaints should alert the examiner to the possibility.

The treatment of depression in the patient with AD or other dementing disorders is basically the same as in depressed patients without dementia, and it rests on a combination of pharmacotherapy and supportive psychotherapy, both tailored to meet the special needs of the dementia patient. As yet, there is no "antidepressant of choice" in the treatment of demented patients with depression. With the growth of available antidepressants, neuropsychiatrists have increasingly used the newer, selective serotonin reuptake inhibitors (fluoxetine, sertraline, and paroxetine) in an effort to avoid the anticholinergic side effects of the tricyclic antidepressants. The new medications are not, of course, without side effects themselves, and they appear especially prone to accentuate anxiety and restlessness or psychotic manifestations in the susceptible patient. Although one should obey the admonition to begin treatment in the demented (and usually elderly) patient with doses of psychotropic medications lower than those used in the nondemented patient, these patients often require as much medication to achieve an effective result as do nondemented patients.

In summary, depression is a common, treatable accompaniment of AD and other dementing disorders. Although it may be disguised in some patients by behavioral manifestations such as agitation, assaultiveness, or persistent screaming, it can usually be recognized and diagnosed easily. Depression in dementia offers the neuropsychiatrist an excellent best opportunity for gratifying therapeutic intervention in the demented patient.

Delusional Ideation and Behavior

Delusions and the resulting behavioral abnormalities are frequent reasons for patients with AD and other dementing disorders to be brought to physicians. There is some evidence suggesting that delusions become less common as the dementia becomes more severe. Confabulations are "false stories" rather than "false beliefs," and the demented person seldom sticks with the same story for very long. A distinction must also be made between delusions and illusions, which are misperceptions or misinterpretations of real events. These distinctions are important from a therapeutic standpoint because we possess no therapeutic measures likely to reduce confabulation and none except calm reassurance likely to influence illusions. Among the many manifestations of dementia, delusions are among those most amenable to treatment.

Treatment of delusions in the demented patient is basically no different from treatment of delusions in nondemented patients—it rests primarily on pharmacologic intervention, although behavioral approaches may play important supplemental roles. Among the many antipsychotic drugs available, no one agent has been proved more effective than another in the demented patient. As in the nondemented patient, choice is usually based on the desire to manage side effects therapeutically (e.g., to give a sedating antipsychotic to the patient who is sleeping poorly) or to avoid a potentially undesirable side effect (e.g., to choose one having relatively few peripheral cardiovascular side effects in the demented patient with diabetes). In the demented patient, doses chosen to initiate therapy are usually smaller than in nondemented patients, although the demented patient may eventually require comparable doses to achieve adequate control.

Although behavioral techniques seldom play a major role, they may nevertheless be very helpful. Reassurance is, of course, the most basic behavioral tool, and it should be applied freely whenever the patient is anxious and fearful. Equally important, although more difficult for most caretakers, is the technique of benign disinterest. It is not therapeutic to question patients repeatedly about their delusions or to discuss the delusions at length with them. Behavioral measures can help even in patients in whom antipsychotics appear to be ineffective.

Hallucinations

Except as a manifestation of delirium, hallucinations are relatively rare in AD and other dementing illnesses. When they occur, they are more likely to be visual than otherwise, but any sensory modality may be involved. As is the case for delusions, hallucinations must be distinguished from illusions. As already written, we have no effective treatment for illusions, but if the patient can be helped by whatever means to become less anxious and fearful, the illusion may be reduced in frequency or at least become less threatening.

Behavioral Problems

Behavioral abnormalities are probably the leading cause for families and caretakers to

TABLE 22-2. *Classification of behavioral problems in dementia*

Distressing repetitive behaviors
Dangerous, careless behaviors
Restless, agitated, hostile, assaultive behaviors
Overelated, overactive, intrusive behaviors
Insomnia and diurnal rhythm reversals
Repetitive screaming and crying out
Inappropriate sexual behaviors
General regression, refusal to eat

seek help from neuropsychiatrists, and they are unquestionably the chief cause for hospitalizing patients who have AD. There is still no generally accepted classification for the various forms taken in the context of "behavioral problems." Table 22-2 provides a classification culled in part from several recent publications.

Distressing Repetitive Behaviors

Although seldom the primary reason for the family or other caretaker to seek medical help, repetitive bizarre behaviors are often among the most stressful for the caretaker to endure. Repetitive behaviors take many forms. Some patients cling to their caretakers, never wanting them out of their sight, following them from room to room, often even into the bathroom if the door is not locked. Often this is associated with incessant questioning, usually the same question repeated continuously. Others occupy themselves taking objects out of closets, drawers, or kitchen cabinets, then replacing them, though never in the accepted order. Another patient may ceaselessly pack and unpack a suitcase or purse. Others hoard, repetitively hiding food, money, or specific objects, often so that they cannot be found. Most of the time no effective intervention will be found to modify any of these behaviors. Ignoring the behavior may, of course, ease the plight of the caretaker even without modifying the behavior, and anxiolytics may reduce the separation anxiety when the patient is locked out of the bathroom, but the basic problem remains unchanged.

Dangerous, Careless Behaviors

There are three behaviors about which the neuropsychiatrist is most likely to be consulted:

- Unsafe driving
- Unsafe use of mechanical equipment (especially stoves and motorized lawn mowers)
- Wandering

The simple truth is that the only certain way for the patient to be protected from these activities is for the automobile, stove, or mower to be made inoperative or for the keys to be safely hidden. Anxiolytic or antipsychotic agents may help reduce the hostility and frustration that often occurs following these prohibitions, but they do little to alter the attempted behavior itself.

Wandering falls into the same category of dangerous behaviors, but it is often even more distressing to caretakers and more difficult to control. It is often the prime reason for a patient's institutionalization. Although wandering seems most difficult to control in the home setting, it remains a serious problem in many institutionalized patients as well. Unless the family or caretaker is willing for the patient to be locked up at home or in a unit in an institution, we have no good measures to stop wandering. Anxiolytics or antipsychotics may reduce restless and exploratory behavior, but they seldom control it entirely, and most patients who are so inclined continue occasionally to wander, although their attempts may be less frequent.

Restless, Agitated, Hostile, Assaultive Behavior

Although all these behaviors may be manifestations of the dementing disease itself, the physician should always seek other explanations. Is the patient taking some medication that might be causing such behavior? Is the patient slipping into and out of delirium? Is the patient experiencing pain that he or she cannot describe but that results in such behaviors? Is the patient delusional and acting on the basis of the delusions? Treatment of these problem behaviors usually requires both behavioral and pharmacologic measures. Even though behavioral techniques are essential in the management of these patients, pharmacologic measures remain the definitive treatment when no specific inciting cause for the agitation and hostility can be found.

Antipsychotics and anxiolytics are the agents most frequently employed, but the process of trial and error often expands to include antidepressants, mood stabilizers, and anticonvulsants. The choice of an initial pharmacologic agent is usually between an antipsychotic or an anxiolytic drug. The two side effects that figure most prominently in the physician's deliberations are sedation and extrapyramidal motility disturbances. In most agitated patients, one cannot deal with the agitation effectively without encountering at least one of these side effects. In practice, it is often easier to deal with sedation than with stiffness and immobility; this means that initial considerations usually include chlorpromazine, thioridazine, loxapine or molindone, and lorazepam or clonazepam. Although one usually begins with small doses of psychotropic medicines, especially in elderly patients with diagnosed brain disease, small doses are often remarkably ineffective in highly agitated demented patients. In many patients, the dangers of prolonged agitation outweigh the dangers of short-term oversedation.

Beyond the antipsychotic and benzodiazepine anxiolytics, a wide variety of psychopharmacologic agents have been reported to be effective in quelling agitation in some patients. These include lithium, carbamazepine, propranolol, buspirone, antidepressants (especially trazodone and fluoxetine), and meprobamate.

For the agitated demented patient, hospitalization may be not only desirable but also essential for the dangerous behavior to be brought under control. Hospitalization offers several distinct advantages and opportunities:

- Observation of the patient's behavior in a new, structured environment
- Observation of the effectiveness of behavioral treatment techniques put into use by well-trained and experienced staff
- Modification of pharmacologic agents on a daily basis based on observations of the physician and those of a well-trained nursing staff
- Use of medications in combinations and dosages that would be avoided were the patient not under constant medical observation

The dividends of hospitalization are greatest if the patient can be hospitalized on a psychiatric unit devoted to the care of demented patients.

Insomnia and Diurnal Rhythm Reversals

Most medical publications dealing with insomnia emphasize the importance of the obvious: keeping the patient awake during the daytime if possible, a program of daily exercise as tolerated, and regular toileting. In practice, effective treatment of insomnia in the demented patient depends on the pharmacologic management. Caution is advised with regard to the side effects that these medications often but not inevitably produce, but these can usually be managed by selection and careful dosage titration. Sometimes, when the insomnia is not too severe, sleep can be improved by bedtime doses of aspirin, acetaminophen, or an antihistamine (such as promethazine). A wide variety of agents may be used to promote sleep, including not only the recognized hypnotics but also antidepressants, anxiolytics, and sometimes even the more sedating neuroleptics. In fact, an antidepressant, specifically trazodone, may be used first, in part because it can foster nighttime sleep without excessive daytime sedation and because it may also reduce daytime restlessness. Small doses of amitriptyline may also be effective and well tolerated. As for the hypnotics, chloral hydrate is often a first choice because even in doses of 1 g or more, it is often tolerated in the elderly demented patient without daytime drowsiness or hangover ensuing. Temazepam too is often surprisingly well tolerated, even for prolonged periods of time. When patients have sleep problems and also require daytime anxiolytics or antipsychotics because of restlessness and agitation, the insomnia can often be handled effectively simply by giving a larger dose of the anxiolytic or neuroleptic agent at bedtime.

Repetitive Screaming and Crying Out

Repetitive unprovoked screaming and crying out are among the most distressing symptoms encountered in patients with dementia and among those most difficult to treat effectively. In these patients, one should always seek a nonpharmacologic remedy first. Does the screaming result from some unsuspected pain or discomfort (a collapsed vertebral body, urinary retention, fecal impaction)? Can the screaming be stopped by some environmental intervention (isolation in a quiet spot, placement among a group of active people, one-to-one personal attention, soothing music)?

The response of the screaming patient to pharmacologic intervention is so unpredictable that a reasonable protocol as to how the physician should proceed cannot even be offered. Anxiolytics, antipsychotics, antidepressants, sedatives, lithium, and antiepileptic drugs all appear to have been helpful periodically.

Inappropriate Sexual Behavior

These behaviors occur most often in men. The most common problems include inappropriate fondling or touching, offensive sexual verbalizations, and masturbation that can be observed by others. Occasionally, it takes the form of continued insistence by the demented patient on sexual congress with the spouse or other convenient sexual object.

Most of the first group of behaviors can be dealt with effectively by behavioral techniques. Calm, quiet, firm statements (made without any sign of irritability or embarrassment) that such behavior is inappropriate and will not be tolerated usually stops troublesome fondling, touching, or verbalizations, but occasionally the caretaker may have to leave the patient alone for a while to prove that it truly will not be tolerated. Masturbation within the view of others is best handled by providing the patient adequate opportunities for masturbation in private. When behavioral techniques are ineffective, hormonal therapy should be tried. In men, conjugated estrogens given on a daily basis may be adequate to reduce these unacceptable behaviors. When this is not effective or if the patient refuses oral medications, regular injections of medroxyprogesterone are often effective.

Incontinence

Urinary incontinence usually precedes fecal incontinence in the progression of these diseases and can often be managed effectively. When urinary incontinence begins in the demented patient, the physician should consider first the possibility of a urinary tract infection. If there is no evidence for infection, and if the patient is still capable of recognizing the urge to urinate and can still find the bathroom, the patient should probably be referred for urologic consultation. For the patient with advanced dementia, escorting the patient to the bathroom on a regular schedule (about every 2 hours during waking hours) is often surprisingly effective. Nocturnal incontinence can often be controlled by waking the patient on schedule once or twice nightly for trips to the bathroom.

In the patient with beginning fecal incontinence, the physician should first be certain that the patient does not have a fecal impaction, a frequent problem in the elderly population. If not, fecal incontinence can often be avoided with a regular toileting schedule as described previously, especially when this is combined with a daily regimen of stool softener plus fiber additives, augmented as needed by judicious use of laxatives.

General Regression and Refusal to Eat

If they live long enough for the disease to progress to very advanced stages, almost all demented patients eventually lose their appetite and quit eating. No remedy is known. At this point, each family confronts a serious problem, one in which no choice appears satisfactory. They must choose either to let the disease take its natural course or to begin artificial feeding (by nasogastric tube, gastrostomy, or total parenteral nutrition).

Respite Needs

From the beginning, the physician should emphasize the caretakers' need for rest and respite from their caretaking responsibilities. Whenever possible, it is best to begin providing the caretaker with regular periods of relief early in the course of the disease, so that it becomes a part of the patient's scheduled care plan. Respite can be provided either in the home or away from home. At home, it is usually not too difficult to set up a plan whereby the principal caretaker is provided regular relief by other family members, close friends, or paid attendants. In most larger cities, a variety of respite services are also available outside the home. As the disease advances, thought should be given to placing the patient, for one or several days each week, in an adult day care program.

CONTEMPORARY ISSUES AND FUTURE DEVELOPMENTS

We are in the midst of an explosion in neuroscientific research related to Alzheimer's disease and other cortical dementias of later life. More than 8000 citations have appeared on these topics in the English language between the years 1995 through 1998. The study of AD remains one of the most active areas of molecular, cellular, and systems neuroscience. The major challenge ahead is to bridge the gap between a molecular characterization of the genetic abnormalities in the disease and the clinical symptoms that we see in patients, so that we can more effectively develop both diagnostic tests and therapeutic interventions. We anticipate a continuing stream of new diagnostic and therapeutic reports through the foreseeable future. The reader will have no alternative to cognizant and ongoing survey of the neuropsychiatric literature.

REFERENCES

1. Ball MJ, Fisman M, Hachinski V, et al. A new definition of Alzheimer's disease: a hippocampal dementia. *Lancet* 1985;1:14–16.
2. Yankner BA, Duffy LK, Kirschner DA. Neurotrophic and neurotoxic effects of amyloid β protein: reversal by tachykinin neuropeptides. *Science* 1990;250:279–282.
3. Goate A, Chartier-Harlin M-C, Mullan M, et al. Segregation of a missense mutation in the amyloid precursor protein gene with familial Alzheimer's disease. *Nature* 1991;349:704–706.
4. Strittmatter WJ, Saunders AM, Schmechel D, et al. Apolipoprotein E: high-avidity binding to β-amyloid and increased frequency of type 4 allele in late-onset familial Alzheimer's disease. *Proc Natl Acad Sci U S A* 1993;90:1977–1981.
5. American Psychiatric Association. *Diagnostic and statistical manual of mental disorders.* 4th ed, revised. Washington, DC: American Psychiatric Association, 1994.
6. McKhann G, Drachman D, Folstein M, Katzman R, Price D, Stadlan EM. Clinical diagnosis of Alzheimer's disease: report of the NINCDS-ADRDA work group under the auspices of Department of Health and Human Services Task Force on Alzheimer's disease. *Neurology* 1984;34:939–944.
7. Rogers SL, Friedhoff, LT. Long-term efficacy and safety of donepezil in the treatment of Alzheimer's disease: an interim analysis of the results of a US multicentre open label extension study. *Eur Neuropsychopharmacol* 1998;8:67—75.

23

Cerebrovascular Disorders

John R. Absher and James F. Toole

OVERVIEW

This chapter reviews the behavioral components of cerebrovascular disease (CVD), including common vascular syndromes and lesion localization, with an emphasis on arterial distribution, clinical diagnosis of focal neurobehavioral syndromes such as vascular dementia, and management considerations. CVD is a leading cause of mortality in the United States. Alterations in behavior resulting from CVD produce substantial morbidity in the elderly population. As more people survive strokes, behavioral problems secondary to CVD become increasingly common.

Stroke is a disorder particularly suited to behavioral study because of its defined limits and its sudden onset with abrupt change in behavior and because the typical survival of the patient is such that the evolution of the process over time can be evaluated. Initially, only clinical pathologic correlations related the neuroanatomic damage to specific clinical phenomena. These anatomic studies can now be coupled with functional mapping by positron emission tomography (PET) and single photon emission computed tomography (SPECT), so that functional changes occurring at the site of the lesion and at distantly interconnected regions can be evaluated and defined anatomically.

Functional neuroimaging has been particularly valuable because behavioral disturbances may occur without an abnormal neurologic examination, yet changes may be seen on magnetic resonance imaging (MRI), PET, SPECT, and specialized cognitive or neurobehavioral tests. Many of these specialized "bedside examination" techniques are discussed in this chapter. When possible, the neurobehavioral features of CVD are described on a vessel-by-vessel basis to illustrate their potential localizing value.

NEUROANATOMIC SYNDROMES CLASSIFIED BY ARTERIAL TERRITORY

The brain is supplied by the two carotid arteries and, in general, by the two vertebral arteries. Most neurobehavioral deficits result from lesions in the distribution of the carotid artery, left hemisphere strokes being slightly more likely to produce noticeable deficits than right hemisphere strokes, owing largely to left hemisphere dominance for language. Neurobehavioral deficits also result from strokes within the vertebral basilar distribution, particularly the posterior cerebral artery territory. The neurobehavioral disorders, therefore, can be divided into carotid or vertebrobasilar syndromes with subdivisions into large surface-conducting arteries, boundary zone (or watershed) disorders, and arteriolar and capillary disorders. The vascular dementias are discussed separately. See Fig. 23-1 for an overview of arterial domains.

Internal Carotid Artery and Branches

Neurobehavioral disorders may result from CVD in most branches of the internal carotid artery. Even ophthalmic artery infarction (or optic nerve damage) can produce, rarely, a classic neurobehavioral disorder: Anton's syndrome, or the denial of blindness.

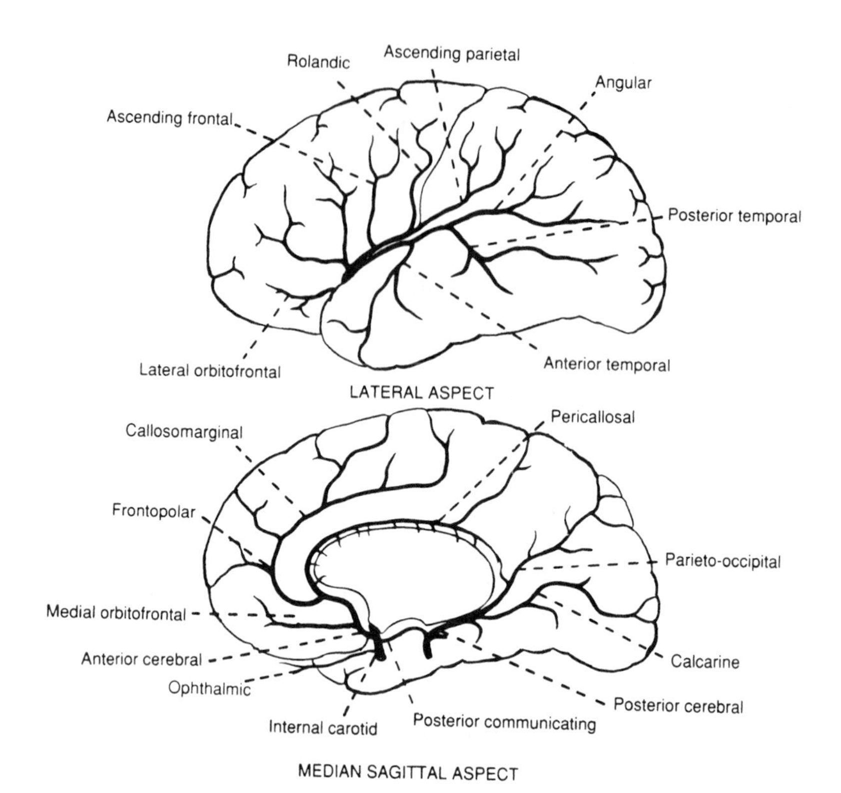

FIG. 23-1. Overview of arterial domain. (From Toole JF. *Cerebrovascular disorders.* 4th ed. New York: Raven, 1990:81.)

Anterior Choroidal Artery

The anterior choroidal artery supplies a portion of the thalamus, globus pallidus, and internal capsule. The anterior choroidal artery syndrome generally results from occlusion due to atherosclerosis and can produce unilateral or bilateral thalamic lesions. Left-hemisphere infarcts can produce a thalamic aphasia characterized by relatively preserved repetition and comprehension ability, with impaired language output. This variety of thalamic aphasia resolves rapidly and has a favorable prognosis compared with most other aphasias. The bilateral anterior choroidal artery syndrome may produce acute pseudobulbar palsy, mutism, abulia, and inappropriate laughter (1).

Anterior Cerebral Artery

The anterior cerebral artery supplies the medial aspects of frontal and parietal lobes and the anterior two thirds of the corpus callosum. This artery joins its opposite through the anterior communicating artery. Its loss, either unilaterally or bilaterally, can result in lesions of the corpus callosum, cingulate gyri, supplementary motor area, or other medial structures.

Callosal infarction can produce a callosal disconnection syndrome. This syndrome includes three essential features:

- Left limb tactile anomia
- Left limb apraxia
- Left limb agraphia

These deficits result from disconnection of the left-hemisphere centers for language from the right hemisphere.

Lesions of the cingulate gyri may cause profound behavioral alterations. Strokes within anterior cerebral artery territory may lead to infarction of the opposite hemisphere through propagation of clot through the anterior communicating artery or because one anterior cerebral artery is "dominant" for blood supply to the other hemisphere. Although unilateral strokes in the cingulate region may not produce substantial behavioral compromise, bilateral damage often leads to *akinetic mutism,* which is "a state of unresponsiveness to the environment with extreme reluctance to perform even elementary motor activities. (2)" The patient lies quietly in bed, immobile, sometimes sleepy but often open-eyed and seemingly alert.

Most patients with cingulate strokes remain vigilant, unlike those with the *paramedian thalamic syndrome,* resulting from mesencephalic artery infarction, which disconnects the ascending reticular activation system from its thalamic and cortical targets, impairing basic arousal. Cingulate damage leaves these pathways for basic arousal intact in akinetic mutism and selectively impairs the will to interact with one's surroundings.

Supplementary motor area strokes in the left hemisphere can produce a transcortical motor aphasia, characterized by preserved repetition, despite naming and verbal fluency deficits. This type of aphasia sometimes is accompanied by failure to initiate motor programs for language (motor neglect) because many of these patients are mute at presentation; bilateral lesions in the region of the supplementary motor area can produce akinetic mutism.

Middle Cerebral Artery

The middle cerebral artery supplies vast areas of the cerebral hemisphere. There are usually two major branches of the middle cerebral: an anterior or frontal branch and a posterior branch. Deep nuclei, such as the caudate, putamen, and globus pallidus, and deep white matter (internal capsule and corona radiata) receive blood supply through this crucial arterial system. Vast areas of the brain may be affected by disease within the middle cerebral distribution, but the occipital lobe, cerebellum, and brain stem are typically spared.

In the left hemisphere, the middle cerebral artery supplies important functional centers for motor behaviors such as language output. The left hemisphere is usually dominant for language, even in non–right-handed people. Executive functions and praxis, language comprehension, reading, and writing are important functions of the dominant left hemisphere. Working memory is probably a dominant function of the left dorsolateral convexity. The left middle cerebral territory is arguably the most important vascular distribution because neurobehavioral deficits within this territory can be devastating. For example, a left middle cerebral artery infarction of its entire territory produces grossly impaired language comprehension, inability to produce meaningful written or spoken language or gestures (global aphasia), and right hemiplegia.

The anterior branch of the middle cerebral artery supplies the dorsolateral frontal lobe. Broca's aphasia results from damage to the left inferior and posterior frontal lobe premotor cortices responsible for coordinating complex linguistic motor activities of the mouth, tongue, palate, and lips. Basic coordination and praxis may be normal, despite language output abnormality, or a facial apraxia may be found in association with the Broca's aphasia. The left premotor association area is important for the integration of language and motor behaviors, and strokes in this same brain area may lead to ideomotor apraxia, in which the left hand is unable to follow verbal commands, despite adequate strength, coordination, and sensory function (sympathetic apraxia). In some cases, there is preserved ability to perform the same maneuvers spontaneously. The left hand is apraxic because of disconnection of the right hemisphere from verbal commands that are normally processed in the left superior temporal gyrus and inferior parietal lobule, transferred to the left motor association areas, and then interpreted transcallosally by the right motor association area.

Left frontal convexity strokes may impair other complex aspects of executive ability, the ability to shift or divide attentional resources, or reason. Depression may result from left dorsolateral frontal lobe strokes, whereas right-sided strokes may produce aprosodias. The anatomic basis and clinical features of the aprosodias are similar to those of the aphasias (e.g., motor, conduction, and transcortical varieties have been described). Damage to the orbital frontal lobe often produces severe personality alterations and an acquired sociopathy, manifested either as inappropriate social behaviors or an incapacity to perform meaningful planning and organizational tasks. Judgment and reasoning are severely affected, and patients demonstrate little concern for the effect their behaviors may have on others.

The posterior branch of the middle cerebral arterial territory supplies a large portion of primary sensory and sensory association areas (including parietal, temporal, and occipital association areas), insula, and deep white and gray matter. Strokes within this territory can produce a number of classic neurobehavioral syndromes, such as alexia *with* agraphia, Gerstmann's syndrome, conduction aphasia, parietal apraxia, transcortical sensory aphasia, hemi-inattention or neglect, and Wernicke's aphasia.

Alexia with agraphia occurs with occlusions of the posterior branches of the left middle cerebral artery that destroy the angular gyrus in the critical brain area for translating written symbols into meaning and vice versa, and for instructing the left motor association areas important for the mechanics of writing.

Gerstmann's syndrome consists of agraphia, acalculia, finger agnosia, and right–left disorientation and typically results from left angular gyrus lesions. Rarely, a right-sided lesion may produce this syndrome if the right hemisphere is dominant for language. Damage to the angular gyrus of the left inferior parietal lobule or nearby regions produces these symptoms. Partial syndromes are common.

Conduction aphasia typically results from disconnection lesions situated between Wernicke's area and Broca's area. Patients cannot repeat, but comprehension and verbal fluency are preserved. The anatomic basis of conduction aphasia has become a controversial topic. See reference 3 for a review of the disconnection syndromes.

Parietal apraxia is a bilateral apraxia without hemiparesis. Damage in the parietal lobe (especially on the left) may disrupt "visuokinesthetic engrams" leading to bilateral apraxia. Parietal apraxia is similar in its pathogenesis to the agraphia and constructional disturbances seen in Gerstmann's syndrome in that the mechanisms for these syndromes probably relate to faulty communications between posterior language areas and the left frontal motor association cortex, which is responsible for relaying verbal commands to the right hemisphere. Patients with transcortical sensory aphasia follow commands poorly with both hands, but comprehension testing documents severe deficits to simple yes–no questions and often an inability to name objects. Preserved repetition allows these patients to repeat commands or questions verbatim (echolalia). Both of these disorders result from damage within the inferior parietal lobule. Lesions deep to the supramarginal gyrus have been associated with parietal apraxia, whereas damage to angular gyrus and adjacent (more posterior) heteromodal association cortex produces transcortical sensory aphasia.

Hemi-inattention and neglect may occur with left middle cerebral territory strokes, particularly the posterior branch, but right-hemisphere damage is probably more likely to produce left hemi-neglect than the reverse.

Wernicke's aphasia results from temporal lobe infarction in the posterior third of the superior temporal gyrus and in small portions of the inferior parietal lobule. The exact boundaries of Wernicke's area vary among patients (e.g., as with right-hemispheric language dominance). Patients with damage in this region cannot comprehend spoken or written language, have rapid and paraphasic speech output, and cannot repeat. Paranoia or violence are occasional manifestations of this aphasia.

A rare vascular syndrome that results from bilateral strokes in this region is pure word deafness or auditory verbal agnosia. The primary auditory cortices or nearby first-order association areas of both temporal lobes are destroyed. Patients complain about being unable to understand anything spoken, as if others are speaking a foreign language. The meaning of familiar sounds (e.g., doorbell) is typically preserved in such patients, and reading, writing, and spontaneous speech are preserved.

Vertebrobasilar Circulation

The vertebral arteries ascend through the transverse foramina of the cervical vertebrae and join to form the basilar artery. There are several branches of the vertebrobasilar system, but most neurobehavioral disturbances due to CVD in the vertebrobasilar (or "posterior") circulation are caused by disease within the posterior cerebral arteries. Thalamoperforating vessels are also important for disturbances such as the paramedian thalamic syndrome, thalamic amnesia, thalamic aphasia, and thalamic neglect. Strokes within small penetrating vessels to the midbrain and pons can also produce neurobehavioral syndromes such as the mesencephalic form of akinetic mutism, peduncular hallucinosis, and amnesia.

Basilar Artery

A large midbrain and pontile infarct may result from CVD of the basilar artery, producing the locked-in syndrome: loss of all motor functions other than eyelid or eye movements. Patients with the locked-in syndrome lack the capacity to communicate except through blinking or moving the eyes, although such movements may indicate intact cortical functions such as language comprehension. These patients, in theory, have normal intellect, normal feelings, and normal emotions.

A milder deficit from damage within this arterial system is the mesencephalic form of the akinetic mute state, which probably should be viewed as a locked-in state with preservation of some limb or head movements. Arousal is generally impaired, in contrast to akinetic mutism from cingulate damage.

Small lesions within the cerebral peduncles, specifically the pars reticulata of the substantia nigra, may rarely produce hallucinosis. Micropsia and kaleidoscopic, highly patterned visual hallucinations occur. Fortunately, this syndrome is extremely rare.

Lesions of the small arteries and arterioles that supply the hypothalamus, the region of the mammillary bodies, and the medial thalamus may result in a devastating amnestic syndrome called *thalamic amnesia.* There is almost always some damage to hypothalamus, mammillary bodies, or fornices.

Posterior Cerebral Artery

Posterior cerebral artery infarction can result in parietal and occipital damage. Four classic neurobehavioral syndromes occur with strokes in the posterior cerebral artery territory: achromatopsia, prosopagnosia, pure alexia, and Anton's syndrome.

Achromatopsia occurs with damage to the ventral visual association areas, especially on the left. Often, there is an associated visual field defect, so that part or all of the contralateral visual field cannot be tested. Right-sided lesions tend to produce only contralateral achromatopsia, whereas left-sided lesions may affect the whole visual field.

Prosopagnosia occurs with damage to both inferior visual association areas and rarely with unilateral right-sided lesions. This disorder is characterized by normal ability to match, sort, or draw faces, as well as preserved ability to recognize gender and emotional facial expressions. What is lacking is the ability to connect this normal visual perception with its meaning. Patients are unable to recognize any familiar face, although other clues (voice, characteristic gait, a distinguishing trait such as facial hair) may lead to correct identification.

Alexia without agraphia may occur after complete left posterior cerebral artery infarction (Fig. 23-2).

All left-hemisphere visual input is disrupted, and right-hemisphere visual input is disconnected from the left-hemisphere center for language. The result is a visual anomia and alexia. Writing is preserved because the

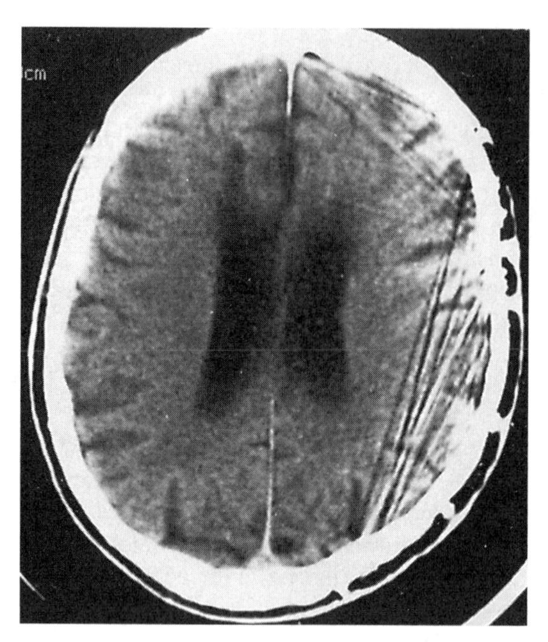

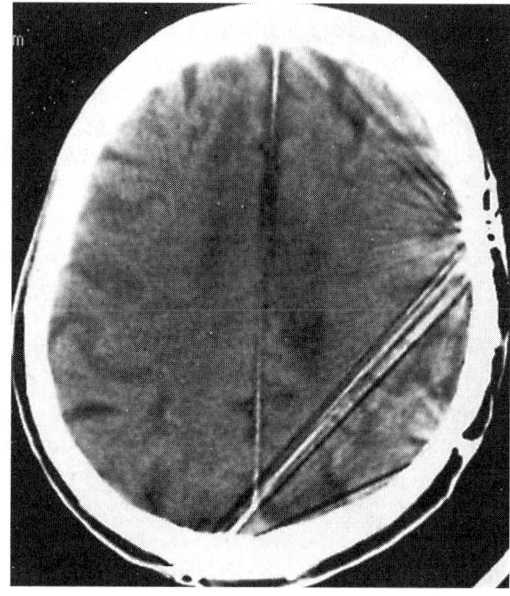

FIG. 23-2. CT of a 60-year-old man who became acutely mute and immobile. During his hospitalization, bilateral anterior cerebral artery infarctions were identified. Note the lesion in the right anterior cingulate region and the larger lucency in the left anterior cerebral artery territory. He has remained essentially akinetic, but is now able to speak in soft, short phrases.

posterior and anterior language areas of the left hemisphere are intact.

Anton's syndrome is a denial of blindness or lack of insight into the fact that one is unable to see. In such cases, the patients may hallucinate and imagine scenes that they describe and, based up hearing and previous experience, act as if they can see. Bilateral occipital lobe infarcts that produce bilateral homonymous hemianopsias (cortical blindness) are often found in patients with Anton's syndrome.

Boundary Zone (Watershed Area)

The major conducting arteries on the surface of the brain result in large infarctions in their vascular distribution. In certain instances, these are confined to the watershed distribution between the major arteries, where there are insufficient collaterals or reduced perfusion pressure in both arteries. This boundary, or watershed zone, extends in a C-shaped pattern from the anterior polar frontal lobe and parasagittally a few centimeters from midline to the posterior parietal-occipital visual association areas. A boundary zone probably also exists in the deep parenchyma, but the term *watershed infarction* is typically reserved for cortical damage. Frontal lobe damage in the boundary zone may produce a transcortical motor aphasia, and transcortical sensory aphasia can result from damage within the posterior cerebral artery territory, or the posterior boundary zone. An isolation aphasia (mixed transcortical aphasia) can result when both the anterior and posterior boundary zones are damaged or when there is a "double disconnection" affecting both anterior and posterior speech areas. When the most posterior aspects of these boundary zones are damaged bilaterally, Balint's syndrome may result.

The mixed transcortical aphasias are a combination of the language output problems of the transcortical motor aphasias and the language comprehension problems of the transcortical sensory aphasias. Patients with this disorder can repeat well and, in fact, may have echolalia as their major speech output. There may be a remarkable preservation of naming ability in some patients with mixed transcortical or atypical aphasias. This observation raises the argument that such disorders should be considered disconnection syndromes rather than aphasias because naming is impaired in virtually all aphasias.

Balint's syndrome consists of three unique visuospatial deficits:

- Inability to direct gaze on command (apraxia of gaze or ocular apraxia)
- Inability to perceive simultaneously all elements of a visual scene (simultananagnosia)
- Paradoxical ability to direct hand movements better without than with visual guidance (optic ataxia)

Arteriolar-Capillary Infarctions

The penetrating arterioles have no anastomoses between them. As a consequence, each arteriolar capillary is isolated. Small infarctions within those vessels can result in microinfarctions, which until the advent of MRI, often went undetected. Now, one can see the results of chronic ischemia or abnormalities in the capillaries as leukoaraiotic changes or lacunar infarctions. The distinction between these two syndromes is traditional but may be artificial. Lacunar infarctions are due to the loss of an arteriole and result in spherical infarctions no larger than 1 to 2 cm. In leukoaraiosis, the extreme example being Binswanger's disease (see later), diffuse damage occurs deep within the substance of the brain near the ventricles. Lacunar infarcts traditionally involve gray matter in the basal ganglia or brain stem, whereas leukoaraiosis affects the white matter. Most of the lacunar syndromes (e.g., pure motor hemiparesis, pure sensory stroke, clumsy-hand dysarthria) are not further discussed in this chapter. Leukoaraiosis is discussed later in the context of the vascular dementias.

DIAGNOSTIC EVALUATION

The diagnostic evaluation of patients for vascular dementia includes an appropriate history to determine whether there is a progressive, acquired cognitive impairment fulfilling the diagnosis of dementia, and vascular risk factors such as hypertension. The Hachinski ischemic score (HIS), or the modified HIS, should be considered as a gauge to the probability that the dementia results from CVD (Table 23-1). The diagnostic evaluation to determine the source of the problem causing vascular dementia includes a full cardiac evaluation of the heart, the extracranial arteries, and the intracranial vessels down to and including the capillary circulation. Acceptable studies would include cardiovascular examination with palpation and auscultation, inspection of the retinal vascular system for disease of the arterioles, a search for cardiac sources for emboli, and constrictions of the outflow tracts. This would include echocardiography, duplex ultrasound of the carotid bifurcations, imaging—preferably with MRI, but with computed tomography (CT) if the former is unavailable—complete blood count, lipid profile, platelet count and differential, and studies of blood oxygenation. One must exclude the possibilities of hypothyroidism, vitamin B_{12} deficiency, toxic encephalopathy, and drug or medication reaction.

TABLE 23-1. *Hachinski ischemic score*

Symptom or sign	Points
Abrupt onset of dementia	2
Stepwise deterioration	1
Fluctuating course	2
Nocturnal confusion	1
Relative preservation of personality	1
Depression	1
Somatic complaints	1
Emotional lability	1
Hypertension	2
History of stroke	2
Focal neurologic symptoms	2
Focal neurologic signs	2
Other signs of arteriosclerosis	1

PET and SPECT techniques can demonstrate the patchy pattern of functional impairment that is characteristic of CVD. Dysfunction at sites distant from a single lesion may be identified, and the subcortical effects of microcirculatory disorders (e.g., Binswanger's disease) can also be demonstrated (4).

SYNDROMES

The discussion of vascular syndromes must begin with a definition of terms. There are many types of vascular dementia, each of which fulfills the essential criteria for a dementia syndrome. Two common definitions of dementia are used clinically. The fourth edition of the *Diagnostic and Statistical Manual of Mental Disorders* (DSM-4) provides a definition that requires impairment in memory and at least one other cognitive domain. Cummings and Benson advocate a slightly different definition of dementia that allows patients without memory impairment to be labeled as demented, but three cognitive domains must be impaired to fulfill the criteria (5). No confusional state can be present that could account for the symptoms, confusion being a reflection of grossly disturbed attention. The description and classification of the vascular dementia syndromes may be based on analyses of cognitive domains, or vascular mechanisms. Most often, mechanistic classification terms are preferred (e.g., multiinfarct dementia [MID], lacunar state, and hypoxic-ischemic encephalopathy), and cognitive domains are analyzed to verify the required number of deficits.

Cognitive Domains Affected by Cerebrovascular Disease

All major spheres (or domains) of higher mental ability may be individually affected by CVD: attention, language, memory, visuospatial ability, and executive ability. These domains of mental functioning represent "cognitive" abilities, whereas mood or affective disorder, personality change, hallucinations, delusions, and anxiety are examples of psychiatric alterations. CVD may produce circumscribed cognitive and psychiatric behavioral alterations.

The major behavioral syndromes related to CVD are listed in Table 23-2 and are discussed in the following sections.

TABLE 23-2. *Neurobehavioral syndromes of significance associated with cerebrovascular disease*

Categorized by arterial supply	Categorized by symptom or sign
Anterior choroidal artery	Attentional disturbances
Thalamic/subcortical aphasia	Acute confusion
Acute pseudobulbar palsy, mutism, abulia, inappropriate laughter	Inattention
Anterior cerebral artery	Neglect
Callosal disconnection syndrome	Motor neglect
Akinetic mutism	Anosognosia
Transcortical motor aphasia	Language disturbances
Middle cerebral artery	Dysarthria
All aphasia types	Dysfluency
All aprosodia types	Comprehension defect
Personality change	Impaired repetition
Alexia with agraphia	Anomia
Gerstmann's syndrome	Reading/writing disturbance
Sympathetic and parietal apraxias	Memory disturbances
Neglect and inattention	Transient global amnesia
Pure word deafness	Procedural memory impairment
Basilar artery	Category-specific amnesias
Locked-in syndrome	Visuospatial dysfunction
Peduncular hallucinosis	Executive dysfunction
Amnesia	Emotional/affective disturbances
Posterior cerebral artery	Pseudobulbar affect
Achromatopsia	Aprosodia
Prosopagnosia	Mania
Pure alexia	Depression
Anton's syndrome	Anxiety
Boundary zone (watershed)	Hallucinations/delusions
Transcortical aphasias	Personality changes
Balint's syndrome	Vascular dementia syndromes
Arteriolar-capillary infarction	Multiinfarct dementia
Leukoaraiosis	Binswanger's disease
Vascular dementia syndromes	Angular gyrus syndrome
	Amnesia

Attentional Disturbances

Immediate Attention

Immediate attention, or attention span, is a basic cognitive process that influences many brain functions. Attention may be disturbed in the acute period after injury from CVD. Mesulam and colleagues noted impaired attention (i.e., "acute confusional state") due to right middle cerebral artery infarction, and a similar syndrome may follow left-hemisphere stroke (6). Even when immediate attention (i.e., forward digit span, or word span) is intact, the ability to direct, shift, and sustain attention can be impaired. Strokes can impair any one of these aspects of attention, without affecting other attentional processes. Attention can be impaired at the input, processing, or output stages of cognition.

Inattention

Inattention is a condition in which a patient fails to report or respond to stimuli presented to one side of his or her body. Visual, tactile, and auditory stimuli are most commonly tested. The involvement of several modalities, the improvement in performance when the patient is asked to attend to the defective side, and "extinction" to double simultaneous stimulation (failure to acknowledge two simultaneous stimuli, although both stimuli, presented individually, are correctly acknowledged) are all suggestive of inattention. Inattention is a partial neglect syndrome in that the patient can be aided to recognize the affected side. Attention to perceptual signals or inputs is disrupted in inattention.

Neglect

Neglect denies relatively normal percepts access to consciousness and is characterized by defective attention to stimuli originating in the spatial environment contralateral to a brain lesion (hemispace). As with inattention, neglect should not be attributable to a primary sensory or perceptual impairment, although minor abnormality is common. Denial of clinical deficits (asomatognosia, or anosognosia) is relatively common and may affect half of the body. It may be impossible to verify intact sensory ability on the neglected side. Clinically, there appears to be a continuum from mild hemi-inattention to severe hemineglect and hemianosognosia (hemisomatognosia).

Motor Neglect

Motor neglect is also called inattention or intentional neglect, defined as the impaired ability to initiate a movement that cannot be attributed to paresis or abnormalities of tone. It is distinguished from apraxia by the normal performance seen when the patient finally initiates the maneuver. It is distinguished from bradykinesia by the relatively normal speed of movement, once initiated. Thus, it is best conceptualized as a subtle form of motor initiation disturbance, with psychomotor retardation and akinesia being extreme forms. Motor neglect can occur as an isolated manifestation of stroke.

Thus, input, processing, and output "attentional" processes may all be selectively disrupted by CVD, but combined syndromes are common. Heilman and co-workers suggested a complex circuit for attention involving the mesencephalic reticular formation, several nonspecific nuclei of the thalamus, portions of the inferior parietal lobule, the frontal lobe, and cingulate gyrus (7). Mesulam recently proposed a cortical network theory for directed attention, language, and memory (8). Dysfunction of the network for attention affects many brain areas, and this may explain the tendency for attentional disturbances to produce impairments in many cognitive and noncognitive functions.

Anosognosia

When attentional impairments eliminate awareness of one's own impairments (e.g., hemiparesis), the hemineglect involves some degree of denial or *anosognosia,* the lack of knowledge of deficits. Somatagnosia is not uncommon and may be detected in patients with hemineglect who deny ownership of the paralyzed limb. Patients may develop anosognosia for blindness (Anton's syndrome), aphasia, hemichorea-hemiballism, prosopagnosia, and apraxia. Thus, lesions in a multitude of different brain areas may produce anosognosia, but the right hemisphere is most often damaged.

Language Disturbances

Speech and Language

It is clinically convenient to conceptualize communication disorders as speech or language problems. Speech disorders consist of defective phonation (sound or "phoneme" generation) or defective articulation (the ability to modify sounds using the tongue, palate, lips, and respiratory musculature). For example, in the syndrome of clumsy-hand dysarthria, oral, facial, and lingual articulatory functions are disrupted, usually by infarction of the basis pontis. Medullary infarcts may selectively impair phonation, when lower cranial nerves are affected. Thus, speech impairments occur in the presence of intact language ability or interfere with language evaluation.

Language is acquired communication ability based on written symbols (graphemes), auditory-verbal symbols (phonemes), or gestures (i.e., as in sign languages) that are combined (e.g., through grammar and syntax) to convey meaning. Aphasias are acquired behavioral syndromes characterized by specific patterns of language abnormality. The aphasias are classified based on the pattern of impairments in naming, repetition, comprehension, and fluency. Reading, writing, pragmatics (the normal give and take of dialogue), discourse (the overall plan or story line), and prosody (the ability to understand and express emotion by alterations in the melody or intonation of speech) may also be impaired in the aphasias (9).

Each language function can be selectively affected by CVD. Table 23-3 lists examples of major language functions along with reported instances of selective impairment due to stroke. Identifying the components of language disorder provides valuable clues to lesion localization.

Fluency

Fluency is a term used to describe the ease of language production. People who hesitate when speaking, truncate sentences, produce language output slowly (usually less than 50 words per minute), exhibit many grammatical errors, and seem to struggle to produce each word are described as nonfluent. Fluent speech flows easily and effortlessly such that many words are produced (e.g., often more than 200 words per minute). Fluent language disorders are far more likely to arise from damage posterior to the central sulcus (or inferior to the sylvian fissure) than from frontal lobe damage, which typically produces nonfluent aphasias.

Comprehension

Comprehension deficits are more likely to result from posterior temporoparietal brain damage than from frontal lobe damage, al-

TABLE 23-3. *Language disturbances and focal brain damage*

Function	Condition	Damaged sites
Naming	Anomia, proper names only	Left temporal lobe
Repetition	Conduction aphasia	Left perisylvian zone
Comprehension	Transcortical sensory aphasia	Left temporal, parietal, occipital lobes; thalamus
Fluency	Transcortical motor aphasia	Left frontal lobe, thalamus
Reading	Alexia	Left occipital lobe
Writing	Agraphia	Left frontal lobe
Prosody	Aprosodia	Right hemisphere
Gesture	Pantomime agnosia	Left occipital

though frontal lobe injury can impair comprehension as well. Simple commands and yes–no questions test basic comprehension, but more complex commands and questions are necessary to identify minor abnormalities of comprehension.

Repetition

The ability to repeat words, phrases, or sentences is mediated by cortical regions surrounding the sylvian fissure, and damage almost anywhere within this large zone may impair repetition ability. Outside this perisylvian zone, cerebral damage usually spares repetition ability.

Naming

Naming requires neural structures that retrieve meaning and mechanisms to translate meaning into "motor programs" that drive the speech apparatus for communication. Anterior and posterior sites of damage may impair naming ability; some degree of anomia is almost always present in aphasia, with rare exceptions in some cases of transcortical or atypical aphasia.

Fig. 23-3 depicts a simplified classification scheme for the aphasias (atypical, jargon, and subcortical aphasias are not included). By testing naming, fluency, repetition, and comprehension, excellent localizing information is gained during the neurologic examination.

Memory Disturbances

Memory impairments can result from brain damage following hypoxic-ischemic injury (e.g., cardiac arrest), hemorrhage, or infarction (e.g., bilateral posterior cerebral artery occlusion). Transient global amnesia may follow a similar process (i.e., a transient ischemic attack with true amnesia as its hallmark), although other etiologic possibilities, such as seizures, have not been excluded. Selective memory problems may result from stroke and may be limited in duration or limited to specific types of information. Detection of memory impairments is enhanced by attention to each remarkable facet of this complex cognitive domain.

Procedural memory is stored knowledge of motor acts or procedures, such as riding a bicycle or walking. Focal strokes can disrupt procedural memory mechanisms, leading to impairment in ability to learn new motor skills. Defective procedural memory is an "anterograde" problem, demonstrated by failure to acquire skill at a normal rate, whereas

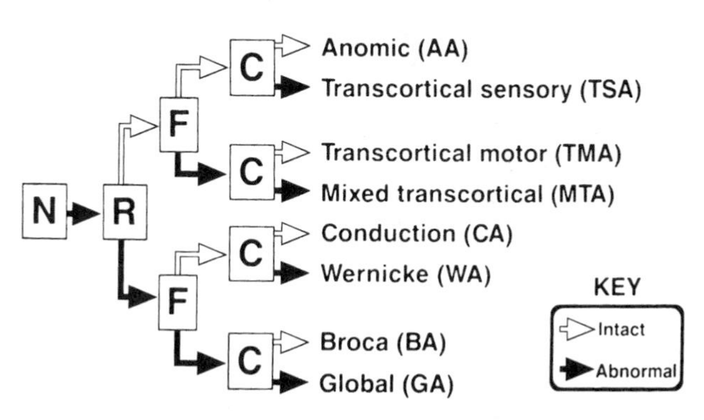

FIG. 23-3. CT of a 70-year-old man who was admitted for slight right-sided weakness and numbness. He was noted on exam to have a right homonymous hemianopsia, hemi-inattention to tactile stimuli on the right (despite only minimally impaired sensory function), and decreased exploratory head and eye movements into the right hemispace. In addition, he could not read but was able to write. A complete left posterior cerebral artery territory stroke prevented linguistic interpretation of visual material perceived by the right hemisphere.

apraxia reflects inability to demonstrate motor skills that have already been learned. It is tempting to speculate that retrograde procedural amnesia is the same thing as ideational apraxia. Ideomotor apraxia would then be a failure to access procedural memories (motor programs) on the basis of auditory or visual commands.

A straightforward approach that tests learning, recall, and recognition of words and pictures is sufficient to identify most patients with recent memory impairment. Typically, the patient is read a list of 8 to 10 words, three or four times. After each presentation, the patient attempts to recall all the words, and the examiner keeps track of his or her responses. Normally, people learn 9 or 10 words after four tries. About 5 to 10 minutes later, the patient is again asked to provide the words. If recall is impaired (i.e., less than 7 or 8 words were recalled), recognition memory should be tested. One way to do this is simply to provide a category clue for each word. Multiple-choice lists should be presented if the category clue does not lead to a correct response. Alternatively, a long list of words, including those on the original list of words and an equal or greater number of foils, can be read to the patient, who must provide a yes or no response indicating recognition of each word. A similar procedure may be used for visuospatial memory. Patients are shown a series of figures, then asked to draw them from memory. A delayed picture recall is also useful, as is having the patient circle one of several figures in a multiple-choice array to see whether he or she can recognize previously drawn pictures that could not be spontaneously recalled. Nonverbal memory can also be tested by hiding objects in the room while the patient observes. Later, the patient is asked to state the names of the objects and where each has been hidden.

Long-term memory testing requires knowledge of several facts about the patient's past. Birth date, schools attended, years of military service, family members, and personal data (e.g., Social Security number, phone number, address) are convenient examples of such facts that may be readily verified. Patients are also asked to name the current president and recent presidents. Most normal adults can name at least the last four to five presidents. These simple bedside memory tests identify most patients with clinically significant memory disturbances.

Perceptual Deficits

There is a continuum of cognitive function between sensation and memory. Sensory impressions blend into percepts. Bits and pieces of a single object may be perceived separately (Balint's syndrome), indicating that some "assembly" or synthesis of individual percepts is required for normal perception. The failure of prosopagnosic patients to access knowledge relevant to correctly perceived faces indicates that access to long-term memories is also necessary for recognition of percepts. This also suggests that recognition requires some sort of emotional or covert recognition that may be either impaired or spared in prosopagnosia.

Visuospatial Dysfunction

Visuospatial abilities encompass a large number of complex brain functions, many of which may be disturbed by CVD. Visuomotor abilities are those skills related to eye–hand coordination, praxis, and constructional ability. Visuoverbal skills relate to language tasks such as reading, or naming or describing things perceived visually.

Naming and reading are typically assessed as part of the language evaluation. Visuomotor ability is continually probed by having the patient copy simple or complex two-dimensional figures. Memory for the drawings can be checked by asking the subject to draw the figures from memory after a 5- to 30-minute delay. Recognition is checked by slowing the subject clusters of three to five similar drawings and asking the patient to identify the one that he or she drew previously. These simple bedside tests identify many patients with visuospatial deficits.

Frontal-Subcortical (Executive) Disturbances

Complex motor behaviors require intact strength, coordination, and praxis. When planning, sequencing, timing, and other integrative functions are required for complex behavior, executive ability is required. Because subcortical-cortical circuits interconnecting the frontal lobe, thalamus, and basal ganglia are probably involved with some of these higher abilities, the term *frontal-subcortical functions* seems an appropriate synonym for executive functions.

Frontal-subcortical circuit impairments may result from CVD. An excellent example of a highly integrated executive ability is the ability to play a musical instrument. Musical execution and appreciation require many frontal-subcortical circuits that play important roles in normal cognition. Neurobehavioral tests, such as the serial hand sequences test of Luria, the multiple loops, the alternating-reciprocal programs, and the go–no-go tasks are used to explore frontal-executive skills at the bedside. The ability to sing, reproduce a rhythmic series of finger taps, and speak with normal melody and prosody are related frontal-executive skills that likely have a right frontal lobe localization.

Psychiatric Types of Behavioral Alteration in Cerebrovascular Disease

Emotional or Affective Disturbances

Pseudobulbar Affect

Mood is an internal experience of emotion that usually corresponds to outward appearance, or affect. Occasionally, CVD produces dissociations between mood and affect so that appearances do not correspond to internal emotional experience. When bilateral brain damage is present, particularly in the frontal operculum, corticobulbar tracts, or basal ganglia, pseudobulbar palsy may result. This disorder is characterized by dysarthria, brisk jaw jerk, and stereotypical emotional facial expressions (e.g., crying, smiling) that are not accompanied by an appropriate degree of mood change.

The right hemisphere may be organized for emotional aspects of language, and its organization partially parallels that of the left hemisphere. Patients with right frontal lobe damage may be unable to modify language output (or facial expression or gesture) to reflect deeply felt emotions. Comprehension of the emotional qualities of speech (e.g., an angry voice, a sad voice) may be selectively impaired by right posterior temporoparietal damage. There may be an emotional language disorder corresponding to each major aphasia type. Emotional comprehension, expression, and affect all appear to be dissociable in some patients with CVD.

Depression

Left frontal lobe damage predisposes to major depressive disorder, with higher risks as the lesion approaches the frontal pole. Right-hemispheric strokes can also predispose to dysthymic mood, or minor depression, and this has been most often found in association with posterior damage. Subcortical strokes, such as within the left thalamus and caudate, can also produce depression. The degree of functional impairment does not predict risk for depressed mood, but depressed mood appears to impair functional recovery.

Anxiety

Anxiety is a feeling of worry or concern, usually without obvious cause. It is differentiated from phobic disorders because the latter result from identifiable precipitants (e.g., snakes, closed spaces). Anxiety may be a direct consequence of a focal brain injury, often accompanies other mood disturbances such as depression, and usually occurs when there is cortical infarction rather than isolated subcortical damage. Anxiety is relatively common among patients with vascular dementia, affecting 30% to 40% of patients. Orbital frontal lobe structures are likely anatomic sites where metabolic abnormalities occur in obsessive-compulsive disorder, and damage in this area may cause some cases of organic anxiety or obsessive-compulsive behavior.

Hallucinations and Delusions

Hallucinations are sensoriperceptual experiences that do not result from tactile, visual, olfactory, gustatory, or auditory stimuli. Illusions are partly based on real sensory impressions, but these stimuli evoke an aberrant perception or interpretation (e.g., one sees a snake when looking at an electrical cord). Delusions are false beliefs that are qualitatively or quantitatively beyond the realm of accepted culturally distinctive religious or metaphysical beliefs. Delusions may or may not be associated with perceptual abnormalities, such as illusions or hallucinations. Delusions are the hallmark of psychosis; hallucinations and illusions may occur without psychosis, as in the visual hallucinations associated with classic migraine.

Visual hallucinations may follow occipital, parietal, or temporal lobe infarctions. Although right-sided temporal, parietal, and occipital strokes have been described in association with visual hallucinations, the site of damage producing hallucinations may be left sided or subcortical. An accumulation of lesions, as in MID, leads to increasing risk for organic hallucinosis due to CVD. Psychosis often accompanies the hallucinations in these patients.

Personality Changes

Personality reflects deeply ingrained, chronic qualities of behavior, some of which may be genetically determined. Many facets of cognitive and behavioral function contribute to personality, perhaps because the term subsumes so many aspects of behavior. Cerebrovascular damage can dramatically alter personality.

Vascular Dementia Syndromes

Multiple infarcts scattered through a variety of locations within the central nervous system, each presumably with a definite effect on brain function, can combine into a syndrome of deficits that exceeds that of individual infarcts (e.g., MID). A crucial exercise in making this diagnosis is identifying the focal deficits that result from the scattered infarcts. Enhanced sensitivity to focal deficits can be achieved by increasing the sophistication of the neurologic examination to include the neurobehavioral examination of cognitive and behavioral functions, as briefly described earlier. Certain clinical clues may also suggest an MID diagnosis.

Among the generally accepted characteristics of MID are its sudden onset (in contrast to the slow decline that is characteristic of Alzheimer's disease) and often stepwise or fluctuating course. Patients with MID often have preserved insight into their deficits, whereas those with Alzheimer's disease generally do not. As a rule, there is a history of one or more strokes in the past, and because of the male propensity to cerebral infarction, the disease tends to be a masculine disorder.

Multiple infarcts are not the only cause of vascular dementia. On occasion, extracranial arterial disease, particularly in the carotid and vertebral arteries, may constrict cerebral blood flow and perfusion pressure, and a global cerebral ischemic syndrome can result. Chronic or repeated hypotension can also lead to an hypoxic-ischemic encephalopathy. Global hypoperfusion may result from arteriolar narrowing from chronic hypertension or from hypotension, producing ischemic demyelination and lacunar infarction. When vast areas of the subcortical white matter surrounding the ventricles become damaged by this process, the patient has Binswanger's disease, a slowly progressive or stepwise form of vascular dementia that is usually severe. Binswanger's disease patients typically have a subcortical dementia syndrome (i.e., a dementia syndrome with subcortical behavioral features, such as dysarthria, mood disturbance, a retrieval deficit on memory testing, and slow movements), and hypertension or other vascular risk factors. Patients with this disorder may have a variety of neurologic signs, such as motor deficits, clumsiness, incoordination of limb movements, vertiginous episodes, gait abnormalities, convulsions, and hemianopic or quadrantanopic defects in vision. They frequently develop pseudobulbar affect, depression, apathy, and psychomotor retardation. In addition, they develop deficits in judgment with lack of insight and, occasionally, hallucinatory episodes and paranoid ideation.

MANAGEMENT

Primary and Secondary Stroke Prevention

The best way to preserve brain functioning is to prevent neuronal damage. The neurochemistry of neuronal injury and death may reveal clues to neuroprotective therapies. For example, calcium-mediated release of excitatory amino acid neurotransmitters may promote neuronal injury. Levels of aspartate and glutamate increase in response to neuronal injury, and these excitotoxins are known to activate specific receptors that mediate cell death. Because calcium transport is mediated by these receptors, their activation can result in calcium accumulation. The extra calcium potentiates lysosomal activity, further excitatory amino acid release, and neuronal injury and death. Inhibitors of *N*-methyl D-aspartate (NMDA) receptors, calcium-channel blockers, and other related strategies are being explored as potential neuroprotective therapies. The ultimate goal of this research is to develop specific drug treatments that can be administered within minutes of stroke onset to salvage or protect portions of the brain that might otherwise die.

Having suffered an initial infarction, all effective means for reducing subsequent infarction must be initiated by addressing the modifiable risk factors. Patients are counseled regarding cessation of tobacco or alcohol abuse, normalization of blood constituents (e.g., cholesterol, platelets, red blood cells) and blood pressure, and the initiation of regular exercise to keep the blood circulating properly and the heart in good condition. For those who have had an initial cerebral infarction, the daily use of aspirin in appropriate dosage is a necessary concomitant to therapy. For those who are aspirin resistant, ticlopidine adds an extra measure of risk reduction. Patients with presumed embolic infarction (e.g., due to cardiac arrhythmia), hypercoagulability, or continued strokes on aspirin or ticlopidine typically receive warfarin. Hemorrheologic agents such as pentoxifylline can be used as well.

Management of Behavioral Complications

In general, psychotropic agents are used to control the various behavioral complications of CVD. Neuroseptic agents, such as haloperidol or chlorpromazine, may be used for hallucinations and delusions. Antidepressants, such as the tricyclic agents, serotonin reuptake inhibitors (e.g., fluoxetine), stimulants (e.g., methylphenidate), or tetracyclic agents (e.g., trazodone), may sometimes be helpful for depression, psychomotor retardation, agitation, sleep disturbance, and episodic aggressiveness. Benzodiazepines (e.g., oxazepam, temazepam), lithium, and propranolol may occasionally be useful for anxiety, agitation, episodic violence, and aggressiveness. Antiepileptics are also useful in some patients with aggressiveness (e.g., carbamazepine) or mood disturbance (e.g., valproic acid).

Basic pharmacologic guidelines for managing patients with CVD should note that pharmacokinetics, pharmacodynamics, and physiologic responsiveness to the effects of psychotherapeutic drugs are likely to be different in patients with CVD than in usual psychiatric settings. Therefore, low doses should be used at first, and all increases should be monitored carefully for evidence of deleterious side effects. The minimum number of drugs necessary to control symptoms should be used. This requires clear discussion with the patient and family about the extent of control that is necessary. Drug management must coincide with consultation to caregivers, so that they consider advanced directives, alternative living arrangements for the patient, and other nonpharmacologic means of support, such as day care centers and support groups.

REFERENCES

1. Hom J. Contributions of the Halstead-Reitan battery in the neuropsychological investigation of stroke. In: Bornstein RA, Brown G, eds. *Neurobehavioral aspects of cerebrovascular disease.* New York: Oxford University Press, 1991:165–181.
2. Segarra JM, Angelo NJN. In: Benton AL, ed. *Behavioral change in cerebrovascular disease.* New York: Harper and Row, 1970:3–14.
3. Geschwind N. Disconnexion syndromes in animals and man. I. *Brain* 1965;88:237–294; II. *Brain* 1965;88:585–644.
4. Hachinski VC, Iliff LD, Zilhka E, et al. Cerebral blood flow in dementia. *Arch Neurol* 1975;32:632–637.
5. Cummings JL, Benson DF. *Dementia: a clinical approach.* 2nd ed. Boston, MA: Butterworth-Heinemann; 1993:1.
6. Mesulam MM, Waxman SG, Geschwind N, Sabin TD. Acute confusional states with right middle cerebral artery infarctions. *J Neurol Neurosurg Psychiatry* 1976; 39:84–89.
7. Heilman KM, Valenstein E. Mechanisms underlying hemispatial neglect. *Ann Neurol* 1979;5:166–170.
8. Mesulam MM. Large-scale neurocognitive networks and distributed processing for attention, language, and memory. *Ann Neurol* 1990;28:597–613.
9. Damasio AR, Geschwind N. The neural basis of language. *Annu Rev Neurosci* 1984;7:127–147.

24

White-Matter Disorders

Christopher M. Filley

OVERVIEW

Behavioral neurology has traditionally concerned itself primarily with the cerebral cortex and the many syndromes related to cortical damage. The phrase *higher cortical function* is often used to describe the behavioral neurologist's area of interest. In recent years, however, it has been repeatedly demonstrated that higher functions depend on many more structures than the cortex alone, and a broader view is required. This chapter addresses an area that has been insufficiently discussed in the examination of brain—behavior relationships: the cerebral white matter. Despite the facts that white matter constitutes a large component of cerebral volume and that it is often selectively damaged by pathology, strong evidence relating neurobehavioral dysfunction to white-matter involvement has been lacking. The advent of magnetic resonance imaging (MRI) in recent years, and its excellent demonstration of cerebral white matter, has played a major role in highlighting this area. In addition, neuropsychological studies have contributed significant information pertaining to the behavioral associations of cerebral white matter. As the various afflictions of cerebral white matter are reviewed, it becomes clear that neurobehavioral disorders are frequent in these conditions.

NEUROBIOLOGY OF CEREBRAL WHITE MATTER

Anatomy

White matter makes up slightly less than half of the volume of the mature human brain. Its characteristic structural feature is myelin, a complex mixture of lipids (70%) and protein (30%) that encases axons. Myelin is laid down in a concentric fashion around axons by oligodendrocytes, cells of glial origin that serve to myelinate central nervous system (CNS) axons in a manner analogous to Schwann's cells of the peripheral nervous system. The myelination of axons proceeds in such a way that leaves short segments of the axonal membrane unmyelinated. These segments, called *Ranvier's nodes,* have special relevance to the speed of impulse propagation.

Myelination of the brain is a gradual process in development that occurs primarily after neuronal differentiation is complete. All of the brain's neurons are formed by the end of the sixth fetal month, and myelination begins in the third month and continues well into postnatal life. CNS structures show great variability in terms of their myelination, with some areas maturing in utero or shortly after birth, and some not becoming fully myelinated until the second or third decade of life. There is a strong likelihood that, in general, myelination parallels functional maturity, so that cerebral areas attain their maximal capabilities only when myelination is complete. This characteristic of white matter has important neurobehavioral implications. In particular, axons in the cerebral commissures and association areas are late to myelinate, a fact that probably relates to the continuing processes of social and intellectual maturation that occur into adolescence and beyond.

In later life, white-matter volume declines to a greater extent than that of gray matter: There is an increase in the hemispheric gray matter–to–white ratio ratio from 1.1 at age 50 years to 1.5 at age 100 years (1). This deterioration may be due to changes in the chemical composition of myelin, to ischemia in the white matter, or to other factors. It is likely that the seemingly ubiquitous designation "cerebral atrophy" that is applied to brain imaging studies in the normal elderly patient may, in many cases, refer more to degeneration of white matter than to cortical cell loss.

In terms of gross anatomy, the large central core of white matter that lies between the lateral ventricles and the cortex is known as the *centrum semiovale.* Within this mass of tissue, three types of cerebral white-matter fibers can be identified:

- *Projection fibers,* which convey impulses to the cortex from distant sites or vice versa
- *Association fibers,* which interconnect regions of the same hemisphere
- *Commissural fibers,* which connect homologous regions in the two hemispheres

The most important projection fiber systems are the corticofugal bundles, which convey motor impulses to effector regions of the neuraxis, and the corticopetal bundles, which carry incoming sensory information to the primary sensory cortices. The major corticofugal tracts lie in the internal capsule, which contains upper motor neuron fibers destined for the brain stem and spinal cord. The primary corticopetal bundles are the optic radiations, which project visual information to the occipital lobes; the auditory radiations to the temporal lobes; and the thalamocortical radiations, which relay somatosensory information to the parietal lobes.

Association systems are of major importance for the interaction of neocortical areas that are crucial to higher functions. These systems include the short association fibers, also known as *arcuate* or *U fibers,* which connect adjacent cortical gyri, and five long association fiber systems. The long tracts are the arcuate (superior longitudinal) fasciculus, the superior occipitofrontal fasciculus, the inferior occipitofrontal fasciculus, the cingulum, and the uncinate fasciculus. On close inspection, it is evident that these fiber systems provide the basis for extensive communication between the frontal lobes and the temporal, parietal, and occipital lobes.

A high concentration of white matter underlies the frontal lobes. The frontal lobes are the largest and most recently evolved areas of the brain, and their connections by way of white-matter pathways to other cerebral regions enable the integration of all aspects of cerebral function that are unique to humans. The ability of the frontal lobes to connect to the rest of the brain implies that white-matter disorders can be positioned to disrupt the executive functions by which behavior is regulated.

Commissural fibers run primarily in the massive corpus callosum, a broad, dense structure of some 300 million myelinated axons that crosses the midline and connects all zones of the cortex on both sides. Smaller commissural connections are made by the anterior commissure, the hippocampal commissure, and in some brains, a bundle connecting the thalami, known as the *massa intermedia.*

Other fiber systems that deserve comment are the ascending neuronal pathways from the brain stem and basal forebrain that project to widespread areas of the cerebral hemispheres. The reticular formation of the brain stem plays a vital role in the maintenance of arousal and alertness. Of equal importance because of their participation in mood, motivation, and memory are the major neurotransmitter systems that innervate the cortex:

- Noradrenergic fibers originating in the locus ceruleus of the rostral pons
- Serotonergic tracts arising from the raphe nuclei of the brain stem
- Dopaminergic systems projecting rostrally from the midbrain ventral tegmental area
- Cholinergic fibers from the nucleus basalis of Meynert

These neuropharmacologic systems course rostrally in the deep subcortical white matter—most notably the median forebrain bundle—to reach their destinations in the cortical mantle. In addition, important cholinergic fiber systems travel within the fornix and the external capsule. These systems are all vulnerable to white-matter lesions because of the long trajectories between their sites of origin and their final destinations.

A final anatomic feature of interest is the observation that the ratio of white matter to gray matter is higher in the right than in the left hemisphere (2). The implications of this asymmetry are as yet uncertain, but it is plausible that a pathologic process that affects white matter diffusely has a disproportionate effect on the right hemisphere and its functions.

Physiology

Communication between neurons ultimately depends on synaptic transmission, but in order for the nerve impulse to reach the synapse at all, axonal conduction is required. The speed of impulse conduction is significantly influenced by the presence of myelin around the axon. Large myelinated fibers conduct impulses as much as 100 times faster than small unmyelinated axons (3). This capacity is made possible in part by Ranvier's nodes, which allow for a "jump" of the impulse from node to node in a process known as *saltatory conduction.* In contrast, conduction in unmyelinated fibers is much slower because the action potential travels in a continuous fashion over the entire length of the axon. Another advantage conferred by myelin is that axonal conduction is much more efficient in terms of energy expenditure; there is less need for ATP to restore the membrane's resting potential if ion conductance takes place only at the widely separated Ranvier's nodes.

Dysfunction of myelin, whether from pathologic states or from changes related to the developmental stage of the brain, can lead to neurobehavioral alterations that reflect a general slowing of neuronal communication. Interestingly, aged laboratory animals show a slowing of conduction velocity between the basal forebrain and the neocortex, an observation that may reflect an age-related decrement in subcortical myelin. It is conceivable that some of the neurobehavioral alterations of normal aging in humans—cognitive slowing, impaired vigilance, forgetfulness, and the like—may be due in part to changes in cerebral white matter.

Clinical assessment of cerebral white matter can be assisted by the use of in vivo neurophysiologic techniques. In recent years, the emergence of evoked potentials has allowed for the measurement of conduction time across certain well-defined CNS white-matter tracts (4). Evoked potentials involve the application of a sensory stimulus (visual, auditory, or somatosensory) in a standardized manner, and the subsequent recording of the electrical response as it occurs along the appropriate CNS pathway. By identifying and localizing areas of white-matter dysfunction in the CNS, evoked potentials have improved the detection of many CNS white-matter lesions that impair primary sensory function.

CEREBRAL WHITE-MATTER DISORDERS

In attempting to address the neurobehavioral consequences of cerebral white-matter disorders, two problems immediately arise. First, the identification of a cerebral white-matter disorder is often difficult. Many illnesses known to damage hemispheric white matter, such as Binswanger's disease, also have lesser effects on other structures such as the basal ganglia. Even multiple sclerosis, the classic CNS demyelinative disease, can damage the cortex because there is some myelin within the cortical gray matter. Second, clinical data describing neurobehavioral impairment in white-matter diseases are often fragmentary or even completely lacking.

Cerebral white-matter disorders are defined as diseases or injuries that primarily, although perhaps not exclusively, damage the white matter of the brain. This criterion is based on valid neuropathologic information. The following section describes cerebral white-matter disorders that commonly affect adults (Table 24-1) and those that affect in-

TABLE 24-1. *Adult cerebral white-matter disorders*

Demyelinative	Multiple sclerosis
Vascular	Binswanger's disease
	Stroke
Toxic	Toluene dementia
	Alcoholic dementia
	Radiation
	Chemotherapy
Metabolic	Cobalamin deficiency
	Hypoxia
	Marchiafava-Bignami disease
	Central pontine myelinolysis
Infectious	Acquired immunodeficiency syndrome
	Progressive multifocal leukoencephalopathy
Traumatic	Traumatic brain injury
	Corpus callosotomy
Hydrocephalic	Normal pressure hydrocephalus
Neoplastic	Gliomatosis cerebri
	Solitary tumors

fants and children (Table 24-2). Some disorders are appropriately included in both categories. It is customary to classify diseases based on demyelination (myelinoclasis) as opposed to dysmyelination (the leukodystrophies), the former being acquired and the latter hereditary. Whereas this distinction is still useful neuropathologically, other types of CNS white-matter disorders are now recognized, including a variety of vascular, toxic, metabolic, infectious, traumatic, congenital, hydrocephalic, and neoplastic causes.

TABLE 24-2. *Childhood cerebral white-matter disorders*

Leukodystrophies	Metachromatic leukodystrophy
	Adrenoleukodystrophy
	Krabbe's disease
	Pelizaeus-Merzbacher disease
	Canavan's disease
	Alexander's disease
Amino acidopathies	Phenylketonuria
	Maple syrup urine disease
Congenital	Hydrocephalus
	Callosal agenesis
Vascular	Periventricular leukomalacia
Infectious	Postinfectious encephalomyelitis
	Acute hemorrhagic leukoencephalitis
Traumatic	Traumatic brain injury
Toxic	Radiation
	Chemotherapy

Dementia is the predominant neurobehavioral sequela of cerebral white-matter involvement (5). Affective disorders and psychoses are also recognized. Focal neurobehavioral syndromes are uncommon in these conditions, but a growing literature has provided documentation of amnesia, aphasia, and other discrete syndromes caused by white-matter disease.

Adult Disorders

Demyelinative Diseases

The demyelinative diseases, the most significant of which is multiple sclerosis (MS), have been known for many years to be capable of significant morbidity, but only recently have their impact on neurobehavioral function been recognized. MS is the most common demyelinative disease and the one subjected to the most detailed neurobehavioral analysis. Schilder's disease is a variant of severe MS. A clinically similar illness is the concentric sclerosis of Balo, the distinguishing feature of which is the presence of alternating bands of myelin preservation and destruction in concentric rings. The variant of MS known as *neuromyelitis optica,* or *Devic's disease,* specifically attacks spinal cord and optic nerves and typically has few neurobehavioral manifestations.

Despite the fact that Charcot did recognize an "enfeeblement of the memory" when he made his detailed observations of MS in the late nineteenth century, little appreciation of such complications was apparent until a century later (6). There is evidence that the memory problem is more related to retrieval than encoding or storage because free recall tasks are more difficult for MS patients than are recognition tasks.

Attentional function has been examined in some detail, and deficits of sustained attention, also known as *vigilance,* have been documented. These disturbances are closely allied with slowing of information processing. Impairments in attention and vigilance may play a key role in the functional limitations experienced by many patients with multiple sclerosis (MS), especially as they contribute to memory dysfunction.

Total lesion area of cerebral MS plaques on MRI is significantly correlated with cognitive impairment (Fig. 24-1). Cerebral white-matter involvement may cause cognitive impairment when a "threshold" level of demyelination is reached. Furthermore, deficits on the Wisconsin Card Sorting Test have been found in MS patients to correlate with frontal lobe demyelination, suggesting that focal lesion areas can cause specific impairments (7). It is likely, in fact, that cognitive and emotional dysfunction are frequently overlooked because even routine screening tests of mental status, such as the Mini-Mental State Examination, are insensitive to the deficits of many MS patients. The pattern of cognitive dysfunction in demyelinative disease has led to MS being variously termed a subcortical or white-matter dementia (8,9).

Emotional disorders are also common in MS. Charcot first made reference to disorders of affect, noting among other changes a "foolish laughter without cause." This phenomenon, which has come to be known as *euphoria,* has long been recognized as a particularly striking feature, but in actuality, it is a rather uncommon late sequela that occurs in severe disease. Euphoria is characterized by the striking appearance of elated unconcern in the presence of severe neurologic disability. Extensive frontal lobe demyelination has been suggested to account for this clinical feature.

More frequent are the mood disorders; depression and bipolar disorder both appear to occur with greater than expected frequency in MS. Depression would not be surprising, of course, in a patient diagnosed with a chronic and potentially disabling illness such as MS, and it is not possible as yet to exclude a reactive cause for a depressive disorder in this setting. However, considerable evidence indicates that the presence of cerebral demyelination is associated with depression.

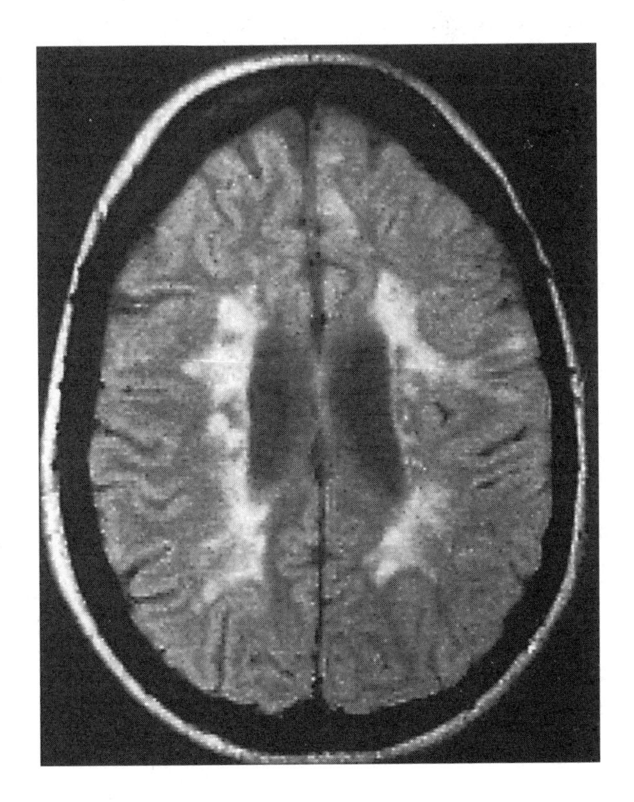

FIG. 24-1. T2-weighted axial magnetic resonance imaging (MRI) scan of a demented patient with multiple sclerosis (MS). (From ref. 9, with permission.)

Another notable syndrome is emotional dysregulation, or pseudobulbar affect, characterized by laughter and weeping without sufficient emotional stimulation and resulting from disinhibition of the facial musculature. Bilateral lesions of the corticobulbar tracts that disinhibit brain-stem motor nuclei are thought to be responsible; patients report that laughter and weeping occur with emotional provocation that would not evoke these reactions typically and that the affective display is more dramatic than that which is suggested by their actual mood at the time.

Vascular Disease

Cerebrovascular disease remains an important clinical problem from many points of view. Although classic stroke syndromes tend to reflect cortical dysfunction, white matter can also be preferentially or even uniquely damaged. A number of older reports describing cerebral white-matter disconnections are pertinent in this context, and a growing body of information is documenting both diffuse and focal syndromes as a result of vascular white-matter disease of the brain. In addition, there is growing interest in frequently noted white-matter changes in the brains of elderly people, which probably reflect more subtle vascular abnormalities as well.

Binswanger's disease is the term given to an old clinical entity that has been studied recently with renewed interest because of the availability of MRI. Originally described neuropathologically as a condition with demyelination of the cerebral white matter, usually with scattered lacunar infarctions, Biswanger's disease (BD) is also known as *subcortical arteriosclerotic encephalopathy.* The disease is thought to be caused by chronic cerebral ischemia that is most pronounced in the subcortical periventricular white matter. Binswanger's disease is a form of vascular dementia that should be distinguished from more widespread cerebral infarcts known as *multiinfarct dementia.* The exact mechanisms leading to the development of Binswanger's disease are still not entirely clear, but possibilities include cerebral atherosclerosis secondary to hypertension, repeated hypotensive episodes in patients with long-standing hypertension, and ischemia related to cerebral amyloid angiopathy. The clinical course is one of slowly progressive dementia, accompanied by variable degrees of psychosis, motor signs, and gait disorder; abrupt changes in neurologic status that imply an episode of infarction have occurred. The disease is rarely encountered in the absence of hypertension. Aphasia and movement disorder are uncommon.

The relationship between Binswanger's disease and the frequent appearance of white-matter changes on the MRI scans of elderly patients is of considerable interest. These "unidentified bright objects" or "UBOs" take the form of focal, sometimes confluent areas of white-matter change that are best seen on T2-weighted images. At first somewhat loosely termed Binswanger's disease by early observers, it was soon appreciated that many patients with these white-matter hyperintensities are cognitively intact. The term *leukoaraiosis* has also been applied to these findings, in an effort to use a purely descriptive label in the absence of a secure neurobiologic explanation. In healthy elderly subjects, there appears to be a threshold effect: the greater the white matter area involved, the greater the likelihood of cognitive impairment. Neuropsychological studies have documented subtle deficits in attention among symptom-free hypertensive men; these cognitive deficits have been associated with degree of white-matter change on MRI (10). It is increasingly plausible that hypertension and other atherosclerotic risk factors contribute to a spectrum of alterations ranging from asymptomatic white-matter change to leukoaraiosis to frank dementia in Binswanger's disease. Studies of elderly patients presenting to psychiatric hospitals have suggested a relationship between white-matter lesions and syndromes of depression and paranoid psychosis. The nature of these associations is not clear, but vascular abnormalities may play a role in the pathogenesis of the emotional disorders.

Stroke syndromes have been associated with discrete white-matter infarcts. Conduction aphasia and ideomotor apraxia occur from lesions of the left arcuate fasciculus, and apraxia of the left hand from a lesion of the anterior corpus callosum is well known (11). Pure word deafness due to a lesion in the left auditory radiation, pure alexia from a lesion of the splenium of the corpus callosum and left occipital cortex, and associative visual agnosia following bilateral lesions of the inferior occipitofrontal fasciculus are also recognized. Verbal memory deficit can occur with an infarct of the posterior limb of the left internal capsule. An infarct confined to the posterior limb of the right internal capsule can produce neglect and topographic disorientation. Spatial delirium, also known as *reduplicative paramnesia,* can result from an infarct in the subcortical white matter of the right frontal lobe. All of these syndromes have also been reported as a result of cortical damage, suggesting that white-matter structures contribute to cerebral networks subserving neurobehavioral functions.

Toxic Disorders

A small number of toxins exert their main effect on the white matter of the brain. These intoxications are incompletely understood because observations have been made primarily with MRI, and there have been few neuropathologic investigations.

Toluene is a widely used organic solvent that is now conclusively associated with cerebral white-matter damage (12). Convincing cases first came to medical attention because of the unfortunate practice of solvent vapor abuse, which involves extraordinary exposure of the CNS to daily inhalation of toluene, found in commonly available spray paints. Dementia has been noted in patients with prolonged exposure, in addition to ataxia, eye-movement disorders, and anosmia. MRI has disclosed diffuse white-matter rarefaction with cerebral atrophy. A careful postmortem study of one demented long-term abuser disclosed cerebral and cerebellar myelin loss. The severity of dementia has been shown to correlate well with the degree of cerebral white-matter change, as displayed by MRI (Fig. 24-2).

Alcohol (ethanol, ethyl alcohol) is a more common intoxicant that has a long history but still unsettled effect on the CNS. The best known syndrome consequent to the long-term

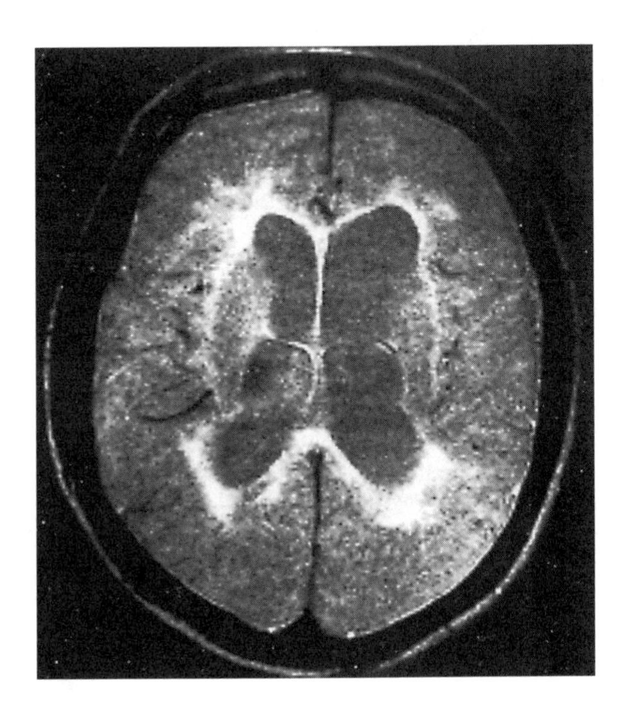

FIG. 24-2. T2-weighted axial MRI scan of a severely demented toluene abuser. (From ref. 9, with permission.)

problem of alcoholism is Wernicke-Korsakoff syndrome, a combination of acute encephalopathy and chronic amnesia due to dietary deficiency of thiamine. This disorder does not have a striking tendency to affect myelin, and in contrast, numerous gray-matter nuclei in the diencephalon and brain stem are affected. Alcoholic dementia, however, is more strongly associated with atrophy of white matter. Dementia in alcoholics is potentially reversible: Computed tomography (CT) scans can show improvement—decrease in size of lateral ventricular volume—with abstinence.

Cancer chemotherapeutic agents constitute another category of cerebral white-matter toxins. Toxic effects of chemotherapeutic agents on the brain were first recognized with CT, but MRI greatly improved the imaging of these effects because they take the form of leukoencephalopathy. The most frequently encountered drug in this regard is methotrexate, a folic acid antagonist that causes either an insidious dementia or, less often, an acute confusional state. Diffuse white-matter involvement has been documented neuropathologically. Another agent capable of causing similar leukoencephalopathy is BCNU. Drugs that are infrequently associated with these changes are arabinosyl cytosine, thiotepa, 5-fluorouracil and its derivatives, and cisplatin.

Finally, another toxic insult to the cerebral white matter is cranial irradiation. This effect is most often noted 12 years after radiation and generally requires 5000 cGy or more. A dementia syndrome resembling subcortical dementia has been described and is presumably caused by leukoencephalopathy that spares the cortex and subcortical gray matter.

Metabolic Disorders

Included in the category of metabolic disorders are four syndromes in which white-matter pathology is firmly documented. Cobalamin (vitamin B_{12}) is an essential nutrient of the nervous system that has long been associated with pernicious anemia and subacute combined degeneration of the spinal cord. In addition to myelopathy, however, cerebral white-matter changes are also pathologically verified, and dementia and psychosis have both been attributed to cobalamin deficiency. Folate, another vitamin routinely assessed in this setting, can be deficient in a manner clinically similar to that of cobalamin deficiency, and a reversible dementia has occasionally been documented. No neuropathologic data on folate deficiency are available, however, and the question of whether this condition damages white matter cannot presently be answered.

Anoxia is a well-known neurologic syndrome in its usual manifestations. These include acute encephalopathy, coma, persistent vegetative state, and death, and the pathology typically affects gray-matter areas such as cerebral cortex, hippocampus, and basal ganglia. An unusual sequel, however, is a demyelinative encephalopathy that supervenes days after an apparent recovery from an anoxic episode. The pathogenesis of postanoxic demyelination is unknown, but dementia and motor abnormalities are common if the patient survives.

Marchiafava-Bignami disease is a poorly understood disorder often associated with alcoholism, in which the corpus callosum is severely damaged. Originally described in Italian drinkers of red wine, it is now evident that Marchiafava-Bignami is not confined to Italians or even to alcoholics. Although the condition is rare, a variety of neurobehavioral features have been described, and dementia may occur.

Central pontine myelinolysis is another white-matter disorder in alcoholism, probably caused by overly rapid correction of hyponatremia. Acute confusional state due to central pontine myelinolysis has been observed and attributed to the pontine lesion interrupting the ascending neurotransmitter systems responsible for the maintenance of arousal and attention. Myelinolysis in the hemispheric white matter has also been observed.

Infectious Diseases

Many infectious diseases attack the cerebrum, but relatively few have their primary effect on the white matter. The dementia associated with acquired immunodeficiency syndrome (AIDS), known as the *AIDS—dementia complex* (ADC), is the best example of a cerebral white-matter infection, and the opportunistic infection progressive multifocal leukoencephalopathy is the other entity in this group. AIDS is now well recognized to be a disease with frequent CNS involvement. The brunt of the pathology appears to involve the white matter of the hemispheres, with multinucleated giant cells and myelin pallor; the basal ganglia, brain stem, and cortex are less affected. Concomitant with these findings, MRI scans demonstrate diffuse white-matter hyperintensity; the dementia syndrome itself has features suggestive of subcortical dementia (13). It is noteworthy, however, that movement disorder is uncommon in ADC, as in other white-matter dementias.

Progressive multifocal leukoencephalopathy is an uncommon viral infection that has come to increased prominence as a result of the AIDS epidemic but that also occurs in patients with malignancies such as leukemia and lymphoma, chronic steroid therapy, systemic lupus erythematosus, and other illnesses. The disease is caused by an infection of oligodendrocytes in cerebral white matter, originating with a member of the papovavirus group known as the *JC virus*. In contrast to the diffuse nature of the ADC, progressive multifocal leukoencephalopathy is a multifocal infection of the white matter. Neurologic deficits and neurobehavioral features are those associated with the areas of white matter involved.

Traumatic Disorders

The investigation of CNS trauma has recently shed new light on mechanisms of brain injury (see Chapter 26), and increasing evidence is accumulating to indicate that cerebral white matter is significantly involved. The cortex can clearly be damaged as well, as in the case of cortical contusion or in hypoxic-ischemic injury, but the entity of diffuse axonal injury has been gaining increasing acceptance as the most common sequel of traumatic brain injury (TBI). Clinical and experimental studies of TBI have demonstrated that diffuse axonal injury commonly occurs in nonpenetrating head injury. This widespread injury involves both myelin and axons and leaves the cortex essentially spared. Depending on the severity of the impact, the clinical picture may be one of concussion, defined as a traumatically induced alteration of the mental state, or of loss of consciousness, with prolonged coma and severe long-term disability. Neurobehavioral deficits in TBI survivors are frequent and range from inattention and memory disturbance in the postconcussion syndrome to severe dementia or persistent vegetative state after severe TBI. Imaging studies, particularly MRI, document the frequent occurrence of white-matter lesions. Associated cortical damage often results in superimposed neurobehavioral impairments.

Lesions of the corpus callosum are not often encountered as isolated neuropathologic findings, but an extensive literature exists on surgically induced corpus callosum lesions. These patients are of considerable interest in view of the still mysterious role of the commissure in the processes of interhemispheric communication. Corpus callosotomy has been performed frequently for the relief of intractable seizure disorders, and studies of neurobehavioral function after callosotomy have indicated that deficits are relatively subtle. The striking phenomenon of intermanual conflict, however, can be seen acutely after callosotomy, and apraxia, agraphia, and tactile agnosia of the left hand can also be demonstrated.

Hydrocephalic Disorders

Normal-pressure hydrocephalus (NPH) is considered a reversible dementia syndrome of adults. The cause of the illness is unexplained in many patients, although cases are known to follow TBI, meningitis, and subarachnoid hemorrhage. NPH presents with the well-known triad of dementia, incontinence, and gait disorder, and some cases respond favorably to shunting procedures. The major neuropathology is in the periventricular white matter, and the dementia syndrome is similar to other disorders of white matter in that there is a subcortical pattern to the neurobehavioral disturbance and no movement disorder.

Neoplastic Disorders

Neoplasms of the brain typically involve widespread areas of the cerebrum or posterior fossa and do not as a rule confine themselves to gray or white matter selectively. In addition, mass effects from a tumor and the often significant edema associated with malignancy add further to the diffuse nature of brain involvement. There is, however, the unusual condition of gliomatosis cerebri, in which cerebral white matter is diffusely invaded by a tumor of the glioma category. This rare neoplasm is characterized clinically by changes in personality and memory loss. Other signs and symptoms, such as hemiparesis, ataxia, papilledema, headache, and seizures, are seen less often.

Childhood Disorders

Leukodystrophies

The leukodystrophies, a group of hereditary diseases of myelin, havw been the subject of much biochemical and genetic study, but detailed knowledge of neurobehavioral features is not as available, in part because there is often a fatal outcome from these diseases in early life. Nevertheless, these disorders offer dramatic evidence that deficiencies in the development of normal mentation can be due to abnormalities in cerebral white matter.

Metachromatic leukodystrophy (MLD) is the best known and most clinically relevant member of this group because of its well-characterized biochemical pathology, relative frequency, and presentation with adult and juvenile as well as infantile neurobehavioral syndromes. MLD is an autosomal recessive disorder of central and peripheral myelin that is caused by a deficiency in the enzyme arylsulfatase A. Typically, the onset is between the first and fourth years of life, with delayed mental and motor development; the combination of dysmyelination of central white matter and dysmyelinating peripheral neuropathy results in a variable mixture of upper and lower motor neuron signs. Onset in childhood and adult life has been well described, and psychosis resembling schizophrenia is frequent. Dementia is also well recognized and often supervenes after a long period of psychosis. MRI scans show diffuse white-matter hyperintensity in the cerebrum (Fig. 24-3) that correlates well with widespread white-matter changes at autopsy.

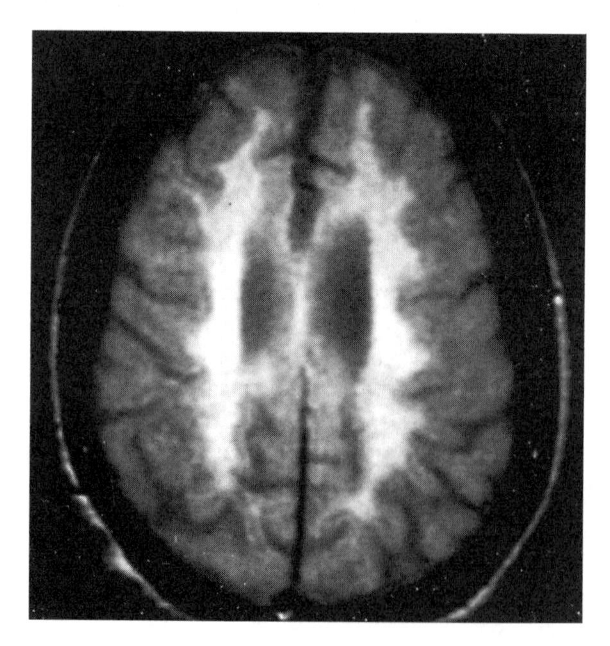

FIG. 24-3. T2-weighted axial MRI scan of a patient with metachromatic leukodystrophy who had psychosis followed later by dementia. (From Erratum. Psychosis with cerebral white matter disease. *Neuropsychiatry Neuropsychol Behav Neurol* 1993;6:142.)

Adrenoleukodystrophy is similar to MLD in many respects but is accompanied by adrenocortical insufficiency. Two forms of the disease are recognized: an X-linked form presenting in boys and rarely in adolescence or adult life and an autosomal recessive type that occurs in the neonatal period. Both varieties are characterized by an inability to degrade very-long-chain fatty acids, and the diagnosis depends on the demonstration of these fatty acids in the blood. Treatment of adrenal insufficiency does not correct the neurologic features of the illness. Dementia and psychosis are prominent in this disorder, and MRI scans demonstrate diffuse white-matter disease.

Four other white-matter diseases are usually listed as members of the leukodystrophy group. Krabbe's disease, or globoid cell leukodystrophy, affects infants most often and usually causes death within 2 years. Onset of dementia in adolescence with survival into adulthood has been observed. An autosomal-recessive disease, this condition is caused by a deficiency of galactocerebrosidase, and cerebral dysmyelination is extensive. Pelizaeus-Merzbacher disease rarely presents in adulthood, and patients may show dementia and psychosis. Islands of intact myelin appear amid abnormal white matter, giving a tigroid pattern in the hemispheres. Canavan's disease, also known as *spongy degeneration* of the white matter, usually begins in early infancy with macrocephaly and arrested development, but onset after 15 years of age has been noted. Alexander's disease typically presents with developmental delay before 2 years of age, but adult-onset cases have been reported to resemble MS.

Aminoacidopathies

The aminoacidopathies, metabolic disorders of infancy, are generally of little relevance to the behavioral neurologist and of greater interest to the pediatric neurologist and biochemist. However, phenylketonuria (PKU) and its rarer relative maple syrup urine disease are marked by striking changes in the cerebral white matter that are likely to account for the neurobehavioral sequelae of these diseases. PKU is the most common of the aminoacidurias and is inherited as an autosomal recessive trait. The fundamental biochemical lesion is a deficiency of the hepatic enzyme phenylalanine hydroxylase, which results in severe white-matter pathology in the cerebrum and mental retardation if untreated with dietary restriction of phenylalanine. Maple syrup urine disease is a less common affliction that is also inherited as an autosomal-recessive trait. Treatment with dietary restriction of branched-chain amino acids may permit normal development, presumably by preventing the white-matter changes in the cerebrum that have been documented.

Congenital Disorders

Hydrocephalus falls into the congenital category and can be a result of aqueductal stenosis, Arnold-Chiari malformation, Dandy-Walker syndrome, germinal matrix hemorrhage, or perinatal infection. Mental retardation is common in untreated cases, although language is relatively spared. Also in this category is callosal agenesis, in which the corpus callosum does not develop normally. Neurobehavioral signs are typically minimal, mainly because of the preservation of noncallosal commissures in this condition.

Vascular Disorders

An important neonatal disease of cerebral white matter is periventricular leukomalacia. This disorder, encountered in newborns who are premature and in cardiorespiratory distress, involves bilateral destruction of periventricular white matter; despite some uncertainty about its pathogenesis, the condition appears to result from circulatory insufficiency. There seems to be a selective vulnerability of the developing white matter to hypoperfusion, so that gray matter is spared; this situation parallels that of adult Binswanger's disease, in which considerable evidence suggests that the cerebral white matter is also selectively damaged by hypotension. Survivors often have significant mental retardation.

Infectious Disorders

Although postinfectious encephalomyelitis is not, strictly speaking, a white-matter infection, it nearly always follows acute infection with measles, mumps, rubella, varicella, or other viruses. The disease may also occur after vaccination. Mental changes, particularly acute confusional states, are common during the illness, and mental retardation occurs in some survivors. The cerebral white matter is significantly demyelinated, as demonstrated on MRI. The disorder is assumed to be a virus-induced immune-mediated condition, and the term *acute disseminated encephalomyelitis* is an alternate designation. Acute hemorrhagic leukoencephalitis is a more severe form of postinfectious encephalomyelitis. This fulminant illness also follows viral infection and is frequently fatal, with survivors having severe mental retardation.

Traumatic Disorders

TBI in children is common. Diffuse axonal injury is again the most consistent pathology, and deficits in cognitive and emotional behavior may be devastating. In particular, impulse control and comportment may be severely disturbed at a time in social development when they are critically needed.

Toxic Disorders

Toxic insults to the brain from radiation and chemotherapy in children are increasingly recognized. Radiation appears to have especially severe effects on neurobehavioral function through its propensity to damage cerebral white matter. Younger children are more vulnerable to both these insults, and nonlanguage skills appear to be more frequently impaired.

PROGNOSIS AND TREATMENT

A summary of the prognosis and treatment of cerebral white-matter disorders presents difficulties because of the impressive variety of pathologic processes involved. The severity of these conditions ranges from mild to catastrophic, and therapy varies greatly, depending on such factors as the type of white-matter disorder, the extent of functional disability, and the age of the patient. Some general principles of white-matter structure and function in health and illness, however, may shed light on these poorly understood questions.

It is well known that neurons in the human nervous system do not develop after birth, and it has been assumed that neurons lost from disease or injury do not regenerate. White-matter lesions, however, even if diffuse, do not primarily injure neuronal cell bodies, and therefore, the cellular sites of neurobehavioral function remain intact. Even those white-matter lesions that damage axons in addition to myelin, such as severe MS plaques, ischemic infarcts, and diffuse axonal injury, may not involve such lasting dysfunction as lesions in which the major target is the neuronal cell body.

Spontaneous Recovery

A substantial body of evidence suggests that myelin can regenerate and restore normal function if the pathology is not overwhelmingly severe. It is a common clinical observation that MS patients typically have a waxing and waning course, and indeed the subtype of relapsing-remitting MS is defined by this fluctuation. Remyelination can occur, and even in axonal segments that remain demyelinated, functional recovery may proceed as a result of an increase in the membrane density of sodium channels. Both of these mechanisms would indicate that specific interventions might be designed that could enhance natural recovery processes.

Another likely mechanism of recovery, most relevant in white-matter infarction, would be the assumption of a lost ability by another cerebral area, similar to the recovery that follows cortical lesions. An interesting report of anterior commissure hypertrophy in callosal agenesis suggests that functional compensation may occur by structural modification of white-matter tracts.

Spontaneous recovery may also occur in alcoholism. As reviewed earlier, alcohol abusers who are abstinent can often show neuropsychological improvement, and imaging studies of the brain demonstrate reversal of cerebral atrophy. There is good reason to believe that the white matter is selectively damaged in alcoholism and that improvement on CT or MRI is due to a restitution of white-matter that eventually results in clinical recovery of neurobehavioral function. Recovery from other toxic disorders, notably dementia from toluene, chemotherapeutic agents, and cranial irradiation, has not been as evident; these observations may imply heavier exposure to the toxin or a different mode of white-matter toxicity. Nevertheless, it is plausible that white matter may possess a resiliency that can be exploited for clinical purposes.

Pharmacotherapy

Pharmacologic treatment of cerebral white-matter disorders could take one of three forms: prevention, restitution of function in damaged areas, and stimulant therapy for bolstering the function of intact cerebral regions. Prevention would be most pertinent in the case of cerebrovascular disease, for which an evolving body of data is accumulating to clarify the value of antiplatelet aggregation drugs, such as aspirin and ticlopidine, and the standard anticoagulants, heparin and warfarin. With the same intent, careful control of arterial blood pressure and other risk factors has an established prophylactic effect for vascular disease. Avoidance of alcohol and solvent vapor abuse has clearly beneficial effects, and TBI is preventable with certain precautions. Restitution of function in diseased areas is exemplified by the ongoing efforts to find an effective immune treatment for MS, cobalamin treatment for vitamin B_{12} deficiency, antiviral drugs for ADC, and dietary adjustment for PKU and maple syrup urine disease. The enhancement of cerebral regions left undamaged by the disease or injury involves consideration of CNS stimulants or other agents designed to improve some form of compensation in view of the primary deficit. Methylphenidate, dextroamphetamine, amantadine, and bromocriptine have been explored in a range of CNS disorders; these drugs have effects on catecholaminergic systems that travel through white-matter tracts. Preliminary studies in MS, AIDS-related complex, and TBI have been encouraging.

Surgical Treatment

Surgical treatment in cerebral white-matter diseases has a fairly limited role. Two instances in which surgery may prove valuable are shunting procedures in ventriculoperitoneal or lumboperitoneal NPH and the bone marrow transplantation procedure being explored in MLD. Shunting for NPH is an established technique but remains controversial because criteria for patient selection are uncertain. Metachromatic leukodystrophy has been reported to be arrested by bone marrow transplantation.

Cognitive Rehabilitation

As an approach to therapy, cognitive rehabilitation has only recently been subjected to critical evaluation. It attempts to improve functional ability in neurobehavioral conditions through various combinations of cognitive retraining (with or without computers), psychotherapy, and vocational counseling. The greatest use thus far for cognitive rehabilitation has been in patients with TBI. There are many theories about why such intervention might be efficacious, but knowledge of this area is very limited. In addition, few data exist on outcome, and it is best to adopt a skeptical attitude until more compelling information appears.

ROLE OF CEREBRAL WHITE MATTER IN HUMAN BEHAVIOR

Having reviewed the neurobiology of cerebral white matter and the many disorders to which it is vulnerable, we now examine how this sizable portion of the cerebral hemispheres participates in neurobehavioral function.

Subcortical Dementia

Subcortical dementia, a concept popularized in the 1970s (14), was first described in patients with progressive supranuclear palsy and Huntington's disease and later came to be associated with the dementia syndromes of Parkinson's disease, Wilson's disease, spinocerebellar degeneration, the lacunar state, depression, and even tardive dyskinesia. A distinct profile of forgetfulness, slowness of thought processes, emotional changes, and an inability to manipulate acquired knowledge was observed in the presence of a movement disorder and without aphasia, apraxia, and agnosia. Because the subcortical dementias are typically associated with diseases of the basal ganglia, it has been argued that the only difference between subcortical and cortical dementias is the presence or absence of a movement disorder. In the past decade, several non-—Alzheimer's disease dementias (MS, Binswanger's disease, and ADC) without significant basal ganglia but with pronounced white-matter involvement, have extended the list of subcortical dementias.

Behavioral Neurology of Cerebral White Matter

White matter, of course, qualifies as another subcortical structure, but at first glance, it is not immediately apparent that it should have any significant role in higher function. Myelin, after all, may be irrelevant to the synaptic events that occur in the gray matter, such as learning, memory, and complex cognitive and emotional processes. Undoubtedly, the extraordinary activities of the cortex play a preeminent role in the elaboration of the human behavioral repertoire. White-matter pathways figure prominently in the multiple computations that produce the final output that is observable as human behavior. Neuronal function is intimately dependent on intact myelin; dramatic slowing of conduction velocity and even conduction block, seen typically in demyelinating peripheral neuropathy, are also thought to result from white-matter lesions in the CNS. Although it is too simplistic to suppose that slowed central conduction in these disorders is solely responsible for neurobehavioral impairments, it is undoubtedly the case that certain aspects of cognitive and emotional processing can be rendered less efficient by a primary disturbance of cerebral white matter.

Lesions that affect the cerebral connections between areas concerned with specific functions act to disconnect these zones from each other and cause identifiable neurobehavioral syndromes. These disconnections may be in either cortical or white-matter regions. Studies using positron emission tomography scans of deep white-matter infarction have demonstrated metabolic abnormalities in cortical regions anatomically connected but pathologically intact (15). The former concept of *diaschisis,* which refers to the remote effects of a focal cerebral lesion, is also illuminated by these considerations; cerebral dysfunction in areas other than the site of the primary pathology must surely implicate white-matter pathways.

The anatomic complexity of the cerebral white matter, as it participates in vast networks of cerebral neurons, is extraordinary. Tables 24-3 and 24-4 list the major neurobehavioral syndromes that have been described. The most frequent neurobehavioral presentation of cerebral white-matter disorder is dementia. In children, the related syndrome of mental retardation is also common. Diseases and injuries of white matter tend to involve diffuse and widespread areas of the cerebral hemispheres, and the lesions disrupt many

TABLE 24-3. *Diffuse neurobehavioral syndromes of cerebral white matter*

Syndrome	Structures involved
Dementia	Hemispheric white matter
Mental retardation	Hemispheric white matter
Depression	Ascending catecholaminergic fibers
Bipolar disorder	Ascending catecholaminergic fibers
Euphoria	Frontal lobe white matter
Psychosis	Frontal-limbic connections

TABLE 24-4. *Focal neurobehavioral syndromes of cerebral white matter*

Syndrome	Structures involved
Amnesia	Fornix
Conduction aphasia	Left arcuate fasciculus
Sympathetic apraxia	Anterior corpus callosum
Ideomotor apraxia	Left arcuate fasciculus
Pure word deafness	Left auditory radiation
Pure alexia	Splenium of corpus callosum[a]
Associative visual agnosia	Bilateral inferior occipitofrontal fasciculus

[a]Left occipital cortex is also damaged.

different networks to produce the multidimensional impairments of dementia. Involvement of frontal lobe white matter is common, and manifestations of frontal lobe dysfunction frequently contribute to the dementia syndrome. In general, the white-matter dementias resemble the subcortical dementias in that inattention, forgetfulness, emotional changes, and absence of aphasia are typical. One key distinction is that movement disorders are not prominent in the white-matter dementias.

Emotional disorders are recognized as additional complications of the cerebral white-matter disorders. Depression, bipolar disorder, euphoria, and emotional dysregulation are all well described, particularly in MS. Disruption of the ascending catecholaminergic systems from the brain stem to the cortex appears to be the most plausible explanation for depression and bipolar disorder, and bifrontal white-matter disease seems to play a role in the genesis of euphoria. Psychosis is another recognized feature of the cerebral white-matter disorders and is particularly striking in MLD. The localization of white-matter areas associated with psychosis is uncertain, but disruption of frontal-limbic connections has been postulated.

Although focal syndromes are apparent in the white-matter disorders that involve limited areas of pathology, diffuse impairment affecting arousal, attention, motivation, and mood is more common. Focal syndromes occur by virtue of discrete damage to association, commissural, or projection fibers and cause instrumental deficits such as aphasia, whereas long ascending fibers and multiple intracortical tracts are diffusely affected in a widespread fashion to result in disturbances of fundamental functions, such as dementia, mood disorder, and psychosis.

Neuropsychological Aspects

Systematic investigation of neuropsychological status in patients with cerebral white-matter disorders has confirmed that the white-matter disorders resemble the subcortical diseases closely. Memory is frequently impaired, and the deficit appears to involve retrieval more than encoding because recognition memory is normal or less impaired than retrieval. Language is typically spared, although dysarthria may be present. Visuospatial abilities have not been carefully studied but appear to be impaired. Complex cognitive tasks such as reasoning and concept formation are often affected, and perseveration can be seen. Alterations in mood, affect, and personality have been frequently noted. Clinical deficits in attention and vigilance are notable in the white-matter disorders and may well relate to the damage in connecting pathways of neural systems devoted to the maintenance of attention. The right hemisphere may be dominant for some aspects of attention, particularly directed attention to the left hemispace, and the larger amount of white-matter in the right hemisphere may be relevant in this regard.

In children, an interesting entity called the *nonverbal learning disabilities syndrome* has been presented by Rourke to characterize the neuropsychological deficits resulting from dysfunction of cerebral white matter (16). This is defined as deficits in visuospatial abilities, complex problem solving, and emotional adjustment in the presence of well-developed language skills. The syndrome has been seen in children with a variety of white-matter disorders.

SUMMARY

The task of defining the role of cerebral white matter in neurobehavioral function is an enormous one. Relatively little systematic effort has been devoted to this inquiry, reflecting a bias in favor of "higher *cortical* function." This review illustrates that the traditional lesion method of behavioral neurology, based on careful clinical observation and supplemented by detailed neuropsychological evaluation, can be equally well applied to the white-matter disorders of the brain, particularly now that MRI provides an elegant and noninvasive means of viewing the areas of interest. The insights gained from functional neuroimaging promise to be of additional utility in delineating the unique contributions of white matter. Promising information is also beginning to appear from clinical neurophysiology, and the measurement of long-latency event-related potentials may be a useful way to measure the role of white-matter tracts in neurobehavioral function. Finally, neuropharmacologists will doubtless make a contribution to the neurochemical aspects of these disorders and their treatment.

REFERENCES

1. Miller AKH, Alston RL, Corsellis JAN. Variation with age in the volumes of grey and white matter in the cerebral hemispheres of man: measurements with an image analyzer. *Neuropathol Appl Neurobiol* 1980;6:119–132.
2. Gur RC, Packer IK, Hungerbuhler JP, et al. Differences in the distribution of gray and white matter in the human cerebral hemispheres. *Science* 1980;207:1226–1228.
3. Martin JH, Jessell TM. Modality coding in the somatic sensory system. In: Kandel ER, Schwartz JH, Jessell TM. *Principles of neural science.* 3rd ed. New York: Elsevier, 1991:351.
4. Chiappa KH. Pattern shift visual, brainstem auditory, and short-latency somatosensory evoked potentials in multiple sclerosis. *Neurology* 1980;30:110–123.
5. Filley CM, Davis KA, Schmitz SP, et al. Neuropsychological performance and magnetic resonance imaging in Alzheimer's disease and normal aging. *Neuropsychiatry Neuropsychol Behav Neurol* 1989;2:81–91.
6. Charcot JM. *Lectures on the diseases of the nervous system delivered at La Salpetriere.* London, England: New Sydenham Society, 1877.
7. Arnett PA, Rao SM, Bernardin L, et al. Relationship between frontal lobe lesions and Wisconsin Card Sorting Test performance in patients with multiple sclerosis. *Neurology* 1994;44:420–425.
8. Rao SM. Neuropsychology of multiple sclerosis: a critical review. *J Clin Exp Neuropsychol* 1986:8: 503–542.
9. Filley CM, Franklin GM, Heaton RK, Rosenberg NL. White matter dementia: clinical disorders and implications. *Neuropsychiatry Neuropsychol Behav Neurol* 1989;1:239–254.
10. van Swieten JC, Geyskes GG, Derix MMA, et al. Hypertension in the elderly is associated with white matter lesions and cognitive decline. *Ann Neurol* 1991;30: 825–830.
11. Geschwind N. Disconnexion syndromes in animals and man. *Brain* 1965;88:237–294, 585–644.
12. Rosenberg NL, Kleinschmidt-DeMasters BK, Davis KA, Dreisbach JN, Hormes JT, Filley CM. Toluene abuse causes diffuse central nervous system white matter changes. *Ann Neurol* 1988;23:611–614.
13. Navia BA, Jordan BD, Price RW. The AIDS dementia complex. I. Clinical features. *Ann Neurol* 1986;19: 517–524.
14. Albert ML, Feldman RG, Willis AL. The "subcortical dementia" of progressive supranuclear palsy. *J Neurol Neurosurg Psychiatry* 1974;37:121–130.
15. Metter EJ, Mazziotta JC, Itabashi HH, Mankovich NJ, Phelps ME, Kuhl DE. Comparison of glucose metabolism, x-ray CT, and post-mortem data in a patient with multiple cerebral infarcts. *Neurology* 1985;35: 1695–1701.
16. Rourke BP. Syndrome of nonverbal learning disabilities: the final common pathway of white-matter disease/ dysfunction? *Clin Neuropsychol* 1987;1:209–234.

25

Epilepsy

Michael R. Trimble, Howard A. Ring, and Bettina Schmitz

OVERVIEW

Classification of Epilepsy and Seizures

The proposed classifications by the International League Against Epilepsy from 1981 and 1989 are based on agreements among international epileptologists and compromises between various viewpoints; they must not be regarded as definitive (1). The 1989 classification does not mention psychiatric criteria in the clinical definitions of epileptic syndromes.

The International Classification of Epileptic Seizures (ICES) from 1981 is based on improved intensive monitoring capabilities that have permitted accurate recognition of seizure symptoms and their longitudinal evolution. Presently, classification of seizures is weighted clinically and gives no clear definitions in terms of seizure origin. The principal feature of the ICES is the distinction between seizures that are generalized from the beginning and those that are partial or focal at onset and may or may not evolve to secondary generalized seizures (Table 25-1). In generalized seizures, there is initial involvement of both hemispheres. Consciousness may be impaired, and this impairment may be the initial manifestation. Motor manifestations are bilateral. The ictal electroencephalogram (EEG) patterns initially are bilateral. Spikes, spike and wave complexes, and polyspike and wave complexes are all typical. Partial seizures are those in which, in general, the first clinical and EEG changes indicate initial activation of a system of neurons limited to part of one cerebral hemisphere. The other important feature of the ICES is the separation between simple and complex partial seizures, depending on whether there is preservation or impairment of consciousness.

Classification of Syndromes

Recurrent epileptic seizures are pathognomonic for all types of epilepsies (Table 25-2). The clinical spectrum of epilepsy, however, is much more complex, and an epileptic syndrome is characterized by a cluster of signs and syndromes customarily occurring together. These include such items as type of seizure, etiology, anatomy, precipitating factors, age of onset, severity, chronicity, diurnal and circadian cycling, and sometimes prognosis.

The ICES distinguishes generalized and localization-related (focal, local, partial) epilepsies. Generalized epilepsies are syndromes characterized by generalized seizures in which there is an involvement of both hemispheres from the beginning of the seizure. Seizures in localization-related epilepsies start in a circumscribed region of the brain. They may be simple or complex focal and may progress to secondary generalized tonic-clonic seizures.

The other important classification criterion refers to etiology. The ICES distinguishes idiopathic, symptomatic, and cryptogenic epilepsies. Idiopathic means that a disease is not preceded or occasioned by another. Symptomatic epilepsies and syndromes are considered the consequence of a known or suspected disorder of the central nervous system. In cryptogenic diseases, a cause is suspected but remains obscure.

TABLE 25-1. *International classification of epileptic seizures (ILAE 1981)*

Clinical seizure type	Ictal electroencephalogram
I. FOCAL (PARTIAL, LOCAL) SEIZURES	
A. Simple partial seizures 1. With motor symptoms a. Focal motor without march b. Focal motor with march (jacksonian) c. Versive d. Postural e. Phonatory (vocalization or arrest of speech) 2. With somatosensory or special-sensorysymptoms (simple hallucinations, e.g., tingling, light flashes, buzzing) a. Somatosensory b. Visual c. Auditory d. Olfactory e. Gustatory f. Vertiginous 3. With autonomic symptoms or signs (including epigastric sensation, pallor, sweating, flushing, piloerection, pupillary dilation) 4. With psychic symptoms (disturbance of higher cortical function). These symptoms rarely occur without impairment of consciousness and are much more commonly experienced as complex partial seizures. a. Dysphasic b. Dysmnesic (e.g. déja vu) c. Cognitive (e.g., dreamy states, distortions of time sense) d. Affective (fear, anger, etc.) e. Ilusions (e.g., macropsia) f. Structured hallucinations (e.g., music, scenes)	Local contralateral discharge starting over corresponding area of cortical representation (not always recorded on the scalp)
B. Complex focal seizures (with impairment of consciousness: may sometimes begin with simple symptomatology) 1. Simple partial onset followed by impairment of consciousness a. With simple partial features (as in A. 1–4) followed by impaired consciousness b. With automatisms 2. With impairment of consciousness at onset a. With impairment of consciousness only b. With automatisms	Unilateral or frequently bilateral discharge, diffuse or focal in temporal or frontotemporal regions
C. Focal seizures evolving to secondarily generalized seizures (this may be generalized tonic-clonic, tonic, or clonic) 1. Simple partial seizures (A) evolving to generalized seizures 2. Complex partial seizures (B) evolving to generalized seizures 3. Simple focal seizures evolving to complex focal seizures evolving to generalized seizures	Above discharge becomes secondarily and rapidly generalized
II. GENERALIZED SEIZURES	
A. 1. Absence seizures a. Impairment of consciousness only b. With mild clonic components c. With atonic components d. With tonic components e. With automatisms f. With autonomic components	Usually regular and symmetric 3-Hz but may be 2- to 4-Hz spike and slow-wave complexes and may have multiple spike and slow-wave complexes. Abnormalities are bilateral.
2. Atypical absence May have: a. Changes in tone that are more pronounced than in A.1 b. Onset and/or cessation that is not abrupt	EEG more heterogeneous; may include irregular spike and slow-wave complexes, fast activity, or other paroxysmal activity. Abnormalities are bilateral but often irregular and asymmetric
B. Myoclonic seizures Myoclonic jerks (single or multiple)	Polyspike and wave, or sometimes spike and wave or sharp and slow waves
C. Clonic seizures	Fast activity (10c/s or more) and slow waves; occasional spike and wave patterns
D. Tonic seizures	Low-voltage, fast activity or a fast rhythm of 9–10c/s or more decreasing in frequency and increasing in amplitude
E. Tonic-clonic seizures	Rhythm at 10c/s or more decreasing in frequency and increasing in amplitude during tonic phase, interrupted by slow waves during the clonic phase
F. Atonic seizures	Polyspikes and wave or flattening or low-voltage fast activity

EEG, electroencephalogram.

TABLE 25-2. *International classification of epilepsies and epileptic syndromes (ILAE 1989)*

1. Localization-related (focal, local, partial) epilepsies and syndromes
   1.1. Idiopathic (with age-related onset)
      At present, the following syndromes are established, but more may be identified in the future:
      Benign childhood epilepsy with centrotemporal spikes
      Childhood epilepsy with occipital paroxysms
      Primary reading epilepsy
   1.2. Symptomatic
      Chronic progressive epilepsia partialis continua of childhood (Kozhevnikov's syndrome)
      Syndromes characterized by seizures with specific modes of precipitation
      Temporal lobe epilepsy
         With amygdala-hippocampal seizures
         With lateral temporal seizures
      Frontal lobe epilepsy
         With supplementary motor seizures
         With cingulate seizures
         With seizures of the anterior frontopolar region
         With orbitofrontal seizures
         With dorsolateral seizures
         With opercular seizures
         With seizures of the motor cortex
         Parietal lobe epilepsies
         Occipital lobe epilepsies
2. Generalized epilepsies and syndromes
   2.1. Idiopathic, with age-related onset, listed in order of age
      Benign neonatal convulsions
      Benign myoclonic epilepsy in infancy
      Childhood absence epilepsy (pyknolepsy)
      Juvenile absence epilepsy
      Juvenile myoclonic epilepsy (impuslive petit mal)
      Epilepsy with grand mal seizures (GTCS) on awakening
      Other generalized idiopathic epilepsies not defined above
      Epilepsies precipitated by specific modes of activation
   2.2. Cryptogenic or symptomatic (in order of age)
      West's syndrome (infantile spasms, Blitz-Nick-Salaam Krämpfe)
      Lennox-Gastaut syndromes
      Epilepsy with myoclonic-astatic seizures
      Epilepsy with myoclonic absences
   2.3. Symptomatic
      2.3.1 Nonspecific etiology
         Early myoclonic encephalopathy
         Early infantile epileptic encephalopathy with suppression burst
         Other symptomatic generalized epilepsies not defined above
      2.3.2 Specific syndromes
         Epileptic seizures may complicate many disease states. Under this heading are included diseases in which seizures are presenting or predominant feature.
3. Epilepsies and syndromes undetermined as to whether they are focal or generalized
   3.1. With both generalized and focal seizures
      Neonatal seizures
      Severe myoclonic epilepsy in infancy
      Epilepsy with continuous spike waves during slow-wave sleep
      Acquired epileptic aphasia (Landau-Kleffner syndrome)
      Other undetermined epilepsies not defined above
   3.2. Without unequivocal generalized or focal features

Localization-Related Epilepsies and Syndromes

Temporal lobe epilepsies are divided into those with lateral temporal seizures with auditory hallucinations, language disorders (in cases of dominant-hemisphere focus), or visual illusions and those with amygdala-hippocampal (mesiobasal limbic or rhinencephalic) seizures. The latter are characterized by simple seizure symptoms, such as rising epigastric discomfort, nausea, marked autonomic signs, and other symptoms, including borborygmi, belching, pallor, fullness of the face, flushing of the face, arrest of respiration, pupillary dilation, fear, panic, and olfactory hallucinations. Complex focal seizures often begin with a motor arrest, typically followed by oroalimentary automatisms.

Frontal lobe epilepsies are characterized by seizures of short duration, minimal or no postictal confusion, rapid secondary generalization, prominent motor manifestations that are tonic or postural, complex gestural automatisms, and frequent falling. Ictal scalp EEGs may show bilateral or multilobar discharges. Accurate localization of frontal lobe epilepsies may be difficult.

Generalized Epilepsies and Syndromes

Idiopathic generalized epilepsies are characterized by an age-related onset. In general, patients are normal between seizures. Radiologic investigations are negative. Frequently, there is an overlap of idiopathic generalized epilepsies, especially of those manifesting in later childhood and adolescence (2).

Symptomatic generalized epilepsies and syndromes usually start in infancy or early childhood. In most children, several seizure types occur. EEG discharges are less rhythmical and less synchronous than in idiopathic generalized epilepsies. There are neurologic, neuropsychological, and radiologic signs of diffuse encephalopathy. The only difference between cryptogenic and symptomatic syndromes is that in cryptogenic syndromes, the presumed cause cannot be identified.

Epilepsies and Syndromes Undetermined as to Whether They Are Focal or Generalized

There are two groups of patients in whom epilepsy cannot be classified as focal or generalized. The first group consists of patients with both generalized and focal seizures (e.g., patients with both focal seizures and absence seizures). The second group includes patients without unequivocal generalized or focal features (e.g., patients with nocturnal grand mal).

NEUROBIOLOGY OF EPILEPSY

Much recent research into the biochemistry of epilepsy has focused on investigation of the effects of inhibitory and excitatory amino acid neurotransmitters on the excitability of neurons, together with the role of calcium in this process (3).

The major inhibitory neurotransmitter in the brain is γ-aminobutyric acid (GABA), a monocarboxylic amino acid. It is distributed throughout the central nervous system and is a potent inhibitor of neuronal discharge. It may be used by as many as 40% of neurons. Two classes of GABA receptors have been characterized, termed $GABA_A$ and $GABA_B$ and distinguished on the basis of their pharmacologic characteristics. The $GABA_A$ receptor is thought to be responsible for mediating CNS neuronal inhibition. The role of GABA-ergic mechanisms in epilepsy has been the focus of much pharmacologic research into the treatment of the condition. In animal models of epilepsy, enhancing GABA-ergic neurotransmission has been shown to protect against seizures of various origins.

Just as a deficit in inhibitory neurotransmitter activity has been implicated in generating seizure activity, so has an excess of excitatory activity. The excitatory amino acids glutamate and aspartate mediate most excitatory transmission in the vertebrate nervous system (4). Modulation of excitatory neurotransmission provides another avenue for pharmacologic manipulation of seizure activity. Indeed, a new anticonvulsant drug, lamotrigine, most likely acts by inhibiting the release of excitatory amino acids.

Experimental models of epilepsy have demonstrated that calcium has an important role both in the induction of epilepsy and in the generation of seizures. The NMDA subtype of glutamate receptors constitutes an important route for influx of extracellular Ca^{2+} and functions as a receptor-operated channel. Studies in experimental models of epilepsy have demonstrated that there is a fall in extracellular Ca^{2+} at the time of seizure onset, probably because of increases in postsynaptic calcium uptake, mediated through NMDA-type glutamate receptors and voltage-operated calcium channels.

EPIDEMIOLOGY OF EPILEPSY

Prevalence and Incidence

Calculated incidence rates of epilepsy range between 20 and 70 per 100,000 population per year. The incidence is age dependent, with a maximum in early childhood and lowest rates in early adulthood. Incidence figures rise again in older age groups, probably because of the higher prevalence of cerebrovascular disease. The overall risk for epilepsy is slightly higher in males than in females. The point prevalence of active epilepsy is about 35 per 1000. The cumulative lifetime prevalence has been estimated to be 3.5%.

Relative Frequency of Epileptic Syndromes

In population-based studies, epilepsies with complex focal and focal seizures secondarily evolving to tonic-clonic seizures are most frequent, occurring in 69% of all patients. This is followed by primary generalized seizures in 30% and absence or myoclonic seizures in less than 5%.

PROGNOSIS OF EPILEPSY

Recurrence Risk After a First Seizure

Recurrence figures in the literature range between 27% and 71%, depending on inclusion criteria, duration of follow-up, and whether patients are treated after a first seizure. The recurrence risk calculated from a population-based study in Rochester was 67% (5).

Prognostic Studies

Recent prospective and population-based studies have challenged the older views that epilepsy is likely to be a chronic disease in as many as 80% of cases. In a population-based survey in Rochester, 20 years after the initial diagnosis of epilepsy, 70% of patients were in 5-year remission, and 50% of patients had successfully withdrawn medication (6).

About 5% to 10% of all epilepsies eventually include intractable seizures, despite optimal medication, most of them occurring in patients with complex partial seizures. Despite the overall favorable prognosis of epilepsy and the good response to treatment, the mortality of patients with epilepsy is 2.3-fold higher than in the general population, being 3.8-fold higher in the first years of the illness. Sudden unexpected death has been estimated to occur in about 1 of 525 epileptic patients.

The prognosis of epilepsy largely depends on the syndromatic diagnosis. Idiopathic localization-related epilepsies, such as rolandic epilepsy, have an excellent prognosis in all respects. Prognosis in terms of seizure remission, social adjustment, and life expectancy, on the other hand, is extremely poor in symptomatic generalized epilepsies, such as West's syndrome, and in progressive myoclonus epilepsies.

DIAGNOSIS OF EPILEPSY

Clinical Diagnosis and Differential Diagnosis

Epilepsy is a clinical diagnosis, defined by recurrent epileptic seizures. Most important for accurate syndromatic classification and optimal application of diagnostic techniques is the clinical interview. This should cover seizure-related information, such as subjective and objective ictal symptoms, precipitation and frequency of seizures, history of seizures in first-degree relatives, and data relevant for etiology, such as complications during pregnancy and birth, early psychomotor development, history of brain injuries, and other disorders of the central nervous system. Other important information that should be obtained include doses, side effects, and efficacy of previous medical or nonmedical treatment, evidence of psychiatric complications in the past, and psychosocial parameters, in-

TABLE 25-3. *Differential diagnosis of epilepsy*

Neurologic disorders
Transient ischemic attacks
Migraine
Paroxysmal dysfunction in multiple sclerosis
Transient global amnesia
Movement disorders (hyperexplexia, tics, myoclonus, dystonia, paroxysmal choreoathetosis)
Drop attacks due to impaired CSF dynamics
Sleep disorders
Physiologic myoclonus
Pavor nocturnus
Somnabulism
Enuresis
Periodic movements in sleep
Sleep-talking
Bruxism
Nightmares
Sleep apnea
Narcolepsy (cataplexy, automatic behavior, sleep attacks, hallucinations)
Psychiatric disorders
Pseudoseizures
Anxiety attacks
Hyperventilation syndrome
Dissociative states, fugues
Episodic dyscontrol, rage attacks
Medical disorders
Cardiac arrhythmias
Syncope (cardiac, orthostatic, reflex)
Metabolic disorders (e.g., hypoglycemia)
Hypertensive crisis
Endocrine disorders (e.g., pheochromocytoma)

CSF, cerebrospinal fluid.

cluding educational and professional status, social independence, and psychosexual history. The neurologic examination may reveal signs of localized or diffuse brain damage.

In most cases, a clinical interview and neurologic examination are sufficient to distinguish between epilepsy and its wide spectrum of differential diagnoses (Table 25-3). There is, however, a substantial problem with pseudoseizures (see later discussion).

Electroencephalography

The interictal surface EEG is still the most important method in the diagnosis and assessment of all types of epilepsy. A routine EEG is recorded over 30 minutes during a relaxed condition, including photic stimulation procedures and 5 minutes of hyperventilation. Paroxysmal discharges strongly suggestive of epilepsy are spikes, spike waves, and sharp waves. These epileptiform patterns are, however, not epilepsy specific. They may be observed in patients suffering from nonepileptic neurologic diseases and even in a small proportion of healthy subjects.

The sensitivity of the routine EEG is limited by restrictions of spatial and temporal sampling. About 50% of patients with epilepsy do not show paroxysmal epileptiform discharges on a single EEG recording. The temporal sensitivity can be increased by repeating the EEG or by carrying out long-term recordings with mobile EEGs. Paroxysmal discharges may furthermore be brought out by performing an EEG after sleep deprivation while the subject is asleep. Simultaneous video EEG recordings of seizures are useful for differentiating between different types of epileptic and nonepileptic seizures. Ictal EEGs are also required for exact localization of the epileptogenic focus when epilepsy surgery is considered.

Surface EEGs record only a portion of the underlying brain activity. Discharges that are restricted to deep structures or to small cortical regions may not be detected. The spatial resolution of the EEG can be improved by special electrode placements, such as pharyngeal and sphenoidal electrodes.

Invasive EEG methods with chronic intracranial electrode placement are necessary for complex analysis in cases with discordant or multifocal results of the ictal surface EEG and imaging techniques. These include foramen ovale electrodes positioned in the subdural space along the amygdala-hippocampal formation, epidural and subdural strip electrodes, and grids to study larger brain areas. Stereotactic depth electrodes provide excellent sensitivity to detect small areas of potentially epileptogenic tissue.

Magnetoencephalography

Multichannel magnetoencephalography (MEG) has recently been introduced in the presurgical assessment as a supplement to EEG. The electric activity, which can be measured by EEG, produces a magnetic field perpendicular to the electric flow. This magnetic signal can be measured by MEG. In contrast to EEG, MEG is not influenced by intervening tissues with the advantage of noninvasive localization of deep electric sources. Disadvantages include high costs and the susceptibility to movement artifacts, making it extremely difficult to do ictal studies (7).

Structural Imaging

Imaging studies should always be performed when a symptomatic etiology is suspected. Cranial computed tomography (CCT) is a quick, easy, and inexpensive technique. Except for a few pathologies, such as calcifications, magnetic resonance imaging (MRI) is superior to CCT in terms of sensitivity and specificity in detecting epilepsy-related lesions, such as malformations, gliosis, and tumors. With optimized MRI technique, including T2-weighted images, inverse recovery sequences, coronal images perpendicular to the hippocampus, and thin sections, the sensitivity in depicting mesial temporal sclerosis reaches 90%.

MR spectroscopy (MRS) is a noninvasive method of measuring chemicals in the body. MRS does not produce images but instead generates numeric values for chemicals. With phosphate spectroscopy, it is possible to study energy metabolism in relation to seizure activity. Proton MRS measures neuronal density, which has been found to be significantly decreased in the mesial temporal lobe of patients with mesial temporal sclerosis (8).

Functional Imaging

Single photon emission computed tomography (SPECT) in epilepsy has mainly been confined to the imaging of cerebral blood flow in focal epilepsy. The tracer most widely used is ^{99m}Tc-HMPAO. Interictally, there is localized hypoperfusion in an area extending beyond the epileptogenic region. Figures on the sensitivity of interictal focus detection of SPECT in the literature range from 40% to 80%. In recent years, there have been major technical developments in instrumentation. Using brain-dedicated multiheaded camera systems, the sensitivity of SPECT is comparable to that of ^{18}F-fluordeoxyglucose (FDG) positron emission tomography (PET). ^{99m}Tc-HMPAO is distributed within a few minutes after injection in the brain, where it remains fixed for about 2 hours. Postictal and ictal SPECT techniques are more sensitive than interictal SPECT and typically show hyperperfusion ipsilateral to the epileptogenic focus.

PET is superior to SPECT in terms of spatial and contrast resolution. PET, however, is expensive and requires an on-site medical cyclotron. Ictal studies are difficult with PET because of the short half-life of positron-emitting radioisotopes. PET has mainly been used to study interictal blood flow with ^{13}N-labeled ammonia and oxygen-15, and glucose metabolism with FDG in focal epilepsy. Localized hypoperfusion and hypometabolism in the epileptogenic area as shown by PET are seen as a reliable confirmatory finding in the presurgical assessment of temporal lobe epilepsy.

Neuropsychological Evaluation

The presurgical assessment neuropsychological evaluation is used to identify localizable deficits that can be related to the epileptogenic lesion. Crucial for lateralizing temporal lobe epilepsies is the function of verbal and nonverbal memory. Another neuropsychological task in the presurgical assessment is forecasting postsurgical cognitive outcome, which sometimes requires the intracarotid sodium amylobarbital procedure (Wada test).

PERSONALITY AND BEHAVIOR DISORDER IN EPILEPSY

Many studies have attempted to assess whether patients with epilepsy, when tested on personality rating scales, show profiles different from those of patients without epilepsy. Generally, the results support the hypothesis that patients with epilepsy have abnormal personalities. This outcome, however, may be attributed to many factors. These include biologic variables, for example, head injury after recurrent seizures or the prescription of long-term anticonvulsant drugs, which may lead to behavioral change. Also, it may relate to psychosocial variables, such as stigmatization, and a low expectancy of achievement by the family or by teachers.

The rating scale most commonly used for scientific investigations has been the Minnesota Multiphasic Personality Inventory. By and large, results using this scale have been mixed, although the scales that tend to be reported as most abnormal are the depression scale (D), the paranoia scale (Pa), the schizophrenia scale (Sc), and the psychesthenia scale (Pt).

The controversial Bear-Fedio Scale was developed by bringing together characteristics described in prior clinical reports of personality in epilepsy. In their original paper, Bear and Fedio compared patients with unilateral temporal lobe epilepsy to healthy subjects and a group with neuromuscular disorders (9). They noted that the epileptic patients had abnormal profiles, as predicted, particularly highlighting such features as humorlessness, dependence, circumstantiality, and an increased sense of personal destiny. Further, some laterality differences were noted; patients with left temporal lobe epilepsy described more anger, paranoia, and dependence, whereas those with right temporal foci reported more elation.

The view that there is indeed an interictal behavior syndrome of epilepsy, associated largely with temporal lobe epilepsy, was most strongly supported by the writings of Waxman and Geschwind (10). These reports highlighted changes in sexual behavior, hypergraphia (a tendency to compulsive and extensive writing), and religiosity. These reports also mentioned "stickiness" or viscosity, patients showing a striking preoccupation with detail and concerns over moral or ethical issues. Estimates of the frequency with which such a profile occurs, from cluster analysis studies using the Minnesota Multiphasic Personality Inventory and other clinical variables, vary from 7% to 21%.

Disagreement about the association between epilepsy and personality disorder persists, but those dealing with patients with chronic epilepsy are familiar with these personality changes. The evidence supporting the view that these abnormalities are more common in patients with temporal lobe epilepsy is not conclusive, but the number of positive studies surpasses greatly the number of negative ones.

EPILEPSY AND DEPRESSION

Prevalence

Although most authors acknowledge that depression is more often seen in epilepsy than in the larger population, the significance of the association depends on the population studied. Edeh and Toone, in a community-based survey of 88 patients, recorded neurotic depression in 22% of their sample (11). In a study of patients with temporal lobe epilepsy, 19% were found to have clinically apparent anxiety, whereas 11% were depressed (12).

The importance of understanding depression in epilepsy is highlighted when the frequency of suicide and parasuicide in this population is considered. Having reviewed 11 previous studies, Barraclough reported a suicide rate five times that in the general population (13). In patients with temporal lobe epilepsy, the relative risk is even greater. There is also an increased risk of parasuicide among these patients, particularly of overdoses.

Mania is rarely associated with epilepsy. Williams described elation in just 3 of 2000 patients (14).

Seizure-Related Factors

The findings of several studies suggest that neither specific seizure types (complex partial with or without secondary generalization or primary generalized seizures), nor frequency, nor duration of epilepsy are usually associated with depression. Most studies that have addressed the issue report an association between interictal depression and a temporal lobe epilepsy. The situation is more equivocal, however, regarding the laterality of the epileptic focus.

The role of the temporal lobes in the generation of affective states may also be investigated by considering the effects of temporal lobe surgery performed for the control of epilepsy. Apparently, in a minority of patients with chronic epilepsy, depression has worsened or developed de novo after temporal lobectomy. The mechanism of this effect is unknown, although the postsurgical increase in depression may in part be an artifact of comparing rates at a single point in time with those obtained over a prolonged follow-up period. It may also be a reflection of "forced normalization" in some cases. This concept refers to patients who develop psychopathology when their seizures are brought under control and their EEG becomes "more normal" than before, when they were psychiatrically well.

Biochemical Factors

One potential biochemical mechanism that is based on human studies is that of folate depletion. This has been observed in patients on polytherapy, and a relationship between mental illness and folate deficiency has been hypothesized. Phenytoin and barbiturates may both lead to falls in serum, red blood cell, and cerebrospinal fluid folate. In epilepsy, a link between low folate and depression has been noticed in several patient groups.

Diagnosis of Depression in Epilepsy

Prodromal Phenomena

The prodrome, occurring hours to days before a seizure, was investigated by Blanchet and Frommer (15). They reported a prospective study in which 27 patients with epilepsy self-rated their moods using Personal Feelings Scales and recorded life events on a daily basis for at least 56 days. During this time, 13 patients had at least one seizure. The authors observed that in these patients, the mean ratings of mood on 8 of the 10 scales showed a decline on the days preceding the seizure and an increase after the seizure.

There are several possible explanations for the association between lowered mood ratings and the subsequent occurrence of seizures. Lower mood may be a symptom of the prodromal phase of seizure activity, initiated by the same biologic processes that bring about the seizure. Alternatively, perhaps the mood change itself precipitates a seizure.

Periictal Phenomena

The aura is the earliest stage of subjective awareness of seizure activity. Many different sensations have been recorded. Williams investigated emotional phenomena in 2000 patients with epilepsy and found that 100 of them reported an emotion as part of the "epileptic experience" (14). The most commonly reported emotion was fear, occurring in Williams' sample in 61% of the 100 patients with emotional phenomena. On some occasions, this fear was quite pervasive, with psychic and somatic features. In contrast to this, depression was reported less often, in just 21% (i.e., in 1% of the whole group). Williams observed that when depression did occur, however, it tended to last far longer than the other periictal phenomena, persisting for up to several days after the ictus.

Although of short duration, postictal depression may be severe, at times including suicidal ideation. Case descriptions give a clear impression that the depressive phase is more than purely an emotional reaction to the advent of a seizure, suggesting instead a biologic link with the seizure process.

Thus, periictal depression, although it does occur, is not common but is characterized by a greater persistence than other postictal emotional phenomena. It is interesting to note that depressive auras appear to be particularly rare, whereas fear is more common.

Interictal Phenomena

Mendez and associates compared 20 depressed epileptic in-patients to 20 nonepileptic depressed subjects (16). Both groups met criteria for major depression according to the third edition of the *Diagnostic and Statistical*

Manual of Mental Disorders (DSM-3). All the patients had endogenous features of depression—anergia, anhedonia, appetite, and sleep disturbance—but the authors concluded that there were other major distinguishing characteristics of the depressed patients with epilepsy compared with those of the nonepileptic group. These factors included a chronic dysthymic background, a relative lack of neurotic traits, such as somatization or self-pity, and a history of periods of agitated periictal psychotic behavior.

EPILEPSY AND PSYCHOSIS

Slater and colleagues published a detailed analysis of 69 patients from two London, England hospitals who suffered from epilepsy and interictal psychoses (17). On the basis of this case series, the authors postulated a positive link between epilepsy and schizophrenia. The temporal lobe hypothesis soon became broadly accepted and stimulated extensive research into the role of temporal lobe pathology in schizophrenia.

Epidemiology

There are few population-based studies on the frequency of mixed psychoses in epilepsy. Krohn, in a population-based survey in Norway, found a 2% prevalence of psychoses with epilepsy (18). In a field study of the Warsaw population, Zielinski found prevalence rates for psychoses in epilepsy of 2% to 3% (19). Gudmundsson, in a study on the frequency of mixed psychoses in epilepsy in the population of Iceland, found prevalence rates for males of 6% and for females of 9% (20). Most figures in the literature on the frequency of psychosis in epilepsy derive from clinical case series and are therefore likely to be biased by unknown selection mechanisms. They cannot be regarded as representative of epilepsy in general.

Classification

There is no internationally accepted syndromic classification of psychoses in epilepsy. Psychiatric aspects are not considered in the international classification of epilepsies, and the use of operational diagnostic systems for psychiatric disorders such as the DSM-4 is limited because, if applied strictly, a diagnosis of functional psychosis is not allowed in the context of epilepsy. “Atypical” syndromes are not unusual, such as ictal and postictal psychoses in clear consciousness. Variations of phenomenology and precipitating factors can also be seen within individual patients who experience recurrent psychotic episodes. For the time being, patients with epilepsy and psychoses should probably receive two separate diagnoses according to both the International Classification of Epilepsy (ICE) and the Diagnostic and Statistical Manual, 4th edition (DSM-4). In addition, the relation among onset of psychosis and seizure activity, antiepileptic therapy, and changes of EEG findings should be noted.

For pragmatic reasons, psychoses in epilepsy are grouped in this chapter according to their temporal relationship to seizures. Table 25-4 provides a comparative list of the syndromes described in more detail in the sections that follow.

TABLE 25-4. *Clinical characteristics of psychoses in relation to seizure activity*

	Ictal psychosis	Postictal psychosis	Periictal psychosis	Alternative psychosis	Interictal psychosis
Consciousness	Impaired	Impaired or normal	Impaired	Normal	Normal
Duration	Hours to days	Days to weeks	Days to weeks	Days and weeks	Months
EEG	Status epilepticus	Increased epileptic and slow activity	Increased epileptic and slow activity	Normalized	Unchanged
Treatment	Antiepileptic drugs (IV)	Spontaneous recovery in many cases	Improvement of seizure control	Reduction of antiepileptic drugs	Neuroleptic drug

EEG, electroencephalogram.

Syndromes of Psychoses in Relation to Seizure Activity

Ictal Psychoses

Prolonged focal and generalized nonconvulsive epileptic activity (occurring for several hours or days) may present with psychotic symptoms. Generalized nonconvulsive status, also called *absence status, petit mal status,* or *spike and wave stupor,* is characterized by altered or narrowed consciousness. Patients are disoriented and apathetic. Contact with the environment is partially preserved, and patients are often able to perform simple tasks. Positive psychotic symptoms, such as delusions and hallucinations, occur only in some patients. The EEG shows generalized bilateral synchronous spike and wave complexes of variable frequency between 1 and 4 Hz. Absence status typically occurs in patients with a known history of generalized epilepsy, but "atypical absence status" may occur as a first manifestation of epilepsy, especially in later life.

Two types of complex focal status (synonyms: *status psychomotoricus, epileptic twilight state)* have been distinguished:

- Continuous form
- Discontinuous or cyclic form

The latter consists of frequently recurring complex partial seizures. Between seizures, patients may or may not experience simple focal seizure symptoms, and consciousness may recover to nearly normal states.

Noncyclic forms of complex partial status consist of prolonged confusional episodes or psychotic behavior. The EEG during complex partial status shows focal or bilateral epileptiform patterns and slowed background activity. Subtle rudiments of motor seizure symptoms, such as eyelid fluttering and bursts of myoclonic jerks in absence status or mild oral activity automatisms in continuous complex partial status, may point to the underlying epileptic activity. Mutism and paucity of speech—or even speech arrest—occur in both absence and complex partial status.

Nonconvulsive status epilepticus requires immediate treatment with intravenous antiepileptic drugs (AEDs).

Simple focal status or aura continua may cause complex hallucinations, thought disorders, and affective symptoms. The continuous epileptic activity is restricted and may escape scalp EEG recordings. Insight is usually maintained, and true psychoses emerging from such a state have not been described.

Sometimes, patients with partial status resemble schizophrenic patients very closely, and without knowledge of a history of epilepsy (e.g., in the emergency room), the ictal origin may be overlooked. Any suspicion of epilepsy should lead to a request for an EEG. If this is unavailable, an intravenous injection of a benzodiazepine may help to resolve the diagnosis.

Postictal Psychoses

Most postictal psychoses are precipitated by a series of generalized tonic-clonic seizures. More rarely, psychoses occur after single grand mal seizures or after a series of complex partial seizures. Postictal psychoses account for about 25% of psychoses in epilepsy. The relation to the type of epilepsy is not clear. In most patients, there is a characteristic lucid interval lasting from 1 to 6 days between the epileptic seizures and onset of psychosis, which may lead to an incorrect diagnosis.

The psychopathology of postictal psychosis is polymorphic, but most patients present with abnormal mood and paranoid delusions. Some patients are confused throughout the episode; others present with fluctuating impairment of consciousness and orientation.

The EEG during postictal psychosis is deteriorated with increased epileptic as well as slow-wave activity.

Psychotic symptoms spontaneously remit within days or weeks, often without need for additional neuroleptic treatment. In some cases, however, chronic psychoses develop from recurrent episodes or even from a single postictal psychosis. The pathophysiology is not known.

Periictal Psychosis

In periictal psychosis, psychotic symptoms develop gradually and parallel to increases in seizure frequency. The relation to seizures is easily overlooked if seizure frequency is not carefully documented over prolonged periods. Impairment of consciousness is more frequent than in postictal psychosis. Treatment of periictal psychoses requires improvement of seizure control.

Interictal Psychoses

Interictal psychoses occur between seizures and cannot be linked directly to the ictus. They are less frequent than periictal psychoses and account for 10% to 30% of diagnoses in unselected case series. Interictal psychoses are, however, clinically more significant in terms of severity and duration than periictal psychoses, which usually are short-lasting and often self-limiting.

Slater stated that in the absence of epilepsy, the psychoses in their study group would have been diagnosed as schizophrenia, but they also mentioned distinct differences between process schizophrenia and the schizophrenia-like psychoses associated with epilepsy. They highlighted the preservation of warm affect and a high frequency of delusions and religious mystical experiences. Other authors stressed the rarity of negative symptoms and the absence of formal thought disorder and catatonic states.

Using the Present State Examination and the CATEGO computer program, which is a semistandardized and validated method for quantifying psychopathology, it has been possible to compare the presentation of psychosis in epilepsy with process schizophrenia. Few significant differences emerged from such studies, suggesting that, assuming the patients were representative, a significant number will have a schizophrenia-like presentation indistinguishable from schizophrenia in the absence of epilepsy (21).

Risk Factors

The literature on risk factors is highly controversial; studies are difficult to compare because of varying definitions of the epilepsy, the psychiatric disorder, and the investigated risk factors. Most studies are restricted to interictal psychoses.

Genetic Predisposition

With few exceptions, most authors do not find any evidence for an increased rate of psychiatric disorders in relatives of epilepsy patients with psychoses.

Sex Distribution

There has been a bias toward female sex in several case series, which has not been confirmed in controlled studies.

Duration of Epilepsy

The interval between age at onset of epilepsy and age at first manifestation of psychosis has been remarkably homogeneous in many series, ranging from 11 to 15 years. This interval has been used to postulate the etiologic significance of the seizure disorder and a kindling-like mechanism.

Type of Epilepsy

There is a clear excess of temporal lobe epilepsy in almost all case series of patients with epilepsy and psychosis. Summarizing the data of 10 studies, 217 of 287 patients (76%) suffered from temporal lobe epilepsy. The preponderance of this type of epilepsy is, however, not a uniform finding; in Gudmundsson's epidemiologic study, for example, only 7% suffered from "psychomotor" epilepsy (20). Unfortunately, most authors have not sufficiently differentiated frontal and temporal lobe epilepsy. The findings are similar regarding rates of psychosis in temporal lobe epilepsy and generalized epilepsies. Several studies demonstrate that psychoses in generalized epilepsies differ from psychoses in temporal lobe epilepsy. The former are more likely to be short-lasting and confusional. Psychoses that develop in generalized epilepsy are usually relatively mild and often remit before paranoid-hallucinatory symptoms are seen. Schneiderian first-rank symptoms and chronicity are more frequent in patients with temporal lobe epilepsy.

Type of Seizures

There is evidence from several studies that focal seizure symptoms that indicate ictal mesial temporal or limbic involvement are overrepresented in patients with psychosis.

Severity of Epilepsy

The strongest risk factors for psychosis in epilepsy are those that correlate with severity of epilepsy. These include long duration of active epilepsy, multiple seizure types, history of status epilepticus, and poor response to drug treatment. Seizure frequency, however, is reported by most authors to be lower in psychotic epilepsy patients than in nonpsychotic patients.

Laterality

Left lateralization of temporal lobe dysfunction or temporal lobe pathology as a risk factor for schizophreniform psychosis was originally suggested by Flor-Henry (22). Studies supporting the laterality hypothesis have been made using computed tomography (CT), neuropathology, neuropsychology, and PET.

Forced Normalization

In the 1950s, Landolt published a series of papers on patients who had epilepsy who became psychotic when their seizures were under control (23). He defined forced normalization as follows:

> Forced normalization is the phenomenon characterized by the fact that, with the recurrence of psychotic states, the EEG becomes more normal, or entirely normal as compared with previous and subsequent EEG findings. These phenomena have now been well documented clinically. Clearly, forced normalization may be provoked by the administration of anticonvulsants, and has been reported with barbiturates, benzodiazepines, ethosuximide, and vigabatrin.

PSEUDOSEIZURES

The term *pseudoseizure* is a misnomer; patients can present with pseudoepileptic seizures, nonepileptic attack disorder, or nonepileptic seizures. The seizures in these patients, however, are very real to patients, third-party observers, and their physicians. Indeed, nearly 20% of patients attending chronic epilepsy clinics do not have epilepsy but have some form of nonepileptic disorder. Alternative terms, including hysterical seizures and psychogenic seizures, are also inadequate, the former being pejorative and the latter defining a different kind of seizure altogether. Thus, psychogenic seizure should logically be used only to imply a form of reflex seizure induced by mental activities.

Differential Diagnosis of Nonepileptic Attacks or Pseudoseizures

The medical and psychiatric conditions to be considered are various. Obvious cardiologic causes (e.g., vasovagal attacks) frequently present in adolescence and often herald the development of pseudoseizures. Metabolic conditions (e.g., hypoglycemia) should always be considered in the differential diagnosis, and attempts made to rule out other common medical conditions should be listed. One of the most common areas of misdiagnosis is with the sleep disorders. Both rapid eye movement (REM) and non-REM disorders can be misdiagnosed as epilepsy, as can some of the hypersomnolent conditions. Night terrors arise out of stage-four sleep, as opposed to the nightmares of REM sleep. They are associated with the sudden onset of restless motor activity, sometimes with a vocalization of fear.

REM behavior disorder, which occurs in patients who have subtle brain-stem damage and no longer have the characteristic REM muscle paralysis, can present with kicking around, self-injury, and aggression. Cataplexy, if the other elements of Gélineau's triad (narcolepsy, sleep paralysis, and either hypnagogic or hypnopompic hallucinations) are absent, can look deceptively like an epileptic drop attack, although the crucial history of the precipitation by emotional events should be solicited by the physician.

Of the psychiatric conditions, the anxiety-related disorders are the most commonly misdiagnosed as epilepsy. In particular, panic attacks, especially if associated with episodes of déja vu, depersonalization, or autoscopy, are deceptive. The onset of the panic may be sudden, not obviously triggered by environmental events, and it may last only for a short time. Some patients have all of the autonomic elements of a panic attack without the subjective sense of fear—panic sine panic.

A depressive illness often underlies a diagnosis of pseudoseizures. Clues in the history include a past history of depression, recent significant life events (e.g., a bereavement that might have precipitated a depressive illness), and the central features of a depression, which, unless solicited, may not be volunteered by a patient.

Conversion disorder, presenting either as dissociative symptoms or as conversion hysteria, is still seen frequently in specialist practice and must always be considered.

Conversion epilepsy involves the mechanism of dissociation, which means that patients are amnesic for their attack. Patients who appear to know absolutely nothing about their seizures, especially after a several-year history of frequent episodes, rarely have epilepsy. Sexual abuse quite possibly may have occurred and has important treatment implications.

Briquet's hysteria, or somatization disorder, is another condition in which patients with pseudoseizures are seen. Such patients are often female and polysymptomatic. They characteristically have a history of an excessive number of surgeries and hospitalizations and present with numerous somatic complaints. Sociopathic personality traits and alcoholism may be noted in these patients' families.

In the differential diagnosis of pseudoseizures, it is essential to take an adequate history, taking note of both psychiatric and neurologic background features. In addition, information regarding such obvious factors as seizure-precipitating events should be sought, paying particular attention to the setting and timing of the very first attack. Seizures occurring many times a day in the presence of a normal interictal EEG are likely to be nonepileptic. In addition, talking, screaming, and displays of emotional behavior immediately after the attack are likely to lead to a similar conclusion.

The presence of paroxysmal abnormalities on the EEG does not necessarily mean a diagnosis of epilepsy. There are several patterns of EEG abnormality seen in psychiatric patients that may be reported as epileptiform. These patterns include small sharp spikes, which may be temporal in distribution, and are seen with affective disorder and rhythmic midtemporal discharges, which arise in drowsiness and may be mistaken for bitemporal I waves. Further, I waves, sometimes paroxysmal, and even sharp waves, are often seen in patients with extreme anxiety disorders and schizophrenia. The diagnosis of epilepsy should never be based on a reporting of "epileptiform features" on an EEG.

Pseudostatus epilepticus also occurs. In fact, if patients present with status epilepticus, and their EEG is normal, including interictal EEGs, it is highly unlikely that they have epilepsy.

IMPAIRMENT OF COGNITIVE FUNCTION AND MEMORY IN EPILEPSY

Several investigators have noted that patients with symptomatic epilepsy are more likely to have impaired intellect than those with epilepsy of no known cause. Preexisting brain damage, although an important variable, does not entirely explain the neuropsychological deficit.

Most investigators report an early age of onset to have a poorer prognosis with regard to intellectual abilities. Patients with generalized seizures tend to show more deficits of attention and concentration compared with patients with focal seizures. The latter, particularly with seizures arising from the temporal lobes, are more likely to show memory impairments. Patients with generalized absence seizures show impaired cognitive function for the duration of the EEG spike and wave abnormality, although generally, patients with this form of seizure show minimal interictal dysfunction. If absences are frequent, impaired performance in the classroom setting may lead to educational underachievement.

In some patients, it is reasonable to consider a dementia of epilepsy. These patients show a cognitive deterioration, which is thus an acquired intellectual deficit, although this is not progressive in the way that parenchymatous degenerative dementias are. Patients halt or arrest in the progression of the intellectual decline. Identifying these patients prospectively is currently impossible, but retrospective studies have suggested that generalized tonic-clonic seizures with recurrent head injury and the prescription of certain anticonvulsant drugs—notably phenytoin and primidone—are somehow associated with this clinical picture.

The effects of AEDs on cognitive function have been an area of intense investigation in recent years. Extensive reviews are available. In patients on polytherapy, rationalization with a diminished burden of anticonvulsant prescriptions improves cognitive function over a wide range of cognitive abilities (24).

There has been a debate regarding possible differences between individual AEDs. Generally, data favor the newer AEDs, carbamazepine, sodium valproate, and vigabatrin, and emphasize more cognitive impairments with phenytoin and phenobarbital. Several new AEDs have recently been introduced to clinical practice, but information on their cognitive effects is rather limited. The only drug to have been investigated in any detail is vigabatrin, and several studies suggest that it appears to have no significant influence on cognitive abilities.

PSYCHIATRIC DISORDERS SECONDARY TO ANTIEPILEPTIC TREATMENTS

The main adverse effect of AEDs is their link with depression. Epileptic patients on therapy, including phenobarbitone, are especially vulnerable to depressive illness. This is in contrast to patients treated with carbamazepine who generally are less depressed.

Another drug recently linked with depression is vigabatrin. Idiosyncratic responses have been described in a group of patients given the drug as add-on therapy. The depression is in some cases related to control of seizure frequency (an example of forced normalization) and is more common in patients with a past history of affective disorder.

Adverse effects of AEDs are often seen in children. Essentially, these include the provocation of conduct disorder, or the development of hyperactivity syndrome, clinically similar to attention deficit hyperactivity disorder. The drug most frequently implicated is phenobarbital, but similar responses have been shown after the use of some benzodiazepines (e.g., clonazepam) and vigabatrin. Interestingly, all of these drugs have some action at the GABA-benzodiazepine receptor.

Throughout the literature on the effects of AEDs on the mental state, two features are noteworthy:

1. Patients on polytherapy experience more problems with personality and cognitive function than do patients on monotherapy.
2. Polytherapy is associated with depression.

Psychiatric Disorders After Surgery

Ever since the early series, the possibility that surgery may be associated with the development of psychiatric disturbance, in particular psychosis, has been discussed. Most surgeons have stopped operating on floridly psychotic patients, based on the observation that psychoses generally do not improve with the operation. The Maudsley series (25) showed that some patients develop new psychosis postoperatively, and there is an increased reporting of depression. Suicide was reported in 2.4% of the sample but accounted for 22% of the postoperative deaths.

Bruton has suggested that the development of postoperative psychoses may be more common with certain pathologies (gangliogliomas), and patients with right-sided temporal lobectomies may be more prone to these psychiatric disturbances (25). In some cases, the sudden relief of seizures that occurs after surgery may suggest a mechanism similar to forced normalization, although no persistent clear relationship emerges between the success of operation and the development of psychotic postoperative states. In the unpublished series from the National Hospitals, analysis of 50 patients assessed psychiatrically preoperatively and then followed for at least 3 months postoperatively revealed that 16% developed depression requiring psychiatric intervention. One patient became psychotic. With potential suicide rates of up to 5%, these data emphasize the need for continuing psychiatric observations in patients who have received temporal lobectomy.

TREATMENT OF EPILEPSY

Most patients with epilepsy are treated with anticonvulsants. There has also been renewed interest in attempts at behavioral and nonpharmacologic approaches to the management of seizures. A minority of patients with drug-resistant epilepsy may proceed to have surgery.

With regard to motor behavior, patients sometimes have a pattern of bradyphrenia, which may in some respects resemble a psychomotor retardation. In some patients, this may be associated with ongoing, subclinical seizures. The latter may be revealed through subtle motor manifestations. These include paroxysms of rhythmic eye blinking, mouth movements, or small frequent jerks of the hands, arms, or fingers, referred to as *minipolymyoclonus.*

The mental state may contain personality changes, including undue religiosity, altered sexual behavior, circumstantiality, rigidity of thinking, and an undertone of paranoia. Hypergraphia should always be asked about. This may be obvious only when, for example, a patient's diary is requested and read. The writing may be pedantic, repetitive, and mystical in content.

In patients with depression, there is often evidence of the underlying neurologic condition. The mood swings may be rapid and severe and may not last long. There is often much anxiety associated with the depression, and the presentation can be further influenced by the presence of an excessive anticonvulsant burden. Suicidal thoughts are frequent and should always be asked about.

Anxiety is also a common problem, and it can sometimes be difficult, without EEG monitoring, to distinguish an aura of fear from a panic attack. The features can seem identical.

The delusions and hallucinations have varying themes, but mystical and religious themes are common. In postictal psychoses, complex visual hallucinations can occur with almost clear consciousness. In patients who present with a postictal twilight state or a partial seizure status, careful testing of the cognitive state reveals impairment of performance. This may fluctuate, and repeated testing may be necessary after a time.

Antiepileptic Agents

Ideally, patients should be managed with a single drug. About 75% of patients with epilepsy can be fully controlled on monotherapy, the choice of agent being determined by seizure type (Table 25-5). Carbamazepine and phenytoin are most effective for complex partial and generalized tonic-clonic seizures. For absence seizures, ethosuximide and sodium valproate are the most useful treatments. These drugs are generally thought of as the first-line treatments. Some patients with myoclonic epilepsy respond well to sodium valproate, but others may require clonazepam. Of the remaining patients not controlled on monotherapy, addition of another first-line drug gains control in a further 15%. Some patients, however, go on to develop chronic seizures unrelieved by these treatments. In such circumstances, adjunctive therapies should be considered. A number of the more recently introduced anticonvulsants, often considered second-line treatments, may be introduced,

TABLE 25-5. *Antiepileptic drugs*

Seizure type	First-line treatment	Second-line treatment
Partial	Carbamazepine Phenytoin	Valproic acid, clobazam, phenobarbital, acetazolamide, clonazepam, vigabatrin, lamotrigine, gabapentin
Generalized	Carbamazepine Phenytoin Valproic acid	(Same as above)
Absence	Ethosuximide Valproic acid	
Myoclonic	Clonazepam Valproic acid	

either alone or in combination with a first-line agent. New anticonvulsants, including vigabatrin, gabapentin, and lamotrigine, have recently been introduced. These drugs have been approved initially as add-on treatments for treatment-resistant partial complex epilepsy.

An important feature in the clinical use of the first-line treatments is therapeutic drug monitoring, which is appropriate for agents without unmeasured active metabolites and for which therapeutic response correlates with and depends on the serum concentration rather than the dose. Sampling should occur once a steady state has been achieved, four to six half-lives after treatment has been modified or introduced.

Behavioral Treatment

It has been suggested that many patients with epilepsy have a mental mechanism they use to attempt to inhibit their seizures. In one study of 70 patients, 36% claimed that they could sometimes stop their seizures. A behavioral approach to the treatment of epilepsy is based on observations that epilepsy can be manipulated in a systematic way through environmental, psychological, and physical changes. The initial stage in this approach is a behavioral analysis of the ways in which environmental and behavioral factors interact with seizure occurrence (26).

Significant reductions in seizure frequency may be achieved by teaching patients a specific contingent relaxation technique that they must be able to employ rapidly when they identify a situation when they are at high risk of having a seizure. Some patients suffer from reflex seizures in that their seizures are precipitated by external stimuli. A number of people can identify specific environmental or affective triggers and may be able to develop specific strategies to abort or delay a seizure.

Primary seizure inhibition describes the direct inhibition of seizures by an act of will. The nature of the successful act varies from person to person, and this treatment approach will not be effective unless it is individually tailored, based on an analysis of each patient's seizures and any actions he or she may already have noticed that modify the seizures.

The term *secondary inhibition* was employed by Fenwick to describe behavioral techniques that effect change in cortical activity in the partially damaged group 2 neurons around the focus, thereby reducing the risk both of a partial seizure discharge and of a generalized seizure discharge (26).

In addition to these seizure-related approaches, more general psychological strategies have been investigated. Several anecdotal reports have been published demonstrating benefit from a reward system that aims to reward seizure-free periods.

Specific biofeedback techniques have also been explored. Measurement of scalp electrical activity has demonstrated that there is an increase in surface-negative slow cortical potentials in the seconds before a seizure occurs. These potentials represent the extent to which apical dendrites of cortical pyramidal cells are depolarized and hence indicate neuronal excitability. Studies using visual feedback of this effect have demonstrated that some patients are able to modulate cortical electrical activity with an associated decrease in seizure frequency.

A quite different biologic approach that has been tried in some centers is that of vagal stimulation using an implanted stimulator. Among the drawbacks to this approach is that while the nerve is being stimulated—usually for 30 seconds every 5 to 10 minutes—the voice changes. More intense stimulation may be associated with throat pain or coughing. Nevertheless, in one series of 130 patients, mean seizure frequency decreased by 30% after 3 months and by 50% after 1 year of therapy.

Surgical Treatment

In patients with persistent epilepsy unrelieved by other treatments, surgical intervention should be considered. Surgery may be effective for the following reasons:

- Surgery *removes* pathologic tissue, including the primary focus
- Surgery *disconnects* a focus from the rest of the brain when the primary focus is incompletely removed
- Surgery *reduces* the mass of neurons that behave abnormally

Treatment centers that engage in routine surgical management of epilepsy generally have a standardized assessment process for patients being considered for such treatment. Although the program may vary a little from place to place, the general procedure is similar. Clinical history and examination focus on looking for etiologic factors, evidence of localizing signs and symptoms, and a witnessed description of the seizures. In addition, psychosocial information must be gathered relating to education, employment, social support, and past and present mental state findings. All patients considered for surgery must undergo EEG investigation. This may include several days of continuous video telemetry. The aim of telemetry is to obtain ictal recordings that give more valuable localizing information than interictal records. Recent advances in structural and functional neuroimaging have made invasive EEG recording less necessary. High-resolution CT of patients with focal epilepsy reveals a lesion in 60% to 70% of cases. MRI is now used routinely and reveals local structural abnormalities not seen on CT. Current research programs are evaluating the relative benefits of functional imaging using SPECT compared with more invasive methods of seizure localization (27).

A number of surgical procedures have been developed. The most frequently performed is the removal of neocortical structures together with deep structures. The nature of potential perioperative complications and postsurgical neurologic, cognitive, and psychiatric sequelae depends in part on the site of surgery.

TREATMENT OF DEPRESSION IN EPILEPSY

Initially, a clinical assessment of the severity of depression should be made because there is an increased risk for suicide. Subsequent actions are determined by the result of this assessment. In addition, epilepsy, anticonvulsant, and psychosocial variables should be examined.

Seizure Status

Seizure control should be optimized. In addition, any change in the nature of seizures should be considered. For instance, a recent increase in seizure frequency or the development of secondarily generalized seizures in a patient with a focal epilepsy may be associated with the development of depression by virtue of the increased disruption to the daily life of the patient. Alternatively, depression may lead to increased seizures by such mechanisms as sleep deprivation or failure of compliance.

Antiepileptic Drug Use

Polypharmacy has been associated with depression by some authors. As described earlier, the use of phenobarbital and vigabatrin have been linked to the development of depression. The introduction of carbamazepine, structurally related to the tricyclic antidepressants, may have antidepressant effects.

Specific Antidepressant Treatments

A potential difficulty in giving antidepressants to patients with epilepsy is that these drugs may lower the seizure threshold. A prospective study of the incidence of antidepressant-induced seizures found that seizures requiring treatment with an AED occurred in less than 1 case in 1000. All these observations were made in patients without epilepsy, and the implications for those who have the condition are not clear. In our practice, we are cautious in giving tricyclic agents to patients with epilepsy.

Although there is limited experience with electroconvulsive therapy for epilepsy, Betts reported that it causes no problems and pointed out that in severe drug-resistant depression, with the attendant risk for suicide, it may be life-saving (28).

In light of the interplay of psychosocial and biologic factors in the genesis of depression, several authors have pointed out the value of including the appropriate type of psychotherapy within the overall management plan.

TREATMENT OF PSYCHOSES OF EPILEPSY

Essentially, management of psychiatric problems in patients with epilepsy is similar to that in patients without epilepsy, with a few caveats. Patients with psychoses should be treated with neuroleptic medications, although these, like most antidepressants, can lower the seizure threshold. To date, all known neuroleptics have this potential, although some more than others. Of the neuroleptics, the phenothiazines are more likely to provoke seizures than the butyrophenones, and of the available drugs, pimozide is perhaps the least likely to precipitate seizures.

When patients with epilepsy have no alteration of the seizure frequency, or the psychosis is occurring in the setting of increased seizure frequency, a neuroleptic less likely to precipitate seizures—such as haloperidol or pimozide—is logical. Sulpiride appears to be a reasonable alternative.

It should be recalled that patients taking AEDs that increase hepatic metabolism show lower serum levels of neuroleptics and may therefore require somewhat higher doses than patients not taking these medications to achieve a similar clinical effect. Occasionally, the addition of an antidepressant or a neuroleptic to a patient's prescription may lead to increases in serum AED concentrations.

As with all psychiatric problems, psychopharmacologic management alone is not sufficient. It is important to acknowledge that epileptic patients with psychosis bear the burden of epilepsy in addition to their psychosis. Patients with intermittent psychotic states are often perplexed and embarrassed about what has happened to them while psychotic. They also fear further continuing bouts, with a descent into insanity. Patients with continuous psychosis require the skills of paramedical intervention, and the full resources of community care may be needed to help them rehabilitate and to assist their families in coping with their difficulties. In many patients with chronic psychoses of epilepsy, the preservation of affect and lack of personality disintegration over years allow them to live in their communities, with their families, and even to marry. Maintaining them and bringing such support to them is important to sustaining them in the community and preventing their recurrent admission into the hospital.

OPTIMIZATION OF PSYCHOSOCIAL POTENTIAL

The concept of quality of life (QOL) has gained popularity in recent years, although its definition is problematic. To date, there is little research on QOL in epilepsy; however, it is important to recognize that epilepsy involves more than seizures. Improving these patients' psychosocial potential requires attention to the many aspects of their lives; their own internal concept of their QOL has thus become an important management goal.

A review of the literature on QOL reveals several life domains or areas thought essential to determining QOL in epilepsy, covering physical, cognitive, affective, social, and economic aspirations of patients. The word *aspiration* is here used to emphasize an important concept in QOL research. Thus, future expectations are a major component of perceived QOL, with actual abilities being less important than a discrepancy between the patients' current position and their expected situation.

The diagnosis of epilepsy brings with it many psychosocial problems, including stigmatization, social isolation, psychological problems, and education and employment difficulties. Societal attitudes play a major role in determining QOL of patients with epilepsy. Discrimination and nonacceptance of patients are still quite common. There are obvious ways to enhance these patients' psychosocial adjustment. At the outset, when patients are diagnosed, they should explore the concept of epilepsy and discuss their fears and myths about the condition. It is important to provide constant support from known individuals who will look after them as they learn to live with epilepsy.

CONTEMPORARY ISSUES

Several new AEDs have been introduced in the last several years:

- *Felbamate* blocks repetitive neuronal firing and NMDA receptor.
- *Gabapentin* may enhance GABA synthesis or release.
- *Lamotrigine* decreases release of excitatory neurotransmitters.
- *Tiagabine* blocks GABA reuptake.
- *Topiramate* blocks repetitive neuronal firing and facilitates GABA inhibition.
- *Vigabatrin* inhibits GABA transaminase.

These agents have been developed in short-term trials, where their value as add-on therapy to treatment-refractory partial and secondarily generalized epilepsy was demonstrated. Some have significant side effects, especially felbamate, which has been associated with aplastic anemia and hepatic failure. Their potential roles in therapy for the neurobehavioral syndromes of epilepsy are not yet known (29).

REFERENCES

1. Commission on Classification and Terminology of the International League Against Epilepsy. 1989 Proposal for revised classification of epilepsies and epileptic syndromes. *Epilepsia* 1989;30:389–399.
2. Roger J, Bureau M, Dravet CH, Dreifuss FE, Perret A, Wolf P. *Epileptic syndromes in infancy, childhood and adolescence.* London, England: John Libbey, 1992.
3. Perlin JB, DeLorenzo RJ. Recent advances in epilepsy. In: Pedley TA, Meldrum BS, eds. *Recent advances in epilepsy.* Edinburgh, Scotland: Churchill Livingstone, 1992:15–36.
4. Geddes JW, Cahan LD, Cooper SM, Kim RC, Choi BH, Cotman CW. Altered distribution of excitatory amino acid receptors in temporal lobe epilepsy. *Exp Neurol* 1990;108:214–220.
5. Hauser W. Anderson VE, Loewnson RB, McRoberts EM. Seizure recurrence after a first unprovoked seizure. *N Engl J Med* 1982;307:522–528.
6. Annegers JF, Hauser WA, Elveback LR. Remission of seizures and relapse in patients with epilepsy. *Epilepsia* 1979;20:729–737.
7. Stefan H. Multichannel magnetoencephalography: recordings of epileptiform discharges. In: Lüders H, ed. *Epilepsy surgery.* New York: Raven, 1991:423–428.
8. Connelly A, Jackson GD, Duncan JD, et al. 1H MRS in the investigation of temporal lobe epilepsy. *Neurology* 1994;44,850.
9. Bear DM, Fedio P. Quantitative analysis of interictal behaviour in temporal lobe epilepsy. *Arch Neurol* 1977; 34:454–467.
10. Waxman SG, Geschwind N. The interictal behaviour syndromes of temporal lobe epilepsy. *Arch Gen Psychiatry* 1975;32:1580–1586.
11. Edeh J, Toone BK. Antiepileptic therapy, folate deficiency, and psychiatric morbidity: a general practice survey. *Epilepsia* 1985;26:434–440.
12. Currie S, Heathfield KWG, Henson RA, Scott DF. Clinical course and prognosis of temporal lobe epilepsy. *Brain* 1971;94:173–190.
13. Barraclough B. Suicide and epilepsy. In: Reynolds EH, Trimble MR, eds. *Epilepsy and psychiatry.* Edinburgh, Scotland: Churchill Livingstone, 1981:72–76.
14. Williams D. The structure of emotions reflected in epileptic experiences. *Brain* 1956;79:29–67.
15. Blanchet P, Frommer GP. Mood change preceding epileptic seizures. *J Nerv Ment Dis* 1986;174:471–476.
16. Mendez MF, Cummings JL, Benson F. Depression in epilepsy: significance and phenomenology. *Arch Neurol* 1986;43:766–770.
17. Slater E, Beard AW, Glithero E. The schizophrenia-like psychoses of epilepsy. V. Discussion and conclusions. *Br J Psychiatry* 1963;109:95–150.
18. Krohn W. A study of epilepsy in Northern Norway, its frequency and character. *Acta Psychiatr Scand* 1961; 150(Suppl):215–225.
19. Zielinski JJ. *Epidemiology and medical-social problems of epilepsy in Warsaw.* (Final report on research pro-

gram No. 19-P-58325-F-01 DHEW, Social and Rehabilitation Services). Washington, DC: U.S. Government Printing Office, 1974.
20. Gudmundsson G. Epilepsy in Iceland. *Acta Neurol Scand* 1966;43(Suppl):E1–124.
21. Perez MM, Trimble MR. Epileptic psychosis: diagnostic comparison with process schizophrenia. *Br J Psychiatry* 1980;137:245–249.
22. Flor-Henry P. Psychosis and temporal lobe epilepsy: a controlled investigation. *Epilepsia* 1969;10:363–395.
23. Landolt H. Serial electroencephalographic investigations during psychotic episodes in epileptic patients and during schizophrenic attacks. In: Lorentz de Haas AM, ed. *Lectures on epilepsy.* Amsterdam, Netherlands: Elsevier, 1958:91–133.
24. Trimble MR, Thompson PJ. Neuropsychological aspects of epilepsy. In: Grant I, Adams KM, eds. *Neuropsychological assessment of neuropsychiatric disorders.* New York: Oxford University Press, 1986:321–346.
25. Bruton CJ. *The neuropathology of temporal lobe epilepsy.* Maudsley Monograph No. 31. Oxford, England: Oxford University Press, 1988.
26. Fenwick P. Evocation and inhibition of seizures: behavioral treatment. In: Smith D, Treiman D, Trimble M, eds. *Neurobehavioral problems in epilepsy.* Vol 55. New York: Raven, 1991:163–183.
27. Polkey CE. Surgical treatment of chronic epilepsy. In: Trimble MR, ed. *Chronic epilepsy:* its prognosis and management. New York: John Wiley & Sons, 1989:189–207.
28. Betts TA. Depression, anxiety and epilepsy. In: Reynolds EH, Trimble MR, eds. *Epilepsy and psychiatry.* Edinburgh, Scotland: Churchill Livingstone, 1981:60–71.
29. Mattson RH. Medical management of epilepsy in adults. *Neurology* 1998;51(Suppl 4):S15—S20.

26
Traumatic Brain Injury

Jordan Grafman and Andres Salazar

OVERVIEW

Traumatic brain injury (TBI) is the leading cause of death and disability in young adults today; every 5 minutes one person dies and another is permanently disabled from TBI. Largely because it affects the young, the total economic cost of TBI has been estimated at more than $25 billion per year. The incidence of TBI requiring hospitalization is about 200 in 100,000 population, yet it has been generally ignored by psychiatrists and neurologists, perhaps as no other subject in these specialties (relative to its incidence). In this chapter, we review the current understanding of the pathogenesis of TBI and its neurobehavioral consequences and then outline a practical management approach to the TBI patient based on that knowledge.

NEUROBIOLOGY OF TRAUMATIC BRAIN INJURY

TBI is a dynamic process. Not only does the pathologic picture continue to evolve over the first few hours and days after trauma, often with devastating secondary injury, but the physiologic and clinical aspects of the recovery process itself can continue for a period of years. Thus, the notion of a "dynamic prognosis" requiring intermittent revision is especially relevant to the head-injured patient both because of the long period of recovery and because the many poorly understood variables involved still make outcome prediction as much of an art as a science. In addition, the TBI victim often manifests a multitude of systemic abnormalities, not only as consequence of concomitant trauma elsewhere in the body but also secondary to the brain injury itself. Changes in nutrition, cardiopulmonary status, circulating catecholamines, and coagulation are among those described.

The pathology of closed head injury (CHI) can be reduced to a four-component classification. Three parallel components were initially identified:

- Focal injury
- Diffuse axonal injury (DAI)
- Superimposed hypoxia and ischemia

Later, diffuse microvascular injury with loss of autoregulation was implicated as playing an important role in the acute stage of moderate and severe head injury. All of these pathologic features have been reproduced in animal models of angular acceleration without impact.

Focal Injury

Focal contusions often occur under the site of impact and thus result in focal neurologic deficits referable to that area (e.g., aphasia, hemiparesis). The most common location for contusions after acceleration-deceleration injury is in the orbitofrontal and anterior temporal lobes, where brain tissue lies next to bony edges. Thus, a relatively typical pathologic picture is often seen in CHI, and the most troubling clinical sequelae are behavioral and cognitive abnormalities that may be referable to the frontal and temporal lobe injury. Subdural hematomas are common occurrences, with rapid decelerations such as occur with impact after a fall, especially in the aged, and are usually the result of rupture of bridging veins. Delays longer than 4 hours in the surgical management of hematomas may significantly worsen prognosis. Delayed hematomas and bleeding into contusions are particularly important in the so-called talk-and-die patient, who may initially appear to be at low risk but then deteriorates unexpectedly.

Diffuse Axonal Injury

Diffuse axonal injury (DAI) is one of the most important causes of persistent, severe neurologic deficit in CHI. Originally described as a "shearing" injury of axons, it was characterized by axonal "retraction" balls microscopically in the hemispheric white matter, corpus callosum, and brainstem (1). *Mild to moderate* fluid-percussion injury in animal models has shown that the typical light microscopic histopathology of DAI may not emerge until 12 to 24 hours after injury. The only early abnormality is a relatively subtle, focal intraaxonal disruption seen on electron microscopy, with an intact axon sheath. This leads to a disturbance of axonal flow, accumulation of transport material with axonal ballooning proximal to the injury, and then eventual *delayed* severing of axons several hours later (2). One obvious clinical implication of these findings is that there may be a potential 12- to 24-hour window of therapeutic opportunity after injury during which future treatments may prevent total axonal disruption. Another important conclusion from these studies is that DAI can be demonstrated after "minor" head injury and occurs even in the absence of morphopathologic change in any other vascular, neural, or glial elements. This axonal damage is the most likely basis for the *postconcussion syndrome* and for the cumulative effects of repeated concussion, as seen in some boxers.

Hypoxia-Ischemia

The classic pathology of hypoxia-ischemia, involving mainly the hippocampus and the vascular border zones of the brain, is all too often superimposed on the other pathologic features that are more specific to TBI. The traumatized brain is particularly sensitive to hypoxia-ischemia, and the relationship is probably more than just additive. When present, such pathology, including the concomitant brain swelling, can become a major determinant of ultimate clinical outcome; the most significant improvements in the management of the TBI patient have resulted from recognition of the importance of this component and its prevention.

Diffuse Microvascular Damage

The finding of diffuse microvascular damage has also been implicated as a major component of both closed and penetrating TBI. Diffuse perivascular damage with astrocytic foot-plate swelling is a prominent feature at both the light and electron microscopic levels within minutes of high-velocity gunshot wound in nonhuman primates. In CHI, the vascular response appears to be biphasic. Depending up the severity of the trauma, early changes include the following:

- Initial transient systemic hypertension (probably related to release of catecholamines)
- Early loss of cerebrovascular autoregulation, with a decreased response to changes in CO_2
- Transient breakdown of the blood–brain barrier (BBB), probably because of endothelial changes (although endothelial tight junctions may remain intact early)

The loss of autoregulation makes the brain particularly susceptible to fluctuations in systemic blood pressure; for example, systemic hypertension can increase the risk for hyperemia and brain swelling seen more commonly in younger patients. The early dysfunction of the blood–brain barrier results in rapid swelling of perivascular astrocytes, which peaks at about 1 hour after injury but begins to recover by 6 hours. Later endothelial changes include formation of intraluminal microvilli or blebs and craters, which peak about 6 hours after injury but can persist as long as 6 days. Although the clinical significance of these changes is still not known, they are probably related to the loss of autoregulation, to the altered vascular sensitivity to circulating neurotransmitters, and to cerebral edema.

Mechanisms of Secondary Tissue Injury

During the past decade, *delayed* secondary injury at the cellular level has come to be recognized as a major contributor to the accumulated tissue loss after TBI. A cascade of physiologic, vascular, and biochemical events is set in motion in injured tissue. This includes changes in arachidonic acid metabolites such as the prostaglandins and the leukotrienes, the formation of oxygen free radicals, and changes in neuropeptides, electrolytes such as calcium and magnesium, excitatory neurotransmitters such as glutamate or acetylcholine, lymphokines such as interleukin-1, or lactic acid (Table 26-1). These products can result in progressive secondary injury to otherwise viable brain tissue through a number of mechanisms, such as the following:

- Producing further ischemia or altering vascular reactivity
- Producing brain swelling (edema or hyperemia)
- Injuring neurons and glia directly or activating macrophages that result in such injury
- Establishing conditions favorable to secondary infection

In other words, much of the ultimate brain loss after TBI may not be due to the injury itself but rather to an uncontrolled vicious cycle of biochemical events set in motion by the trauma.

TABLE 26-1. *Secondary injury in traumatic brain injury*

Hypoxia/ischemia
Mass effect
  Delayed hematoma
  Brain swelling
    Cerebral edema
    Hyperemia
  Hydrocephalus
Infection
Potential cellular mechanisms
  Phospholipid metabolism
    Lipid peroxidation (arachidonic acid chain)
    Prostaglandins, leukotrienes
    Platelet-activating factor (PAF)
  Oxygen-free radicals
    Free iron catalysis
  Excitotoxic mechanisms
    Glutamate (NMDA) receptors
    Acetylcholine
  Neuropeptides
    Endorphins (dynorphin)
    Thyrotropin-releasing hormone (TRH)
  Calcium and magnesium metabolism abnormalities
  Central nervous system (CNS) lactic acidosis
  Axonal flow abnormalities

Arachidonic Acid Metabolites

As one of the breakdown products of phospholipids, arachidonic acid is particularly plentiful in the brain, and its metabolites are likely to play a role in secondary brain injury. Arachidonic acid is metabolized through two major pathways: the cyclooxygenase path, leading to the formation of prostaglandins; and the lipoxygenase path, leading to the formation of leukotrienes. Both pathways may play a role, but the cyclooxygenase pathway appears to be the most important in TBI; among the metabolites that may be most active are prostaglandin E_2 and thromboxane. Marked elevations in cerebrospinal fluid and brain prostaglandins occur within minutes of injury. Theoretically, these metabolites could produce secondary injury by inducing vasospasm, thrombosis, and edema.

Oxygen Free Radicals

Oxygen free radicals are also active species biologically and are produced early in ischemic and traumatic tissue injury, in both the central nervous system and elsewhere. The superoxide radical (O^*) is formed through a variety of mechanisms, including both the xanthine oxidase and the cyclooxygenase pathways, and results in tissue injury in its own right by combining directly with cellular elements. When combined with its own breakdown product, hydrogen peroxide, in the presence of free iron, however, it forms the hydroxyl radical, (OH^*), which is even more destructive. The affinity of the hydroxyl radical for the abundant lipids in brain results in lipid peroxidation, with further release of arachidonic acid. A vicious cycle ensues in which more free radicals are produced through the cyclooxygenase pathway, along with prostaglandins, overwhelming natural superoxide scavenging mechanisms. The continued presence of free iron is essential for this vicious cycle, thus providing one likely explanation for the toxicity of free blood in TBI patients, including its possible relationship to posttraumatic epilepsy (PTE). Pharmacologic intervention to reduce the formation of such radicals or to scavenge those already formed would be expected to reduce ultimate tissue injury, and the complexity of the biochemical events involved provides several potential therapeutic avenues.

Excitotoxins

Another potential mechanism of secondary injury that has received increasing attention is the role of excitotoxins, especially of agonists of the NMDA subclass of glutamate receptors. Theoretically, the sustained release of excess amounts of such naturally occurring neurotransmitters after an injury can lead to eventual neuronal death. The mechanisms for this effect are not yet clear, but they may involve alterations in calcium and magnesium metabolism and activation of various enzyme systems. Among these might be phospholipase A, with consequent release of arachidonic acid and activation of the superoxide cycle discussed earlier. Experimental therapeutic interventions aimed at the excitotoxin mechanism include NMDA receptor antagonists, such as dextromethorphan or MK-801, and acetylcholine antagonists, such as scopolamine. Other endogenous agents that have received attention in recent years are various neuropeptides, such as the endorphins (particularly dynorphin) and thyrotropin-releasing hormone.

ASSESSMENT AND TREATMENT

Acute Care

Acute management of TBI is primarily directed at the prevention of secondary injury, especially that related to hypoxia-ischemia or to expanding mass lesions. An organized team approach is essential to accomplish this goal, from prehospital through intensive care unit (ICU) and postacute care. The cornerstone of early neuropsychiatric evaluation is the use of the Glasgow Coma Scale score (GCS), along with checks of lateralization, brain-stem function, and pupillary response; a system for easily recording sequential changes in these and other vital parameters is an integral part of trauma care (3). Although the GCS has been criticized for its simplicity, it has proved to be very reproducible across individual examiners and institutions and serves as a solid basis for evaluation of potential deterioration over time. More detailed neuropsychiatric examinations are probably not warranted until the patient is well stabilized in the ICU. A history from witnesses, particularly with regard to the onset of coma, is important not only for decisions on acute care but also for long-term prognosis. For example, the presence of an initial "lucid" or "semilucid" interval in a now-comatose patient suggests a possible hematoma requiring prompt surgery. At the same time, it makes severe DAI unlikely and points to a relatively favorable prognosis, provided there is no further secondary damage.

The importance of cardiopulmonary resuscitation and management in TBI care cannot be overstated; airway and shock management should be the top priority in any trauma patient. Superimposed hypoxia-ischemia can be the single most important determinant of ultimate outcome in severely head-injured patients. The traumatized brain is particularly sensitive to hypoxia-ischemia. In addition, levels of hypercarbia tolerated by the normal brain can lead to critical marginal increases in intracranial pressure (ICP) after TBI. Among the most important improvements in TBI patient care over the past decade has been the introduction of emergency care and transport systems that include paramedic training in early, on-site fluid resuscitation and intubation.

The comatose TBI patient is often hypoxic or hypercarbic, even though he or she may appear to be ventilating normally. Patients in coma (GCS > 8) should thus be intubated and hyperventilated, if possible, to a pCO_2 of 25 to 30 (but preferably not below that level for more than brief periods). Sedation with morphine, 4 to 12 mg intravenous every 2 to 4 hours, to prevent systemic hypertension, or paralysis with pancuronium bromide, 4 mg every 2 to 4 hours, should be used as needed. The stomach should be emptied to prevent aspiration. Immobilization of the head in the plane of the body is advisable not only because of the possibility of associated cervical fracture (about 5%) but also for airway maintenance and prevention of venous occlusion, which might raise ICP. Elevation of the head further facilitates craniovenous return.

Shock should suggest the possibility of hemorrhage elsewhere in the body. Fluid resuscitation should rely on normal saline or Ringer's lactate solution, but TBI patients should not be overly hydrated; central venous pressure monitoring can be helpful in this regard. Dextrose and water should be avoided not only because it is hypotonic but also because of the potential for increased lactic acidosis and cerebral necrosis in the hypoxic-ischemic patient with elevated blood sugars. Although formal studies have not been done on this latter issue in head-injured patients, we prefer to avoid maintenance with any dextrose solutions in the early acute phase. In addition to standard laboratory tests, evaluation for coagulopathies with platelet count, prothrombin and partial thromboplastin times, thrombin time, fibrinogen, and fibrinogen split products may also be indicated.

Radiologic Examination

Computed tomography (CT) has become standard in the management of mass lesions in the head-injured patient and should be used when available to assess all patients with a GCS of less than 12 ("does not obey commands") or when focal signs accompany a mild head injury. This should be done as soon as possible after the patient has been resuscitated and stabilized. As noted earlier, delays of more than 4 hours after injury in evacuation of hematomas have been associated with significant deterioration in outcome. Comatose patients, however, must remain accompanied by a physician or critical care nurse; often a "stabilized" patient arrests or suffers irreversible brain damage because of a simple airway problem in the elevator on the way to the CT suite. The usefulness of magnetic resonance imaging (MRI) in the acute situation is limited in part by the difficulty of managing the comatose patient in most scanners. In conscious patients with mild confusion and no lateralizing signs, a skull radiograph and observation may be sufficient. A fracture seen on the radiograph, however, increases markedly the risk for a surgical lesion and is indication for a CT scan even in the alert TBI patient. In any case, a high index of suspicion for delayed hematomas is imperative.

Intensive Care Unit

After a surgical mass lesion has been treated or excluded, the comatose patient should be managed in the ICU. The avoidance of secondary insults to the brain remains the principal goal of therapy. The same principles of care and treatment just given for earlier stages of care are generally continued, and, as before, organization, training, and adherence to fairly simple principles are the mainstay of care.

Intracranial Pressure Monitoring

Although many physicians, particularly neurologists, may be reluctant to give up the neurologic examination as the principal measure of patient progress, recent studies suggest that the ICP, which is one determinant of cerebral perfusion pressure, is a more sensitive parameter. For example, the classic Cushing's triad has been shown to occur less than 25% of the time in patients with ICP greater than 30 mm Hg, a concentration that almost invariably proves fatal if not controlled. It is much easier to prevent a rise to that level by treating when the patient is at 15 mm Hg than it is to bring ICP down from a level of 25 to 30 mm Hg. ICP has been shown repeatedly to correlate significantly with outcome, and its monitoring is increasingly used in the care of the comatose TBI patient.

A relatively simple algorithm in use for treating ICP elevations is the therapeutic intensity level (TIL). It has the advantage of also providing a standard measure of severity of ICP elevations in TBI patients. The TIL outlines an orderly increase in therapeutic vigor from simple sedation through barbiturate coma. Although treatment is always individualized for each patient, a new level of therapy is generally instituted when the previous level has failed to control ICP below 20 mm Hg. Each specific therapy is assigned a point value; the TIL score at a specific time is the sum of points for the interventions in use at that time. Thus, a patient with an ICP of 15 mm Hg at a TIL of 12 is quite different from a patient with the same ICP at a TIL of 3. It should be emphasized that using the TIL algorithm must not replace entertaining the possibility that progressive ICP elevations may also occur because of surgical lesions, such as delayed hematoma, or hydrocephalus. Similarly, seizures, hyponatremia, and airway problems increase the ICP.

Barbiturate coma is the last step in the recommended nonsurgical control of ICP and has been shown to improve outcome in patients with otherwise uncontrolled ICP. Barbiturate coma is induced with pentobarbital at an initial loading dose of 10 mg/kg intravenously over 30 minutes. An additional 5 mg/kg is then given every hour for three doses, always with close monitoring of blood pressure. Serum levels should then be maintained at 3 to 4 mg/100 mL with doses of about 1 mg/kg/h.

Medical Therapy for Prevention of Secondary Injury

Specific medical therapy aimed at minimizing secondary injury is still in its infancy. Ideally, in this context, active clinical research begs to be integrated with clinical care. Such treatments generally should be started as soon as possible after the injury, preferably even before the patient goes to surgery or the ICU. Mannitol is the most valuable of the agents presently available, perhaps because in addition to its osmotic effects, it is also an oxygen free radical scavenger. The usual initial dose is 1 g/kg in adults. Patients are then maintained on 0.25-g boluses every 4 hours as necessary to control ICP (see earlier) as long as serum osmolarity is less than 310. Some surgeons advocate the continued use of low-dose mannitol. Other diuretics, such as furosemide, are still used by some practitioners in specific situations. Corticosteroids for acute TBI are still overprescribed, despite several well-controlled studies that show that steroids either show no benefit or have a deleterious effect on the metabolism of the TBI patient.

Posttraumatic Epilepsy

The overall risk for epilepsy in patients with CHI is relatively small: 2% to 5% overall and about 11% for patients with severe CHI (4). A higher incidence is observed in patients with depressed skull fracture (15%), hematoma (31%), or penetrating brain wounds (50%). In all cases, the risk decreases markedly as time passes. Although the relative risk of developing epilepsy after penetrating head injury (PHI) is still 25 times higher than the normal age-matched population 10 to 15 years after injury, most patients with PHI can be 95% certain of remaining seizure free if they have no seizures for the first 3 years after injury (5).

The ongoing debate over the use of prophylactic antiepileptic drugs (AEDs) in head-injured patients must be separated into two questions:

1. Are AEDs indicated in a patient with PTE?
2. Do prophylactic AEDs prevent the onset of PTE?

In light of data suggesting that most patients with one posttraumatic seizure will have recurrent seizures for some time, most clinicians reply yes to the first question. The use of prophylactic anticonvulsants to prevent the onset of PTE is the more controversial issue. It has been demonstrated that phenytoin, even when given under carefully monitored conditions with maintenance of adequate blood levels, does not prevent the development of PTE beyond the first week after injury (6). Prophylactic phenobarbital is theoretically preferable because of its suppressant effect on the kindling phenomenon and its reported superoxide radical scavenging effect, but further controlled studies of this and other agents are clearly needed. In any case, we recommend routine acute use of phenytoin or phenobarbital in high-risk CHI and in PHI patients for a period of 2 to 4 weeks only. Because of the sometimes subtle cognitive effects of phenytoin and phenobarbital, however, carbamazepine may be the agent of choice for longer-term therapy in patients who have manifested PTE with one or more seizures.

Assessing the Severity of a Head Injury

Kraus and Sorenson have described a variety of approaches to classify the severity of a traumatic CHI injury (7). Most commonly, severity of injury is defined on the basis of a patient's GCS score on entrance to an emergency room. Typically, on the GCS, a *mild* head injury is defined by a score that falls into the 13 to 15 range; a *moderate* head injury falls into the 9 to 12 range; and a *severe* head injury falls into the 0 to 8 range. The GCS is used most frequently to record the severity of a head injury because it tends to be a relatively objective scale, whereas other indices of head injury severity, such as duration of loss of consciousness or posttraumatic amnesia, may be more difficult to document. Greater severity of head injury, as measured by the GCS score, correlates well with a poorer outcome as measured by global scales such as the Glasgow Outcome Scale or more detailed neuropsychological tests.

Postacute and Long-Term Rehabilitation

The field of TBI rehabilitation has grown in recent years. Multiple therapies, including coma stimulation, reality orientation, cognitive rehabilitation, speech therapy, occupational therapy, and recreation therapy, among others, have been applied to the TBI patient. Yet their use has been largely empirical, and there has been a paucity of scientific validation for these sometimes expensive interventions (including comparison with minimal care, supportive models). If progress is to be made in this area, rehabilitation modalities must be subject to the same scrutiny for indications, dosage, duration of treatment, and efficacy as are other medical treatments, such as drugs. The most pressing challenge in the field is the development of reproducible, universally accepted measures of function and ultimate outcome with which to compare the value of various interventions.

One of the most encouraging aspects of TBI rehabilitation is the amazing ability of the young adult brain to *compensate* for many aspects of injury naturally. This is particularly apparent in head injury, as opposed to progressive conditions such as MS, or even stroke in older people. Disabilities such as hemiparesis, seizures, and certain language disorders may appear more dramatic initially, but the most devastating long-term impairments are the cognitive and especially the attentional and behavioral deficits that often persist after TBI. The goal of therapy should be the independence and community reintegration of the patient within his or her limits, rather than the specialized treatment of specific deficits simply because "they are there." Scarce resources available to the patient are often depleted in the early acute and postacute phases on evaluation and therapy of deficits that will improve anyway or that have little effect on the ultimate goal of independence. Some therapies may actually be counterproductive by fostering continued dependence. Interventions that may be more cost-effective, like training in specific community reintegration skills such as decision making and certain forms of behavioral modification, may end up being omitted for lack of funds.

MRI may be especially useful at this stage in identifying clinically significant focal contusions, but electrophysiologic studies such as electroencephalogram and evoked responses have not proved to be particularly helpful. Evaluation at this stage should include particular attention to input from family and attendants who spend considerable time with the patient. Neuropsychological testing is an important part of the evaluation but is of limited value in the confused patient. When performed, these tests should focus on measurement of expected deficits for guiding therapy and evaluating progress (attentional deficits, posttraumatic amnesia) rather than on standard batteries that seek to confirm anatomic deficits already identified on MRI or CT, or that investigate in detail cognitive domains of limited practical interest to the case.

The use of pharmacologic agents (and particularly psychotropic medications) in TBI rehabilitation continues to hold much promise, but treatment still remains largely empirical. The sensitivity of the traumatized brain or the confused patient to medication must always be considered, and treatment must be tailored to the individual patient. Overmedication with AEDs, sedatives, or stimulants is a frequent problem; paradoxical responses to sedation in confused patients are especially common. Nevertheless, judicious use of adequate sedation can help reestablish sleep–wake cycles; and methylphenidate, dextroamphetamine, or bromocryptine may be useful as adjuncts in the management of the lethargic or apathetic patient. Carbamazepine is also a possible useful adjunct in the management of certain behavioral problems.

OUTCOME

A surprisingly good overall outcome can be seen in many young, moderately severe to severely injured patients, a finding that probably reflects compensation for lost functions more so than recovery of the injured tissue itself. Thus, "floating" or dynamic end points can be identified in the post-TBI course:

- Resolution of coma
- Return to orientation
- Resolution of posttraumatic amnesia
- Resolution and stable duration of retrograde amnesia
- Number of significant deficits identified on initial neuropsychological testing
- Number of significant deficits identified on the last neuropsychological evaluation
- Steepness of the recovery slopes

Final outcome is a composite of a number of elements, including preinjury, neurologic, cognitive, behavioral, and psychosocial functions, all of which may interact differently in each individual patient. When evaluating efficacy of a given therapy, one must study all of these elements in the context of outcome as a whole; any evaluation battery should thus include at least some measure of each. Return to gainful employment may be a practical one in most TBI populations. Patients with aphasia, posttraumatic epilepsy, hemiparesis, visual field loss, verbal memory loss, visual memory loss, psychological problems, or violent behavior are less likely to be employed.

Cognitive and Behavioral Aspects of Outcome

Given the mix of pathologies that could affect outcome in TBI, it may seem surprising that a typical syndrome of cognitive and behavioral change can be described. Some TBI patients may present striking deficits in language, perception, and visuospatial processing, but the more typical picture after TBI emphasizes deficits in memory, attention, personality, and social cognition.

Memory problems include a notable loss of explicit retrieval of new information presented after injury (posttraumatic amnesia) along with a more modest period of retrograde amnesia for events preceding the injury (8). Immediate recall and older memories, however, are generally intact, as are implicit or automatic memory processes. Temporal order judgment may be affected. The deficit thus appears to be primarily in the encoding and "consolidation" of new episodic memories and may reflect hippocampal, reticular, thalamic, or even basal forebrain damage. In the case of story recall, head injury patients may have difficulty not only in recalling the story but also in conceptualizing its thematic elements. This may give them particular difficulty in selectively recalling the most important story elements. Head injury patients may be able to use semantic information in encoding and retrieving information, but they are less successful at this than controls. Of course, the longer the duration of posttraumatic amnesia and the more severe the initial head injury as judged by the emergency room GCS score, the more likely the patient is to experience persistent memory problems. These same predictive factors appear important for predicting other persistent cognitive deficits.

Memory problems may be accompanied by slowed information processing (identified by slowed response times or increased latencies in the late event—-related brain potentials such as the P300 or N400). This slowing of information processing may be due to a combination of diffuse axonal shearing, damage to projection systems from the brain stem, and focal frontal lesions. Slowed information processing can affect the quality of memory encoding and rehearsal, as well as fluency and response times independent of motor control deficits. Often, access to a source of knowledge is slowed even though the search through the semantic network where that knowledge is stored may be normal. Dramatically slowed information processing usually recovers after several months, except in the case of the most severely injured patient.

Attentional deficits are frequently reported after CHI, and they usually take the form of a deficit in sustained effortful attention. The patient may also have difficulty refocusing his or her attention after a period of delay. This deficit may not be as apparent if the task is of short duration. It is not always clear whether the subject is distracted by other stimuli on the screen or in the room or by internalized thought. It is usually characterized by omitted responses and increased variance in response times to targets. Patients may also have difficulty inhibiting responses to previously associated stimuli, even when the association is currently irrelevant. These deficits are associated with damage to the prefrontal cortex. Attentional control deficits may also result in a diminished ability to divide available cognitive resources to handle a multifaceted task.

A deficit in concept formation is frequently reported by head-injured patients and their families. This problem is usually coupled with an inability to shift mental set, manifested by perseveration on tasks such as the Wisconsin Card Sorting Test. Patients may appear to use concrete problem-solving strategies. They may also have trouble initiating any problem-solving strategies unless they are encouraged to do so by the environment or another person. These deficits in executive functions almost always follow prefrontal lobe lesions (most frequently with dorsolateral lesions) but may even appear after reticular thalamic lesions. Patients may demonstrate reasonably intact memory, perception, language, and even attention yet still have major executive function deficits. A subtler form of this problem may even appear after so-called minor closed head injury.

Perhaps the most impressive problem in TBI patients (usually reported by family members) is a change in mood and personality (9). The patient may be disinhibited, behave irrationally, have mood swings, ignore social convention, and not care about the future consequences of his or her current actions. A few may also manifest aggressive or even violent behavior for months or years after the injury. This aggressive behavior can be either directed or nondirected and may be accompanied by an often explosive autonomic and emotional response. Some of these "episodic dyscontrol" events are thought by some to be a subtle form of temporal lobe epilepsy, particularly because they often appear to respond to AEDs such as carbamazepine. The most likely neuropathologic basis for these social cognitive and behavioral disorders is damage to the prefrontal cortex, including the orbitofrontal region. Family members may benefit from neuropsychiatric intervention. Early in the course of recovery, family members may appear relieved simply to be told that the patient will survive and recover. As time passes, however, significant others can become overwhelmed in their effort to adjust to the permanent neurobehavioral changes in a loved one.

Neuropsychological Assessment

The neuropsychological assessment of TBI can begin at a very early stage. The Galveston Orientation and Amnesia Test can be used with the disoriented patient and is designed for repeated administration (8). It allows the investigator to evaluate the duration and severity of posttraumatic amnesia as well as the duration of retrograde amnesia. This evaluation is important because the duration of posttraumatic amnesia is a powerful predictor of eventual outcome. Other tests that make minimal demands on the patient, such as choice reaction time tasks (to measure speed of information processing), picture naming (to measure visual recognition and name retrieval), verbal fluency (to measure strategic memory search and name retrieval), and letter cancellation (to measure spatial attention), can aid the investigator in establishing the acute recovery slope (over days or weeks early in the course of recovery) and in determining when more extensive neuropsychological testing can be attempted.

The more formal neuropsychological evaluation should incorporate standard clinical measures, such as the Wechsler Adult Intelligence Scale—Revised, the Wechsler Memory Scale—Revised, and the neurobehavioral rating scale. Other, more experimental tests of memory, attention, personality, and social cognition are used in TBI, but none has emerged as specific for TBI, and a full review is beyond the scope of this chapter. The specific battery used in a given hospital is often dependent on the particular interests of the attending neuropsychologist.

TBI patients are usually evaluated at least three times in the postacute recovery period. To chart the slope of cognitive recovery, it is important to include some repeatable tests in the evaluation that are relatively resistant to test–retest or "practice" artifact. Repeatable tests may be clinical tests that have different versions equated for difficulty, or information processing tests on which stimuli can be randomly selected for presentation. In these tests, the dependent measure reflects the subject strategy rather than the particular stimuli used. Examples are various versions of the selective reminding test and of choice reaction time tests.

Long-term recovery is usually assessed by evaluating the rate of improvement, preferably with at least three testing points spread over at least 1 year after the injury. In addition to the neuropsychological measures, other indices of recovery, such as functional independence in activities of daily living, employability, school performance, family and community adjustment, and other social functions, become more important at this stage.

"Minor" Head Injury

A group of patients who have been frequently mismanaged in the past are those with so-called minor head injury (10). Not only has axonal damage been demonstrated in animal models of concussion but also MRI as well as positron emission tomography have repeatedly shown structural and metabolic changes in humans with minor head injury as well. The most important element in the management of these cases is the recognition that there is usually an organic, pathologic basis for their complaints, at least in the early, postinjury period and that it usually resolves over a few months. If mishandled, however, these patients often develop an overlying neurosis that makes evaluation and management infinitely more difficult.

There is nothing more frustrating to the intelligent minor head injury victim than to be told there is "nothing wrong" by his or her physician, family, and employer. Proper counseling should thus include not only the patient but also the family, school, or employer. MRI, auditory evoked potentials, and specific neuropsychological tests such as choice reaction time early in the course can help delineate the deficits. The basic elements of the postconcussion syndrome are cognitive, somatic, and affective. Clinically significant neuropsychological impairments have been documented repeatedly even after minor "dings" without loss of consciousness. The most frequent somatic complaints are headache (71%), decreased energy or fatigue (60%), and dizziness (53%); these had all markedly improved at 3 months (11). The proper management of the fatigue element (which may relate to orbitofrontal injury) is a major factor in recovery and requires the cooperation of the school or employer. We suggest a graded return to full workload over a period of 4 to 8 weeks.

SUMMARY

Head trauma is a common cause of neuropsychiatric impairment. The development of therapeutic agents that can reduce the severity of the pathologic changes caused by the trauma should have a major effect on survival rates and the quality of that survival. The neuropsychiatric sequelae of head trauma include deficits in attention, executive functions, social cognition and personality, and memory. Moderate and severe head injuries are likely to lead to persistent problems in these areas. Mild head injury can often result in subtle but significant cognitive impairment that is often difficult to discriminate from a neurotic reaction to an experience of trauma. Although rehabilitation techniques may be of benefit to patients recovering from TBI, there are few controlled studies that can reliably document these benefits. The neuropsychiatrist is encouraged to assume an important new role in contributing to the evaluation and management of the cognitive and social-personality problems that emerge after TBI.

REFERENCES

1. Gennarelli T, Thibault L, Adams J, Graham D, Thompson C, Marcinin R. Diffuse axonal injury and traumatic coma in the primate. *Ann Neurol* 1982;12:564–574.
2. Povlishock J, Coburn T. Morphopathological^ change associated with mild head injury. In: Levin H, Eisenberg H, Benton A, eds. *Mild head injury.* New York: Oxford University Press, 1989:37–53.
3. Jennett B, Teasdale G. *Management of head injuries.* Philadelphia, PA: FA Davis, 1981.
4. Annegers J, Grabow J, Groover R, Laws EJ, Elveback L, Kurland L. Seizures after head trauma: a population study. *Neurology* 1980;30:683–689.
5. Weiss G, Salazar A, Vance S, Grafman J, Jabbari B. Predicting posttraumatic epilepsy in penetrating head injury. *Arch Neurol* 1986;43:771–773.
6. Salazar A, Jabbari B, Vance S, Grafman J, Amin^ D, Dillon J. Epilepsy after penetrating head injury. I. Clinical correlates. *Neurology* 1985;35:1406–1414.
7. Kraus JF, Sorenson SB. Epidemiology. In: Silver JM, Yudofsky SC, Hales RE, eds. *Neuropsychiatry of traumatic brain injury.* Washington, DC: American Psychiatric Press, 1994:3–42.
8. Levin H, Goldstein F. Neurobehavioral aspects of traumatic brain injury. In: Bach-Y-Rita P, ed. *Traumatic brain injury.* New York: Demos, 1989:53–72.
9. Brooks N, Campsie L, Symington C, Beattie A, McKinlay W. The five year outcome of severe blunt head injury: a relative's view. *J Neurol Neurosurg Psychiatry* 1986;49:764–770.
10. Levin HS, Eisenberg HM, Benton AL, eds. *Mild head injury.* New York: Oxford University Press, 1989.
11. Levin H, Mattis S, Ruff R, et al. Neurobehavioral outcome following minor head injury: a three-center study. *J Neurosurg* 1987;66:234–243.

27

Neurotoxicology

Roberta F. White, Robert G. Feldman, and Susan P. Proctor

OVERVIEW

Exposure to neurotoxicants often produces effects on cerebral structures and functions. Acute exposure may manifest obvious symptoms. Insidious and chronic exposures result in long-standing, undiagnosed toxic states, often characterized by more subtle and often permanent functional impairments. Attention has been focused on the need for earliest detection of these "subclinical" toxic states, especially those behavioral syndromes associated with exposure to particular neurotoxicants (1).

This chapter describes an approach to the evaluation of toxin-related neurologic dysfunction and clinical differential diagnosis. Emphasis is given to methods of assessing behavioral syndromes associated with neurotoxic exposures.

NEUROLOGIC EVALUATION AND DIAGNOSIS

Diagnosis of neurologic abnormalities secondary to exposure requires evaluation to ascertain that abnormal observed findings are not due to any other primary neurologic disease and can be attributed only to the exposure. Many neurologic symptoms and examination findings are similar in toxic encephalopathy and other disorders. The strategies used to reach a diagnosis in occupational and environmental neurology include both a detailed exposure history and symptom review, a careful neurologic examination, and the use of selected sensitive neurophysiologic and neuropsychological test batteries. To evaluate alterations in central nervous system functioning and to identify structural versus functional disturbances, certain laboratory tests are useful:

Computerized axial tomography (CT) can indicate ventricular size, symmetry, and presence of mass lesions. Experience with certain toxic exposures suggests that they are sometimes associated with the finding of cortical atrophy on CT scan.

Magnetic resonance imaging (MRI) provides essentially the same information as the CT scan except that white and gray matter can be differentiated more clearly. Our experience with MRI in toxicant-exposed patients also suggests that atrophy is a consistent finding. In addition, we are seeing evidence of small white-matter lesions in some patients, some of whom seem to improve in the absence of exposure. Other investigators have reported seeing white-matter lesions after toluene exposure (2).

Electroencephalography (EEG) records the electrical activity of the brain during waking and sleeping. It is useful in demonstrating asymmetries of electrical activity due to mass lesions, identifying focal disturbances, showing epileptic tendencies, and demonstrating slowing due to toxic or metabolic conditions.

NEUROPSYCHOLOGICAL TESTING

Formal neuropsychological testing offers reliable, standardized procedures for objectively evaluating specific aspects of change in cognitive function. There are specific, standardized rules for administering and scoring these tests. Results may be analyzed objectively through

TABLE 27-1. *Neuropsychological assessment of possible toxic encephalopathy*

Domain	Description	Implications
General Intellect		
Wechsler IQ tests (WAIS-R, WISC, WPPSI)	IQ measures	Overall level of cognitive function compared with population norms
Peabody Picture Vocabulary Test	Single-word comprehension	Measure of verbal intelligence in adults; can be sensitive to exposure in children
Stanford-Binet	IQ measure	Similar to Wechsler tests
Wide Range Achievement Test	Academic skills in arithmetic, spelling, reading	Estimate of premorbid ability patterns in adults; can be sensitive to exposure in children
Attention, Executive Functioning		
Digit Span (WAIS-R)	Digits forward and backward	Measures simple attention and cognitive tracking
Arithmetic (Wechsler tests)	Oral calculations	Assesses attention, tracking, and calculation
Trail Making Test	Connect-a-dot task requiring sequencing and alternating sequences	Measures attention, sequencing, visual scanning, speed of processing
Continuous Performance Test	Acknowledgment of occurrence of critical stimuli in a series of orally or visually presented stimuli	Assesses attention
Paced Auditory Serial Addition	Serial calculation test	Sensitive measure of attention and tracking speed
Wisconsin Card Sorting Test	Requires subject to infer decision-making rules	Tests ability to think flexibly
Verbal, Language		
Information (Wechsler tests)	Information usually learned in school	Estimate of native abilities in adults
Vocabulary (Wechsler tests)	Verbal vocabulary definitions	Estimate of verbal intelligence; sensitive to concreteness associated with brain damage (including toxic encephalopathy)
Comprehension (Wechsler tests)	Proverb definitions, social judgment, problem solving	Sensitive to reasoning skills; can be impaired after exposure to neurotoxicants
Similarities (Wechsler tests)	Inference of similarities between nominative words	Sensitive to reasoning skills; can be impaired after exposure to neurotoxicants
Controlled Oral Word Association	Word list generation within alphabetical or semantic categories	Assesses flexibility, planning, arousal, processing speed, ability to generate strategies, somewhat sensitive to exposure
Boston Naming Test	Naming of objects depicted in line drawings	Sensitive to aphasia; also sensitive to native verbal processing deficits or those acquired through childhood exposure
Reading Comprehension (Boston Diagnostic Aphasia Exam)	A direct screening test of simple reading comprehension	Sensitive to moderate-to-severe dyslexia, usually insensitive to toxic exposure in adults
Writing Sample	Patient writes to dictation or describes a picture	Assesses graphomotor skills, spelling
Visuospatial, Visuomotor		
Picture Completion (Wechsler tests)	Identification of missing details in line drawing	Measures perceptual analysis
Santa Ana Formboard Test	Knobs in a formboard are turned 180 degrees with each hand individually and both hands together	Measures motor speed and coordination
Finger tapping	Speed of tapping with each index finger	Sensitive to lateralized manual motor speed
Memory		
Logical Memories-Immediate and Delayed Recall (IR, DR) (Wechsler Memory Scales)	Recall of paragraph information read orally on an immediate and 20-minute delayed recall	Sensitive to new learning and retention of newly learned information
Verbal Paired Associate Learning DR (Wechsler Memory Scales)	Two paired words are presented in a list of pairs; subject must recall second word; test is presented on immediate and delayed recall	Measures abstract verbal list IR, learning, retention
Figural Memory (Wechsler Memory Scales)	Multiple choice recognition of using recognition (not recall) performance measures	Assesses visual recognition memory
Digit Symbol (Wechsler tests)	Coding task requiring matching symbols to digits	Complex task assessing motor speed, visual scanning, working memory
Picture Arrangement (Wechsler tests)	Sequencing of cartoon frames to represent meaningful stories	Measures visual sequencing, ability to infer relationships from visuospatial/social stimuli
Block Design (Wechsler tests)	Assembly of 3-D blocks to replicate 2-D representation of designs	Assesses abstract visual construction ability and planning
Object Assembly (Wechsler tests)	Assembly of puzzles	Measure of concrete visual construction skills, gestalt recognition
Boston Visuospatial Quantitative Battery	Drawings of common objects spontaneously and to copy	Measures constructional abilities, motor functioning

TABLE 27-1. *Continued.*

Domain	Description	Implications
Hooper Visual Organization Test	Identification of correct outline of drawings of cut up objects	Sensitive to gestalt integration processing
Rey-Osterreith Complex Figure (copy condition)	Drawing of a complicated abstract visual design	Sensitive to deficits in visuospatial planning and construction
Visual Paired Associate Learning, IR, DR (Wechsler Memory Scales)	Six visual designs are paired with six colors; recognition memory is tested immediately after the six are presented on learning trials and at delayed recall	Test of abstract visual learning using recognition (not recall) performance measures
Visual Reproductions, IR, DR (Wechsler Memory Scales)	Visual designs are drawn immediately after presentation and on delayed recall	Measures visual learning and retention
Delayed Recognition Span Test	Based on delayed nonmatching to sample paradigm, discs are moved about on a board to assess recognition memory for words, color, spatial locations	Assesses new learning
Peterson Task	Words or consonants are presented and must be recalled after a period of distraction	Measures sensitivity to interference in new learning
California Verbal Learning Test	Subject is presented with list of 16 words (which can be semantically related) over multiple learning trials and with an interference list	Provides multiple measures of new learning recall, recognition memory, use of strategies and sensitivity to interference
Rey-Osterreith (IR, DR)	Complex design is drawn from IR immediately after it has been copied and at a 20-minute delayed recall	Assesses memory for visual information that is difficult to encode verbally
Personality, Mood		
Profile of Mood states	65 single word descriptors of affective symptoms are endorsed by degree of severity on six scales	Sensitive to clinical mood disturbance and to affective changes secondary to toxicant exposure
Minnesota Multiphasic Personality Inventory (R)	True–false responses provided on personality inventory summarized on multiple clinical dimensions	Provides description of current personality function; some scales sensitive to exposure; screening for inconsistency and malingering

Modified from White RF, Proctor SP. Research and clinical criteria for the development of neurobehavioral test batteries. *J Occup Med* 1992;34:140–148.

published normative data on the expected ranges of scores based on age, sex, and education. In addition, the tests have been validated in both research and clinical settings on patients with known brain damage, allowing clear interpretation of results by independent neuropsychologists. Reliability of the tests (test and retest) is often established. Finally, the tests are used widely, and results from them may be applied in many diagnostic and treatment settings.

In clinical situations, it is generally more appropriate to select neuropsychological tests that assess as many functions as possible to provide a profile of abilities and deficits in a given patient. Perhaps the most difficult situation arises with respect to estimating the severity of past exposure. Evaluation of deficits often occurs after exposure has ceased, at a time when the body burden of suspected substances is no longer elevated or environmental levels can no longer be ascertained. Obtaining an occupational and environmental history to estimate the extent of exposure (duration and intensity) is of the utmost importance. Essential to this inquiry are questions about the patient's current and previous occupations, job tasks, places of residence and employment, and hobbies. To elicit information about exposures to known neurotoxic substances, a checklist of neurotoxicants can be used for patient response.

An extensive neuropsychological test battery for clinical assessment has evolved (Table 27-1). The battery is designed to cover a wide range of functions and to analyze as many manifestations of neurotoxicant-induced encephalopathy as possible. The battery is extensive enough to allow localization of cerebral dysfunction in cases with specific constellations of deficits, and to aid in the differential diagnosis of toxic encephalopathy if cognitive deficits are observed.

BEHAVIORAL EFFECTS OF EXPOSURE TO NEUROTOXICANTS

Metals

Inorganic Lead

Lead neurotoxicity has been associated with signs of encephalopathy such as restlessness, irritability, tremor, poor memory, drowsiness, and stupor, progressing to seizures, coma, and death in severe forms of lead poisoning. As cases of severe acute toxicity have become less frequent, increasing attention has been focused on the milder, subclinical manifestations of lead exposure. Evidence suggests that changes in behavioral and cognitive functioning are observable in patients with blood lead levels greater than 70 mg/100 dL. Although results in studies of adults have not always been consistent, there is good reason to suggest that lead exposure can lead to impairments in affect, attention, psychomotor function, verbal concept formation, short-term memory, and visuospatial abilities.

In lead-exposed workers evaluated in our Environmental and Occupational Neurology program, we have observed dysfunction in affect, attention and cognitive tracking, short-term memory, verbal reasoning, and concept formation as well as loss of motor coordination and speed and visuospatial abilities. Blood-lead levels in such patients are generally highly elevated (60 to 90 mg/100 dL). Mood disorders are common, with subjective complaints of apathy, irritability, and diminished ability to control anger. Attentional deficits may be observed on simple tasks, such as the Continuous Performance Test, and on repeating digits forward. These deficits become even more pronounced on tasks requiring holding and manipulating information (e.g., reciting digits backward, counting by threes beginning with the number 1) and sequencing of material. Difficulties in accurately completing oral arithmetic problems are common among lead-exposed patients and appear to reflect attention deficits.

Despite the aforementioned areas of dysfunction, certain well-retained abilities are apparent in *adult* patients with lead levels of less than 90 mg/100 dL. Language skills, long-term memory, recall of information learned in school, and reading appear to be unaffected in the adults tested. This is not true in children, however. Needleman and colleagues, controlling for several sources of error that may have contributed to these discrepancies, found significant impairment in neuropsychological test function in children whose teeth had the highest dentin lead levels (3). Other studies have also demonstrated significant effects of lead exposure on test performance in children (4–6). Again, deficits were observed in the areas of psychomotor speed, verbal concept formation, visuospatial abilities, and attention.

Mercury

Elemental and inorganic mercury compounds are used in industry, agriculture, and medicine. Organic mercury compounds are also found in fungicides, preservatives, denaturants for ethyl alcohol, antiseptics, and herbicides.

Accidental absorption of organic mercury occurs as a result of inhalation, percutaneous absorption, or ingestion. Acute intoxication has reportedly resulted in abdominal pain, nausea, vomiting, diarrhea, headaches, and chills. Behavioral disturbances manifested by mood changes, depression, irritability, and changes in personality may result from chronic mercury intoxication. Behavioral effects of mercurialism have been classified into three major areas (7):

- Disturbances of the motor system, manifested by fine muscle tremor
- Deterioration of intellectual capacities (memory, concentration, and logical reasoning)
- Alterations of emotional state with associated symptoms of depression, fatigue, listlessness, irritability, and social sensitivity

Kurland and coauthors described massive destruction of cerebellum, calcarine cortex, and basal ganglia that resulted from the ingestion of methyl mercury-containing fish by people and animals living in the vicinity of Minamata Bay, Japan. The onset of illness in this outbreak began with a progressive peripheral neuropathy, followed by ataxia, dysarthria, deafness, blindness, the development of spasticity, and intellectual impairment (8).

In contrast to these severe neurologic effects, subtle changes, such as those seen in "micromercurialism," may be the result of chronic low-level exposure. The affective symptoms of anxiety and depression have been observed.

Arsenic

Arsenic, a constituent of numerous minerals, is found in soil, water, and foods, particularly seafood. This element exists in various chemical states, each of which has its own toxicologic potential. Arsenic trioxide is recovered as a byproduct from smelting copper-, lead-, and gold-bearing ores. Arsenic compounds are used in insecticides, herbicides, and wood preservatives; in solder and steel as an alloying ingredient; as oxidizing and refining agents in glass manufacture; as preservatives in tanning and taxidermy; and as a component of semiconductors.

Inhalation and ingestion are the main routes of entry of arsenic into the body. Once absorbed, arsenic is widely distributed in the tissues, including the liver, abdominal viscera, bone, skin, hair, and nails. Remarkably, arsenic can be detected in hair and nails months after it has disappeared from urine and feces. Both peripheral nervous system and central nervous system effects have been documented after acute and chronic exposures to arsenic. Electrophysiologic studies have provided objective quantification of functional impairment in the peripheral nervous system, including both sensory and motor dysfunction, after exposure to arsenic. Only case reports exist to link arsenic exposure with neurobehavioral syndromes, however, such as lassitude, dizziness, fatigue, and emotional lability.

Manganese

Manganese is especially known for its alloying properties with both ferrous and nonferrous metals. In the form of manganese oxide, it is used in the manufacture of paints, disinfectants, fertilizers, varnishes, and pyrotechnics and as a depolarizer in cell batteries. Principal sources of industrial exposure occur in mining, transporting, crushing, and sieving of ore. Exposure primarily results from inhalation of metal dust or fumes and can occur in industrial workers as well as in people living in the areas surrounding manganese smelters.

The patterns of neurologic deficit in severe manganese intoxication are characteristic and are marked by a behavioral and postural (dystonic) syndrome. The neurotoxicity of manganese was first recognized after the appearance of neurologic disorders in manganese miners in Chile and India. Early neurologic patterns were suggestive of an acute disorder and included irritability, nervousness, and emotional instability. For some, visual and auditory hallucinations appeared in addition to compulsive, repetitive, uncontrollable actions.

The second stage of manganese intoxication, sometimes referred to as *Mn psychosis,* may, after months of continued exposure, progress into a stage characterized by speech disorder, gait disturbance, slowness and clumsiness of movement, and postural imbalance. Signs of dystonia appear in the third stage. These signs include an awkward high-stepping gait ("cock walk"), as well as tremor and chorea.

Solvents

Carbon Disulfide

Carbon disulfide, a heavy volatile liquid, is a solvent for lipids, sulfur, rubber, oils, waxes, industrial chemicals, phosphorus, and resins. Primarily, it is used in the agricultural industry in various insecticides, in the rubber industry, and in the rayon industry, where it is used for the preparation of rayon viscose fibers. Inhalation of the vapor is the major route of entry, although percutaneous absorption of both liquid or vapor may also occur. Perhaps the best-known manifestation of advanced carbon disulfide intoxication is polyneuropathy. Parkinsonism has also been described after carbon disulfide exposure, as have CT scan abnormalities. Hanninen and associates demonstrated a broad range of psychological effects that could be differentiated in acute and latent carbon disulfide poisoning (9).

Most investigations of the effects of carbon disulfide have focused on visuomotor function, attention, and intelligence, which are motor and cognitive processes primarily dependent on the integrity of the central nervous system. Other behavioral disturbances, called *psychoneurasthenic* difficulties, have also been related to carbon disulfide exposure. These difficulties may represent an organic affective disturbance and include changes in mood or personality; excessive irritability; increased physical complaints, such as headache, dizziness, weakness, and fatigue; and memory loss. These complaints usually precede the onset of more obvious neurologic involvement, that is, peripheral neuropathy. Frequently, depressive symptoms have been reported.

Carbon disulfide has been shown convincingly to cause impaired psychomotor function, especially affecting dexterity and speed. Tests that measure these changes reliably include the Santa Ana Dexterity Test and tests of simple reaction time. Higher cortical functions, especially visuomotor abilities and concentration, are also affected by carbon disulfide exposure and may be assessed by using the Block Design, Digit Symbol, and Digit Span subtests of the Wechsler Adult Intelligence Scale-Revised (WAIS-R). Subtle symptoms, such as mood changes, irritability, and increased systemic complaints, may precede overt neurologic illness from carbon disulfide exposure and can be assessed by the use of standard interview questionnaires.

Trichloroethylene

Trichloroethylene (TCE), a colorless aliphatic hydrocarbon liquid, is a widely used industrial solvent, particularly in metal degreasing and extraction processes. It is also used in many chemical processes, in the cleaning of optical lenses and photographic plates, in painting and enameling, and as an adhesive in the shoe industry. In these processes, TCE may be heated and is easily volatilized, increasing the danger of inhalation, the most significant route of exposure. In the presence of light, flame, or alkaline substances, TCE can produce decomposition products, namely dichloroacetylene, chlorine, hydrochloric acid, carbon monoxide, and phosgene.

The toxicology of TCE has been extensively described. Occurrences of undesirable effects after exposure to TCE are frequent in industrial settings, with central nervous system depression as the predominant physiologic response to exposure. Typical manifestations include visual disturbances, mental confusion, fatigue, and impaired concentration. Of particular note is the predilection of TCE for selective neurotoxic action on the trigeminal nerve and the lasting depression after significant exposure (10).

The behavioral aspects of TCE intoxication have been reported to include effects on reaction time and motor dexterity.

Toluene

Toluene, a colorless liquid derived from coal tar, is used as a solvent and thinner in the printing, rubber, lumber, furniture, and chemical industries. It is a basic component in many glues, lacquers, inks, cleaning liquids, paints, and adhesives. Atmospheric toluene largely results from motor vehicle vapor emissions because toluene is a constituent of gasoline. As with other solvents, inhalation is a major route of entry; to a lesser extent, percutaneous absorption occurs. Exposure to intensely high levels may occur through accidental inhalation at the workplace or intentional inhalation as in the case of glue sniffing for "recreational" purposes.

The principal health effect of toluene is on the central nervous system. Low levels produce reduced performance and perceptions of fatigue and dizziness, whereas higher levels have excitatory effects such as euphoria, exhilaration, and agitation. Although the inhaled vapors usually induce a temporary euphoria, addiction can result, leading to serious damage to the nervous system. Caution must be used in generalization of results from epidemiologic studies because exact exposure levels and knowledge of other concomitant exposures are not always known, and toluene is often used in combination with other solvents. Several case reports exist describing a constellation of symptoms for toluene abusers. These include mental confusion, inappropriate laughter, suicidal tendencies, dyscoordination, and emotional lability. Toluene has been associated with cerebellar dysfunction and with lesions in cerebral white matter, with associated neuropsychological deficits.

Perchloroethylene

Perchloroethylene (PCE, tetrachloroethylene), a colorless liquid, is another major solvent used particularly in dry cleaning, fabric finishing, metal degreasing, and other applications. Exposure to PCE may occur as a result of inhalation of vapors and secondarily through direct skin contact. Symptoms attributable to central nervous system (CNS) depression have been reported and include vertigo, impaired memory, confusion, fatigue, drowsiness, irritability, loss of appetite, and lack of coordination.

SUMMARY

In 1983, a Joint World Health Organization and National Institute for Occupational Safety and Health Workshop on Neurotoxic Illness addressed the importance of international collaboration in the development, validation, and application of neurobehavioral tests for detecting neurotoxic illness. In developing neuropsychological test batteries to assess occupational exposures for both research and clinical settings, we have used the following guidelines:

- Examination of a broad range of cognitive functions
- Inclusion of tests that are ecologically valid, that is, tests that will generate predictions about the patient's functioning in daily life
- Administration of tests known to be sensitive to neurotoxicant exposures
- Inclusion of tests with demonstrated clinical utility in other settings, that is, tests available and familiar to practicing clinicians
- Inclusion of tests that have been validated on patients with specific types of brain damage

In completing occupational studies, we have found it necessary to edit carefully the test battery because of the limited time that is usually available for testing. Thus, in addition to the guidelines just listed, we consider brevity of testing as essential.

Test batteries for clinical patients with suspected neurotoxicant-related encephalopathy can generally include greater numbers of tests and time spent in testing. In addition, the patient is usually self-motivated to participate fully and attentively in the testing process. However, several problems are encountered in working with these patients.

One such problem is that of *hypochondriacal response* to occupational or environmental exposures. We have tested a number of patients who are convinced that they have suffered intellectual changes and memory disturbances as a result of exposures to toxicants. These claims are not verifiable on formal cognitive testing. When no encephalopathy is identified with intellectual testing, the use of the personality tests (e.g., Minnesota Multiphasic Personality Inventory—see Table 27-1), qualitative data, and interview responses may be necessary to form alternative psychiatric and social hypotheses.

Another problem frequently mentioned in occupational settings is that of *malingering*; however, in our population, we have found many fewer instances of malingering than we have of *hysterical* or *somatoform disorders*. Malingering tends to show up in a large battery of tests as highly uneven and inconsistent test performance, with exaggeration of subjective symptoms on interview.

Other areas of difficulty in neuropsychological testing involve background patient characteristics. Again, *psychiatric disorders*, manifesting typically as somatoform responses to occupational disorder, are frequent. In addition, a patient may have a long-term psychiatric disorder upon which neurotoxicant exposure is superimposed. Psychological testing and a psychiatric interview are included to assess these factors. Many patients with occupational or environmental exposures have histories of *learning disability*, and many of them are unaware of this fact. Very rarely do they have prior testing or prior diagnostic information that is helpful in differentiating the effects of neurotoxicant exposure versus that of early "minimal brain dysfunction." However, with sufficiently broad testing of cognitive and academic skills, the patterns associated with learning disability often can be identified and taken into account when other diagnoses are being considered.

Another area of difficulty is *cultural background* of patients. Frequently, patients whose second language is English are tested; in these cases, greater emphasis is placed on the motor, visuospatial, visual memory, and mood tests than on language and other highly culturally bound tasks for diagnostic purposes. We have, through interpreters, used translation in completing testing, but still find that less language-dependent tests are the most reliable.

Finally, a patient may have concurrent *neurologically based impairments* arising from head injuries, alcohol abuse, or other primary neurologic disorders. These are assessed by including tests known to be sensitive to the patient's neurologic disorder, though differentiation of etiologies can become quite difficult.

REFERENCES

1. Hernberg S. Neurotoxic effects of long-term exposure to organic hydrocarbon solvents: epidemiologic aspects. In: Holmstedt B, Lauwerys R, Mercier M, Roberfroid M, eds. *Mechanisms of toxicity and hazard evaluation.* Amsterdam, Netherlands: Elsevier, 1980:307–317.
2. Filley CM, Heaton RK, Rosenberg NL. White matter dementia in chronic toluene abuse. *Neurology* 1990;40: 532–534.
3. Needleman HL, Gunnoe C, Leviton A, et al. Deficits in psychological and classroom performance of children with elevated dentine lead levels. *N Engl J* Med 1979; 300:689–695.
4. Needleman HL, Schell A, Bellinger D, et al. The long-term effects of exposure to low doses of lead in childhood. *N Engl J Med* 1990;322:83–88.
5. McMichael AJ, Baghurst PA, Wigg NR, et al. Port Prairie cohort study: environmental exposure to lead and children's abilities at the age of four years. *N Engl J Med* 1988;319:1037–1043.
6. Dietrich KN, Succop PA, Berger OG, et al. Lead exposure and the cognitive development of urban preschool children: the Cincinnati lead study cohort at age 4 years. *Neurotoxicol Teratol* 1991;13:203–211.
7. Feldman RG. Neurological manifestations of mercury intoxication. *Acta Neurol Scand* 1982;66:201–209.
8. Kurland LT, Faro SN, Siedler H. Minamata disease. *World Neurol* 1960;1:370–391.
9. Hanninen H, Nurminen M, Tolonen M, Martelin T. Psychological tests as indicators of excessive exposure to carbon disulfide. *Scand J Psychiatry* 1978;19:163–174.
10. Feldman RG. Trichloroethylene. In: Vinken PJ, Bruyn GW, eds. *Handbook of clinical neurology.* Amsterdam, Netherlands: North-Holland 1979:457–464.

Subject Index

Q

R